High-speed CMOS Logic Data Book

TEXAS INSTRUMENTS

IMPORTANT NOTICE

Texas Instruments reserves the right to make changes at any time in order to improve design and to supply the best product possible.

TI cannot assume any responsibility for any circuits shown or represent that they are free from patent infringement.

Information contained herein supersedes data published in the *High-Speed CMOS Logic Data Book*, 1983, SCLD001.

ISBN 0-89512-114-X
Library of Congress No. 82-074480

INTRODUCTION

The high-speed silicon-gate CMOS logic family (SN54HC/SN74HC) from Texas Instruments offers a broad range of functions: from basic gates and flip-flops to bus-compatible complex devices. These devices are pin-for-pin and functionally compatible (but not necessarily interchangeable) with the corresponding devices in the popular LSTTL family. Also, many of the metal-gate CMOS devices (4000 series) and TTL-voltage-compatible functions ('HCT) are available in the high-speed CMOS logic family from Texas Instruments.

The original CMOS devices were used in applications where the main concerns were low power consumption, wide power supply range, and high noise immunity. These requirements were satisfied by the metal-gate CMOS family. However, metal-gate CMOS could not satisfy system designs that required high speeds such as those imposed by microprocessor-based applications. For such designs, the system designers used faster families (STTL and LSTTL), and thus traded the advantages of CMOS for faster switching speeds. With the introduction of the high-speed CMOS family, Texas Instruments now provides the system designer with the best of both TTL and CMOS; fast switching speeds (comparable to LSTTL) and most of the advantages of CMOS.

The drawbacks of metal-gate CMOS arise because the source and drain areas are diffused before the gate is defined (Figure 1), and therefore the metal gate needs to overlap the source and drain to allow for misalignment, resulting in higher gate capacitances. Junction capacitance is increased by the deep diffusions required for the source and drain. The slow switching speeds are a result of the combined gate and junction capacitances.

New generations of CMOS technology (high-speed CMOS or HCMOS) have now evolved through improvements in process technology. High speeds and low power consumption have been made possible by the 3-μm, self-aligned poly-silicon-gate CMOS process. In this process, poly-silicon gates are deposited over the gate oxide before the source and drain implants are made (Figure 2). Then the gate itself is used as a mask for the source and drain implants. This self-aligning process results in reduced gate capacitance. Junction capacitance, a function of the junction area, is also minimized on a per gate basis through shallower implants and minimal sideways diffusion. The net result is an increase in the switching speeds. An added benefit of the self-aligning feature is that it permits smaller channel lengths, hence smaller gates and less gate capacitance. This corresponds to higher gate densities and further reduction in power consumption.

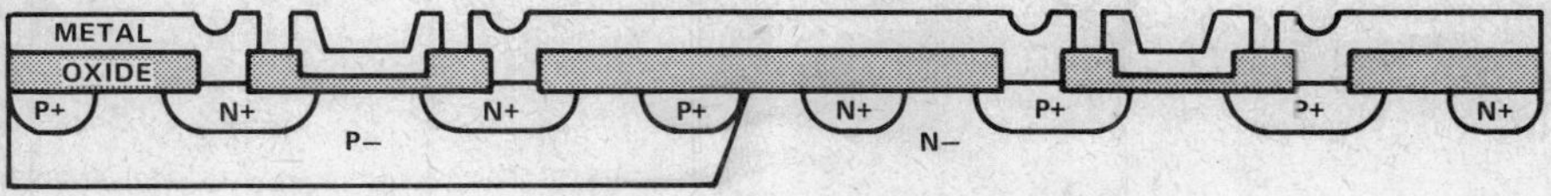

FIGURE 1. METAL-GATE CMOS

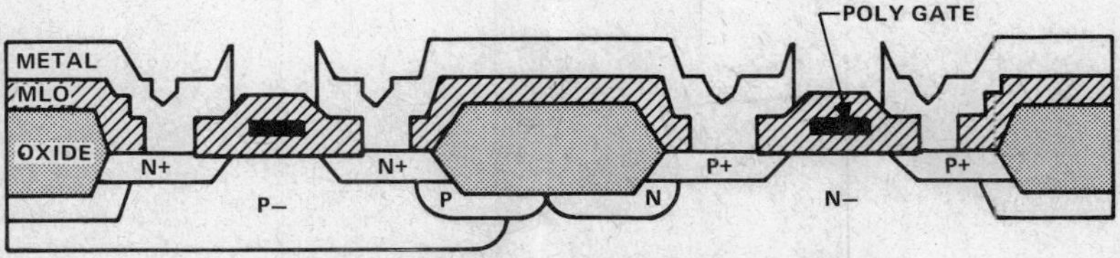

FIGURE 2. HIGH-SPEED SILICON-GATE CMOS

Designers' Information (Section 7) provides detailed discussion of interchangeability, electrostatic discharge (ESD) protection, latch-up circuitry, design considerations, interfacing, and other pertinent subjects regarding this family.

ATTENTION

These devices contain circuits to protect the inputs and outputs against damage due to high static voltages or electrostatic fields, however, it is advised that precautions be taken to avoid application of any voltage higher than maximum-rated voltages to these high-impedance circuits.

Unused inputs must always be connected to an appropriate logic voltage level, preferably either V_{CC} or ground.

ATTENTION

These devices contain circuits to protect the inputs and outputs against damage due to high static voltages or electrostatic fields, however, it is advised that precautions be taken to avoid application of any voltage higher than maximum-rated voltages to these high-impedance circuits.

Unused inputs must always be connected to an appropriate logic voltage level, preferably either V_{CC} or ground.

†See these pages for absolute maximum ratings, recommended operating conditions, and electrical characteristics.
‡See these pages for description, pin assignments, timing requirements, and switching characteristics.

TYPE NUMBER	RATINGS AND CHARACTERISTICS† TABLE	RATINGS AND CHARACTERISTICS† PAGE	DESCRIPTIVE INFORMATION‡ PAGE
'HC490	IV	2-10	3-281
'HC533	III	2-8	4-61
'HCT533	VII	2-14	4-65
'HC534	III	2-8	4-69
'HCT534	VII	2-14	5-23
'HC540	III	2-8	4-73
'HCT540	VII	2-14	4-77
'HC541	III	2-8	4-73
'HCT541	VII	2-14	4-77
'HC563	III	2-8	3-285
'HCT563	VII	2-14	3-289
'HC564	III	2-8	3-293
'HCT564	VII	2-14	3-297
'HC573	III	2-8	3-301
'HCT573	VII	2-14	3-305
'HC574	III	2-8	3-309
'HCT574	VII	2-14	3-313
'HC589	III	2-8	4-81
'HC590	III	2-8	4-87
'HC592	IV	2-10	5-27
'HC593	III	2-8	5-27
'HC594	IV	2-10	5-31
'HC595	III	2-8	5-33
'HC597	IV	2-10	5-35
'HC598	III	2-8	5-35
'HC604	III	2-8	3-317
'HC620	III	2-8	3-321
'HCT620	VII	2-14	4-89
'HC623	III	2-8	3-321
'HCT623	VII	2-14	4-89
'HC630	III	2-8	5-41
'HC632	III	2-8	5-45
'HC640	III	2-8	4-93
'HCT640	VII	2-14	4-97
'HC643	III	2-8	4-93
'HCT643	VII	2-14	4-97
'HC645	III	2-8	4-93
'HCT645	VII	2-14	4-97
'HC646	III	2-8	3-325
'HCT646	VII	2-14	3-331
'HC648	III	2-8	3-325
'HCT648	VII	2-14	3-331
'HC651	III	2-8	3-337
'HCT651	VII	2-14	3-343
'HC652	III	2-8	3-337
'HCT652	VII	2-14	3-343
'HC658	III	2-8	3-349
'HCT658	VII	2-14	3-357
'HC659	III	2-8	3-349
'HCT659	VII	2-14	3-357
'HC664	III	2-8	3-363
'HCT664	VII	2-14	3-371
'HC665	III	2-8	3-363
'HCT665	VII	2-14	3-371
'HC670	III	2-8	5-55
'HC673	III	2-8	4-103
'HC674	III	2-8	4-103
'HC677	III	2-8	4-109
'HC678	III	2-8	4-109
'HC679	III	2-8	4-115
'HC680	III	2-8	4-115
'HC682	IV	2-10	4-121
'HC684	IV	2-10	4-121
'HC686	IV	2-10	4-121
'HC688	IV	2-10	4-127
'HC690	III	2-8	4-129
'HC691	III	2-8	4-129
'HC692	III	2-8	4-129
'HC693	III	2-8	4-129
'HC696	III	2-8	4-137
'HC697	III	2-8	4-137
'HC698	III	2-8	4-137
'HC699	III	2-8	4-137
'HC804	III	2-8	3-377
'HC805	III	2-8	3-379
'HC808	III	2-8	3-381
'HC832	III	2-8	3-383
'HC881	IV	2-10	5-57
'HC882	IV	2-10	5-59
'HC4002	I	2-4	3-385
'HC4016	—	—	5-61
'HC4017	IV	2-10	4-145
'HC4020	IV	2-10	3-387
'HC4022	IV	2-10	4-149
'HC4024	IV	2-10	3-391
'HC4040	IV	2-10	3-395
'HC4049	—	—	5-63
'HC4050	—	—	5-63
'HC4051	—	—	5-65
'HC4052	—	—	5-65
'HC4053	—	—	5-65
'HC4060	IV	2-10	3-399
'HC4061	I	2-4	3-403
'HC4075	I	2-4	3-407
'HC4078A	I	2-4	3-409
'HC4316	—	—	5-67
'HC4511	IV	2-10	5-69
'HC4514	IV	2-10	5-71
'HC4515	IV	2-10	5-71
'HC4724	IV	2-10	3-411
'HC7001	I	2-4	5-75
'HC7002	I	2-4	5-77
'HC7003	I	2-4	5-79
'HC7006	I	2-4	5-81
'HC7022	IV	2-10	4-153
'HC7032	I	2-4	5-85
'HC7074	II	2-6	4-157
'HC7266	I	2-4	3-415
'HC7340	III	2-8	4-163

†See these pages for absolute maximum ratings, recommended operating conditions, and electrical characteristics.
‡See these pages for description, pin assignments, timing requirements, and switching characteristics.

FUNCTIONS	PAGE

AND, NAND GATES, BUFFERS, AND INVERTERS

DESCRIPTION	OUTPUT TYPE	DEVICE TYPE	RATINGS AND CHARACTERISTICS TABLE	RATINGS AND CHARACTERISTICS PAGE	DESCRIPTIVE INFORMATION
Hex Inverters	Totem-pole	'HC04	I	2-4	3-9
	Open-drain	'HC05			3-13
	Totem-pole	'HC4049	—	—	5-63
Hex Unbuffered Inverters	Totem-pole	'HCU04	IX	2-16	3-11
Hex Buffers		'HC4050	—	—	5-63
		'HC4061	I	2-4	3-403
Quad 2-Input NAND Gates	Totem-pole	'HC00			3-3
	Open-drain	'HC03			3-7
Quad 2-Input AND Gates	Totem-Pole	'HC08			3-15
	Open-drain	'HC09			3-17
Hex 2-Input NAND Drivers	Totem-pole	'HC804	III	2-8	3-377
Hex 2-Input AND Drivers		'HC808			3-381
Triple 3-Input NAND Gates		'HC10	I	2-4	3-19
Triple 3-Input AND Gates		'HC11			3-21
Dual 4-Input NAND Gates		'HC20			3-25
Dual 4-Input AND Gates		'HC21			3-27
8-Input NAND Gates		'HC30			3-31
13-Input AND Gates		'HC133			3-73

OR, NOR, EXCLUSIVE-OR, AND AND-OR-INVERT GATES

DESCRIPTION	OUTPUT TYPE	DEVICE TYPE	RATINGS AND CHARACTERISTICS TABLE	RATINGS AND CHARACTERISTICS PAGE	DESCRIPTIVE INFORMATION
Quad 2-Input NOR Gates	Totem-pole	'HC02	I	2-4	3-5
		'HC36			3-35
Quad 2-Input OR Gates		'HC32			3-33
Quad 2-Input Exclusive-NOR Gates		'HC7266			3-415
	Open-drain	'HC266			3-237
Quad 2-Input Exclusive-OR Gates	Totem-pole	'HC86			3-49
		'HC386			3-273
Hex 2-Input NOR Drivers		'HC805	III	2-8	3-379
Hex 2-Input OR Drivers		'HC832			3-383
Dual 2-Wide 2-Input AND-OR-Invert Gates		'HC51	I	2-4	3-39
Triple 3-Input NOR Gates		'HC27			3-29
Triple 3-Input OR Gates		'HC4075			3-407
Dual 4-Input NOR Gates		'HC4002			3-385
8-Input OR/NOR Gates		'HC4078A			3-409

GATE BUFFER/DRIVERS

DESCRIPTION	DEVICE TYPE	RATINGS AND CHARACTERISTICS TABLE	RATINGS AND CHARACTERISTICS PAGE	DESCRIPTIVE INFORMATION
Hex 2-Input NAND Drivers	'HC804	III	2-8	3-377
Hex 2-Input AND Drivers	'HC808			3-381
Hex 2-Input NOR Drivers	'HC805			3-379
Hex 2-Input OR Drivers	'HC832			3-383

SCHMITT-TRIGGER GATES AND INVERTERS

DESCRIPTION	OUTPUT TYPE	DEVICE TYPE	RATINGS AND CHARACTERISTICS TABLE	RATINGS AND CHARACTERISTICS PAGE	DESCRIPTIVE INFORMATION
Hex Inverters	Totem-pole	'HC14			3-23
Quad 2-Input NAND Gates		'HC132			5-5
	Open-drain	'HC7003	I	2-4	5-79
Quad 2-Input AND Gates	Totem-pole	'HC7001			5-75
Quad 2-Input NOR Gates		'HC7002			5-77
Quad 2-Input OR Gates		'HC7032			5-85

MULTI-FUNCTION CIRCUITS

DESCRIPTION	DEVICE TYPE	RATINGS AND CHARACTERISTICS TABLE	RATINGS AND CHARACTERISTICS PAGE	DESCRIPTIVE INFORMATION
Inverter, 3-/4-Input NAND/NOR Combination	'HC7006	I	2-4	5-81
Dual D-Type Flip-Flop, Inverter 2-Input NAND/NOR Combination	'HC7074	II	2-6	4-157

SHIFT REGISTERS

DESCRIPTION	INPUTS	OUTPUTS	DEVICE TYPE	RATINGS AND CHARACTERISTICS TABLE	RATINGS AND CHARACTERISTICS PAGE	DESCRIPTIVE INFORMATION
4-Bit Shift Registers with Clear	J-$\overline{K}$/Parallel	Parallel	'HC195			3-175
4-Bit Bidirectional Shift Registers with Clear	Serial/Parallel	Parallel	'HC194			3-171
8-Bit Shift Registers	Serial/Parallel, Clear, Clock Inhibit, Shift/Load	2 Serial	'HC165			3-131
	2 Serial, Clear	Parallel	'HC164	IV	2-10	3-127
	Serial/Parallel, Clear, Clock Inhibit, Shift/Load	Serial	'HC166			3-135
8-Bit Shift Registers with Input Registers	Serial/Parallel	Serial	'HC597			5-35
		3-State Parallel (Multiplexed I/O)	'HC598	III	2-8	5-35
		3-State Serial	'HC589			4-81
8-Bit Shift Registers with Output Registers	Serial	Parallel	'HC594	IV	2-10	5-31
		3-State Parallel	'HC595			5-33
	Serial/Parallel	3-State Parallel (Multiplexed I/O)	'HC299			4-25
			'HC322	III	2-8	4-31
			'HC323			4-37
16-Bit Shift Registers	Serial	3-State Serial, Parallel	'HC673			4-103
			'HC674			4-103

REGISTER FILES

DESCRIPTION	FEATURES	DEVICE TYPE	RATINGS AND CHARACTERISTICS TABLE	RATINGS AND CHARACTERISTICS PAGE	DESCRIPTIVE INFORMATION
4-by-4 Register Files	3-State Outputs	'HC670	III	2-8	5-55

LATCHES AND REGISTERS

DESCRIPTION	OUTPUT CONFIGURATION	DEVICE TYPE	RATINGS AND CHARACTERISTICS TABLE	RATINGS AND CHARACTERISTICS PAGE	DESCRIPTIVE INFORMATION
Quad D-type Latches	Complementary	'HC75	II	2-6	3-43
		'HC375			4-59
	Q only	'HC77			3-47
Quad D-type Registers	Q only, 3-State	'HC173	III	2-8	3-139
Octal D-type Latches	Q only, 3-State	'HC373			3-261
		'HC573			3-301
	$\overline{Q}$ only, 3-State	'HC533			4-65
		'HC563			3-285
Octal D-type Latches with TTL-Compatible Inputs	Q only, 3-State	'HCT373	VII	2-14	4-51
		'HCT573			3-305
	$\overline{Q}$ only, 3-State	'HCT533			4-61
		'HCT563			3-289
8-Bit Addressable Latches	Q only	'HC4724	IV	2-10	3-411
		'HC259			3-233

MONOSTABLE MULTIVIBRATORS

DESCRIPTION	FEATURES		DEVICE TYPE	RATINGS AND CHARACTERISTICS TABLE	RATINGS AND CHARACTERISTICS PAGE	DESCRIPTIVE INFORMATION
Dual Monostable Multivibrators with Direct Clear, Positive and Negative Inputs, and Complementary Outputs			'HC221	IV	2-10	5-9
	Retriggerable		'HC123			5-3
		Will not Trigger from Clear	'HC423			5-21

D-TYPE FLIP-FLOPS

DESCRIPTION	OUTPUT CONFIGURATION	OTHER FEATURES	DEVICE TYPE	RATINGS AND CHARACTERISTICS TABLE	RATINGS AND CHARACTERISTICS PAGE	DESCRIPTIVE INFORMATION
Dual D-type Flip-Flops with Preset and Clear	Complementary	Independent clocks, Preset, and Clear	'HC74	II	2-6	3-41
Dual D-type Flip-Flops with 2-Input NAND/NOR Gates	Complementary	Independent clocks, Preset, and Clear	'HC7074			4-157
Quad D-type Flip-Flops with Common Clocks	Complementary	Common Clear	'HC175	IV	2-10	3-143
		Output Enable	'HC379			3-269
Hex D-type Flip-Flops with Common Clocks	Q only	Common Clear	'HC174			3-143
		Output Enable	'HC378			3-269
Octal D-type Flip-Flops with Common Clocks	Q only	Common Clear	'HC273			3-239
		Output Enable	'HC377			3-269
	3-State, Q only	Output control	'HC374	III	2-8	3-265
			'HC574			3-309
	3-State, $\overline{Q}$ only	Output control	'HC534			4-69
			'HC564			3-293
Octal D-type Flip-Flops with Common Clocks and TTL-Compatible Inputs	3-State, Q only	Output control	'HCT374	VII	2-14	4-55
			'HCT574			3-313
	3-State, $\overline{Q}$ only	Output control	'HCT534			5-23
			'HCT564			3-297

DUAL J-K FLIP-FLOPS

DESCRIPTION	DEVICE TYPE	RATINGS AND CHARACTERISTICS TABLE	RATINGS AND CHARACTERISTICS PAGE	DESCRIPTIVE INFORMATION
Dual J-K Flip-Flops with Clear	'HC73	II	2-6	4-3
	'HC107			3-51
Dual J-K Flip-Flops with Preset	'HC113			3-63
Dual J-K Flip-Flops with Preset, Common Clock, and Common Clear	'HC78			4-5
	'HC114			3-65
Dual J-K Flip-Flops with Preset and Clear	'HC76			3-45
	'HC112			3-59
Dual J-$\overline{K}$ Flip-Flops with Preset and Clear	'HC109			3-55

BUS DRIVERS AND TRANSCEIVERS WITH 3-STATE OUTPUTS

DESCRIPTION	OUTPUT DATA	CONTROL INPUTS	DEVICE TYPE	RATINGS AND CHARACTERISTICS TABLE	RATINGS AND CHARACTERISTICS PAGE	DESCRIPTIVE INFORMATION
Quad Bus Drivers/Receivers	True	Individual Enables	'HC125	III	2-8	3-69
			'HC126			3-69
Quad Bus Transceivers	Inverting	Independent Enables for A and B Buses	'HC242			3-207
	True		'HC243			3-207
Hex Bus Drivers/Receivers	True	Common Enables	'HC365			3-257
	Inverting		'HC366			3-257
	True	Symmetrical Enables	'HC367			3-257
	Inverting		'HC368			3-257
Octal Bus Drivers/Receivers	Inverting	Symmetrical Enables	'HC240			3-197
		2 Enables	'HC540			4-73
	True	Complementary Enables	'HC241			3-197
		Symmetrical Enables	'HC244			3-213
		2 Enables	'HC541			4-73
Octal Bus Transceivers	Inverting	Independent Enables for A and B Buses	'HC620			3-321
	True		'HC623			3-321
	Inverting	Enable and Direction Control	'HC640			4-93
	True and Inverting		'HC643			4-93
	True		'HC645			4-93
			'HC245			3-219
Octal Bus Transceivers with Registers	True	Enable and Direction Control	'HC646			3-325
	Inverting		'HC648			3-325
	Inverting	Independent Enables for A and B Buses	'HC651			3-337
	True		'HC652			3-337
Octal Bus Drivers with Registers	Inverting	Independent Enables for A and B Buses	'HC7340			4-163
8-/9-Bit Bus Transceivers with Parity Checker/Generator	True	Enable and Direction Control	'HC659			3-349
	Inverting		'HC658			3-349
	True	Independent Enables for A and B Buses	'HC665			3-363
	Inverting		'HC664			3-363

1 GENERAL INFORMATION

BUS DRIVERS AND TRANSCEIVERS WITH 3-STATE OUTPUTS AND TTL-COMPATIBLE INPUTS

DESCRIPTION	OUTPUT DATA	CONTROL INPUTS	DEVICE TYPE	RATINGS AND CHARACTERISTICS TABLE	RATINGS AND CHARACTERISTICS PAGE	DESCRIPTIVE INFORMATION
Quad Bus Transceivers	Inverting	Individual Enables for A and B Buses	'HCT242	VII	2-14	3-209
	True		'HCT243			3-209
Octal Bus Drivers/Receivers	Inverting	Symmetrical Enables	'HCT240			3-201
		2 Enables	'HCT540			4-77
	True	Complementary Enables	'HCT241			3-201
		Symmetrical Enables	'HCT244			3-217
		2 Enables	'HCT541			4-77
Octal Bus Transceivers	Inverting	Independent Enables for A and B Buses	'HCT620			4-89
	True		'HCT623			4-89
	Inverting	Enable and Direction Control	'HCT640			4-97
	True and Inverting		'HCT643			4-97
	True		'HCT645			4-97
			'HCT245			4-19
Octal Bus Transceivers with Registers	True	Enable and Direction Control	'HCT646			3-331
	Inverting		'HCT648			3-331
	Inverting	Independent Enables for A and B Buses	'HCT651			3-343
	True		'HCT652			3-343
8-/9-Bit Bus Transceivers with Parity Checker/Generator	True	Enable and Direction Control	'HCT659			3-357
	Inverting		'HCT658			3-357
	True	Independent Enables for A and B Buses	'HCT665			3-371
	Inverting		'HCT664			3-371

ASYNCHRONOUS (RIPPLE-CLOCK) COUNTERS

DESCRIPTION	FEATURES	DEVICE TYPE	RATINGS AND CHARACTERISTICS TABLE	RATINGS AND CHARACTERISTICS PAGE	DESCRIPTIVE INFORMATION
7-Bit Binary Counters		'HC4024	IV	2-10	3-391
12-Bit Binary Counters		'HC4040			3-395
14-Bit Binary Counters		'HC4020			3-387
	On-Chip Oscillator	'HC4060			3-399
Dual Decade Counters	Biquinary or BCD	'HC390			3-275
	Set-to-9 input	'HC490			3-281
Dual 4-Bit Binary Counters		'HC393			3-275

PROGRAMMABLE FREQUENCY DIVIDERS/TIMERS

DESCRIPTION	FEATURES	DEVICE TYPE	RATINGS AND CHARACTERISTICS TABLE	RATINGS AND CHARACTERISTICS PAGE	DESCRIPTIVE INFORMATION
Programmable Frequency Dividers/Digital Timers	Programming range 2^2 to 2^{15}	'HC294	IV	2-10	5-11
	Programming range 2^2 to 2^{31}	'HC292			5-11

SYNCHRONOUS COUNTERS

DESCRIPTION	FEATURES		DEVICE TYPE	RATINGS AND CHARACTERISTICS TABLE	RATINGS AND CHARACTERISTICS PAGE	DESCRIPTIVE INFORMATION
Decade	Async Clear	Synchronous Load	'HC160	IV	2-10	3-115
	Sync Clear		'HC162			3-115
Decade Counters with Output Registers	Sync Clear	Multiplexed 3-State Outputs	'HC692	III	2-8	4-129
	Async Clear		'HC690			4-129
Decade Up/Down	Clock Inhibit	Asynchronous Load	'HC190	IV	2-10	3-155
	Async Clear		'HC192			3-163
Decade Up/Down Counters with Output Registers	Sync Clear	Multiplexed 3-State Outputs	'HC698	III	2-8	4-137
	Async Clear		'HC696			4-137
Divide-by-8 Johnson Counter	Sync Clear		'HC4022	IV	2-10	4-149
			'HC7022			4-153
Divide-by-10 Johnson Counter	Async Clear		'HC4017			4-145
4-Bit Binary	Async Clear	Synchronous Load	'HC161			3-115
	Sync Clear		'HC163			3-115
4-Bit Binary Counters with Output Registers	Sync Clear	Multiplexed 3-State Outputs	'HC693	III	2-8	4-129
	Async Clear		'HC691			4-129
4-Bit Binary Up/Down	Clock Inhibit	Asynchronous Load	'HC191	IV	2-10	3-155
	Async Clear		'HC193			3-163
4-Bit Binary Up/Down Counters with Output Registers	Sync Clear	Multiplexed 3-State Outputs	'HC699	III	2-8	4-137
	Async Clear		'HC697			4-137
8-Bit Binary with Input Registers	Sync Clear		'HC592	IV	2-10	5-27
		Multiplexed 3-State I/O	'HC593	III	2-8	5-27
8-Bit Binary with Output Registers	Sync Clear	3-State Outputs	'HC590			4-87

MAGNITUDE COMPARATORS, PARITY GENERATORS/CHECKERS, AND PARITY ENCODERS

DESCRIPTION	FEATURES		DEVICE TYPE	RATINGS AND CHARACTERISTICS TABLE	RATINGS AND CHARACTERISTICS PAGE	DESCRIPTIVE INFORMATION
4-Bit Magnitude Comparators			'HC85	IV	2-10	4-7
8-Bit Magnitude Comparators	$\overline{P=Q}$, $\overline{P>Q}$ Outputs		'HC682			4-121
			'HC684			4-121
		Enable Inputs	'HC686			4-121
	$\overline{P=Q}$ Outputs		'HC688			4-127
9-Bit Odd/Even Parity Generator/Checkers	Even, Odd Inputs		'HC180			3-147
			'HC280			3-243
8-/9-Bit Bus Transceivers with Parity Generator/Checkers	True Outputs	Enable and Direction Control	'HC659	III	2-8	3-349
			'HCT659	VII	2-14	3-357
	Inverting Outputs		'HC658	III	2-8	3-349
			'HCT658	VII	2-14	3-357
	True Outputs	Independent Enables for A and B Buses	'HC665	III	2-8	3-363
			'HCT665	VII	2-14	3-371
	Inverting Outputs		'HC664	III	2-8	3-363
			'HCT664	VII	2-14	3-371
8-Line to 3-Line Priority Encoders	Enable Inputs and Outputs		'HC148	IV	2-10	3-93
10-Line Decimal to 4-Line BCD Priority Encoders			'HC147			3-93

TEXAS INSTRUMENTS
POST OFFICE BOX 225012 • DALLAS, TEXAS 75265

ADDRESS COMPARATORS

DESCRIPTION	FEATURES	DEVICE TYPE	RATINGS AND CHARACTERISTICS TABLE	RATINGS AND CHARACTERISTICS PAGE	DESCRIPTIVE INFORMATION
16-Bit to 4-Bit Address Comparators	Output Enable	'HC677	III	2-8	4-109
	Latched Output	'HC678			4-109
12-Bit to 4-Bit Address Comparators	Output Enable	'HC679			4-115
	Latched Output	'HC680			4-115

ARITHMETIC CIRCUITS

DESCRIPTION		DEVICE TYPE	RATINGS AND CHARACTERISTICS TABLE	RATINGS AND CHARACTERISTICS PAGE	DESCRIPTIVE INFORMATION
4-Bit Arithmetic Logic Units/Function Generators	16 Functions	'HC181	IV	2-10	3-149
	8 Functions	'HC381			5-17
	16 Functions	'HC881			5-57
4-Bit ALU with Ripple Carry		'HC382			5-17
4-Bit Adders		'HC283			4-21
Look-Ahead Carry Generators	16-Bit	'HC182			5-7
	32-Bit	'HC882			5-59

ERROR DETECTORS/CORRECTORS

DESCRIPTION	DEVICE TYPE	RATINGS AND CHARACTERISTICS TABLE	RATINGS AND CHARACTERISTICS PAGE	DESCRIPTIVE INFORMATION
16-Bit Parallel Error Detection and Correction	'HC630	III	2-8	5-41
32-Bit Parallel Error Detection and Correction	'HC632			5-45

DATA SELECTORS/MULTIPLEXERS

DESCRIPTION	INPUTS	OUTPUTS	DEVICE TYPE	RATINGS AND CHARACTERISTICS TABLE	RATINGS AND CHARACTERISTICS PAGE	DESCRIPTIVE INFORMATION
8-Line to 1-Line		Inverting	'HC152	III	2-8	3-103
	Enable	Complementary	'HC151			3-99
		Complementary 3-State	'HC251			3-221
	Transparent Latches, Enable		'HC354			4-43
	Registers, Enable		'HC356			4-47
Dual 4-Line to 1-Line	Independent Enables	True, 3-State	'HC253			3-225
		Inverting, 3-State	'HC353			3-253
		True	'HC153			3-107
		Inverting	'HC352			3-249
Quad 2-Line to 1-Line	Common Enable	True	'HC157			3-111
		Inverting	'HC158			3-111
		True, 3-State	'HC257			3-229
		Inverting, 3-State	'HC258			3-229
Quad 2-Line to 1-Line with Storage		True	'HC298	IV	2-10	3-245
Octal 2-Line to 1-Line	Input Registers	True, 3-State	'HC604	III	2-8	3-317

DECODERS/DEMULTIPLEXERS

DESCRIPTION	FEATURES	OUTPUTS	DEVICE TYPE	RATINGS AND CHARACTERISTICS TABLE	RATINGS AND CHARACTERISTICS PAGE	DESCRIPTIVE INFORMATION
4-Line to 16-Line	2 Enables	Inverting	'HC154	IV	2-10	4-9
	Input Latches, Output Enable	True	'HC4514			5-71
		Inverting	'HC4515			5-71
4-Line to 10-Line BCD-to-Decimal			'HC42			3-37
3-Line to 8-Line	3 Enables	True	'HC238			3-187
			'HCT238	VIII	2-15	3-191
		Inverting	'HC138	IV	2-10	3-83
			'HCT138	VIII	2-15	3-87
	3 Enables, Address Latches	True	'HC237	IV	2-10	3-179
			'HCT237	VIII	2-15	3-183
		Inverting	'HC137	IV	2-10	3-75
			'HCT137	VIII	2-15	3-79
Dual 2-Line to 4-Line	Independent Enables	Inverting	'HC139	IV	2-10	3-91
		True	'HC239			3-195

DISPLAY DECODERS/DRIVERS

DESCRIPTION	DEVICE TYPE	RATINGS AND CHARACTERISTICS TABLE	RATINGS AND CHARACTERISTICS PAGE	DESCRIPTIVE INFORMATION
BCD-to-7-Segment Decoders/Drivers with Input Latches	'HC4511	IV	2-10	5-69

ANALOG SWITCHES/MULTIPLEXERS/DEMULTIPLEXERS

DESCRIPTION		DEVICE TYPE	RATINGS AND CHARACTERISTICS TABLE	RATINGS AND CHARACTERISTICS PAGE	DESCRIPTIVE INFORMATION
Quad Analog Switch/Transmission Gate		'HC4016	—	—	5-61
	Enable, Level Translators	'HC4316	—	—	5-67
8-Channel Analog Multiplexer/Demultiplexer		'HC4051	—	—	5-65
Dual 4-Channel Analog Multiplexer/Demultiplexer		'HC4052	—	—	5-65
Triple 2-Channel Analog Multiplexer/Demultiplexer		'HC4053	—	—	5-65

RANDOM ACCESS MEMORIES

DESCRIPTION	ORGANIZATION	FEATURES	DEVICE TYPE	RATINGS AND CHARACTERISTICS TABLE	RATINGS AND CHARACTERISTICS PAGE	DESCRIPTIVE INFORMATION
64-Bit	16 × 4	3-State Inverting Outputs	'HC189	III	2-8	4-11
			'HCT189	VII	2-14	4-15
		3-State Noninverting Outputs	'HC219	III	2-8	4-11
			'HCT219	VII	2-14	4-15

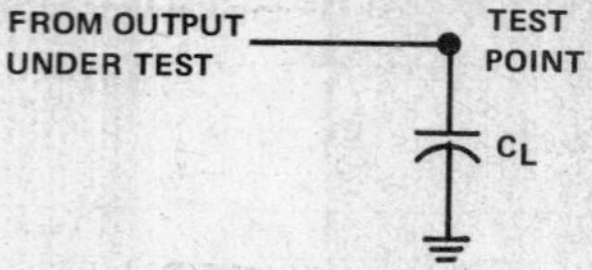

LOAD CIRCUIT

PARAMETER		C_L†
t_{pd} or t_t	Standard outputs	50 pF
	High-current outputs‡	50 pF or 150 pF

†C_L includes probe and test fixture capacitance.
‡High-current outputs are indicated by the ▷ in the logic symbol.

FIGURE 1. TOTEM-POLE OUTPUTS

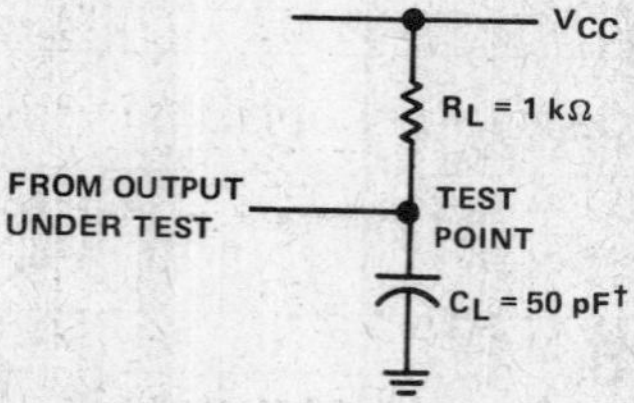

LOAD CIRCUIT

†C_L includes probe and test fixture capacitance.

FIGURE 2. OPEN-DRAIN OUTPUTS

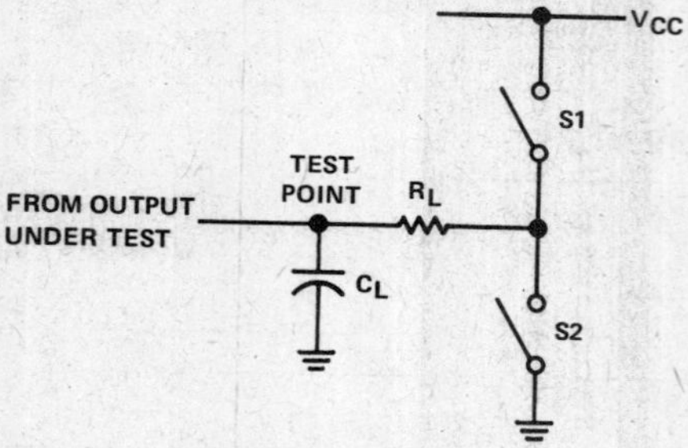

LOAD CIRCUIT

PARAMETER		R_L	C_L†	S_1	S_2
t_{en}	t_{PZH}	1 kΩ	50 pF or 150 pF	OPEN	CLOSED
	t_{PZL}			CLOSED	OPEN
t_{dis}	t_{PHZ}	1 kΩ	50 pF	OPEN	CLOSED
	t_{PLZ}			CLOSED	OPEN
t_{pd} or t_t		—	50 pF or 150 pF	OPEN	OPEN

†C_L includes probe and test fixture capacitance.

FIGURE 3. 3-STATE OUTPUTS

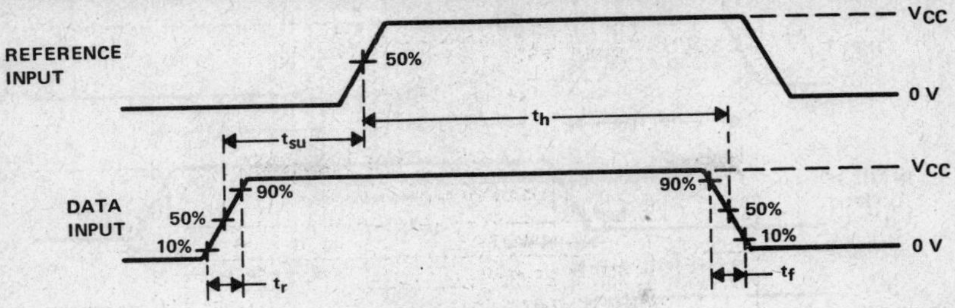

VOLTAGE WAVEFORMS

NOTE 1: Phase relationships between waveforms were chosen arbitrarily. All input pulses are supplied by generators having the following characteristics: PRR ≤ 1 MHz, Z_{out} ≈ 50 Ω, t_r = 6 ns, t_f = 6 ns.

FIGURE 4. HC AND HCU — SETUP AND HOLD TIMES, AND INPUT RISE AND FALL TIMES

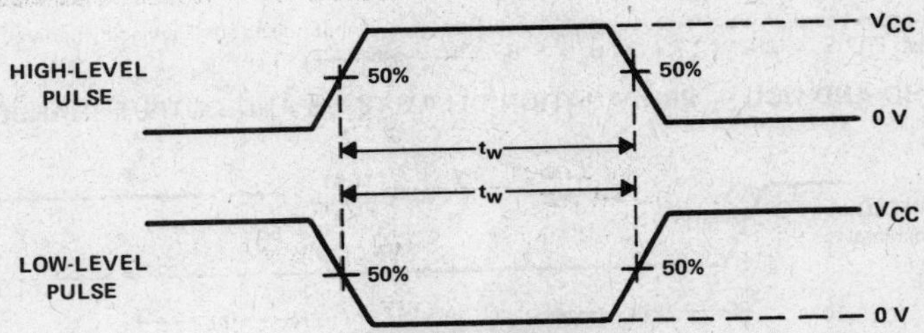

VOLTAGE WAVEFORMS

NOTES: 1. Phase relationships between waveforms were chosen arbitrarily. All input pulses are supplied by generators having the following characteristics: PRR ≤ 1 MHz, Z_{out} ≈ 50 Ω, t_r = 6 ns, t_f = 6 ns.

2. For clock inputs, f_{max} is measured when the input duty cycle is 50%.

FIGURE 5. HC AND HCU — PULSE DURATIONS

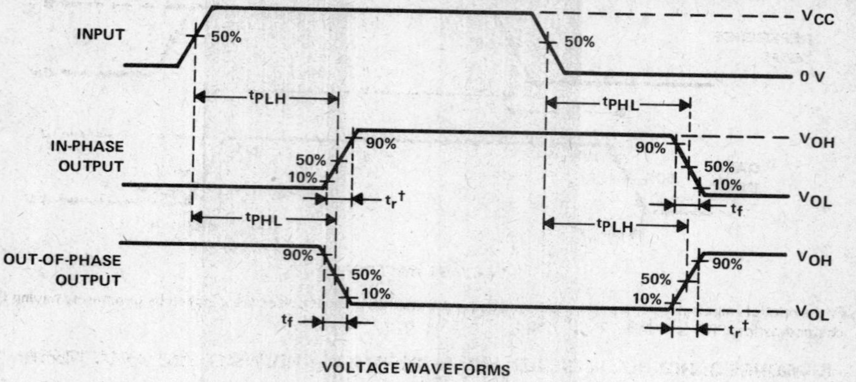

† t_r is not applicable to SN54/74HCU' devices.

NOTE 1: Phase relationships between waveforms were chosen arbitrarily. All input pulses are supplied by generators having the following characteristics: PRR $\leq$ 1 MHz, $Z_{out} \approx 50\ \Omega$, t_r = 6 ns, t_f = 6 ns.

FIGURE 6. HC AND HCU — PROPAGATION DELAY TIMES AND OUTPUT TRANSITION TIMES

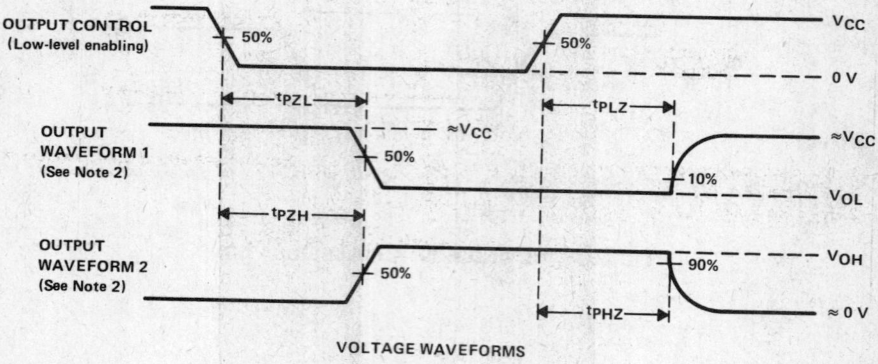

NOTES: 1. Phase relationships between waveforms were chosen arbitrarily. All input pulses are supplied by generators having the following characteristics: PRR $\leq$ MHz, $Z_{out} \approx 50\ \Omega$, t_r = 6 ns, t_f = 6 ns.

2. Waveform 1 is for an output with internal conditions such that the output is low except when disabled by the output control. Waveform 2 is for an output with internal conditions such that the output is high except when disabled by the output control.

FIGURE 7. HC AND HCU — ENABLE AND DISABLE TIMES, 3-STATE OUTPUTS

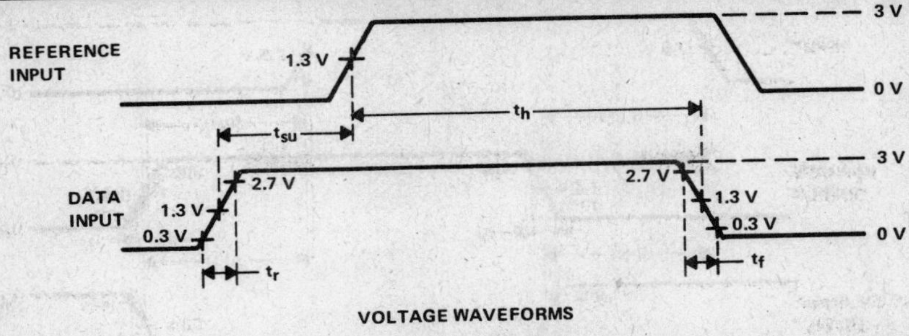

NOTE 1: Phase relationships between waveforms were chosen arbitrarily. All input pulses are supplied by generators having the following characteristics: PRR ≤ 1 MHz, Z_{out} ≈ 50 Ω, t_r = 6 ns, t_f = 6 ns.

FIGURE 8. HCT — SETUP AND HOLD TIMES, AND INPUT RISE AND FALL TIMES

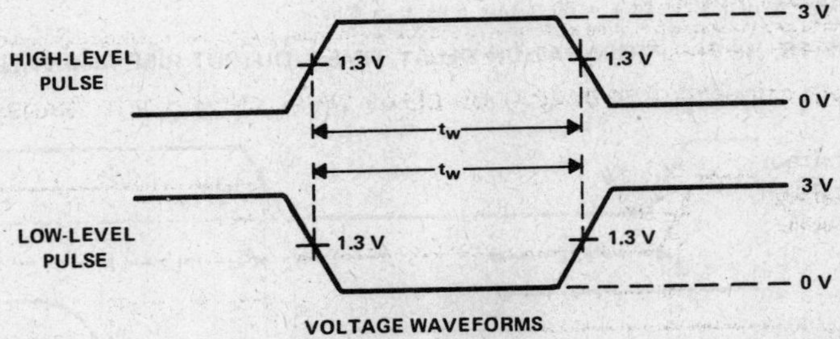

NOTES: 1. Phase relationships between waveforms were chosen arbitrarily. All input pulses are supplied by generators having the following characteristics: PRR ≤ 1 MHz, Z_{out} ≈ 50 Ω, t_r = 6 ns, t_f = 6 ns.
2. For clock inputs, f_{max} is measured when the input duty cycle is 50%.

FIGURE 9. HCT — PULSE DURATIONS

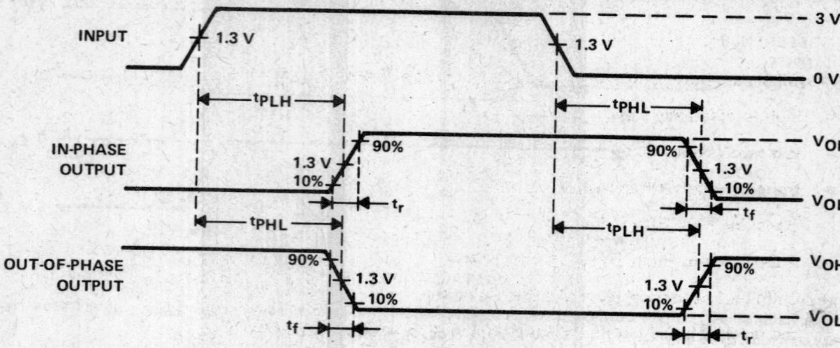

VOLTAGE WAVEFORMS

NOTE 1: Phase relationships between waveforms were chosen arbitrarily. All input pulses are supplied by generators having the following characteristics: PRR ≤ 1 MHz, Z_{out} ≈ 50 Ω, t_r = 6 ns, t_f = 6 ns.

FIGURE 10. HCT — PROPAGATION DELAY TIMES, OUTPUT RISE AND FALL TIMES

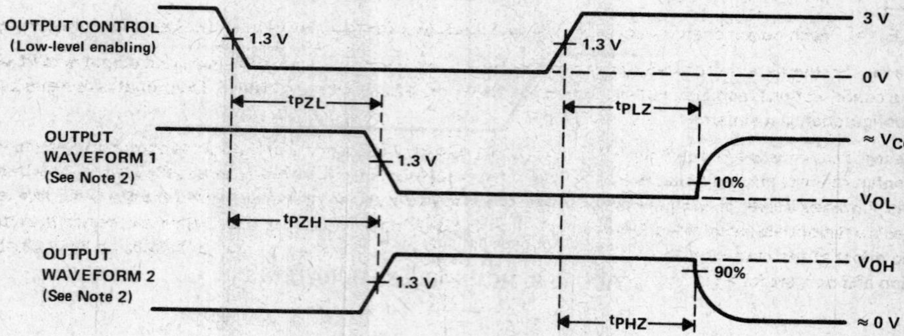

VOLTAGE WAVEFORMS

NOTES: 1. Phase relationships between waveforms were chosen arbitrarily. All input pulses are supplied by generators having the following characteristics: PRR ≤ 1 MHz, Z_{out} ≈ 50 Ω, t_r = 6 ns, t_f = 6 ns.
2. Waveform 1 is for an output with internal conditions such that the output is low except when disabled by the output control. Waveform 2 is for an output with internal conditions such that the output is high except when disabled by the output control.

FIGURE 11. HCT — ENABLE AND DISABLE TIMES FOR 3-STATE OUTPUTS

The following symbols are now being used in function tables on TI data sheets:

H	=	high level (steady state)
L	=	low level (steady state)
↑	=	transition from low to high level
↓	=	transition from high to low level
X	=	irrelevant (any input, including transitions)
Z	=	off (high-impedance) state of a 3-state output
a . . h	=	the level of steady-state inputs at inputs A through H, respectively
Q_0	=	level of Q before the indicated steady-state input conditions were established
$\overline{Q}_0$	=	complement of Q_0 or level of $\overline{Q}$ before the indicated steady-state input conditions were established
Q_n	=	level of Q before the most recent active transition indicated by ↑ or ↓
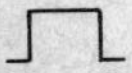	=	one high-level pulse
	=	one low-level pulse
TOGGLE	=	each output changes to the complement of its previous level on each active transition indicated by ↑ or ↓.

If, in the input columns, a row contains only the symbols H, L, and/or X, this means the indicated output is valid whenever the input configuration is achieved and regardless of the sequence in which it is achieved. The output persists so long as the input configuration is maintained.

If, in the input columns, a row contains, H, L, and/or X together with ↑ and/or ↓, this means the output is valid whenever the input configuration is achieved but the transition(s) must occur following the achievement of the steady-state levels. If the output is shown as a level (H, L, Q_0, or $\overline{Q}_0$), it persists so long as the steady-state input levels and the levels that terminate indicated transitions are maintained. Unless otherwise indicated, input transitions in the opposite direction to those shown have no effect at the output. (If the output is shown as a pulse,  or 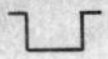, the pulse follows the indicated input transition and persists for an interval dependent on the circuit.)

EXPLANATION OF FUNCTION TABLES

Among the most complex function tables in this book are those of the shift registers. These embody most of the symbols used in any of the function tables, plus more. Below is the function table of a 4-bit bidirectional universal shift register, e.g., type SN74HC194.

FUNCTION TABLE

INPUTS										OUTPUTS			
CLEAR	MODE		CLOCK	SERIAL		PARALLEL				Q_A	Q_B	Q_C	Q_D
	S1	S0		LEFT	RIGHT	A	B	C	D				
L	X	X	X	X	X	X	X	X	X	L	L	L	L
H	X	X	L	X	X	X	X	X	X	Q_{A0}	Q_{B0}	Q_{C0}	Q_{D0}
H	H	H	↑	X	X	a	b	c	d	a	b	c	d
H	L	H	↑	X	H	X	X	X	X	H	Q_{An}	Q_{Bn}	Q_{Cn}
H	L	H	↑	X	L	X	X	X	X	L	Q_{An}	Q_{Bn}	Q_{Cn}
H	H	L	↑	H	X	X	X	X	X	Q_{Bn}	Q_{Cn}	Q_{Dn}	H
H	H	L	↑	L	X	X	X	X	X	Q_{Bn}	Q_{Cn}	Q_{Dn}	L
H	L	L	X	X	X	X	X	X	X	Q_{An}	Q_{Bn}	Q_{Cn}	Q_{D0}

The first line of the table represents a synchronous clearing of the register and says that if clear is low, all four outputs will be reset low regardless of the other inputs. In the following lines, clear is inactive (high) and so has no effect.

The second line shows that so long as the clock input remains low (while clear is high), no other input has any effect and the outputs maintain the levels they assumed before the steady-state combination of clear high and clock low was established. Since on other lines of the table only the rising transition of the clock is shown to be active, the second line implicitly shows that no further change in the outputs will occur while the clock remains high or on the high-to-low transition of the clock.

The third line of the table represents synchronous parallel loading of the register and says that if S1 and S0 are both high then, without regard to the serial input, the data entered at A will be at output Q_A, data entered at B will be at Q_B, and so forth, following a low-to-high clock transition.

The fourth and fifth lines represent the loading of high- and low-level data, respectively, from the shift-right serial input and the shifting of previously entered data one bit; data previously at Q_A is now at Q_B, the previous levels of Q_B and Q_C are now at Q_C and Q_D respectively, and the data previously at Q_D is no longer in the register. This entry of serial data and shift takes place on the low-to-high transition of the clock when S1 is low and S0 is high and the levels at inputs A through D have no effect.

The sixth and seventh lines represent the loading of high- and low-level data, respectively, from the shift-left serial input and the shifting of previously entered data one bit; data previously at Q_B is now at Q_A, the previous levels of Q_C and Q_D are now at Q_B and Q_C, respectively, and the data previously at Q_A is no longer in the register. This entry of serial data and shift takes place on the low-to-high transition of the clock when S1 is high and S0 is low and the levels at inputs A through D have no effect.

The last line shows that as long as both mode inputs are low, no other input has any effect and, as in the second line, the outputs maintain the levels they assumed before the steady-state combination of clear high and both mode inputs low was established.

INTRODUCTION

These symbols, terms, and definitions are in accordance with those currently agreed upon by the JEDEC Council of the Electronic Industries Association (EIA) for use in the USA and by the International Electrotechnical Commission (IEC) for international use.

OPERATING CONDITIONS AND CHARACTERISTICS (IN SEQUENCE BY LETTER SYMBOLS)

C_{pd} **Power dissipation capacitance**
Used to determine the no-load dynamic power dissipation per logic function (see individual circuit pages): $P_D = C_{pd} V_{CC}^2 f + I_{CC} V_{CC}$.

f_{max} **Maximum clock frequency**
The highest rate at which the clock input of a bistable circuit can be driven through its required sequence while maintaining stable transitions of logic level at the output with input conditions established that should cause changes of output logic level in accordance with the specification.

I_{CC} **Supply current**
The current into* the V_{CC} supply terminal of an integrated circuit.

I_{IH} **High-level input current**
The current into* an input when a high-level voltage is applied to that input.

I_{IL} **Low-level input current**
The current into* an input when a low-level voltage is applied to that input.

I_{OH} **High-level output current**
The current into* an output with input conditions applied that, according to the product specification, will establish a high level at the output.

I_{OL} **Low-level output current**
The current into* an output with input conditions applied that, according to the product specification, will establish a low level at the output.

I_{OZ} **Off-state (high-impedance-state) output current (of a three-state output)**
The current flowing into* an output having three-state capability with input conditions established that, according to the production specification, will establish the high-impedance state at the output.

V_{IH} **High-level input voltage**
An input voltage within the more positive (less negative) of the two ranges of values used to represent the binary variables.
NOTE: A minimum is specified that is the least-positive value of high-level input voltage for which operation of the logic element within specification limits is guaranteed.

V_{IL} **Low-level input voltage**
An input voltage level within the less positive (more negative) of the two ranges of values used to represent the binary variables.
NOTE: A minimum is specified that is the most-positive value of low-level input voltage for which operation of the logic element within specification limits is guaranteed.

*Current out of a terminal is given as a negative value.

V_{OH} **High-level output voltage**
The voltage at an output terminal with input conditions applied that, according to product specification, will establish a high level at the output.

V_{OL} **Low-level output voltage**
The voltage at an output terminal with input conditions applied that, according to product specification, will establish a low level at the output.

V_{T+} **Positive-going threshold level**
The voltage level at a transition-operated input that causes operation of the logic element according to specification as the input voltage rises from a level below the negative-going threshold voltage, V_{T-}.

V_{T-} **Negative-going threshold level**
The voltage level at a transition-operated input that causes operation of the logic element according to specification as the input voltage falls from a level above the positive-going threshold voltage, V_{T+}.

t_a **Access time**
The time interval between the application of a specified input pulse and the availability of valid signals at an output.

t_{dis} **Disable time (of a three-state output)**
The time interval between the specified reference points on the input and output voltage waveforms, with the three-state output changing from either of the defined active levels (high or low) to a high-impedance (off) state. (t_{dis} = t_{PHZ} or t_{PLZ}).

t_{en} **Enable time (of a three-state output)**
The time interval between the specified reference points on the input and output voltage waveforms, with the three-state output changing from a high-impedance (off) state to either of the defined active levels (high or low). (t_{en} = t_{PZH} or t_{PZL}).

t_f **Fall time**
The time interval between two reference points (90% and 10% unless otherwise specified) on a waveform that is changing from the defined high level to the defined low level.

t_h **Hold time**
The time interval during which a signal is retained at a specified input terminal after an active transition occurs at another specified input terminal.

NOTES: 1. The hold time is the actual time interval between two signal events and is determined by the system in which the digital circuit operates. A minimum value is specified that is the shortest interval for which correct operation of the digital circuit is guaranteed.

2. The hold time may have a negative value in which case the minimum limit defines the longest interval (between the release of the signal and the active transition) for which correct operation of the digital circuit is guaranteed.

t_{pd} **Propagation delay time**
The time between the specified reference points on the input and output voltage waveforms with the output changing from one defined level (high or low) to the other defined level. (t_{pd} = t_{PHL} or t_{PLH}).

t_{PHL} **Propagation delay time, high-to-low level output**
The time between the specified reference points on the input and output voltage waveforms with the output changing from the defined high level to the defined low level.

t_{PHZ} **Disable time (of a three-state output) from high level**
The time interval between the specified reference points on the input and the output voltage waveforms with the three-state output changing from the defined high level to a high-impedance (off) state.

t_{PLH} **Propagation delay time, low-to-high-level output**
The time between the specified reference points on the input and output voltage waveforms with the output changing from the defined low level to the defined high level.

t_{PLZ} **Disable time (of a three-state output) from low level**
The time interval between the specified reference points on the input and output voltage waveforms with the three-state output changing from the defined low level to a high-impedance (off) state.

t_{PZH} **Enable time (of a three-state output) to high level**
The time interval between the specified reference points on the input and output voltage waveforms with the three-state output changing from a high-impedance (off) state to the defined high level.

t_{PZL} **Enable time (of a three-state output) to low level**
The time interval between the specified reference points on the input and output voltage waveforms with the three-state output changing from a high-impedance (off) state to the defined low level.

t_r **Rise time**
The time interval between two reference points (10% and 90% unless otherwise specified) on a waveform that is changing from the defined low level to the defined high level.

t_{sr} **Sense recovery time**
The time interval needed to switch a memory from a write mode to a read mode and to obtain valid data signals at the output.

t_{su} **Setup time**
The time interval between the application of a signal at a specified input terminal and a subsequent active transition at another specified input terminal.

NOTES:
1. The setup time is the actual time interval between two signal events and is determined by the system in which the digital circuit operates. A minimum value is specified that is the shortest interval for which correct operation of the digital circuit is guaranteed.
2. The setup time may have a negative value in which case the minimum limit defines the longest interval (between the active transition and the application of the other signal) for which correct operation of the digital circuit is guaranteed.

t_t **Transition time (general)**
The time interval between two reference points (10% and 90% unless otherwise specified) on a waveform that is changing from the defined low level to the defined high level (rise time) or from the defined high level to the defined low level (fall time).

t_w **Pulse duration (width)**
The time interval between specified reference points on the leading and trailing edges of the pulse waveform.

These devices contain circuits to protect the inputs and outputs against damage due to high static voltages or electrostatic fields, however, it is advised that precautions be taken to avoid application of any voltage higher than maximum-rated voltages to these high-impedance circuits.

Unused inputs must always be connected to an appropriate logic voltage level, preferably either V_{CC} or ground.

LIST OF TABLES

D2804, DECEMBER 1982—REVISED MARCH 1984

absolute maximum ratings over operating free-air temperature range†

Supply voltage range, V_{CC} .... −0.5 V to 7 V
Input diode current, I_{IK} ($V_I < 0$ or $V_I > V_{CC}$) .... ±20 mA
Output diode current, I_{OK} ($V_O < 0$ or $V_O > V_{CC}$) .... ±20 mA
Continuous output current, I_O ($V_O = 0$ to V_{CC}) .... ±25 mA
Continuous current through V_{CC} or GND pins .... ±50 mA
Lead temperature 1,6 mm (1/16 inch) from case for 60 seconds: FH, FK, or J package .... 300°C
Lead temperature 1,6 mm (1/16 inch) from case for 10 seconds: FN or N package .... 260°C
Storage temperature range .... −65°C to 150°C

† Stresses beyond those listed under "absolute maximum ratings" may cause permanent damage to the device. These are stress ratings only and functional operation of the device at these or any other conditions beyond those indicated under "recommended operating conditions" is not implied. Exposure to absolute-maximum-rated conditions for extended periods may affect device reliability.

recommended operating conditions

			SN54HC'			SN74HC'			UNIT
			MIN	NOM	MAX	MIN	NOM	MAX	
V_{CC}	Supply voltage		2	5	6	2	5	6	V
V_{IH}	High-level input voltage	V_{CC} = 2 V	1.5			1.5			V
		V_{CC} = 4.5 V	3.15			3.15			
		V_{CC} = 6 V	4.2			4.2			
V_{IL}	Low-level input voltage	V_{CC} = 2 V	0		0.3	0		0.3	V
		V_{CC} = 4.5 V	0		0.9	0		0.9	
		V_{CC} = 6 V	0		1.2	0		1.2	
V_I	Input voltage		0		V_{CC}	0		V_{CC}	V
V_O	Output voltage		0		V_{CC}	0		V_{CC}	V
t_t	Input transition (rise and fall) times (except Schmitt-trigger inputs)	V_{CC} = 2 V	0		1000	0		1000	ns
		V_{CC} = 4.5 V	0		500	0		500	
		V_{CC} = 6 V	0		400	0		400	
T_A	Operating free-air temperature		−55		125	−40		85	°C

electrical characteristics over recommended operating free-air temperature range (unless otherwise noted)

PARAMETER	TEST CONDITIONS	V_{CC}	T_A = 25°C MIN	TYP	MAX	SN54HC' MIN	MAX	SN74HC' MIN	MAX	UNIT
V_{OH} (Totem-pole outputs)	$V_I = V_{IH}$ or V_{IL}, $I_{OH} = -20\ \mu A$	2 V	1.9	1.998		1.9		1.9		V
		4.5 V	4.4	4.499		4.4		4.4		
		6 V	5.9	5.999		5.9		5.9		
	$V_I = V_{IH}$ or V_{IL}, $I_{OH} = -4$ mA	4.5 V	3.98	4.30		3.7		3.84		
	$V_I = V_{IH}$ or V_{IL}, $I_{OH} = -5.2$ mA	6 V	5.48	5.80		5.2		5.34		
I_{OH} (Open-drain outputs)	$V_I = V_{IH}$ or V_{IL}, $V_O = V_{CC}$	6 V		0.01	0.5		10		5	μA
V_{OL}	$V_I = V_{IH}$ or V_{IL}, $I_{OL} = 20\ \mu A$	2 V		0.002	0.1		0.1		0.1	V
		4.5 V		0.001	0.1		0.1		0.1	
		6 V		0.001	0.1		0.1		0.1	
	$V_I = V_{IH}$ or V_{IL}, $I_{OL} = 4$ mA	4.5 V		0.17	0.26		0.4		0.33	
	$V_I = V_{IH}$ or V_{IL}, $I_{OL} = 5.2$ mA	6 V		0.15	0.26		0.4		0.33	
V_{T+}†		2 V	0.8	1.2	1.5					V
		4.5 V	2	2.5	3.15					
		6 V	2.5	3.3	4.2					
V_{T-}†		2 V	0.3	0.6	0.8					V
		4.5 V	0.9	1.6	2					
		6 V	1.2	2	2.5					
$V_{T+} - V_{T-}$†		2 V	0.2	0.6	1					V
		4.5 V	0.4	0.9	1.4					
		6 V	0.5	1.3	1.7					
I_I	V_I = 0 to V_{CC}	6 V		±0.1	±100		±1000		±1000	nA
I_{CC}	$V_I = V_{CC}$ or 0, $I_O = 0$	6 V			2		40		20	μA
C_I		2 to 6 V		3	10		10		10	pF

†This parameter applies only for Schmitt-trigger inputs.

switching characteristics

See individual circuit pages.

TABLE II SPECIFICATIONS FOR HC DUAL AND QUAD FLIP-FLOPS AND LATCHES

D2804, DECEMBER 1982—REVISED MARCH 1984

absolute maximum ratings over operating free-air temperature range†

Supply voltage range, V_{CC} .... −0.5 V to 7 V
Input diode current, I_{IK} ($V_I < 0$ or $V_I > V_{CC}$) .... ±20 mA
Output diode current, I_{OK} ($V_O < 0$ or $V_O > V_{CC}$) .... ±20 mA
Continuous output current, I_O ($V_O = 0$ to V_{CC}) .... ±25 mA
Continuous current through V_{CC} or GND pins .... ±50 mA
Lead temperature 1,6 mm (1/16 inch) from case for 60 seconds: FH, FK, or J package .... 300°C
Lead temperature 1,6 mm (1/16 inch) from case for 10 seconds: FN or N package .... 260°C
Storage temperature range .... −65°C to 150°C

† Stresses beyond those listed under "absolute maximum ratings" may cause permanent damage to the device. These are stress ratings only and functional operation of the device at these or any other conditions beyond those indicated under "recommended operating conditions" is not implied. Exposure to absolute-maximum-rated conditions for extended periods may affect device reliability.

2

RATINGS AND CHARACTERISTICS

recommended operating conditions

			SN54HC'			SN74HC'			UNIT
			MIN	NOM	MAX	MIN	NOM	MAX	
V_{CC}	Supply voltage		2	5	6	2	5	6	V
V_{IH}	High-level input voltage	V_{CC} = 2 V	1.5			1.5			V
		V_{CC} = 4.5 V	3.15			3.15			
		V_{CC} = 6 V	4.2			4.2			
V_{IL}	Low-level input voltage	V_{CC} = 2 V	0		0.3	0		0.3	V
		V_{CC} = 4.5 V	0		0.9	0		0.9	
		V_{CC} = 6 V	0		1.2	0		1.2	
V_I	Input voltage		0		V_{CC}	0		V_{CC}	V
V_O	Output voltage		0		V_{CC}	0		V_{CC}	V
t_t	Input transition (rise and fall) times	V_{CC} = 2 V	0		1000	0		1000	ns
		V_{CC} = 4.5 V	0		500	0		500	
		V_{CC} = 6 V	0		400	0		400	
T_A	Operating free-air temperature		−55		125	−40		85	°C

See individual circuits for additional timing requirements.

electrical characteristics over recommended operating free-air temperature range (unless otherwise noted)

PARAMETER	TEST CONDITIONS	V_{CC}	T_A = 25°C			SN54HC′		SN74HC′		UNIT
			MIN	TYP	MAX	MIN	MAX	MIN	MAX	
V_{OH}	$V_I = V_{IH}$ or V_{IL}, $I_{OH} = -20\ \mu A$	2 V	1.9	1.998		1.9		1.9		V
		4.5 V	4.4	4.499		4.4		4.4		
		6 V	5.9	5.999		5.9		5.9		
	$V_I = V_{IH}$ or V_{IL}, $I_{OH} = -4$ mA	4.5 V	3.98	4.30		3.7		3.84		
	$V_I = V_{IH}$ or V_{IL}, $I_{OH} = -5.2$ mA	6 V	5.48	5.80		5.2		5.34		
V_{OL}	$V_I = V_{IH}$ or V_{IL}, $I_{OL} = 20\ \mu A$	2 V		0.002	0.1		0.1		0.1	V
		4.5 V		0.001	0.1		0.1		0.1	
		6 V		0.001	0.1		0.1		0.1	
	$V_I = V_{IH}$ or V_{IL}, $I_{OL} = 4$ mA	4.5 V		0.17	0.26		0.4		0.33	
	$V_I = V_{IH}$ or V_{IL}, $I_{OL} = 5.2$ mA	6 V		0.15	0.26		0.4		0.33	
I_I	$V_I = 0$ to V_{CC}	6 V		±0.1	±100		±1000		±1000	nA
I_{CC}	$V_I = V_{CC}$ or 0, $I_O = 0$	6 V			4		80		40	μA
C_I		2 to 6 V		3	10		10		10	pF

switching characteristics

See individual circuit pages.

TABLE III
SPECIFICATIONS FOR HC CIRCUITS WITH HIGH-CURRENT OUTPUTS

D2804, DECEMBER 1982—REVISED MARCH 1984

absolute maximum ratings over operating free-air temperature range†

Supply voltage range, V_{CC} . . . . −0.5 V to 7 V
Input diode current, I_{IK} ($V_I < 0$ or $V_I > V_{CC}$) . . . . ±20 mA
Output diode current, I_{OK} ($V_O < 0$ or $V_O > V_{CC}$) . . . . ±20 mA
Continuous output current, I_O ($V_O = 0$ to V_{CC}) . . . . ±35 mA
Continuous current through V_{CC} or GND pins . . . . ±70 mA
Lead temperature 1,6 mm (1/16 inch) from case for 60 seconds: FH, FK, or J package . . . . 300°C
Lead temperature 1,6 mm (1/16 inch) from case for 10 seconds: FN or N package . . . . 260°C
Storage temperature range . . . . −65°C to 150°C

† Stresses beyond those listed under "absolute maximum ratings" may cause permanent damage to the device. These are stress ratings only and functional operation of the device at these or any other conditions beyond those indicated under "recommended operating conditions" is not implied. Exposure to absolute-maximum-rated conditions for extended periods may affect device reliability.

recommended operating conditions

			SN54HC′			SN74HC′			UNIT
			MIN	NOM	MAX	MIN	NOM	MAX	
V_{CC}	Supply voltage		2	5	6	2	5	6	V
V_{IH}	High-level input voltage	$V_{CC} = 2$ V	1.5			1.5			V
		$V_{CC} = 4.5$ V	3.15			3.15			
		$V_{CC} = 6$ V	4.2			4.2			
V_{IL}	Low-level input voltage	$V_{CC} = 2$ V	0		0.3	0		0.3	V
		$V_{CC} = 4.5$ V	0		0.9	0		0.9	
		$V_{CC} = 6$ V	0		1.2	0		1.2	
V_I	Input voltage		0		V_{CC}	0		V_{CC}	V
V_O	Output voltage		0		V_{CC}	0		V_{CC}	V
t_t	Input transition (rise and fall) times (except Schmitt-trigger inputs)	$V_{CC} = 2$ V	0		1000	0		1000	ns
		$V_{CC} = 4.5$ V	0		500	0		500	
		$V_{CC} = 6$ V	0		400	0		400	
T_A	Operating free-air temperature		−55		125	−40		85	°C

TEXAS INSTRUMENTS
POST OFFICE BOX 225012 • DALLAS, TEXAS 75265

SPECIFICATIONS FOR HC CIRCUITS WITH HIGH-CURRENT OUTPUTS

electrical characteristics over recommended operating free-air temperature range (unless otherwise noted)

PARAMETER	TEST CONDITIONS	V_{CC}	$T_A = 25°C$			SN54HC′		SN74HC′		UNIT
			MIN	TYP	MAX	MIN	MAX	MIN	MAX	
V_{OH}	$V_I = V_{IH}$ or V_{IL}, $I_{OH} = -20\ \mu A$	2 V	1.9	1.998		1.9		1.9		V
		4.5 V	4.4	4.499		4.4		4.4		
		6 V	5.9	5.999		5.9		5.9		
	$V_I = V_{IH}$ or V_{IL}, I_{OH} = See Notes 1 and 5	4.5 V	3.98	4.30		3.7		3.84		
	$V_I = V_{IH}$ or V_{IL}, I_{OH} = See Notes 2 and 5	6 V	5.48	5.80		5.2		5.34		
V_{OL}	$V_I = V_{IH}$ or V_{IL}, $I_{OL} = 20\ \mu A$	2 V		0.002	0.1		0.1		0.1	V
		4.5 V		0.001	0.1		0.1		0.1	
		6 V		0.001	0.1		0.1		0.1	
	$V_I = V_{IH}$ or V_{IL}, V_{OL} = See Notes 3 and 5	4.5 V		0.17	0.26		0.4		0.33	
	$V_I = V_{IH}$ or V_{IL}, I_{OL} = See Notes 4 and 5	6 V		0.15	0.26		0.4		0.33	
I_I	V_I = 0 to V_{CC}	6 V		±0.1	±100		±1000		±1000	nA
I_{OZ} †	$V_O = V_{CC}$ or 0, $V_I = V_{IH}$ or V_{IL}	6 V		±0.01	±0.5		±10		±5	μA
I_{CC}	$V_I = V_{CC}$ or 0, $I_O = 0$	6 V			8		160		80	μA
C_I ‡		2 to 6 V		3	10		10		10	pF

†This parameter, I_{OZ}, the high-impedance-state output current, applies only to three-state outputs and transceiver I/O pins.

‡This parameter, C_I, does not apply to transceiver I/O ports.

NOTES:
1. $I_{OH} = -4$ mA for standard outputs and -6 mA for high-current outputs.
2. $I_{OH} = -5.2$ mA for standard outputs and -7.8 mA for high-current outputs.
3. $I_{OL} = 4$ mA for standard outputs and 6 mA for high-current outputs.
4. $I_{OL} = 5.2$ mA for standard outputs and 7.8 mA for high-current outputs.
5. High-current outputs are indicated by the ▷ in the logic symbol. All 3-state outputs (indicated by the ▽ in the logic symbol) are also high-current outputs.

switching characteristics

See individual circuit pages.

D2804, DECEMBER 1982—REVISED MARCH 1984

absolute maximum ratings over operating free-air temperature range†

Supply voltage range, V_{CC} . . . . −0.5 V to 7 V
Input diode current, I_{IK}(V_I < 0 or V_I > V_{CC}) . . . . ±20 mA
Output diode current, I_{OK}(V_O < 0 or V_O > V_{CC}) . . . . ±20 mA
Continuous output current, I_O (V_O = 0 to V_{CC}) . . . . ±25 mA
Continuous current through V_{CC} or GND pins . . . . ±50 mA
Lead temperature 1,6 mm (1/16 inch) from case for 60 seconds: FH, FK, or J package . . . . 300°C
Lead temperature 1,6 mm (1/16 inch) from case for 10 seconds: FN or N package . . . . 260°C
Storage temperature range . . . . −65°C to 150°C

† Stresses beyond those listed under "absolute maximum ratings" may cause permanent damage to the device. These are stress ratings only and functional operation of the device at these or any other conditions beyond those indicated under "recommended operating conditions" is not implied. Exposure to absolute-maximum-rated conditions for extended periods may affect device reliability.

2

RATINGS AND CHARACTERISTICS

recommended operating conditions

			SN54HC′			SN74HC′			UNIT
			MIN	NOM	MAX	MIN	NOM	MAX	
V_{CC}	Supply voltage		2	5	6	2	5	6	V
V_{IH}	High-level input voltage	V_{CC} = 2 V	1.5			1.5			V
		V_{CC} = 4.5 V	3.15			3.15			
		V_{CC} = 6 V	4.2			4.2			
V_{IL}	Low-level input voltage	V_{CC} = 2 V	0		0.3	0		0.3	V
		V_{CC} = 4.5 V	0		0.9	0		0.9	
		V_{CC} = 6 V	0		1.2	0		1.2	
V_I	Input voltage		0		V_{CC}	0		V_{CC}	V
V_O	Output voltage		0		V_{CC}	0		V_{CC}	V
t_t	Input transition (rise and fall) times (except Schmitt-trigger inputs)	V_{CC} = 2 V	0		1000	0		1000	ns
		V_{CC} = 4.5 V	0		500	0		500	
		V_{CC} = 6 V	0		400	0		400	
T_A	Operating free-air temperature		−55		125	−40		85	°C

electrical characteristics over recommended operating free-air temperature range (unless otherwise noted)

PARAMETER	TEST CONDITIONS	V_{CC}	$T_A = 25°C$			SN54HC'		SN74HC'		UNIT
			MIN	TYP	MAX	MIN	MAX	MIN	MAX	
V_{OH}	$V_I = V_{IH}$ or V_{IL}, $I_{OH} = -20\ \mu A$	2 V	1.9	1.998		1.9		1.9		V
		4.5 V	4.4	4.499		4.4		4.4		
		6 V	5.9	5.999		5.9		5.9		
	$V_I = V_{IH}$ or V_{IL}, $I_{OH} = -4$ mA	4.5 V	3.98	4.30		3.7		3.84		
	$V_I = V_{IH}$ or V_{IL}, $I_{OH} = -5.2$ mA	6 V	5.48	5.80		5.2		5.34		
V_{OL}	$V_I = V_{IH}$ or V_{IL}, $I_{OL} = 20\ \mu A$	2 V		0.002	0.1		0.1		0.1	V
		4.5 V		0.001	0.1		0.1		0.1	
		6 V		0.001	0.1		0.1		0.1	
	$V_I = V_{IH}$ or V_{IL}, $I_{OL} = 4$ mA	4.5 V		0.17	0.26		0.4		0.33	
	$V_I = V_{IH}$ or V_{IL}, $I_{OL} = 5.2$ mA	6 V		0.15	0.26		0.4		0.33	
I_I	$V_I = 0$ to V_{CC}	6 V		±0.1	±100		±1000		±1000	nA
$I_{S(off)}$ †	$V_I = V_{CC}$ or 0, $V_S = \pm V_{CC}$	6 V			±0.1		±1		±1	μA
V_{T+} ‡		2 V	0.8	1.2	1.5					V
		4.5 V	2	2.5	3.15					
		6 V	2.5	3.3	4.2					
V_{T-} ‡		2 V	0.3	0.6	0.8					V
		4.5 V	0.9	1.6	2					
		6 V	1.2	2	2.5					
$V_{T+} - V_{T-}$ ‡		2 V	0.2	0.6	1					V
		4.5 V	0.4	0.9	1.4					
		6 V	0.5	1.3	1.7					
I_{CC}	$V_I = V_{CC}$ or 0, $I_O = 0$	6 V			8		160		80	μA
C_I		2 to 6 V		3	10		10		10	pF

† This parameter, $I_{S(off)}$, is for analog switches only.
‡ These threshold parameters apply only to Schmitt-trigger inputs.

switching characteristics

See individual circuit pages.

absolute maximum ratings over operating free-air temperature range†

Supply voltage range, V_{CC} . . . . −0.5 V to 7 V
Input diode current, I_{IK} ($V_I < 0$ or $V_I > V_{CC}$) . . . . ±20 mA
Output diode current, I_{OK} ($V_O < 0$ or $V_O > V_{CC}$) . . . . ±20 mA
Continuous output current, I_O ($V_O = 0$ to V_{CC}) . . . . ±25 mA
Continuous current through V_{CC} or GND pins . . . . ±50 mA
Lead temperature 1,6 mm (1/16 inch) from case for 60 seconds: FH, FK, or J package . . . . 300°C
Lead temperature 1,6 mm (1/16 inch) from case for 10 seconds: FN or N package . . . . 260°C
Storage temperature range . . . . −65°C to 150°C

† Stresses beyond those listed under "absolute maximum ratings" may cause permanent damage to the device. These are stress ratings only and functional operation of the device at these or any other conditions beyond those indicated under "recommended operating conditions" is not implied. Exposure to absolute-maximum-rated conditions for extended periods may affect device reliability.

recommended operating conditions

			SN54HCT′			SN74HCT′			UNIT
			MIN	NOM	MAX	MIN	NOM	MAX	
V_{CC}	Supply voltage		4.5	5	5.5	4.5	5	5.5	V
V_{IH}	High-level input voltage	V_{CC} = 4.5 V to 5.5 V	2			2			V
V_{IL}	Low-level input voltage	V_{CC} = 4.5 V to 5.5 V	0		0.8	0		0.8	V
V_I	Input voltage		0		V_{CC}	0		V_{CC}	V
V_O	Output voltage		0		V_{CC}	0		V_{CC}	V
t_t	Input transition (rise and fall) times		0		500	0		500	ns
T_A	Operating free-air temperature		−55		125	−40		85	°C

electrical characteristics over recommended operating free-air temperature range (unless otherwise noted)

PARAMETER	TEST CONDITIONS	V_{CC}	T_A = 25°C			SN54HCT′		SN74HCT′		UNIT
			MIN	TYP	MAX	MIN	MAX	MIN	MAX	
V_{OH}	$V_I = V_{IH}$ or V_{IL}, $I_{OH} = -20\ \mu A$	4.5 V	4.4	4.499		4.4		4.4		V
	$V_I = V_{IL}$, $I_{OH} = -4$ mA	4.5 V	3.98	4.30		3.7		3.84		
V_{OL}	V_I or V_{IL}, $I_{OL} = 20\ \mu A$	4.5 V		0.001	0.1		0.1		0.1	V
	$V_I = V_{IH}$ or V_{IL}, $I_{OL} = 4$ mA	4.5 V		0.17	0.26		0.4		0.33	
I_I	$V_I = 0$ to V_{CC}	5.5 V		±0.1	±100		±1000		±1000	nA
I_{CC}	$V_I = V_{CC}$ or 0, $I_O = 0$	5.5 V			2		40		20	µA
ΔI_{CC}‡	One input at 0.5 V or 2.4 V, Other inputs at 0 V or V_{CC}	5.5 V		1.4	2.4		2.9		3	mA
C_I		4.5 to 5.5 V		3	10		10		10	pF

‡This is the increase in supply current for each input that is at one of the specified TTL voltage levels rather than 0 V or V_{CC}.

switching characteristics

See individual circuit pages.

TABLE VI
SPECIFICATIONS FOR HCT DUAL AND QUAD FLIP-FLOPS AND LATCHES

D2804, MARCH 1984

absolute maximum ratings over operating free-air temperature range†

Supply voltage range, V_{CC} .... −0.5 V to 7 V
Input diode current, I_{IK} ($V_I < 0$ or $V_I > V_{CC}$) .... ±20 mA
Output diode current, I_{OK} ($V_O < 0$ or $V_O > V_{CC}$) .... ±20 mA
Continuous output current, I_O ($V_O = 0$ to V_{CC}) .... ±25 mA
Continuous current through V_{CC} or GND pins .... ±50 mA
Lead temperature 1,6 mm (1/16 inch) from case for 60 seconds: FH, FK, or J package .... 300°C
Lead temperature 1,6 mm (1/16 inch) from case for 10 seconds: FN or N package .... 260°C
Storage temperature range .... −65°C to 150°C

† Stresses beyond those listed under "absolute maximum ratings" may cause permanent damage to the device. These are stress ratings only and functional operation of the device at these or any other conditions beyond those indicated under "recommended operating conditions" is not implied. Exposure to absolute-maximum-rated conditions for extended periods may affect device reliability.

recommended operating conditions

			SN54HCT'			SN74HCT'			UNIT
			MIN	NOM	MAX	MIN	NOM	MAX	
V_{CC}	Supply voltage		4.5	5	5.5	4.5	5	5.5	V
V_{IH}	High-level input voltage	V_{CC} = 4.5 V to 5.5 V	2			2			V
V_{IL}	Low-level input voltage	V_{CC} = 4.5 V to 5.5 V	0		0.8	0		0.8	V
V_I	Input voltage		0		V_{CC}	0		V_{CC}	V
V_O	Output voltage		0		V_{CC}	0		V_{CC}	V
t_t	Input transition (rise and fall) times		0		500	0		500	ns
T_A	Operating free-air temperature		−55		125	−40		85	°C

electrical characteristics over recommended operating free-air temperature range (unless otherwise noted)

PARAMETER	TEST CONDITIONS	V_{CC}	T_A = 25°C			SN54HCT'		SN74HCT'		UNIT
			MIN	TYP	MAX	MIN	MAX	MIN	MAX	
V_{OH}	$V_I = V_{IH}$ or V_{IL}, $I_{OH} = -20\ \mu A$	4.5 V	4.4	4.499		4.4		4.4		V
	$V_I = V_{IH}$ or V_{IL}, $I_{OH} = -4$ mA	4.5 V	3.98	4.30		3.7		3.84		
V_{OL}	$V_I = V_{IH}$ or V_{IL}, $I_{OH} = 20\ \mu A$	4.5 V		0.001	0.1		0.1		0.1	V
	$V_I = V_{IH}$ or V_{IL}, $I_{OL} = 4$ mA	4.5 V		0.17	0.26		0.4		0.33	
I_I	$V_I = 0$ to V_{CC}	5.5 V		±0.1	±100		±1000		±1000	nA
I_{CC}	$V_I = V_{CC}$ or 0, $I_O = 0$	5.5 V			4		80		40	μA
ΔI_{CC}‡	One input at 0.5 V or 2.4 V, Other inputs at 0 V or V_{CC}	5.5 V		1.4	2.4		2.9		3	mA
C_I		4.5 to 5.5 V		3	10		10		10	pF

‡This is the increase in supply current for each input that is at one of the specified TTL voltage levels rather than 0 V or V_{CC}.

switching characteristics

See individual circuit pages.

TABLE VII SPECIFICATIONS FOR HCT DEVICES WITH HIGH-CURRENT OUTPUTS

D2804, MARCH 1984

absolute maximum ratings over operating free-air temperature range†

Supply voltage range, V_{CC} . . . . −0.5 V to 7 V
Input diode current, I_{IK} ($V_I < 0$ or $V_I > V_{CC}$) . . . . ±20 mA
Output diode current, I_{OK} ($V_O < 0$ or $V_O > V_{CC}$) . . . . ±20 mA
Continuous output current, I_O ($V_O = 0$ to V_{CC}) . . . . ±35 mA
Continuous current through V_{CC} or GND pins . . . . ±70 mA
Lead temperature 1,6 mm (1/16 inch) from case for 60 seconds: FH, FK, or J package . . . . 300°C
Lead temperature 1,6 mm (1/16 inch) from case for 10 seconds: FN or N package . . . . 260°C
Storage temperature range . . . . −65°C to 150°C

† Stresses beyond those listed under "absolute maximum ratings" may cause permanent damage to the device. These are stress ratings only and functional operation of the device at these or any other conditions beyond those indicated under "recommended operating conditions" is not implied. Exposure to absolute-maximum-rated conditions for extended periods may affect device reliability.

recommended operating conditions

			SN54HCT'			SN74HCT'			UNIT
			MIN	NOM	MAX	MIN	NOM	MAX	
V_{CC}	Supply voltage		4.5	5	5.5	4.5	5	5.5	V
V_{IH}	High-level input voltage	V_{CC} = 4.5 V to 5.5 V	2			2			V
V_{IL}	Low-level input voltage	V_{CC} = 4.5 V to 5.5 V	0		0.8	0		0.8	V
V_I	Input voltage		0		V_{CC}	0		V_{CC}	V
V_O	Output voltage		0		V_{CC}	0		V_{CC}	V
t_t	Input transition (rise and fall) times		0		500	0		500	ns
T_A	Operating free-air temperature		−55		125	−40		85	°C

electrical characteristics over recommended operating free-air temperature range (unless otherwise noted)

PARAMETER	TEST CONDITIONS	V_{CC}	T_A = 25°C			SN54HCT'		SN74HCT'		UNIT
			MIN	TYP	MAX	MIN	MAX	MIN	MAX	
V_{OH}	$V_I = V_{IH}$ or V_{IL}, $I_{OH} = -20\ \mu A$	4.5 V	4.4	4.499		4.4		4.4		V
	$V_I = V_{IH}$ or V_{IL}, I_{OH} = See Notes 1 and 3	4.5 V	3.98	4.30		3.7		3.84		
V_{OL}	$V_I = V_{IH}$ or V_{IL}, $I_{OH} = 20\ \mu A$	4.5 V		0.001	0.1		0.1		0.1	V
	$V_I = V_{IH}$ or V_{IL}, I_{OL} = See Notes 2 and 3	4.5 V		0.17	0.26		0.4		0.33	
I_I	V_I = 0 to V_{CC}	5.5 V		±0.1	±100		±1000		±1000	nA
I_{OZ}‡	$V_O = V_{CC}$ or 0, $V_I = V_{IH}$ or V_{IL}	5.5 V		±0.01	±0.5		±10		±5	μA
I_{CC}	$V_I = V_{CC}$ or 0, $I_O = 0$	5.5 V			8		160		80	μA
ΔI_{CC}§	One input at 0.5 V or 2.4 V Other inputs at 0 V or V_{CC}	5.5 V		1.4	2.4		2.9		3	mA
C_I¶		4.5 to 5.5 V		3	10		10		10	pF

‡This parameter, I_{OZ}, the high impedance-state output current, applies only for three-state outputs and transceiver I/O pins.
§This is the increase in supply current for each input that is at one of the specified TTL voltage levels rather than 0 V or V_{CC}.
¶This parameter, C_I, does not apply to transceiver I/O ports.

NOTES: 1. I_{OH} = −4 mA for standard outputs and −6 mA for high-current outputs.
2. I_{OL} = 4 mA for standard outputs and 6 mA for high-current outputs.
3. High-current outputs are indicated by the ▷ in the logic symbol. All 3-state outputs (indicated by the ▽ in the logic symbol) are also high-current outputs.

switching characteristics

See individual circuit pages.

D2804, MARCH 1984

absolute maximum ratings over operating free-air temperature range†

Supply voltage range, V_{CC} .... −0.5 V to 7 V
Input diode current, I_{IK} ($V_I < 0$ or $V_I > V_{CC}$) .... ±20 mA
Output diode current, I_{OK} ($V_O < 0$ or $V_O > V_{CC}$) .... ±20 mA
Continuous output current, I_O ($V_O = 0$ to V_{CC}) .... ±25 mA
Continuous current through V_{CC} or GND pins .... ±50 mA
Lead temperature 1,6 mm (1/16 inch) from case for 60 seconds: FH, FK, or J package .... 300°C
Lead temperature 1,6 mm (1/16 inch) from case for 10 seconds: FN or N package .... 260°C
Storage temperature range .... −65°C to 150°C

† Stresses beyond those listed under "absolute maximum ratings" may cause permanent damage to the device. These are stress ratings only and functional operation of the device at these or any other conditions beyond those indicated under "recommended operating conditions" is not implied. Exposure to absolute-maximum-rated conditions for extended periods may affect device reliability.

recommended operating conditions

			SN54HCT'			SN74HCT'			UNIT
			MIN	NOM	MAX	MIN	NOM	MAX	
V_{CC}	Supply voltage		4.5	5	5.5	4.5	5	5.5	V
V_{IH}	High-level input voltage	V_{CC} = 4.5 V to 5.5 V	2			2			V
V_{IL}	Low-level input voltage	V_{CC} = 4.5 V to 5.5 V	0		0.8	0		0.8	V
V_I	Input voltage		0		V_{CC}	0		V_{CC}	V
V_O	Output voltage		0		V_{CC}	0		V_{CC}	V
t_t	Input transition (rise and fall) times		0		500	0		500	ns
T_A	Operating free-air temperature		−55		125	−40		85	°C

electrical characteristics over recommended operating free-air temperature range (unless otherwise noted)

PARAMETER	TEST CONDITIONS	V_{CC}	T_A = 25°C MIN	TYP	MAX	SN54HCT' MIN	MAX	SN74HCT' MIN	MAX	UNIT
V_{OH}	$V_I = V_{IH}$ or V_{IL}, $I_{OH} = -20\ \mu A$	4.5 V	4.4	4.499		4.4		4.4		V
	$V_I = V_{IH}$ or V_{IL}, $I_{OH} = -4$ mA	4.5 V	3.98	4.30		3.7		3.84		
V_{OL}	$V_I = V_{IH}$ or V_{IL}, $I_{OH} = 20\ \mu A$	4.5 V		0.001	0.1		0.1		0.1	V
	$V_I = V_{IH}$ or V_{IL}, $I_{OL} = 4$ mA	4.5 V		0.17	0.26		0.4		0.33	
I_I	$V_I = 0$ to V_{CC}	5.5 V		±0.1	±100		±1000		±1000	nA
$I_{S(off)}$‡	$V_I = V_{IH}$ or V_{IL}, $V_S = \pm V_{CC}$	5.5 V			±0.1		±1		±1	μA
I_{CC}	$V_I = V_{CC}$ or 0, $I_O = 0$	5.5 V			8		160		80	μA
ΔI_{CC}§	One input at 0.5 V or 2.4 V, Other inputs at 0 V or V_{CC}	5.5 V		1.4	2.4		2.9		3	mA
C_I		4.5 to 5.5 V		3	10		10		10	pF

‡This parameter, $I_{S(off)}$, is for analog switches only.
§This is the increase in supply current for each input that is at one of the specified TTL voltage levels rather than 0 V or V_{CC}.

switching characteristics

See individual circuit pages.

TABLE IX
SPECIFICATIONS FOR HCU CIRCUITS

D2804, MARCH 1984

absolute maximum ratings over operating free-air temperature range†

Supply voltage range, V_{CC} . . . . −0.5 V to 7 V
Input diode current, I_{IK} ($V_I < 0$ or $V_I > V_{CC}$) . . . . ±20 mA
Output diode current, I_{OK} ($V_O < 0$ or $V_O > V_{CC}$) . . . . ±20 mA
Continuous output current, I_O ($V_O = 0$ to V_{CC}) . . . . ±25 mA
Continuous current through V_{CC} or GND pins . . . . ±50 mA
Lead temperature 1,6 mm (1/16 inch) from case for 60 seconds: FH, FK, or J package . . . . 300°C
Lead temperature 1,6 mm (1/16 inch) from case for 10 seconds: FN or N package . . . . 260°C
Storage temperature range . . . . −65°C to 150°C

† Stresses beyond those listed under "absolute maximum ratings" may cause permanent damage to the device. These are stress ratings only and functional operation of the device at these or any other conditions beyond those indicated under "recommended operating conditions" is not implied. Exposure to absolute-maximum-rated conditions for extended periods may affect device reliability.

recommended operating conditions

			SN54HCU'			SN74HCU'			UNIT
			MIN	NOM	MAX	MIN	NOM	MAX	
V_{CC}	Supply voltage		2	5	6	2	5	6	V
V_{IH}	High-level input voltage	V_{CC} = 2 V	1.7			1.7			V
		V_{CC} = 4.5 V	3.6			3.6			
		V_{CC} = 6 V	4.8			4.8			
V_{IL}	Low-level input voltage	V_{CC} = 2 V	0		0.3	0		0.3	V
		V_{CC} = 4.5 V	0		0.8	0		0.8	
		V_{CC} = 6 V	0		1.1	0		1.1	
V_I	Input voltage		0		V_{CC}	0		V_{CC}	V
V_O	Output voltage		0		V_{CC}	0		V_{CC}	V
t_t	Input transition (rise and fall) times	V_{CC} = 2 V	0		1000	0		1000	ns
		V_{CC} = 4.5 V	0		500	0		500	
		V_{CC} = 6 V	0		400	0		400	
T_A	Operating free-air temperature		−55		125	−40		85	°C

electrical characteristics over recommended operating free-air temperature range (unless otherwise noted)

PARAMETER	TEST CONDITIONS	V_{CC}	T_A = 25°C			SN54HCU'		SN74HCU'		UNIT
			MIN	TYP	MAX	MIN	MAX	MIN	MAX	
V_{OH}	$V_I = V_{IH}$ or V_{IL}, $I_{OH} = -20\ \mu A$	2 V	1.8			1.8		1.8		V
		4.5 V	4			4		4		
		6 V	5.5			5.5		5.5		
	$V_I = V_{IH}$ or V_{IL}, $I_{OH} = -4$ mA	4.5 V	3.98			3.7		3.84		
	$V_I = V_{IH}$ or V_{IL}, $I_{OH} = -5.2$ mA	6 V	5.48			5.2		5.34		
V_{OL}	$V_I = V_{IH}$ or V_{IL}, $I_{OL} = 20\ \mu A$	2 V			0.2		0.2		0.2	V
		4.5 V			0.5		0.5		0.5	
		6 V			0.5		0.5		0.5	
	$V_I = V_{IH}$ or V_{IL}, $I_{OL} = 4$ mA	4.5 V			0.26		0.4		0.33	
	$V_I = V_{IH}$ or V_{IL}, $I_{OL} = 5.2$ mA	6 V			0.26		0.4		0.33	
I_I	$V_I = 0$ to V_{CC}	6 V			±100		±1000		±1000	nA
I_{CC}	$V_I = V_{CC}$ or 0, $I_O = 0$	6 V			2		40		20	μA
C_I		2 to 6 V		3	10		10		10	pF

switching characteristics

See individual circuit pages.

ATTENTION

These devices contain circuits to protect the inputs and outputs against damage due to high static voltages or electrostatic fields, however, it is advised that precautions be taken to avoid application of any voltage higher than maximum-rated voltages to these high-impedance circuits.

Unused inputs must always be connected to an appropriate logic voltage level, preferably either V_{CC} or ground.

TYPES SN54HC00, SN74HC00 QUADRUPLE 2-INPUT POSITIVE-NAND GATES

D2684, DECEMBER 1982—REVISED MARCH 1984

- **Package Options Include Both Plastic and Ceramic Chip Carriers in Addition to Plastic and Ceramic DIPs**
- **Dependable Texas Instruments Quality and Reliability**

SN54HC00 . . . J PACKAGE
SN74HC00 . . . J OR N PACKAGE
(TOP VIEW)

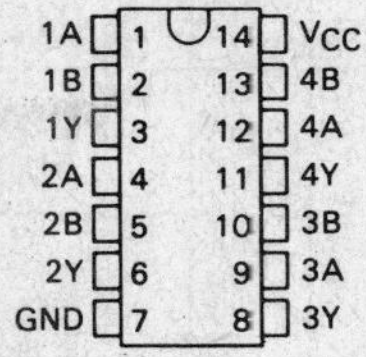

SN54HC00 . . . FH OR FK PACKAGE
SN74HC00 . . . FH OR FN PACKAGE
(TOP VIEW)

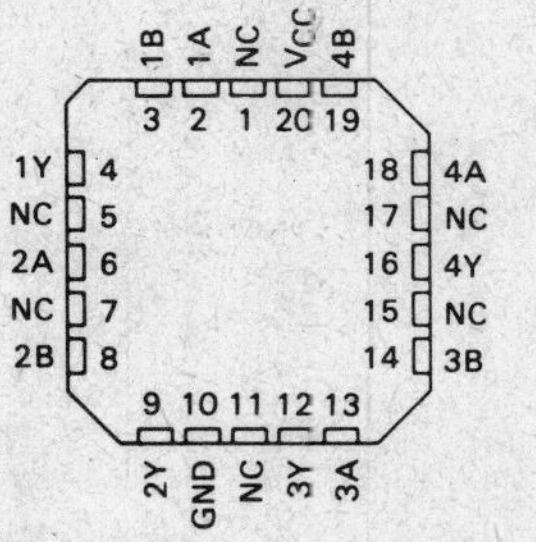

NC—No internal connection

description

These devices contain four independent 2-input NAND gates. They perform the Boolean functions $Y = \overline{A \cdot B}$ or $Y = \overline{A} + \overline{B}$ in positive logic.

The SN54HC00 is characterized for operation over the full military temperature range of −55°C to 125°C. The SN74HC00 is characterized for operation from −40°C to 85°C.

FUNCTION TABLE (each gate)

INPUTS		OUTPUT
A	B	Y
H	H	L
L	X	H
X	L	H

logic symbol

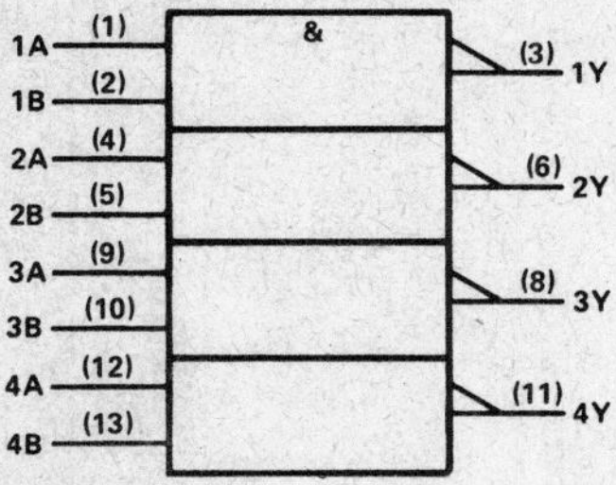

Pin numbers shown are for J and N packages.

maximum ratings, recommended operating conditions, and electrical characteristics

See Table I, page 2-4.

switching characteristics over recommended operating free-air temperature range (unless otherwise noted), C_L = 50 pF (see Note 1)

PARAMETER	FROM (INPUT)	TO (OUTPUT)	V_{CC}	T_A = 25°C			SN54HC00		SN74HC00		UNIT
				MIN	TYP	MAX	MIN	MAX	MIN	MAX	
t_{pd}	A or B	Y	2 V		45	90		135		115	ns
			4.5 V		9	18		27		23	
			6 V		8	15		23		20	
t_t		Y	2 V		38	75		110		95	ns
			4.5 V		8	15		22		19	
			6 V		6	13		19		16	

C_{pd}	Power dissipation capacitance per gate	No load, T_A = 25°C	20 pF typ

NOTE 1: For load circuit and voltage waveforms, see page 1-14.

3

HCMOS DEVICES

TYPES SN54HC02, SN74HC02
QUADRUPLE 2-INPUT POSITIVE-NOR GATES

D2684, DECEMBER 1982—REVISED MARCH 1984

- **Package Options Include Both Plastic and Ceramic Chip Carriers in Addition to Plastic and Ceramic DIPs**
- **Dependable Texas Instruments Quality and Reliability**

SN54HC02 . . . J PACKAGE
SN74HC02 . . . J OR N PACKAGE
(TOP VIEW)

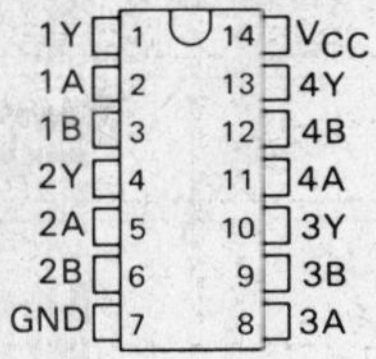

description

These devices contain four independent 2-input NOR gates. They perform the Boolean functions $Y = \overline{A + B}$ or $Y = \overline{A} \cdot \overline{B}$ in positive logic.

The SN54HC02 is characterized for operation over the full military temperature range of −55°C to 125°C. The SN74HC02 is characterized for operation from −40°C to 85°C.

SN54HC02 . . . FH OR FK PACKAGE
SN74HC02 . . . FH OR FN PACKAGE
(TOP VIEW)

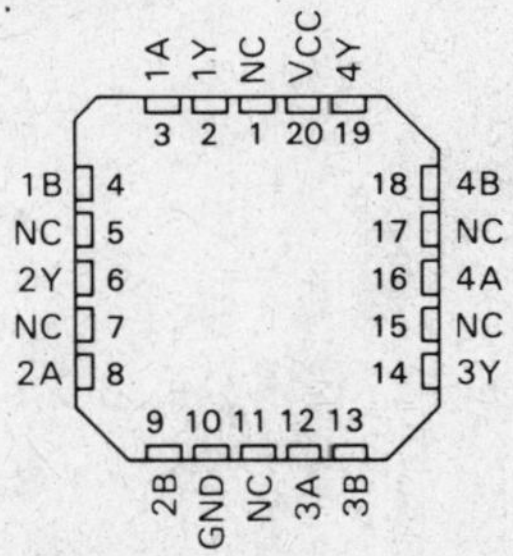

NC — No internal connection

logic symbol

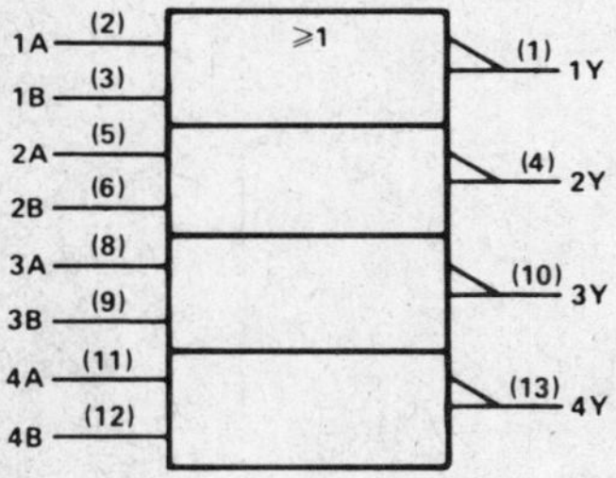

Pin numbers shown are for J and N packages.

FUNCTION TABLE (each gate)

INPUTS		OUTPUT
A	B	Y
H	X	L
X	H	L
L	L	H

maximum ratings, recommended operating conditions, and electrical characteristics

See Table I, page 2-3.

switching characteristics over recommended operating free-air temperature range (unless otherwise noted), C_L = 50 pF (see Note 1)

PARAMETER	FROM (INPUT)	TO (OUTPUT)	V_{CC}	T_A = 25°C MIN	TYP	MAX	SN54HC02 MIN	MAX	SN74HC02 MIN	MAX	UNIT
t_{pd}	A or B	Y	2 V		45	90		135		115	ns
			4.5 V		9	18		27		23	
			6 V		8	15		23		20	
t_t		Y	2 V		38	75		110		95	ns
			4.5 V		8	15		22		19	
			6 V		6	13		19		16	

C_{pd}	Power dissipation capacitance per gate	No load, T_A = 25°C	22 pF typ

NOTE 1: For load circuit and voltage waveforms, see page 1-14.

TYPES SN54HC03, SN74HC03 QUADRUPLE 2-INPUT POSITIVE-NAND GATES WITH OPEN-DRAIN OUTPUTS

D2804, MARCH 1984

- **Package Options Include Both Plastic and Ceramic Chip Carriers in Addition to Plastic and Ceramic DIPs**
- **Dependable Texas Instruments Quality and Reliability**

SN54HC03 . . . J PACKAGE
SN74HC03 . . . J OR N PACKAGE
(TOP VIEW)

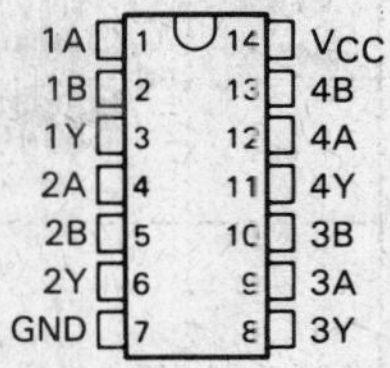

SN54HC03 . . . FH OR FK PACKAGE
SN74HC03 . . . FH OR FN PACKAGE
(TOP VIEW)

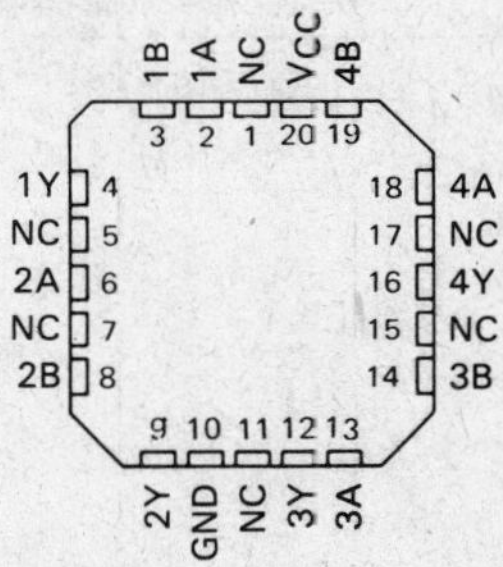

NC—No internal connection

description

These devices contain four independent 2-input NAND gates. They perform the Boolean functions $Y = \overline{A \cdot B}$ or $Y = \overline{A} + \overline{B}$ in positive logic. The open-drain outputs require pull-up resistors to perform correctly. They may be connected to other open-drain outputs to implement active-low wired-OR or active-high wired-AND functions.

The SN54HC03 is characterized for operation over the full military temperature range of −55°C to 125°C. The SN74HC03 is characterized for operation from −40°C to 85°C.

FUNCTION TABLE (each gate)

INPUTS		OUTPUT
A	B	Y
H	H	L
L	X	H
X	L	H

logic symbol

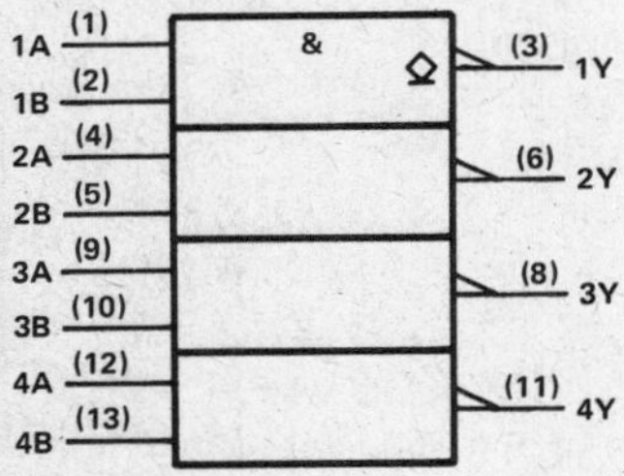

Pin numbers shown are for J and N packages.

maximum ratings, recommended operating conditions, and electrical characteristics

See Table I, page 2-4.

switching characteristics over recommended operating free-air temperature range (unless otherwise noted), R_L = 1 kΩ, C_L = 50 pF (see Note 1)

PARAMETER	FROM (INPUT)	TO (OUTPUT)	V_{CC}	T_A = 25°C MIN	TYP	MAX	SN54HC03 MIN	MAX	SN74HC03 MIN	MAX	UNIT
t_{PLH}	A or B	Y	2 V		60	105		155		131	ns
			4.5 V		13	25		36		31	
			6 V		10	23		31		27	
t_{PHL}			2 V		50	100		150		125	
			4.5 V		10	20		30		25	
			6 V		8	17		25		21	
t_t		Y	2 V		38	75		110		95	ns
			4.5 V		8	15		22		19	
			6 V		6	13		19		16	

C_{pd}	Power dissipation capacitance per gate	No load, T_A = 25°C	20 pF typ

NOTE 1: For load circuit and voltage waveforms, see page 1-14.

TEXAS INSTRUMENTS
POST OFFICE BOX 225012 • DALLAS, TEXAS 75265

D2684, DECEMBER 1982 — REVISED MARCH 1984

- Package Options Include Both Plastic and Ceramic Chip Carriers in Addition to Plastic and Ceramic DIPs
- Dependable Texas Instruments Quality and Reliability

SN54HC04 . . . J PACKAGE
SN74HC04 . . . J OR N PACKAGE
(TOP VIEW)

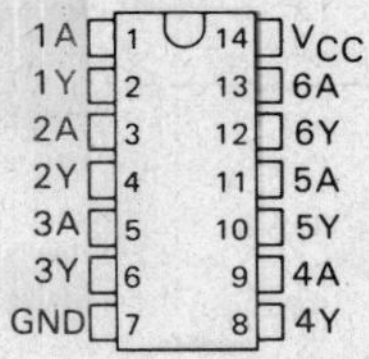

SN54HC04 . . . FH OR FK PACKAGE
SN74HC04 . . . FH OR FN PACKAGE
(TOP VIEW)

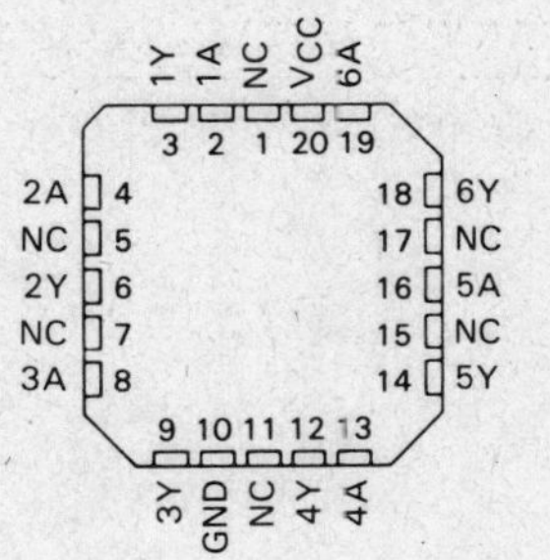

NC — No internal connection

description

These devices contain six independent inverters. They perform the Boolean function $Y = \overline{A}$.

The SN54HC04 is characterized for operation over the full military temperature range of −55°C to 125°C. The SN74HC04 is characterized for operation from −40°C to 85°C.

FUNCTION TABLE
(each inverter)

INPUT A	OUTPUT Y
H	L
L	H

logic symbol

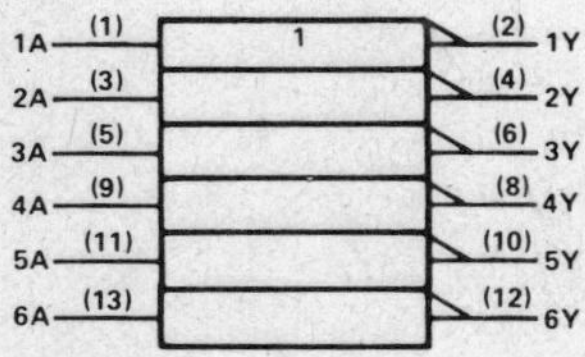

Pin numbers shown are for J and N packages.

maximum ratings, recommended operating conditions, and electrical characteristics

See Table I, page 2-4.

switching characteristics over recommended operating free-air temperature range (unless otherwise noted), C_L = 50 pF (see Note 1)

PARAMETER	FROM (INPUT)	TO (OUTPUT)	V_{CC}	T_A = 25°C MIN	TYP	MAX	SN54HC04 MIN	MAX	SN74HC04 MIN	MAX	UNIT
t_{pd}	A	Y	2 V		45	95		145		120	ns
			4.5 V		9	19		29		24	
			6 V		8	16		25		20	
t_t		Y	2 V		38	75		110		95	ns
			4.5 V		8	15		22		19	
			6 V		6	13		19		16	

C_{pd}	Power dissipation capacitance per inverter	No load, T_A = 25°C	20 pF typ

NOTE 1: For load circuit and voltage waveforms, see page 1-14.

D2804, MARCH 1984

- Package Options Include Both Plastic and Ceramic Chip Carriers in Addition to Plastic and Ceramic DIPs
- Unbuffered Outputs
- Dependable Texas Instruments Quality and Reliability

SN54HCU04 . . . J PACKAGE
SN74HCU04 . . . J OR N PACKAGE
(TOP VIEW)

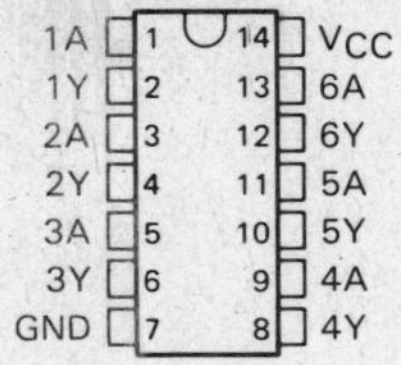

description

These devices contain six independent inverters. They perform the Boolean function $Y = \overline{A}$.

The SN54HCU04 is characterized for operation over the full military temperature range of −55°C to 125°C. The SN74HCU04 is characterized for operation from −40°C to 85°C.

SN54HCU04 . . . FH OR FK PACKAGE
SN74HCU04 . . . FH OR FN PACKAGE
(TOP VIEW)

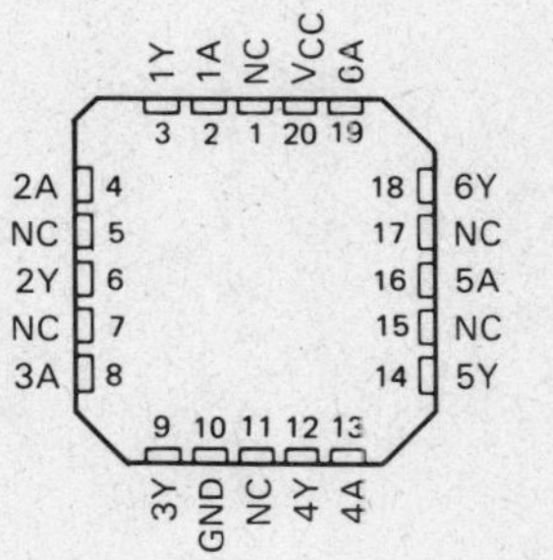

NC – No internal connection

FUNCTION TABLE
(each inverter)

INPUT A	OUTPUT Y
H	L
L	H

logic symbol

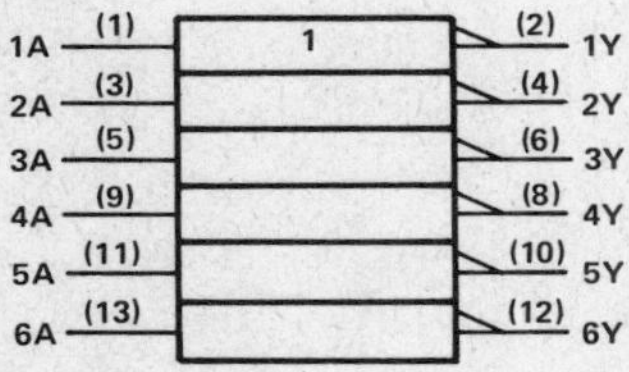

Pin numbers shown are for J and N packages.

maximum ratings, recommended operating conditions, and electrical characteristics

See Table IX, page 2-16.

switching characteristics over recommended operating free-air temperature range (unless otherwise noted), C_L = 50 pF (see Note 1)

PARAMETER	FROM (INPUT)	TO (OUTPUT)	V_{CC}	T_A = 25°C MIN	T_A = 25°C TYP	T_A = 25°C MAX	SN54HCU04 MIN	SN54HCU04 MAX	SN74HCU04 MIN	SN74HCU04 MAX	UNIT
t_{pd}	A	Y	2 V		40	80		120		100	ns
			4.5 V		8	16		24		20	
			6 V		7	14		20		17	
t_t		Y	2 V		38	75		110		95	ns
			4.5 V		8	15		22		19	
			6 V		6	13		19		16	

C_{pd}	Power dissipation capacitance per inverter	No load, T_A = 25°C	20 pF typ

NOTE 1: For load circuit and voltage waveforms, see page 1-14.

D2804, MARCH 1984

- Package Options Include Both Plastic and Ceramic Chip Carriers in Addition to Plastic and Ceramic DIPs
- Dependable Texas Instruments Quality and Reliability

SN54HC05 . . . J PACKAGE
SN74HC05 . . . J OR N PACKAGE
(TOP VIEW)

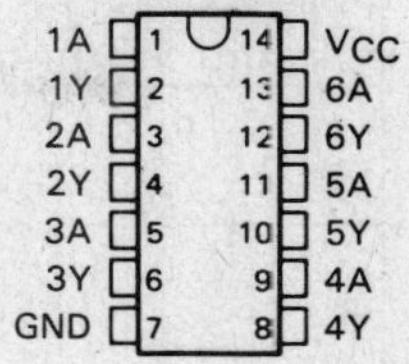

SN54HC05 . . . FH OR FK PACKAGE
SN74HC05 . . . FH OR FN PACKAGE
(TOP VIEW)

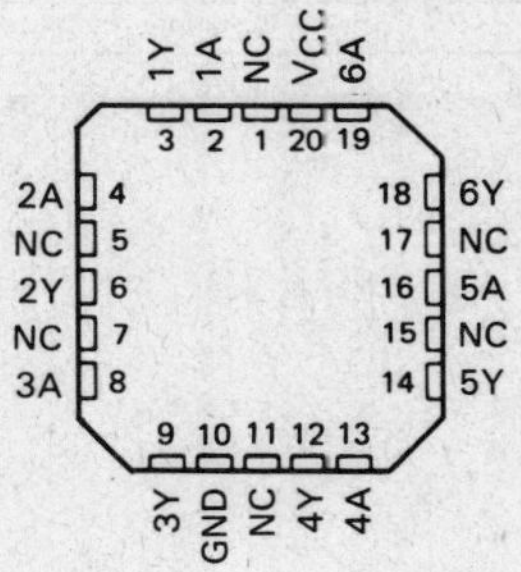

NC—No internal connection

description

These devices contain six independent inverters. They perform the Boolean function $Y = \bar{A}$. The open-drain outputs require pull-up resistors to perform correctly. They may be connected to other open-drain outputs to implement active-low wired-OR or active-high wired-AND functions.

The SN54HC05 is characterized for operation over the full military temperature range of −55°C to 125°C. The SN74HC05 is characterized for operation from −40°C to 85°C.

FUNCTION TABLE (each inverter)

INPUT A	OUTPUT Y
H	L
L	H

logic symbol

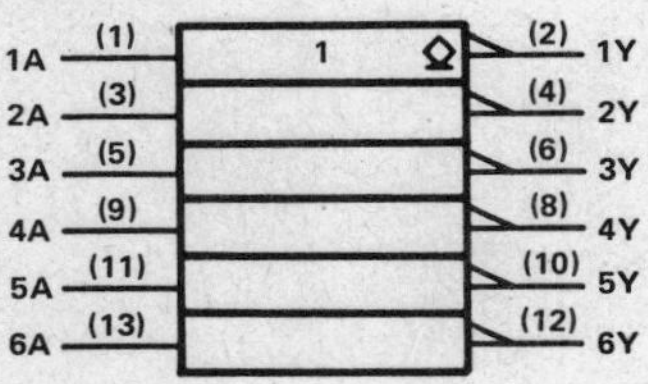

Pin numbers shown are for J and N packages.

maximum ratings, recommended operating conditions, and electrical characteristics

See Table I, page 2-4.

switching characteristics over recommended operating free-air temperature range (unless otherwise noted), C_L = 50 pF (see Note 1)

PARAMETER	FROM (INPUT)	TO (OUTPUT)	V_{CC}	T_A = 25°C			SN54HC05		SN74HC05		UNIT
				MIN	TYP	MAX	MIN	MAX	MIN	MAX	
t_{PLH}	A	Y	2 V		60	115		175		145	ns
			4.5 V		13	23		35		29	
			6 V		10	20		30		25	
t_{PHL}	A	Y	2 V		45	85		130		105	ns
			4.5 V		9	17		26		21	
			6 V		8	14		22		18	
t_f		Y	2 V		38	75		110		95	ns
			4.5 V		8	15		22		19	
			6 V		6	13		19		16	

C_{pd}	Power dissipation capacitance per inverter	No load, T_A = 25°C	20 pF typ

NOTE 1: For load circuit and voltage waveforms, see page 1-14.

3

HCMOS DEVICES

TYPES SN54HC08, SN74HC08 QUADRUPLE 2-INPUT POSITIVE-AND GATES

D2684, DECEMBER 1982—REVISED MARCH 1984

- **Package Options Include Both Plastic and Ceramic Chip Carriers in Addition to Plastic and Ceramic DIPs**
- **Dependable Texas Instruments Quality and Reliability**

SN54HC08 . . . J PACKAGE
SN74HC08 . . . J OR N PACKAGE
(TOP VIEW)

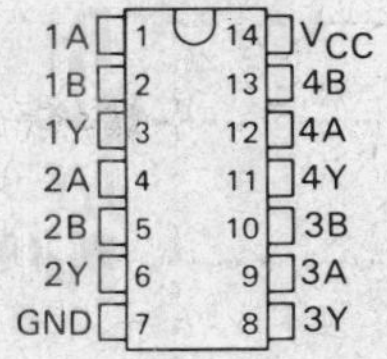

description

These devices contain four independent 2-input AND gates. They perform the Boolean functions $Y = A \cdot B$ or $Y = \overline{\overline{A} + \overline{B}}$ in positive logic.

The SN54HC08 is characterized for operation over the full military temperature range of −55 °C to 125 °C. The SN74HC08 is characterized for operation from −40 °C to 85 °C.

SN54HC08 . . . FH OR FK PACKAGE
SN74HC08 . . . FH OR FN PACKAGE
(TOP VIEW)

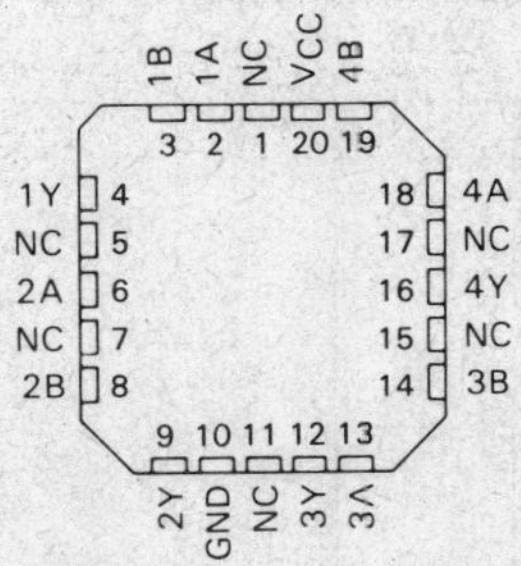

NC—No internal connection

logic symbol

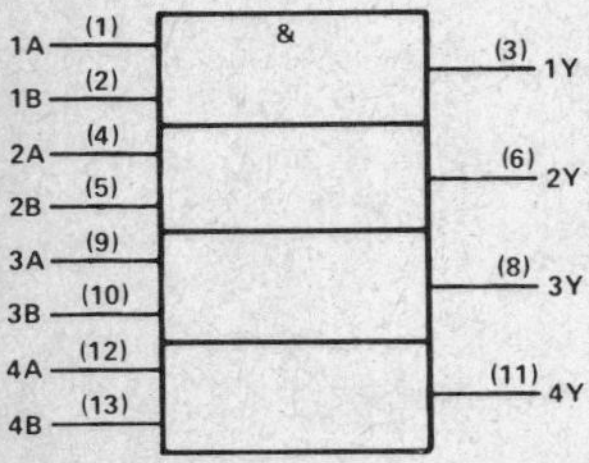

Pin numbers shown are for J and N packages.

FUNCTION TABLE
(each gate)

INPUTS		OUTPUT
A	B	Y
H	H	H
L	X	L
X	L	L

maximum ratings, recommended operating conditions, and electrical characteristics

See Table I, page 2-4.

switching characteristics over recommended operating free-air temperature range (unless otherwise noted), C_L = 50 pF (see Note 1)

PARAMETER	FROM (INPUT)	TO (OUTPUT)	V_{CC}	T_A = 25 °C			SN54HC08		SN74HC08		UNIT
				MIN	TYP	MAX	MIN	MAX	MIN	MAX	
t_{pd}	A or B	Y	2 V		50	100		150		125	ns
			4.5 V		10	20		30		25	
			6 V		8	17		25		21	
t_t		Y	2 V		38	75		110		95	ns
			4.5 V		8	15		22		19	
			6 V		6	13		19		16	

C_{pd}	Power dissipation capacitance per gate	No load, T_A = 25 °C	20 pF typ

NOTE 1: For load circuit and voltage waveforms, see page 1-14.

TYPES SN54HC09, SN74HC09 QUADRUPLE 2-INPUT POSITIVE-AND GATES WITH OPEN-DRAIN OUTPUTS

D2804, MARCH 1984

- Package Options Include Both Plastic and Ceramic Chip Carriers in Addition to Plastic and Ceramic DIPs
- Dependable Texas Instruments Quality and Reliability

SN54HC09 . . . J PACKAGE
SN74HC09 . . . J OR N PACKAGE
(TOP VIEW)

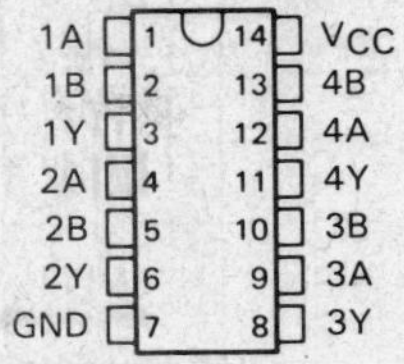

SN54HC09 . . . FH OR FK PACKAGE
SN74HC09 . . . FH OR FN PACKAGE
(TOP VIEW)

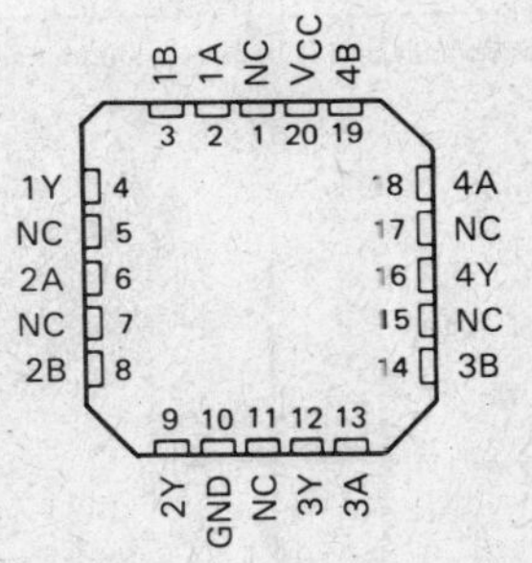

NC—No internal connection

description

These devices contain four independent 2-input AND gates. They perform the Boolean functions $Y = A \cdot B$ or $Y = \overline{\overline{A} + \overline{B}}$ in positive logic. The open-drain outputs require pull-up resistors to perform correctly. They may be connected to other open-drain outputs to implement active-low wired-OR or active-high wired-AND functions.

The SN54HC09 is characterized for operation over the full military temperature range of $-55\,°C$ to $125\,°C$. The SN74HC09 is characterized for operation from $-40\,°C$ to $85\,°C$.

FUNCTION TABLE (each gate)

INPUTS		OUTPUT
A	B	Y
H	H	L
L	X	H
X	L	H

logic symbol

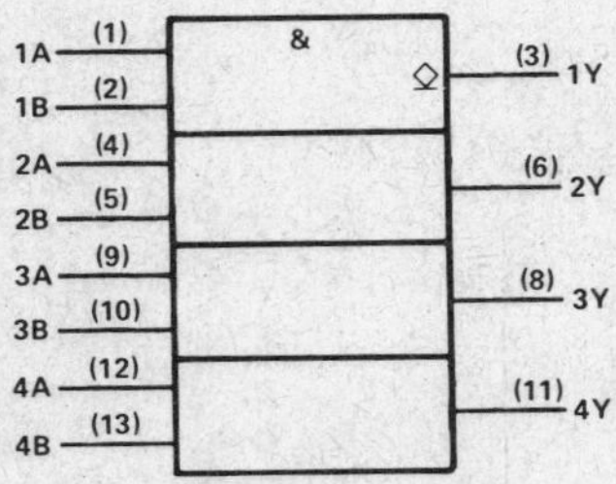

Pin numbers shown are for J and N packages.

maximum ratings, recommended operating conditions, and electrical characteristics

See Table I, page 2-4.

switching characteristics over recommended operating free-air temperature range (unless otherwise noted), R_L = 1 kΩ, C_L = 50 pF (see Note 1)

PARAMETER	FROM (INPUT)	TO (OUTPUT)	V_{CC}	T_A = 25°C MIN	T_A = 25°C TYP	T_A = 25°C MAX	SN54HC09 MIN	SN54HC09 MAX	SN74HC09 MIN	SN74HC09 MAX	UNIT
t_{PLH}	A or B	Y	2 V		60	105		155		131	ns
			4.5 V		13	25		36		31	
			6 V		10	23		31		27	
t_{PHL}	A or B	Y	2 V		50	100		150		125	ns
			4.5 V		10	20		30		25	
			6 V		8	17		25		21	
t_f		Y	2 V		38	75		110		95	ns
			4.5 V		8	15		22		19	
			6 V		6	13		19		16	

C_{pd}	Power dissipation capacitance per gate	No load, T_A = 25°C	20 pF typ

NOTE 1: For load circuit and voltage waveforms, see page 1-14.

TYPES SN54HC10, SN74HC10 TRIPLE 3-INPUT POSITIVE-NAND GATES

D2684, DECEMBER 1982—REVISED MARCH 1984

- **Package Options Include Both Plastic and Ceramic Chip Carriers in Addition to Plastic and Ceramic DIPs**
- **Dependable Texas Instruments Quality and Reliability**

SN54HC10 . . . J PACKAGE
SN74HC10 . . . J OR N PACKAGE
(TOP VIEW)

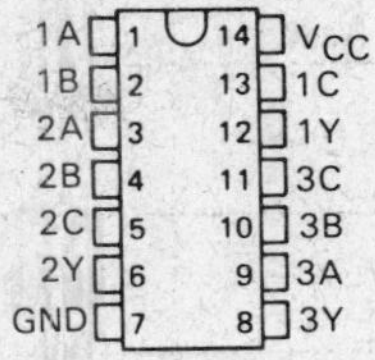

description

These devices contain three independent 3-input NAND gates. They perform the Boolean functions $Y = \overline{A \cdot B \cdot C}$ or $Y = \overline{A} + \overline{B} + \overline{C}$ in positive logic.

The SN54HC10 is characterized for operation over the full military temperature range of −55°C to 125°C. The SN74HC10 is characterized for operation from −40°C to 85°C.

SN54HC10 . . . FH OR FK PACKAGE
SN74HC10 . . . FH OR FN PACKAGE
(TOP VIEW)

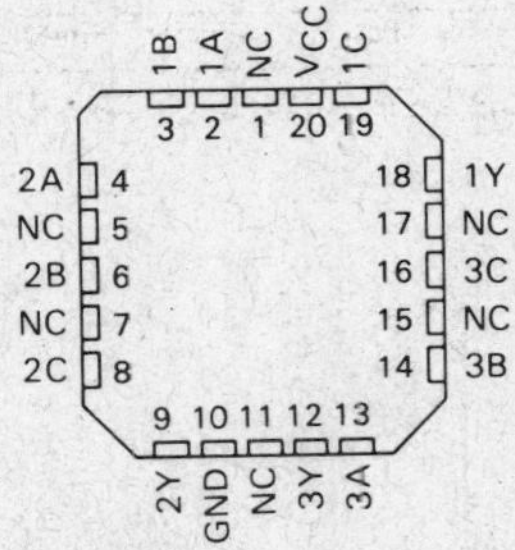

NC—No internal connection

logic symbol

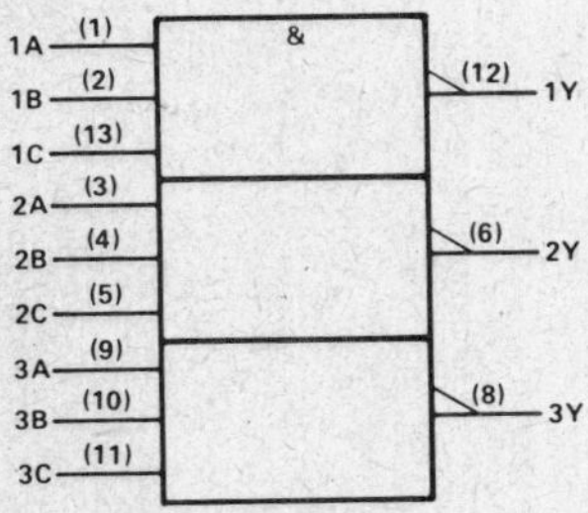

Pin numbers shown are for J and N packages.

FUNCTION TABLE (each gate)

INPUTS			OUTPUT
A	B	C	Y
H	H	H	L
L	X	X	H
X	L	X	H
X	X	L	H

maximum ratings, recommended operating conditions, and electrical characteristics

See Table I, page 2-4.

switching characteristics over recommended operating free-air temperature range (unless otherwise noted), C_L = 50 pF (see Note 1)

PARAMETER	FROM (INPUT)	TO (OUTPUT)	V_{CC}	T_A = 25 °C MIN	TYP	MAX	SN54HC10 MIN	MAX	SN74HC10 MIN	MAX	UNIT
t_{pd}	A, B, or C	Y	2 V		35	95		145		120	ns
			4.5 V		10	19		29		24	
			6 V		9	16		25		20	
t_t		Y	2 V		23	75		110		95	ns
			4.5 V		6	15		22		19	
			6 V		5	13		19		16	

C_{pd}	Power dissipation capacitance per gate	No load, T_A = 25 °C	25 pF typ

NOTE 1: For load circuit and voltage waveforms, see page 1-14.

TYPES SN54HC11, SN74HC11 TRIPLE 3-INPUT POSITIVE-AND GATES

D2684, DECEMBER 1982—REVISED MARCH 1984

- **Package Options Include Both Plastic and Ceramic Chip Carriers in Addition to Plastic and Ceramic DIPs**
- **Dependable Texas Instruments Quality and Reliability**

SN54HC11 . . . J PACKAGE
SN74HC11 . . . J OR N PACKAGE
(TOP VIEW)

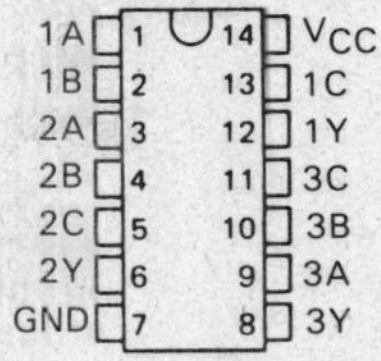

SN54HC11 . . . FH OR FK PACKAGE
SN74HC11 . . . FH OR FN PACKAGE
(TOP VIEW)

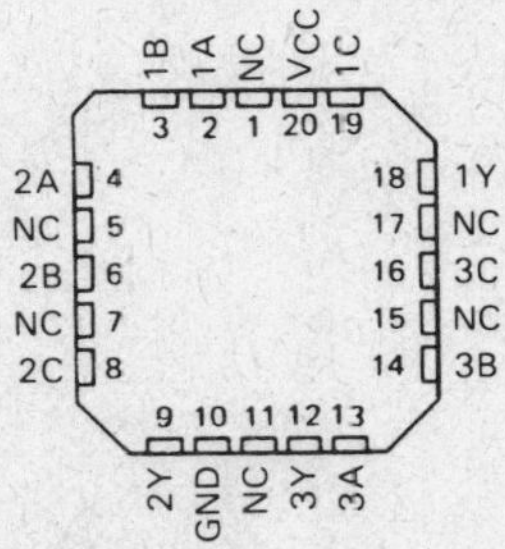

NC — No internal connection

description

These devices contain three independent 3-input AND gates. They perform the Boolean functions $Y = A \cdot B \cdot C$ or $Y = \overline{\overline{A} + \overline{B} + \overline{C}}$ in positive logic.

The SN54HC11 is characterized for operation over the full military temperature range of −55°C to 125°C. The SN74HC11 is characterized for operation from −40°C to 85°C.

logic symbol

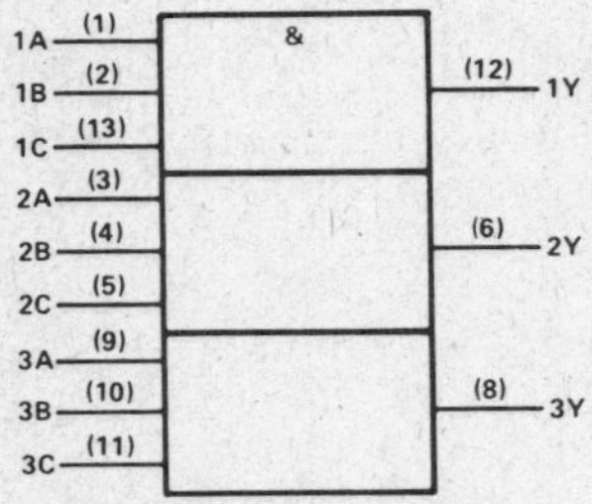

Pin numbers shown are for J and N packages.

FUNCTION TABLE (each gate)

INPUTS			OUTPUT
A	B	C	Y
H	H	H	H
L	X	X	L
X	L	X	L
X	X	L	L

maximum ratings, recommended operating conditions, and electrical characteristics

See Table I, page 2-4.

switching characteristics over recommended operating free-air temperature range (unless otherwise noted), C_L = 50 pF (see Note 1)

PARAMETER	FROM (INPUT)	TO (OUTPUT)	V_{CC}	T_A = 25°C MIN	T_A = 25°C TYP	T_A = 25°C MAX	SN54HC11 MIN	SN54HC11 MAX	SN74HC11 MIN	SN74HC11 MAX	UNIT
t_{pd}	A, B, or C	Y	2 V		35	100		150		125	ns
			4.5 V		10	20		30		25	
			6 V		8	17		25		21	
t_t		Y	2 V		25	75		110		95	ns
			4.5 V		7	15		22		19	
			6 V		5	13		19		16	

C_{pd}	Power dissipation capacitance per gate	No load, T_A = 25°C	25 pF typ

NOTE 1: For load circuit and voltage waveforms, see page 1-14.

TYPES SN54HC14, SN74HC14 HEX SCHMITT-TRIGGER INVERTERS

D2684, DECEMBER 1982—REVISED MARCH 1984

- Package Options Include Both Plastic and Ceramic Chip Carriers in Addition to Plastic and Ceramic DIPs
- Dependable Texas Instruments Quality and Reliability

SN54HC14 . . . J PACKAGE
SN74HC14 . . . J OR N PACKAGE
(TOP VIEW)

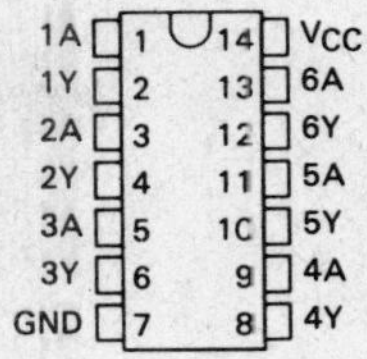

SN54HC14 . . . FH OR FK PACKAGE
SN74HC14 . . . FH OR FN PACKAGE
(TOP VIEW)

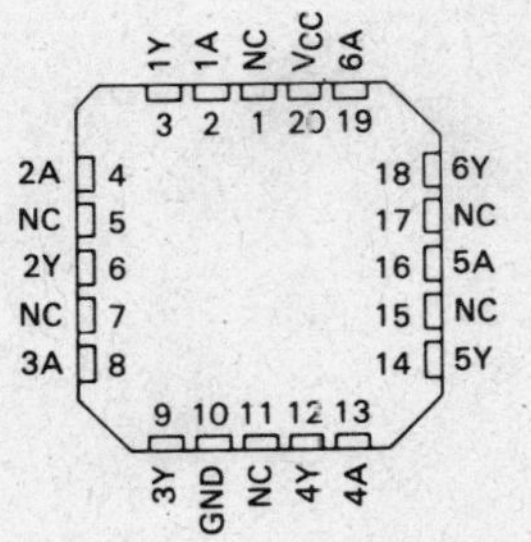

NC—No internal connection

description

These Schmitt-trigger devices contain six independent inverters. They perform the Boolean function $Y = \overline{A}$.

The SN54HC14 is characterized for operation over the full military temperature range of −55°C to 125°C. The SN74HC14 is characterized for operation from −40°C to 85°C.

FUNCTION TABLE
(each inverter)

INPUT A	OUTPUT Y
H	L
L	H

logic symbol

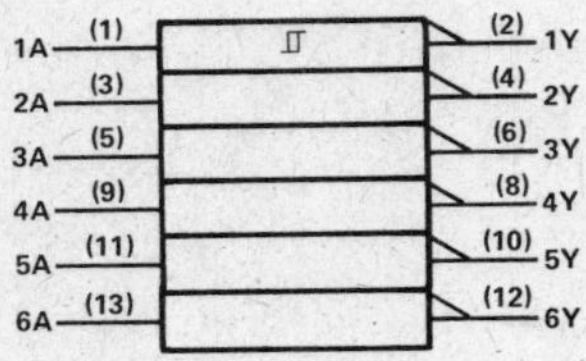

Pin numbers shown are for J and N packages.

maximum ratings, recommended operating conditions, and electrical characteristics

See Table I, page 2-4.

switching characteristics over recommended operating free-air temperature range (unless otherwise noted), C_L = 50 pF (see Note 1)

PARAMETER	FROM (INPUT)	TO (OUTPUT)	V_{CC}	T_A = 25°C MIN	T_A = 25°C TYP	T_A = 25°C MAX	SN54HC14 MIN	SN54HC14 MAX	SN74HC14 MIN	SN74HC14 MAX	UNIT
t_{pd}	A	Y	2 V		55	125		190		155	ns
			4.5 V		12	25		38		31	
			6 V		11	21		32		26	
t_t		Y	2 V		38	75		110		95	ns
			4.5 V		8	15		22		19	
			6 V		6	13		19		16	

C_{pd}	Power dissipation capacitance per inverter	No load, T_A = 25°C	20 pF typ

NOTE 1: For load circuit and voltage waveforms, see page 1-14.

3

HCMOS DEVICES

TYPES SN54HC20, SN74HC20 DUAL 4-INPUT POSITIVE-NAND GATES

D2684, DECEMBER 1982—REVISED MARCH 1984

- Package Options Include Both Plastic and Ceramic Chip Carriers in Addition to Plastic and Ceramic DIPs
- Dependable Texas Instruments Quality and Reliability

SN54HC20 . . . J PACKAGE
SN74HC20 . . . J OR N PACKAGE
(TOP VIEW)

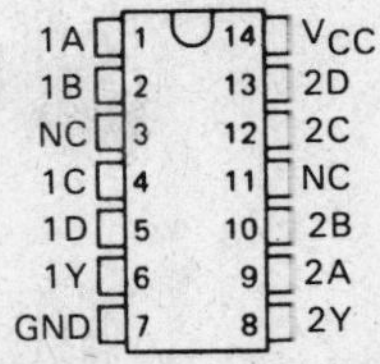

description

These devices contain two independent 4-input NAND gates. They perform the Boolean functions $Y=\overline{A \cdot B \cdot C \cdot D}$ or $Y=\overline{A}+\overline{B}+\overline{C}+\overline{D}$ in positive logic.

The SN54HC20 is characterized for operation over the full military temperature range of −55 °C to 125 °C. The SN74HC20 is characterized for operation from −40 °C to 85 °C.

SN54HC20 . . . FH OR FK PACKAGE
SN74HC20 . . . FH OR FN PACKAGE
(TOP VIEW)

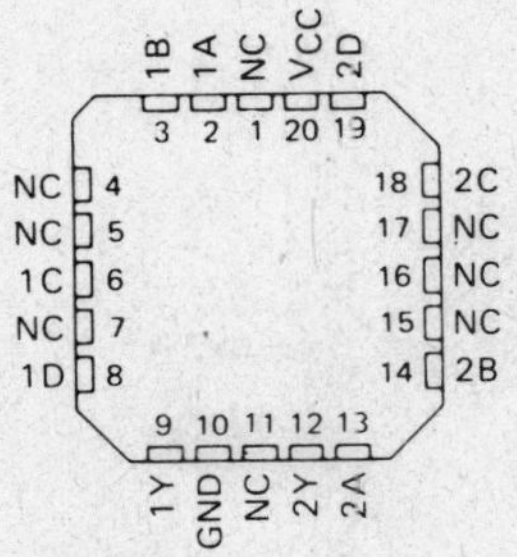

NC—No internal connection

logic symbol

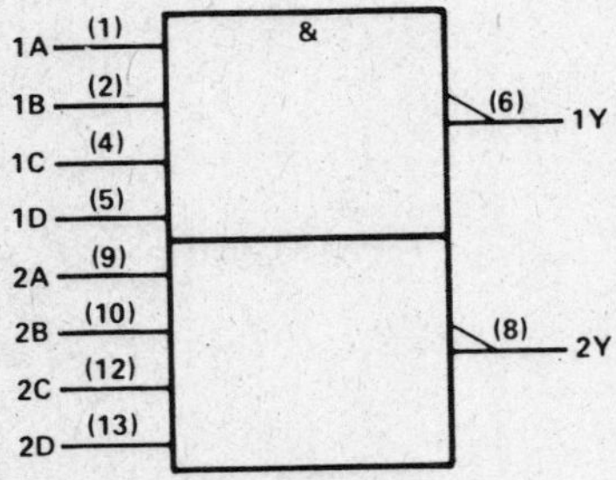

Pin numbers shown are for J and N packages.

FUNCTION TABLE (each gate)

INPUTS				OUTPUT
A	B	C	D	Y
H	H	H	H	L
L	X	X	X	H
X	L	X	X	H
X	X	L	X	H
X	X	X	L	H

maximum ratings, recommended operating conditions, and electrical characteristics

See Table I, page 2-4.

switching characteristics over recommended operating free-air temperature range (unless otherwise noted), C_L = 50 pF (see Note 1)

PARAMETER	FROM (INPUT)	TO (OUTPUT)	V_{CC}	T_A = 25 °C			SN54HC20		SN74HC20		UNIT
				MIN	TYP	MAX	MIN	MAX	MIN	MAX	
t_{pd}	A, B, C, or D	Y	2 V		45	110		165		140	ns
			4.5 V		14	22		33		28	
			6 V		11	19		28		24	
t_t		Y	2 V		27	75		110		95	ns
			4.5 V		9	15		22		19	
			6 V		7	13		19		16	

C_{pd}	Power dissipation capacitance per gate	No load, T_A = 25 °C	25 pF typ

NOTE 1: For load circuit and voltage waveforms, see page 1-14.

TEXAS INSTRUMENTS
POST OFFICE BOX 225012 • DALLAS, TEXAS 75265

TYPES SN54HC21, SN74HC21 DUAL 4-INPUT POSITIVE-AND GATES

D2684, DECEMBER 1982—REVISED MARCH 1984

- **Package Options Include Both Plastic and Ceramic Chip Carriers in Addition to Plastic and Ceramic DIPs**
- **Dependable Texas Instruments Quality and Reliability**

SN54HC21 . . . J PACKAGE
SN74HC21 . . . J OR N PACKAGE
(TOP VIEW)

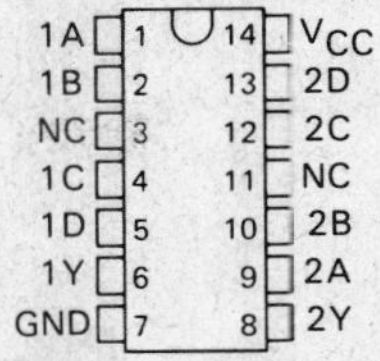

SN54HC21 . . . FH OR FK PACKAGE
SN74HC21 . . . FH OR FN PACKAGE
(TOP VIEW)

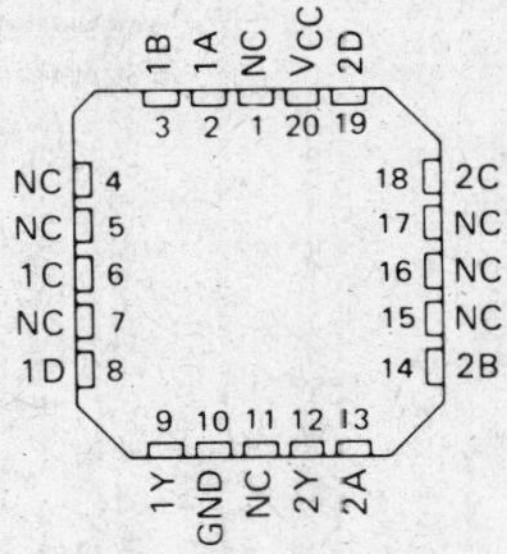

NC—No internal connection

description

These devices contain two independent 4-input AND gates. They perform the Boolean functions $Y = A \bullet B \bullet C \bullet D$ or $Y = \overline{\overline{A} + \overline{B} + \overline{C} + \overline{D}}$ in positive logic.

The SN54HC21 is characterized for operation over the full military temperature range of −55 °C to 125 °C. The SN74HC21 is characterized for operation from −40 °C to 85 °C.

logic symbol

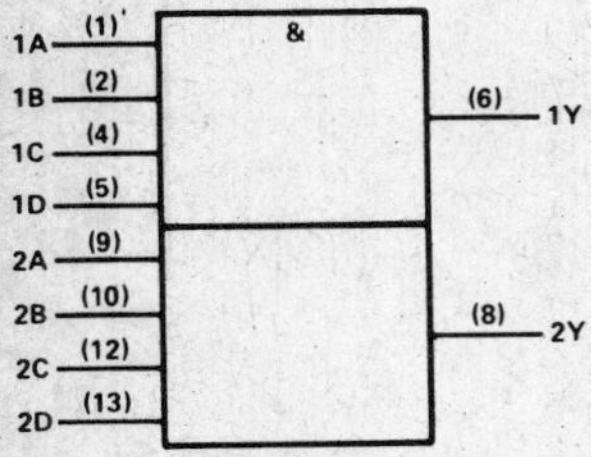

Pin numbers shown are for J and N packages.

FUNCTION TABLE (each gate)

INPUTS				OUTPUT
A	B	C	D	Y
H	H	H	H	H
L	X	X	X	L
X	L	X	X	L
X	X	L	X	L
X	X	X	L	L

maximum ratings, recommended operating conditions, and electrical characteristics

See Table I, page 2-4.

switching characteristics over recommended operating free-air temperature range (unless otherwise noted), C_L = 50 pF (see Note 1)

PARAMETER	FROM (INPUT)	TO (OUTPUT)	V_{CC}	T_A = 25 °C MIN	TYP	MAX	SN54HC21 MIN	MAX	SN74HC21 MIN	MAX	UNIT
t_{pd}	A, B, C, or D	Y	2 V		44	110		165		140	ns
			4.5 V		14	22		33		28	
			6 V		11	19		28		24	
t_t		Y	2 V		29	75		110		95	ns
			4.5 V		10	15		22		19	
			6 V		8	13		19		16	

C_{pd}	Power dissipation capacitance per gate	No load, T_A = 25 °C	25 pF typ

NOTE 1: For load circuit and voltage waveforms, see page 1-14.

TYPES SN54HC27, SN74HC27 TRIPLE 3-INPUT POSITIVE-NOR GATES

D2684, DECEMBER 1982—REVISED MARCH 1984

- **Package Options Include Both Plastic and Ceramic Chip Carriers in Addition to Plastic and Ceramic DIPs**
- **Dependable Texas Instruments Quality and Reliability**

SN54HC27 . . . J PACKAGE
SN74HC27 . . . J OR N PACKAGE
(TOP VIEW)

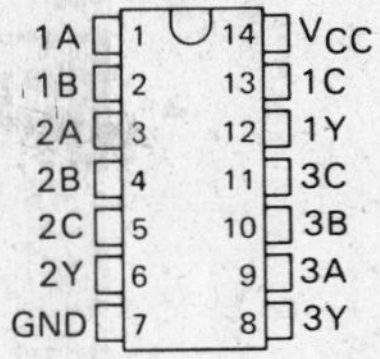

description

These devices contain three independent 3-input NOR gates. They perform the Boolean functions $Y=\overline{A+B+C}$ or $Y=\overline{A}\cdot\overline{B}\cdot\overline{C}$ in positive logic.

The SN54HC27 is characterized for operation over the full military temperature range of −55°C to 125°C. The SN74HC27 is characterized for operation from −40°C to 85°C.

SN54HC27 . . . FH OR FK PACKAGE
SN74HC27 . . . FH OR FN PACKAGE
(TOP VIEW)

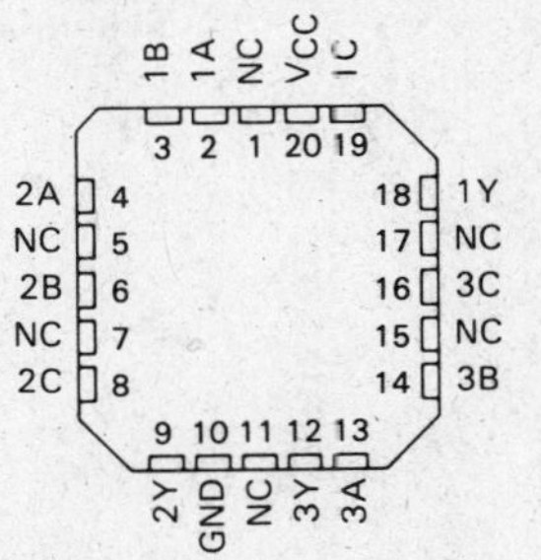

NC—No internal connection

logic symbol

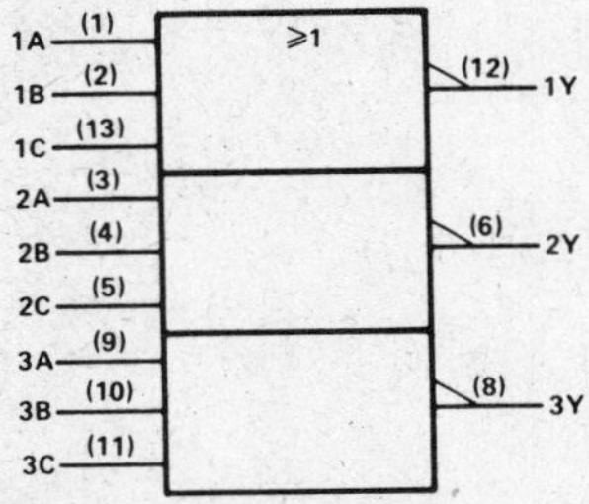

Pin numbers shown are for J and N packages.

FUNCTION TABLE (each gate)

INPUTS			OUTPUT
A	B	C	Y
H	X	X	L
X	H	X	L
X	X	H	L
L	L	L	H

maximum ratings, recommended operating conditions, and electrical characteristics

See Table I, page 2-4.

switching characteristics over recommended operating free-air temperature range (unless otherwise noted), C_L = 50 pF (see Note 1)

PARAMETER	FROM (INPUT)	TO (OUTPUT)	V_{CC}	T_A = 25°C MIN	TYP	MAX	SN54HC27 MIN	MAX	SN74HC27 MIN	MAX	UNIT
t_{pd}	A, B, or C	Y	2 V		35	90		135		115	ns
			4.5 V		10	18		27		23	
			6 V		9	15		23		20	
t_t		Y	2 V		27	75		110		95	ns
			4.5 V		7	15		22		19	
			6 V		6	13		19		16	

C_{pd}	Power dissipation capacitance per gate	No load, T_A = 25°C	25 pF typ

NOTE 1: For load circuit and voltage waveforms, see page 1-14.

D2684, DECEMBER 1982—REVISED MARCH 1984

- **Package Options Include Both Plastic and Ceramic Chip Carriers in Addition to Plastic and Ceramic DIPs**
- **Dependable Texas Instruments Quality and Reliability**

SN54HC30 . . . J PACKAGE
SN74HC30 . . . J OR N PACKAGE
(TOP VIEW)

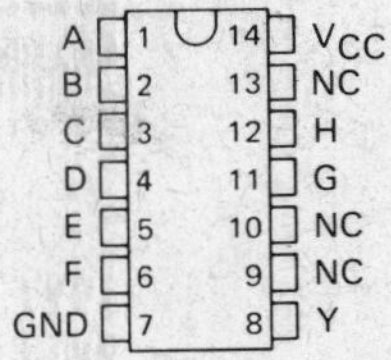

SN54HC30 . . . FH OR FK PACKAGE
SN74HC30 . . . FH OR FN PACKAGE
(TOP VIEW)

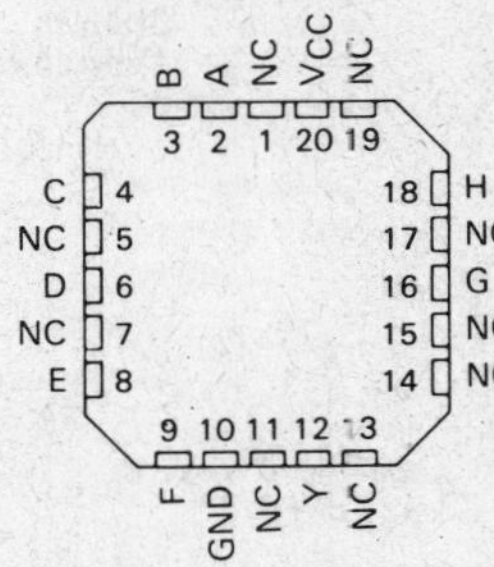

NC—No internal connection

description

These devices contain a single 8-input NAND gate and perform the following Boolean functions in positive logic:

$$Y = \overline{A \cdot B \cdot C \cdot D \cdot E \cdot F \cdot G \cdot H}$$

or

$$Y = \overline{A} + \overline{B} + \overline{C} + \overline{D} + \overline{E} + \overline{F} + \overline{G} + \overline{H}$$

The SN54HC30 is characterized for operation over the full military temperature range of −55 °C to 125 °C. The SN74HC30 is characterized for operation from −40 °C to 85 °C.

logic symbol

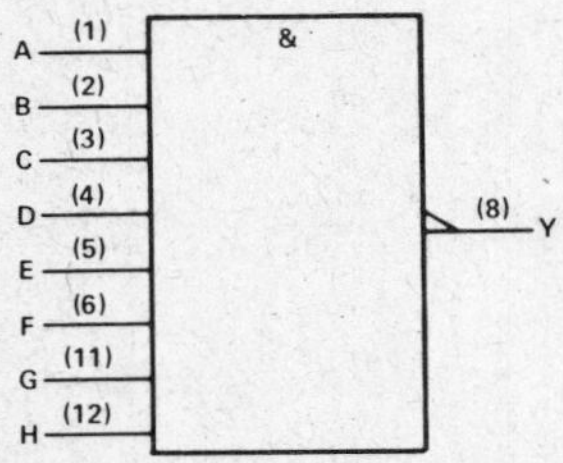

Pin numbers shown are for J and N packages.

FUNCTION TABLE

INPUTS A THRU H	OUTPUT Y
All inputs H	L
One or more inputs L	H

maximum ratings, recommended operating conditions, and electrical characteristics

See Table I, page 2-4.

switching characteristics over recommended operating free-air temperature range (unless otherwise noted), C_L = 50 pF (see Note 1)

PARAMETER	FROM (INPUT)	TO (OUTPUT)	V_{CC}	T_A = 25 °C MIN	TYP	MAX	SN54HC30 MIN	MAX	SN74HC30 MIN	MAX	UNIT
t_{pd}	A thru H	Y	2 V		51	130		195		165	ns
			4.5 V		15	26		39		33	
			6 V		12	22		33		28	
t_t		Y	2 V		28	75		110		95	ns
			4.5 V		8	15		22		19	
			6 V		6	13		19		16	

C_{pd}	Power dissipation capacitance per gate	No load, T_A = 25 °C	22 pF typ

NOTE 1: For load circuit and voltage waveforms, see page 1-14.

TYPES SN54HC32, SN74HC32
QUADRUPLE 2-INPUT POSITIVE-OR GATES

D2684, DECEMBER 1982—REVISED MARCH 1984

- **Package Options Include Both Plastic and Ceramic Chip Carriers in Addition to Plastic and Ceramic DIPs**
- **Dependable Texas Instruments Quality and Reliability**

SN54HC32 . . . J PACKAGE
SN74HC32 . . . J OR N PACKAGE
(TOP VIEW)

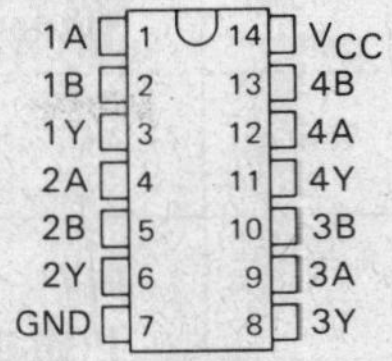

SN54HC32 . . . FH OR FK PACKAGE
SN74HC32 . . . FH OR FN PACKAGE
(TOP VIEW)

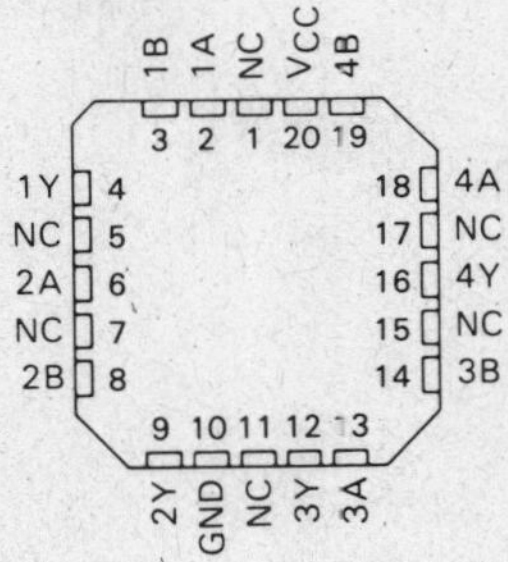

NC — No internal connection

description

These devices contain four independent 2-input OR gates. They perform the Boolean functions $Y = A + B$ or $Y = \overline{\overline{A} \cdot \overline{B}}$ in positive logic.

The SN54HC32 is characterized for operation over the full military temperature range of −55 °C to 125 °C. The SN74HC32 is characterized for operation from −40 °C to 85 °C.

FUNCTION TABLE
(each gate)

INPUTS		OUTPUT
A	B	Y
H	X	H
X	H	H
L	L	L

logic symbol

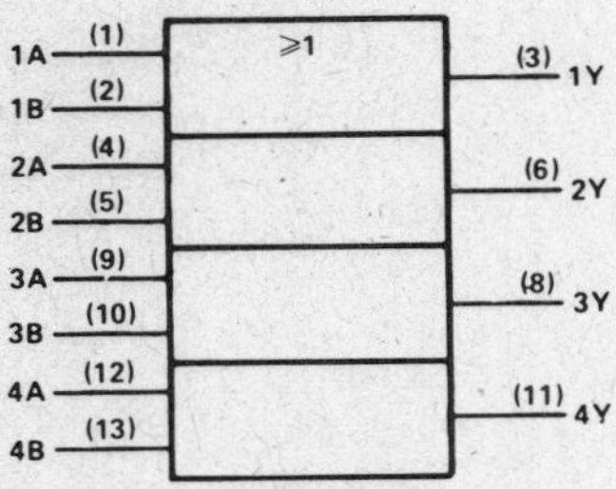

Pin numbers shown are for J and N packages.

maximum ratings, recommended operating conditions, and electrical characteristics

See Table I, page 2-4.

switching characteristics over recommended operating free-air temperature range (unless otherwise noted), C_L = 50 pF (see Note 1)

PARAMETER	FROM (INPUT)	TO (OUTPUT)	V_{CC}	T_A = 25°C			SN54HC32		SN74HC32		UNIT
				MIN	TYP	MAX	MIN	MAX	MIN	MAX	
t_{pd}	A or B	Y	2 V		50	100		150		125	ns
			4.5 V		10	20		30		25	
			6 V		8	17		25		21	
t_t		Y	2 V		38	75		110		95	ns
			4.5 V		8	15		22		19	
			6 V		6	13		19		16	

C_{pd}	Power dissipation capacitance per gate	No load, T_A = 25°C	20 pF typ

NOTE 1: For load circuit and voltage waveforms, see page 1-14.

TYPES SN54HC36, SN74HC36 QUADRUPLE 2-INPUT POSITIVE-NOR GATES

D2684, DECEMBER 1982—REVISED MARCH 1984

- **Package Options Include Both Plastic and Ceramic Chip Carriers in Addition to Plastic and Ceramic DIPs**
- **Dependable Texas Instruments Quality and Reliability**

SN54HC36 . . . J PACKAGE
SN74HC36 . . . J OR N PACKAGE
(TOP VIEW)

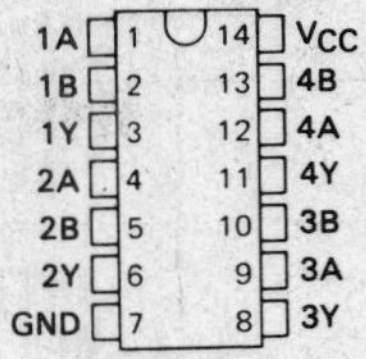

SN54HC36 . . . FH OR FK PACKAGE
SN74HC36 . . . FH OR FN PACKAGE
(TOP VIEW)

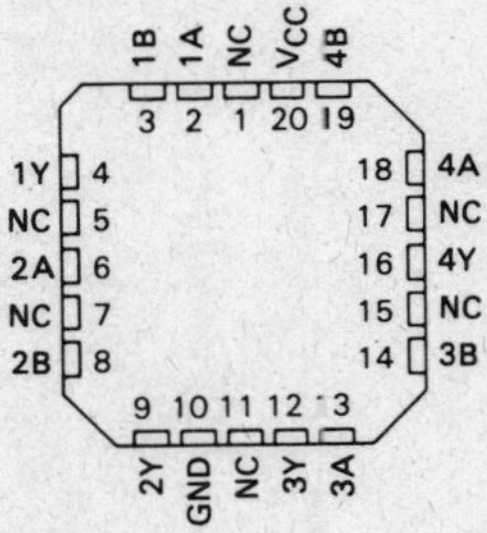

NC—No internal connection

description

These devices contain four independent 2-input NOR gates. They perform the Boolean functions $Y = \overline{A+B}$ or $Y = \overline{A} \cdot \overline{B}$ in positive logic.

The SN54HC36 is characterized for operation over the full military temperature range of −55 °C to 125 °C. The SN74HC36 is characterized for operation from −40 °C to 85 °C.

logic symbol

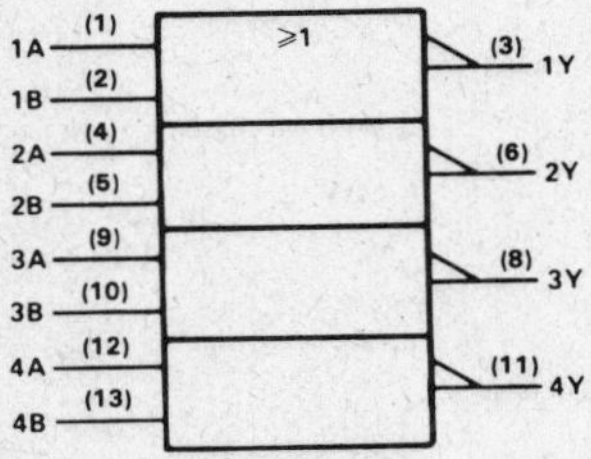

Pin numbers shown are for J and N packages.

FUNCTION TABLE (each gate)

INPUTS		OUTPUT
A	B	Y
H	X	L
X	H	L
L	L	H

maximum ratings, recommended operating conditions, and electrical characteristics

See Table I, page 2-4.

switching characteristics over recommended operating free-air temperature range (unless otherwise noted), C_L = 50 pF (see Note 1)

PARAMETER	FROM (INPUT)	TO (OUTPUT)	V_{CC}	T_A = 25 °C MIN	TYP	MAX	SN54HC36 MIN	MAX	SN74HC36 MIN	MAX	UNIT
t_{pd}	A or B	Y	2 V		50	100		150		125	ns
			4.5 V		10	20		30		25	
			6 V		8	17		25		21	
t_t		Y	2 V		38	75		110		95	ns
			4.5 V		8	15		22		19	
			6 V		6	13		19		16	

C_{pd}	Power dissipation capacitance per gate	No load, T_A = 25 °C	20 pF typ

NOTE 1: For load circuit and voltage waveforms, see page 1-14.

TYPES SN54HC42, SN74HC42 4-LINE TO 10-LINE DECODERS (1-of-10)

D2684, DECEMBER 1982—REVISED MARCH 1984

- **Full Decoding of Input Logic**
- **All Outputs Are High for Invalid BCD Conditions**
- **Also for Application as 3-Line to 8-Line Decoders**
- **Package Options Include Both Plastic and Ceramic Chip Carriers in Addition to Plastic and Ceramic DIPs**
- **Dependable Texas Instruments Quality and Reliability**

SN54HC42 . . . J PACKAGE
SN74HC42 . . . J OR N PACKAGE
(TOP VIEW)

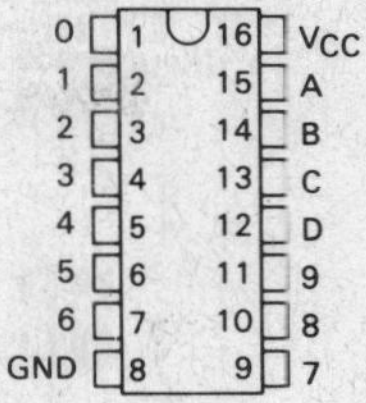

SN54HC42 . . . FH OR FK PACKAGE
SN74HC42 . . . FH OR FN PACKAGE
(TOP VIEW)

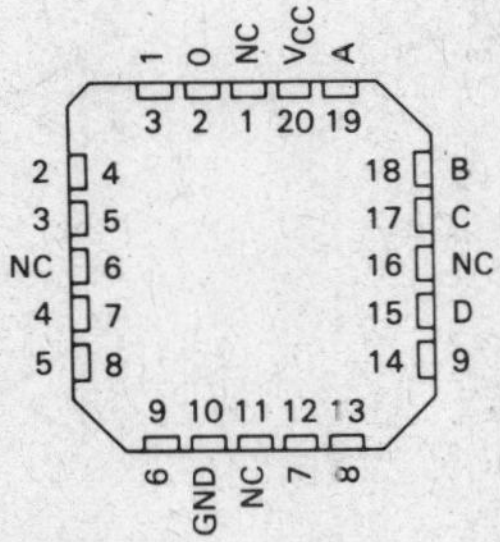

NC—No internal connection

description

These monolithic decimal decoders consist of eight inverters and ten four-input NAND gates. The inverters are connected in pairs to make BCD input data available for decoding by the NAND gates. Full decoding of valid input logic ensures that all inputs remain off for all invalid input conditions.

The SN54HC42 is characterized for operation over the full military temperature range of −55°C to 125°C. The SN74HC42 is characterized for operation from −40°C to 85°C.

FUNCTION TABLE

NO.	INPUTS				OUTPUTS									
	D	C	B	A	0	1	2	3	4	5	6	7	8	9
0	L	L	L	L	L	H	H	H	H	H	H	H	H	H
1	L	L	L	H	H	L	H	H	H	H	H	H	H	H
2	L	L	H	L	H	H	L	H	H	H	H	H	H	H
3	L	L	H	H	H	H	H	L	H	H	H	H	H	H
4	L	H	L	L	H	H	H	H	L	H	H	H	H	H
5	L	H	L	H	H	H	H	H	H	L	H	H	H	H
6	L	H	H	L	H	H	H	H	H	H	L	H	H	H
7	L	H	H	H	H	H	H	H	H	H	H	L	H	H
8	H	L	L	L	H	H	H	H	H	H	H	H	L	H
9	H	L	L	H	H	H	H	H	H	H	H	H	H	L
INVALID	H	L	H	L	H	H	H	H	H	H	H	H	H	H
	H	L	H	H	H	H	H	H	H	H	H	H	H	H
	H	H	L	L	H	H	H	H	H	H	H	H	H	H
	H	H	L	H	H	H	H	H	H	H	H	H	H	H
	H	H	H	L	H	H	H	H	H	H	H	H	H	H
	H	H	H	H	H	H	H	H	H	H	H	H	H	H

logic symbol

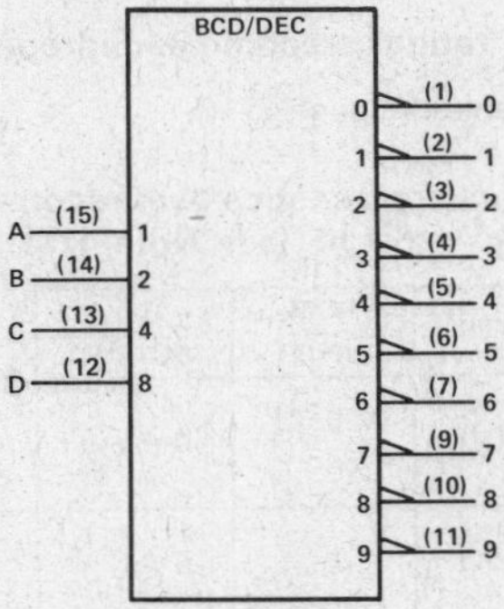

Pin numbers shown are for J and N packages.

logic diagram (positive logic)

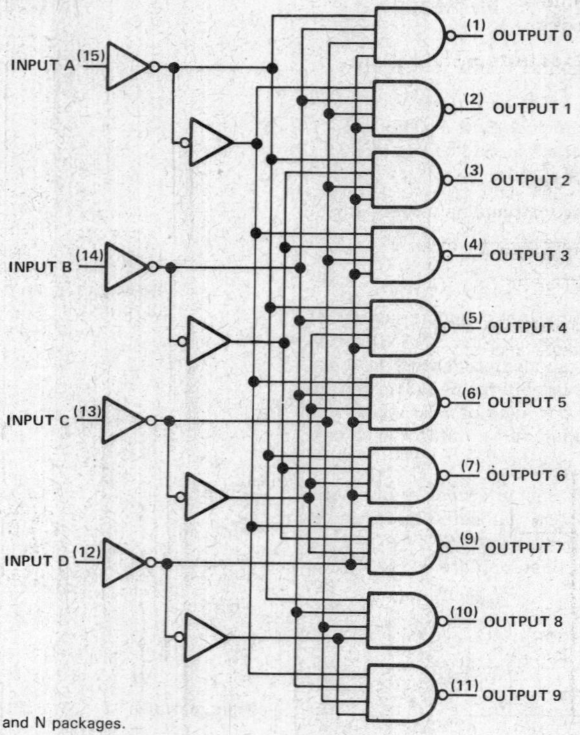

Pin numbers shown are for J and N packages.

maximum ratings, recommended operating conditions, and electrical characteristics

See Table IV, page 2-10.

switching characteristics over recommended operating free-air temperature range (unless otherwise noted), C_L = 50 pF (see Note 1)

PARAMETER	FROM (INPUT)	TO (OUTPUT)	V_{CC}	T_A = 25°C MIN	T_A = 25°C TYP	T_A = 25°C MAX	SN54HC42 MIN	SN54HC42 MAX	SN74HC42 MIN	SN74HC42 MAX	UNIT
t_{pd}	A, B, C, or D	0 thru 9	2 V		65	150		225		190	ns
			4.5 V		18	30		45		38	
			6 V		14	26		38		32	
t_t		Any	2 V		28	75		110		95	ns
			4.5 V		8	15		22		19	
			6 V		7	13		19		16	

C_{pd}	Power dissipation capacitance	No load, T_A = 25°C	39 pF typ

NOTE 1: For load circuit and voltage waveforms, see page 1-14.

TEXAS INSTRUMENTS
POST OFFICE BOX 225012 • DALLAS, TEXAS 75265

D2684, DECEMBER 1982—REVISED MARCH 1984

- **Package Options Include Both Plastic and Ceramic Chip Carriers in Addition to Plastic and Ceramic DIPs**
- **Dependable Texas Instruments Quality and Reliability**

SN54HC51 . . . J PACKAGE
SN74HC51 . . . J OR N PACKAGE
(TOP VIEW)

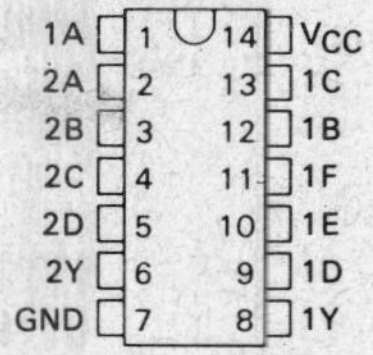

SN54HC51 . . . FH OR FK PACKAGE
SN74HC51 . . . FH OR FN PACKAGE
(TOP VIEW)

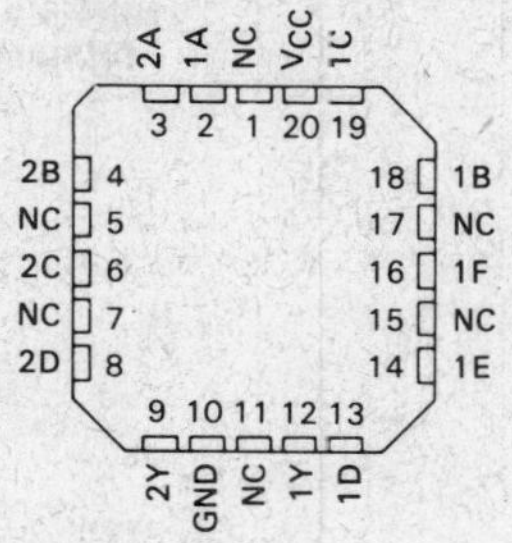

NC—No internal connection

description

The 'HC51 provides 2-wide, 2-input, and 2-wide, 3-input AND-OR-INVERT gates. The device performs the following Boolean functions:

$$1Y = \overline{(1A \cdot 1B \cdot 1C) + (1D \cdot 1E \cdot 1F)}$$

$$2Y = \overline{(2A \cdot 2B) + (2C \cdot 2D)}$$

The SN54HC51 is characterized for operation over the full military temperature range of −55°C to 125°C. The SN74HC51 is characterized for operation from −40°C to 85°C.

logic symbol

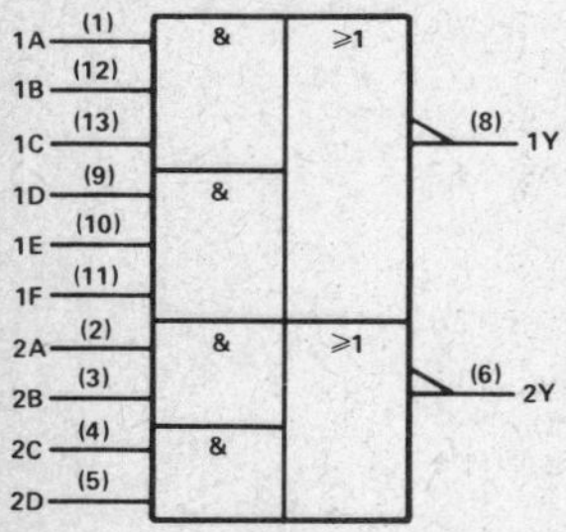

Pin numbers shown are for J and N packages.

FUNCTION TABLES

INPUTS						OUTPUT
1A	1B	1C	1D	1E	1F	1Y
H	H	H	X	X	X	L
X	X	X	H	H	H	L
Any other combination						H

INPUTS				OUTPUT
2A	2B	2C	2D	2Y
H	H	X	X	L
X	X	H	H	L
Any other combination				H

maximum ratings, recommended operating conditions, and electrical characteristics

See Table I, page 2-4.

switching characteristics over recommended operating free-air temperature range (unless otherwise noted), C_L = 50 pF (see Note 1)

PARAMETER	FROM (INPUT)	TO (OUTPUT)	V_{CC}	T_A = 25°C MIN	TYP	MAX	SN54HC51 MIN	MAX	SN74HC51 MIN	MAX	UNIT
t_{pd}	Any	Y	2 V		54	140		210		175	ns
			4.5 V		15	28		42		35	
			6 V		12	24		36		30	
t_t		Y	2 V		28	75		110		95	ns
			4.5 V		9	15		22		19	
			6 V		8	13		19		16	

C_{pd}	Power dissipation capacitance per AOI gate	No load, T_A = 25°C	25 pF typ

NOTE 1: For load circuit and voltage waveforms, see page 1-14.

TYPES SN54HC74, SN74HC74 DUAL D-TYPE POSITIVE-EDGE-TRIGGERED FLIP-FLOPS WITH CLEAR AND PRESET

D2684, DECEMBER 1982—REVISED JUNE 1984

- Package Options Include Both Plastic and Ceramic Chip Carriers in Addition to Plastic and Ceramic DIPs
- Dependable Texas Instruments Quality and Reliability

SN54HC74 . . . J PACKAGE
SN74HC74 . . . J OR N PACKAGE
(TOP VIEW)

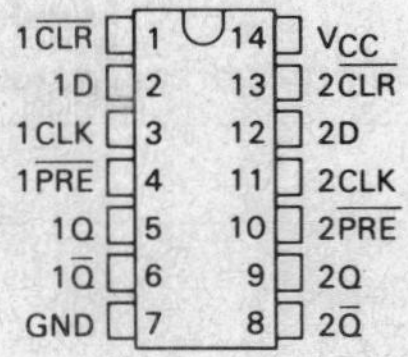

description

These devices contain two independent D-type positive-edge-triggered flip-flops. A low level at the Preset or Clear inputs sets or resets the outputs regardless of the levels of the other inputs. When Preset and Clear are inactive (high), data at the D input meeting the setup time requirements are transferred to the outputs on the positive-going edge of the clock pulse. Clock triggering occurs at a voltage level and is not directly related to the rise time of the clock pulse. Following the hold time interval, data at the D input may be changed without affecting the levels at the outputs.

The SN54HC74 is characterized for operation over the full military temperature range −55°C to 125°C. The SN74HC74 is characterized for operation from −40°C to 85°C.

SN54HC74 . . . FH OR FK PACKAGE
SN74HC74 . . . FH OR FN PACKAGE
(TOP VIEW)

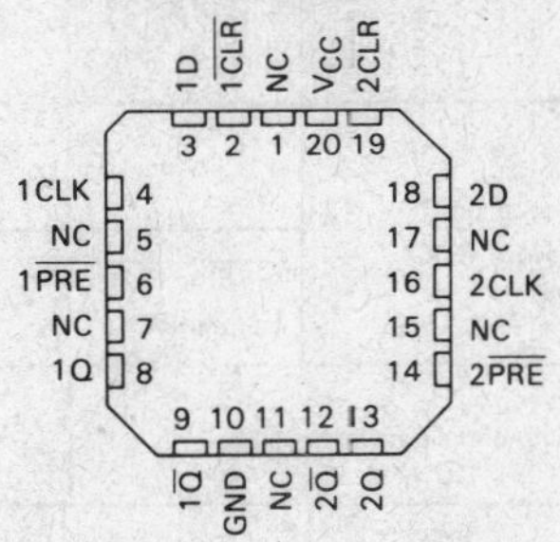

NC—No internal connection

FUNCTION TABLE

INPUTS				OUTPUTS	
$\overline{PRE}$	$\overline{CLR}$	CLK	D	Q	$\overline{Q}$
L	H	X	X	H	L
H	L	X	X	L	H
L	L	X	X	H†	H†
H	H	↑	H	H	L
H	H	↑	L	L	H
H	H	L	X	Q_0	$\overline{Q}_0$

†This configuration is nonstable; that is, it will not persist when Preset or Clear returns to its inactive (high) level.

logic symbol

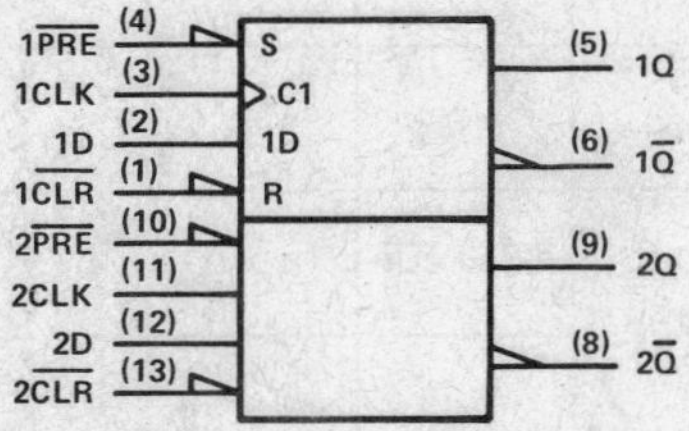

Pin numbers shown are for J and N packages.

logic diagram, each flip-flop (positive logic)

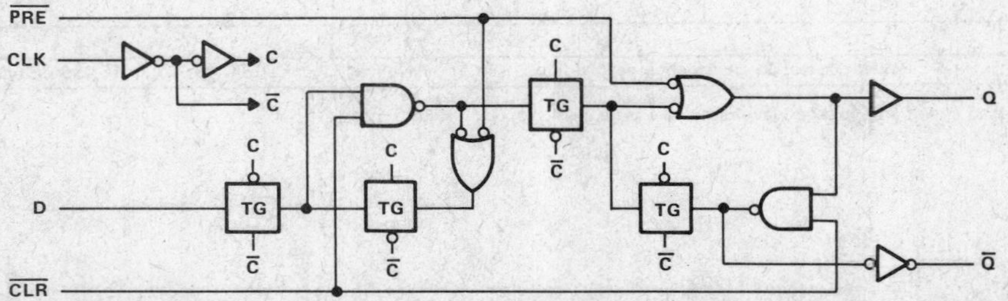

maximum ratings, recommended operating conditions, and electrical characteristics

See Table II, page 2-6.

timing requirements over recommended operating free-air temperature range (unless otherwise noted)

			V_{CC}	T_A = 25°C		SN54HC74		SN74HC74		UNIT
				MIN	MAX	MIN	MAX	MIN	MAX	
f_{clock}	Clock frequency		2 V	0	6	0	4.2	0	5	MHz
			4.5 V	0	31	0	21	0	25	
			6 V	0	36	0	25	0	29	
t_w	Pulse duration	$\overline{PRE}$ or $\overline{CLR}$ low	2 V	100		150		125		ns
			4.5 V	20		30		25		
			6 V	17		25		21		
		CLK high or low	2 V	80		120		100		
			4.5 V	16		24		20		
			6 V	14		20		17		
t_{su}	Setup time before CLK↑	Data	2 V	100		150		125		ns
			4.5 V	20		30		25		
			6 V	17		25		21		
		$\overline{PRE}$ or $\overline{CLR}$ inactive	2 V	25		40		30		
			4.5 V	5		8		6		
			6 V	4		7		5		
t_h	Hold time data after CLK↑		2 V	0		0		0		ns
			4.5 V	0		0		0		
			6 V	0		0		0		

switching characteristics over recommended operating free-air temperature range (unless otherwise noted), C_L = 50 pF (see Note 1)

PARAMETER	FROM (INPUT)	TO (OUTPUT)	V_{CC}	T_A = 25°C			SN54HC74		SN74HC74		UNIT
				MIN	TYP	MAX	MIN	MAX	MIN	MAX	
f_{max}			2 V	6	10		4.2		5		MHz
			4.5 V	31	50		20		25		
			6 V	36	60		25		29		
t_{pd}	$\overline{PRE}$ or $\overline{CLR}$	Q or $\overline{Q}$	2 V		70	230		345		290	ns
			4.5 V		20	46		69		58	
			6 V		15	39		59		49	
	CLK	Q or $\overline{Q}$	2 V		70	175		250		220	
			4.5 V		20	35		50		44	
			6 V		15	30		42		37	
t_t		Q or $\overline{Q}$	2 V		28	75		110		95	ns
			4.5 V		8	15		22		19	
			6 V		6	13		19		16	

C_{pd}	Power dissipation capacitance per flip-flop	No load, T_A = 25°C	35 pF typ

NOTE 1: For load circuit and voltage waveforms, see page 1-14.

TEXAS INSTRUMENTS
POST OFFICE BOX 225012 • DALLAS, TEXAS 75265

D2684, DECEMBER 1982—REVISED MARCH 1984

- Complementary Q and $\overline{Q}$ Outputs
- Package Options Include Both Plastic and Ceramic Chip Carriers in Addition to Plastic and Ceramic DIPs
- Dependable Texas Instruments Quality and Reliability

SN54HC75 . . . J PACKAGE
SN74HC75 . . . J OR N PACKAGE
(TOP VIEW)

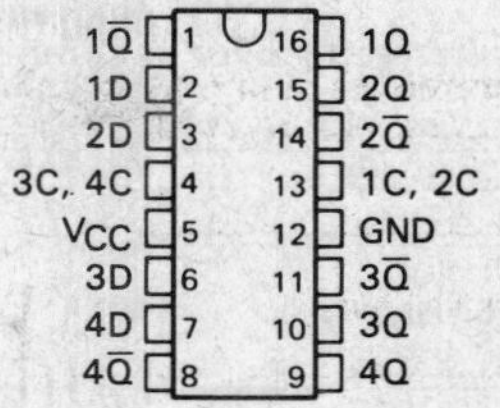

For functionally and electrically identical parts in chip carrier packages, see SN54HC375 and SN74HC375.

description

These latches are ideally suited for use as temporary storage for binary information between processing units and input/output or indicator units. Information present at a data (D) input is transferred to the Q output when the enable (C) is high and the Q output will follow the data input as long as the enable remains high. When the enable goes low, the information (that was present at the data input at the time the transition occurred) is retained at the Q output until the enable is permitted to go high.

The SN54HC75 is characterized for operation over the full military temperature range of −55°C to 125°C. The SN74HC75 is characterized for operation from −40°C to 85°C.

logic symbol

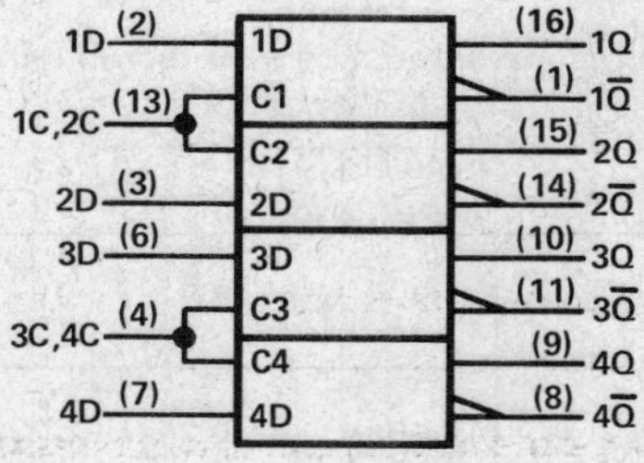

Pin numbers shown are for J and N packages.

FUNCTION TABLE
(Each Latch)

INPUTS		OUTPUTS	
D	C	Q	$\overline{Q}$
L	H	L	H
H	H	H	L
X	L	Q_0	$\overline{Q}_0$

logic diagram, each latch (positive logic)

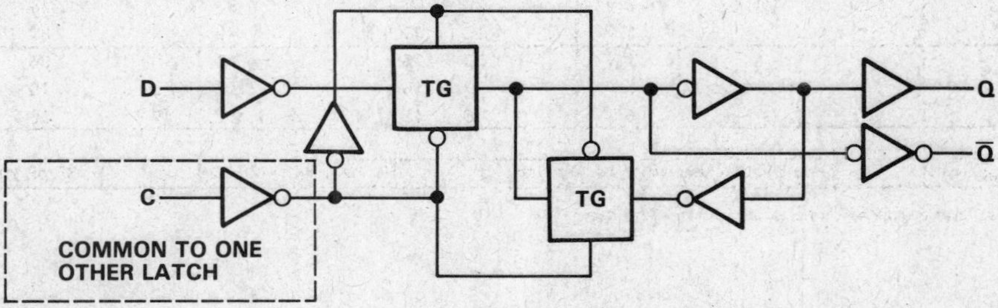

maximum ratings, recommended operating conditions, and electrical characteristics

See Table II, page 2-6.

timing requirements over recommended operating free-air temperature range (unless otherwise noted)

		V_{CC}	T_A = 25°C		SN54HC75		SN74HC75		UNIT
			MIN	MAX	MIN	MAX	MIN	MAX	
t_w	Pulse duration, C high	2 V	80		120		100		ns
		4.5 V	16		24		20		
		6 V	14		20		17		
t_{su}	Setup time, data before C↓	2 V	100		150		125		ns
		4.5 V	20		30		25		
		6 V	17		26		21		
t_h	Hold time, data after C↓	2 V	5		5		5		ns
		4.5 V	5		5		5		
		6 V	5		5		5		

switching characteristics over recommended operating free-air temperature range (unless otherwise noted), C_L = 50 pF (see Note 1)

PARAMETER	FROM (INPUT)	TO (OUTPUT)	V_{CC}	T_A = 25°C			SN54HC75		SN74HC75		UNIT
				MIN	TYP	MAX	MIN	MAX	MIN	MAX	
t_{pd}	D	Q or $\overline{Q}$	2 V		40	120		180		150	ns
			4.5 V		14	24		36		30	
			6 V		11	20		31		26	
t_{pd}	C	Q or $\overline{Q}$	2 V		44	130		195		165	ns
			4.5 V		15	26		39		33	
			6 V		12	22		33		28	
t_t		Any	2 V		38	75		110		95	ns
			4.5 V		8	15		22		19	
			6 V		6	13		19		16	

C_{pd}	Power dissipation capacitance per latch	No load, T_A = 25°C	46 pF typ

NOTE 1: For load circuits and voltage waveforms, see page 1-14.

3 HCMOS DEVICES

D2684, DECEMBER 1982—REVISED MARCH 1984

- **Package Options Include Both Plastic and Ceramic Chip Carriers in Addition to Plastic and Ceramic DIPs**
- **Dependable Texas Instruments Quality and Reliability**

SN54HC76 . . . J PACKAGE
SN74HC76 . . . J OR N PACKAGE
(TOP VIEW)

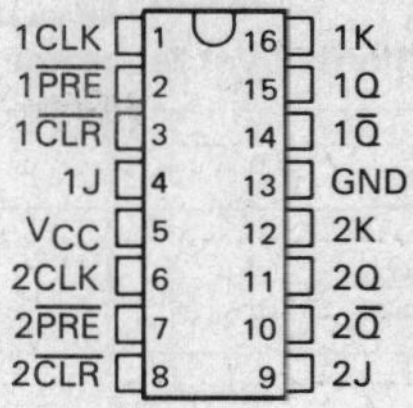

For functionally and electrically identical parts in chip carrier packages, see SN54HC112 and SN74HC112.

description

These devices contain two independent J-K negative-edge-triggered flip-flops. A low level at the Preset or Clear input sets or resets the outputs regardless of the levels of the other inputs. When Preset and Clear are inactive (high), data at the J and K inputs meeting the setup time requirements are transferred to the outputs on the negative-going edge of the clock pulse. Clock triggering occurs at a voltage level and is not directly related to the rise time of the clock pulse. Following the hold time interval, data at the J and K inputs may be changed without affecting the levels at the outputs. These versatile flip-flops can also perform as toggle flip-flops by tying J and K high.

FUNCTION TABLE
(EACH FLIP-FLOP)

INPUTS					OUTPUTS	
$\overline{PRE}$	$\overline{CLR}$	CLK	J	K	Q	$\overline{Q}$
L	H	X	X	X	H	L
H	L	X	X	X	L	H
L	L	X	X	X	H*	H*
H	H	↓	L	L	Q_0	$\overline{Q}_0$
H	H	↓	H	L	H	L
H	H	↓	L	H	L	H
H	H	↓	H	H	TOGGLE	
H	H	H	X	X	Q_0	$\overline{Q}_0$

*This configuration is nonstable; that is, it will not persist when either Preset or Clear returns to its inactive (high) level.

logic symbol

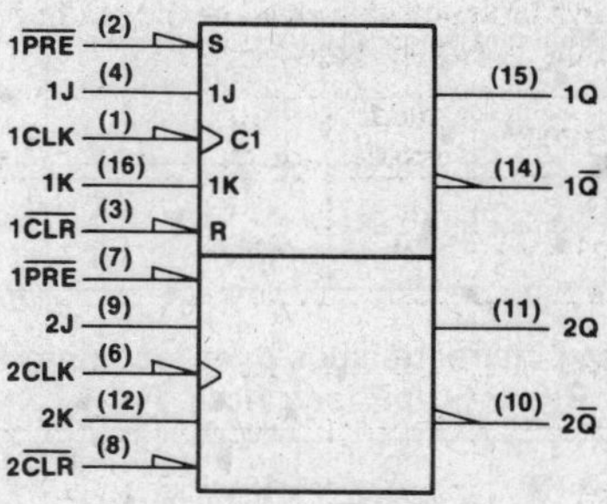

Pin numbers shown are for J and N packages.

logic diagram, each flip-flop (positive logic)

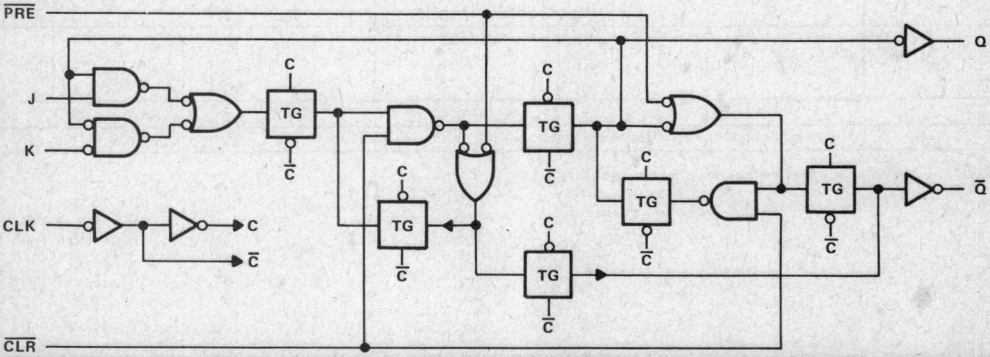

TEXAS INSTRUMENTS
POST OFFICE BOX 225012 • DALLAS, TEXAS 75265

maximum ratings, recommended operating conditions, and electrical characteristics

See Table II, page 2-6.

timing requirements over recommended operating free-air temperature range (unless otherwise noted)

			V_{CC}	$T_A = 25°C$ MIN	$T_A = 25°C$ MAX	SN54HC76 MIN	SN54HC76 MAX	SN74HC76 MIN	SN74HC76 MAX	UNIT
f_{clock}	Clock frequency		2 V	0	6	0	4.2	0	5	MHz
			4.5 V	0	31	0	21	0	25	
			6 V	0	36	0	25	0	29	
t_w	Pulse duration	$\overline{PRE}$ or $\overline{CLR}$ low	2 V	100		150		125		ns
			4.5 V	20		30		25		
			6 V	17		25		21		
		CLK high or low	2 V	80		120		100		
			4.5 V	16		24		20		
			6 V	14		20		17		
t_{su}	Setup time before CLK↓	Data	2 V	150		225		190		ns
			4.5 V	30		45		38		
			6 V	25		38		32		
		$\overline{PRE}$ or $\overline{CLR}$ inactive	2 V	100		150		125		
			4.5 V	20		30		25		
			6 V	17		25		21		
t_h	Hold time after CLK↓		2 V	0		0		0		ns
			4.5 V	0		0		0		
			6 V	0		0		0		

switching characteristics over recommended operating free-air temperture range (unless otherwise noted), C_L = 50 pF (see Note 1)

PARAMETER	FROM (INPUT)	TO (OUTPUT)	V_{CC}	$T_A = 25°C$ MIN	$T_A = 25°C$ TYP	$T_A = 25°C$ MAX	SN54HC76 MIN	SN54HC76 MAX	SN74HC76 MIN	SN74HC76 MAX	UNIT
f_{max}			2 V	6	9		4.2		5		MHz
			4.5 V	31	41		21		25		
			6 V	36	50		25		29		
t_{pd}	$\overline{PRE}$ or $\overline{CLR}$	Q or $\overline{Q}$	2 V		65	155		250		190	ns
			4.5 V		16	31		47		39	
			6 V		15	26		40		33	
t_{pd}	CLK	Q or $\overline{Q}$	2 V		70	145		220		180	ns
			4.5 V		19	29		44		36	
			6 V		16	25		37		31	
t_t		Q or $\overline{Q}$	2 V		38	75		110		95	ns
			4.5 V		8	15		22		19	
			6 V		6	13		19		16	

C_{pd}	Power dissipation capacitance per flip-flop	No load, $T_A = 25°C$	36 pF typ

NOTE 1: For load circuit and voltage waveforms, see page 1-14.

D2684, DECEMBER 1982—REVISED MARCH 1984

- Package Options Include Both Plastic and Ceramic DIPs
- Dependable Texas Instruments Quality and Reliability

SN54HC77 . . . J PACKAGE
SN74HC77 . . . J OR N PACKAGE
(TOP VIEW)

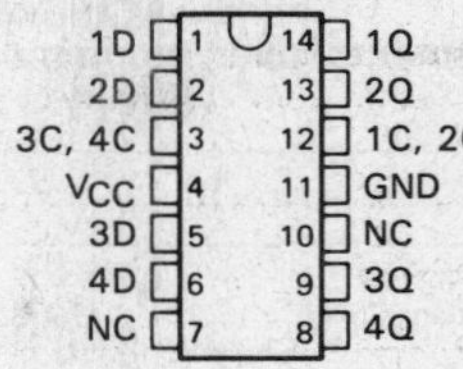

NC—No internal connection

Not available in chip carrier package with JEDEC-Standard pin-out. For chip carrier information, contact the factory.

description

These latches are ideally suited for use as temporary storage for binary information between processing units and input/output or indicator units. Information present at a data (D) input is transferred to the Q output when the enable (C) is high and the Q output will follow the data input as long as the enable remains high. When the enable goes low, the information (that was present at the data input at the time the transition occurred) is retained at the Q output until the enable is permitted to go high.

The SN54HC77 is characterized for operation over the full military temperature range of −55°C to 125°C. The SN74HC77 is characterized for operation from −40°C to 85°C.

logic symbol

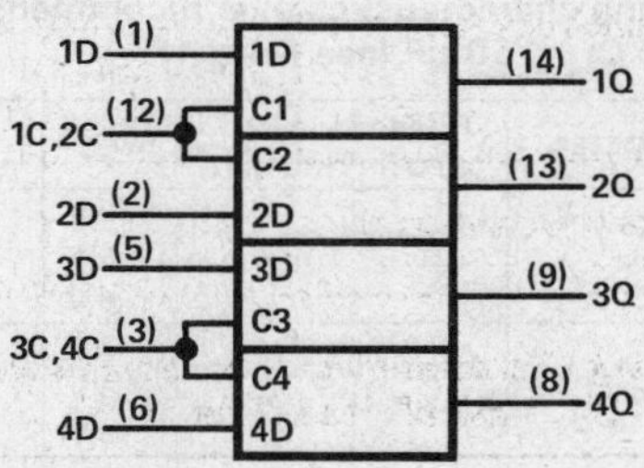

Pin numbers shown are for J and N packages.

FUNCTION TABLE
(Each Latch)

INPUTS		OUTPUT
D	C	Q
L	H	L
H	H	H
X	L	Q_0

logic diagram, each latch (positive logic)

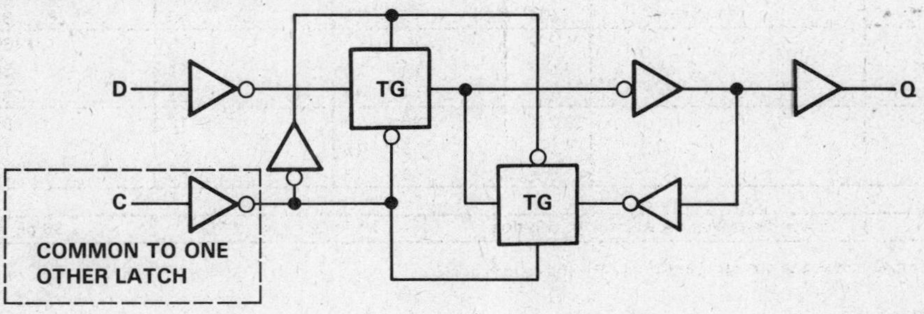

maximum ratings, recommended operating conditions, and electrical characteristics

See Table II, page 2-6.

timing requirements over recommended operating free-air temperature range (unless otherwise noted)

		V_{CC}	T_A = 25°C		SN54HC77		SN74HC77		UNIT
			MIN	MAX	MIN	MAX	MIN	MAX	
t_w	Pulse duration, C high	2 V	80		120		100		ns
		4.5 V	16		24		20		
		6 V	14		20		17		
t_{su}	Setup time, data before C↓	2 V	100		150		125		ns
		4.5 V	20		30		25		
		6 V	17		26		21		
t_h	Hold time, data after C↓	2 V	5		5		5		ns
		4.5 V	5		5		5		
		6 V	5		5		5		

switching characteristics over recommended operating free-air temperature range (unless otherwise noted), C_L = 50 pF (see Note 1)

PARAMETER	FROM (INPUT)	TO (OUTPUT)	V_{CC}	T_A = 25°C			SN54HC77		SN74HC77		UNIT
				MIN	TYP	MAX	MIN	MAX	MIN	MAX	
t_{pd}	D	Q	2 V		40	120		180		150	ns
			4.5 V		12	24		36		30	
			6 V		10	20		31		26	
t_{pd}	C	Q	2 V		45	130		195		165	ns
			4.5 V		14	26		39		33	
			6 V		11	22		33		28	
t_t		Any	2 V		28	75		110		95	ns
			4.5 V		8	15		22		19	
			6 V		6	13		19		16	

NOTE 1: For load circuit and voltage waveforms, see page 1-14.

TEXAS INSTRUMENTS
POST OFFICE BOX 225012 • DALLAS, TEXAS 75265

- Package Options Include Both Plastic and Ceramic Chip Carriers in Addition to Plastic and Ceramic DIPs
- Dependable Texas Instruments Quality and Reliability

SN54HC86 . . . J PACKAGE
SN74HC86 . . . J OR N PACKAGE
(TOP VIEW)

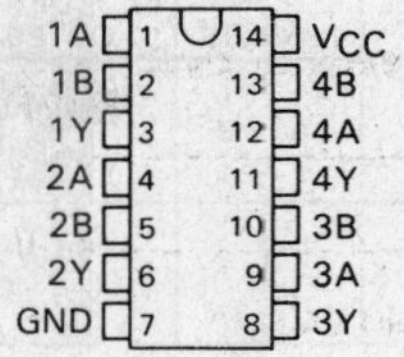

SN54HC86 . . . FH OR FK PACKAGE
SN74HC86 . . . FH OR FN PACKAGE
(TOP VIEW)

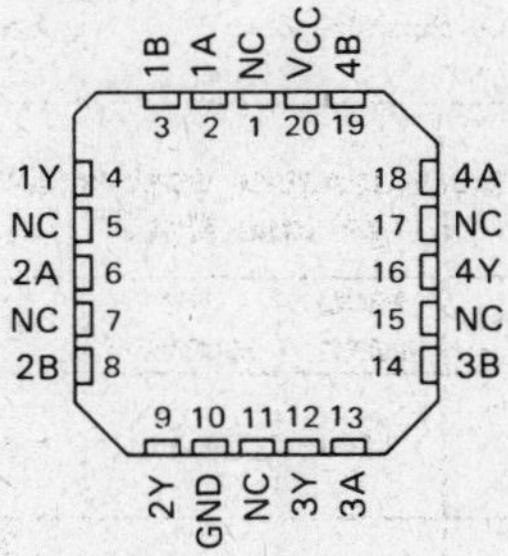

NC—No internal connection

description

These devices contain four independent 2-input Exclusive-OR gates. They perform the Boolean functions $Y = A \oplus B = \overline{A}B + A\overline{B}$ in positive logic.

A common application is as a true/complement element. If one of the inputs is low, the other input will be reproduced in true form at the output. If one of the inputs is high, the signal on the other input will be reproduced inverted at the output.

The SN54HC86 is characterized for operation over the full military temperature range of −55 °C to 125 °C. The SN74HC86 is characterized for operation from −40 °C to 85 °C.

logic symbol

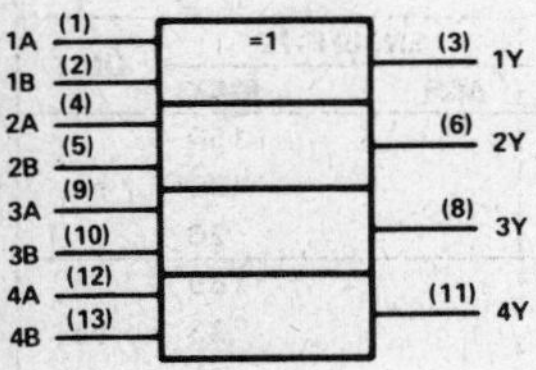

Pin numbers shown are for J and N packages.

FUNCTION TABLE
(each gate)

INPUTS		OUTPUT
A	B	Y
L	L	L
L	H	H
H	L	H
H	H	L

exclusive-OR logic

An exclusive-OR gate has many applications, some of which can be represented better by alternative logic symbols.

EXCLUSIVE-OR

These are five equivalent Exclusive-OR symbols valid for an 'HC86 gate in positive logic; negation may be shown at any two ports.

LOGIC IDENTITY ELEMENT

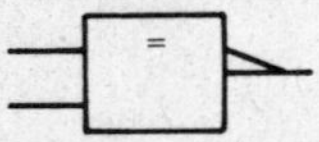

The output is active (low) if all inputs stand at the same logic level (i.e., A = B).

EVEN-PARITY

The output is active (low) if an even number of inputs (i.e., 0 or 2) are active.

ODD-PARITY ELEMENT

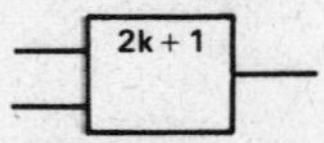

The output is active (high) if an odd number of inputs (i.e., only 1 of the 2) are active.

maximum ratings, recommended operating conditions, and electrical characteristics

See Table I, page 2-4.

switching characteristics over recommended operating free-air temperature range (unless otherwise noted), C_L = 50 pF (see Note 1)

PARAMETER	FROM (INPUT)	TO (OUTPUT)	V_{CC}	T_A = 25°C			SN54HC86		SN74HC86		UNIT
				MIN	TYP	MAX	MIN	MAX	MIN	MAX	
t_{pd}	A or B	Y	2 V		40	100		150		125	ns
			4.5 V		12	20		30		25	
			6 V		10	17		25		21	
t_t		Y	2 V		28	75		110		95	ns
			4.5 V		8	15		22		19	
			6 V		6	13		19		16	

C_{pd}	Power dissipation capacitance per gate	No load, T_A = 25°C	35 pF typ

NOTE 1: For load circuit and voltage waveforms, see page 1-14.

TEXAS INSTRUMENTS
POST OFFICE BOX 225012 • DALLAS, TEXAS 75265

TYPES SN54HC107, SN74HC107 DUAL J-K NEGATIVE-EDGE-TRIGGERED FLIP-FLOPS WITH CLEAR

D2684, DECEMBER 1982—REVISED MARCH 1984

- **Package Options Include Both Plastic and Ceramic Chip Carriers in Addition to Plastic and Ceramic DIPs**
- **Dependable Texas Instruments Quality and Reliability**

SN54HC107 . . . J PACKAGE
SN74HC107 . . . J OR N PACKAGE
(TOP VIEW)

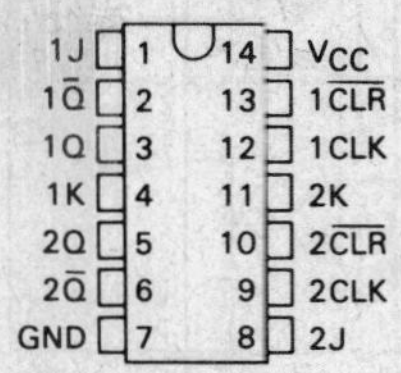

SN54HC107 . . . FH OR FK PACKAGE
SN74HC107 . . . FH OR FN PACKAGE
(TOP VIEW)

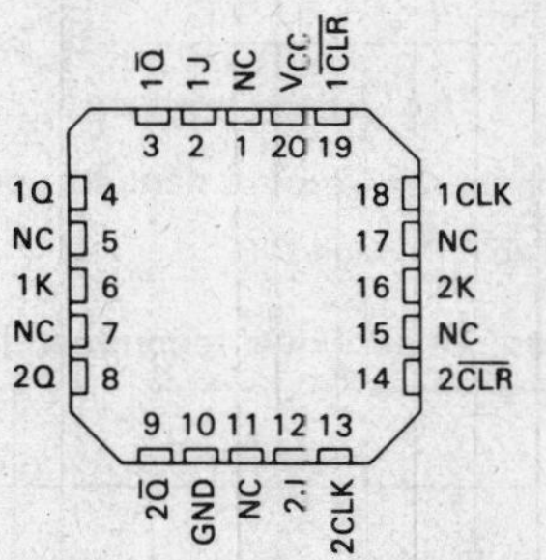

NC—No internal connection

description

These devices contain two independent J-K negative-edge-triggered flip-flops. A low level at the $\overline{\text{CLR}}$ input resets the outputs regardless of the levels of the other inputs. When $\overline{\text{CLR}}$ is inactive (high), data at the J and K inputs meeting the setup time requirements are transferred to the outputs on the negative-going edge of the clock pulse. Clock triggering occurs at a voltage level and is not directly related to the rise time of the clock pulse. Following the hold time interval, data at the J and K inputs may be changed without affecting the levels at the outputs. These versatile flip-flops can perform as toggle flip-flops by tying J and K high.

The SN54HC107 is characterized for operation over the full military temperature range of −55°C to 125°C. The SN74HC107 is characterized for operation from −40°C to 85°C.

FUNCTION TABLE

INPUTS				OUTPUTS	
$\overline{\text{CLR}}$	CLK	J	K	Q	$\overline{\text{Q}}$
L	X	X	X	L	H
H	↓	L	L	Q_0	$\overline{Q}_0$
H	↓	H	L	H	L
H	↓	L	H	L	H
H	↓	H	H	TOGGLE	
H	H	X	X	Q_0	$\overline{Q}_0$

logic symbol

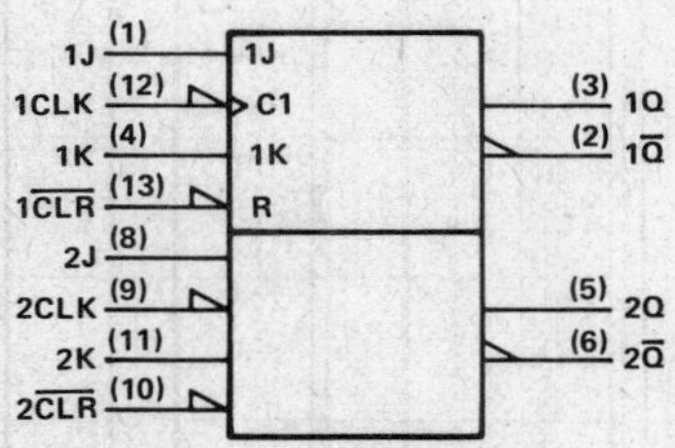

Pin numbers shown are for J and N packages.

logic diagram, each flip-flop (positive logic)

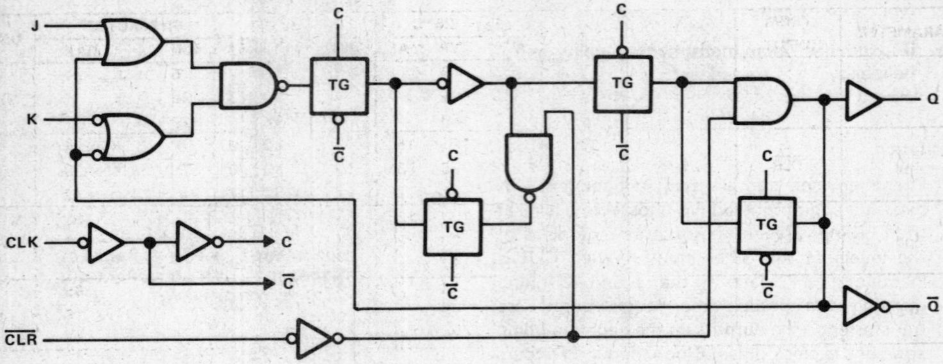

maximum ratings, recommended operating conditions, and electrical characteristics

See Table II, page 2-6.

timing requirements over recommended operating free-air temperature range (unless otherwise noted)

			V_{CC}	T_A = 25°C MIN	T_A = 25°C MAX	SN54HC107 MIN	SN54HC107 MAX	SN74HC107 MIN	SN74HC107 MAX	UNIT
f_{clock}	Clock frequency		2 V	0	6	0	4.2	0	5	MHz
			4.5 V	0	31	0	21	0	25	
			6 V	0	36	0	25	0	29	
t_w	Pulse duration	$\overline{CLR}$ low	2 V	100		150		125		ns
			4.5 V	20		30		25		
			6 V	17		25		21		
		CLK high or low	2 V	80		120		100		
			4.5 V	16		24		20		
			6 V	14		20		17		
t_{su}	Setup time before CLK↓	Data (J, K)	2 V	100		150		125		ns
			4.5 V	20		30		25		
			6 V	17		25		21		
		$\overline{CLR}$ inactive	2 V	100		150		125		
			4.5 V	20		30		25		
			6 V	17		25		21		
t_h	Hold time, data after CLK↓		2 V	0		0		0		ns
			4.5 V	0		0		0		
			6 V	0		0		0		

TEXAS INSTRUMENTS
POST OFFICE BOX 225012 • DALLAS, TEXAS 75265

switching characteristics over recommended operating free-air temperature range (unless otherwise noted), C_L = 50 pF (see Note 1)

PARAMETER	FROM (INPUT)	TO (OUTPUT)	V_{CC}	T_A = 25°C			SN54HC107		SN74HC107		UNIT
				MIN	TYP	MAX	MIN	MAX	MIN	MAX	
f_{max}			2 V	6	9		4.2		5		MHz
			4.5 V	31	45		21		25		
			6 V	36	53		25		29		
t_{pd}	$\overline{CLR}$	Q or $\overline{Q}$	2 V		126	155		235		195	ns
			4.5 V		25	31		47		39	
			6 V		21	26		40		32	
t_{pd}	CLK	Q or $\overline{Q}$	2 V		100	125		185		160	ns
			4.5 V		20	25		37		32	
			6 V		17	21		32		27	
t_t		Q or $\overline{Q}$	2 V		38	75		110		95	ns
			4.5 V		8	15		22		19	
			6 V		6	13		19		16	

C_{pd}	Power dissipation capacitance per flip-flop	No load, T_A = 25°C	35 pF typ

NOTE 1: For load circuit and voltage waveforms, see page 1-14.

TYPES SN54HC109, SN74HC109 DUAL J-$\overline{K}$ POSITIVE-EDGE-TRIGGERED FLIP-FLOPS WITH CLEAR AND PRESET

D2684, DECEMBER 1982—REVISED MARCH 1984

- **Package Options Include Both Plastic and Ceramic Chip Carriers in Addition to Plastic and Ceramic DIPs**
- **Dependable Texas Instruments Quality and Reliability**

SN54HC109 . . . J PACKAGE
SN74HC109 . . . J OR N PACKAGE
(TOP VIEW)

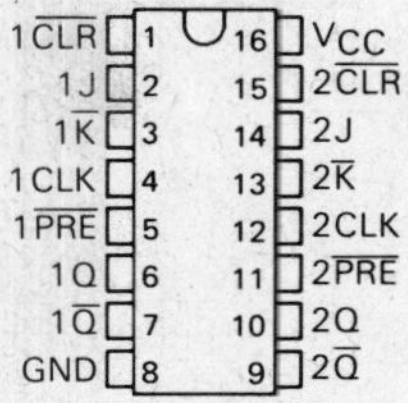

SN54HC109 . . . FH OR FK PACKAGE
SN74HC109 . . . FH OR FN PACKAGE
(TOP VIEW)

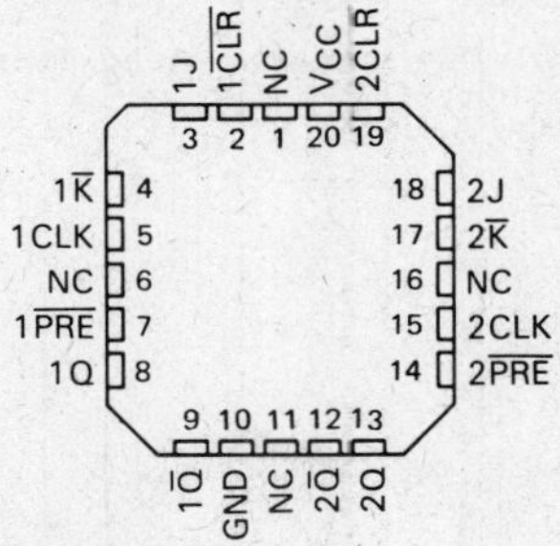

NC—No internal connection

description

These devices contain two independent J-$\overline{K}$ positive-edge-triggered flip-flops. A low level at the Preset or Clear inputs sets or resets the outputs regardless of the levels of the other inputs. When Preset and Clear are inactive (high), data at the J and $\overline{K}$ inputs meeting the setup time requirements are transferred to the outputs on the positive-going edge of the clock pulse. Clock triggering occurs at a voltage level and is not directly related to the rise time of the clock pulse. Following the hold time interval, data at the J and $\overline{K}$ inputs may be changed without affecting the levels at the outputs. These versatile flip-flops can perform as toggle flip-flops by grounding $\overline{K}$ and tying J high. They also can perform as D-type flip-flops if J and $\overline{K}$ are tied together.

The SN54HC109 is characterized for operation over the full military temperature range of −55°C to 125°C. The SN74HC109 is characterized for operation from −40°C to 85°C.

FUNCTION TABLE

INPUTS					OUTPUTS	
$\overline{PRE}$	$\overline{CLR}$	CLK	J	$\overline{K}$	Q	$\overline{Q}$
L	H	X	X	X	H	L
H	L	X	X	X	L	H
L	L	X	X	X	H*	H*
H	H	↑	L	L	L	H
H	H	↑	H	L	TOGGLE	
H	H	↑	L	H	Q_0	$\overline{Q}_0$
H	H	↑	H	H	H	L
H	H	L	X	X	Q_0	$\overline{Q}_0$

* This configuration is nonstable; that is, it will not persist when Preset or Clear return to their inactive (high) level.

logic symbol

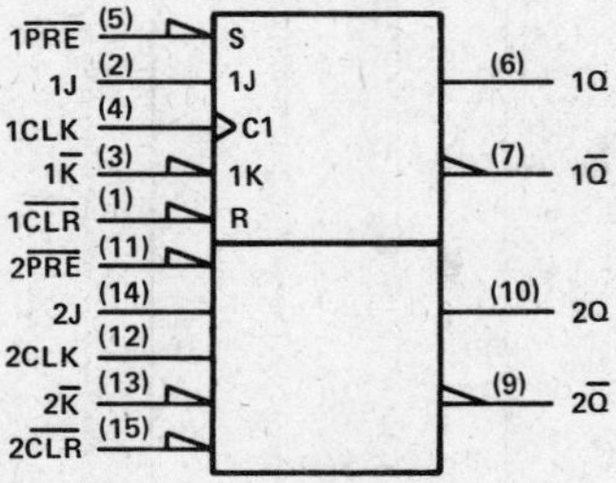

Pin numbers shown are for J and N packages.

logic diagram, each flip-flop (positive logic)

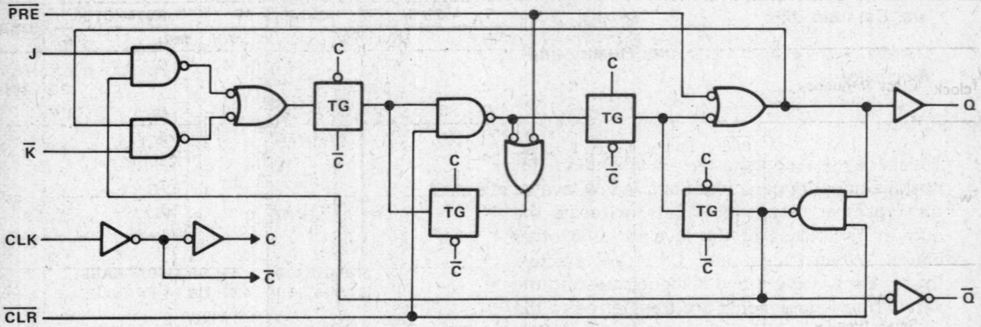

maximum ratings, recommended operating conditions, and electrical characteristics

See Table II, page 2-6.

timing requirements over recommended operating free-air temperature range (unless otherwise noted)

			V_{CC}	T_A = 25°C MIN	T_A = 25°C MAX	SN54HC109 MIN	SN54HC109 MAX	SN74HC109 MIN	SN74HC109 MAX	UNIT
f_{clock}	Clock frequency		2 V	0	6	0	4.2	0	5	MHz
			4.5 V	0	31	0	21	0	25	
			6 V	0	36	0	25	0	29	
t_w	Pulse duration	$\overline{PRE}$ or $\overline{CLR}$ low	2 V	100		150		125		ns
			4.5 V	20		30		25		
			6 V	17		25		21		
		CLK high or low	2 V	80		120		100		
			4.5 V	16		24		20		
			6 V	14		20		17		
t_{su}	Setup time before CLK↑	Data (J, K)	2 V	100		150		125		ns
			4.5 V	20		30		25		
			6 V	17		25		21		
		$\overline{PRE}$ or $\overline{CLR}$ inactive	2 V	25		40		30		
			4.5 V	5		8		6		
			6 V	4		7		5		
t_h	Hold time, data after CLK↑		2 V	0		0		0		ns
			4.5 V	0		0		0		
			6 V	0		0		0		

switching characteristics over recommended operating free-air temperature range (unless otherwise noted), C_L = 50 pF (see Note 1)

PARAMETER	FROM (INPUT)	TO (OUTPUT)	V_{CC}	T_A = 25°C MIN	T_A = 25°C TYP	T_A = 25°C MAX	SN54HC109 MIN	SN54HC109 MAX	SN74HC109 MIN	SN74HC109 MAX	UNIT
f_{max}			2 V	6	10		4.2		5		MHz
			4.5 V	31	50		21		25		
			6 V	36	60		25		29		
t_{pd}	$\overline{PRE}$ or $\overline{CLR}$	Q or $\overline{Q}$	2 V		60	230		345		290	ns
			4.5 V		15	46		69		58	
			6 V		12	39		59		49	
t_{pd}	CLK	Q or $\overline{Q}$	2 V		50	175		250		220	ns
			4.5 V		15	35		50		44	
			6 V		12	30		42		37	
t_t		Q or $\overline{Q}$	2 V		28	75		110		95	ns
			4.5 V		8	15		22		19	
			6 V		6	13		19		16	

C_{pd}	Power dissipation capacitance per flip-flop	No load, T_A = 25°C	35 pF typ

NOTE 1: For load circuit and voltage waveforms, see page 1-14.

TYPES SN54HC112, SN74HC112 DUAL J-K NEGATIVE-EDGE-TRIGGERED FLIP-FLOPS WITH CLEAR AND PRESET

D2684, DECEMBER 1982—REVISED MARCH 1984

- **Package Options Include Both Plastic and Ceramic Chip Carriers in Addition to Plastic and Ceramic DIPs**
- **Dependable Texas Instruments Quality and Reliability**

description

These devices contain two independent J-K negative-edge-triggered flip-flops. A low level at the Preset or Clear inputs sets or resets the outputs regardless of the levels of the other inputs. When Preset and Clear are inactive (high), data at the J and K inputs meeting the setup time requirements are transferred to the outputs on the negative-going edge of the clock pulse. Clock triggering occurs at a voltage level and is not directly related to the rise time of the clock pulse. Following the hold time interval, data at the J and K inputs may be changed without affecting the levels at the outputs. These versatile flip-flops can perform as toggle flip-flops by tying J and K high.

The SN54HC112 is characterized for operation over the full military temperature range of −55°C to 125°C. The SN74HC112 is characterized for operation from −40°C to 85°C.

FUNCTION TABLE

INPUTS					OUTPUTS	
$\overline{PRE}$	$\overline{CLR}$	CLK	J	K	Q	$\overline{Q}$
L	H	X	X	X	H	L
H	L	X	X	X	L	H
L	L	X	X	X	H*	H*
H	H	↓	L	L	Q_0	$\overline{Q}_0$
H	H	↓	H	L	H	L
H	H	↓	L	H	L	H
H	H	↓	H	H	TOGGLE	
H	H	H	X	X	Q_0	$\overline{Q}_0$

*This configuration is nonstable; that is, it will not persist when either Preset or Clear returns to its inactive (high) level.

SN54HC112 . . . J PACKAGE
SN74HC112 . . . J OR N PACKAGE
(TOP VIEW)

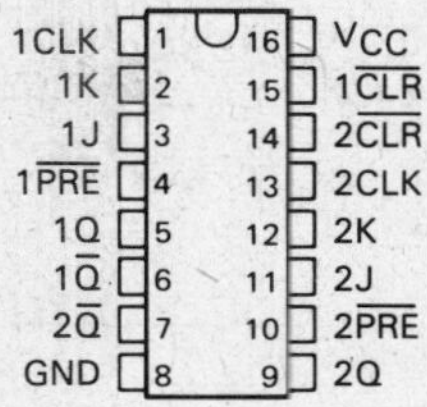

SN54HC112 . . . FH OR FK PACKAGE
SN74HC112 . . . FH OR FN PACKAGE
(TOP VIEW)

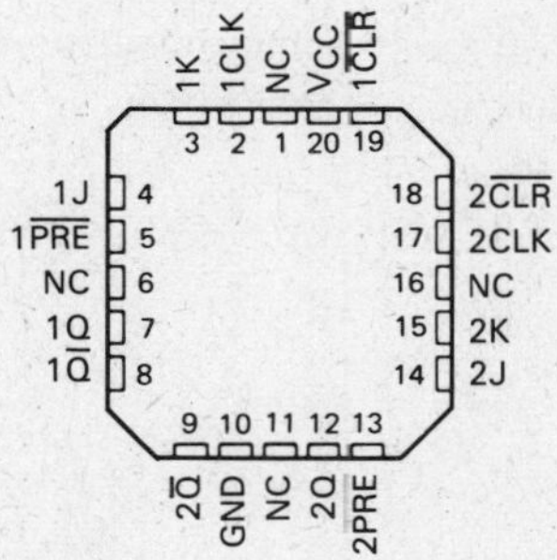

NC—No internal connection

logic symbol

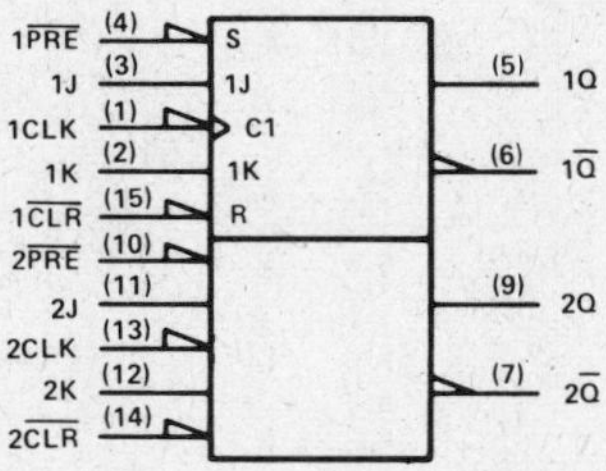

Pin numbers shown are for J and N packages.

logic diagram, each flip-flop (positive logic)

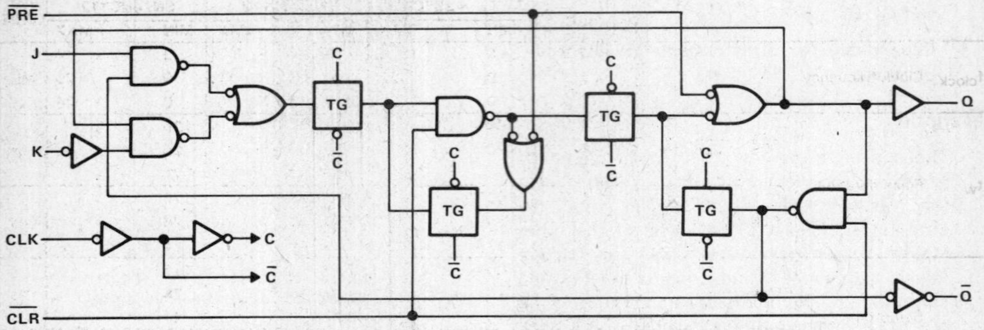

maximum ratings, recommended operating conditions, and electrical characteristics

See Table II, page 2-6.

TEXAS INSTRUMENTS
POST OFFICE BOX 225012 • DALLAS, TEXAS 75265

timing requirements over recommended operating free-air temperature range (unless otherwise noted)

			V_{CC}	T_A = 25°C MIN	T_A = 25°C MAX	SN54HC112 MIN	SN54HC112 MAX	SN74HC112 MIN	SN74HC112 MAX	UNIT
f_{clock}	Clock frequency		2 V	0	5	0	3.3	0	4	MHz
			4.5 V	0	25	0	17	0	20	
			6 V	0	29	0	20	0	24	
t_w	Pulse duration	$\overline{PRE}$ or $\overline{CLR}$ low	2 V	100		150		125		ns
			4.5 V	20		30		25		
			6 V	17		25		21		
		CLK high or low	2 V	100		150		125		ns
			4.5 V	20		30		25		
			6 V	17		25		21		
t_{su}	Setup time before CLK↓	Data (J,K)	2 V	100		150		125		ns
			4.5 V	20		30		25		
			6 V	17		25		21		
		$\overline{PRE}$ or $\overline{CLR}$ inactive	2 V	100		150		125		ns
			4.5 V	20		30		25		
			6 V	17		25		21		
t_h	Hold time, data after CLK↓		2 V	0		0		0		ns
			4.5 V	0		0		0		
			6 V	0		0		0		

switching characteristics over recommended operating free-air temperature range (unless otherwise noted), C_L = 50 pF (see Note 1)

PARAMETER	FROM (INPUT)	TO (OUTPUT)	V_{CC}	T_A = 25°C MIN	T_A = 25°C TYP	T_A = 25°C MAX	SN54HC112 MIN	SN54HC112 MAX	SN74HC112 MIN	SN74HC112 MAX	UNIT
f_{max}			2 V	5	10		3.3		4		MHz
			4.5 V	25	50		17		20		
			6 V	29	60		20		24		
t_{pd}	$\overline{PRE}$ or $\overline{CLR}$	Q or $\overline{Q}$	2 V		54	165		245		205	ns
			4.5 V		16	33		49		41	
			6 V		13	28		42		35	
t_{pd}	CLK	Q or $\overline{Q}$	2 V		56	125		185		155	ns
			4.5 V		16	25		37		31	
			6 V		13	21		31		26	
t_t		Q or $\overline{Q}$	2 V		29	75		110		95	ns
			4.5 V		9	15		22		19	
			6 V		8	13		19		16	

C_{pd}	Power dissipation capacitance per flip-flop	No load, T_A = 25°C	35 pF typ

NOTE 1: For load circuit and voltage waveforms, see page 1-14.

3 HCMOS DEVICES

TYPES SN54HC113, SN74HC113 DUAL J-K NEGATIVE-EDGE-TRIGGERED FLIP-FLOPS WITH PRESET

D2684, DECEMBER 1982—REVISED MARCH 1984

- **Package Options Include Both Plastic and Ceramic Chip Carriers in Addition to Plastic and Ceramic DIPs**
- **Dependable Texas Instruments Quality and Reliability**

SN54HC113 . . . J PACKAGE
SN74HC113 . . . J OR N PACKAGE
(TOP VIEW)

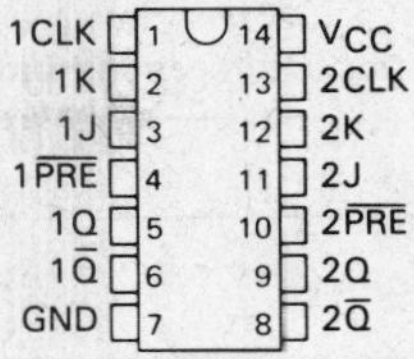

SN54HC113 . . . FH OR FK PACKAGE
SN74HC113 . . . FH OR FN PACKAGE
(TOP VIEW)

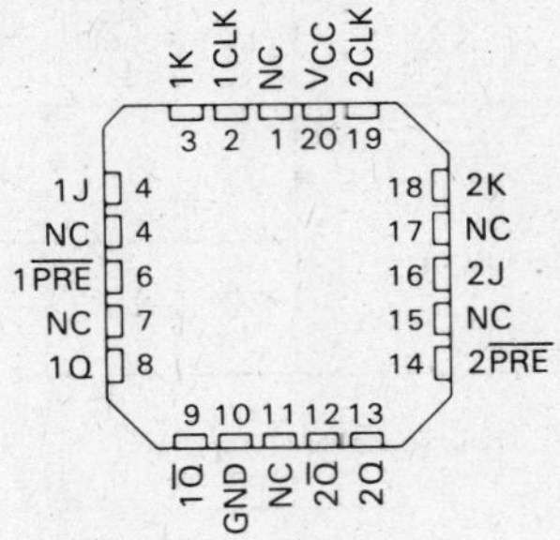

NC—No internal connection

description

These devices contain two independent J-K negative-edge-triggered flip-flops. A low level at the Preset input sets the outputs regardless of the levels of the other inputs. When Preset ($\overline{PRE}$) is inactive (high), data at the J and K inputs meeting the setup time requirements are transferred to the outputs on the negative-going edge of the clock pulse. Clock triggering occurs at a voltage level and is not directly related to the rise time of the clock pulse. Following the hold time interval, data at the J and K inputs may be changed without affecting the outputs. These versatile flip-flops can perform as toggle flip-flops by tying J and K high.

The SN54HC113 is characterized for operation over the full military temperature range of −55°C to 125°C. The SN74HC113 is characterized for operation from −40°C to 85°C.

FUNCTION TABLE

INPUTS				OUTPUTS	
$\overline{PRE}$	CLK	J	K	Q	$\overline{Q}$
L	X	X	X	H	L
H	↓	L	L	Q_0	$\overline{Q}_0$
H	↓	H	L	H	L
H	↓	L	H	L	H
H	↓	H	H	TOGGLE	
H	H	X	X	Q_0	$\overline{Q}_0$

logic symbol

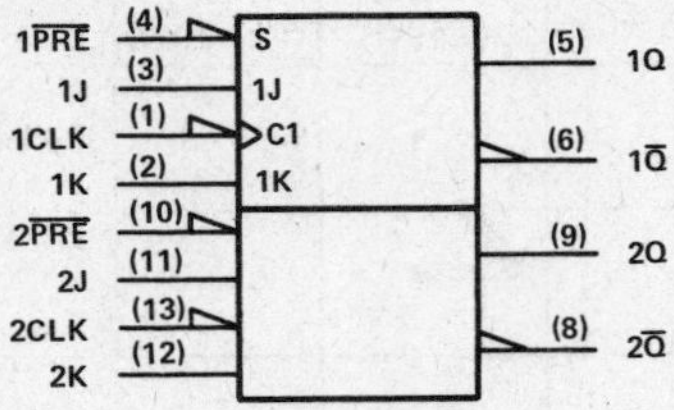

Pin numbers shown are for J and N packages.

logic diagram, each flip-flop (positive logic)

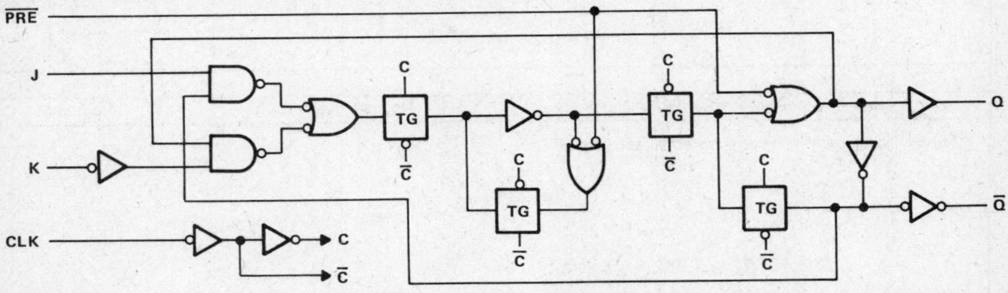

maximum ratings, recommended operating conditions, and electrical characteristics

See Table II, page 2-6.

timing requirements over recommended operating free-air temperature range (unless otherwise noted)

			V_{CC}	T_A = 25°C MIN	T_A = 25°C MAX	SN54HC113 MIN	SN54HC113 MAX	SN74HC113 MIN	SN74HC113 MAX	UNIT
f_{clock}	Clock frequency		2 V		6		4.2		5	MHz
			4.5 V		31		21		25	
			6 V		36		25		29	
t_w	Pulse duration	$\overline{PRE}$ low	2 V	100		150		125		ns
			4.5 V	20		30		25		
			6 V	17		25		21		
		CLK high or low	2 V	80		120		100		ns
			4.5 V	16		24		20		
			6 V	14		20		17		
t_{su}	Setup time before CLK↓	Data (J, K)	2 V	100		150		125		ns
			4.5 V	20		30		25		
			6 V	17		25		21		
		$\overline{PRE}$ inactive	2 V	25		40		30		ns
			4.5 V	5		8		6		
			6 V	4		7		5		
t_h	Hold time, data after CLK↓		2 V	0		0		0		ns
			4.5 V	0		0		0		
			6 V	0		0		0		

switching characteristics over recommended operating free-air temperature range (unless otherwise noted), C_L = 50 pF (see Note 1)

PARAMETER	FROM (INPUT)	TO (OUTPUT)	V_{CC}	T_A = 25°C MIN	T_A = 25°C TYP	T_A = 25°C MAX	SN54HC113 MIN	SN54HC113 MAX	SN74HC113 MIN	SN74HC113 MAX	UNIT
f_{max}			2 V	6	10		4.2		5		MHz
			4.5 V	31	50		21		25		
			6 V	36	60		25		29		
t_{pd}	$\overline{PRE}$	Q or $\overline{Q}$	2 V		60	165		250		205	ns
			4.5 V		18	33		50		41	
			6 V		15	28		43		35	
t_{pd}	CLK	Q or $\overline{Q}$	2 V		85	140		210		175	ns
			4.5 V		19	28		42		35	
			6 V		16	24		36		30	
t_t		Q or $\overline{Q}$	2 V		28	75		110		95	ns
			4.5 V		8	15		22		19	
			6 V		6	13		19		16	

C_{pd}	Power dissipation capacitance per flip-flop	No load, T_A = 25°C	35 pF typ

NOTE 1: For load circuit and voltage waveforms, see page 1-14.

TEXAS INSTRUMENTS
POST OFFICE BOX 225012 • DALLAS, TEXAS 75265

TYPES SN54HC114, SN74HC114 DUAL J-K NEGATIVE-EDGE-TRIGGERED FLIP-FLOPS WITH PRESET, COMMON CLEAR, AND COMMON CLOCK

D2684, DECEMBER 1982—REVISED MARCH 1984

- **Package Options Include Both Plastic and Ceramic Chip Carriers in Addition to Plastic and Ceramic DIPs**
- **Dependable Texas Instruments Quality and Reliability**

SN54HC114 . . . J PACKAGE
SN74HC114 . . . J OR N PACKAGE
(TOP VIEW)

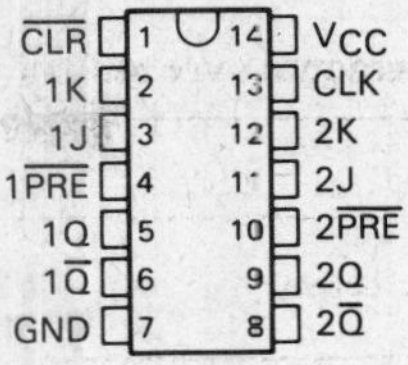

SN54HC114 . . . FH OR FK PACKAGE
SN74HC114 . . . FH OR FN PACKAGE
(TOP VIEW)

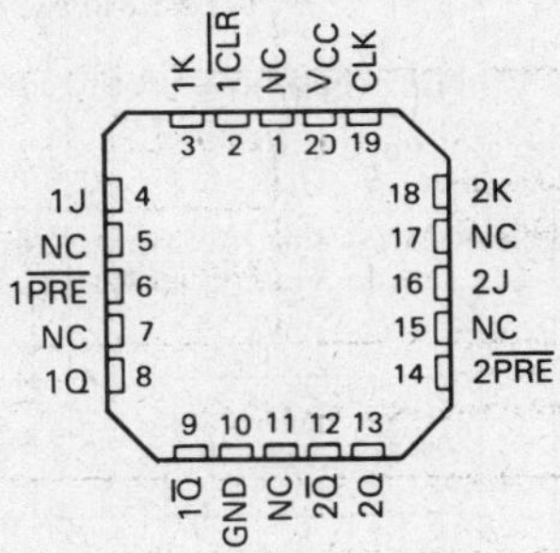

NC—No internal connection

description

These devices contain two independent J-K negative-edge-triggered flip-flops. a low level at the Preset or Clear inputs sets or resets the outputs regardless of the levels of the other inputs. When the Preset and Clear are inactive (high), data at the J and K inputs meeting the setup time requirements are transferred to the outputs on the negative-going edge of the clock pulse. Clock triggering occurs at a voltage level and is not directly related to the rise time of the clock pulse. Following the hold time interval, data at the J and K inputs may be changed without affecting the levels at the outputs. These versatile flip-flops can perform as toggle flip-flops by tying J and K high.

The SN54HC114 is characterized for operation over the full military temperature range of −55°C to 125°C. The SN74HC114 is characterized for operation from −40°C to 85°C.

FUNCTION TABLE

INPUTS					OUTPUTS	
$\overline{PRE}$	$\overline{CLR}$	CLK	J	K	Q	$\overline{Q}$
L	H	X	X	X	H	L
H	L	X	X	X	L	H
L	L	X	X	X	H*	H*
H	H	↓	L	L	Q_0	$\overline{Q}_0$
H	H	↓	H	L	H	L
H	H	↓	L	H	L	H
H	H	↓	H	H	TOGGLE	
H	H	H	X	X	Q_0	$\overline{Q}_0$

* This configuration is nonstable; that is, it will not persist when either Preset of Clear returns to its inactive (high) level.

logic symbol

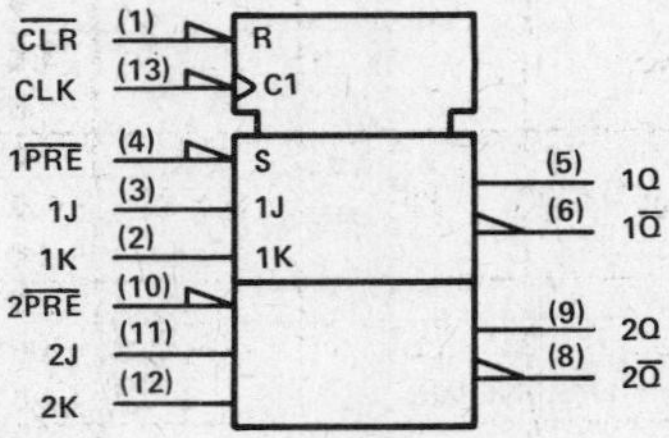

Pin numbers shown are for J and N packages.

logic diagram, each flip-flop (positive logic)

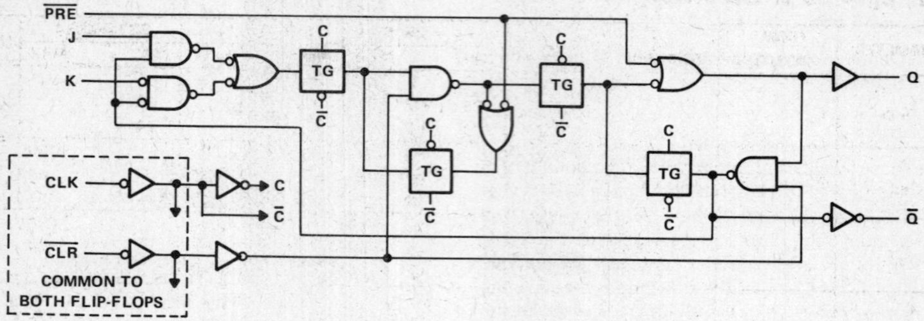

maximum ratings, recommended operating conditions, and electrical characteristics

See Table II, page 2-6.

timing requirements over recommended operating free-air temperature range (unless otherwise noted)

			V_{CC}	$T_A = 25°C$ MIN	$T_A = 25°C$ MAX	SN54HC114 MIN	SN54HC114 MAX	SN74HC114 MIN	SN74HC114 MAX	UNIT
f_{clock}	Clock frequency		2 V	0	5	0	3.3	0	4	MHz
			4.5 V	0	25	0	17	0	20	
			6 V	0	29	0	20	0	24	
t_w	Pulse duration	$\overline{PRE}$ or $\overline{CLR}$ low	2 V	100		150		125		ns
			4.5 V	20		30		25		
			6 V	17		25		21		
		CLK high or low	2 V	100		150		125		
			4.5 V	20		30		25		
			6 V	17		25		21		
t_{su}	Setup time before CLK↓	Data (J,K)	2 V	100		150		125		ns
			4.5 V	20		30		25		
			6 V	17		25		21		
		$\overline{PRE}$ or $\overline{CLR}$ inactive	2 V	100		150		125		
			4.5 V	20		30		25		
			6 V	17		25		21		
t_h	Hold time, data after CLK↓		2 V	0		0		0		ns
			4.5 V	0		0		0		
			6 V	0		0		0		

switching characteristics over recommended operating free-air temperature range (unless otherwise noted), C_L = 50 pF (see Note 1)

PARAMETER	FROM (INPUT)	TO (OUTPUT)	V_{CC}	T_A = 25°C MIN	T_A = 25°C MAX	SN54HC114 MIN	SN54HC114 MAX	SN74HC114 MIN	SN74HC114 MAX	UNIT
f_{max}			2 V	5	9	3.3		4		MHz
			4.5 V	25	45	17		20		
			6 V	29	50	20		24		
t_{pd}	$\overline{PRE}$ or $\overline{CLR}$	Q or $\overline{Q}$	2 V		75 175		250		220	ns
			4.5 V		20 35		50		44	
			6 V		17 30		42		37	
t_{pd}	CLK	Q or $\overline{Q}$	2 V		63 175		250		220	ns
			4.5 V		19 35		50		44	
			6 V		16 30		42		37	
t_t		Q or $\overline{Q}$	2 V		28 75		110		95	ns
			4.5 V		8 15		22		19	
			6 V		6 13		19		16	

C_{pd}	Power dissipation capacitance	No load, T_A = 25°C	50 pF typ

NOTE 1: For load circuit and voltage waveforms, see page 1-14.

3

HCMOS DEVICES

TYPES SN54HC125, SN54HC126 SN74HC125, SN74HC126 QUADRUPLE BUS BUFFER GATES WITH 3-STATE OUTPUTS

D2804, MARCH 1984

- High-Current 3-State Outputs Interface Directly with System Bus or Can Drive Up to 15 LSTTL Loads
- Package Options Include Both Plastic and Ceramic Chip Carriers in Addition to Plastic and Ceramic DIPs
- Dependable Texas Instruments Quality and Reliability

SN54HC125, SN54HC126 . . . J PACKAGE
SN74HC125, SN74HC126 . . . J OR N PACKAGE
(TOP VIEW)

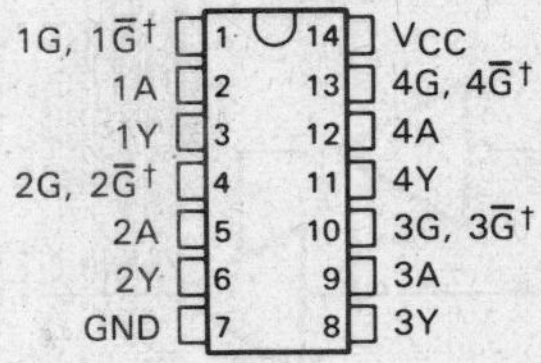

description

These bus buffers feature independent line drivers with three-state outputs. Each 'HC125 output is disabled when the associated $\overline{G}$ is high, and each 'HC126 output is disabled when the associated G is low.

The SN54HC125 and SN54HC126 are characterized for operation over the full military temperature range of −55°C to 125°C. The SN74HC125 and SN74HC126 are characterized for operation from −40°C to 85°C.

SN54HC125, SN54HC126 . . . FH OR FK PACKAGE
SN74HC125, SN74HC126 . . . FH OR FN PACKAGE
(TOP VIEW)

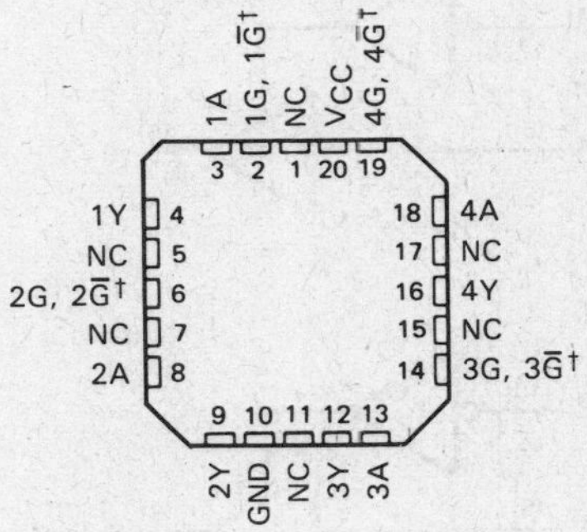

†$\overline{G}$ on 'HC125; G on 'HC126

NC—No internal connection

FUNCTION TABLES

'HC125
(EACH BUFFER)

INPUTS		OUTPUT
$\overline{G}$	A	Y
L	H	H
L	L	L
H	X	Z

'HC126
(EACH BUFFER)

INPUTS		OUTPUT
G	A	Y
H	H	H
H	L	L
L	X	Z

logic symbols

'HC125

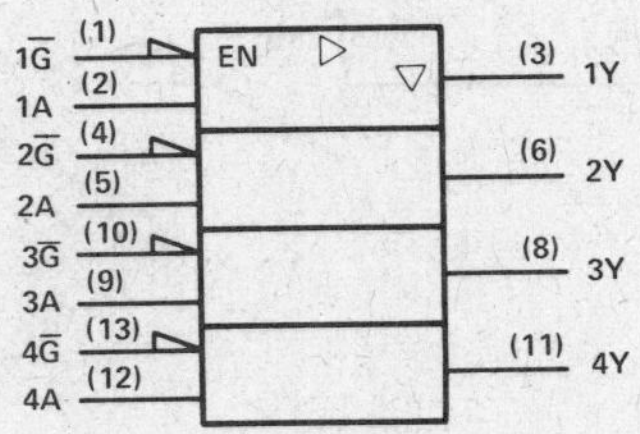

'HC126

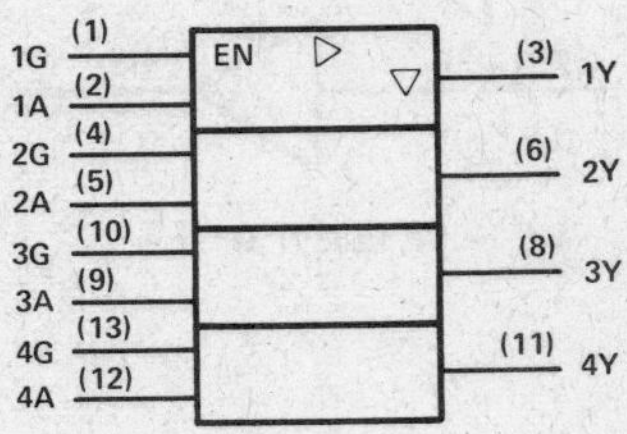

Pin numbers shown are for J and N packages.

logic diagrams (positive logic)

'HC125

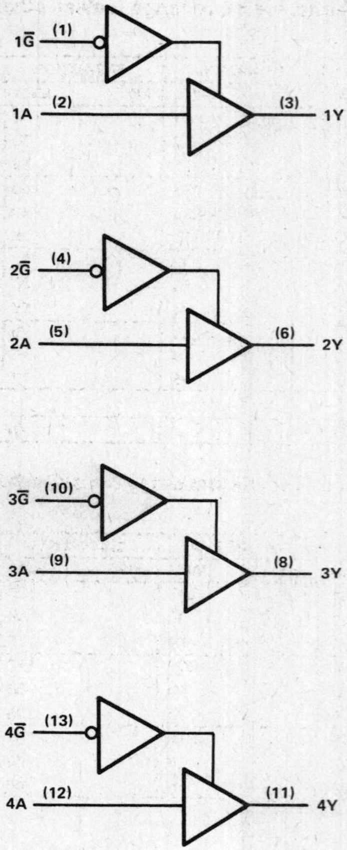

'HC126

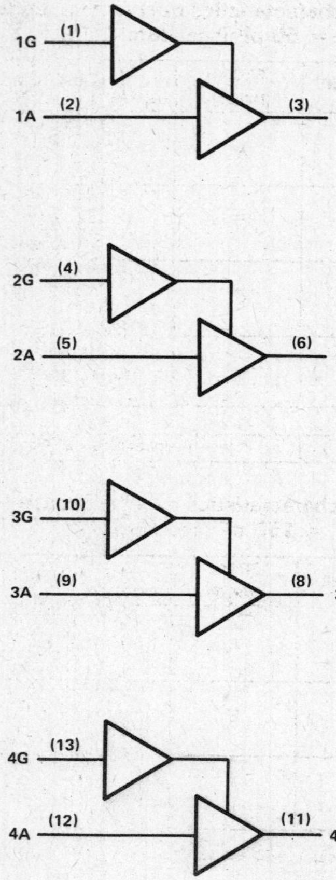

Pin numbers shown are for J and N packages.

maximum ratings, recommended operating conditions, and electrical characteristics

See Table III, page 2-8.

switching characteristics over recommended operating free-air temperature range (unless otherwise noted), C_L = 50 pF (see Note 1)

PARAMETER	FROM (INPUT)	TO (OUTPUT)	V_{CC}	T_A = 25°C MIN	TYP	MAX	SN54HC125 MIN	MAX	SN74HC125 MIN	MAX	UNIT
t_{pd}	A	Y	2 V		48	120		180		150	ns
			4.5 V		14	24		36		30	
			6 V		11	20		31		26	
t_{en}	$\overline{G}$	Y	2 V		53	120		180		150	ns
			4.5 V		14	24		36		30	
			6 V		11	20		31		26	
t_{dis}	$\overline{G}$	Y	2 V		30	120		180		150	ns
			4.5 V		15	24		36		30	
			6 V		14	20		31		26	
t_t		Any	2 V		28	60		90		75	ns
			4.5 V		8	12		18		15	
			6 V		6	10		15		13	

C_{pd}	Power dissipation capacitance per gate	No load, T_A = 25°C	45 pF typ

switching characteristics over recommended operating free-air temperature range (unless otherwise noted), C_L = 150 pF (see Note 1)

PARAMETER	FROM (INPUT)	TO (OUTPUT)	V_{CC}	T_A = 25°C MIN	TYP	MAX	SN54HC125 MIN	MAX	SN74HC125 MIN	MAX	UNIT
t_{pd}	A	Y	2 V		67	150		225		190	ns
			4.5 V		19	30		45		38	
			6 V		15	25		39		32	
t_{en}	$\overline{G}$	Y	2 V		100	135		200		170	ns
			4.5 V		20	27		40		34	
			6 V		17	23		34		29	
t_t		Any	2 V		45	210		315		265	ns
			4.5 V		17	42		63		53	
			6 V		13	36		53		45	

NOTE 1: For load circuit and voltage waveforms, see page 1-14.

maximum ratings, recommended operating conditions, and electrical characteristics

See Table III, page 2-8.

switching characteristics over recommended operating free-air temperature range (unless otherwise noted), C_L = 50 pF (see Note 1)

PARAMETER	FROM (INPUT)	TO (OUTPUT)	V_{CC}	T_A = 25°C MIN	TYP	MAX	SN54HC126 MIN	MAX	SN74HC126 MIN	MAX	UNIT
t_{pd}	A	Y	2 V		47	120		180		150	ns
			4.5 V		14	24		36		30	
			6 V		11	20		31		26	
t_{en}	G	Y	2 V		57	120		180		150	ns
			4.5 V		16	24		36		30	
			6 V		12	20		31		26	
t_{dis}	G	Y	2 V		35	120		180		150	ns
			4.5 V		17	24		36		30	
			6 V		15	20		31		26	
t_t		Any	2 V		28	60		90		75	ns
			4.5 V		8	12		18		15	
			6 V		6	10		15		13	

C_{pd}	Power dissipation capacitance per gate	No load, T_A = 25°C	45 pF typ

switching characteristics over recommended operating free-air temperature range (unless otherwise noted), C_L = 150 pF (see Note 1)

PARAMETER	FROM (INPUT)	TO (OUTPUT)	V_{CC}	T_A = 25°C MIN	TYP	MAX	SN54HC126 MIN	MAX	SN74HC126 MIN	MAX	UNIT
t_{pd}	A	Y	2 V		67	150		225		188	ns
			4.5 V		19	30		45		38	
			6 V		15	25		39		33	
t_{en}	G	Y	2 V		100	135		202		169	ns
			4.5 V		20	27		40		36	
			6 V		17	23		36		30	
t_t		Any	2 V		45	210		315		265	ns
			4.5 V		17	42		63		53	
			6 V		13	36		53		45	

NOTE 1: For load circuit and voltage waveforms, see page 1-14.

TYPES SN54HC133, SN74HC133
13-INPUT POSITIVE-NAND GATES

D2684, DECEMBER 1982—REVISED MARCH 1984

- **Package Options Include Both Plastic and Ceramic Chip Carriers in Addition to Plastic and Ceramic DIPs**
- **Dependable Texas Instruments Quality and Reliability**

SN54HC133 . . . J PACKAGE
SN74HC133 . . . J OR N PACKAGE
(TOP VIEW)

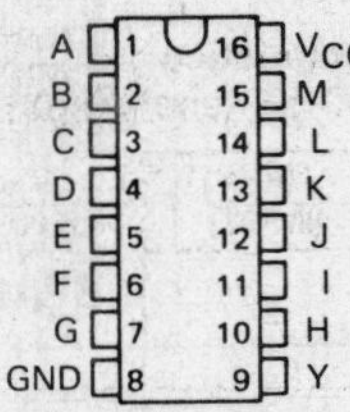

description

These devices contain a single 13-input NAND gate. They perform the Boolean functions in postive logic:

$$Y = \overline{A \cdot B \cdot C \cdot D \cdot E \cdot F \cdot G \cdot H \cdot I \cdot J \cdot K \cdot L \cdot M} \quad \text{or}$$

$$Y = \overline{A} + \overline{B} + \overline{C} + \overline{D} + \overline{E} + \overline{F} + \overline{G} + \overline{H} + \overline{I} + \overline{J} + \overline{K} + \overline{L} + \overline{M}$$

The SN54HC133 is characterized for operation over the full military temperature range of −55°C to 125°C. The SN74HC133 is characterized for operation from −40°C to 85°C.

SN54HC133 . . . FH OR FK PACKAGE
SN74HC133 . . . FH OR FN PACKAGE
(TOP VIEW)

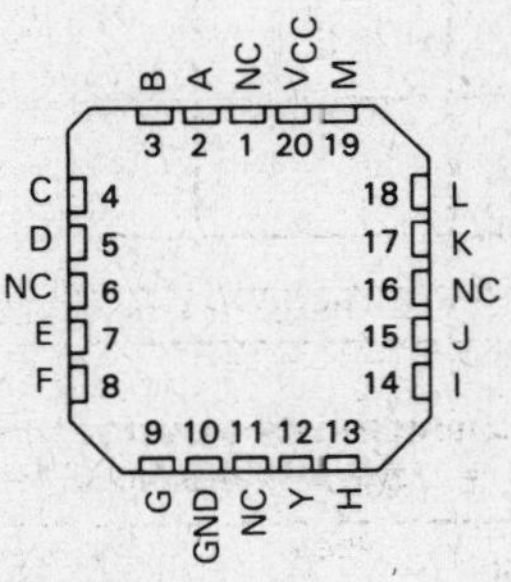

NC—No internal connection

logic symbol

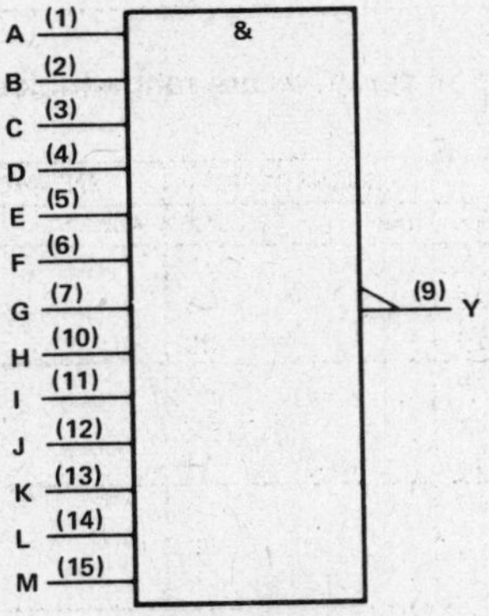

Pin numbers shown are for J and N packages.

FUNCTION TABLE

INPUTS A THRU M	OUTPUT Y
All inputs H	L
One or more inputs L	H

maximum ratings, recommended operating conditions, and electrical characteristics

See Table I, page 2-4.

switching characteristics over recommended operating free-air temperature range (unless otherwise noted), C_L = 50 pF (see Note 1)

PARAMETER	FROM (INPUT)	TO (OUTPUT)	V_{CC}	T_A = 25°C MIN	T_A = 25°C TYP	T_A = 25°C MAX	SN54HC133 MIN	SN54HC133 MAX	SN74HC133 MIN	SN74HC133 MAX	UNIT
t_{pd}	Any	Y	2 V		70	150		225		190	ns
			4.5 V		16	30		45		38	
			6 V		13	26		38		33	
t_t		Y	2 V		38	75		110		95	ns
			4.5 V		8	15		22		19	
			6 V		6	13		19		16	

C_{pd}	Power dissipation capacitance	No load, T_A = 25°C	24 pF typ

NOTE 1: For load circuit and voltage waveforms, see page 1-14.

TYPES SN54HC137, SN74HC137 3-LINE TO 8-LINE DECODERS/DEMULTIPLEXERS WITH ADDRESS LATCHES

D2684, DECEMBER 1982—REVISED MARCH 1984

- Combines Decoder and 3-Bit Address Latch
- Incorporates 2 Output Enables to Simplify Cascading
- Package Options Include Both Plastic and Ceramic Chip Carriers in Addition to Plastic and Ceramic DIPs
- Dependable Texas Instruments Quality and Reliability

SN54HC137 . . . J PACKAGE
SN74HC137 . . . J OR N PACKAGE
(TOP VIEW)

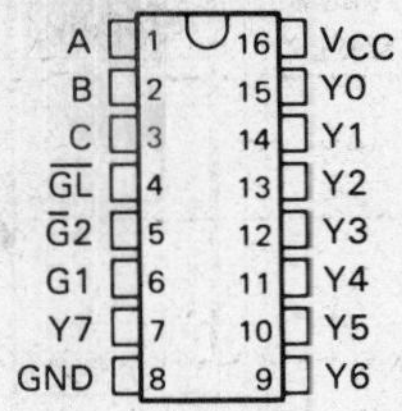

SN54HC137 . . . FH OR FK PACKAGE
SN74HC137 . . . FH OR FN PACKAGE
(TOP VIEW)

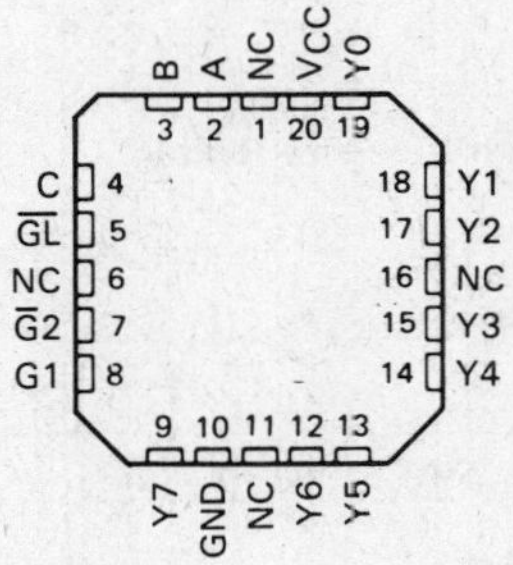

NC—No internal connection

description

The 'HC137 is a three-line to eight-line decoder/demultiplexer with latches on the three address inputs. When the latch-enable input ($\overline{GL}$) is low, the 'HC137 acts as a decoder/demultiplexer. When $\overline{GL}$ goes from low to high, the address present at the select inputs (A, B, and C) is stored in the latches. Further address changes are ignored as long as $\overline{GL}$ remains high. The output enable controls, G1 and $\overline{G}2$, control the outputs independently of the select or latch-enable inputs. All of the outputs are forced high if G1 is low or $\overline{G}2$ is high. The 'HC137 is ideally suited for implementing glitch-free decoders in strobed (stored-address) applications in bus-oriented systems.

The SN54HC137 is characterized for operation over the full military temperature range of −55°C to 125°C. The SN74HC137 is characterized for operation from −40°C to 85°C.

logic symbols (alternatives)

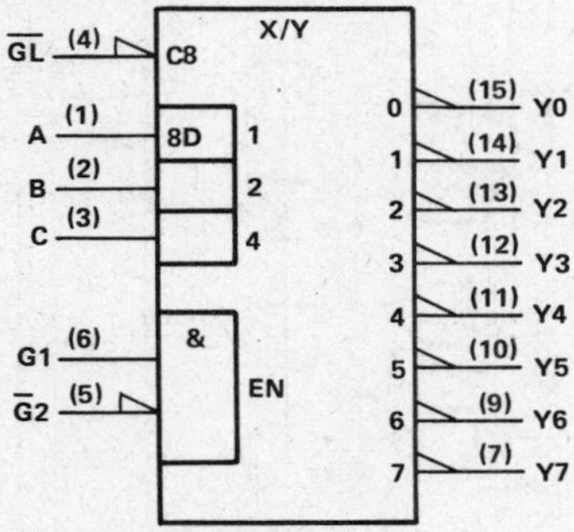

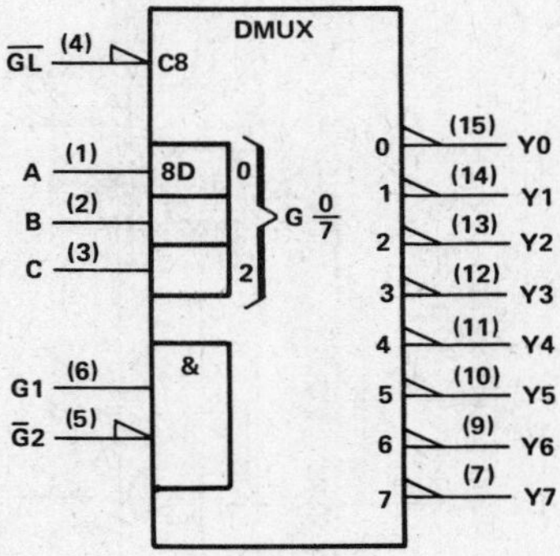

Pin numbers shown are for J and N packages.

FUNCTION TABLE

INPUTS						OUTPUTS							
ENABLE			SELECT										
$\overline{GL}$	G1	$\overline{G2}$	C	B	A	Y0	Y1	Y2	Y3	Y4	Y5	Y6	Y7
X	X	H	X	X	X	H	H	H	H	H	H	H	H
X	L	X	X	X	X	H	H	H	H	H	H	H	H
L	H	L	L	L	L	L	H	H	H	H	H	H	H
L	H	L	L	L	H	H	L	H	H	H	H	H	H
L	H	L	L	H	L	H	H	L	H	H	H	H	H
L	H	L	L	H	H	H	H	H	L	H	H	H	H
L	H	L	H	L	L	H	H	H	H	L	H	H	H
L	H	L	H	L	H	H	H	H	H	H	L	H	H
L	H	L	H	H	L	H	H	H	H	H	H	L	H
L	H	L	H	H	H	H	H	H	H	H	H	H	L
H	H	L	X	X	X	Output corresponding to stored address, L; all others, H							

logic diagram (positive logic)

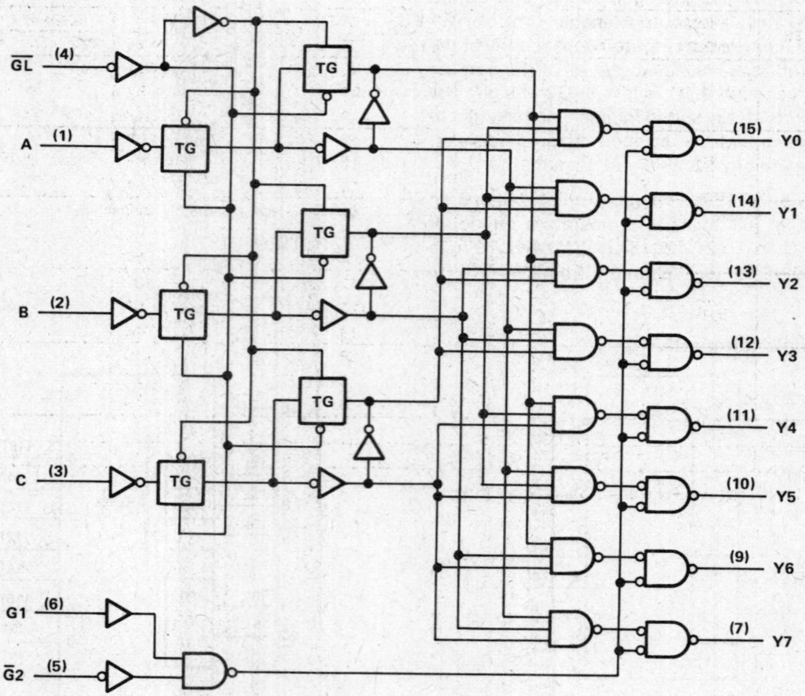

Pin numbers shown are for J and N packages.

TEXAS INSTRUMENTS
POST OFFICE BOX 225012 • DALLAS, TEXAS 75265

absolute maximum ratings, recommended operating conditions, and electrical characteristics

See Table IV, page 2-10.

timing requirements over recommended operating free-air temperature range (unless otherwise noted)

		V_{CC}	T_A = 25°C		SN54HC137		SN74HC137		UNIT
			MIN	MAX	MIN	MAX	MIN	MAX	
t_w	Pulse duration, $\overline{GL}$ low	2 V	80		120		100		ns
		4.5 V	16		24		20		
		6 V	14		20		17		
t_{su}	Setup time, A, B, and C before $\overline{GL}$↑	2 V	75		115		95		ns
		4.5 V	15		23		19		
		6 V	13		20		16		
t_h	Hold time, A, B, and C after $\overline{GL}$↑	2 V	5		5		5		ns
		4.5 V	5		5		5		
		6 V	5		5		5		

switching characteristics over recommended operating free-air temperature range (unless otherwise noted), C_L = 50 pF (see Note 1)

PARAMETER	FROM (INPUT)	TO (OUTPUT)	V_{CC}	T_A = 25°C			SN54HC137		SN74HC137		UNIT
				MIN	TYP	MAX	MIN	MAX	MIN	MAX	
t_{pd}	A, B, C	Y	2 V		82	190		285		240	ns
			4.5 V		23	38		57		48	
			6 V		19	32		48		41	
t_{pd}	$\overline{G}$2	Y	2 V		59	145		220		180	ns
			4.5 V		17	29		44		36	
			6 V		14	25		37		31	
t_{pd}	G1	Y	2 V		61	145		220		180	ns
			4.5 V		17	29		44		36	
			6 V		14	25		37		31	
t_{pd}	$\overline{GL}$	Y	2 V		77	190		285		240	ns
			4.5 V		22	38		57		48	
			6 V		19	32		48		41	
t_t		Y	2 V		38	75		110		95	ns
			4.5 V		8	15		22		19	
			6 V		6	13		19		16	

C_{pd}	Power dissipation capacitance	No load, T_A = 25°C	85 pF typ

NOTE 1: For load circuit and voltage waveforms, see page 1-14.

3

HCMOS DEVICES

TYPES SN54HCT137, SN74HCT137
3-LINE TO 8-LINE DECODERS/DEMULTIPLEXERS WITH ADDRESS LATCHES

D2804, MARCH 1984

- **Inputs are TTL-Voltage Compatible**
- **Combines Decoder and 3-Bit Address Latch**
- **Incorporates 2 Output Enables to Simplify Cascading**
- **Package Options Include Both Plastic and Ceramic Chip Carriers in Addition to Plastic and Ceramic DIPs**
- **Dependable Texas Instruments Quality and Reliability**

SN54HCT137 . . . J PACKAGE
SN74HCT137 . . . J OR N PACKAGE
(TOP VIEW)

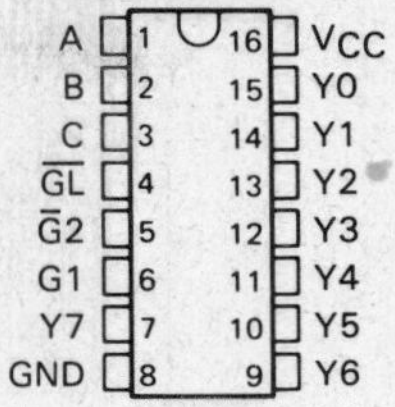

SN54HCT137 . . . FH OR FK PACKAGE
SN74HCT137 . . . FH OR FN PACKAGE
(TOP VIEW)

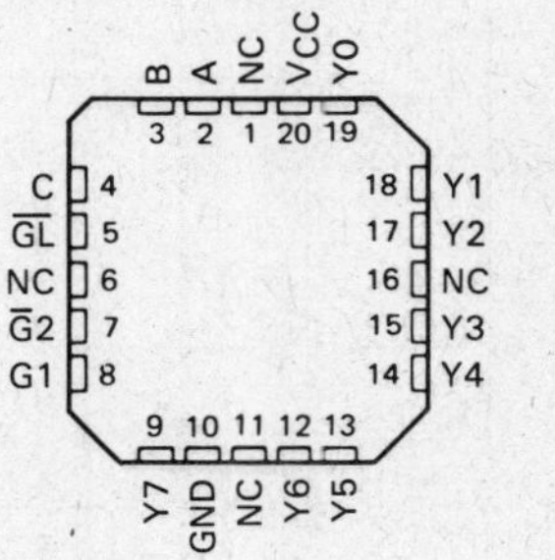

NC—No internal connection

description

The 'HCT137 is a three-line to eight-line decoder/demultiplexer with latches on the three address inputs. When the latch-enable input ($\overline{GL}$) is low, the 'HCT137 acts as a decoder/demultiplexer. When $\overline{GL}$ goes from low to high, the address present at the select inputs (A, B, and C) is stored in the latches. Further address changes are ignored as long as $\overline{GL}$ remains high. The output enable controls, G1 and $\overline{G}2$, control the outputs independently of the select or latch-enable inputs. All of the outputs are forced high if G1 is low or $\overline{G}2$ is high. The 'HCT137 is ideally suited for implementing glitch-free decoders in strobed (stored-address) applications in bus-oriented systems.

The SN54HCT137 is characterized for operation over the full military temperature range of −55°C to 125°C. The SN74HCT137 is characterized for operation from −40°C to 85°C.

logic symbols (alternatives)

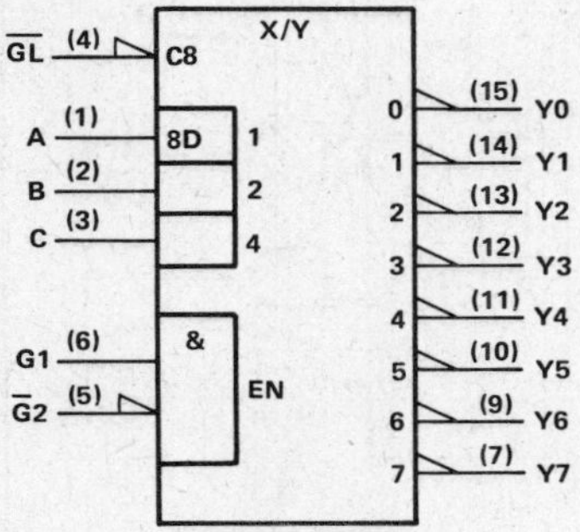

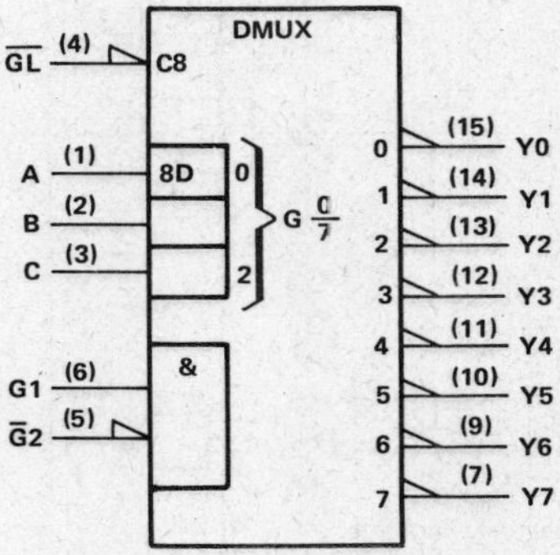

Pin numbers shown are for J and N packages.

FUNCTION TABLE

INPUTS						OUTPUTS							
ENABLE			SELECT										
$\overline{GL}$	G1	$\overline{G2}$	C	B	A	Y0	Y1	Y2	Y3	Y4	Y5	Y6	Y7
X	X	H	X	X	X	H	H	H	H	H	H	H	H
X	L	X	X	X	X	H	H	H	H	H	H	H	H
L	H	L	L	L	L	L	H	H	H	H	H	H	H
L	H	L	L	L	H	H	L	H	H	H	H	H	H
L	H	L	L	H	L	H	H	L	H	H	H	H	H
L	H	L	L	H	H	H	H	H	L	H	H	H	H
L	H	L	H	L	L	H	H	H	H	L	H	H	H
L	H	L	H	L	H	H	H	H	H	H	L	H	H
L	H	L	H	H	L	H	H	H	H	H	H	L	H
L	H	L	H	H	H	H	H	H	H	H	H	H	L
H	H	L	X	X	X	Output corresponding to stored address, L; all others, H							

logic diagram (positive logic)

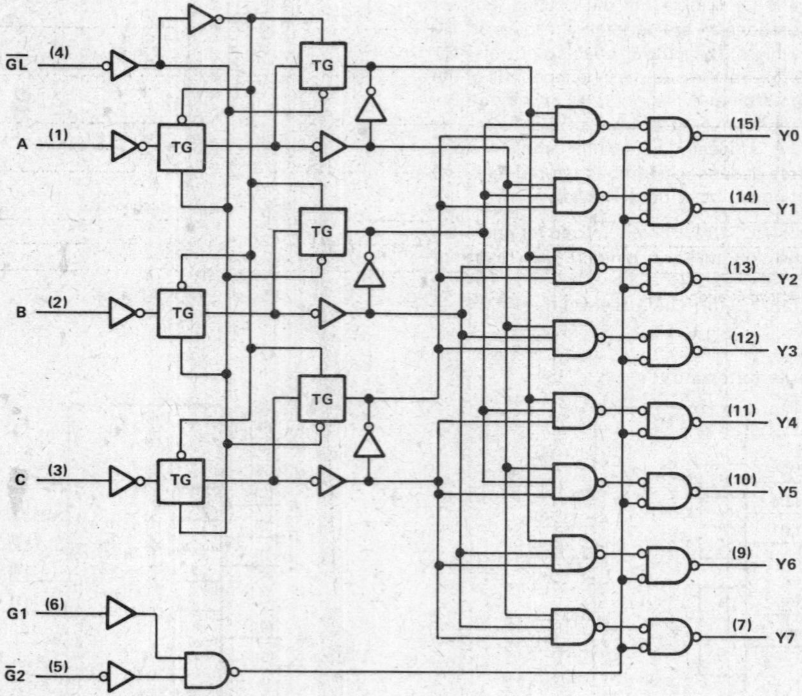

Pin numbers shown are for J and N packages.

TEXAS INSTRUMENTS
POST OFFICE BOX 225012 • DALLAS, TEXAS 75265

absolute maximum ratings, recommended operating conditions, and electrical characteristics

See Table VIII, page 2-15.

timing requirements over recommended operating free-air temperature range (unless otherwise noted)

PARAMETER		V_{CC}	T_A = 25°C			SN54HCT137		SN74HCT137		UNIT
			MIN	TYP	MAX	MIN	MAX	MIN	MAX	
t_w	Pulse duration, $\overline{GL}$ low	4.5 V	26			39		33		ns
		5.5 V	23			35		30		
t_{su}	Setup time, A, B, and C before $\overline{GL}$↑	4.5 V	15			23		19		ns
		5.5 V	14			21		17		
t_h	Hold time, A, B, and C after $\overline{GL}$↑	4.5 V	5			5		5		ns
		5.5 V	5			5		5		

switching characteristics over recommended operating free-air temperature range (unless otherwise noted), C_L = 50 pF (see Note 1)

PARAMETER	FROM (INPUT)	TO (OUTPUT)	V_{CC}	T_A = 25°C			SN54HCT137		SN74HCT137		UNIT
				MIN	TYP	MAX	MIN	MAX	MIN	MAX	
t_{pd}	A, B, C	Y	4.5 V		25	38		57		48	ns
			5.5 V		20	34		51		43	
t_{pd}	$\overline{G}2$	Y	4.5 V		20	29		44		36	ns
			5.5 V		17	25		40		32	
t_{pd}	G1	Y	4.5 V		20	29		44		36	ns
			5.5 V		17	25		40		32	
t_{pd}	$\overline{GL}$	Y	4.5 V		32	42		63		52	ns
			5.5 V		25	36		57		47	
t_t		Any	4.5 V		12	15		22		19	ns
			5.5 V		11	14		20		17	

C_{pd}	Power dissipation capacitance	No load, T_A = 25°C	85 pF typ

NOTE 1: For load circuit and voltage waveforms, see page 1-14.

D2684, DECEMBER 1982—REVISED MARCH 1984

- **Designed Specifically for High-Speed Memory Decoders and Data Transmission Systems**
- **Incorporates 3 Enable Inputs to Simplify Cascading and/or Data Reception**
- **Package Options Include Both Plastic and Ceramic Chip Carriers in Addition to Plastic and Ceramic DIPs**
- **Dependable Texas Instruments Quality and Reliability**

SN54HC138 . . . J PACKAGE
SN74HC138 . . . J OR N PACKAGE
(TOP VIEW)

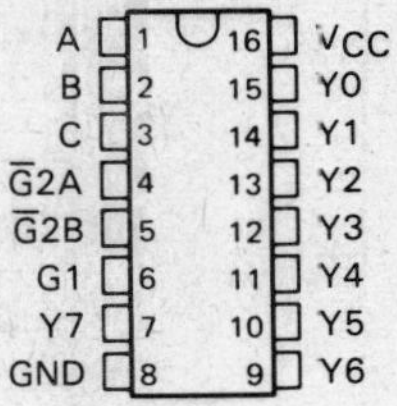

SN54HC138 . . . FH OR FK PACKAGE
SN74HC138 . . . FH OR FN PACKAGE
(TOP VIEW)

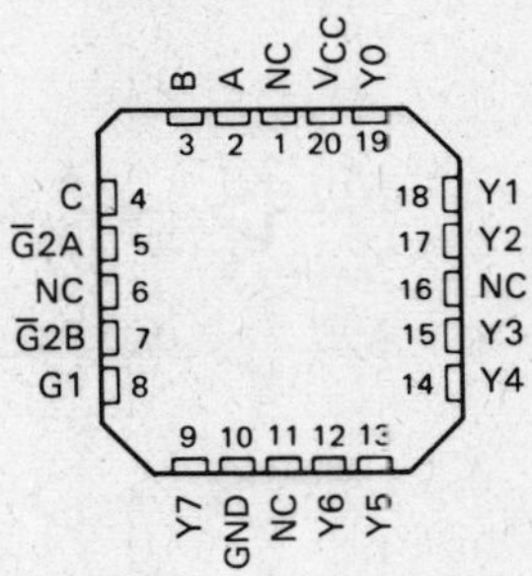

NC—No internal connection

description

The 'HC138 circuit is designed to be used in high-performance memory-decoding or data-routing applications requiring very short propagation delay times. In high-performance memory systems this decoder can be used to minimize the effects of system decoding. When employed with high-speed memories utilizing a fast enable circuit, the delay times of this decoder and the enable time of the memory are usually less than the typical access time of the memory. This means that the effective system delay introduced by the decoder is negligible.

The conditions at the binary select inputs at the three enable inputs select one of eight input lines. Two active-low and one active-high enable inputs reduce the need for external gates or inverters when expanding. A 24-line decoder can be implemented without external inverters and a 32-line decoder requires only one inverter. An enable input can be used as a data input for demultiplexing applications.

The SN54HC138 is characterized for operation over the full military temperature range of −55°C to 125°C. The SN74HC138 is characterized for operation from −40°C to 85°C.

logic symbols (alternatives)

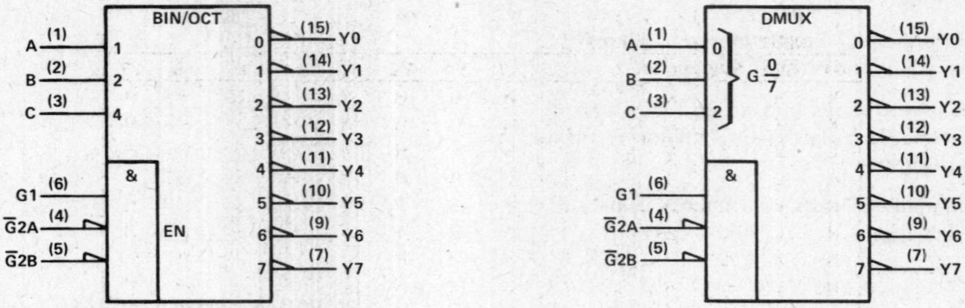

logic diagram (positive logic)

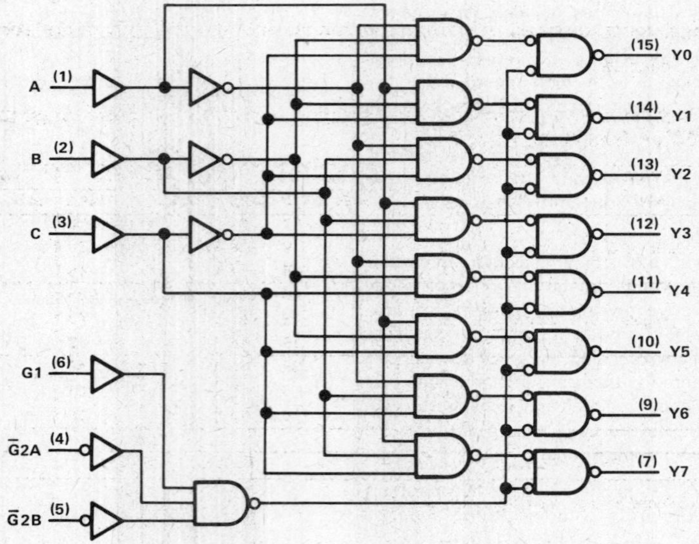

Pin numbers shown are for J and N packages.

FUNCTION TABLE

ENABLE INPUTS			SELECT INPUTS			OUTPUTS							
G1	$\overline{G}$2A	$\overline{G}$2B	C	B	A	Y0	Y1	Y2	Y3	Y4	Y5	Y6	Y7
X	H	X	X	X	X	H	H	H	H	H	H	H	H
X	X	H	X	X	X	H	H	H	H	H	H	H	H
L	X	X	X	X	X	H	H	H	H	H	H	H	H
H	L	L	L	L	L	L	H	H	H	H	H	H	H
H	L	L	L	L	H	H	L	H	H	H	H	H	H
H	L	L	L	H	L	H	H	L	H	H	H	H	H
H	L	L	L	H	H	H	H	H	L	H	H	H	H
H	L	L	H	L	L	H	H	H	H	L	H	H	H
H	L	L	H	L	H	H	H	H	H	H	L	H	H
H	L	L	H	H	L	H	H	H	H	H	H	L	H
H	L	L	H	H	H	H	H	H	H	H	H	H	L

maximum ratings, recommended operating conditions, and electrical characteristics

See Table IV, page 2-10.

switching characteristics over recommended operating free-air temperature range (unless otherwise noted), C_L = 50 pF (see Note 1)

PARAMETER	FROM (INPUT)	TO (OUTPUT)	V_{CC}	T_A = 25°C MIN	T_A = 25°C TYP	T_A = 25°C MAX	SN54HC138 MIN	SN54HC138 MAX	SN74HC138 MIN	SN74HC138 MAX	UNIT
t_{pd}	A, B, or C	Any Y	2 V		67	180		270		225	ns
			4.5 V		18	36		54		45	
			6 V		15	31		46		38	
t_{pd}	Enable	Any Y	2 V		66	155		235		195	ns
			4.5 V		18	31		47		39	
			6 V		15	26		40		33	
t_t		Any	2 V		38	75		110		95	ns
			4.5 V		8	15		22		19	
			6 V		6	13		19		16	

C_{pd}	Power dissipation capacitance	No load, T_A = 25°C	85 pF typ

NOTE 1: For load circuit and voltage waveforms, see page 1-14.

- Inputs Are TTL-Voltage Compatible
- Designed Specifically for High-Speed memory Decoders and Data Transmission Systems
- Incorporates 3 Enable Inputs to Simplify Cascading and/or Data Reception
- Package Options Include Both Plastic and Ceramic Chip Carriers in Addition to Plastic and Ceramic DIPs
- Dependable Texas instruments Quality and Reliability

SN54HCT138 . . . J PACKAGE
SN74HCT138 . . . J OR N PACKAGE
(TOP VIEW)

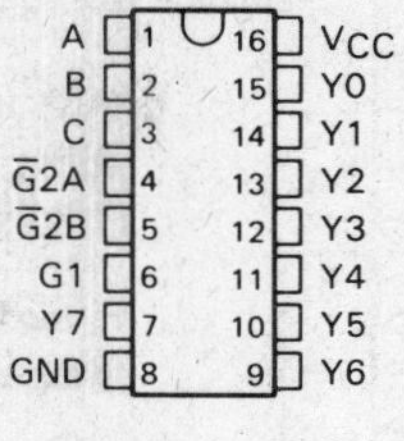

SN54HCT138 . . . FH OR FK PACKAGE
SN74HCT138 . . . FH OR FN PACKAGE
(TOP VIEW)

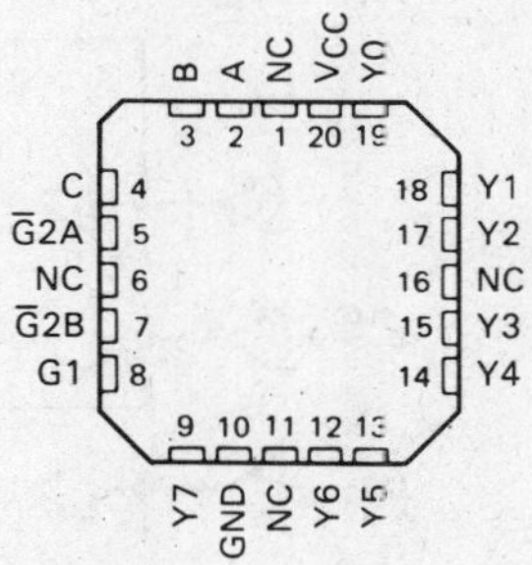

NC—No internal connection

description

The 'HCT138 circuit is designed to be used in high-performance memory-decoding or data-routing applications requiring very short propagation delay times. In high-performance memory systems, this decoder can be used to minimize the effects of system decoding. When employed with high-speed memories utilizing a fast enable circuit, the delay times of this decoder and the enable time of the memory are usually less than typical access time of the memory. This means that the effective system delay introduced by the decoder is negligible.

The conditions at the binary select inputs and the three enable inputs select one of eight input lines. Two active-low and one active-high enable inputs reduce the need for external gates or inverters when expanding. A 24-line decoder can be implemented without external inverters and a 32-line decoder requires only one inverter. An enable input can be used as a data input for demultiplexing applications.

The SN54HCT138 is characterized for operation over the full military temperature range of −55°C to 125°C. The SN74HCT138 is characterized for operation from −40°C to 85°C.

logic symbols (alternatives)

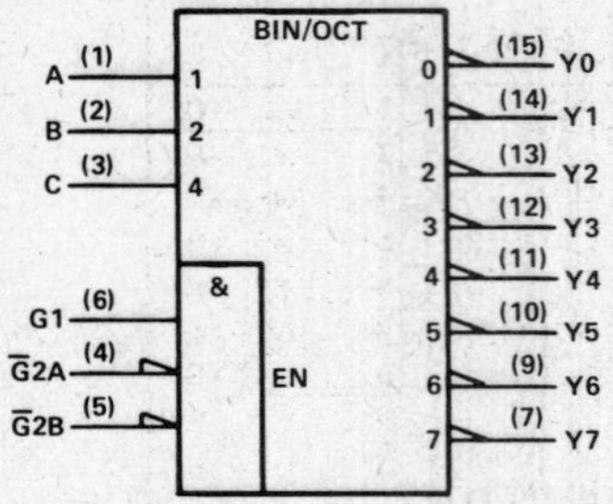

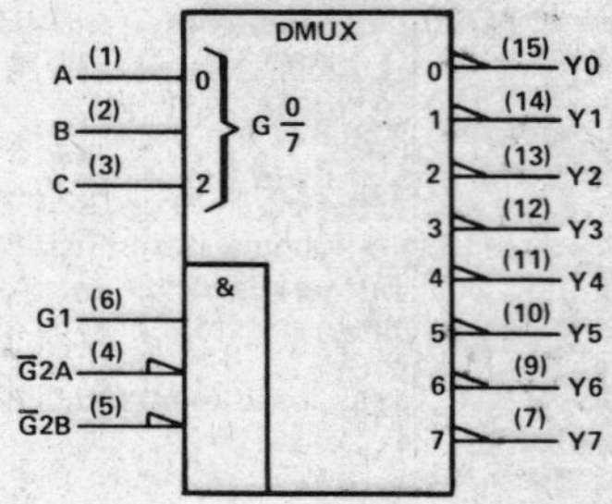

logic diagram (positive logic)

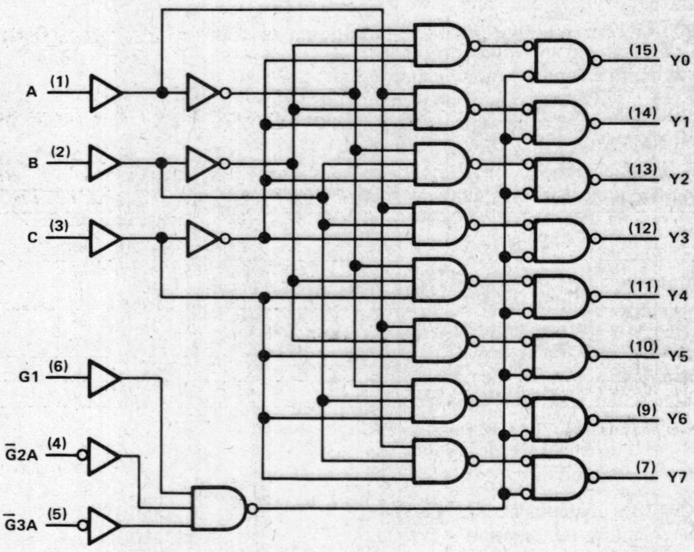

Pin numbers shown are for J and N packages.

TEXAS INSTRUMENTS
POST OFFICE BOX 225012 • DALLAS, TEXAS 75265

FUNCTION TABLE

ENABLE INPUTS			SELECT INPUTS			OUTPUTS							
G1	$\overline{G}$2A	$\overline{G}$2B	C	B	A	Y0	Y1	Y2	Y3	Y4	Y5	Y6	Y7
X	H	X	X	X	X	H	H	H	H	H	H	H	H
X	X	H	X	X	X	H	H	H	H	H	H	H	H
L	X	X	X	X	X	H	H	H	H	H	H	H	H
H	L	L	L	L	L	L	H	H	H	H	H	H	H
H	L	L	L	L	H	H	L	H	H	H	H	H	H
H	L	L	L	H	L	H	H	L	H	H	H	H	H
H	L	L	L	H	H	H	H	H	L	H	H	H	H
H	L	L	H	L	L	H	H	H	H	L	H	H	H
H	L	L	H	L	H	H	H	H	H	H	L	H	H
H	L	L	H	H	L	H	H	H	H	H	H	L	H
H	L	L	H	H	H	H	H	H	H	H	H	H	L

maximum ratings, recommended operating conditions, and electrical characteristics

See Table VIII, page 2-15.

switching characteristics over recommended operating free-air temperature range (unless otherwise noted), C_L = 50 pF (see Note 1)

PARAMETER	FROM (INPUT)	TO (OUTPUT)	V_{CC}	T_A = 25°C MIN	TYP	MAX	SN54HCT138 MIN	MAX	SN74HCT138 MIN	MAX	UNIT
t_{pd}	A, B, or C	Any Y	4.5 V		23	36		54		45	ns
			5.5 V		17	32		49		34	
t_{pd}	Enable	Any Y	4.5 V		22	33		50		42	ns
			5.5 V		18	30		45		38	
t_t		Any	4.5 V		11	15		22		19	ns
			5.5 V		11	14		20		17	

C_{pd}	Power dissipation capacitance	No load, T_A = 25°C	85 pF typ

NOTE 1: For load circuit and voltage waveforms, see page 1-14.

D2684, DECEMBER 1982—REVISED MARCH 1984

- Designed Specifically for High-Speed Memory Decoders and Data Transmission Systems
- Incorporates 2 Enable Inputs to Simplify Cascading and/or Data Reception
- Package Options Include Both Plastic and Ceramic Chip Carriers in Addition to Plastic and Ceramic DIPs
- Dependable Texas Instruments Quality and Reliability

SN54HC139 . . . J PACKAGE
SN74HC139 . . . J OR N PACKAGE
(TOP VIEW)

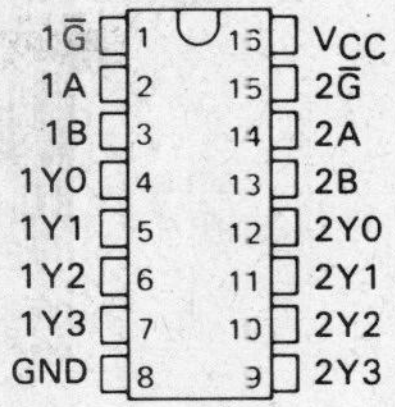

description

The ’HC139 circuit is designed to be used in high-performance memory-decoding or data-routing applications requiring very short propagation delay times. In high-performance memory systems, this decoder can be used to minimize the effects of system decoding. When employed with high-speed memories utilizing a fast-enable circuit, the delay times of this decoder and the enable time of the memory are usually less than the typical access time of the memory. This means that the effective system delay introduced by the decoder is negligible.

The ’HC139 is comprised of two individual two-line to four-line decoders in a single package. The active-low enable input can be used as a data line in demultiplexing applications. These decoders/demultiplexers feature fully buffered inputs, each of which represents only one normalized load to its driving circuit.

The SN54HC139 is characterized for operation over the full military temperature range of −55°C to 125°C. The SN74HC139 is characterized for operation from −40°C to 85°C.

SN54HC139 . . . FH OR FK PACKAGE
SN74HC139 . . . FH OR FN PACKAGE
(TOP VIEW)

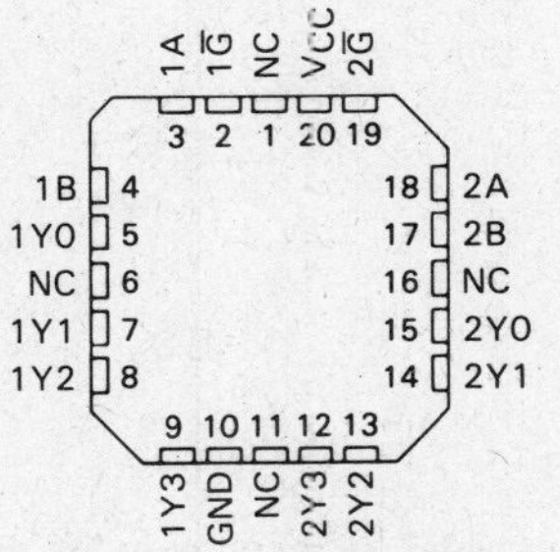

NC—No internal connection

logic symbols (alternatives)

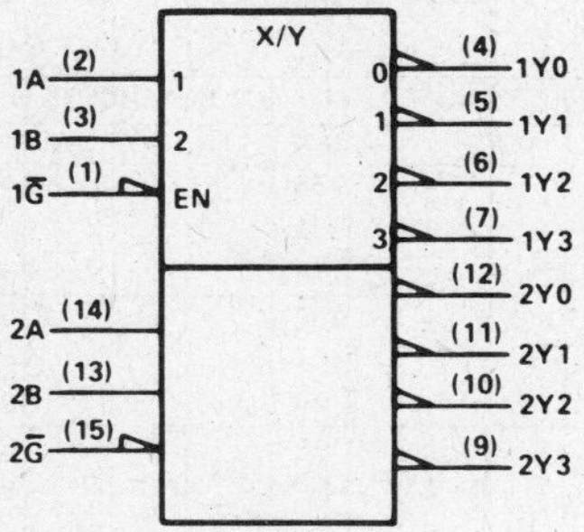

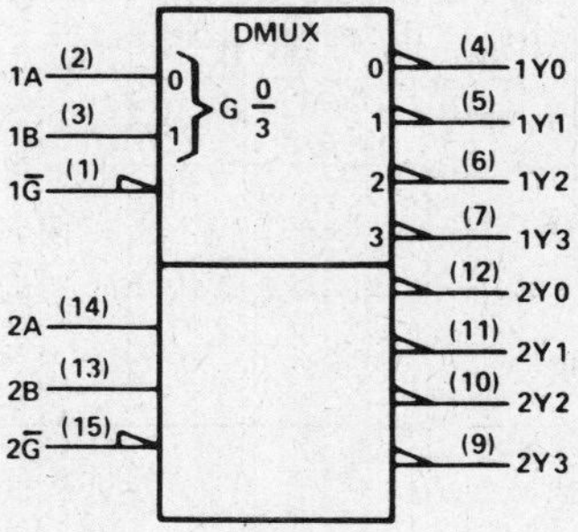

Pin numbers shown are for J and N packages.

logic diagram (positive logic)

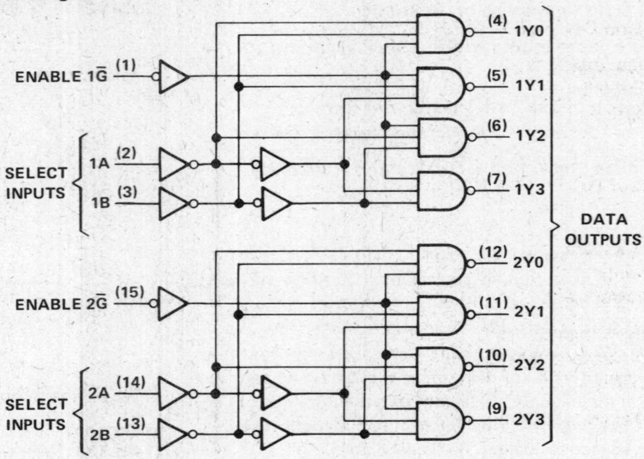

Pin numbers shown are for J and N packages.

FUNCTION TABLE

INPUTS			OUTPUTS			
ENABLE	SELECT					
$\overline{G}$	B	A	Y0	Y1	Y2	Y3
H	X	X	H	H	H	H
L	L	L	L	H	H	H
L	L	H	H	L	H	H
L	H	L	H	H	L	H
L	H	H	H	H	H	L

maximum ratings, recommended operating conditions, and electrical characteristics

See Table IV, page 2-10.

switching characteristics over recommended operating free-air temperature range (unless otherwise noted), C_L = 50 pF (see Note 1)

PARAMETER	FROM (INPUT)	TO (OUTPUT)	V_{CC}	T_A = 25°C MIN	TYP	MAX	SN54HC139 MIN	MAX	SN74HC139 MIN	MAX	UNIT
t_{pd}	A or B	Y	2 V		47	175		255		220	ns
			4.5 V		14	35		51		44	
			6 V		12	30		44		38	
t_{pd}	$\overline{G}$	Y	2 V		39	175		255		220	ns
			4.5 V		11	35		51		44	
			6 V		10	30		44		38	
t_t		Y	2 V		38	75		110		95	ns
			4.5 V		8	15		22		19	
			6 V		6	13		19		16	

C_{pd}	Power dissipation capacitance per decoder	No load, T_A = 25°C	25 pF typ

NOTE 1: For load circuit and voltage waveforms, see page 1-14.

TEXAS INSTRUMENTS
POST OFFICE BOX 225012 • DALLAS, TEXAS 75265

TYPES SN54HC147, SN54HC148 SN74HC147, SN74HC148 10-LINE TO 4-LINE AND 8-LINE TO 3-LINE PRIORITY ENCODERS

D2844, MARCH 1984

'HC147

- Encodes 10-Line Decimal to 4-Line BCD
- Applications Include:
  Keyboard Encoding
  Range Selection

'HC148

- Encodes 8 Data Lines to 3-Line Binary (Octal)
- Applications Include:
  N-Bit Encoding
  Code Converters and Generators
- Package Options Include Both Plastic and Ceramic Chip Carriers in Addition to Plastic and Ceramic DIPs
- Dependable Texas Instruments Quality and Reliability

description

These encoders feature priority decoding of the inputs to ensure that only the highest-order data line is encoded. The 'HC147 encodes nine data lines to four-line (8-4-2-1) BCD. The implied decimal zero condition requires no input condition as zero is encoded when all nine data lines are at a high logic level. The 'HC148 encodes eight data lines to three-line (4-2-1) binary (octal). Cascading circuitry (enable input EI and enable output EO) has been provided to allow octal expansion without the need for external circuitry. For all types, data inputs and outputs are active at the low logic level.

The SN54HC147 and SN54HC148 are characterized for operation over the full military temperature range of −55°C to 125°C. The SN74HC147 and SN74HC148 are characterized for operation from −40°C to 85°C.

SN54HC147 . . . J PACKAGE
SN74HC147 . . . J OR N PACKAGE
(TOP VIEW)

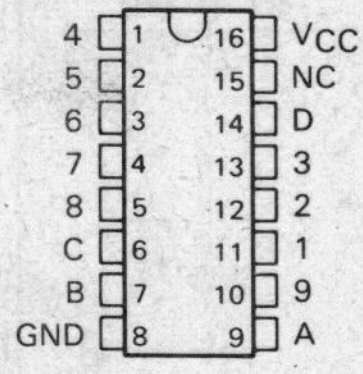

SN54HC147 . . . FH OR FK PACKAGE
SN74HC147 . . . FH OR FN PACKAGE
(TOP VIEW)

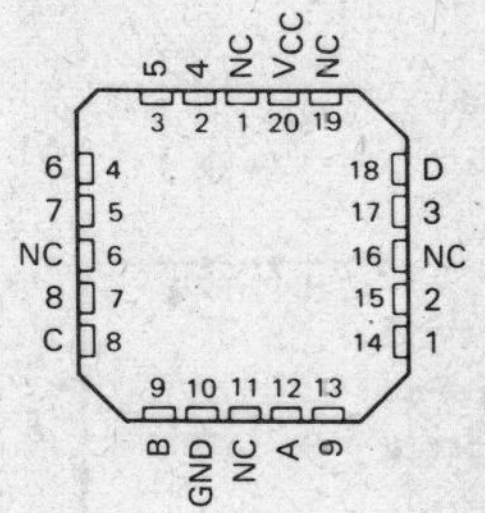

SN54HC148 . . . J PACKAGE
SN74HC148 . . . J OR K PACKAGE
(TOP VIEW)

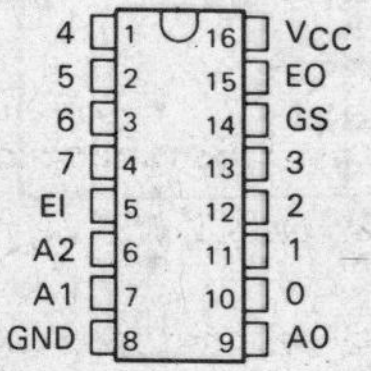

SN54HC148 . . . FH OR FK PACKAGE
SN74HC148 . . . FH OR FN PACKAGE
(TOP VIEW)

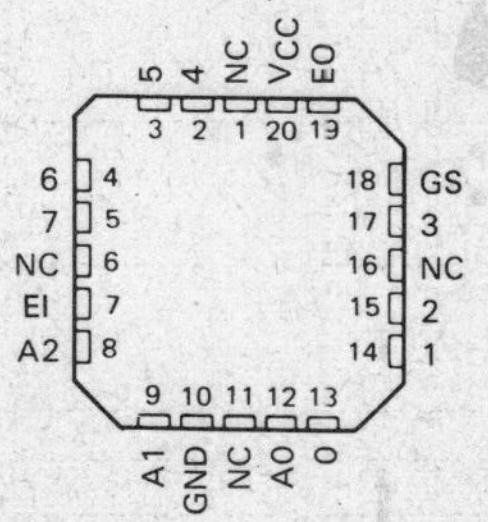

NC—No internal connection

TEXAS INSTRUMENTS
POST OFFICE BOX 225012 • DALLAS, TEXAS 75265

'HC147
FUNCTION TABLE

INPUTS									OUTPUTS			
1	2	3	4	5	6	7	8	9	D	C	B	A
H	H	H	H	H	H	H	H	H	H	H	H	H
X	X	X	X	X	X	X	X	L	L	H	H	L
X	X	X	X	X	X	X	L	H	L	H	H	H
X	X	X	X	X	X	L	H	H	H	L	L	L
X	X	X	X	X	L	H	H	H	H	L	L	H
X	X	X	X	L	H	H	H	H	H	L	H	L
X	X	X	L	H	H	H	H	H	H	L	H	H
X	X	L	H	H	H	H	H	H	H	H	L	L
X	L	H	H	H	H	H	H	H	H	H	L	H
L	H	H	H	H	H	H	H	H	H	H	H	L

'HC148
FUNCTION TABLE

INPUTS									OUTPUTS				
EI	0	1	2	3	4	5	6	7	A2	A1	A0	GS	EO
H	X	X	X	X	X	X	X	X	H	H	H	H	H
L	H	H	H	H	H	H	H	H	H	H	H	H	L
L	X	X	X	X	X	X	X	L	L	L	L	L	H
L	X	X	X	X	X	X	L	H	L	L	H	L	H
L	X	X	X	X	X	L	H	H	L	H	L	L	H
L	X	X	X	X	L	H	H	H	L	H	H	L	H
L	X	X	X	L	H	H	H	H	H	L	L	L	H
L	X	X	L	H	H	H	H	H	H	L	H	L	H
L	X	L	H	H	H	H	H	H	H	H	L	L	H
L	L	H	H	H	H	H	H	H	H	H	H	L	H

H = high logic level, L = low logic level, X = irrelevant

logic symbols

'HC147

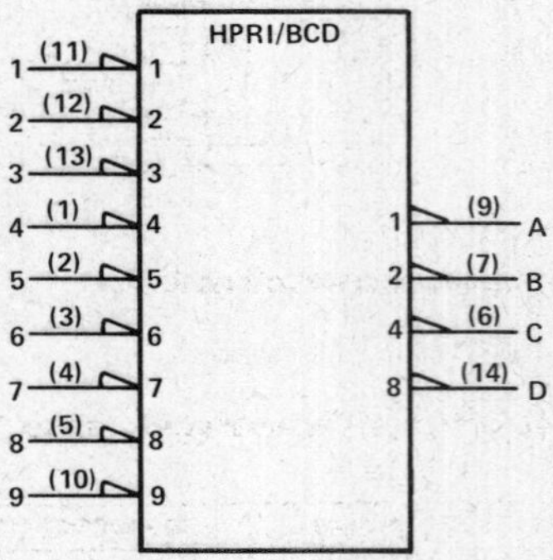

'HC148

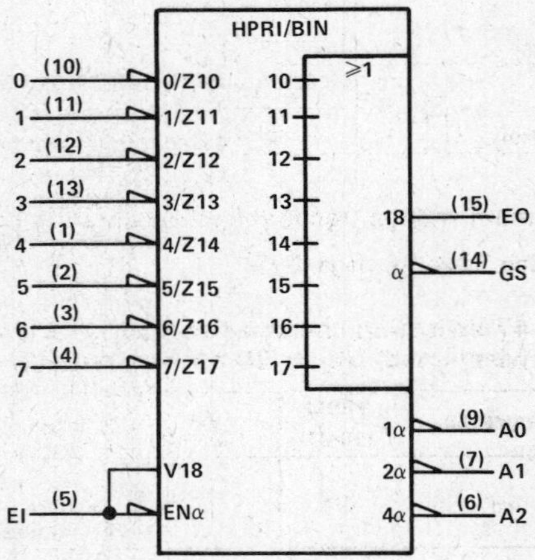

Pin numbers shown are for J and N packages.

logic diagrams

'HC147

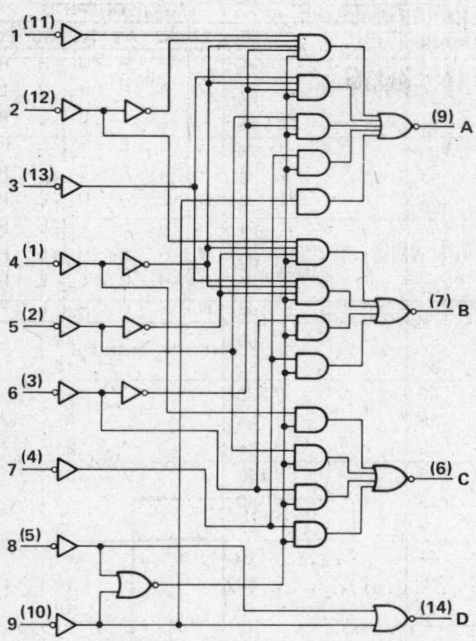

'HC148

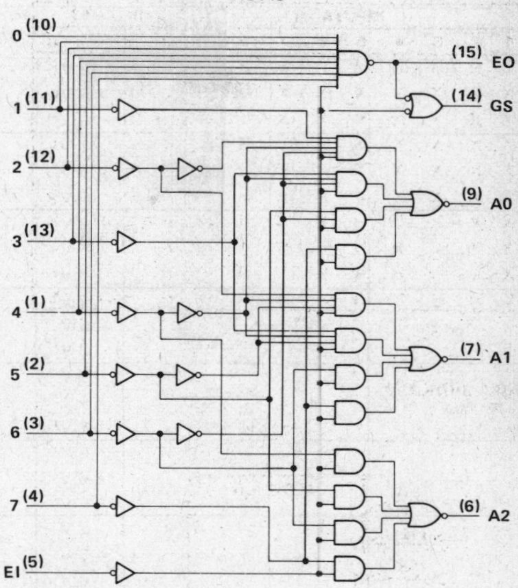

maximum ratings, recommended operating conditions, and electrical characteristics

See Table IV, page 2-10.

'HC147 switching characteristics over recommended operating free-air temperature range (unless otherwise noted), C_L = 50 pF (see Note 1)

PARAMETER	FROM (INPUT)	TO (OUTPUT)	V_{CC}	T_A = 25°C MIN	TYP	MAX	SN54HC147 MIN	MAX	SN74HC147 MIN	MAX	UNIT
t_{pd}	Any	Any	2 V		75	190		285		240	ns
			4.5 V		25	38		57		48	
			6 V		21	32		48		41	
t_t		Any	2 V		28	75		110		95	ns
			4.5 V		8	15		22		19	
			6 V		6	13		19		16	

NOTE 1: For load circuit and voltage waveforms, see page 1-14.

'HC148 switching characteristics over recommended operating free-air temperature range (unless otherwise noted), C_L = 50 pF (see Note 1)

PARAMETER	FROM (INPUT)	TO (OUTPUT)	V_{CC}	T_A = 25°C MIN	TYP	MAX	SN54HC148 MIN	MAX	SN74HC148 MIN	MAX	UNIT
t_{pd}	1–7	A0, A1, or A2	2 V		69	180		270		225	ns
			4.5 V		23	36		54		45	
			6 V		21	31		46		38	
t_{pd}	0–7	EO	2 V		60	150		225		190	ns
			4.5 V		20	30		45		38	
			6 V		17	26		38		33	
t_{pd}	0–7	GS	2 V		75	190		285		240	ns
			4.5 V		25	38		57		48	
			6 V		21	32		48		41	
t_{pd}	EI	A0, A1 or A2	2 V		78	195		295		245	ns
			4.5 V		26	39		59		49	
			6 V		22	33		50		42	
t_{pd}	EI	GS	2 V		57	145		220		180	ns
			4.5 V		19	29		44		36	
			6 V		16	25		38		31	
t_{pd}	EI	EO	2 V		66	165		250		205	ns
			4.5 V		22	33		50		41	
			6 V		19	28		43		35	
t_t		Any	2 V		28	75		110		95	ns
			4.5 V		8	15		22		19	
			6 V		6	13		19		16	

NOTE 1: For load circuit and voltage waveforms, see page 1-14.

TYPICAL APPLICATION DATA

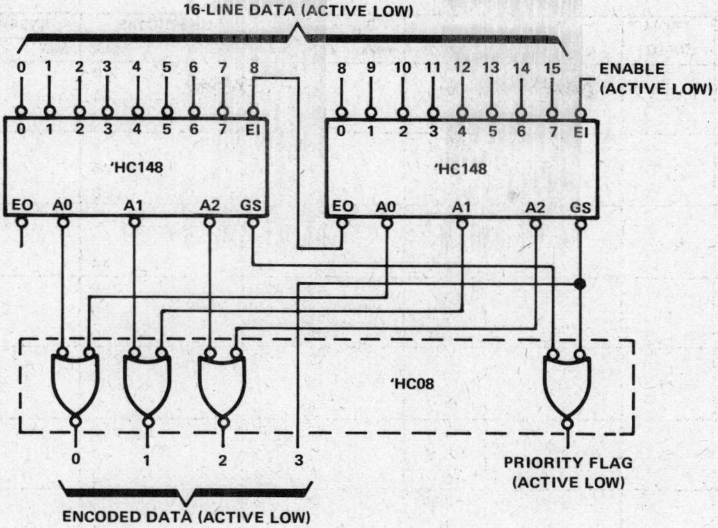

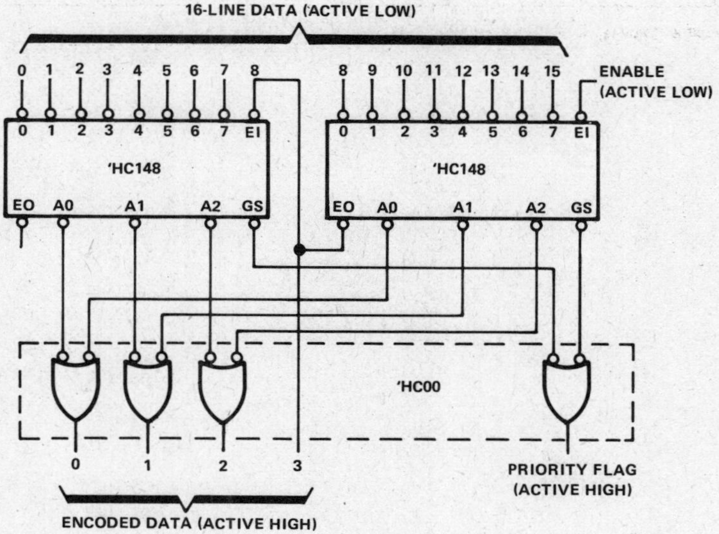

PRIORITY ENCODER FOR 16 BITS

Since the 'HC147 and 'HC148 are combinational logic circuits, wrong addresses can appear during input transients. Moreover, for the 'HC148, a change from high to low at input EI can cause a transient low on the GS output when all inputs are high. This must be considered when strobing the outputs.

D2684, DECEMBER 1982—REVISED MARCH 1984

- **8-Line to 1-Line Multiplexers Can Perform As:**
  - **Boolean Function Generators**
  - **Parallel-to-Serial Converters**
  - **Data Source Selectors**
- **Package Options Include Both Plastic and Ceramic Chip Carriers in Addition to Plastic and Ceramic DIPs**
- **Dependable Texas Instruments Quality and Reliability**

SN54HC151 . . . J PACKAGE
SN74HC151 . . . J OR N PACKAGE
(TOP VIEW)

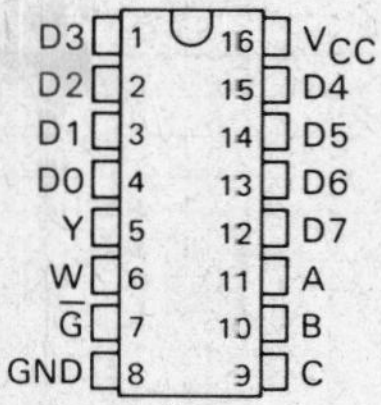

SN54HC151 . . . FH OR FK PACKAGE
SN74HC151 . . . FH OR FN PACKAGE
(TOP VIEW)

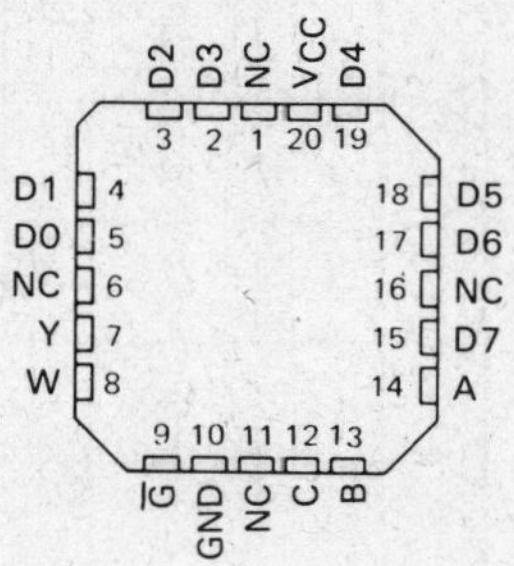

NC—No internal connection

description

These monolithic data selectors/multiplexers provide full binary decoding to select one of eight data sources. The strobe input ($\overline{G}$) must be at a low logic level to enable the inputs. A high level at the strobe terminal forces the W output high and the Y output low.

The SN54HC151 is characterized for operation over the full military temperature range of −55°C to 125°C. The SN74HC151 is characterized for operation from −40°C to 85°C.

FUNCTION TABLE

INPUTS				OUTPUTS	
SELECT			STROBE		
C	B	A	$\overline{G}$	Y	W
X	X	X	H	L	H
L	L	L	L	D0	$\overline{D0}$
L	L	H	L	D1	$\overline{D1}$
L	H	L	L	D2	$\overline{D2}$
L	H	H	L	D3	$\overline{D3}$
H	L	L	L	D4	$\overline{D4}$
H	L	H	L	D5	$\overline{D5}$
H	H	L	L	D6	$\overline{D6}$
H	H	H	L	D7	$\overline{D7}$

H = high level, L = low level, X = irrelevant
D0, D1 . . . D7 = the level of the D respective input

logic symbols

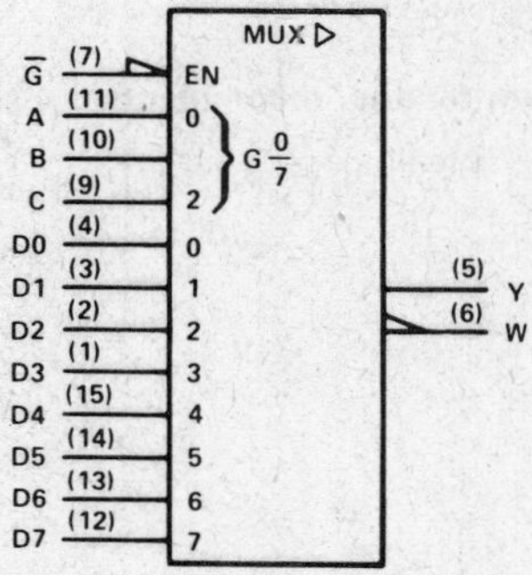

Pin numbers shown are for J and N packages

logic diagram (positive logic)

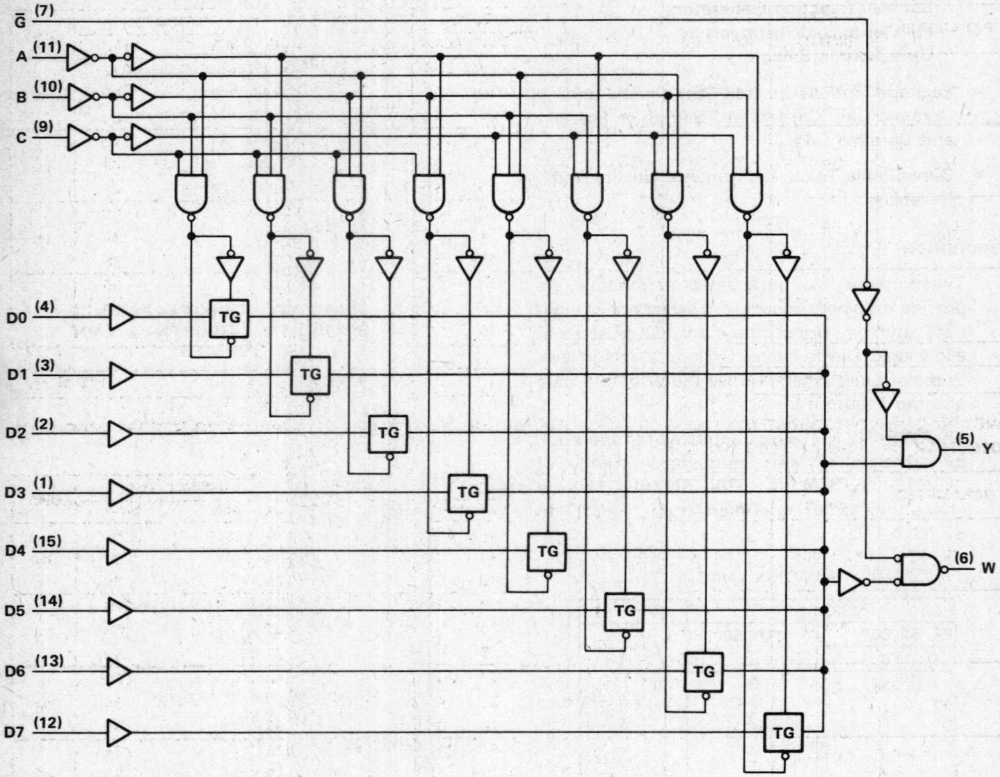

Pin numbers shown are for J and N packages.

maximum ratings, recommended operating conditions, and electrical characteristics

See Table III, page 2-8.

switching characteristics over recommended operating free-air temperature range (unless otherwise noted), C_L = 50 pF (see Note 1)

PARAMETER	FROM (INPUT)	TO (OUTPUT)	V_{CC}	T_A = 25°C			SN54HC151		SN74HC151		UNIT
				MIN	TYP	MAX	MIN	MAX	MIN	MAX	
t_{pd}	A, B, or C	Y or W	2 V		94	250		360		312	ns
			4.5 V		30	50		73		63	
			6 V		25	43		62		54	
t_{pd}	Any D	Y or W	2 V		74	195		283		244	ns
			4.5 V		23	39		57		49	
			6 V		20	33		48		41	
t_{pd}	$\overline{G}$	Y or W	2 V		49	127		185		159	ns
			4.5 V		15	25		37		32	
			6 V		13	22		32		28	
t_t			2 V		22	75		110		95	ns
			4.5 V		9	15		22		19	
			6 V		8	13		19		16	

C_{pd}	Power dissipation capacitance	No load, T_A = 25°C	70 pF typ

switching characteristics over recommended operating free-air temperature range (unless otherwise noted), C_L = 150 pF (see Note 1)

PARAMETER	FROM (INPUT)	TO (OUTPUT)	V_{CC}	T_A = 25°C			SN54HC151		SN74HC151		UNIT
				MIN	TYP	MAX	MIN	MAX	MIN	MAX	
t_{pd}	A, B, or C	Y or W	2 V		107	350		525		440	ns
			4.5 V		33	70		105		88	
			6 V		30	59		89		76	
t_{pd}	Any D	Y or W	2 V		90	275		415		345	ns
			4.5 V		29	51		83		69	
			6 V		25	47		72		59	
t_{pd}	$\overline{G}$	Y or W	2 V		67	205		310		255	ns
			4.5 V		21	41		62		51	
			6 V		18	35		53		43	
t_t			2 V		51	210		315		265	ns
			4.5 V		16	42		63		53	
			6 V		14	36		53		45	

NOTE 1: For load circuit and voltage waveforms, see page 1-14.

TYPES SN54HC152, SN74HC152
8-LINE TO 1-LINE DATA SELECTORS/MULTIPLEXERS

D2684, DECEMBER 1982—REVISED MARCH 1984

- **Selects One-of-Eight Data Sources**
- **Performs Parallel-to-Serial Conversion**
- **Package Options Include Both Plastic and Ceramic Chip Carriers in Addition to Plastic and Ceramic DIPs**
- **Dependable Texas Instruments Quality and Reliability**

SN54HC152 . . . J PACKAGE
SN74HC152 . . . J OR N PACKAGE
(TOP VIEW)

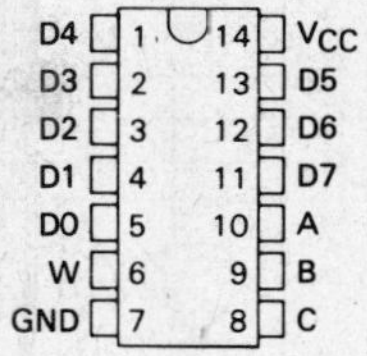

SN54HC152 . . . FH OR FK PACKAGE
SN74HC152 . . . FH OR FN PACKAGE
(TOP VIEW)

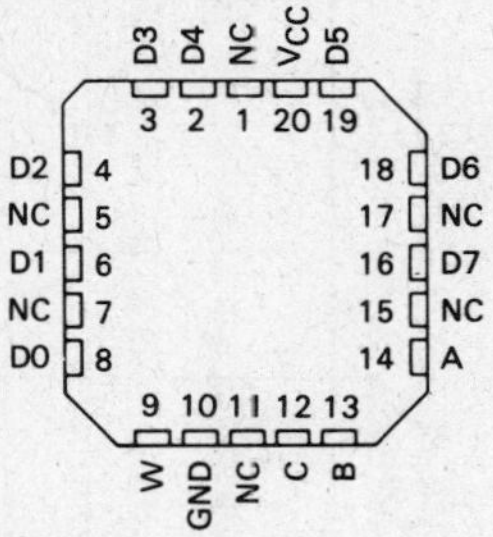

NC—No internal connection

description

These monolithic data selectors/multiplexers contain full on-chip binary decoding to select the desired one-of-eight data sources.

The SN54HC152 is characterized for operation over the full military temperature range of −55°C to 125°C. The SN74HC152 is characterized for operation from −40°C to 85°C.

FUNCTION TABLE

SELECT INPUTS			OUTPUT W
C	B	A	
L	L	L	$\overline{D0}$
L	L	H	$\overline{D1}$
L	H	L	$\overline{D2}$
L	H	H	$\overline{D3}$
H	L	L	$\overline{D4}$
H	L	H	$\overline{D5}$
H	H	L	$\overline{D6}$
H	H	H	$\overline{D7}$

logic symbol

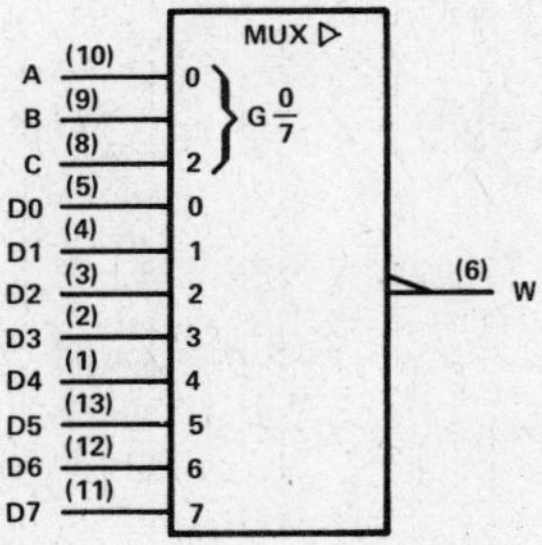

Pin numbers shown are for J and N packages.

logic diagram (positive logic)

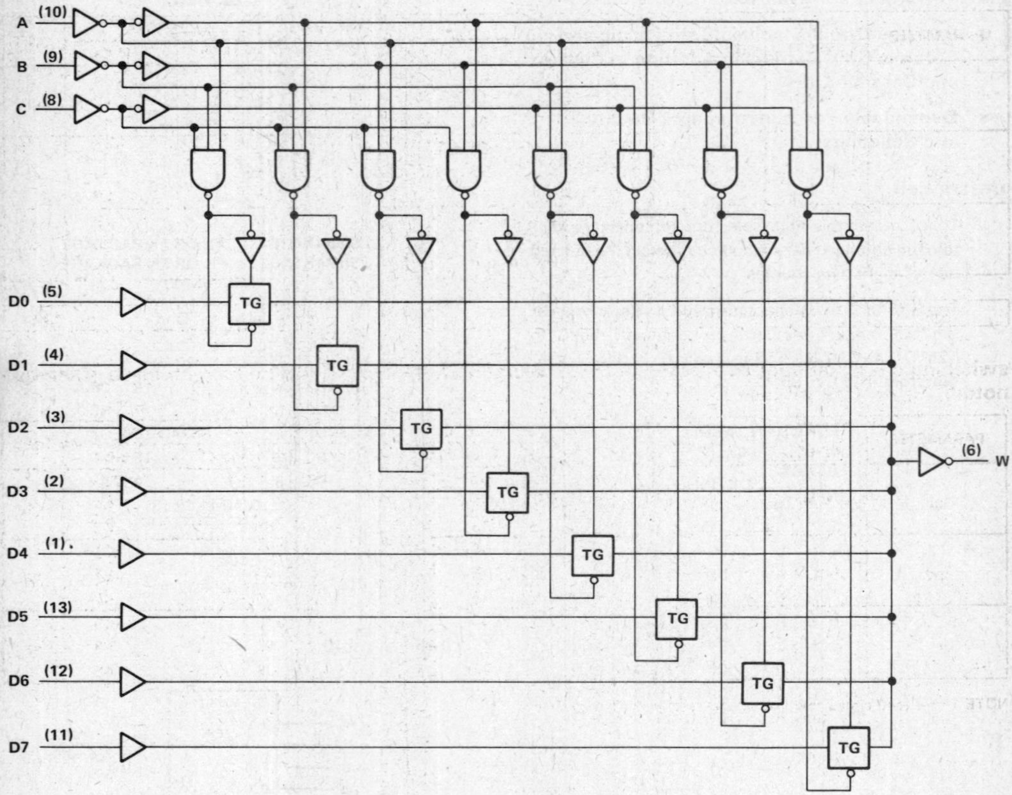

Pin numbers shown are for J and N packages.

maximum ratings, recommended operating conditions, and electrical characteristics

See Table III, page 2-8.

TEXAS INSTRUMENTS
POST OFFICE BOX 225012 • DALLAS, TEXAS 75265

switching characteristics over recommended operating free-air temperature range (unless otherwise noted), C_L = 50 pF (see Note 1)

PARAMETER	FROM (INPUT)	TO (OUTPUT)	V_{CC}	T_A = 25°C MIN	TYP	MAX	SN54HC152 MIN	MAX	SN74HC152 MIN	MAX	UNIT
t_{pd}	A B, or C	W	2 V		50	170		255		213	ns
			4.5 V		18	34		51		43	
			6 V		16	29		44		36	
t_{pd}	Any D	W	2 V		38	130		195		163	ns
			4.5 V		14	26		39		33	
			6 V		12	22		33		28	
t_t		W	2 V		20	60		90		75	ns
			4.5 V		8	12		18		15	
			6 V		6	10		15		13	

C_{pd}	Power dissipation capacitance	No load, T_A = 25°C	70 pF typ

switching characteristics over recommended operating free-air temperature range (unless otherwise noted), C_L = 150 pF (see Note 1)

PARAMETER	FROM (INPUT)	TO (OUTPUT)	V_{CC}	T_A = 25°C MIN	TYP	MAX	SN54HC152 MIN	MAX	SN74HC152 MIN	MAX	UNIT
t_{pd}	A, B, or C	W	2 V		63	225		385		318	ns
			4.5 V		22	51		77		64	
			6 V		19	44		66		55	
t_{pd}	Any D	W	2 V		52	215		325		268	ns
			4.5 V		18	43		65		54	
			6 V		16	37		55		47	
t_t		W	2 V		45	210		315		265	ns
			4.5 V		17	42		63		53	
			6 V		13	36		53		45	

NOTE 1: For load circuit and voltage waveforms, see page 1-14.

TYPES SN54HC153, SN74HC153
DUAL 4-LINE TO 1-LINE DATA SELECTORS/MULTIPLEXERS

D2684, DECEMBER 1982—REVISED MARCH 1984

- **Permits Multiplexing from N Lines to 1 Line**
- **Performs Parallel-to-Serial Conversion**
- **Strobe (Enable) Line Provided for Cascading (N lines to n lines)**
- **Package Options Include Both Plastic and Ceramic Chip Carriers in Addition to Plastic and Ceramic DIPs**
- **Dependable Texas Instruments Quality and Reliability**

SN54HC153 . . . J PACKAGE
SN74HC153 . . . J OR N PACKAGE
(TOP VIEW)

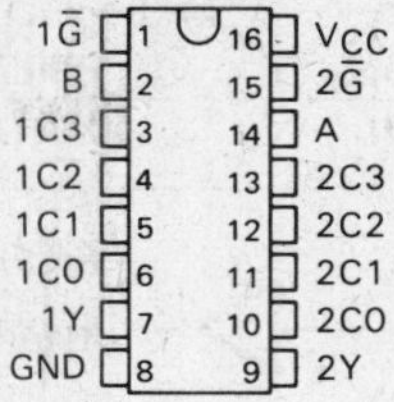

SN54HC153 . . . FH OR FK PACKAGE
SN74HC153 . . . FH OR FN PACKAGE
(TOP VIEW)

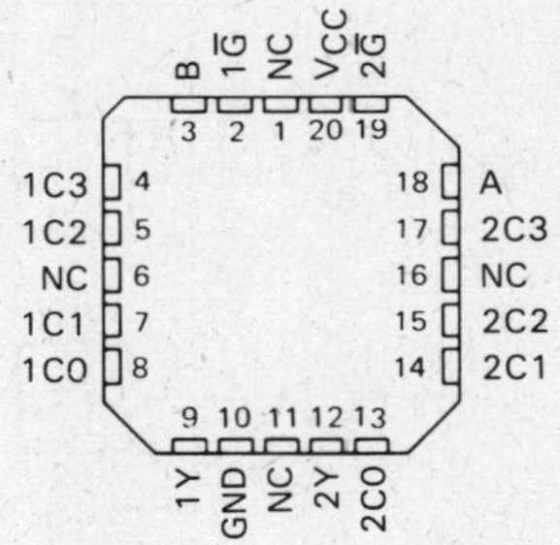

NC—No internal connection

description

Each of these data selectors/multiplexers contains inverters and drivers to supply full binary decoding data selection to the AND-OR gates. Separate strobe inputs (G̅) are provided for each of the two four-line sections.

The SN54HC153 is characterized for operation over the full military temperature range of −55°C to 125°C. The SN74HC153 is characterized for operation from −40°C to 85°C.

FUNCTION TABLE

SELECT INPUTS		DATA INPUTS				STROBE	OUTPUT
B	A	C0	C1	C2	C3	G̅	Y
X	X	X	X	X	X	H	L
L	L	L	X	X	X	L	L
L	L	H	X	X	X	L	H
L	H	X	L	X	X	L	L
L	H	X	H	X	X	L	H
H	L	X	X	L	X	L	L
H	L	X	X	H	X	L	H
H	H	X	X	X	L	L	L
H	H	X	X	X	H	L	H

Select inputs A and B are common to both sections.

logic symbol

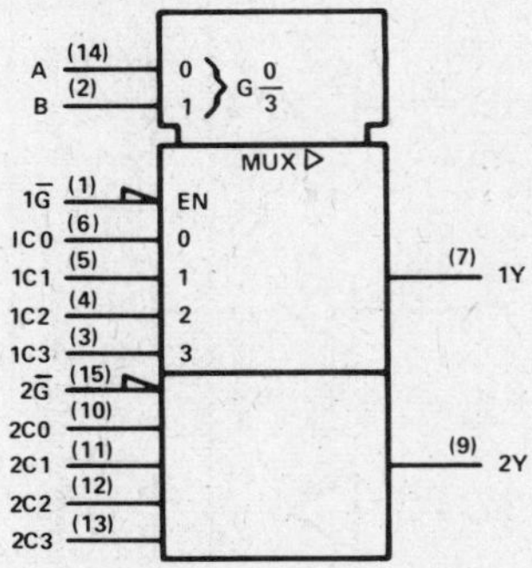

Pin numbers shown are for J and N packages.

logic diagram (positive logic)

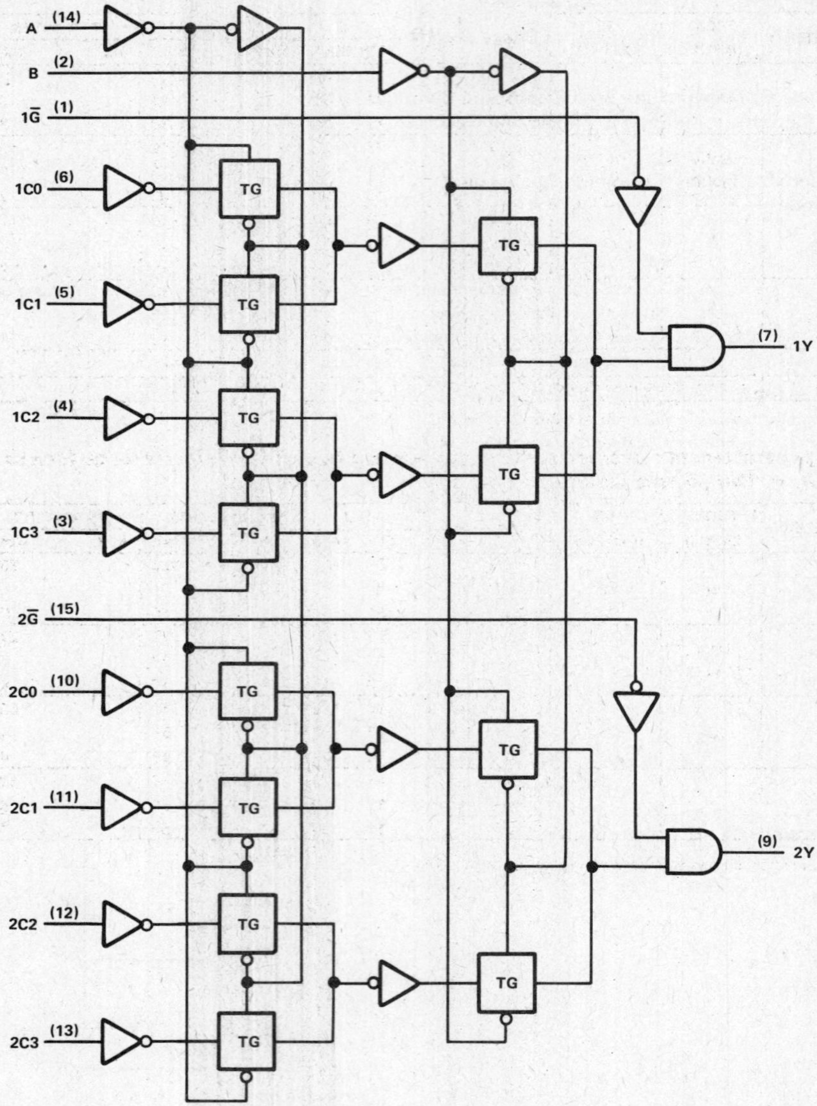

maximum ratings, recommended operating conditions, and electrical characteristics

See Table III, page 2-8.

switching characteristics over recommended operating free-air temperature range (unless otherwise noted), C_L = 50 pF (see Note 1)

PARAMETER	FROM (INPUT)	TO (OUTPUT)	V_{CC}	T_A = 25°C MIN	TYP	MAX	SN54HC153 MIN	MAX	SN74HC153 MIN	MAX	UNIT
t_{pd}	A or B	Y	2 V		90	150		225		190	ns
			4.5 V		21	30		45		38	
			6 V		17	26		38		32	
t_{pd}	Data (Any C)	Y	2 V		73	126		189		158	ns
			4.5 V		17	28		42		35	
			6 V		14	23		35		29	
t_{pd}	G	Y	2 V		38	95		150		125	ns
			4.5 V		11	19		28		24	
			6 V		9	16		24		20	
t_t		Y	2 V		20	60		90		75	ns
			4.5 V		8	12		18		15	
			6 V		6	10		15		13	

C_{pd}	Power dissipation capacitance per multiplexer	No load, T_A = 25°C	40 pF typ

switching characteristics over recommended operating free-air temperature range (unless otherwise noted), C_L = 150 pF (see Note 1)

PARAMETER	FROM (INPUT)	TO (OUTPUT)	V_{CC}	T_A = 25°C MIN	TYP	MAX	SN54HC153 MIN	MAX	SN74HC153 MIN	MAX	UNIT
t_{pd}	A or B	Y	2 V		105	235		355		295	ns
			4.5 V		27	47		71		59	
			6 V		21	41		60		51	
t_{pd}	Data (Any C)	Y	2 V		93	220		335		274	ns
			4.5 V		23	44		67		55	
			6 V		19	38		57		48	
t_{pd}	G	Y	2 V		60	185		280		230	ns
			4.5 V		17	37		56		46	
			6 V		14	32		48		40	
t_t		Y	2 V		45	210		315		265	ns
			4.5 V		17	42		63		53	
			6 V		13	36		53		45	

NOTE 1: For load circuit and voltage waveforms, see page 1-14.

TYPES SN54HC157, SN54HC158, SN74HC157, SN74HC158 QUADRUPLE 2-LINE TO 1-LINE DATA SELECTORS/MULTIPLEXERS

D2684, DECEMBER 1982—REVISED MARCH 1984

- **Package Options Include Both Plastic and Ceramic Chip Carriers in Addition to Plastic and Ceramic DIPs**
- **Dependable Texas Instruments Quality and Reliability**

SN54HC157, SN54HC158 . . . J PACKAGE
SN74HC157, SN74HC158 . . . J OR N PACKAGE
(TOP VIEW)

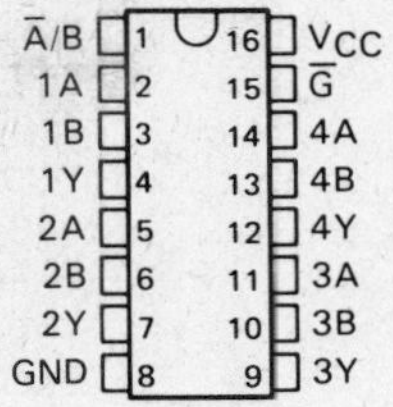

description

These monolithic data selectors/multiplexers contain inverters and drivers to supply full data selection to the four output gates. A separate strobe input (G̅) is provided. A 4-bit word is selected from one of two sources and is routed to the four outputs. The 'HC157 presents true data whereas the 'HC158 presents inverted data.

The SN54HC157 and SN54HC158 are characterized for operation over the full military temperature range of −55°C to 125°C. The SN74HC157 and SN74HC158 are characterized for operation from −40°C to 85°C.

SN54HC157, SN54HC158 . . . FH OR FK PACKAGE
SN74HC157, SN74HC158 . . . FH OR FN PACKAGE
(TOP VIEW)

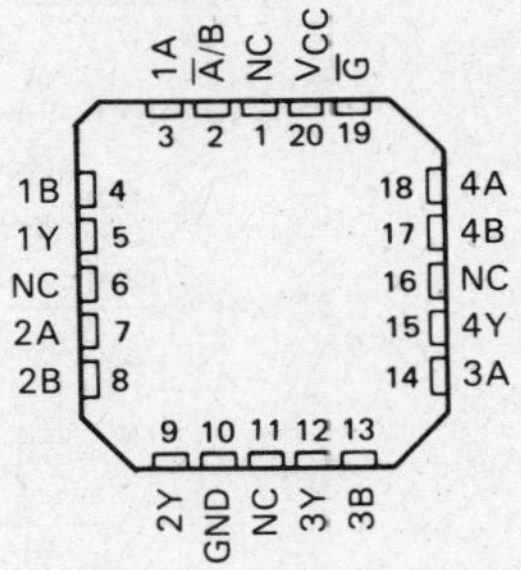

NC—No internal connection

FUNCTION TABLE

INPUTS				OUTPUT Y	
STROBE G̅	SELECT A̅/B	DATA A	DATA B	'HC157	'HC158
H	X	X	X	L	H
L	L	L	X	L	H
L	L	H	X	H	L
L	H	X	L	L	H
L	H	X	H	H	L

logic symbols

'HC157

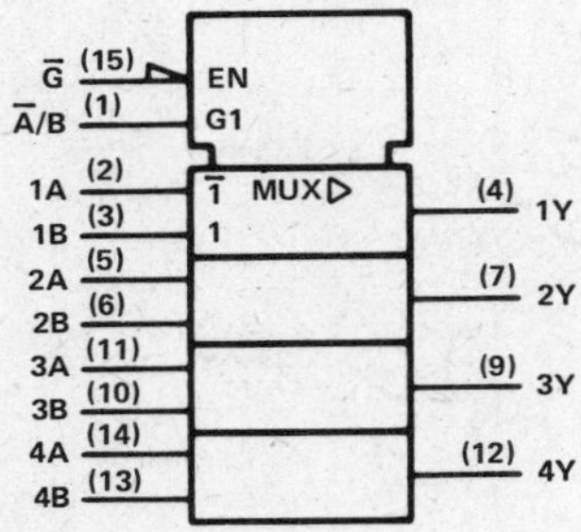

'HC158

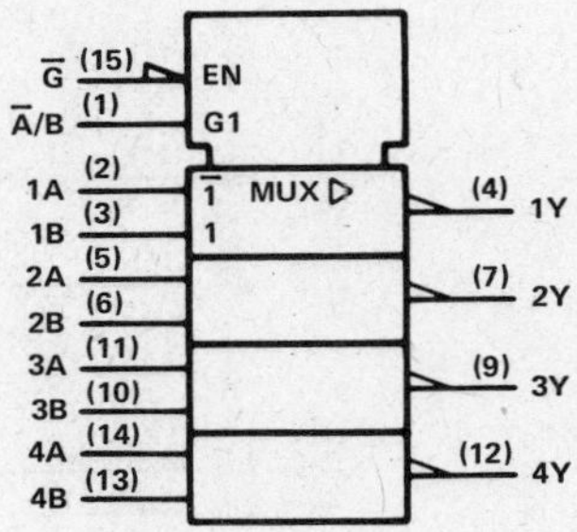

Pin numbers shown are for J and N packages.

logic diagrams (positive logic)

'HC157

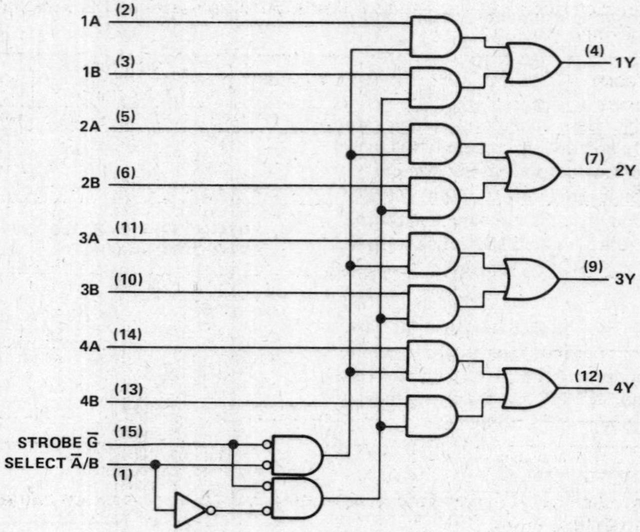

'HC158

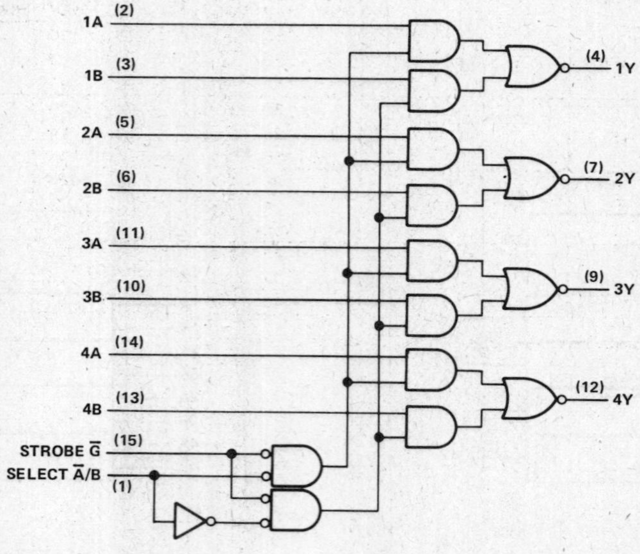

Pin numbers shown are for J and N packages.

3 HCMOS DEVICES

maximum ratings, recommended operating conditions, and electrical characteristics

See Table III, page 2-8.

switching characteristics over recommended operating free-air temperature range (unless otherwise noted), C_L = 50 pF (see Note 1)

PARAMETER	FROM (INPUT)	TO (OUTPUT)	V_{CC}	T_A = 25°C MIN	TYP	MAX	SN54HC157 SN54HC158 MIN	MAX	SN74HC157 SN74HC158 MIN	MAX	UNIT
t_{pd}	A or B	Y	2 V		63	125		190		160	ns
			4.5 V		13	25		38		32	
			6 V		11	21		32		27	
t_{pd}	$\overline{A}$/B	Y	2 V		67	125		190		160	ns
			4.5 V		18	25		38		31	
			6 V		14	21		32		27	
t_{pd}	$\overline{G}$	Y	2 V		59	115		170		145	ns
			4.5 V		16	23		34		29	
			6 V		13	20		29		25	
t_t		Y	2 V		28	60		90		75	ns
			4.5 V		8	12		18		15	
			6 V		6	10		15		13	

C_{pd}	Power dissipation capacitance	No load, T_A = 25°C	40 pF typ

switching characteristics over recommended operating free-air temperature range (unless otherwise noted), C_L = 150 pF (see Note 1)

PARAMETER	FROM (INPUT)	TO (OUTPUT)	V_{CC}	T_A = 25°C MIN	TYP	MAX	SN54HC157 SN54HC158 MIN	MAX	SN74HC157 SN74HC158 MIN	MAX	UNIT
t_{pd}	A or B	Y	2 V		81	190		290		235	ns
			4.5 V		23	38		58		47	
			6 V		18	33		49		41	
t_{pd}	$\overline{A}$/B	Y	2 V		81	210		320		260	ns
			4.5 V		23	42		64		52	
			6 V		18	36		54		45	
t_{pd}	$\overline{G}$	Y	2 V		91	190		290		235	ns
			4.5 V		24	38		58		47	
			6 V		18	33		49		41	
t_t		Y	2 V		45	210		315		265	ns
			4.5 V		17	42		63		53	
			6 V		13	36		53		45	

NOTE 1: For load circuit and voltage waveforms, see page 1-14.

TYPES SN54HC160 THRU SN54HC163 SN74HC160 THRU SN74HC163 SYNCHRONOUS 4-BIT DECADE AND BINARY COUNTERS

D2684, DECEMBER 1982—REVISED MARCH 1984

- **Internal Look Ahead for Fast Counting**
- **Carry Output for N-Bit Cascading**
- **Synchronous Counting**
- **Synchronously Programmable**
- **Package Options Include Both Plastic and Ceramic Chip Carriers in Addition to Plastic and Ceramic DIPs**
- **Dependable Texas Instruments Quality and Reliability**

SN54HC' . . . J PACKAGE
SN74HC' . . . J OR N PACKAGE
(TOP VIEW)

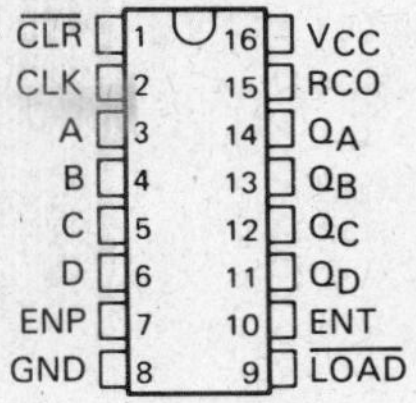

SN54HC' . . . FH OR FK PACKAGE
SN74HC' . . . FH OR FN PACKAGE
(TOP VIEW)

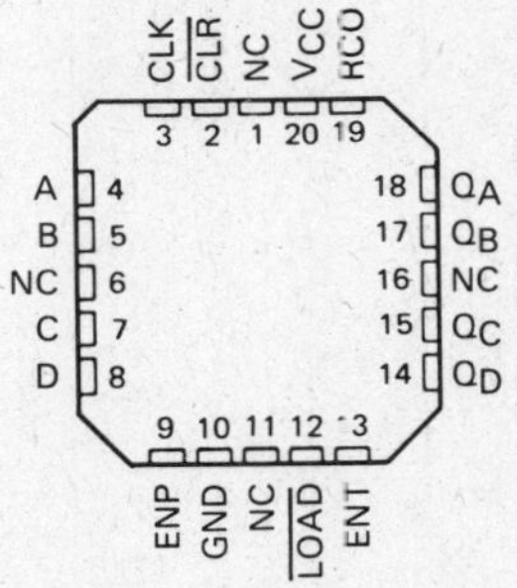

NC—No internal connection

description

These synchronous, presettable counters feature an internal carry look-ahead for application in high-speed counting designs. The 'HC160 and 'HC162 are decade counters, and the 'HC161 and 'HC163 are 4-bit binary counters. Synchronous operation is provided by having all flip-flops clocked simultaneously so that the outputs change coincident with each other when so instructed by the count-enable inputs and internal gating. This mode of operation eliminates the output counting spikes that are normally associated with synchronous (ripple clock) counters. A buffered clock input triggers the four flip-flops on the rising (positive-going) edge of the clock input waveform.

These counters are fully programmable; that is, the outputs may be preset to either level. As presetting is synchronous, setting up a low level at the load input disables the counter and causes the outputs to agree with the setup data after the next clock pulse regardless of the levels of the enable inputs.

The clear function for the 'HC160 and 'HC161 is asynchronous and a low level at the clear input sets all four of the flip-flop outputs low regardless of the levels of the clock, load, or enable inputs.

The clear function for the 'HC162 and 'HC163 is synchronous and a low level at the clear input sets all four of the flip-flop outputs low after the next low-to-high transition of the clock input, regardless of the levels of the enable inputs. This synchronous clear allows the count length to be modified easily as decoding the maximum count desired can be accomplished with one external NAND gate. The gate output is connected to the clear input to synchronously clear the counter to 0000 (LLLL).

The carry look-ahead circuitry provides for cascading counters for n-bits synchronous applications without additional gating. Instrumental in accomplishing this function are two count-enable inputs and a ripple carry output. Both count-enable inputs (ENP and ENT) must be high to count, and ENT is fed forward to enable the ripple carry output. The ripple carry output (RCO) thus enabled will produce a high-level pulse while the count is maximum (9 or 15 with Q_A high). This high-level overflow ripple carry pulse can be used to enable successive cascaded stages. Transitions at the ENP or ENT are allowed regardless of the level of the clock input.

These counters feature a fully independent clock circuit. Changes at control inputs (ENP, ENT, or $\overline{\text{LOAD}}$) that will modify the operating mode have no effect on the contents of the counter until clocking occurs. The function of the counter (whether enabled, disabled, loading, or counting) will be dictated solely by the conditions meeting the stable setup and hold times.

The SN54HC160 through SN54HC163 are characterized for operation over the full military temperature range of −55 °C to 125 °C. The SN74HC160 through SN74HC163 are characterized for operation from −40 °C to 85 °C.

logic symbols

'HC160 DECADE COUNTER WITH DIRECT CLEAR

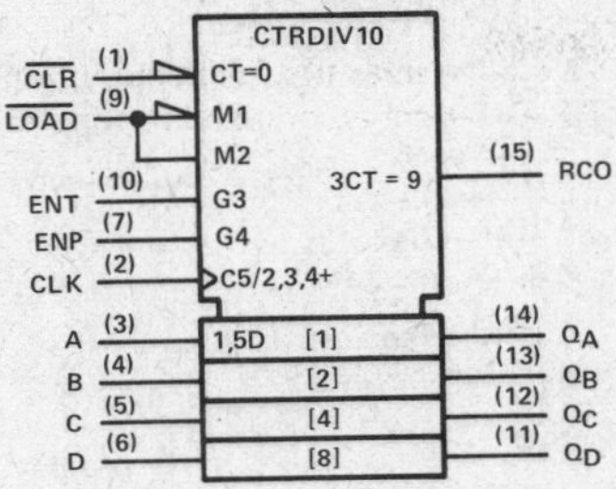

'HC162 DECADE COUNTER WITH SYNCHRONOUS CLEAR

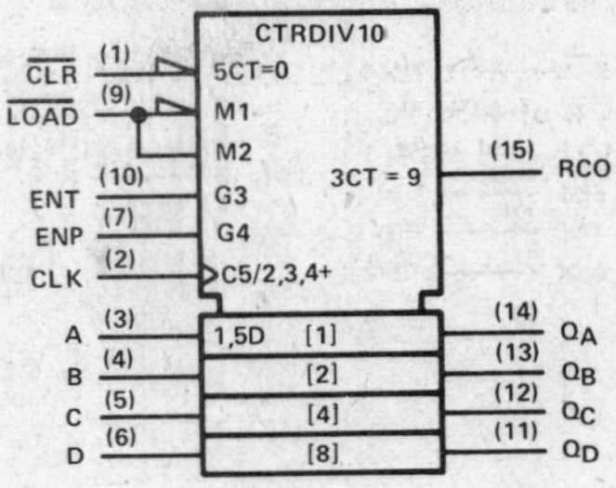

'HC160 and 'HC162 logic diagram (positive logic)

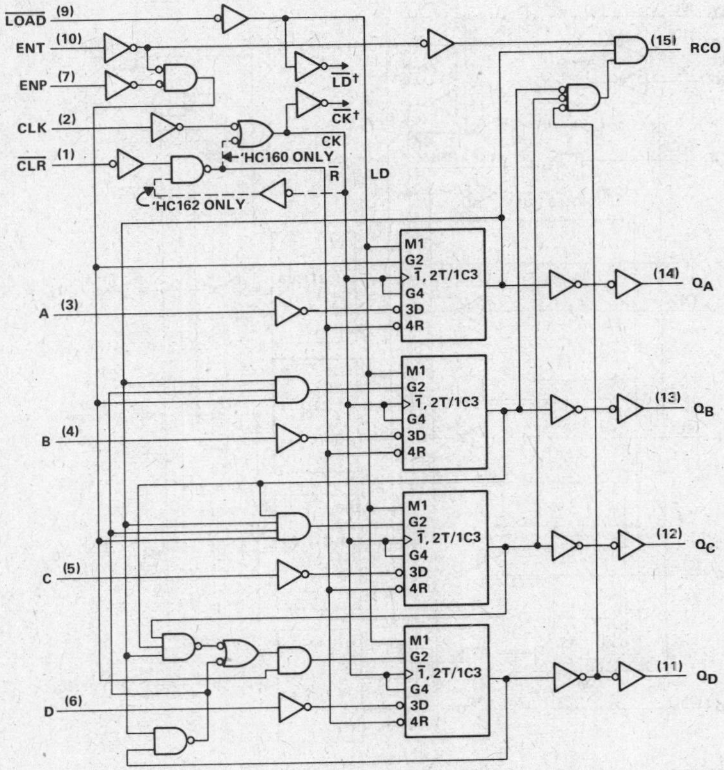

† For the sake of simplicity, the routing of the complementary signals $\overline{LD}$ and $\overline{CK}$ is not shown on this overall logic diagram. The uses of these signals are shown on the logic diagram of the D/T flip-flops.

Pin numbers shown are for J and N packges.

logic symbols

'HC161 BINARY COUNTER
WITH DIRECT CLEAR

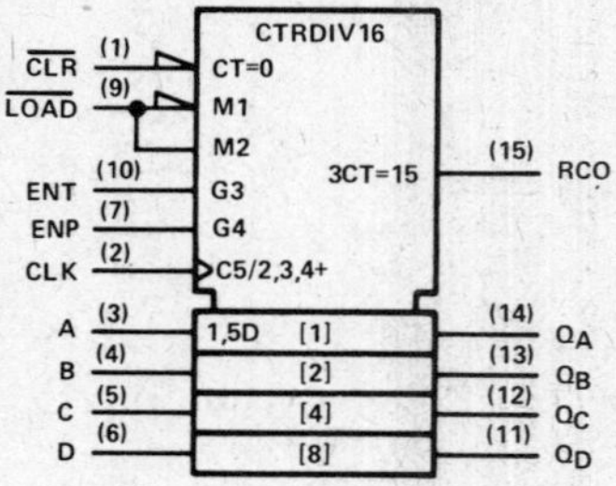

'HC163 BINARY COUNTER
WITH SYNCHRONOUS CLEAR

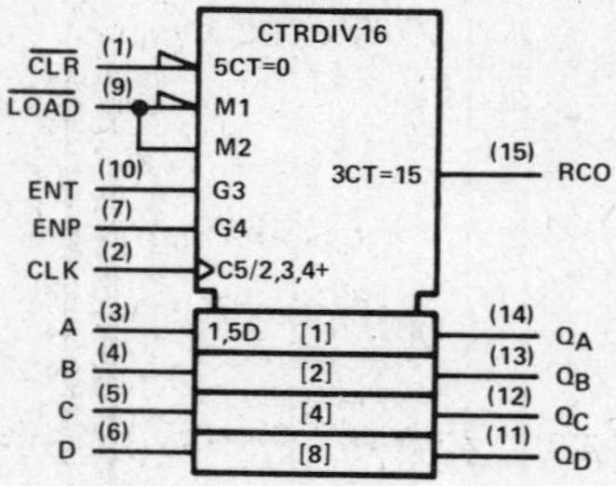

'HC161 and 'HC163 logic diagram (positive logic)

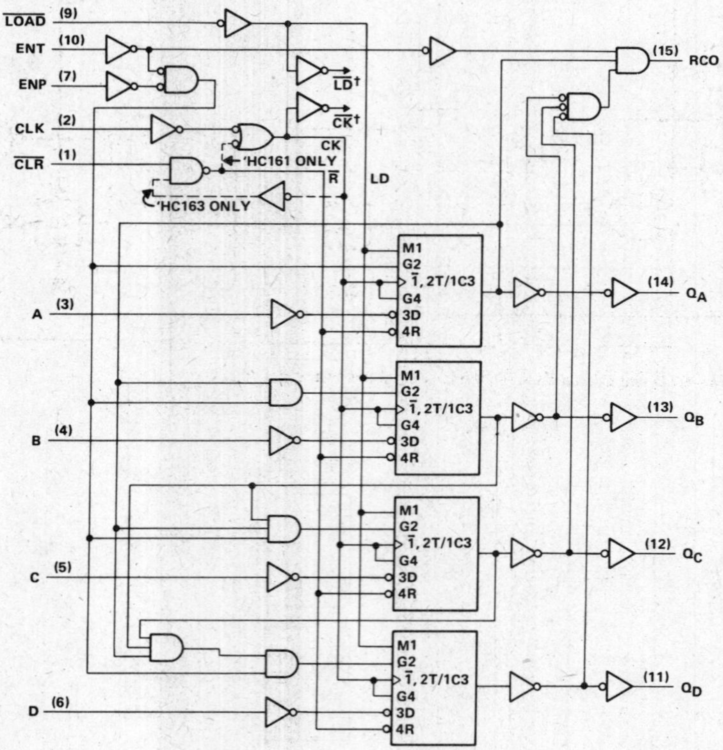

† For the sake of simplicity, the routing of the complementary signals $\overline{LD}$ and $\overline{CK}$ is not shown on this overall logic diagram. The uses of these signals are shown on the logic diagram of the D/T flip-flops.

Pin numbers shown are for J and N packages.

logic symbol, each D/T flip-flop (positive logic)

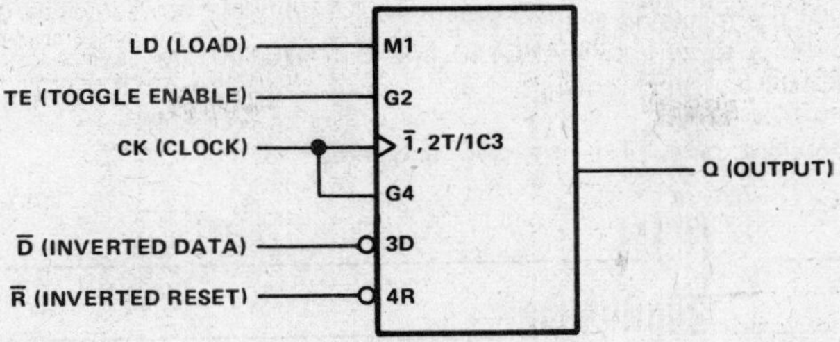

logic diagram, each D/T flip-flop (positive logic)

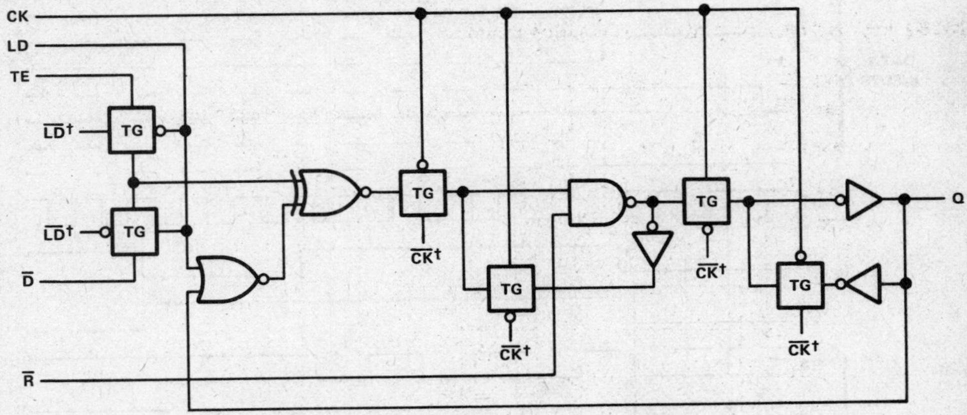

†The origins of the signals LD̄ and CK̄ are shown in the logic diagrams of the overall devices.

'HC160 and 'HC162 output sequence

Illustrated below is the following sequence:

1. Clear outputs to zero (SN54HC160 and SN74HC160 are asynchronous; SN54HC162 and SN74HC162 are synchronous)
2. Preset to BCD seven
3. Count to eight, nine, zero, one, two, and three
4. Inhibit

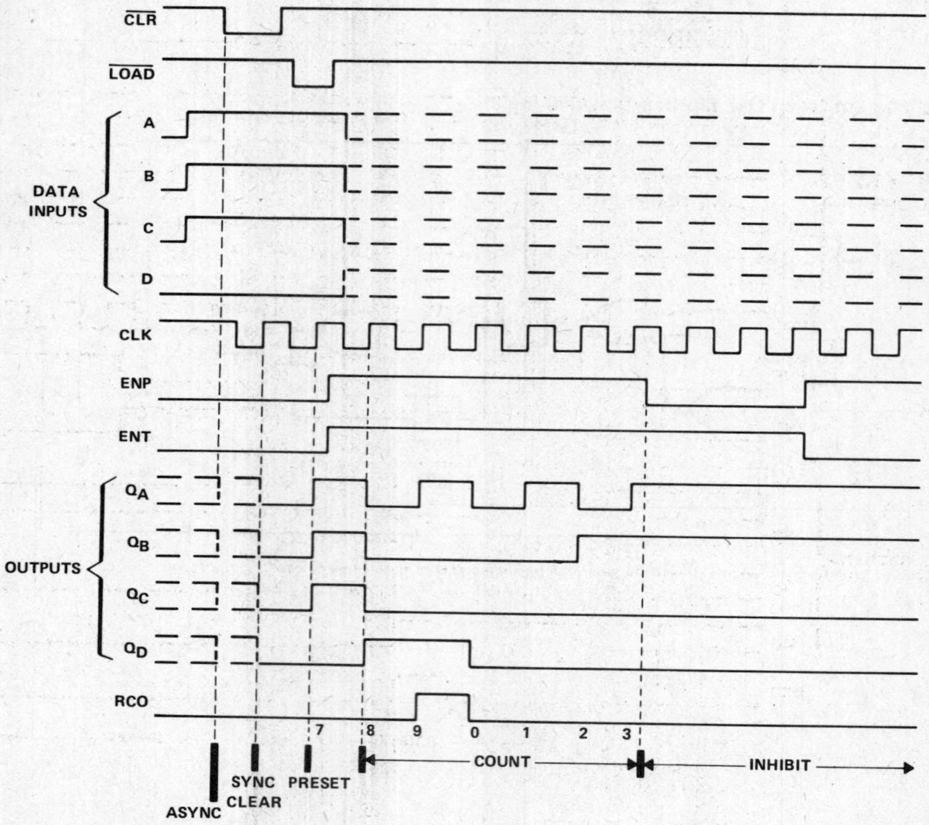

TEXAS INSTRUMENTS
POST OFFICE BOX 225012 • DALLAS, TEXAS 75265

'HC161 and 'HC163 output sequence

Illustrated below is the following sequence:

1. Clear outputs to zero (SN54HC161 and SN74HC161 are asynchronous; SN54HC163 and SN74HC163 are synchronous)
2. Preset to binary twelve
3. Count to thirteen, fourteen, zero, one, and two
4. Inhibit

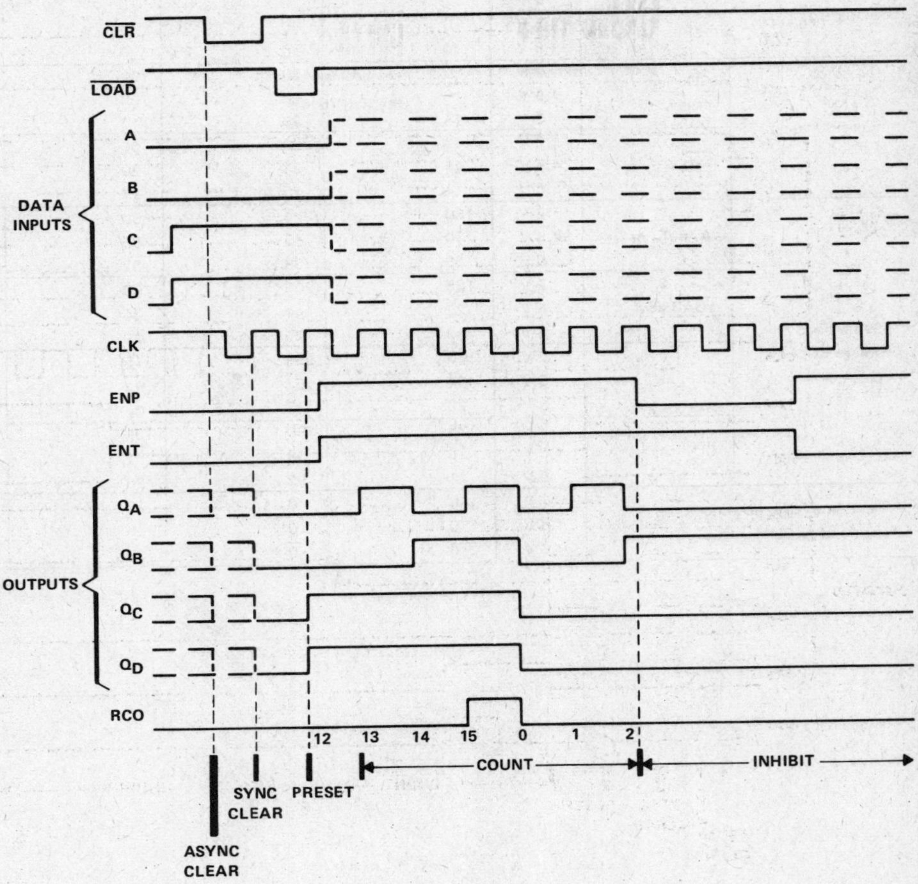

absolute maximum ratings, recommended operating conditions, and electrical characteristics

See Table IV, page 2-10.

timing requirements over recommended operating free-air temperature range (unless otherwise noted)

PARAMETER			V_{CC}	T_A = 25°C		SN54HC160 SN54HC161		SN74HC160 SN74HC161		UNIT
				MIN	MAX	MIN	MAX	MIN	MAX	
f_{clock}	Clock frequency		2 V	0	6	0	4.2	0	5	MHz
			4.5 V	0	31	0	21	0	25	
			6 V	0	36	0	25	0	29	
t_w	Pulse duration	CLK high or low	2 V	80		120		100		ns
			4.5 V	16		24		20		
			6 V	14		20		17		
		$\overline{CLR}$ low	2 V	80		120		100		
			4.5 V	16		24		20		
			6 V	14		20		17		
t_{su}	Setup time, before CLK↑	A, B, C, or D	2 V	150		225		190		ns
			4.5 V	30		45		38		
			6 V	26		38		32		
		$\overline{LOAD}$ low	2 V	135		205		170		
			4.5 V	27		41		34		
			6 V	23		35		29		
		ENP, ENT	2 V	170		255		215		
			4.5 V	34		51		43		
			6 V	29		43		37		
		$\overline{CLR}$ inactive	2 V	125		190		155		
			4.5 V	25		38		31		
			6 V	21		32		26		
t_h	Hold time, all synchronous inputs after CLK↑		2 V	0		0		0		ns
			4.5 V	0		0		0		
			6 V	0		0		0		

switching characteristics over recommended operating free-air temperature range (unless otherwise noted), C_L = 50 pF (see Note 1)

PARAMETER	FROM (INPUT)	TO (OUTPUT)	V_{CC}	T_A = 25°C MIN	TYP	MAX	SN54HC160 SN54HC161 MIN	MAX	SN74HC160 SN74HC161 MIN	MAX	UNIT
f_{max}			2 V	6	14		4.2		5		MHz
			4.5 V	31	40		21		25		
			6 V	36	44		25		29		
t_{pd}	CLK	RCO	2 V		83	215		325		270	ns
			4.5 V		24	43		65		54	
			6 V		20	37		55		46	
t_{pd}	CLK	Any Q	2 V		80	205		310		255	ns
			4.5 V		25	41		62		51	
			6 V		21	35		53		43	
t_{pd}	ENT	RCO	2 V		62	195		295		245	ns
			4.5 V		17	39		59		49	
			6 V		14	33		50		42	
t_{PHL}	$\overline{CLR}$	Any Q	2 V		105	210		315		265	ns
			4.5 V		21	42		63		53	
			6 V		18	36		54		45	
t_{PHL}	$\overline{CLR}$	RCO	2 V		110	220		330		275	ns
			4.5 V		22	44		66		55	
			6 V		19	37		56		47	
t_t		Any	2 V		38	75		110		95	ns
			4.5 V		8	15		22		19	
			6 V		6	13		19		16	

C_{pd}	Power dissipation capacitance	No load, T_A = 25°C	60 pF typ

NOTE 1: For load circuit and voltage waveforms, see page 1-14.

3

HCMOS DEVICES

maximum ratings, recommended operating conditions, and electrical characteristics

See Table IV, page 2-10.

timing requirements over recommended operating free-air temperature range (unless otherwise noted)

PARAMETER			V_{CC}	T_A = 25°C MIN	T_A = 25°C MAX	SN54HC162 SN54HC163 MIN	SN54HC162 SN54HC163 MAX	SN74HC162 SN74HC163 MIN	SN74HC162 SN74HC163 MAX	UNIT
f_{clock}	Clock frequency		2 V	0	6	0	4.2	0	5	MHz
			4.5 V	0	31	0	21	0	25	
			6 V	0	36	0	25	0	29	
t_w	Pulse duration, CLK high or low		2 V	80		120		100		ns
			4.5 V	16		24		20		
			6 V	14		20		17		
t_{su}	Setup time, before CLK↑	A, B, C, or D	2 V	150		225		190		ns
			4.5 V	30		45		38		
			6 V	26		38		32		
		$\overline{LOAD}$ low	2 V	135		205		170		
			4.5 V	27		41		34		
			6 V	23		35		29		
		ENP, ENT	2 V	170		225		215		
			4.5 V	34		51		43		
			6 V	29		43		37		
		$\overline{CLR}$ low	2 V	160		240		200		
			4.5 V	32		48		40		
			6 V	27		41		34		
		$\overline{CLR}$ inactive	2 V	160		240		200		
			4.5 V	32		48		40		
			6 V	27		41		34		
t_h	Hold time, all synchronous inputs after CLK↑		2 V	0		0		0		ns
			4.5 V	0		0		0		
			6 V	0		0		0		

switching characteristics over recommended operating free-air temperature range (unless otherwise noted), C_L = 50 pF (see Note 1)

PARAMETER	FROM (INPUT)	TO (OUTPUT)	V_{CC}	T_A = 25°C MIN	T_A = 25°C TYP	T_A = 25°C MAX	SN54HC162 SN54HC163 MIN	SN54HC162 SN54HC163 MAX	SN74HC162 SN74HC163 MIN	SN74HC162 SN74HC163 MAX	UNIT
f_{max}			2 V	6	14		4.2		5		MHz
			4.5 V	31	40		21		25		
			6 V	36	44		25		29		
t_{pd}	CLK	RCO	2 V		83	215		325		270	ns
			4.5 V		24	43		65		54	
			6 V		20	37		55		46	
t_{pd}	CLK	Any Q	2 V		80	205		310		255	ns
			4.5 V		25	41		62		51	
			6 V		21	35		53		43	
t_{pd}	ENT	RCO	2 V		62	195		295		245	ns
			4.5 V		17	39		59		49	
			6 V		14	33		50		42	
t_t		Any	2 V		38	75		110		95	ns
			4.5 V		8	15		22		19	
			6 V		6	13		19		16	

C_{pd}	Power dissipation capacitance	No load, T_A = 25°C	60 pF typ

NOTE 1: For load circuit and voltage waveforms, see page 1-14.

TYPICAL APPLICATION DATA

N-BIT SYNCHRONOUS COUNTERS

This application demonstrates how the look-ahead carry circuit can be used to implement a high-speed n-bit counter. The 'HC160 and 'HC162 will count in BCD, and the 'HC161 and 'HC163 will count in binary. Virtually any count mode (modulo-N, N_1-to-N_2, N_1-to-maximum) can be used with this fast look-ahead circuit.

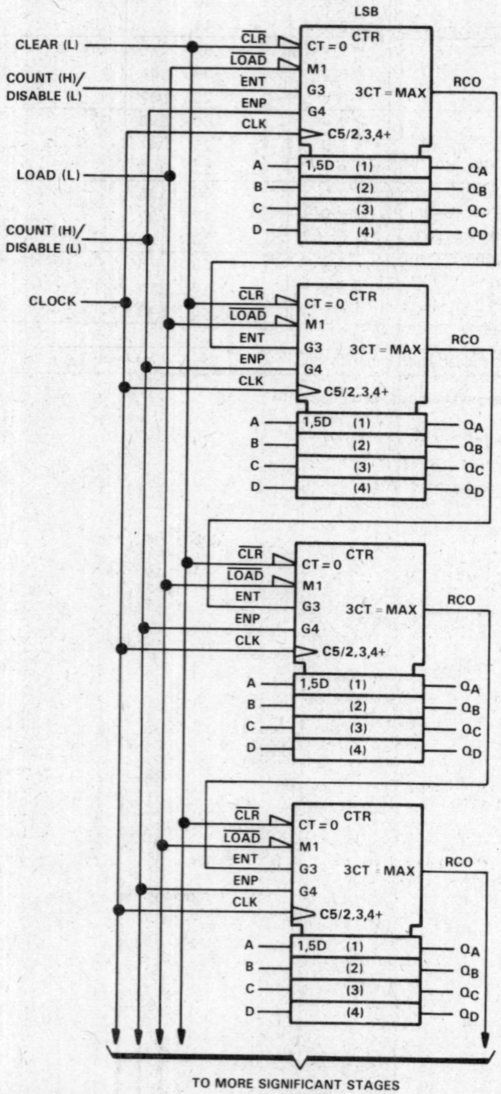

TEXAS INSTRUMENTS
POST OFFICE BOX 225012 • DALLAS, TEXAS 75265

D2684, DECEMBER 1982—REVISED MARCH 1984

- **AND-Gated (Enable/Disable) Serial Inputs**
- **Fully Buffered Clock and Serial Inputs**
- **Direct Clear**
- **Package Options Include Both Plastic and Ceramic Chip Carriers in Addition to Plastic and Ceramic DIPs**
- **Dependable Texas Instruments Quality and Reliability**

SN54HC164 . . . J PACKAGE
SN74HC164 . . . J OR N PACKAGE
(TOP VIEW)

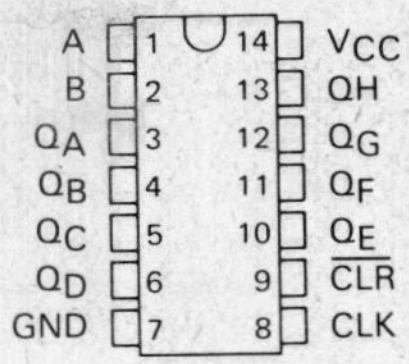

SN54HC164 . . . FH OR FK PACKAGE
SN74HC164 . . . FH OR FN PACKAGE
(TOP VIEW)

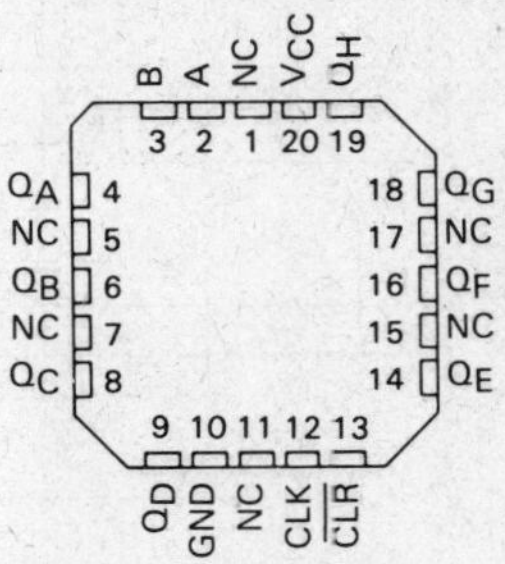

NC—No internal connection

description

These 8-bit shift registers feature AND-gated serial inputs and an asynchronous clear. The gated serial inputs (A and B) permit complete control over incoming data as a low at either input inhibits entry of the new data and resets the first flip-flop to the low level at the next clock pulse. A high-level input enables the other input, which will then determine the state of the first flip-flop. Data at the serial inputs may be changed while the clock is high or low, provided the minimum setup time requirements are met. Clocking occurs on the low-to-high-level transition of the clock input.

The SN54HC164 is characterized for operation over the full military temperature range of −55°C to 125°C. The SN74HC164 is characterized for operation from −40°C to 85°C.

logic symbol

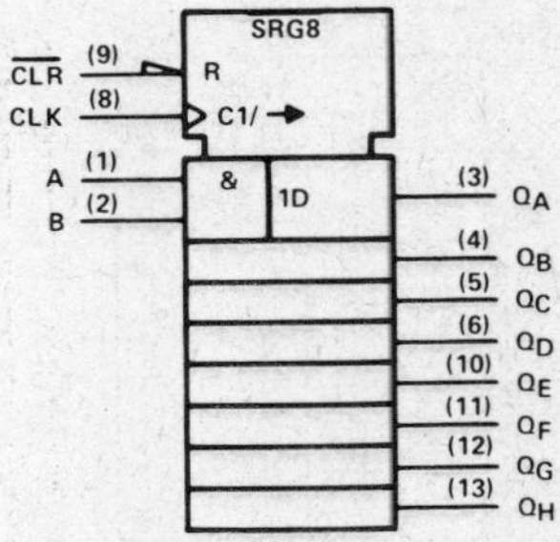

Pin numbers shown are for J and N packages.

FUNCTION TABLE

INPUTS				OUTPUTS		
$\overline{CLR}$	CLK	A	B	Q_A	Q_B . . .	Q_H
L	X	X	X	L	L	L
H	L	X	X	Q_{A0}	Q_{B0}	Q_{H0}
H	↑	H	H	H	Q_{An}	Q_{Gn}
H	↑	L	X	L	Q_{An}	Q_{Gn}
H	↑	X	L	L	Q_{An}	Q_{Gn}

H = high level (steady state), L = low level (steady state)
X = irrelevant (any input, including transitions)
↑ = transition from low to high level.
Q_{A0}, Q_{B0}, Q_{H0} = the level of Q_A, Q_B, or Q_H, respectively, before the indicated steady-state input conditions were established.
Q_{An}, Q_{Gn} = the level of Q_A or Q_G before the most-recent ↑ transition of the clock; indicates a one-bit shift.

logic diagram (positive logic)

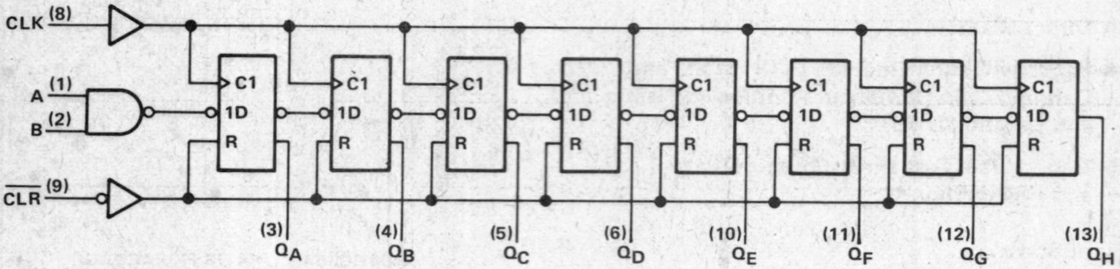

typical clear, shift, and clear sequences

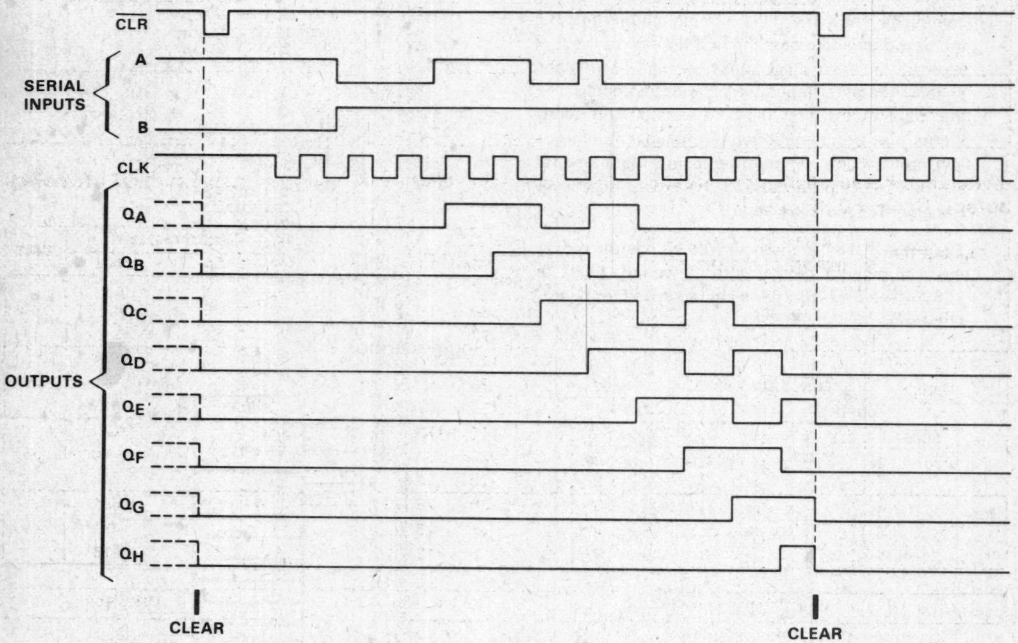

maximum ratings, recommended operating conditions, and electrical characteristics

See Table IV, page 2-10.

TEXAS INSTRUMENTS
POST OFFICE BOX 225012 • DALLAS, TEXAS 75265

timing requirements over recommended operating free-air temperature range (unless otherwise noted)

			V_{CC}	T_A = 25°C MIN	T_A = 25°C MAX	SN54HC164 MIN	SN54HC164 MAX	SN74HC164 MIN	SN74HC164 MAX	UNIT
f_{clock}	Clock frequency		2 V	0	6	0	4.2	0	5	MHz
			4.5 V	0	31	0	21	0	25	
			6 V	0	36	0	25	0	28	
t_w	Pulse duration	$\overline{CLR}$ low	2 V	100		150		125		ns
			4.5 V	20		30		25		
			6 V	17		25		21		
		CLK high or low	2 V	80		120		100		
			4.5 V	16		24		20		
			6 V	14		20		18		
t_{su}	Setup time before CLK↑	Data	2 V	100		150		125		ns
			4.5 V	20		30		25		
			6 V	27		25		21		
		$\overline{CLR}$ inactive	2 V	100		150		125		
			4.5 V	20		30		25		
			6 V	27		25		21		
t_h	Hold time, data after CLK↑		2 V	5		5		5		ns
			4.6 V	5		5		5		
			6 V	5		5		5		

switching characteristics over recommended operating free-air temperature range (unless otherwise noted), C_L = 50 pF (see Note 1)

PARAMETER	FROM (INPUT)	TO (OUTPUT)	V_{CC}	T_A = 25°C MIN	T_A = 25°C TYP	T_A = 25°C MAX	SN54HC164 MIN	SN54HC164 MAX	SN74HC164 MIN	SN74HC164 MAX	UNIT
f_{max}			2 V	6	10		4.2		5		MHz
			4.5 V	31	54		21		25		
			6 V	36	62		25		28		
t_{PHL}	$\overline{CLR}$	Any Q	2 V		140	205		295		255	ns
			4.5 V		28	41		59		51	
			6 V		24	35		51		46	
t_{pd}	CLK	Any Q	2 V		115	175		265		220	ns
			4.5 V		23	35		53		44	
			6 V		20	30		45		38	
t_t			2 V		38	75		110		95	ns
			4.5 V		8	15		22		19	
			6 V		6	13		19		16	

C_{pd}	Power dissipation capacitance	No load, T_A = 25°C	135 pF typ

NOTE 1: For load circuit and voltage waveforms, see page 1-14.

TYPES SN54HC165, SN74HC165 PARALLEL-LOAD 8-BIT SHIFT REGISTERS

D2684, DECEMBER 1982—REVISED MARCH 1984

- **Complementary Outputs**
- **Direct Overriding Load (Data) Inputs**
- **Gated Clock Inputs**
- **Parallel-to-Serial Data Conversion**
- **Package Options Include Both Plastic and Ceramic Chip Carriers in Addition to Plastic and Ceramic DIPs**
- **Dependable Texas Instruments Quality and Reliability**

SN54HC165 . . . J PACKAGE
SN74HC165 . . . J OR N PACKAGE
(TOP VIEW)

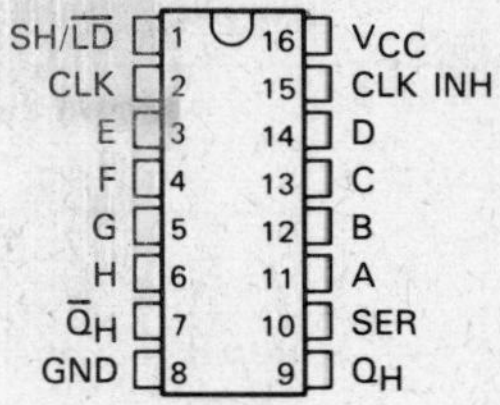

SN54HC165 . . . FH OR FK PACKAGE
SN74HC165 . . . FH OR FN PACKAGE
(TOP VIEW)

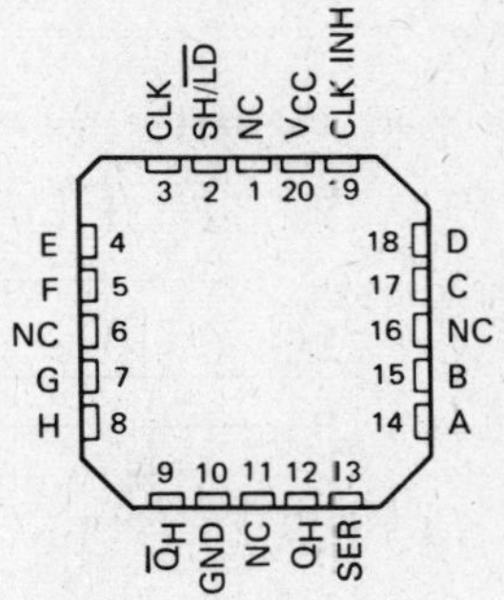

NC—No internal connection

description

The 'HC165 is an 8-bit serial shift register that, when clocked, shifts the data toward serial output Q_H. Parallel-in access to each stage is provided by eight individual direct data inputs that are enabled by a low level at the SH/$\overline{LD}$ input. The 'HC165 also features a clock inhibit function and a complementary serial output $\overline{Q}_H$.

Clocking is accomplished by a low-to-high transition of the CLK input while SH/$\overline{LD}$ is held high and CLK INH is held low. The functions of the CLK and CLK INH (clock inhibit) inputs are interchangeable. Since a low CLK input and a low-to-high transition of CLK INH will also accomplish clocking, CLK INH should be changed to the high level only while the CLK input is high. Parallel loading is inhibited when SH/$\overline{LD}$ is held high. The parallel inputs to the register are enabled while SH/$\overline{LD}$ is low independently of the levels of CLK, CLK INH, or SER inputs.

The SN54HC165 is characterized for operation over the full military temperature range of −55 °C to 125 °C. The SN74HC165 is characterized for operation from −40 °C to 85 °C.

FUNCTION TABLE

INPUTS			FUNCTION
SH/$\overline{LD}$	CLK	CLK INH	
L	X	X	Parallel load
H	H	X	No change
H	X	H	No change
H	L	↑	Shift
H	↑	L	Shift

Shift — content of each internal register shifts toward serial output Q_H. Data at serial input is shifted into first register.

logic symbol

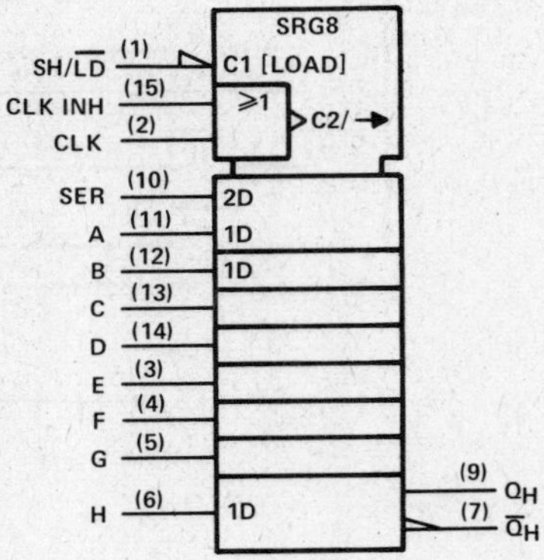

Pin numbers shown are for J and N packages.

logic diagram (positive logic)

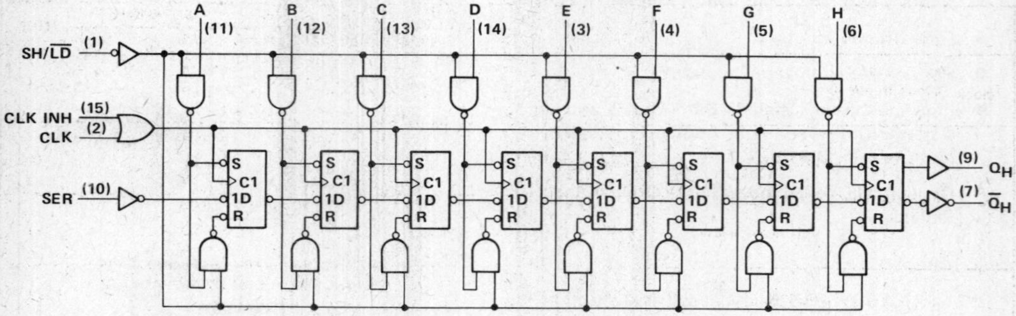

Pin numbers shown are for J and N packages.

typical shift, load, and inhibit sequences

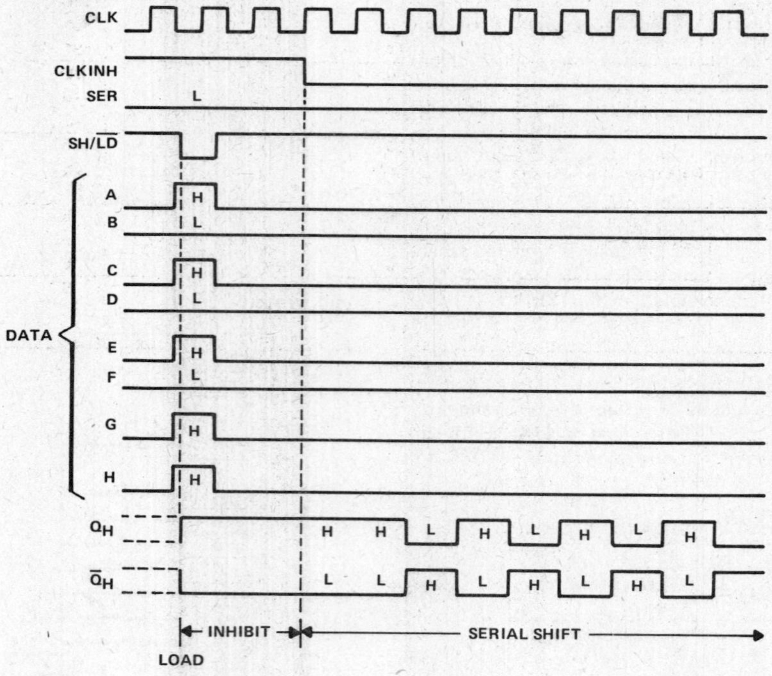

TEXAS INSTRUMENTS
POST OFFICE BOX 225012 • DALLAS, TEXAS 75265

timing requirements over recommended operating free-air temperature range (unless otherwise noted)

			V_{CC}	T_A = 25°C MIN	T_A = 25°C MAX	SN54HC165 MIN	SN54HC165 MAX	SN74HC165 MIN	SN74HC165 MAX	UNIT
f_{clock}	Clock frequency		2 V	0	6	0	4.2	0	5	MHz
			4.5 V	0	31	0	21	0	25	
			6 V	0	36	0	25	0	29	
t_w	Pulse duration	SH/$\overline{LD}$ low	2 V	80		120		100		ns
			4.5 V	16		24		20		
			6 V	14		20		17		
		CLK high or low	2 V	80		120		100		ns
			4.5 V	16		24		20		
			6 V	14		20		17		
t_{su}	Setup time	SH/$\overline{LD}$ high before CLK↑	2 V	80		120		100		ns
			4.5 V	16		24		20		
			6 V	14		20		17		
		SER before CLK↑	2 V	40		60		50		ns
			4.5 V	8		12		10		
			6 V	7		10		9		
		CLK INH low before CLK↑	2 V	100		150		125		ns
			4.5 V	20		30		25		
			6 V	17		25		21		
		CLK INH high before CLK↓	2 V	40		60		50		ns
			4.5 V	8		12		10		
			6 V	7		10		9		
		Data before SH/$\overline{LD}$↑	2 V	100		150		125		ns
			4.5 V	20		30		25		
			6 V	17		26		21		
t_h	Hold time	SER data after CLK↑	2 V	5		5		5		ns
			4.5 V	5		5		5		
			6 V	5		5		5		
		PAR data after SH/$\overline{LD}$↑	2 V	5		5		5		ns
			4.5 V	5		5		5		
			6 V	5		5		5		

switching characteristics over recommended operating free-air temperature range (unless otherwise noted), C_L = 50 pF (see Note 1)

PARAMETER	FROM (INPUT)	TO (OUTPUT)	V_{CC}	T_A = 25°C MIN	TYP	MAX	SN54HC165 MIN	MAX	SN74HC165 MIN	MAX	UNIT
f_{max}			2 V	6	13		4.2		5		MHz
			4.5 V	31	50		21		25		
			6 V	36	62		25		29		
t_{pd}	SH/$\overline{LD}$	Q_H or $\overline{Q}_H$	2 V		80	150		225		190	ns
			4.5 V		20	30		45		38	
			6 V		16	26		38		32	
t_{pd}	CLK	Q_H or $\overline{Q}_H$	2 V		75	150		225		190	ns
			4.5 V		15	30		45		38	
			6 V		13	26		38		32	
t_{pd}	H	Q_H or $\overline{Q}_H$	2 V		75	150		225		190	ns
			4.5 V		15	30		45		38	
			6 V		13	26		38		32	
t_t		Any	2 V		38	75		110		95	ns
			4.5 V		8	15		22		19	
			6 V		6	3		19		16	

C_{pd}	Power dissipation capacitance	No load, T_A = 25°C	75 pF typ

NOTE 1: For load circuit and voltage waveforms, see page 1-14.

TEXAS INSTRUMENTS
POST OFFICE BOX 225012 • DALLAS, TEXAS 75265

TYPES SN54HC166, SN74HC166 PARALLEL-LOAD 8-BIT SHIFT REGISTERS

D2684, DECEMBER 1982—REVISED MARCH 1984

- Synchronous Load
- Direct Overriding Clear
- Parallel to Serial Conversion
- Package Options Include Both Plastic and Ceramic Chip Carriers in Addition to Plastic and Ceramic DIPs
- Dependable Texas Instruments Quality and Reliability

SN54HC166 . . . J PACKAGE
SN74HC166 . . . J OR N PACKAGE
(TOP VIEW)

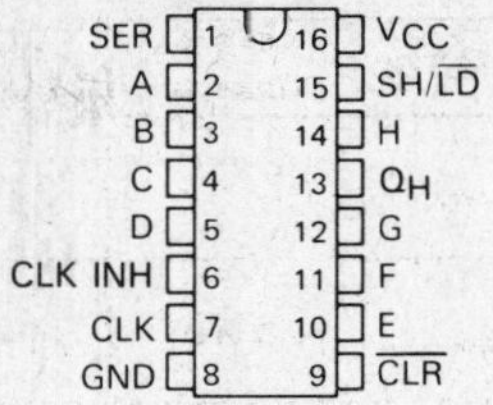

SN54HC166 . . . FH OR FK PACKAGE
SN74HC166 . . . FH OR FN PACKAGE
(TOP VIEW)

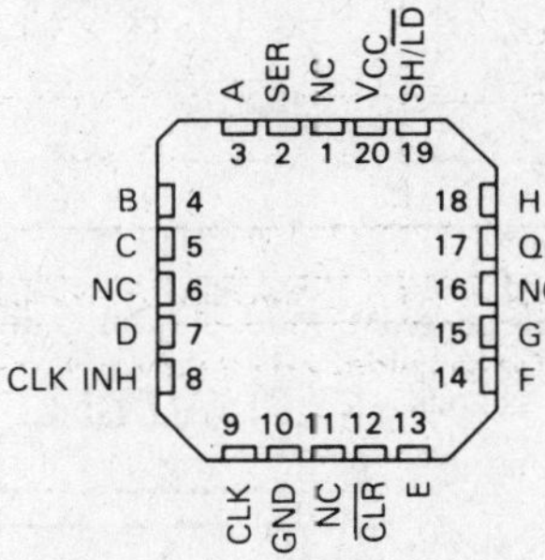

NC—No internal connection

description

The 'HC166 parallel-in or serial-in, serial-out registers feature gated clock inputs and an overriding clear input. The parallel-in or serial-in modes are established by the shift/load input. When high, this input enables the serial data input and couples the eight flip-flops for serial shifting with each clock pulse. When low, the parallel (broadside) data inputs are enabled and synchronous loading occurs on the next clock pulse. During parallel loading, serial data flow is inhibited. Clocking is accomplished on the low-to-high-level edge of the clock pulse through a two-input positive NOR gate permitting one input to be used as a clock-enable or clock-inhibit function. Holding either of the clock inputs high inhibits clocking; holding either low enables the other clock input. This, of course, allows the system clock to be free-running and the register can be stopped on command with the other clock input. The clock-inhibit input should be changed to the high level only when the clock input is high. A direct clear input overrides all other inputs, including the clock, and resets all flip-flops to zero.

The SN54HC166 is characterized for operation over the full military temperature range of −55°C to 125°C. The SN74HC166 is characterized for operation from −40°C to 85°C.

logic symbol

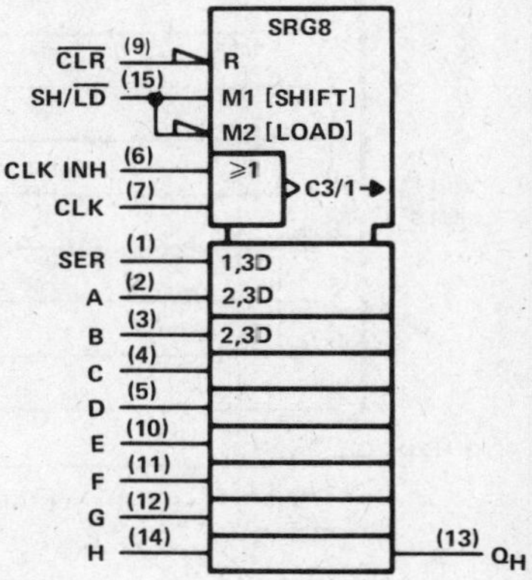

Pin numbers shown are for J and N packages.

FUNCTION TABLE

INPUTS						INTERNAL OUTPUTS		OUTPUT
CLEAR	SHIFT/ LOAD	CLOCK INHIBIT	CLOCK	SERIAL	PARALLEL A . . . H	Q_A	Q_B	Q_H
L	X	X	X	X	X	L	L	L
H	X	L	L	X	X	Q_{A0}	Q_{B0}	Q_{H0}
H	L	L	↑	X	a . . . h	a	b	h
H	H	L	↑	H	X	H	Q_{An}	Q_{Gn}
H	H	L	↑	L	X	L	Q_{An}	Q_{Gn}
H	X	H	↑	X	X	Q_{A0}	Q_{B0}	Q_{H0}

logic diagram (positive logic)

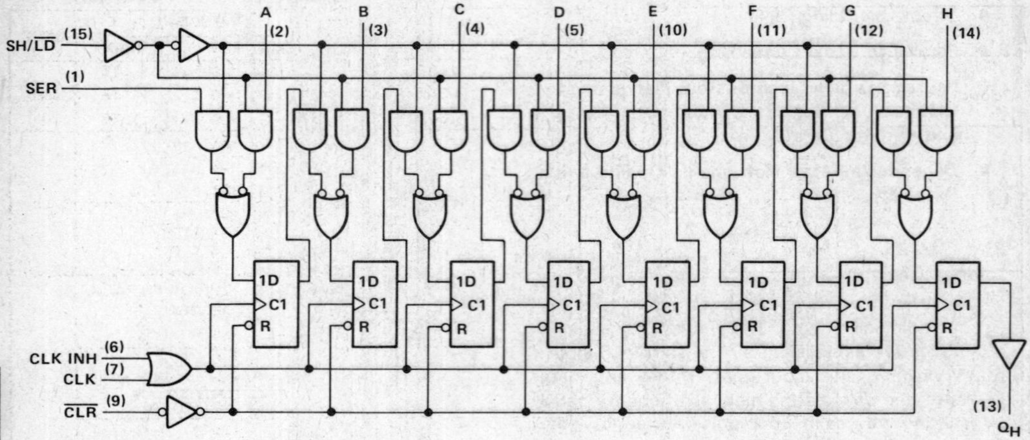

Pin numbers shown are for J and N packages.

typical clear, shift, load, inhibit, and shift sequences

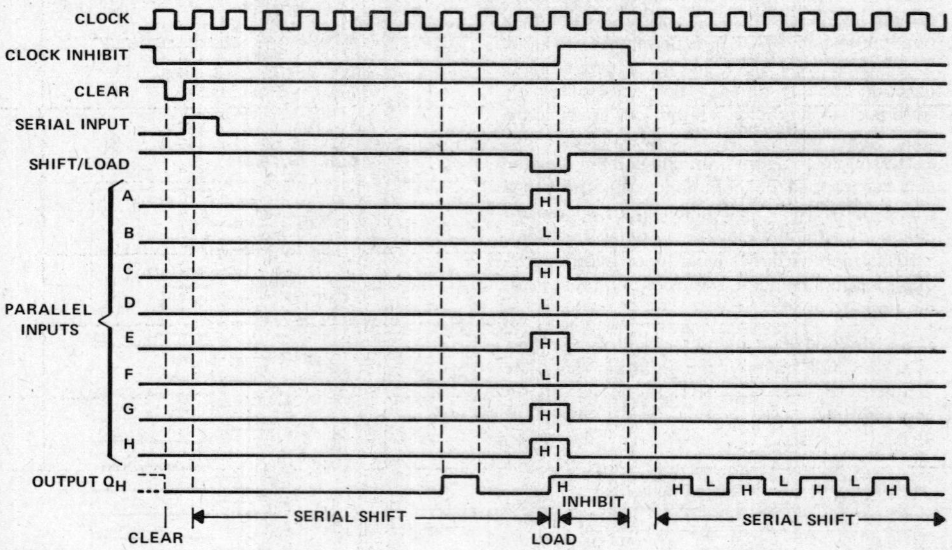

maximum ratings, recommended operating conditions, and eletrical characteristics

See Table IV, page 2-10.

TEXAS INSTRUMENTS
POST OFFICE BOX 225012 • DALLAS, TEXAS 75265

timing requirements over recommended operating free-air temperature range (unless otherwise noted)

			V_{CC}	T_A = 25°C		SN54HC166		SN74HC166		UNIT
				MIN	MAX	MIN	MAX	MIN	MAX	
f_{clock}	Clock frequency		2 V	0	6	0	4.2	0	5	MHz
			4.5 V	0	31	0	21	0	25	
			6 V	0	36	0	25	0	29	
t_w	Pulse duration	$\overline{CLR}$ low	2 V	100		150		125		ns
			4.5 V	20		30		25		
			6 V	17		26		21		
		CLK high or low	2 V	80		120		100		
			4.5 V	16		24		20		
			6 V	14		20		17		
t_{su}	Setup time	SH/$\overline{LD}$ high before CLK↑	2 V	145		220		180		ns
			4.5 V	29		44		36		
			6 V	25		38		31		
		SER before CLK↑	2 V	80		120		100		
			4.5 V	16		24		20		
			6 V	14		20		17		
		CLK INH before CLK↑	2 V	100		150		125		
			4.5 V	20		30		25		
			6 V	17		26		21		
		Data before SH/$\overline{LD}$↑	2 V	80		120		100		
			4.5 V	16		24		20		
			6 V	14		20		17		
		$\overline{CLR}$ inactive before CLK↑	2 V	40		60		50		
			4.5 V	8		12		10		
			6 V	7		10		9		
t_h	Hold time	SH/$\overline{LD}$ high after CLK↑	2 V	0		0		0		ns
			4.5 V	0		0		0		
			6 V	0		0		0		
		SER after CLK↑	2 V	5		5		5		
			4.5 V	5		5		5		
			6 V	5		5		5		
		CLK INH after CLK↑	2 V	0		0		0		
			4.5 V	0		0		0		
			6 V	0		0		0		
		Data after SH/$\overline{LD}$↑	2 V	5		5		5		
			4.5 V	5		5		5		
			6 V	5		5		5		
		$\overline{CLR}$ active after CLK↑	2 V	0		0		0		
			4.5 V	0		0		0		
			6 V	0		0		0		

switching characteristics over recommended operating free-air temperature range (unless otherwise noted), C_L = 50 pF (see Note 1)

PARAMETER	FROM	TO	V_{CC}	T_A = 25°C			SN54HC166		SN74HC166		UNIT
				MIN	TYP	MAX	MIN	MAX	MIN	MAX	
f_{max}			2 V	6	11		4.2		5		MHz
			4.5 V	31	36		21		25		
			6 V	36	45		25		29		
t_{PHL}	$\overline{CLR}$	Q_H	2 V		62	120		180		150	ns
			4.5 V		18	24		36		30	
			6 V		13	20		31		26	
t_{pd}	CLK	Q_H	2 V		75	150		225		190	ns
			4.5 V		15	30		45		38	
			6 V		13	26		38		32	
t_t		Any	2 V		38	75		110		95	ns
			4.5 V		8	15		22		19	
			6 V		6	13		19		16	

C_{pd}	Power dissipation capacitance	No load, T_A = 25°C	50 pF typ

NOTE 1: For load circuit and voltage waveforms, see page 1-14.

3 HCMOS DEVICES

TEXAS INSTRUMENTS
POST OFFICE BOX 225012 • DALLAS, TEXAS 75265

TYPES SN54HC173, SN74HC173
4-BIT D-TYPE REGISTERS WITH 3-STATE OUTPUTS

D2684, DECEMBER 1982—REVISED MARCH 1984

- **High-Current 3-State Outputs Interface Directly with System Bus or Can Drive up to 15 LSTTL Loads**
- **Gated Output-Control Lines for Enabling or Disabling the Outputs**
- **Fully Independent Clock Virtually Eliminates Restrictions for Operating in One of Two Modes**
- **Package Options Include Both Plastic and Ceramic Chip Carriers in Addition to Plastic and Ceramic DIPs**
- **Dependable Texas Instruments Quality and Reliability**

SN54HC173 . . . J PACKAGE
SN74HC173 . . . J OR N PACKAGE
(TOP VIEW)

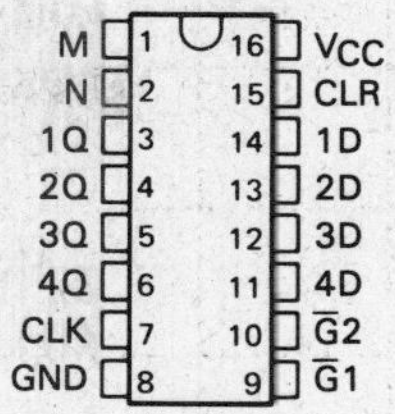

SN54HC173 . . . FH OR FK PACKAGE
SN74HC173 . . . FH OR FN PACKAGE
(TOP VIEW)

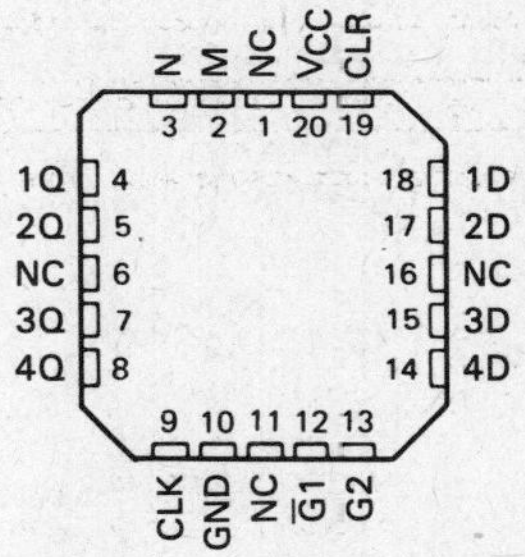

NC—No internal connection

description

The 'HC173 four-bit registers include D-type flip-flops featuring totem-pole three-state outputs capable of driving highly capacitive or relatively low-impedance loads. The high-impedance third state and increased drive provide these flip-flops with the capability of being connected directly to and driving the lines in a bus-organized system without need for interface or pull-up components.

Gated enable inputs are provided on these devices for controlling the entry of data into the flip-flops. When both data-enable inputs are low, data at the D inputs are loaded into their respective flip-flops on the next positive transition of the clock input. Gate output control inputs are also provided. When both are low, the normal logic states (high or low levels) of the four outputs are available for driving the loads or bus lines. The outputs are disabled independently from the level of the clock by a high logic level at either output control input. The outputs then present a high impedance and neither load nor drive the bus line. Detailed operation is given in the function table.

The SN54HC173 is characterized for operation over the full military temperature range of −55°C to 125°C. The SN74HC173 is characterized for operation from −40°C to 85°C.

FUNCTION TABLE

INPUTS					OUTPUT
CLEAR	CLOCK	DATA ENABLE $\overline{G}1$	DATA ENABLE $\overline{G}2$	DATA D	Q
H	X	X	X	X	L
L	L	X	X	X	Q_0
L	↑	H	X	X	Q_0
L	↑	X	H	X	Q_0
L	↑	L	L	L	L
L	↑	L	L	H	H

When either M or N (or both) is (are) high the output is disabled to the high-impedance state; however sequential operation of the flip-flops is not affected.

logic symbol

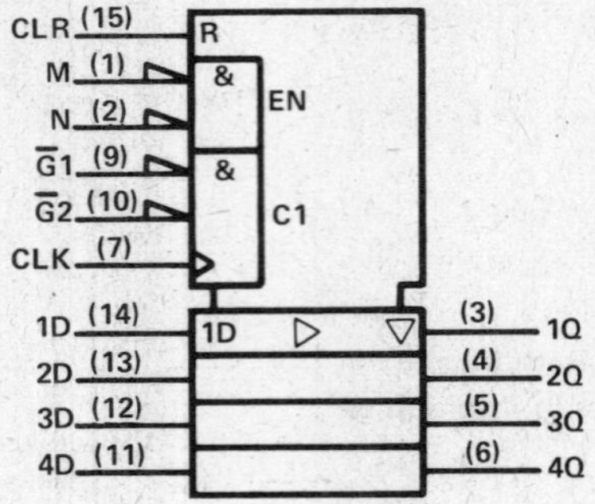

logic diagram (positive logic)

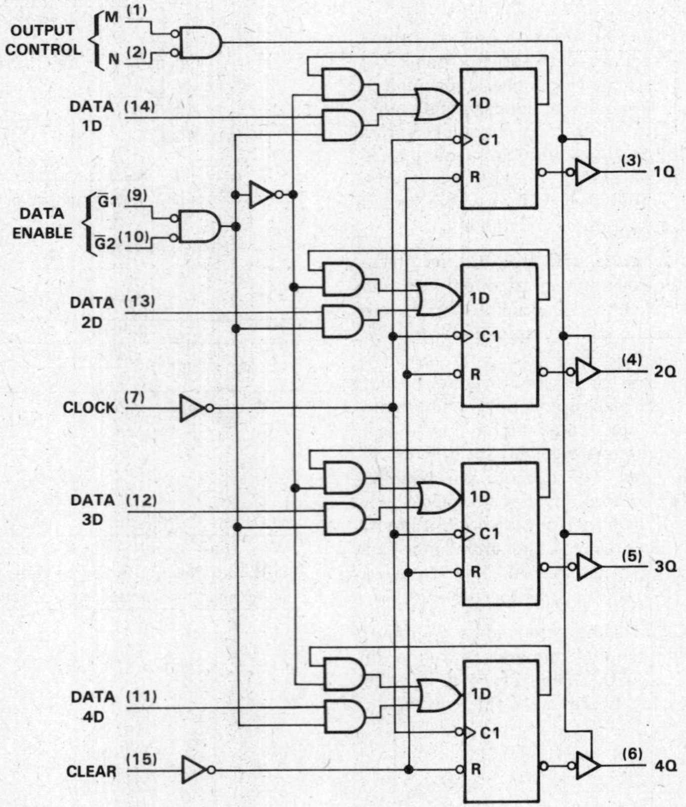

Pin numbers shown are for J and N packages.

TEXAS INSTRUMENTS
POST OFFICE BOX 225012 • DALLAS, TEXAS 75265

maximum ratings, recommended operating conditions, and electrical characteristics

See Table III, page 2-8.

timing requirements over recommended operating free-air temperature range (unless otherwise noted)

		V_{CC}	T_A = 25°C		SN54HC173		SN74HC173		UNIT
			MIN	MAX	MIN	MAX	MIN	MAX	
f_{clock} Input clock frequency		2 V	0	6	0	4.2	0	5	MHz
		4.5 V	0	31	0	21	0	25	
		6 V	0	36	0	25	0	29	
t_w Pulse duration	CLK high or low	2 V	80		120		100		ns
		4.5 V	16		24		20		
		6 V	14		20		17		
	CLR high	2 V	80		120		100		ns
		4.5 V	16		24		20		
		6 V	14		20		17		
t_{su} Setup time before CLK↑	$\overline{G}1$ and $\overline{G}2$	2 V	100		150		125		ns
		4.5 V	20		30		25		
		6 V	17		25		21		
	Data	2 V	100		150		125		ns
		4.5 V	20		30		25		
		6 V	17		25		21		
	CLR inactive	2 V	90		135		115		ns
		4.5 V	18		27		23		
		6 V	15		23		19		
t_h Hold time after CLK↑	$\overline{G}1$ and $\overline{G}2$	2 V	0		0		0		ns
		4.5 V	0		0		0		
		6 V	0		0		0		
	Data	2 V	0		0		0		ns
		4.5 V	0		0		0		
		6 V	0		0		0		

switching characteristics over recommended operating free-air temperature range (unless otherwise noted), C_L = 50 pF (see Note 1)

PARAMETER	FROM (INPUT)	TO (OUTPUT)	V_{CC}	T_A = 25°C MIN	T_A = 25°C TYP	T_A = 25°C MAX	SN54HC173 MIN	SN54HC173 MAX	SN74HC173 MIN	SN74HC173 MAX	UNIT
f_{max}			2 V	6	8		4.2		5		MHz
			4.5 V	31	46		21		25		
			6 V	36	55		25		29		
t_{PHL}	CLR	Any	2 V		78	150		225		190	ns
			4.5 V		21	30		45		38	
			6 V		20	26		38		32	
t_{pd}	CLK	Any	2 V		78	150		225		190	ns
			4.5 V		21	30		45		38	
			6 V		20	26		38		32	
t_{en}	M or N	Any	2 V		78	150		225		190	ns
			4.5 V		20	30		45		38	
			6 V		15	26		38		32	
t_{dis}	M or N	Any	2 V		40	150		225		190	ns
			4.5 V		18	30		45		38	
			6 V		16	26		38		32	
t_t		Any	2 V		20	60		90		75	ns
			4.5 V		8	12		18		15	
			6 V		6	10		15		13	

C_{pd}	Power dissipation capacitance	No load, T_A = 25°C	29 pF typ

switching characteristics over recommended operating free-air temperature range (unless otherwise noted), C_L = 150 pF (see Note 1)

PARAMETER	FROM (INPUT)	TO (OUTPUT)	V_{CC}	T_A = 25°C MIN	T_A = 25°C TYP	T_A = 25°C MAX	SN54HC173 MIN	SN54HC173 MAX	SN74HC173 MIN	SN74HC173 MAX	UNIT
t_{PHL}	CLR	Any	2 V		100	200		300		250	ns
			4.5 V		28	40		60		50	
			6 V		21	34		51		43	
t_{pd}	CLK	Any	2 V		100	200		300		250	ns
			4.5 V		28	40		60		50	
			6 V		21	34		51		43	
t_{en}	M or N	Any	2 V		100	200		300		250	ns
			4.5 V		28	40		60		50	
			6 V		21	34		51		43	
t_t		Any	2 V		45	210		315		265	ns
			4.5 V		17	42		63		53	
			6 V		13	36		53		45	

NOTE 1: For load circuit and voltage waveforms, see page 1-14.

TYPES SN54HC174, SN54HC175 SN74HC174, SN74HC175 HEX/QUADRUPLE D-TYPE FLIP-FLOPS WITH CLEAR

D2684, DECEMBER 1982—REVISED MARCH 1984

- **'HC174 Contains Six Flip-Flops with Single-Rail Outputs**
- **'HC175 Contains Four Flip-Flops with Double-Rail Outputs**
- **Applications Include:**
  - **Buffer/Storage Registers**
  - **Shift Registers**
  - **Pattern Generators**
- **Package Options Include Both Plastic and Ceramic Chip Carriers in Addition to Plastic and Ceramic DIPs**
- **Dependable Texas Instruments Quality and Reliability**

SN54HC174 . . . J PACKAGE
SN74HC174 . . . J OR N PACKAGE
(TOP VIEW)

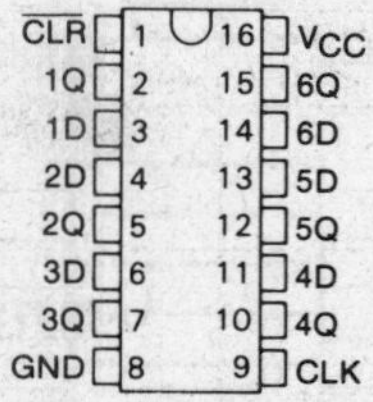

SN54HC174 . . . FH OR FK PACKAGE
SN74HC174 . . . FH OR FN PACKAGE
(TOP VIEW)

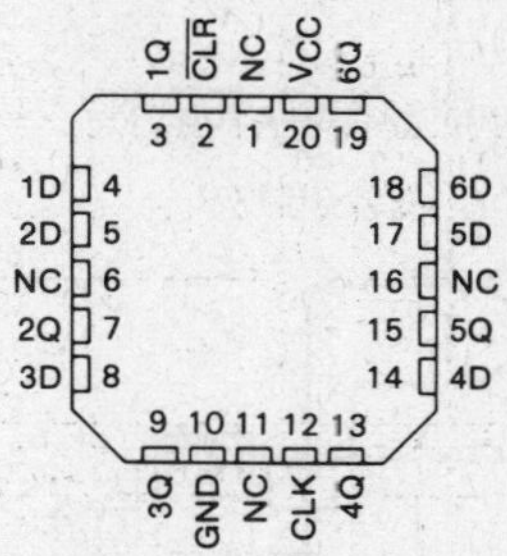

description

These monolithic, positive-edge triggered D-type flip-flops have a direct clear input and the 'HC175 features complementary outputs from each flip-flop.

Information at the D inputs meeting the setup time requirements is transferred to the outputs on the positive-going edge of the clock pulse. Clock triggering occurs at a particular voltage level and is not directly related to the transition time of the positive-going edge of the clock pulse. When the clock input is at either the high or low level, the D input signal has no effect at the output.

The SN54HC174 and SN54HC175 are characterized for operation over the full military temperature range of −55°C to 125°C. The SN74HC174 and SN74HC175 are characterized for operation from −40°C to 85°C.

SN54HC175 . . . J PACKAGE
SN74HC175 . . . J OR N PACKAGE
(TOP VIEW)

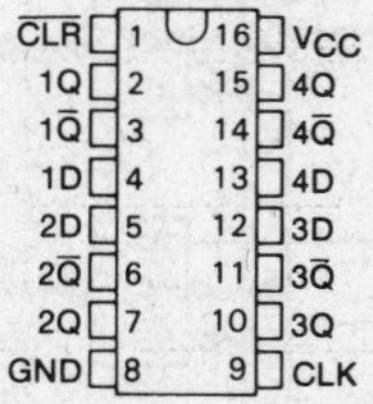

FUNCTION TABLE
(EACH FLIP-FLOP)

INPUTS			OUTPUTS	
$\overline{CLR}$	CLK	D	Q	$\overline{Q}$†
L	X	X	L	H
H	↑	H	H	L
H	↑	L	L	H
H	L	X	Q_0	$\overline{Q}_0$

†'HC175 only

SN54HC175 . . . FH OR FK PACKAGE
SN74HC175 . . . FH OR FN PACKAGE
(TOP VIEW)

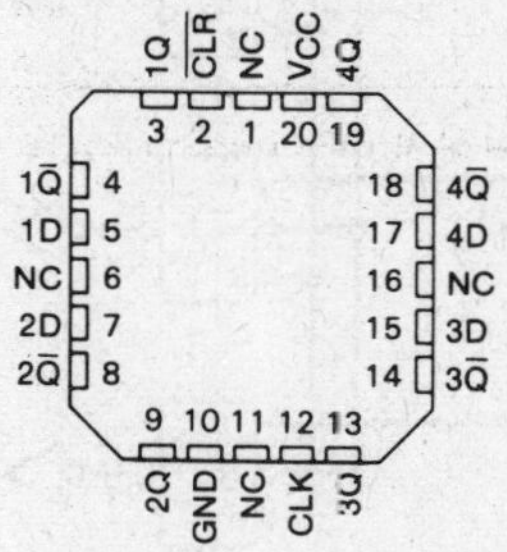

NC—No internal connection

logic symbols

'HC174

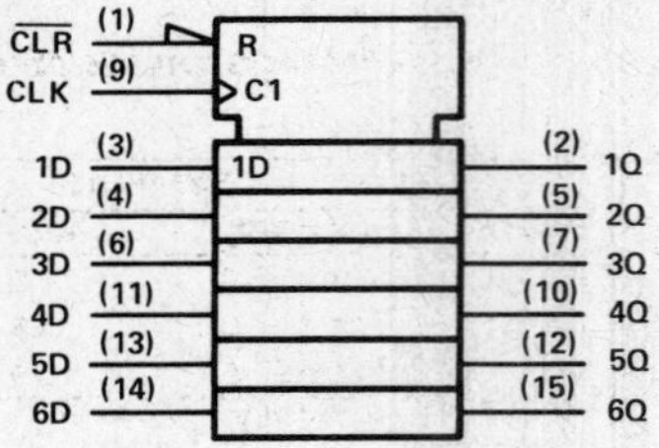

'HC175

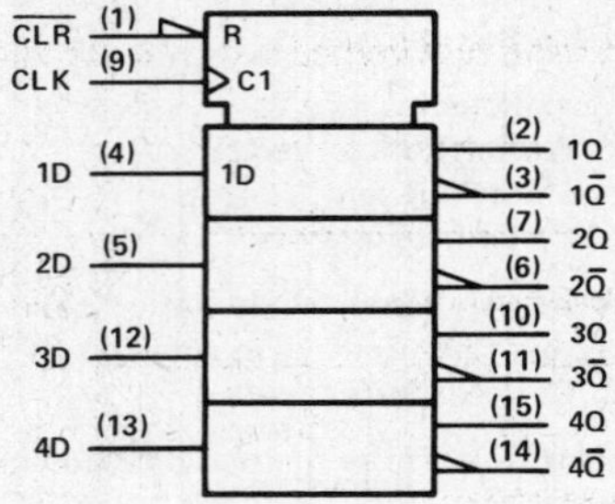

logic diagrams (positive logic)

'HC174

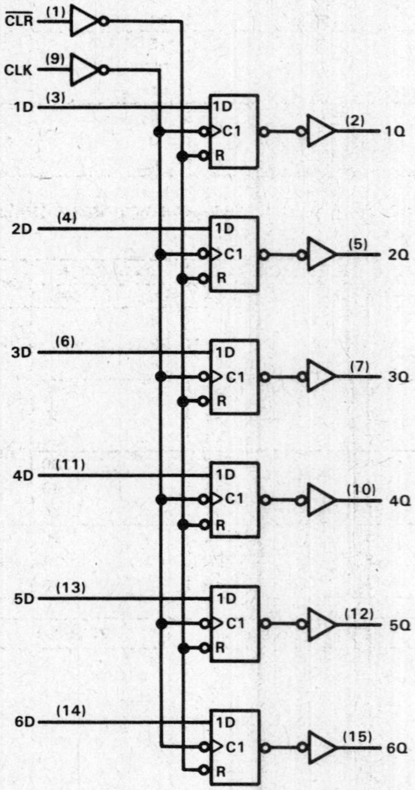

'HC175

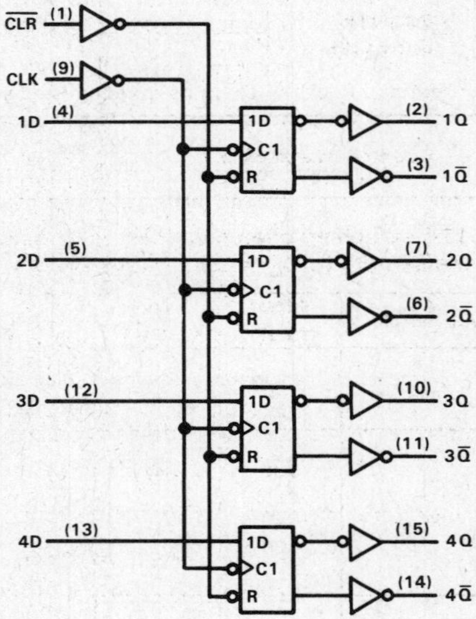

Pin numbers shown are for J and N packages.

maximum ratings, recommended operating conditions, and electrical characteristics

See Table IV, page 2-10.

'HC174 timing requirements over recommended operating free-air temperature range (unless otherwise noted)

			V_{CC}	T_A = 25°C MIN	T_A = 25°C MAX	SN54HC174 MIN	SN54HC174 MAX	SN74HC174 MIN	SN74HC174 MAX	UNIT
f_{clock}	Clock Frequency		2 V	0	6	0	4.2	0	5	MHz
			4.5 V	0	31	0	21	0	25	
			6 V	0	36	0	25	0	29	
t_w	Pulse duration	$\overline{CLR}$ low	2 V	80		120		100		ns
			4.5 V	16		24		20		
			6 V	14		20		17		
		CLK high or low	2 V	80		120		100		
			4.5 V	16		24		20		
			6 V	14		20		17		
t_{su}	Setup time before CLK↑	Data	2 V	100		150		125		ns
			4.5 V	20		30		25		
			6 V	17		25		21		
		$\overline{CLR}$ inactive	2 V	100		150		125		
			4.5 V	20		30		25		
			6 V	17		25		21		
t_h	Hold time, data after CLK↑		2 V	0		0		0		ns
			4.5 V	0		0		0		
			6 V	0		0		0		

'HC174 switching characteristics over recommended operating free-air temperature range (unless otherwise noted), C_L = 50 pF (see Note 1)

PARAMETER	FROM (INPUT)	TO (OUTPUT)	V_{CC}	T_A = 25°C MIN	T_A = 25°C TYP	T_A = 25°C MAX	SN54HC174 MIN	SN54HC174 MAX	SN74HC174 MIN	SN74HC174 MAX	UNIT
f_{max}			2 V	6	9		4.2		5		MHz
			4.5 V	31	44		21		25		
			6 V	36	50		25		29		
t_{pd}	$\overline{CLR}$	Any	2 V		58	160		240		200	ns
			4.5 V		17	32		48		40	
			6 V		14	27		41		34	
	CLK	Any	2 V		58	160		240		200	
			4.5 V		17	32		48		40	
			6 V		14	27		41		34	
t_t		Any	2 V		38	75		110		90	ns
			4.5 V		8	15		22		19	
			6 V		6	13		19		16	

C_{pd}	Power dissipation capacitance per flip-flop	No load, T_A = 25°C	27 pF typ

NOTE 1: For load circuit and voltage waveforms, see page 1-14.

maximum ratings, recommended operating conditions, and electrical characteristics

See Table II, page 2-6.

'HC175 timing requirements over recommended operating free-air temperature range (unless otherwise noted)

			V_{CC}	T_A = 25°C		SN54HC175		SN74HC175		UNIT
				MIN	MAX	MIN	MAX	MIN	MAX	
f_{clock}			2 V	0	6	0	4.2	0	5	MHz
			4.5 V	0	31	0	21	0	25	
			6 V	0	36	0	25	0	29	
t_w	Pulse duration	$\overline{CLR}$ low	2 V	80		120		100		ns
			4.5 V	16		24		20		
			6 V	14		20		17		
		CLK high or low	2 V	80		120		100		
			4.5 V	16		24		20		
			6 V	14		20		17		
t_{su}	Setup time before CLK↑	Data	2 V	100		150		125		ns
			4.5 V	20		30		25		
			6 V	17		25		21		
		$\overline{CLR}$ inactive	2 V	100		150		125		
			4.5 V	20		30		25		
			6 V	17		25		21		
t_h	Hold time, data after CLK↑		2 V	0		0		0		ns
			4.5 V	0		0		0		
			6 V	0		0		0		

'HC175 switching characteristics over recommended operating free-air temperature range (unless otherwise noted), C_L = 50 pF (see Note 1)

PARAMETER	FROM (INPUT)	TO (OUTPUT)	V_{CC}	T_A = 25°C			SN54HC175		SN74HC175		UNIT
				MIN	TYP	MAX	MIN	MAX	MIN	MAX	
f_{max}			2 V	6	12		4.2		5		MHz
			4.5 V	31	50		21		25		
			6 V	36	60		25		29		
t_{pd}	$\overline{CLR}$	Any	2 V		52	150		255		190	ns
			4.5 V		15	30		45		38	
			6 V		13	26		38		32	
	CLK	Any	2 V		58	150		255		190	
			4.5 V		16	30		45		38	
			6 V		13	26		38		32	
t_t		Any	2 V		38	75		110		90	ns
			4.5 V		8	15		22		19	
			6 V		6	13		19		16	

C_{pd}	Power dissipation capacitance per flip-flop	No load, T_A = 25°C	30 pF typ

NOTE 1: For load circuit and voltage waveforms, see page 1-14.

TYPES SN54HC180, SN74HC180
9-BIT ODD/EVEN PARITY GENERATORS/CHECKERS

D2484, MARCH 1984

- Package Options Include Both Plastic and Ceramic Chip Carriers in Addition to Plastic and Ceramic DIPs
- Dependable Texas Instruments Quality and Reliability

SN54HC180 . . . J PACKAGE
SN74HC180 . . . J OR N PACKAGE
(TOP VIEW)

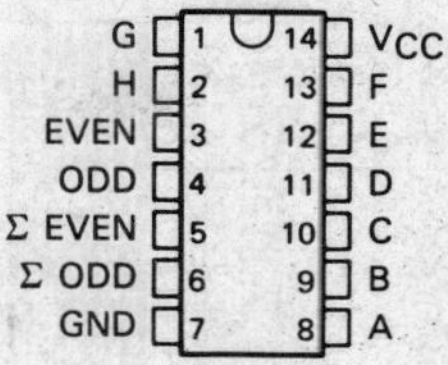

SN54HC180 . . . FH OR FK PACKAGE
SN74HC180 . . . FH OR FN PACKAGE
(TOP VIEW)

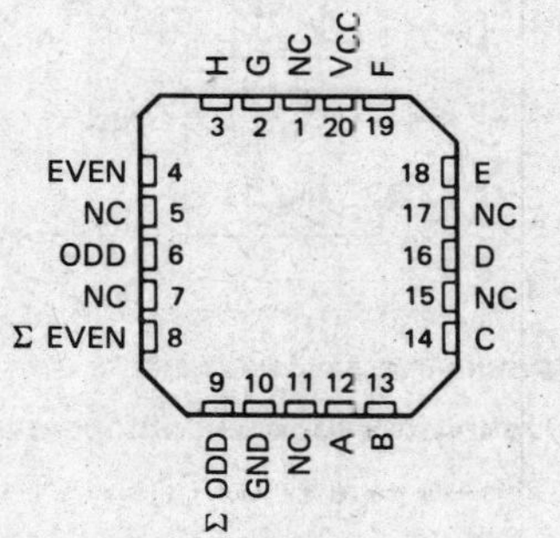

NC—No internal connection

description

These universal, monolithic, 9-bit (8 data bits plus 1 parity bit) parity generators/checkers, feature odd/even outputs and control inputs to facilitate operation in either odd- or even-parity applications. Depending on whether even or odd parity is being generated or checked, the even or odd inputs can be utilized as the parity or 9th-bit input. The word-length capability is easily expanded by cascading.

The SN54HC180 is characterized for operation over the full military temperature range of −55°C to 125°C; and the SN74HC180 is characterized for operation from −40°C to 85°C.

FUNCTION TABLE

INPUTS			OUTPUTS	
Σ OF H's AT A THRU H	EVEN	ODD	Σ EVEN	Σ ODD
EVEN	H	L	H	L
ODD	H	L	L	H
EVEN	L	H	L	H
ODD	L	H	H	L
X	H	H	L	L
X	L	L	H	H

H = high level, L = low level, X = irrelevant

logic symbol

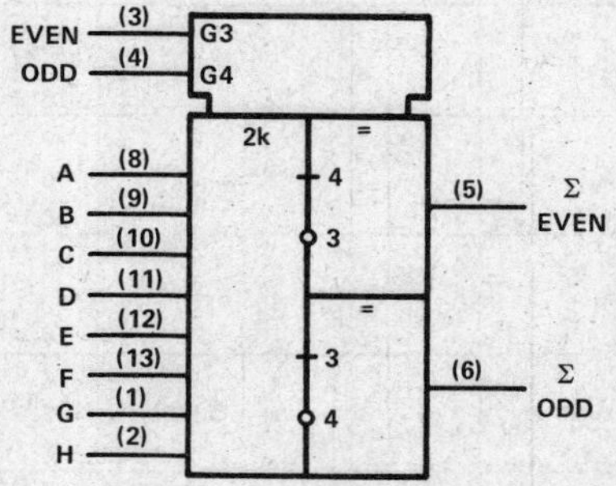

Pin numbers shown are for J and N packages.

TEXAS INSTRUMENTS
POST OFFICE BOX 225012 • DALLAS, TEXAS 75265

logic diagram (positive logic)

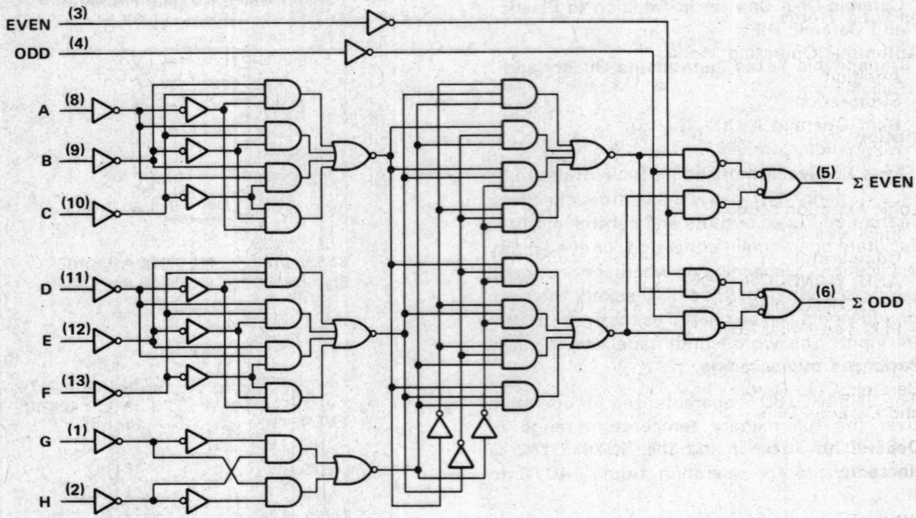

Pin numbers shown are for J and N packages.

maximum ratings, recommended operating conditions, and electrical characteristics

See Table IV, page 2-10.

switching characteristics over recommended operating free-air temperature range (unless otherwise noted), C_L = 50 pF (see Note 1)

PARAMETER	FROM (INPUT)	TO (OUTPUT)	V_{CC}	T_A = 25°C MIN	T_A = 25°C TYP	T_A = 25°C MAX	SN54HC180 MIN	SN54HC180 MAX	SN74HC180 MIN	SN74HC180 MAX	UNIT
t_{pd}	Data (odd = 0)	Even	2 V		119	260		390		325	ns
			4.5 V		36	52		78		65	
			6 V		32	44		66		55	
t_{pd}	Data (odd = 0)	Odd	2 V		113	245		370		305	ns
			4.5 V		33	49		74		61	
			6 V		24	42		63		52	
t_{pd}	Data (even = 0)	Even	2 V		119	260		390		325	ns
			4.5 V		36	52		78		65	
			6 V		32	44		66		55	
t_{pd}	Data (even = 0)	Odd	2 V		113	245		370		305	ns
			4.5 V		33	49		74		61	
			6 V		24	42		63		52	
t_{pd}	Even or Odd	Even or Odd	2 V		49	110		165		140	ns
			4.5 V		15	22		33		28	
			6 V		12	19		28		24	
t_t		Any	2 V		38	75		110		95	ns
			4.5 V		8	15		22		19	
			6 V		6	13		19		16	

C_{pd}	Power dissipation capacitance	No load, T_A = 25°C	60 pF typ

NOTE 1: For load circuit and voltage waveforms, see page 1-14.

TEXAS INSTRUMENTS
POST OFFICE BOX 225012 • DALLAS, TEXAS 75265

TYPES SN54HC181, SN54HC881 SN74HC181, SN74HC881 ARITHMETIC LOGIC UNITS/FUNCTION GENERATORS

D2804, MARCH 1984

- **Full Look-Ahead for High-Speed Operations on Long Words**
- **Arithmetic Operating Modes**
  - **Addition**
  - **Subtraction**
  - **Shift Operand A One Position**
  - **Magnitude Comparison**
  - **Plus Twelve Other Arithmetic Operations**
- **Logic Function Modes**
  - **Exclusive-OR**
  - **Comparator**
  - **AND, NAND, OR, NOR**
  - **'HC881 Provides Status Register Checks**
  - **Plus Ten Other Logic Operations**
- **Package Options Include Both Plastic and Ceramic Chip Carriers in Addition to Plastic and Ceramic DIPs**
- **Dependable Texas Instruments Quality and Reliability**

SN54HC181, SN54HC881 . . . JT PACKAGE
SN74HC181, SN74HC881 . . . JT OR NT PACKAGE
(TOP VIEW)

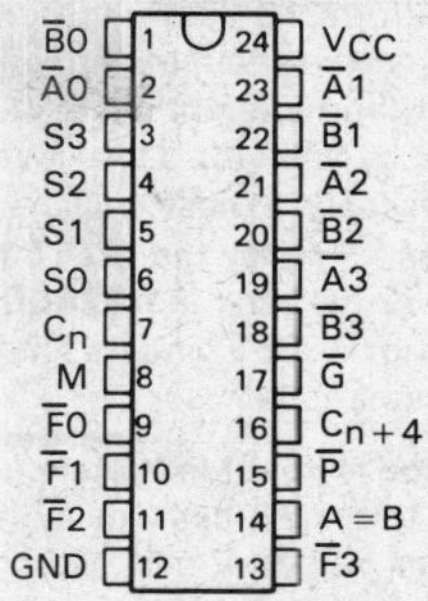

SN54HC181, SN54HC881 . . . FH OR FK PACKAGE
SN74HC181, SN74HC881 . . . FH OR FN PACKAGE
(TOP VIEW)

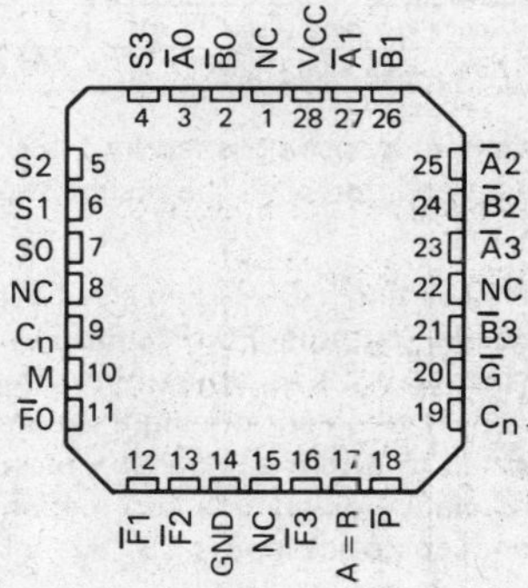

NC—No internal connection

logic symbol

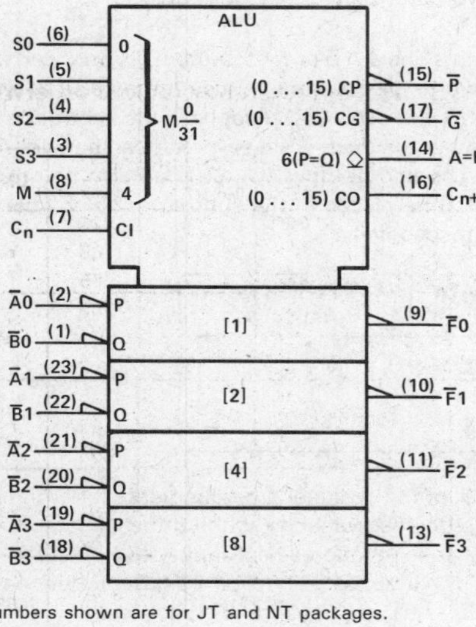

Pin numbers shown are for JT and NT packages.

3

HCMOS DEVICES

PRODUCT PREVIEW

This document contains information on a product under development. Texas Instruments reserves the right to change or discontinue this product without notice.

description

The 'HC181 and 'HC881 are arithmetic logic units (ALU)/function generators on a monolithic chip. These circuits perform 16 binary arithmetic operations on two 4-bit words as shown in Tables I and II. These operations are selected by the four function-select lines (S0, S1, S2, S3) and include addition, subtraction, decrement, and straight transfer. When performing arithmetic manipulations, the internal carriers must be enabled by applying a low-level voltage to the mode control input (M). A full carry look-ahead scheme is made available in these devices for fast, simultaneous carry generation by means of two cascade-outputs (pins 15 and 17) for the four bits in the package. When used in conjunction with the SN54HC882 or SN74HC882 full carry look-ahead circuits, high-speed arithmetic operations can be performed. The method of cascading 'HC882 circuits with these ALUs to provide multilevel full carry look-ahead is illustrated under signal designations.

If high speed is not of importance, a ripple-carry input (C_n) and a ripple-carry output (C_{n+4}) are available. However, the ripple-carry delay has also been minimized so that arithmetic manipulations for small word lengths can be performed without external circuitry.

The 'HC181 and 'HC881 will accommodate active-high or active-low data if the pin designations are interpreted as follows:

PIN NUMBER	2	1	23	22	21	20	19	18	9	10	11	13	7	16	15	17
Active-low data (Table I)	$\overline{A0}$	$\overline{B0}$	$\overline{A1}$	$\overline{B1}$	$\overline{A2}$	$\overline{B2}$	$\overline{A3}$	$\overline{B3}$	$\overline{F0}$	$\overline{F1}$	$\overline{F2}$	$\overline{F3}$	C_n	C_{n+4}	$\overline{P}$	$\overline{G}$
Active-high data (Table II)	A0	B0	A1	B1	A2	B2	A3	B3	F0	F1	F2	F3	$\overline{C_n}$	$\overline{C_{n+4}}$	X	Y

Subtraction is accomplished by 1's complement addition where the 1's complement of the subtrahend is generated internally. The resultant output is A − B − 1, which requires an end-around or forced carry to provide A − B.

The 'HC181 and 'HC881 can also be utilized as a comparator. The A = B output is internally decoded from the function outputs (F0, F1, F2, F3) so that when two words of equal magnitude are applied at the A and B inputs, it will assume a high level to indicate equality (A = B). The ALU must be in the subtract mode with C_n = H when performing this comparison. The A = B output is open-drain so that it can be wire-AND connected to give a comparison for more than four bits. The carry output (C_{n+4}) can also be used to supply relative magnitude information. Again, the ALU must be placed in the subtract mode by placing the function select inputs S3, S2, S1, S0 at L, H, H, L, respectively.

INPUT C_n	OUTPUT C_{n+4}	ACTIVE-LOW DATA (FIGURE 1)	ACTIVE-HIGH DATA (FIGURE 2)
H	H	A ≥ B	A ≤ B
H	L	A < B	A > B
L	H	A > B	A < B
L	L	A ≤ B	A ≥ B

These circuits have been designed to not only incorporate all of the designer's requirements for arithmetic operations, but also to provide 16 possible functions of two Boolean variables without the use of external circuitry. These logic functions are selected by use of the four function-select inputs (S0, S1, S2, S3) with the mode-control input (M) at a high level to disable the internal carry. The 16 logic functions are detailed in Tables I and II and include Exclusive-OR, NAND, AND, NOR, and OR functions.

The 'HC881 has the same pinout and same functionality as the 'HC181 except for the $\overline{P}$, $\overline{G}$, and C_{n+4} outputs when the device is in the logic mode (M = H).

In the logic mode the 'HC881 provides the user with a status check on the input words, A and B, and the output word F. While in the logic mode the $\overline{P}$, $\overline{G}$, and C_{n+4} outputs supply status information based upon the following logical combinations:

$$\overline{P} = F0 + F1 + F2 + F3$$
$$\overline{G} = H$$
$$C_{n+4} = PC_n$$

FUNCTION TABLE FOR INPUT BITS EQUAL/NOT EQUAL

S0 = S3 = H, S1 = S2 = L, and M = H

C_n	DATA INPUTS				OUTPUTS		
					$\overline{G}$	$\overline{P}$	C_{n+4}
H	$\overline{A0}=\overline{B0}$	$\overline{A1}=\overline{B1}$	$\overline{A2}=\overline{B2}$	$\overline{A3}=\overline{B3}$	H	L	H
L	$\overline{A0}=\overline{B0}$	$\overline{A1}=\overline{B1}$	$\overline{A2}=\overline{B2}$	$\overline{A3}=\overline{B3}$	H	L	L
X	$\overline{A0}\neq\overline{B0}$	X	X	X	H	H	L
X	X	$\overline{A1}\neq\overline{B1}$	X	X	H	H	L
X	X	X	$\overline{A2}\neq\overline{B2}$	X	H	H	L
X	X	X	X	$\overline{A3}\neq\overline{B3}$	H	H	L

FUNCTION TABLE FOR INPUT PAIRS HIGH/NOT HIGH

S0 = S1 = S3 = L, S2 = H, and M = H

C_n	DATA INPUTS				OUTPUTS		
					$\overline{G}$	$\overline{P}$	C_{n+4}
H	$\overline{A0}$ or $\overline{B0}$ = L	$\overline{A1}$ or $\overline{B1}$ = L	$\overline{A2}$ or $\overline{B2}$ = L	$\overline{A3}$ or $\overline{B3}$ = L	H	L	H
L	$\overline{A0}$ or $\overline{B0}$ = L	$\overline{A1}$ or $\overline{B1}$ = L	$\overline{A2}$ or $\overline{B2}$ = L	$\overline{A3}$ or $\overline{B3}$ = L	H	L	L
X	$\overline{A0}=\overline{B0}=H$	X	X	X	H	H	L
X	X	$\overline{A1}=\overline{B1}=H$	X	X	H	H	L
X	X	X	$\overline{A2}=\overline{B2}=H$	X	H	H	L
X	X	X	X	$\overline{A3}=\overline{B3}=H$	H	H	L

The combination of signals on the S3 through S0 control lines determine the operation performed on the data words to generate the output bits $\overline{F}i$. By monitoring the $\overline{P}$ and C_{n+4} outputs, the user can determine if all pairs of input bits are equal (see table above) or if any pair of inputs are both high (see table above). The 'HC881 has the unique feature of providing an A = B status while the Exclusive-OR ($\oplus$) function is being utilized. When the control inputs (S3, S2, S1, S0) equal H, L, L, H; a status check is generated to determine whether all pairs ($\overline{A}i$, $\overline{B}i$) are equal in the following manner: $\overline{P} = (A0 \oplus B0) + (A1 \oplus B1) + (A2 \oplus B2) + (A3 \oplus B3)$. This unique bit-by-bit comparison of the data words, which is available on the totem-pole $\overline{P}$ output, is particularly useful when cascading 'HC881's. As the A = B condition is sensed in the first stage, the signal is propagated through the same ports used for carry generation in the arithmetic mode ($\overline{P}$ and $\overline{G}$). Thus the A = B status is transmitted to the second stage more quickly without the need for external multiplexing logic. The A = B open-drain output allows the user to check the validity of the bit-by-bit result by comparing the two signals for parity.

If the user wishes to check for any pair of data inputs ($\overline{A}i$, $\overline{B}i$) being high, it is necessary to set the control lines (S3, S2, S1, S0) to L, H, L, L. The data pairs will then be ANDed together and the results ORed in the following manner: $\overline{P} = \overline{A0}\overline{B0} + \overline{A1}\overline{B1} + \overline{A2}\overline{B2} + \overline{A3}\overline{B3}$.

S3	S2	S1	S0	M	$\overline{P} = F0 + F1 + F2 + F3$
L	H	L	L	H	$\overline{A0}\overline{B0} + \overline{A1}\overline{B1} + \overline{A2}\overline{B2} + \overline{A3}\overline{B3}$
H	L	L	H	H	$(A0 \oplus B0) + (A1 \oplus B1) + (A2 \oplus B2) + (A3 \oplus B3)$

The SN54HC181 and SN54HC881 are characterized for operation over the full military temperature range of −55 °C to 125 °C. The SN74HC181 and SN74HC881 are characterized for operation from −40 °C to 85 °C.

signal designations

In both Figures 1 and 2, the polarity indicators (⊵) indicate that the associated input or output is active-low with respect to the function shown inside the symbol and the symbols are the same in both figures. The signal designations in Figure 1 agree with the indicated internal functions based on active-low data, and are for use with the logic functions and arithmetic operations shown in Table I. The signal designations have been changed in Figure 2 to accommodate the logic functions and arithmetic operations for the active-high data given in Table II. The 'HC181 and 'HC881 together with the 'HC182 and 'HC882 can be used with the signal designation of either Figure 1 or Figure 2.

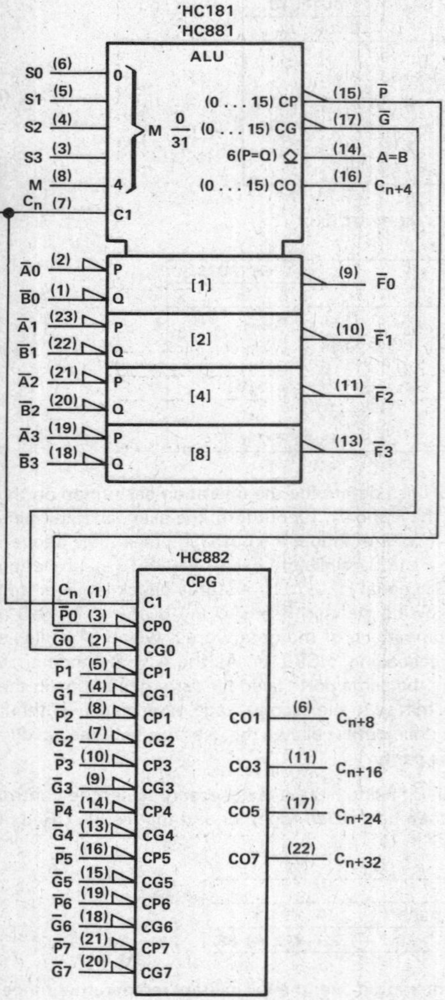

FIGURE 1
(USE WITH TABLE I)

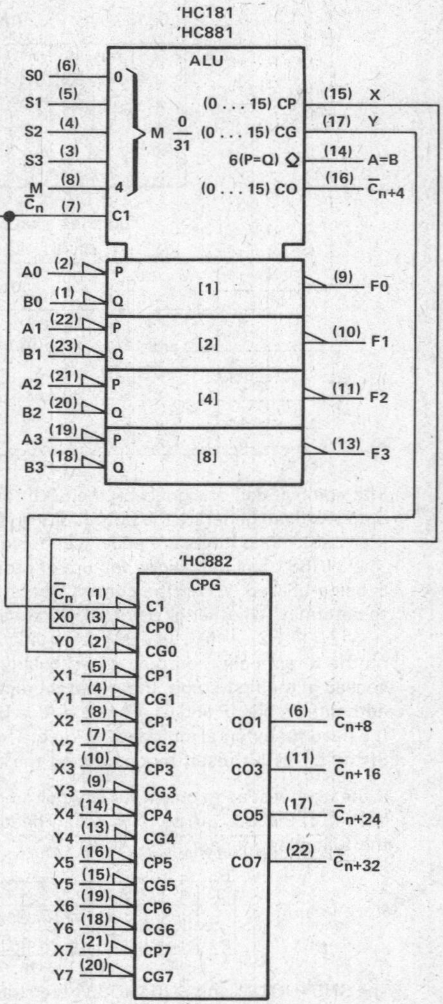

FIGURE 2
(USE WITH TABLE II)

TEXAS INSTRUMENTS
POST OFFICE BOX 225012 • DALLAS, TEXAS 75265

TABLE I

SELECTION				ACTIVE-LOW DATA		
				M = H	M = L; ARITHMETIC OPERATIONS	
S3	S2	S1	S0	LOGIC FUNCTIONS	C_n = L (no carry)	C_n = H (with carry)
L	L	L	L	F = $\overline{A}$	F = A MINUS 1	F = A
L	L	L	H	F = $\overline{AB}$	F = AB MINUS 1	F = AB
L	L	H	L	F = $\overline{A}$ + B	F = A$\overline{B}$ MINUS 1	F = A$\overline{B}$
L	L	H	H	F = 1	F = MINUS 1 (2's COMP)	F = ZERO
L	H	L	L	F = $\overline{A + B}$	F = A PLUS (A + $\overline{B}$)	F = A PLUS (A + $\overline{B}$) PLUS 1
L	H	L	H	F = B	F = AB PLUS (A + $\overline{B}$)	F = AB PLUS (A + $\overline{B}$) PLUS 1
L	H	H	L	F = $\overline{A \oplus B}$	F = A MINUS B MINUS 1	F = A MINUS B
L	H	H	H	F = A + $\overline{B}$	F = A + $\overline{B}$	F = (A + $\overline{B}$) PLUS 1
H	L	L	L	F = $\overline{A}$B	F = A PLUS (A + B)	F = A PLUS (A + B) PLUS 1
H	L	L	H	F = A $\oplus$ B	F = A PLUS B	F = A PLUS B PLUS 1
H	L	H	L	F = B	F = A$\overline{B}$ PLUS (A + B)	F = A$\overline{B}$ PLUS (A + B) PLUS 1
H	L	H	H	F = A + B	F = (A + B)	F = (A + B) PLUS 1
H	H	L	L	F = 0	F = A PLUS A*	F = A PLUS A PLUS 1
H	H	L	H	F = A$\overline{B}$	F = AB PLUS A	F = AB PLUS A PLUS 1
H	H	H	L	F = AB	F = A$\overline{B}$ PLUS A	F = A$\overline{B}$ PLUS A PLUS 1
H	H	H	H	F = A	F = A	F = A PLUS 1

TABLE II

SELECTION				ACTIVE-HIGH DATA		
				M = H	M = L; ARITHMETIC OPERATIONS	
S3	S2	S1	S0	LOGIC FUNCTIONS	$\overline{C}_n$ = H (no carry)	$\overline{C}_n$ = L (with carry)
L	L	L	L	F = $\overline{A}$	F = A	F = A PLUS 1
L	L	L	H	F = $\overline{A + B}$	F = A + B	F = (A + B) PLUS 1
L	L	H	L	F = $\overline{A}$B	F = A + $\overline{B}$	F = (A + $\overline{B}$) PLUS 1
L	L	H	H	F = 0	F = MINUS 1 (2's COMP)	F = ZERO
L	H	L	L	F = $\overline{AB}$	F = A PLUS A$\overline{B}$	F = A PLUS A$\overline{B}$ PLUS 1
L	H	L	H	F = $\overline{B}$	F = (A + B) PLUS A$\overline{B}$	F = (A + B) PLUS A$\overline{B}$ PLUS 1
L	H	H	L	F = A $\oplus$ B	F = A MINUS B MINUS 1	F = A MINUS B
L	H	H	H	F = A$\overline{B}$	F = A$\overline{B}$ MINUS 1	F = A$\overline{B}$
H	L	L	L	F = $\overline{A}$ + B	F = A PLUS AB	F = A PLUS AB PLUS 1
H	L	L	H	F = $\overline{A \oplus B}$	F = A PLUS B	F = A PLUS B PLUS 1
H	L	H	L	F = B	F = (A + $\overline{B}$) PLUS AB	F = (A + $\overline{B}$) PLUS AE PLUS 1
H	L	H	H	F = AB	F = AB MINUS 1	F = AB
H	H	L	L	F = 1	F = A PLUS A*	F = A PLUS A PLUS 1
H	H	L	H	F = A + $\overline{B}$	F = (A + B) PLUS A	A = (A + B) PLUS A PLUS 1
H	H	H	L	F = A + B	F = (A + $\overline{B}$) PLUS A	F = (A + $\overline{B}$) PLUS A PLUS 1
H	H	H	H	F = A	F = A MINUS 1	F = A

*Each bit is shifted to the next more significant position.

3

HCMOS DEVICES

TYPES SN54HC190, SN54HC191, SN74HC190, SN74HC191 SYNCHRONOUS 4-BIT UP/DOWN DECADE AND BINARY COUNTERS

D2684, DECEMBER 1982—REVISED MARCH 1984

- **Single Down/Up Count Control Line**
- **Look-Ahead Circuitry Enhances Speed of Cascaded Counters**
- **Fully Synchronous in Count Modes**
- **Asynchronously Presettable with Load Control**
- **Package Options Include Both Plastic and Ceramic Chip Carriers in Addition to Plastic and Ceramic DIPs**
- **Dependable Texas Instruments Quality and Reliability**

SN54HC190, SN54HC191 . . . J PACKAGE
SN74HC190, SN74HC191 . . . J OR N PACKAGE
(TOP VIEW)

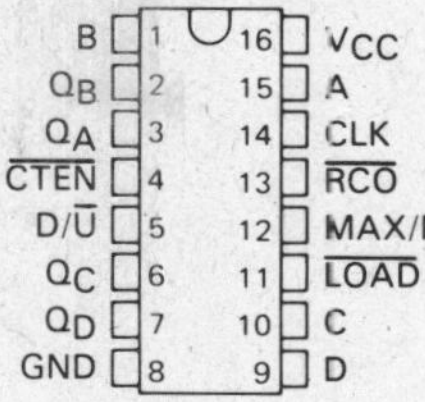

SN54HC190, SN54HC191 . . . FH OR FK PACKAGE
SN74HC190, SN74HC191 . . . FH OR FN PACKAGE
(TOP VIEW)

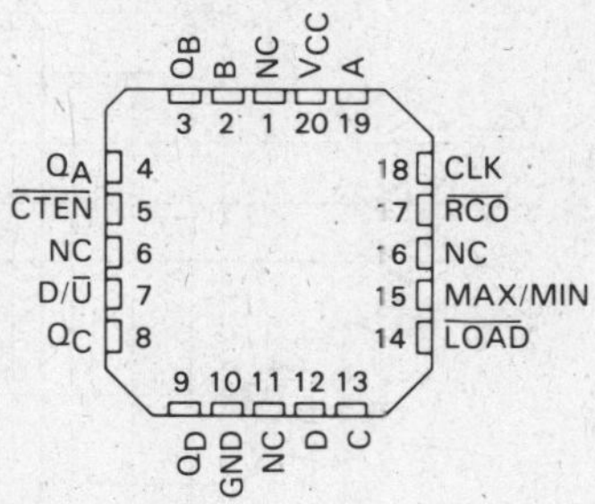

NC — No internal connection

description

The 'HC190 and 'HC191 are synchronous, reversible up/down counters. The 'HC190 is a 4-bit decade counter and the 'HC191 is a 4-bit binary counter. Synchronous counting operation is provided by having all flip-flops clocked simultaneously so that the outputs change coincident with each other when so instructed by the steering logic. This mode of operation eliminates the output counting spikes normally associated with asynchronous (ripple clock) counters.

The outputs of the four flip-flops are triggered on a low-to-high-level transition of the clock input if the enable input ($\overline{CTEN}$) is low. A high at $\overline{CTEN}$ inhibits counting. The direction of the count is determined by the level of the down/up (D/$\overline{U}$) input. When D/$\overline{U}$ is low, the counter counts up and when D/$\overline{U}$ is high, it counts down.

These counters feature a fully independent clock circuit. Changes at the control inputs ($\overline{CTEN}$ and D/$\overline{U}$) that will modify the operating mode have no effect on the contents of the counter until clocking occurs. The function of the counter will be dictated solely by the condition meeting the stable setup and hold times.

These counters are fully programmable; that is, the outputs may each be preset to either level by placing a low on the load input and entering the desired data at the data inputs. The output will change to agree with the data inputs independently of the level of the clock input. This feature allows the counters to be used as modulo-N dividers by simply modifying the count length with the preset inputs.

Two outputs have been made available to perform the cascading function: ripple clock and maximum/minimum count. The latter output produces a high-level output pulse with a duration approximately equal to one complete cycle of the clock while the count is zero (all outputs low) counting down or maximum (9 or 15) counting up. The ripple clock output produces a low-level output pulse under those same conditions but only while the clock input is low. The counters can be easily cascaded by feeding the ripple clock output to the enable input of the succeeding counter if parallel clocking is used, or to the clock input if parallel enabling is used. The maximum/minimum count output can be used to accomplish look-ahead for high-speed operation.

The SN54HC190 and SN54HC191 are characterized for operation over the full military temperature range of −55 °C to 125 °C. The SN74HC190 and SN74HC191 are characterized for operation from −40 °C to 85 °C.

'HC190 logic symbol

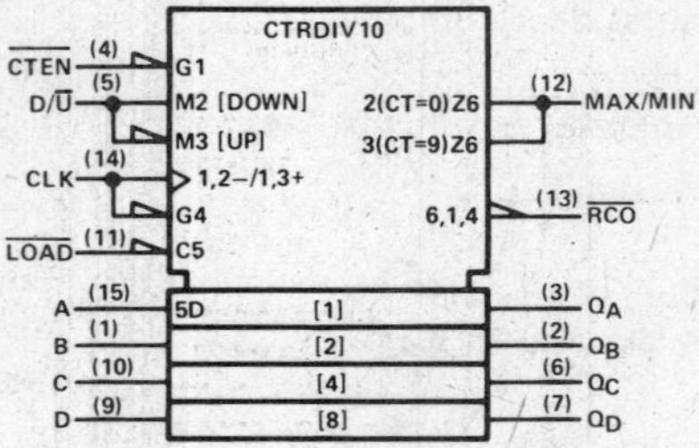

'HC190 logic diagram (positive logic)

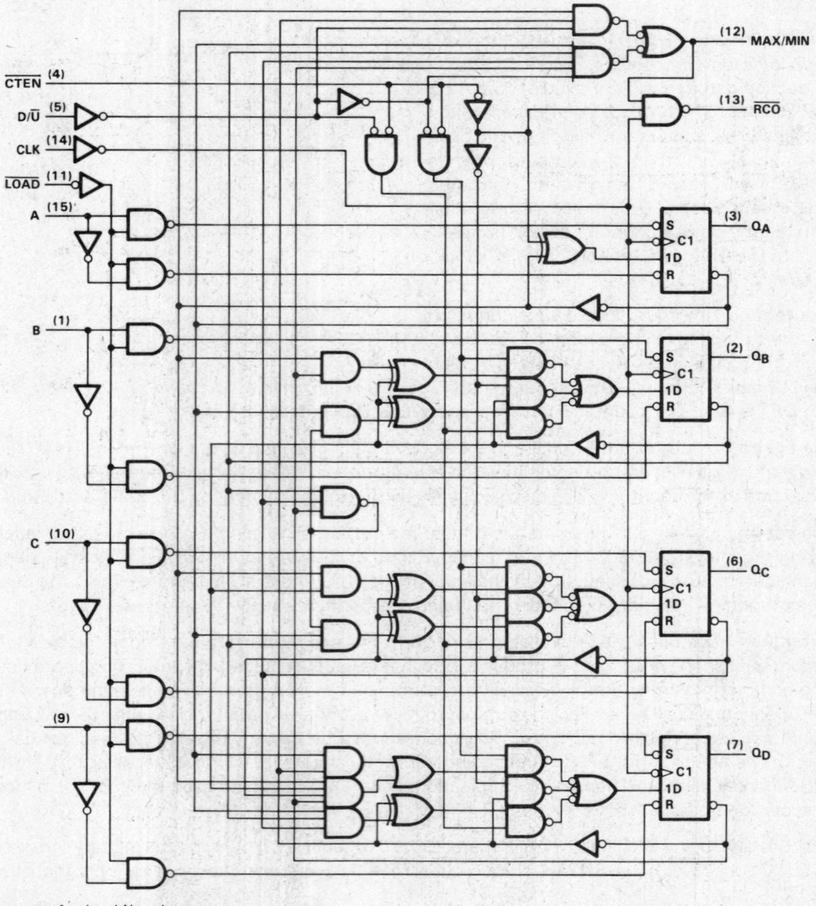

Pin numbers shown are for J and N packages.

TEXAS INSTRUMENTS
POST OFFICE BOX 225012 • DALLAS, TEXAS 75265

'HC191 logic symbol

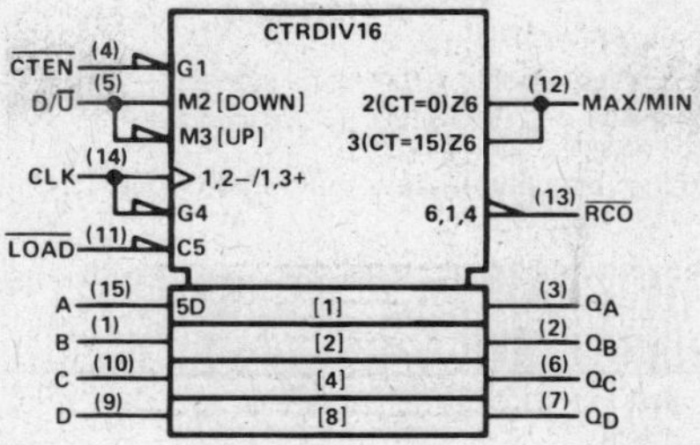

'HC191 logic diagram (positive logic)

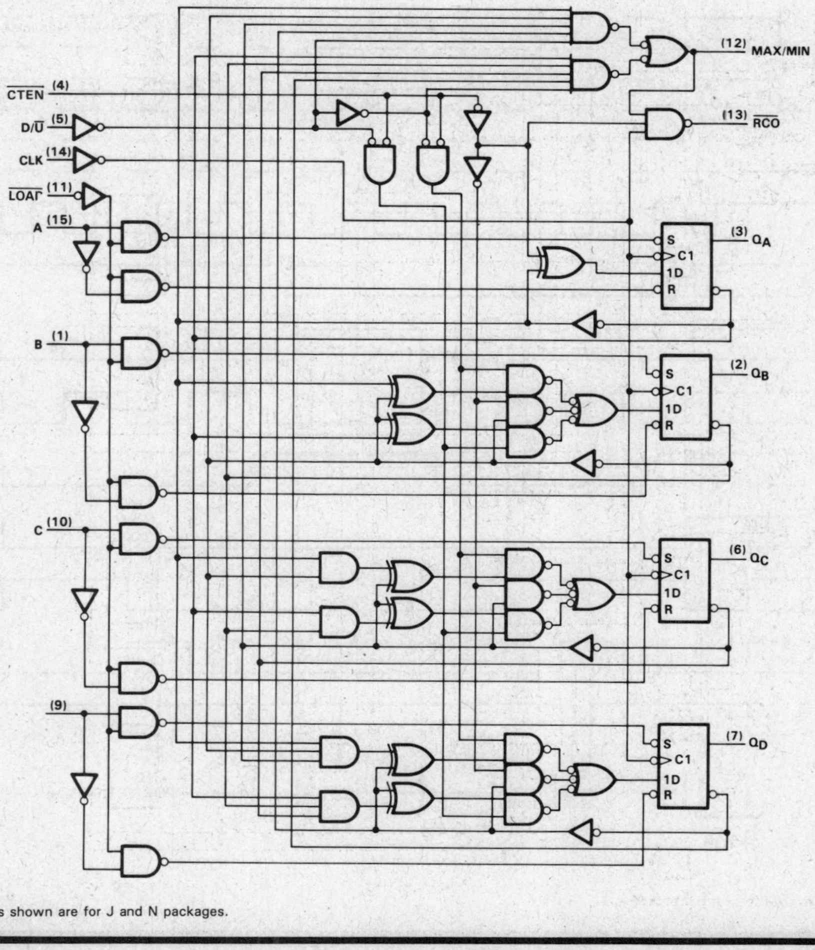

Pin numbers shown are for J and N packages.

typical load, count, and inhibit sequences

'HC190

Illustrated below is the following sequence:

1. Load (preset) to BCD seven.
2. Count up to eight, nine (maximum), zero, one, and two.
3. Inhibit.
4. Count down to one, zero (minimum), nine, eight, and seven.

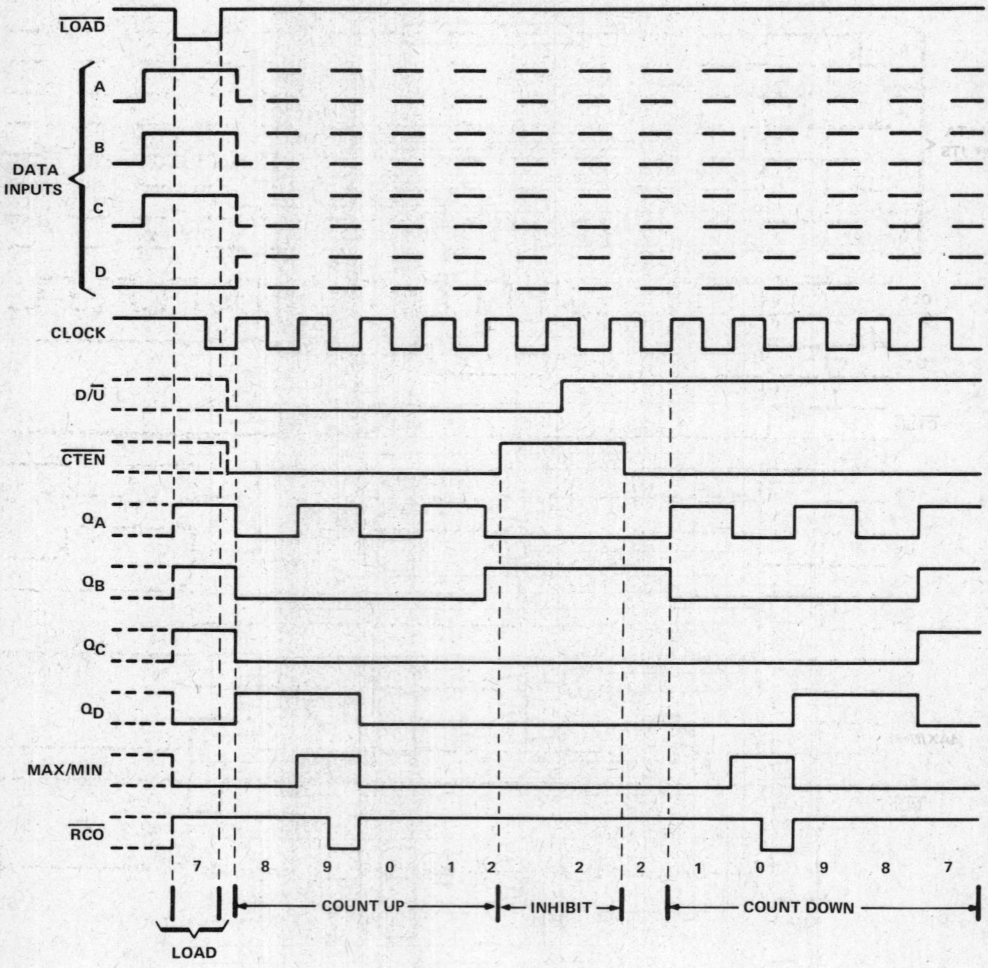

typical load, count, and inhibit sequences

'HC191

Illustrated below is the following sequence:

1. Load (preset) to binary thirteen.
2. Count up to fourteen, fifteen (maximum), zero, one, and two.
3. Inhibit.
4. Count down to one, zero (minimum), fifteen, fourteen, and thirteen.

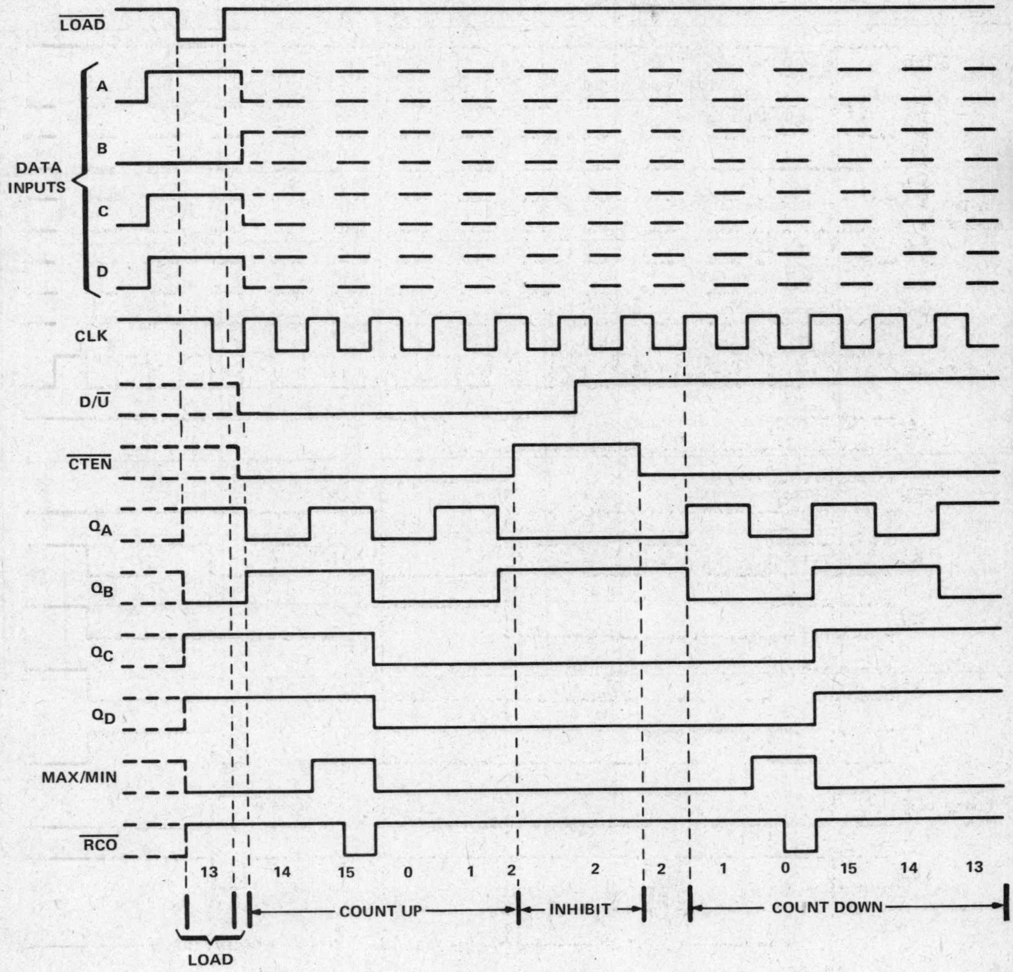

TEXAS INSTRUMENTS
POST OFFICE BOX 225012 • DALLAS, TEXAS 75265

maximum ratings, recommended operating conditions, and electrical characteristics

See Table IV, page 2-10.

timing requirements over recommended operating free-air temperature range (unless otherwise noted)

			V_{CC}	T_A = 25°C MIN	T_A = 25°C MAX	SN54HC190 SN54HC191 MIN	SN54HC190 SN54HC191 MAX	SN74HC190 SN74HC191 MIN	SN74HC190 SN74HC191 MAX	UNIT
f_{clock}	Clock frequency		2 V	0	4.2	0	2.8	0	3.3	MHz
			4.5 V	0	21	0	14	0	17	
			6 V	0	24	0	16	0	19	
t_w	Pulse duration	$\overline{\text{LOAD}}$ low	2 V	120		180		150		ns
			4.5 V	24		36		30		
			6 V	21		31		26		
		CLK high or low	2 V	120		180		150		
			4.5 V	24		36		30		
			6 V	21		31		26		
t_{su}	Setup time	Data before $\overline{\text{LOAD}}$↑	2 V	150		256		188		ns
			4.5 V	30		46		38		
			6 V	25		38		32		
		$\overline{\text{CTEN}}$ before CLK↑	2 V	205		306		255		
			4.5 V	41		61		51		
			6 V	35		53		44		
		D/$\overline{\text{U}}$ before CLK↑	2 V	205		306		255		
			4.5 V	41		61		51		
			6 V	35		53		44		
		$\overline{\text{LOAD}}$ inactive before CLK↑	2 V	150		250		190		
			4.5 V	30		45		38		
			6 V	25		38		32		
t_h	Hold time	Data after $\overline{\text{LOAD}}$↑	2 V	5		5		5		ns
			4.5 V	5		5		5		
			6 V	5		5		5		
		$\overline{\text{CTEN}}$ after CLK↑	2 V	5		5		5		
			4.5 V	5		5		5		
			6 V	5		5		5		
		D/$\overline{\text{U}}$ after CLK↑	2 V	5		5		5		
			4.5 V	5		5		5		
			6 V	5		5		5		

switching characteristics over recommended operating free-air temperature range (unless otherwise noted), C_L = 50 pF (see Note 1)

PARAMETER	FROM (INPUT)	TO (OUTPUT)	V_{CC}	T_A = 25°C MIN	TYP	MAX	SN54HC190 SN54HC191 MIN	MAX	SN74HC190 SN74HC191 MIN	MAX	UNIT
f_{max}			2 V	4.2	8		2.8		3.3		MHz
			4.5 V	21	42		14		17		
			6 V	24	48		16		19		
t_{pd}	$\overline{\text{LOAD}}$	Any Q	2 V		130	264		396		330	ns
			4.5 V		40	53		79		66	
			6 V		33	45		67		56	
t_{pd}	A, B, C, or D	Q_A, Q_B Q_C, or Q_D	2 V		135	240		360		300	ns
			4.5 V		36	48		72		60	
			6 V		30	41		61		51	
t_{pd}	CLK	$\overline{\text{RCO}}$	2 V		58	120		180		150	ns
			4.5 V		17	24		36		30	
			6 V		14	21		31		26	
t_{pd}	CLK	Any Q	2 V		107	192		288		240	ns
			4.5 V		31	38		58		48	
			6 V		26	32		49		41	
t_{pd}	CLK	MAX/MIN	2 V		123	252		378		315	ns
			4.5 V		39	50		76		63	
			6 V		32	43		65		54	
t_{pd}	D/$\overline{\text{U}}$	$\overline{\text{RCO}}$	2 V		102	228		342		285	ns
			4.5 V		29	46		68		57	
			6 V		24	38		59		49	
t_{pd}	D/U	MAX/MIN	2 V		86	192		288		240	ns
			4.5 V		24	38		58		48	
			6 V		20	32		49		41	
t_{pd}	$\overline{\text{CTEN}}$	$\overline{\text{RCO}}$	2 V		50	132		198		165	ns
			4.5 V		15	26		40		33	
			6 V		13	23		34		28	
t_t		Any	2 V		38	75		110		95	ns
			4.5 V		8	15		22		19	
			6 V		6	13		19		16	

C_{pd}	Power dissipation capacitance	No load, T_A = 25°C	50 pF typ

NOTE 1: For load circuit and voltage waveforms, see page 1-14.

3 HCMOS DEVICES

TYPES SN54HC192, SN54HC193 SN74HC192, SN74HC193 SYNCHRONOUS 4-BIT UP/DOWN COUNTERS (DUAL CLOCK WITH CLEAR)

D2684, DECEMBER 1982—REVISED MARCH 1984

- **Look-Ahead Circuitry Enhances Cascaded Counters**
- **Fully Synchronous in Count Modes**
- **Parallel Asynchronous Load for Modulo-N Count Lengths**
- **Asynchronous Clear**
- **Package Options Include Both Plastic and Ceramic Chip Carriers in Addition to Plastic and Ceramic DIPs**
- **Dependable Texas Instruments Quality and Reliability**

SN54HC192, SN54HC193 . . . J PACKAGE
SN74HC192, SN74HC193 . . . J OR N PACKAGE
(TOP VIEW)

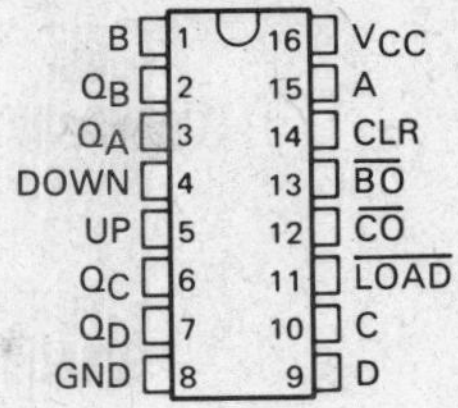

SN54HC192, SN54HC193 . . . FH OR FK PACKAGE
SN74HC192, SN74HC193 . . . FH OR FN PACKAGE
(TOP VIEW)

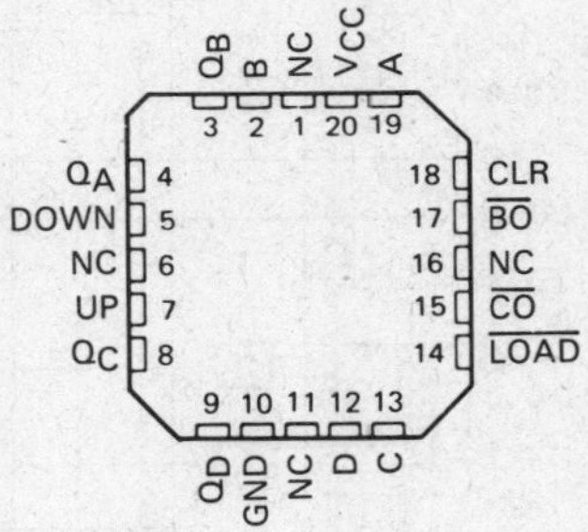

NC—No internal connection

description

The 'HC192 and 'HC193 are synchronous, reversible up/down counters. The 'HC192 is a 4-bit decade counter and the 'HC193 is a 4-bit binary counter. Synchronous operation is provided by having all flip-flops clocked simultaneously so that the outputs change coincidently with each other when so instructed by the steering logic. This mode of operation eliminates the output counting spikes normally associated with asynchronous (ripple clock) counters.

The outputs of the four flip-flops are triggered by a low-to-high-level transition of either count (clock) input (Up or Down). The direction of counting is determined by which count input is pulsed while the other count input is high.

All four counters are fully programmable; that is, each output may be preset to either level by placing a low on the load input and entering the desired data at the data inputs. The output will change to agree with the data inputs independently of the count pulses. This feature allows the counters to be used as modulo-N dividers by simply modifying the count length with the preset inputs.

A clear input has been provided that forces all outputs to the low level when a high level is applied. The clear function is independent of the count and the load inputs.

These counters were designed to be cascaded without the need for external circuitry. The borrow output ($\overline{BO}$) produces a low-level pulse while the count is zero (all outputs low) and the count-down is low. Similarly, the carry output ($\overline{CO}$) produces a low-level pulse while the count is maximum (9 or 15) and the count-up input is low. The counters can then be easily cascaded by feeding the borrow and carry outputs to the count-down and count-up inputs, respectively, of the succeeding counter.

The SN54HC192 and SN54HC193 are characterized for operation over the full military temperature range of −55°C to 125°C. The SN74HC192 and SN74HC193 are characterized for operation from −40°C to 85°C.

'HC192 logic symbol

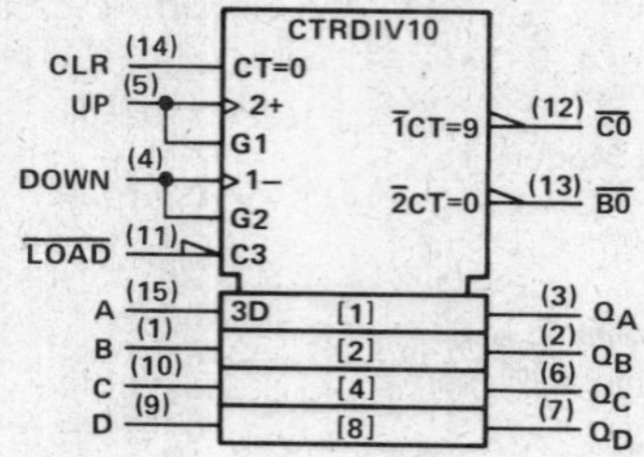

'HC192 logic diagram (positive logic)

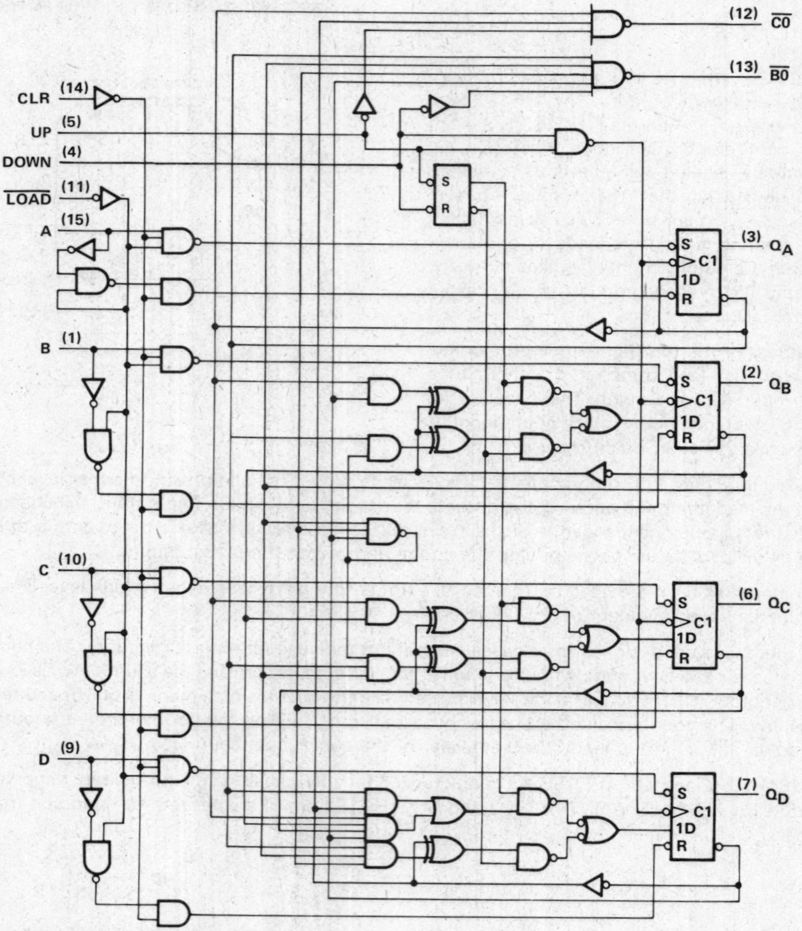

Pin numbers shown are for J and N packages.

TEXAS INSTRUMENTS
POST OFFICE BOX 225012 • DALLAS, TEXAS 75265

'HC193 logic symbol

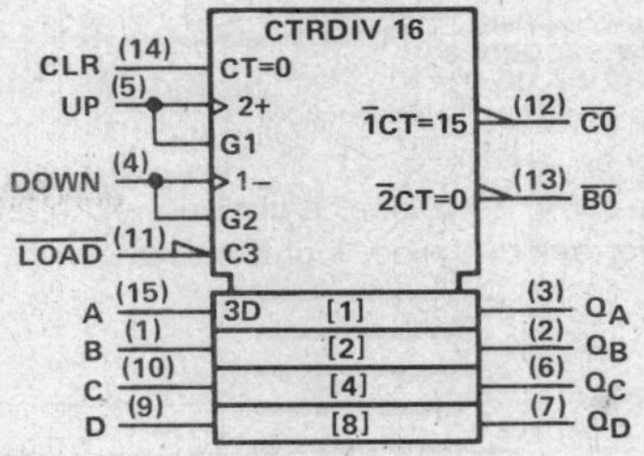

'HC193 logic diagrams (positive logic)

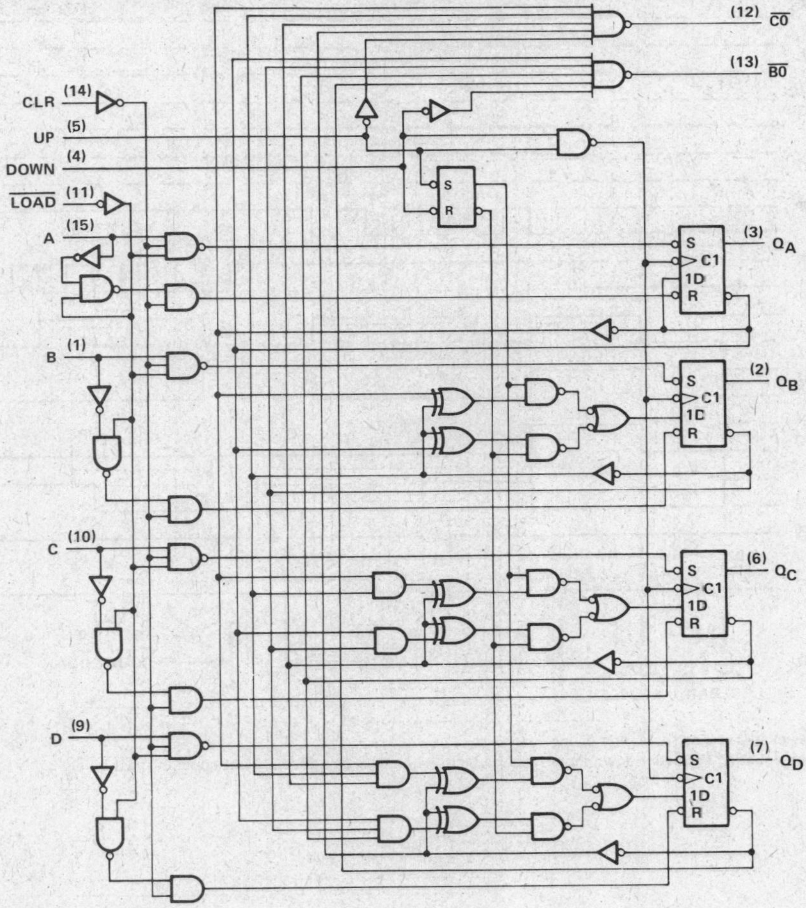

Pin numbers shown are for J and N packages.

'HC192 typical clear, load, and count sequence

Illustrated below is the following sequence:

1. Clear outputs to zero.
2. Load (preset) to BCD seven.
3. Count up to eight, nine, carry, zero, one, and two.
4. Count down to one, zero, borrow, nine, eight, and seven.

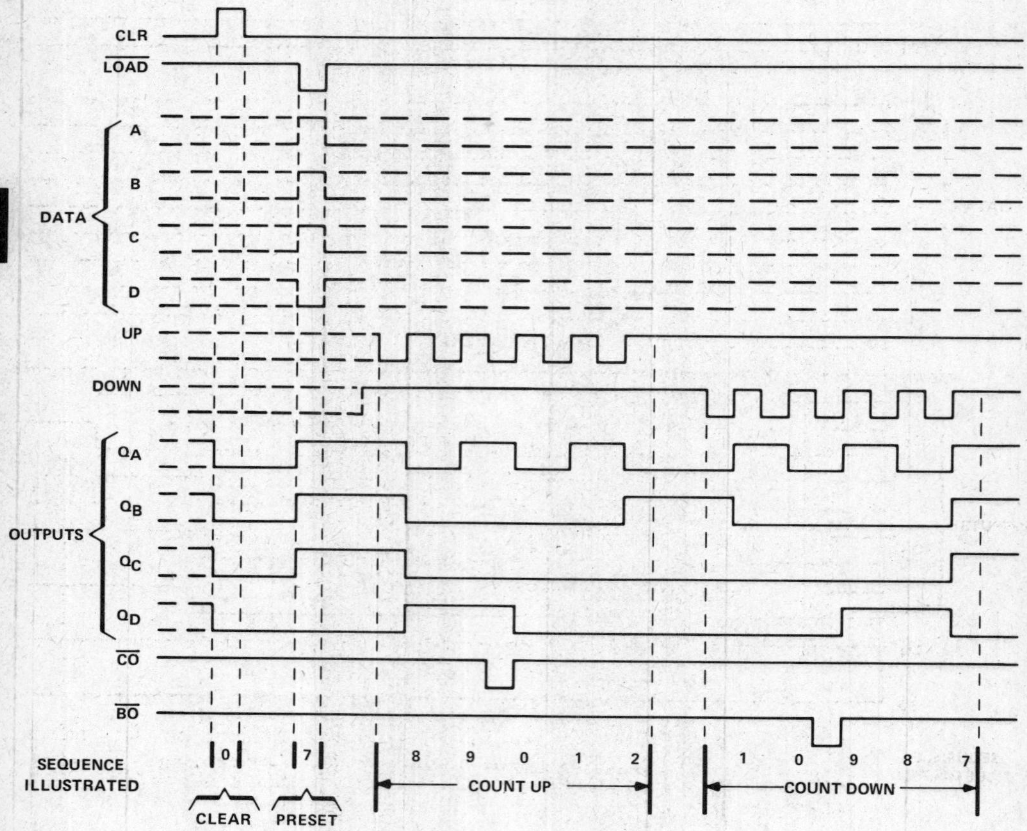

NOTES: A. Clear overrides load, data, and count inputs.
B. When counting up, count-down input must be high; when counting down, count-up input must be high.

TEXAS INSTRUMENTS
POST OFFICE BOX 225012 • DALLAS, TEXAS 75265

'HC193 typical clear, load, and count sequences

Illustrated below is the following sequence:

1. Clear outputs to zero.
2. Load (preset) to binary thirteen.
3. Count up to fourteen, fifteen, carry, zero, one, and two.
4. Count down to one, zero, borrow, fifteen, fourteen, and thirteen.

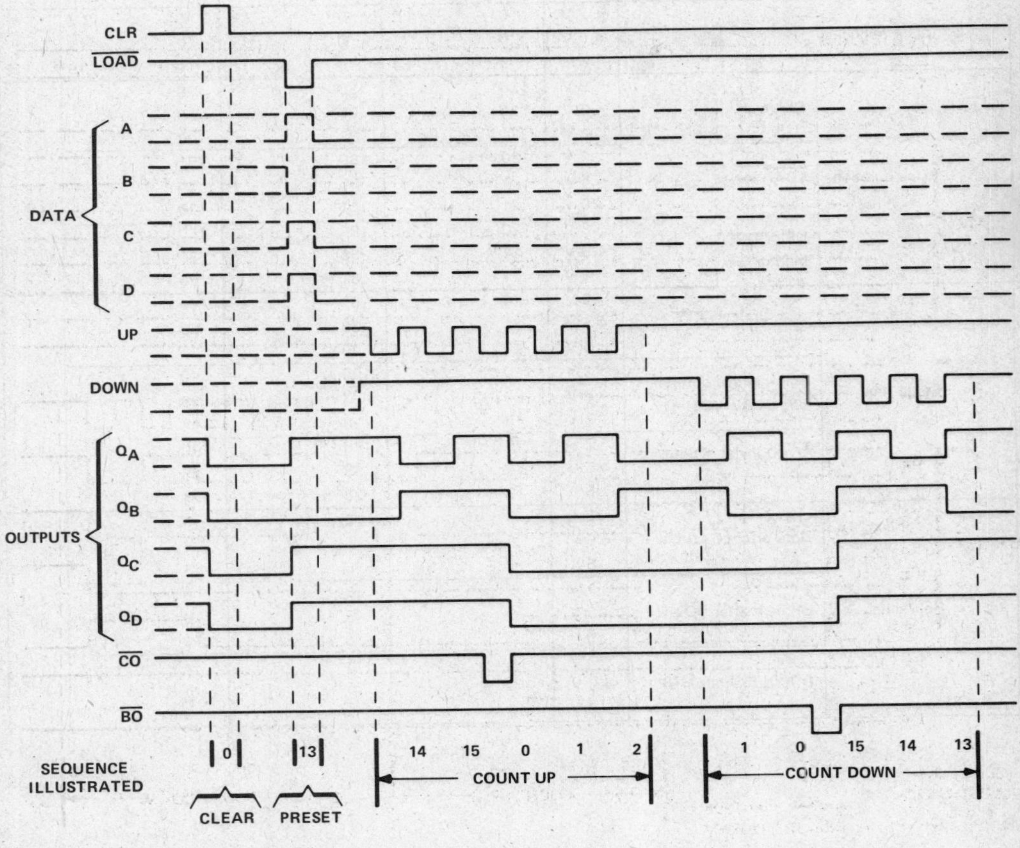

NOTES: A. Clear overrides load, data, and count inputs.
B. When counting up, count-down input must be high; when counting down, count-up input must be high.

maximum ratings, recommended operating conditions, and electrical characteristics

See Table IV, page 2-10.

timing requirements over recommended operating free-air temperature range (unless otherwise noted)

			V_{CC}	T_A = 25°C MIN	T_A = 25°C MAX	SN54HC192 SN54HC193 MIN	SN54HC192 SN54HC193 MAX	SN74HC192 SN74HC193 MIN	SN74HC192 SN74HC193 MAX	UNIT
f_{clock}	Clock frequency		2 V	0	4.2	0	2.8	0	3.3	MHz
			4.5 V	0	21	0	14	0	17	
			6 V	0	24	0	16	0	19	
t_w	Pulse duration	CLR high	2 V	120		180		150		ns
			4.5 V	24		36		30		
			6 V	21		31		26		
		$\overline{\text{LOAD}}$ low	2 V	120		180		150		
			4.5 V	24		36		30		
			6 V	21		31		26		
		UP or DOWN high or low	2 V	120		180		150		
			4.5 V	24		36		30		
			6 V	21		31		26		
t_{su}	Setup time	Data before $\overline{\text{LOAD}}$↑	2 V	110		165		140		ns
			4.5 V	22		33		28		
			6 V	19		28		24		
		CLR inactive before UP↑ or DOWN↑	2 V	110		165		140		
			4.5 V	22		33		28		
			6 V	19		28		24		
		$\overline{\text{LOAD}}$ inactive before UP↑ or DOWN↑	2 V	110		165		140		
			4.5 V	22		33		28		
			6 V	19		28		24		
t_h	Hold time	Data after $\overline{\text{LOAD}}$↑	2 V	5		5		5		ns
			4.5 V	5		5		5		
			6 V	5		5		5		
		UP high after DOWN↑	2 V	5		5		5		
			4.5 V	5		5		5		
			6 V	5		5		5		
		DOWN high after UP↑	2 V	5		5		5		
			4.5 V	5		5		5		
			6 V	5		5		5		

switching characteristics over recommended operating free-air temperature range (unless otherwise noted), C_L = 50 pF (see Note 1)

PARAMETER	FROM (INPUT)	TO (OUTPUT)	V_{CC}	T_A = 25°C MIN	TYP	MAX	SN54HC192 SN54HC193 MIN	MAX	SN74HC192 SN74HC193 MIN	MAX	UNIT
f_{max}			2 V	4.2	8		2.8		3.3		MHz
			4.5 V	21	55		14		17		
			6 V	24	60		16		19		
t_{pd}	UP	$\overline{CO}$	2 V		75	165		250		205	ns
			4.5 V		24	33		50		41	
			6 V		20	28		43		35	
t_{pd}	DOWN	$\overline{BO}$	2 V		75	165		250		205	ns
			4.5 V		24	33		50		41	
			6 V		20	28		43		35	
t_{pd}	UP or DOWN	Any Q	2 V		190	250		375		315	ns
			4.5 V		40	50		75		63	
			6 V		35	43		64		54	
t_{pd}	$\overline{LOAD}$	Any Q	2 V		190	260		390		325	ns
			4.5 V		40	52		78		65	
			6 V		35	44		66		55	
t_{PHL}	CLR	Any Q	2 V		170	240		360		300	ns
			4.5 V		36	48		72		60	
			6 V		31	41		61		51	
t_t		Any	2 V		38	75		110		95	ns
			4.5 V		8	15		22		19	
			6 V		6	13		19		16	

C_{pd}	Power dissipation capacitance	No load, T_A = 25°C	50 pF typ

NOTE 1: For load circuit and voltage waveforms, see page 1-14.

TYPES SN54HC194, SN74HC194
4-BIT BIDIRECTIONAL UNIVERSAL SHIFT REGISTERS

D2684, DECEMBER 1982—REVISED MARCH 1984

- **Parallel Inputs and Outputs**
- **Four Operating Modes:**
  - **Synchronous Parallel Load**
  - **Right Shift**
  - **Left Shift**
  - **Do Nothing**
- **Positive Edge-Triggered Clocking**
- **Direct Overriding Clear**
- **Package Options Include Both Plastic and Ceramic Chip Carriers in Addition to Plastic and Ceramic DIPs**
- **Dependable Texas Instruments Quality and Reliability**

SN54HC194 . . . J PACKAGE
SN74HC194 . . . J OR N PACKAGE
(TOP VIEW)

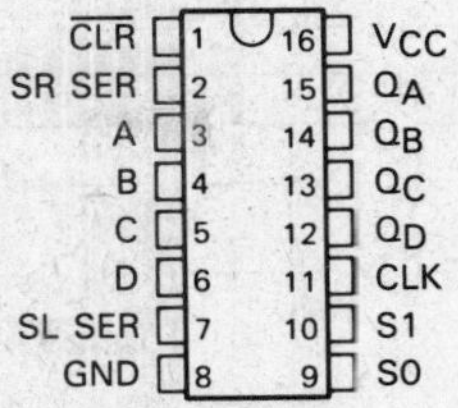

SN54HC194 . . . FH OR FK PACKAGE
SN74HC194 . . . FH OR FN PACKAGE
(TOP VIEW)

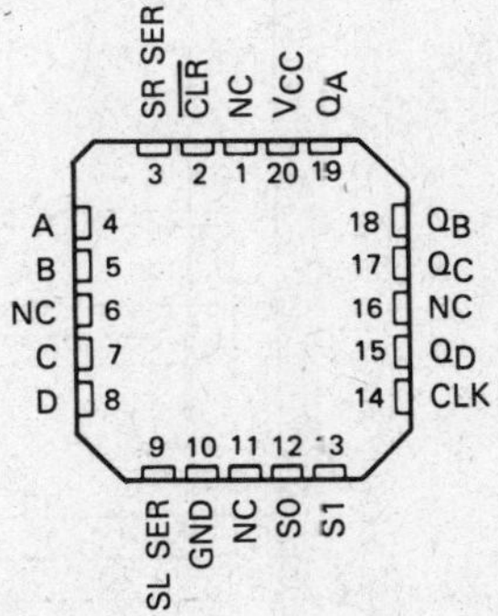

NC—No internal connection

description

These bidirectional shift registers are designed to incorporate virtually all of the features a system designer may want in a shift register. The circuit features parallel inputs, parallel outputs, right-shift and left-shift inputs, operating-mode-control inputs, and a direct overriding clear line. The register has four distinct modes of operation, namely:

Parallel (broadside) load
Shift right (in the direction Q_A toward Q_D)
Shift left (in the direction Q_D toward Q_A)
Inhibit clocking (do nothing)

Synchronous parallel loading is accomplished by applying the four bits of data and taking both mode control inputs, S0 and S1, high. The data are loaded into the associated flip-flops and appear at the outputs after the positive transition of the clock input. During loading, serial data flow is inhibited.

Shift right is accomplished synchronously with the rising edge of the clock pulse when S0 is high and S1 is low. Serial data for this mode is entered at the shift-right data input. When S0 is low and S1 is high, data shifts left synchronously and new data is entered at the shift-left serial input.

Clocking of the shift register is inhibited when both mode control inputs are low.

The SN54HC194 is characterized for operation over the full military temperature range of −55°C to 125°C. The SN74HC194 is characterized for operation from −40°C to 85°C.

logic symbol

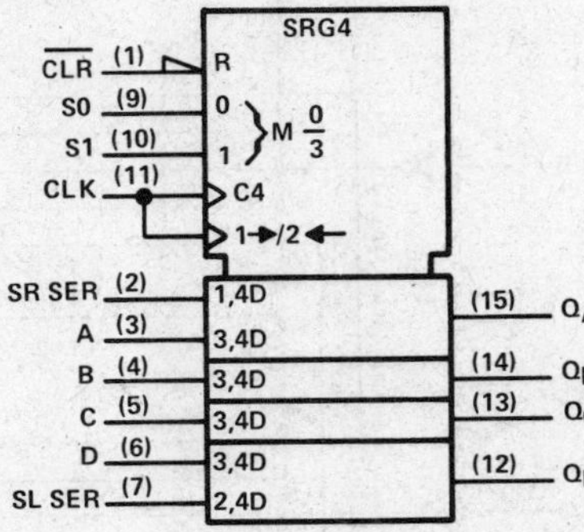

Pin numbers shown are for J and N packages.

TEXAS INSTRUMENTS
POST OFFICE BOX 225012 • DALLAS, TEXAS 75265

logic diagram (positive logic)

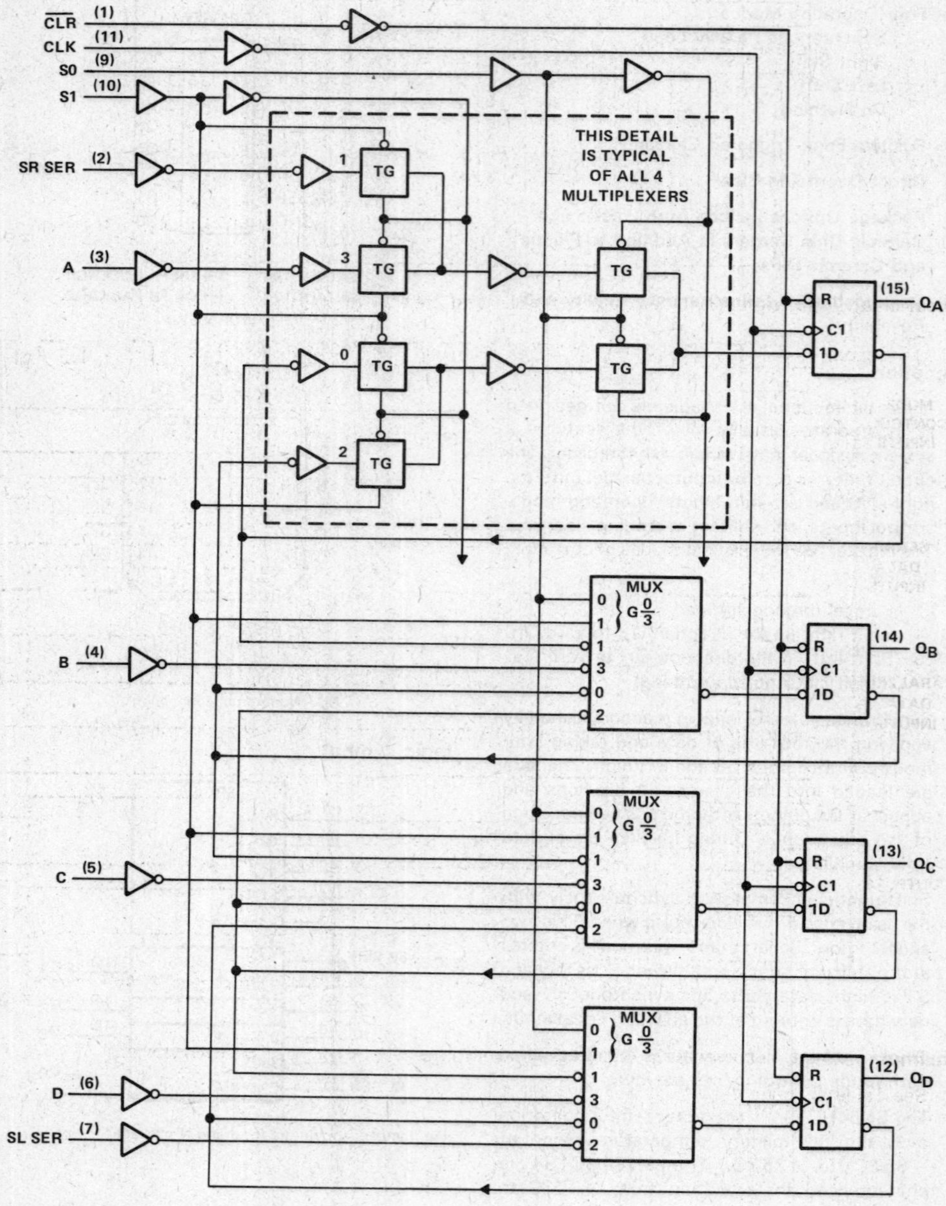

Pin numbers shown are for J and N packages.

TEXAS INSTRUMENTS
POST OFFICE BOX 225012 • DALLAS, TEXAS 75265

FUNCTION TABLE

INPUTS										OUTPUTS			
CLEAR	MODE		CLOCK	SERIAL		PARALLEL				Q_A	Q_B	Q_C	Q_D
	S1	S0		LEFT	RIGHT	A	B	C	D				
L	X	X	X	X	X	X	X	X	X	L	L	L	L
H	X	X	L	X	X	X	X	X	X	Q_{A0}	Q_{B0}	Q_{C0}	Q_{D0}
H	H	H	↑	X	X	a	b	c	d	a	b	c	d
H	L	H	↑	X	H	X	X	X	X	H	Q_{An}	Q_{Bn}	Q_{Cn}
H	L	H	↑	X	L	X	X	X	X	L	Q_{An}	Q_{Bn}	Q_{Cn}
H	H	L	↑	H	X	X	X	X	X	Q_{Bn}	Q_{Cn}	Q_{Dn}	H
H	H	L	↑	L	X	X	X	X	X	Q_{Bn}	Q_{Cn}	Q_{Dn}	L
H	L	L	X	X	X	X	X	X	X	Q_{An}	Q_{Bn}	Q_{Cn}	Q_{D0}

typical clear, load, right-shift, left-shift, inhibit, and clear sequences

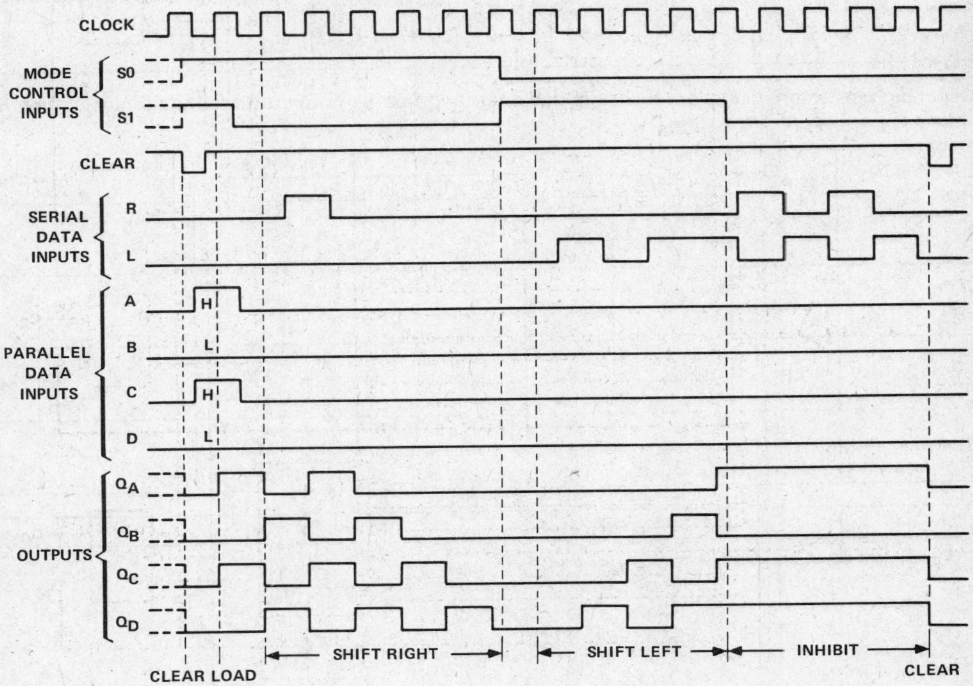

maximum ratings, recommended operating conditions, and electrical characteristics

See Table IV, page 2-10.

timing requirements over recommended operating free-air temperature range (unless otherwise noted)

			V_{CC}	$T_A = 25°C$ MIN	$T_A = 25°C$ MAX	SN54HC194 MIN	SN54HC194 MAX	SN74HC194 MIN	SN74HC194 MAX	UNIT
f_{clock}	Clock frequency		2 V	0	6	0	4.2	0	5	MHz
			4.5 V	0	31	0	21	0	25	
			6 V	0	36	0	25	0	29	
t_w	Pulse duration	CLK high or low	2 V	80		120		100		ns
			4.5 V	16		24		20		
			6 V	14		20		17		
		$\overline{CLR}$ low	2 V	80		120		100		
			4.5 V	16		24		20		
			6 V	14		20		17		
t_{su}	Setup time, any input before CLK↑		2 V	100		150		125		ns
			4.5 V	20		30		25		
			6 V	17		26		21		
t_h	Hold time, data after CLK↑		2 V	0		0		0		ns
			4.5 V	0		0		0		
			6 V	0		0		0		

switching characteristics over recommended operating free-air temperature range (unless otherwise noted), C_L = 50 pF (see Note 1)

PARAMETER	FROM (INPUT)	TO (OUTPUT)	V_{CC}	$T_A = 25°C$ MIN	$T_A = 25°C$ TYP	$T_A = 25°C$ MAX	SN54HC194 MIN	SN54HC194 MAX	SN74HC194 MIN	SN74HC194 MAX	UNIT
f_{max}			2 V	6			4.2		5		MHz
			4.5 V	31			21		25		
			6 V	36			25		29		
t_{PHL}	$\overline{CLR}$	Any	2 V		67	150		225		190	ns
			4.5 V		17	30		45		38	
			6 V		14	26		37		32	
t_{pd}	CLK	Any	2 V		67	145		220		180	ns
			4.5 V		17	29		44		36	
			6 V		14	25		37		31	
t_t		Any	2 V		28	75		110		95	ns
			4.5 V		8	15		22		19	
			6 V		6	13		19		16	

C_{pd}	Power dissipation capacitance	No load, $T_A = 25°C$	65 pF typ

NOTE 1: For load circuit and voltage waveforms, see page 1-14.

TYPES SN54HC195, SN74HC195
4-BIT PARALLEL-ACCESS SHIFT REGISTERS

D2684, DECEMBER 1982—REVISED MARCH 1984

- Synchronous Parallel Load
- Positive-Edge-Triggered Clocking
- J and $\overline{K}$ Inputs to First Stage
- Complementary Outputs from Last Stage
- Package Options Include Both Plastic and Ceramic Chip Carriers in Addition to Plastic and Ceramic DIPs
- Dependable Texas Instruments Quality and Reliability

SN54HC195 . . . J PACKAGE
SN74HC195 . . . J OR N PACKAGE
(TOP VIEW)

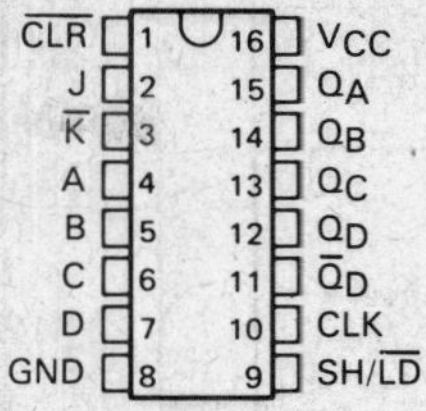

SN54HC195 . . . FH OR FK PACKAGE
SN74HC195 . . . FH OR FN PACKAGE
(TOP VIEW)

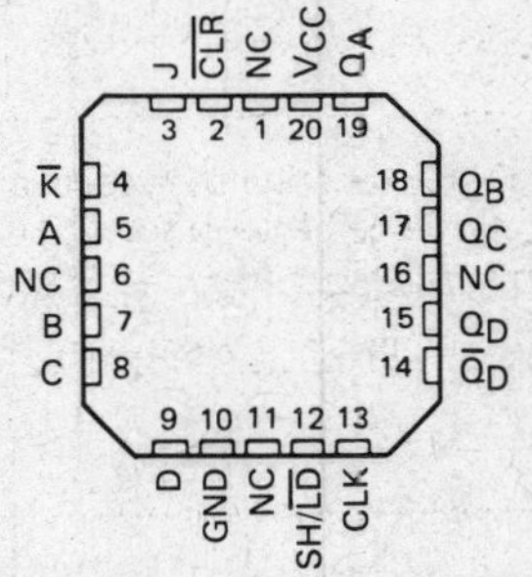

NC—No internal connection

description

These 4-bit registers feature parallel inputs, parallel outputs, J-$\overline{K}$ serial inputs, shift/load control input, and a direct overriding clear. The registers have two modes of operation: parallel (broadside) load, and shift (in the direction Q_A and Q_D).

Parallel loading is accomplished by applying the four bits of data and taking the shift/load control input low. The data is loaded into the associated flip-flop and appears at the outputs after the positive transition of the clock input. During loading, serial data flow is inhibited.

Shifting is accomplished synchronously when the shift/load control input is high. Serial data for this mode is entered at the J-$\overline{K}$ inputs. These inputs permit the first stage to perform as a J-$\overline{K}$-, D-, or T-type flip-flop as shown in the function table.

The SN54HC195 is characterized for operation over the full military temperature range of −55°C to 125°C. The SN74HC195 is characterized for operation from −40°C to 85°C.

logic symbol

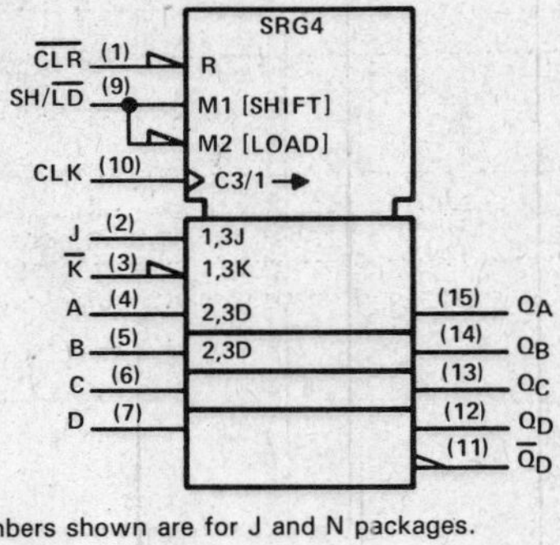

Pin numbers shown are for J and N packages.

logic diagram (positive logic)

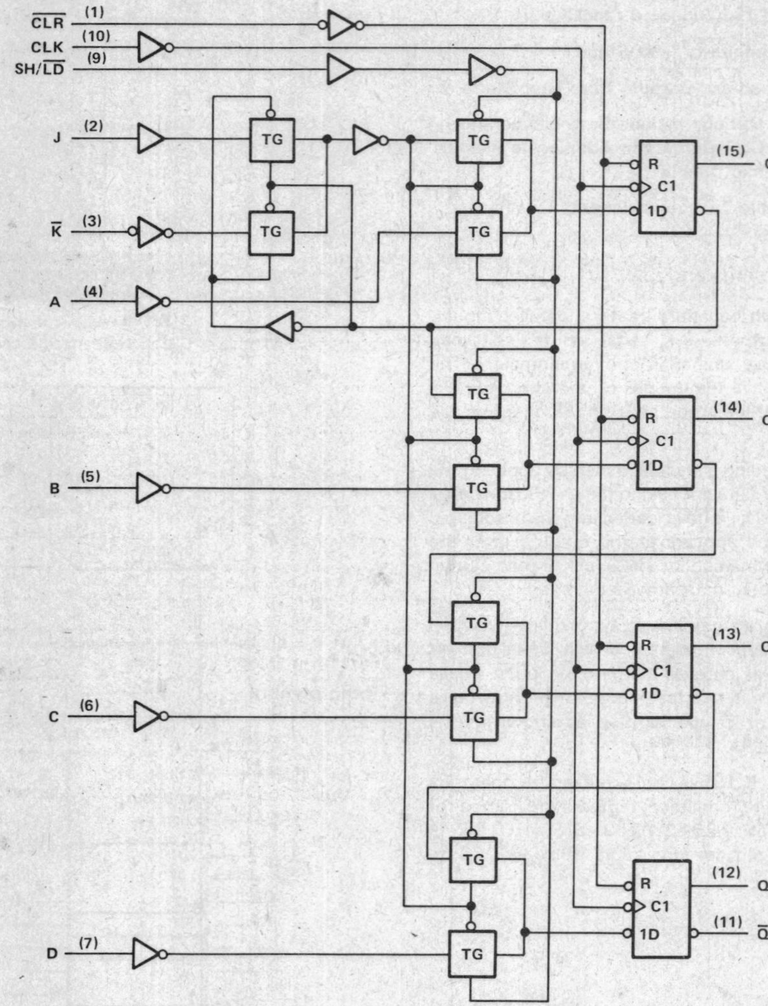

Pin numbers shown are for J and N packages.

TEXAS INSTRUMENTS
POST OFFICE BOX 225012 • DALLAS, TEXAS 75265

FUNCTION TABLE

INPUTS									OUTPUTS				
$\overline{CLR}$	SH/$\overline{LD}$	CLK	SERIAL J	SERIAL $\overline{K}$	PARALLEL A	PARALLEL B	PARALLEL C	PARALLEL D	Q_A	Q_B	Q_C	Q_D	$\overline{Q}_D$
L	X	X	X	X	X	X	X	X	L	L	L	L	H
H	L	↑	X	X	a	b	c	d	a	b	c	d	$\overline{d}$
H	H	L	X	X	X	X	X	X	Q_{A0}	Q_{B0}	Q_{C0}	Q_{D0}	$\overline{Q}_{D0}$
H	H	↑	L	H	X	X	X	X	Q_{A0}	Q_{A0}	Q_{Bn}	Q_{Cn}	$\overline{Q}_{Cn}$
H	H	↑	L	L	X	X	X	X	L	Q_{An}	Q_{Bn}	Q_{Cn}	$\overline{Q}_{Cn}$
H	H	↑	H	H	X	X	X	X	H	Q_{An}	Q_{Bn}	Q_{Cn}	$\overline{Q}_{Cn}$
H	H	↑	H	L	X	X	X	X	$\overline{Q}_{An}$	Q_{An}	Q_{Bn}	Q_{Cn}	$\overline{Q}_{Cn}$

typical clear, shift, and load sequences

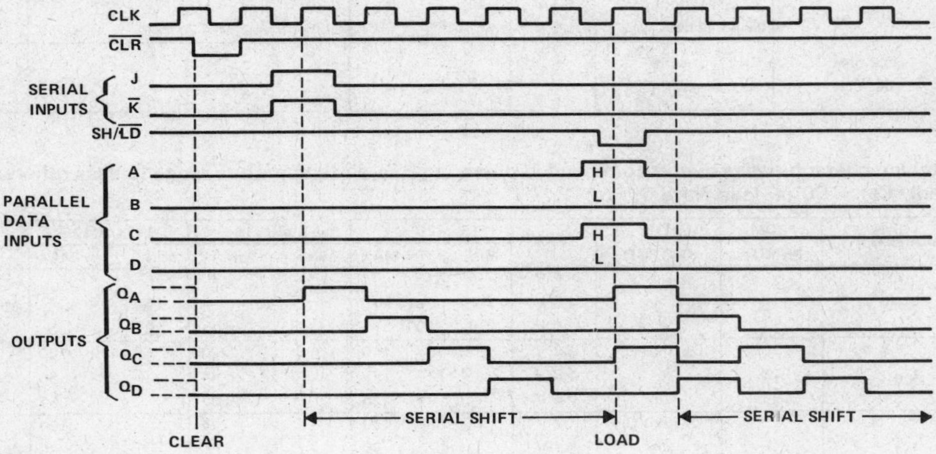

absolute maximum ratings, recommended operating conditions, and electrical characteristics

See Table IV, page 2-10.

timing requirements over recommended operating free-air temperature range (unless otherwise noted)

PARAMETER			V_{CC}	T_A = 25°C		SN54HC195		SN74HC195		UNIT
				MIN	MAX	MIN	MAX	MIN	MAX	
f_{clock}	Clock frequency		2 V	0	6	0	4.2	0	5	MHz
			4.5 V	0	31	0	21	0	25	
			6 V	0	36	0	25	0	29	
t_w	Pulse duration	CLK high or low	2 V	80		120		100		ns
			4.5 V	16		24		20		
			6 V	14		20		17		
		$\overline{CLR}$ low	2 V	80		120		100		
			4.5 V	16		24		20		
			6 V	14		20		17		
t_{su}	Setup time, before CLK↑	SH/$\overline{LD}$, or serial and parallel data, or $\overline{CLR}$ inactive	2 V	100		150		125		ns
			4.5 V	20		30		25		
			6 V	17		26		21		
t_h	Hold time, after CLK↑	SH/$\overline{LD}$ or serial and parallel data	2 V	0		0		0		ns
			4.5 V	0		0		0		
			6 V	0		0		0		

switching characteristics over recommended operating free-air temperature range (unless otherwise noted), C_L = 50 pF (see Note 1)

PARAMETER	FROM (INPUT)	TO (OUTPUT)	V_{CC}	T_A = 25°C			SN54HC195		SN74HC195		UNIT
				MIN	TYP	MAX	MIN	MAX	MIN	MAX	
f_{max}			2 V	6	12		4.2		5		MHz
			4.5 V	31	50		21		25		
			6 V	36	60		25		29		
t_{pd}	CLK	Q_A thru Q_D or $\overline{Q}_D$	2 V		67	145		220		180	ns
			4.5 V		17	29		44		36	
			6 V		14	25		37		31	
t_{pd}	$\overline{CLR}$	Q_A thru Q_D or $\overline{Q}_D$	2 V		67	150		225		190	ns
			4.5 V		17	30		45		38	
			6 V		13	26		38		32	
t_t		Any	2 V		28	75		110		95	ns
			4.5 V		8	15		22		19	
			6 V		6	13		19		16	

C_{pd}	Power dissipation capacitance	No load, T_A = 25°C	65 pF typ

NOTE 1: For load circuit and voltage waveforms, see page 1-14.

TEXAS INSTRUMENTS
POST OFFICE BOX 225012 • DALLAS, TEXAS 75265

TYPES SN54HC237, SN74HC237 3-LINE TO 8-LINE DECODERS/DEMULTIPLEXERS WITH ADDRESS LATCHES

D2804, MARCH 1984

- Combines Decoder and 3-Bit Address Latch
- Incorporates 2 Output Enables to Simplify Cascading
- Package Options Include Both Plastic and Ceramic Chip Carriers in Addition to Plastic and Ceramic DIPs
- Dependable Texas Instruments Quality and Reliability

SN54HC237 . . . J PACKAGE
SN74HC237 . . . J OR N PACKAGE
(TOP VIEW)

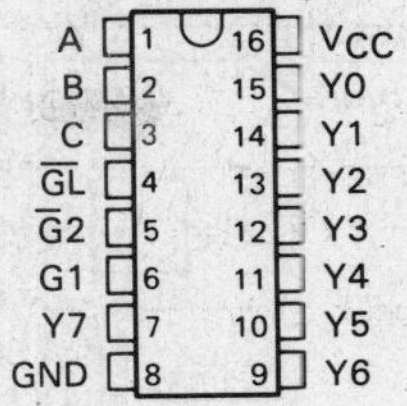

SN54HC237 . . . FH OR FK PACKAGE
SN74HC237 . . . FH OR FN PACKAGE
(TOP VIEW)

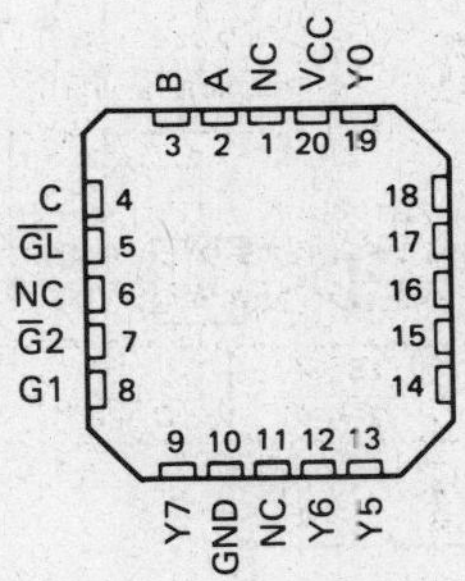

NC—No internal connection

description

The 'HC237 is a three-line to eight-line decoder/demultiplexer with latches on the three address inputs. When the latch-enable ($\overline{GL}$) is low, the 'HC237 acts as a decoder/demultiplexer. When $\overline{GL}$ goes from low to high, the address present at the select inputs (A, B, and C) is stored in the latches. Further address changes are ignored as long as $\overline{GL}$ remains high. The output enable controls, G1 and $\overline{G2}$, control the outputs independently of the select or latch-enable inputs. All of the outputs are forced high if G1 is low or $\overline{G2}$ is high. The 'HC237 is ideally suited for implementing glitch-free decoders in strobed (stored-address) applications in bus-oriented systems.

The SN54HC237 is characterized for operation over the full military temperature range of −55°C to 125°C. The SN74HC237 is characterized for operation from −40°C to 85°C.

logic symbols (alternatives)

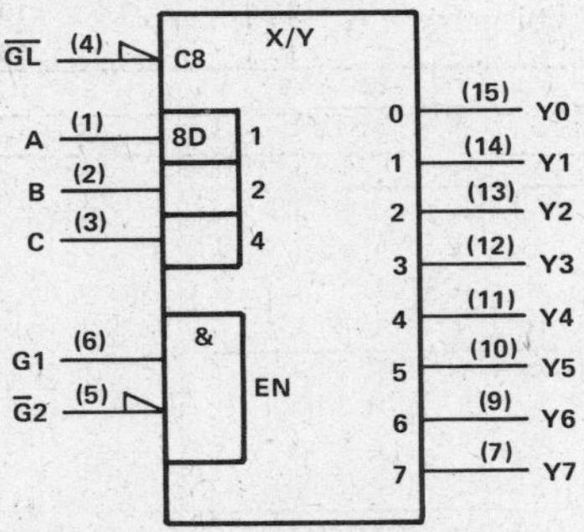

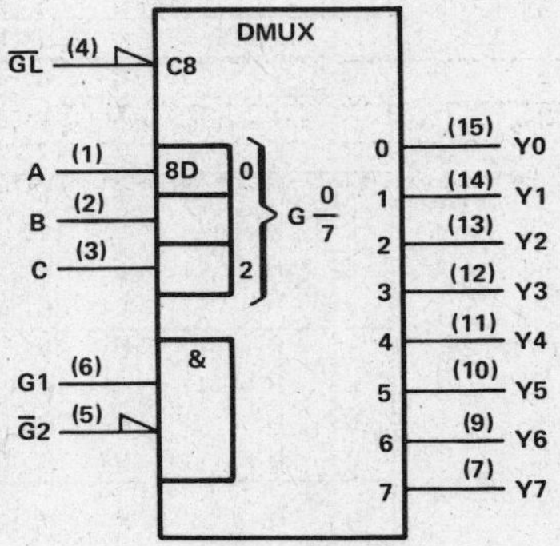

Pin numbers shown are for J and N packages.

logic diagram (positive logic)

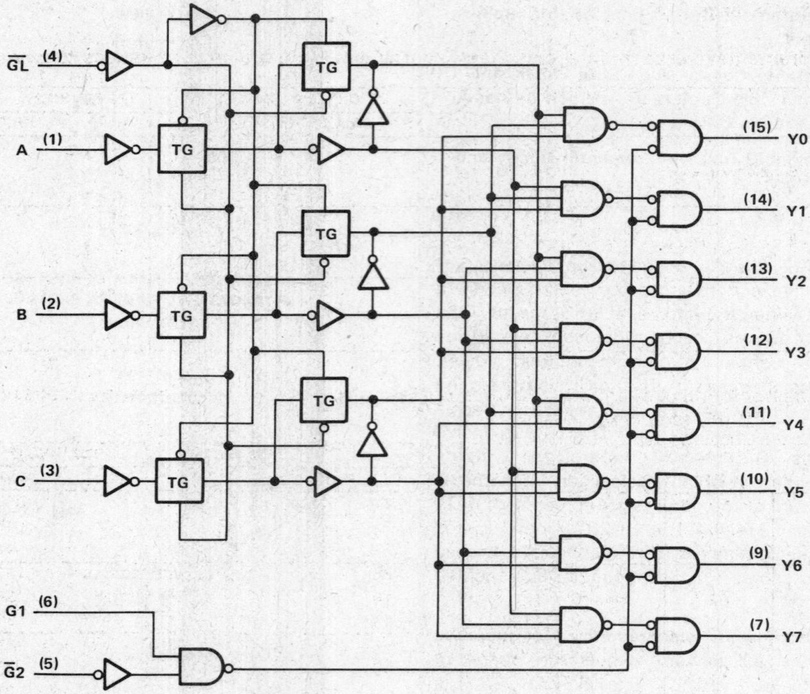

Pin numbers shown are for J and N packages.

FUNCTION TABLE

INPUTS						OUTPUTS							
ENABLE			SELECT										
$\overline{GL}$	G1	$\overline{G2}$	C	B	A	Y0	Y1	Y2	Y3	Y4	Y5	Y6	Y7
X	X	H	X	X	X	L	L	L	L	L	L	L	L
X	L	X	X	X	X	L	L	L	L	L	L	L	L
L	H	L	L	L	L	H	L	L	L	L	L	L	L
L	H	L	L	L	H	L	H	L	L	L	L	L	L
L	H	L	L	H	L	L	L	H	L	L	L	L	L
L	H	L	L	H	H	L	L	L	H	L	L	L	L
L	H	L	H	L	L	L	L	L	L	H	L	L	L
L	H	L	H	L	H	L	L	L	L	L	H	L	L
L	H	L	H	H	L	L	L	L	L	L	L	H	L
L	H	L	H	H	H	L	L	L	L	L	L	L	H
H	H	L	X	X	X	Outputs corresponding to stored address, L; all others, H							

maximum ratings, recommended operating conditions, and electrical characteristics

See Table IV, page 2-10.

timing requirements over recommended operating free-air temperature range (unless otherwise noted)

		V_{CC}	T_A = 25°C		SN54HC237		SN74HC237		UNIT
			MIN	MAX	MIN	MAX	MIN	MAX	
t_w	Pulse duration, $\overline{GL}$ low	2 V	80		120		100		ns
		4.5 V	16		24		20		
		6 V	14		20		17		
t_{su}	Setup time, A, B, and C before $\overline{GL}$↑	2 V	75		115		95		ns
		4.5 V	15		23		19		
		6 V	13		20		16		
t_h	Hold time, A, B, and C after $\overline{GL}$↑	2 V	5		5		5		ns
		4.5 V	5		5		5		
		6 V	5		5		5		

switching characteristics over recommended operating free-air temperature range (unless otherwise noted), C_L = 50 pF, (see Note 1)

PARAMETER	FROM (INPUT)	TO (OUTPUT)	V_{CC}	T_A = 25°C			SN54HC237		SN74HC237		UNIT
				MIN	TYP	MAX	MIN	MAX	MIN	MAX	
t_{pd}	A, B, C	Any	2 V		91	190		285		240	ns
			4.5 V		23	38		57		48	
			6 V		17	32		48		41	
t_{pd}	$\overline{G}2$	Any	2 V		66	145		220		181	ns
			4.5 V		18	29		44		36	
			6 V		13	25		37		31	
t_{pd}	G1	Any	2 V		68	145		220		181	ns
			4.5 V		18	29		44		36	
			6 V		14	25		37		31	
t_{pd}	$\overline{GL}$	Any	2 V		92	190		285		240	ns
			4.5 V		24	38		57		48	
			6 V		19	32		48		41	
t_t		Any	2 V		38	75		110		95	ns
			4.5 V		8	15		22		19	
			6 V		6	13		19		16	

C_{pd}	Power dissipation capacitance	No load, T_A = 25°C	85 pF typ

NOTE 1: For load circuit and voltage waveforms, see page 1-14.

3 HCMOS DEVICES

TYPES SN54HCT237, SN74HCT237 3-LINE TO 8-LINE DECODERS/DEMULTIPLEXERS WITH ADDRESS LATCHES

D2804, MARCH 1984

- **Inputs are TTL-Voltage Compatible**
- **Combines Decoder and 3-Bit Address Latch**
- **Incorporates 2 Output Enables to Simplify Cascading**
- **Package Options Include Both Plastic and Ceramic Chip Carriers in Addition to Plastic and Ceramic DIPs**
- **Dependable Texas Instruments Quality and Reliability**

SN54HCT237 . . . J PACKAGE
SN74HCT237 . . . J OR N PACKAGE
(TOP VIEW)

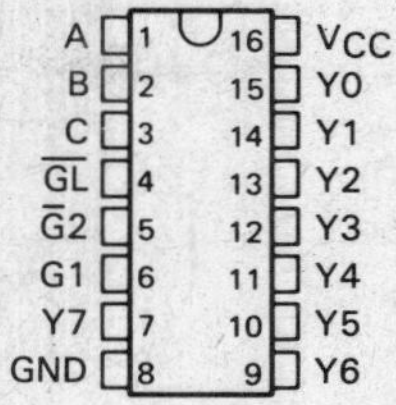

SN54HCT237 . . . FH OR FK PACKAGE
SN74HCT237 . . . FH OR FN PACKAGE
(TOP VIEW)

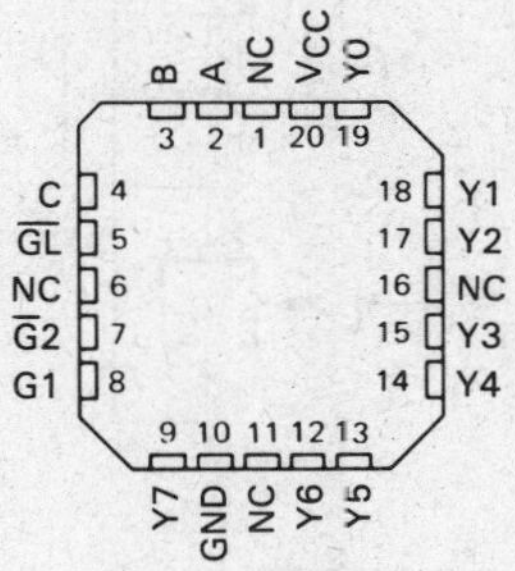

NC—No internal connection

description

The 'HCT237 is a three-line to eight-line decoder/demultiplexer with latches on the three address inputs. When the latch-enable input ($\overline{GL}$) is low, the 'HCT237 acts as a decoder/demultiplexer. When $\overline{GL}$ goes from low to high, the address present at the select inputs (A, B, and C) is stored in the latches. Further address changes are ignored as long as $\overline{GL}$ remains high. The output enable controls, G1 and $\overline{G2}$, control the outputs independently of the select or latch-enable inputs. All of the outputs are forced high if G1 is low or $\overline{G2}$ is high. The 'HCT237 is ideally suited for implementing glitch-free decoders in strobed (stored-address) applications in bus-oriented systems.

The SN54HCT237 is characterized for operation over the full military temperature range of −55°C to 125°C. The SN74HCT237 is characterized for operation from −40°C to 85°C.

logic symbols (alternatives)

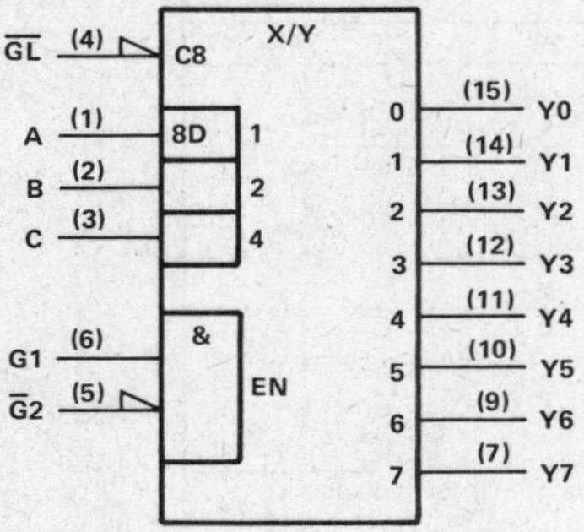

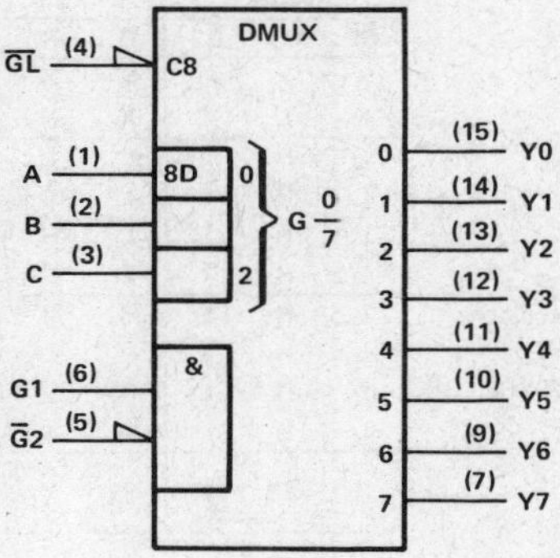

Pin numbers shown are for J and N packages.

logic diagram (positive logic)

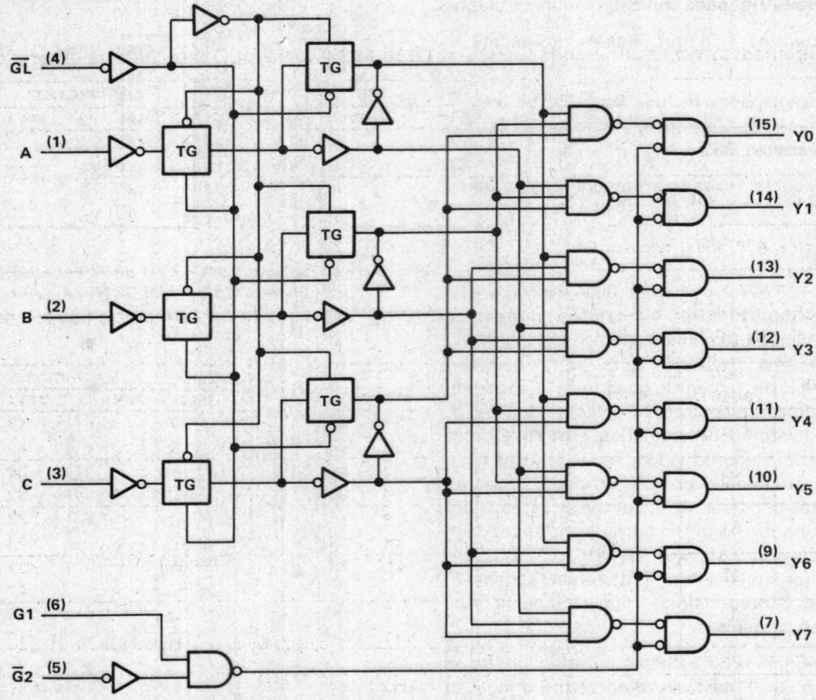

Pin numbers shown are for J and N packages.

FUNCTION TABLE

INPUTS						OUTPUTS							
ENABLE			SELECT										
$\overline{GL}$	G1	$\overline{G2}$	C	B	A	Y0	Y1	Y2	Y3	Y4	Y5	Y6	Y7
X	X	H	X	X	X	L	L	L	L	L	L	L	L
X	L	X	X	X	X	L	L	L	L	L	L	L	L
L	H	L	L	L	L	H	L	L	L	L	L	L	L
L	H	L	L	L	H	L	H	L	L	L	L	L	L
L	H	L	L	H	L	L	L	H	L	L	L	L	L
L	H	L	L	H	H	L	L	L	H	L	L	L	L
L	H	L	H	L	L	L	L	L	L	H	L	L	L
L	H	L	H	L	H	L	L	L	L	L	H	L	L
L	H	L	H	H	L	L	L	L	L	L	L	H	L
L	H	L	H	H	H	L	L	L	L	L	L	L	H
H	H	L	X	X	X	Output corresponding to stored address, L; all others, H							

maximum ratings, recommended operating conditions, and electrical characteristics

See Table VIII, page 2-15.

timing requirements over recommended operating free-air temperature range (unless otherwise noted)

		V_{CC}	T_A = 25°C		SN54HCT237		SN74HCT237		UNIT
			MIN	MAX	MIN	MAX	MIN	MAX	
t_w	Pulse duration, $\overline{GL}$ low	4.5 V	26		39		33		ns
		5.5 V	23		35		30		
t_{su}	Setup time, A, B, and C before $\overline{GL}\uparrow$	4.5 V	15		23		19		ns
		5.5 V	14		21		17		
t_h	Hold time, A, B, and C before $\overline{GL}\uparrow$	4.5 V	5		5		5		ns
		5.5 V	5		5		5		

switching characteristics over recommended operating free-air temperature range (unless otherwise noted), C_L = 50 pF (see Note 1)

PARAMETER	FROM (INPUT)	TO (OUTPUT)	V_{CC}	T_A = 25°C			SN54HCT237		SN74HCT237		UNIT
				MIN	TYP	MAX	MIN	MAX	MIN	MAX	
t_{pd}	A, B, C	Any	4.5 V		24	38		57		48	ns
			5.5 V		20	34		51		43	
t_{pd}	$\overline{G}2$	Any	4.5 V		19	29		44		36	ns
			5.5 V		16	26		40		32	
t_{pd}	G1	Any	4.5 V		19	29		44		36	ns
			5.5 V		16	26		40		32	
t_{pd}	$\overline{GL}$	Any	4.5 V		29	42		63		52	ns
			5.5 V		25	36		57		47	
t_t		Any	4.5 V		12	15		22		19	ns
			5.5 V		11	14		20		17	

C_{pd}	Power dissipation capacitance	No load, T_A = 25°C	85 pF typ

NOTE 1: For load circuit and voltage waveforms, see page 1-14.

TYPES SN54HC238, SN74HC238
3-LINE TO 8-LINE DECODERS/DEMULTIPLEXERS

D2804, MARCH 1984

- **Designed Specifically for High-Speed Memory Decoders and Data Transmission Systems**
- **Incorporates 3 Enable Inputs to Simplify Cascading and/or Data Reception**
- **Package Options Include Both Plastic and Ceramic Chip Carriers in Addition to Plastic and Ceramic DIPs**
- **Dependable Texas Instruments Quality and Reliability**

SN54HC238 . . . J PACKAGE
SN74HC238 . . . J OR N PACKAGE
(TOP VIEW)

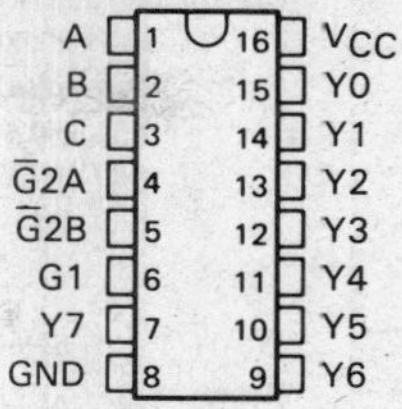

SN54HC238 . . . FH OR FK PACKAGE
SN74HC238 . . . FH OR FN PACKAGE
(TOP VIEW)

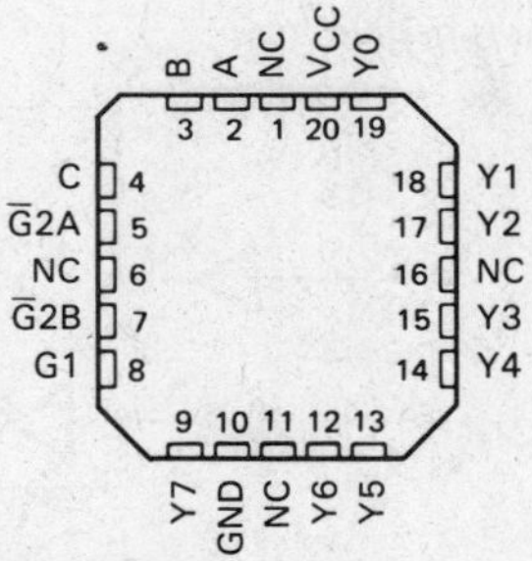

NC—No internal connection

description

The 'HC238 circuit is designed to be used in high-performance memory-decoding or data-routing applications requiring very short propagation delay times. In high-performance memory systems this decoder can be used to minimize the effects of system decoding. When employed with high-speed memories utilizing a fast enable circuit, the delay times of this decoder and the enable time of the memory are usually less than the typical access time of the memory. This means that the effective system delay introduced by the decoder is negligible.

The conditions at the binary select inputs and the three enable inputs select one of eight input lines. Two active-low and one active-high enable inputs reduce the need for external gates or inverters when expanding. A 24-line decoder can be implemented without external inverters and a 32-line decoder requires only one inverter. An enable input can be used as a data input for demultiplexing applications.

The SN54HC238 is characterized for operation over the full military temperature range of −55 °C to 125 °C. The SN74HC238 is characterized for operation from −40 °C to 85 °C.

logic symbols (alternatives)

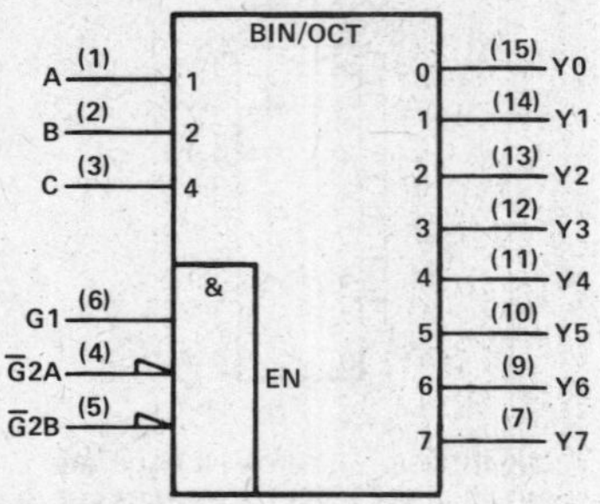

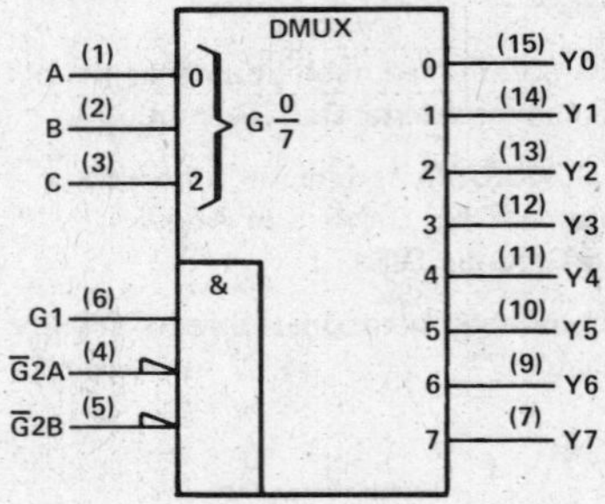

Pin numbers shown are for J and N packages.

logic diagram (positive logic)

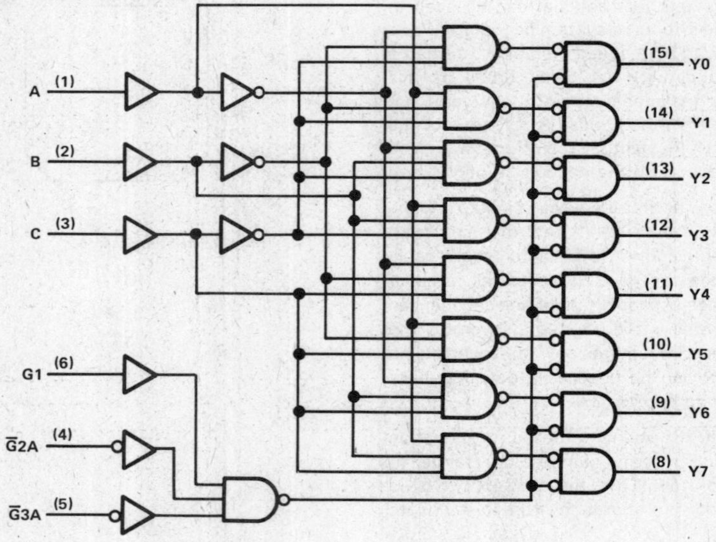

Pin numbers shown are for J and N packages.

FUNCTION TABLE

ENABLE INPUTS			SELECT INPUTS			OUTPUTS							
G1	$\overline{G}$2A	$\overline{G}$2B	C	B	A	Y0	Y1	Y2	Y3	Y4	Y5	Y6	Y7
X	H	X	X	X	X	L	L	L	L	L	L	L	L
X	X	H	X	X	X	L	L	L	L	L	L	L	L
L	X	X	X	X	X	L	L	L	L	L	L	L	L
H	L	L	L	L	L	H	L	L	L	L	L	L	L
H	L	L	L	L	H	L	H	L	L	L	L	L	L
H	L	L	L	H	L	L	L	H	L	L	L	L	L
H	L	L	L	H	H	L	L	L	H	L	L	L	L
H	L	L	H	L	L	L	L	L	L	H	L	L	L
H	L	L	H	L	H	L	L	L	L	L	H	L	L
H	L	L	H	H	L	L	L	L	L	L	L	H	L
H	L	L	H	H	H	L	L	L	L	L	L	L	H

maximum ratings, recommended operating conditions, and electrical characteristics

See Table IV, page 2-10.

switching characteristics over recommended operating free-air temperature range (unless otherwise noted), C_L = 50 pF (see Note 1)

PARAMETER	FROM (INPUT)	TO (OUTPUT)	V_{CC}	T_A = 25°C MIN	T_A = 25°C TYP	T_A = 25°C MAX	SN54HC238 MIN	SN54HC238 MAX	SN74HC238 MIN	SN74HC238 MAX	UNIT
t_{pd}	A, B, or C	Any	2 V		67	180		270		225	ns
			4.5 V		20	36		54		45	
			6 V		15	31		46		38	
t_{pd}	Enable	Any	2 V		60	155		235		195	ns
			4.5 V		17	31		47		39	
			6 V		13	26		40		33	
t_t		Any	2 V		38	75		110		95	ns
			4.5 V		8	15		22		19	
			6 V		6	13		19		16	

C_{pd}	Power dissipation capacitance	No load, T_A = 25°C	85 pF typ

NOTE 1: For load circuit and voltage waveforms, see page 1-14.

TYPES SN54HCT238, SN74HCT238 3-LINE TO 8-LINE DECODERS/DEMULTIPLEXERS

D2804, MARCH 1984

- Inputs are TTL-Voltage Compatible
- Designed Specifically for High-Speed Memory Decoders and Data Transmission Systems
- Incorporates 3 Enable Inputs to Simplify Cascading and/or Data Reception
- Package Options Include Both Plastic and Ceramic Chip Carriers in Addition to Plastic and Ceramic DIPs
- Dependable Texas Instruments Quality and Reliability

SN54HCT238 . . . J PACKAGE
SN74HCT238 . . . J OR N PACKAGE
(TOP VIEW)

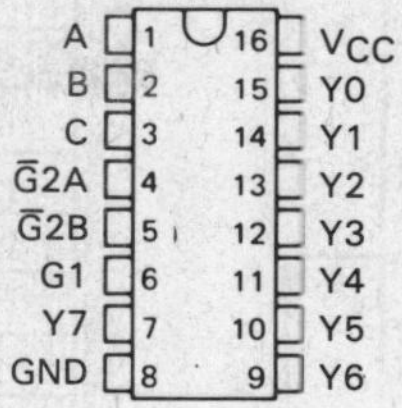

SN54HCT238 . . . FH OR FK PACKAGE
SN74HCT238 . . . FH OR FN PACKAGE
(TOP VIEW)

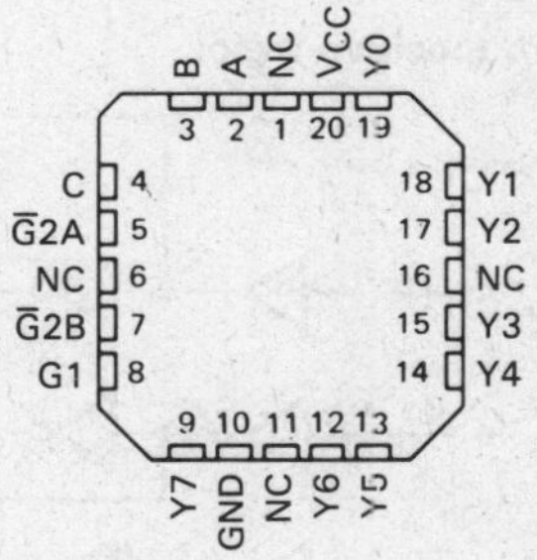

NC—No internal connection

description

The 'HCT238 circuit is designed to be used in high-performance memory-decoding or data-routing applications requiring very short propagation delay times. In high-performance memory systems this decoder can be used to minimize the effects of system decoding. When employed with high-speed memories utilizing a fast enable circuit, the delay times of this decoder and the enable time of the memory are usually less than the typical access time of the memory. This means that the effective system delay introduced by the decoder is negligible.

The conditions at the binary select inputs and the three enable inputs select one of eight input lines. Two active-low and one active-high enable inputs reduce the need for external gates or inverters when expanding. A 24-line decoder can be implemented without external inverters and a 32-line decoder requires only one inverter. An enable input can be used as a data input for demultiplexing applications.

The SN54HCT238 is characterized for operation over the full military temperature range of −55 °C to 125 °C. The SN74HCT238 is characterized for operation from −40 °C to 85 °C.

logic symbols (alternatives)

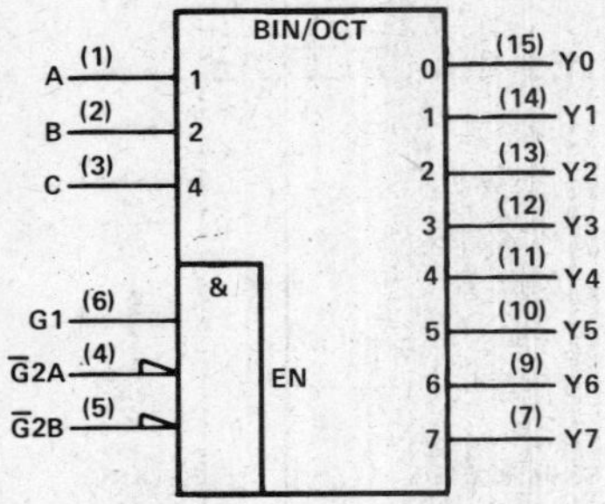

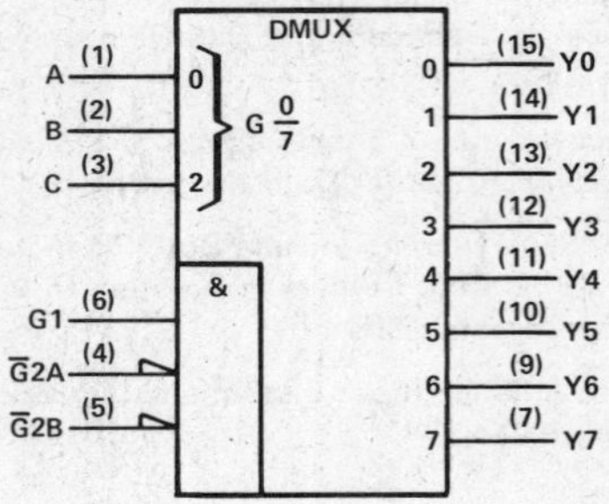

Pin numbers shown are for J and N packages.

logic diagram (positive logic)

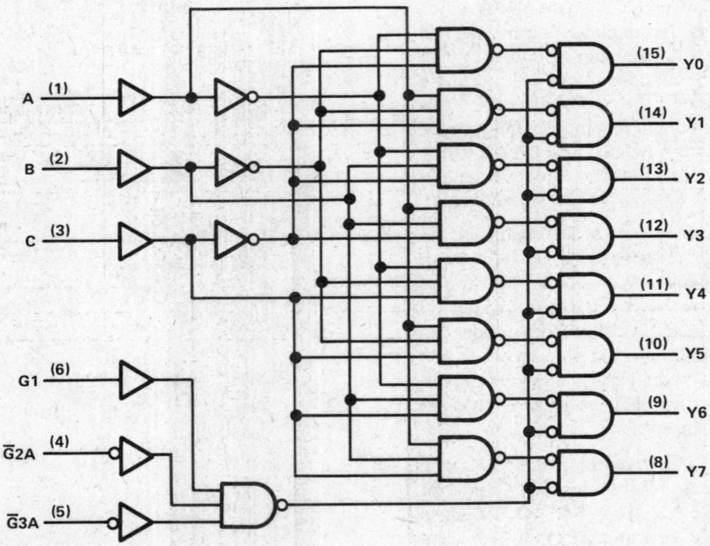

Pin numbers shown are for J and N packages.

TEXAS INSTRUMENTS
POST OFFICE BOX 225012 • DALLAS, TEXAS 75265

FUNCTION TABLE

ENABLE INPUTS			SELECT INPUTS			OUTPUTS							
G1	$\overline{G}$2A	$\overline{G}$2B	C	B	A	Y0	Y1	Y2	Y3	Y4	Y5	Y6	Y7
X	H	X	X	X	X	L	L	L	L	L	L	L	L
X	X	H	X	X	X	L	L	L	L	L	L	L	L
L	X	X	X	X	X	L	L	L	L	L	L	L	L
H	L	L	L	L	L	H	L	L	L	L	L	L	L
H	L	L	L	L	H	L	H	L	L	L	L	L	L
H	L	L	L	H	L	L	L	H	L	L	L	L	L
H	L	L	L	H	H	L	L	L	H	L	L	L	L
H	L	L	H	L	L	L	L	L	L	H	L	L	L
H	L	L	H	L	H	L	L	L	L	L	H	L	L
H	L	L	H	H	L	L	L	L	L	L	L	H	L
H	L	L	H	H	H	L	L	L	L	L	L	L	H

maximum ratings, recommended operating conditions, and electrical characteristics

See Table VIII, page 2-15.

switching characteristics over recommended operating free-air temperature range (unless otherwise noted), C_L = 50 pF, (see Note 1)

PARAMETER	FROM (INPUT)	TO (OUTPUT)	V_{CC}	T_A = 25°C MIN	TYP	MAX	SN54HCT238 MIN	MAX	SN74HCT238 MIN	MAX	UNIT
t_{pd}	A, B, or C	Any	4.5 V		21	36		54		45	ns
			5.5 V		18	32		49		41	
t_{pd}	Enable	Any	4.5 V		21	33		50		42	ns
			5.5 V		17	30		45		38	
t_t		Any	4.5 V		11	15		22		19	ns
			5.5 V		9	14		20		17	

C_{pd}	Power dissipation capacitance	No load, T_A = 25°C	85 pF typ

NOTE 1: For load circuit and voltage waveforms, see page 1-14.

3

HCMOS DEVICES

TYPES SN54HC239, SN74HC239
DUAL 2-LINE TO 4-LINE DECODERS/DEMULTIPLEXERS

D2804, MARCH 1984

- **Designed Specifically for High-Speed Memory Decoders and Data Transmission Systems**
- **Incorporates 2 Enable Inputs to Simplify Cascading and/or Data Reception**
- **Package Options Include Both Plastic and Ceramic Chip Carriers in Addition to Plastic and Ceramic DIPs**
- **Dependable Texas Instrumentss Quality and Reliability**

SN54HC239 . . . J PACKAGE
SN74HC239 . . . J OR N PACKAGE
(TOP VIEW)

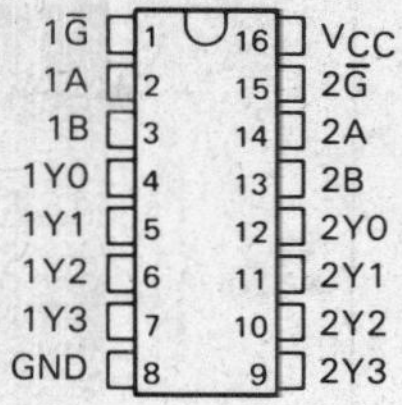

SN54HC239 . . . FH OR FK PACKAGE
SN74HC239 . . . FH OR FN PACKAGE
(TOP VIEW)

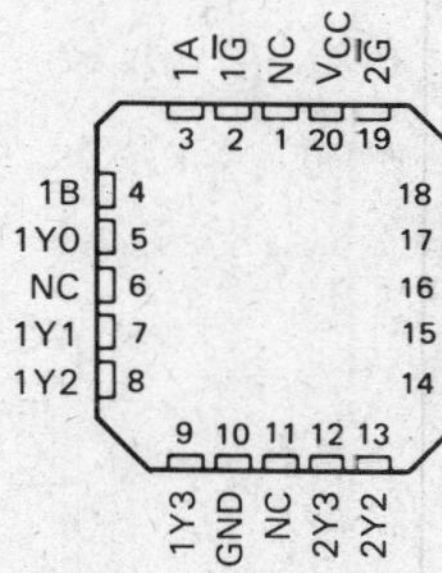

NC—No internal connection

description

The 'HC239 circuit is designed to be used in high-performance memory-decoding or data-routing applications requiring very short propagation delay times. In high-performance memory systems, this decoder can be used to minimize the effects of system decoding. When employed with high-speed memories utilizing a fast-enable circuit, the delay times of this decoder and the enable time of the memory are usually less than the typical access time of the memory. This means that the effective system delay introduced by the decoder is negligible.

The 'HC239 is comprised of two individual two-line to four-line decoders in a single package. The active-low enable input can be used as a data line in demultiplexing applications. These decoders/demultiplexers feature fully buffered inputs, each of which represents only one normalized load to its driving circuit.

The SN54HC239 is characterized for operation over the full military temperature range of −55°C to 125°C. The SN74HC239 is characterized for operation from −40°C to 85°C.

logic symbols (alternatives)

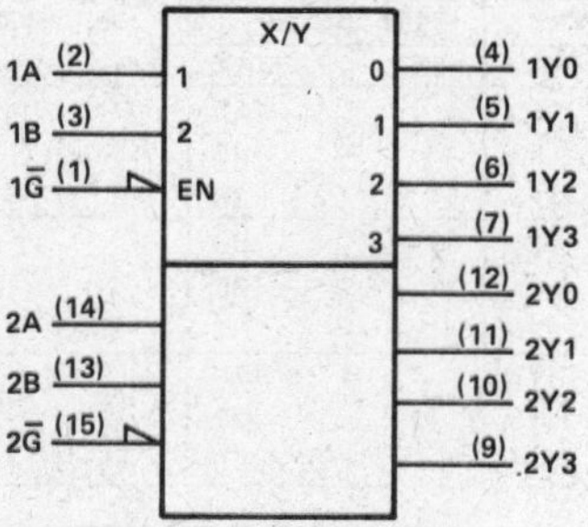

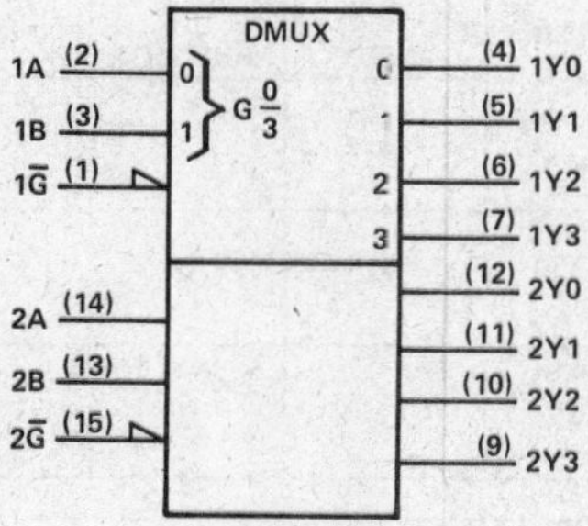

Pin numbers shown are for J and N packages.

logic diagram (positive logic)

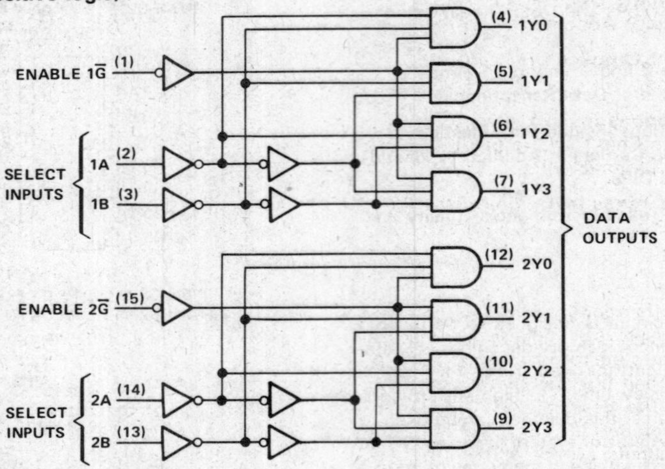

Pin numbers shown are for J and N packages.

FUNCTION TABLE

INPUTS			OUTPUTS			
ENABLE	SELECT					
$\overline{G}$	B	A	Y0	Y1	Y2	Y3
H	X	X	L	L	L	L
L	L	L	H	L	L	L
L	L	H	L	H	L	L
L	H	L	L	L	H	L
L	H	H	L	L	L	H

maximum ratings, recommended operating conditions, and electrical characteristics

See Table IV, page 2-10.

switching characteristics over recommended operating free-air temperature range (unless otherwise noted), C_L = 50 pF (see Note 1)

PARAMETER	FROM (INPUT)	TO (OUTPUT)	V_{CC}	T_A = 25°C MIN	T_A = 25°C TYP	T_A = 25°C MAX	SN54HC239 MIN	SN54HC239 MAX	SN74HC239 MIN	SN74HC239 MAX	UNIT
t_{pd}	A or B	Y	2 V		62	150		225		190	ns
			4.5 V		18	30		45		38	
			6 V		14	26		38		32	
t_{pd}	$\overline{G}$	Y	2 V		53	120		180		150	ns
			4.5 V		14	24		36		30	
			6 V		11	20		31		26	
t_{pd}		Y	2 V		38	75		110		95	ns
			4.5 V		8	15		22		19	
			6 V		6	13		19		16	

C_{pd}	Power dissipation capacitance per decoder	No load, T_A = 25°C	25 pF typ

NOTE 1: For load circuit and voltage waveforms, see page 1-14.

TYPES SN54HC240, SN54HC241, SN74HC240, SN74HC241 OCTAL BUFFERS AND LINE DRIVERS WITH 3-STATE OUTPUTS

D2684, DECEMBER 1982—REVISED MARCH 1984

- **3-State Outputs Drive Bus Lines or Buffer Memory Address Retgisters**
- **High-Current Outputs Drive up to 15 LSTTL Loads**
- **Package Options Include Both Plastic and Ceramic Chip Carriers in Addition to Plastic and Ceramic DIPs**
- **Dependable Texas Instruments Quality and Reliability**

SN54HC' . . . J PACKAGE
SN74HC' . . . J OR N PACKAGE
(TOP VIEW)

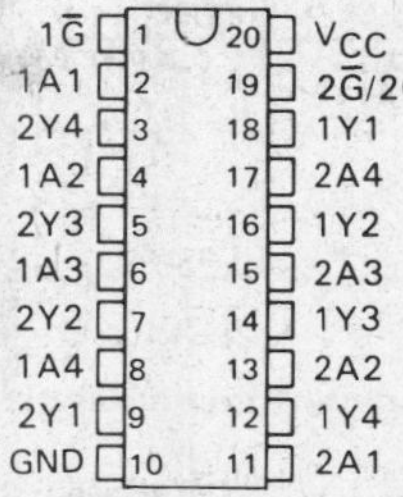

SN54HC' . . . FH OR FK PACKAGE
SN74HC' . . . FH OR FN PACKAGE
(TOP VIEW)

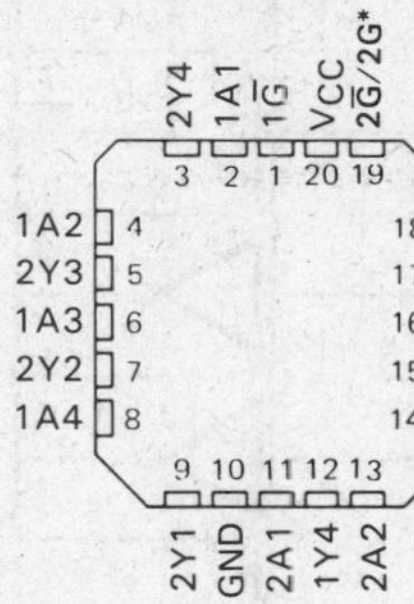

*2G̅ for 'HC240, or 2G for 'HC241

description

These octal buffers and line drivers are designed specifically to improve both the performance and density of three-state memory address drivers, clock drivers, and bus-oriented receivers and transmitters. The designer has a choice of selected combinations of inverting and noninverting outputs, symmetrical G̅ (active-low output control) inputs, and complementary G and G̅ inputs. These devices feature high fan-out.

The SN54HC' family is characterized for operation over the full military temperature range of −55°C to 125°C. The SN74HC' family is characterized for operation from −40°C to 85°C.

logic symbol

'HC240

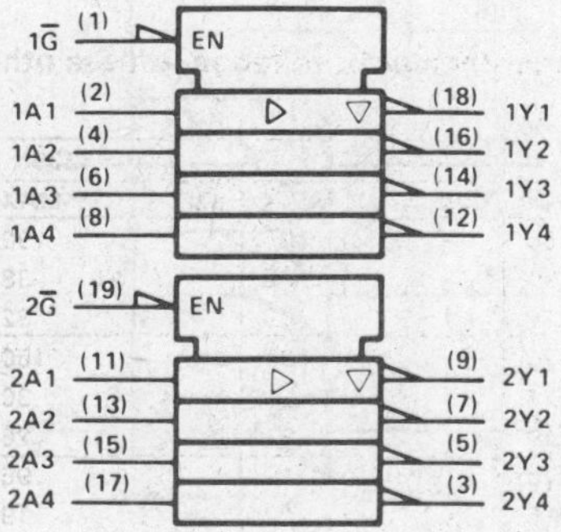

'HC241

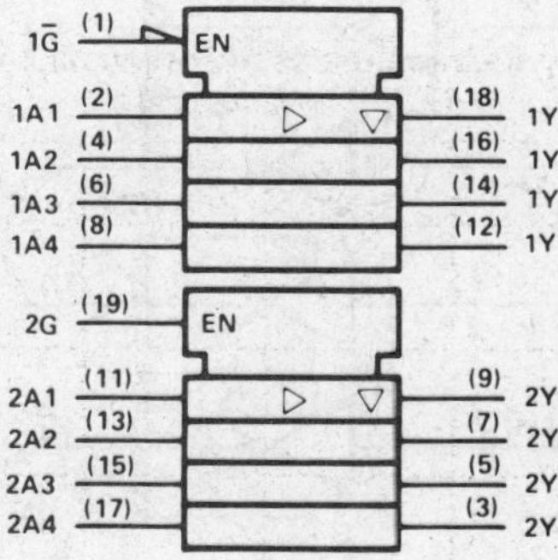

FUNCTION TABLES

'HC240
(EACH BUFFER)

INPUTS		OUTPUT
$\overline{G}$	A	Y
L	H	L
L	L	H
H	X	Z

'HC241
(EACH BUFFER IN FIRST SET)

INPUTS		OUTPUT
1$\overline{G}$	1A	1Y
L	H	H
L	L	L
H	X	Z

'HC241
(EACH BUFFER IN SECOND SET)

INPUTS		OUTPUT
2G	2A	2Y
H	H	H
H	L	L
L	X	Z

logic diagrams (positive logic)

'HC240

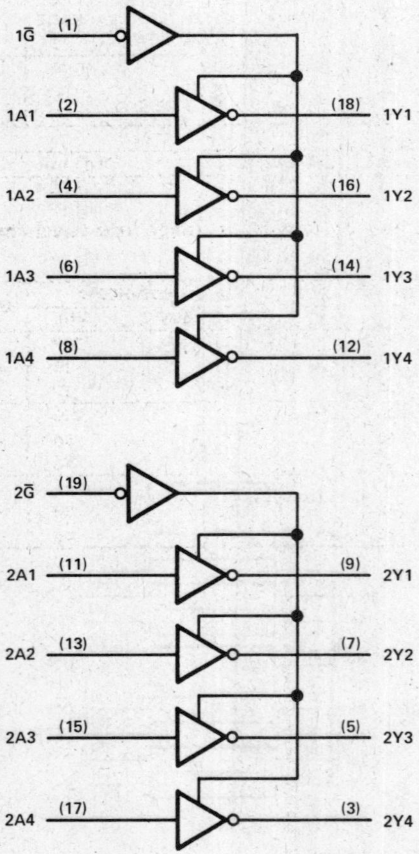

'HC241

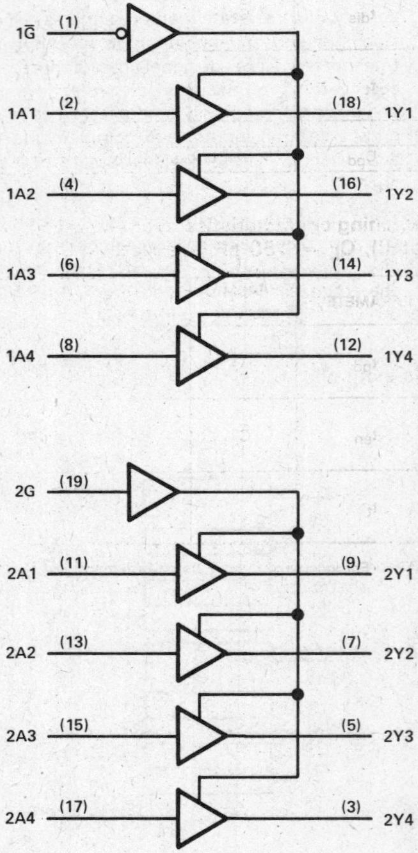

maximum ratings, recommended operating conditions, and electrical characteristics

See Table III, page 2-8.

switching characteristics over recommended operating free-air temperature range (unless otherwise noted), C_L = 50 pF (see Note 1)

PARAMETER	FROM (INPUT)	TO (OUTPUT)	V_{CC}	T_A = 25°C MIN	T_A = 25°C TYP	T_A = 25°C MAX	SN54HC240 MIN	SN54HC240 MAX	SN74HC240 MIN	SN74HC240 MAX	UNIT
t_{pd}	A	Y	2 V		50	100		150		125	ns
			4.5 V		10	20		30		25	
			6 V		9	17		25		21	
t_{en}	$\overline{G}$	Y	2 V		75	150		225		190	ns
			4.5 V		15	30		45		38	
			6 V		13	26		38		32	
t_{dis}	$\overline{G}$	Y	2 V		44	150		225		190	ns
			4.5 V		22	30		45		38	
			6 V		21	26		38		32	
t_t		Y	2 V		28	60		90		75	ns
			4.5 V		8	12		18		15	
			6 V		6	10		15		13	

C_{pd}	Power dissipation capacitance per buffer	No load, T_A = 25°C	35 pF typ

switching characteristics over recommended operating free-air temperature range (unless otherwise noted), C_L = 150 pF (see Note 1)

PARAMETER	FROM (INPUT)	TO (OUTPUT)	V_{CC}	T_A = 25°C MIN	T_A = 25°C TYP	T_A = 25°C MAX	SN54HC240 MIN	SN54HC240 MAX	SN74HC240 MIN	SN74HC240 MAX	UNIT
t_{pd}	A	Y	2 V		75	150		225		190	ns
			4.5 V		15	30		45		38	
			6 V		13	26		38		32	
t_{en}	$\overline{G}$	Y	2 V		100	200		300		250	ns
			4.5 V		20	40		60		50	
			6 V		17	34		51		43	
t_t		Y	2 V		45	210		315		265	ns
			4.5 V		17	42		63		53	
			6 V		13	36		53		45	

NOTE 1: For load circuit and voltage waveforms, see page 1-14.

maximum ratings, recommended operating conditions, and electrical characteristics

See Table III, page 2-8.

switching characteristics over recommended operating free-air temperature range (unless otherwise noted), C_L = 50 pF (see Note 1)

PARAMETER	FROM (INPUT)	TO (OUTPUT)	V_{CC}	T_A = 25°C MIN	TYP	MAX	SN54HC241 MIN	MAX	SN74HC241 MIN	MAX	UNIT
t_{pd}	A	Y	2 V		39	115		170		145	ns
			4.5 V		12	23		34		29	
			6 V		11	20		29		25	
t_{en}	$\overline{G}$ or G	Y	2 V		60	150		225		190	ns
			4.5 V		17	30		45		38	
			6 V		15	26		38		32	
t_{dis}	$\overline{G}$ or G	Y	2 V		40	150		225		190	ns
			4.5 V		18	30		45		38	
			6 V		17	26		38		32	
t_t		Y	2 V		28	60		90		75	ns
			4.5 V		8	12		18		15	
			6 V		6	10		15		13	

C_{pd}	Power dissipation capacitance per buffer	No load, T_A = 25°C	35 pF typ

switching characteristics over recommended operating free-air temperature range (unless otherwise noted), C_L = 150 pF (see Note 1)

PARAMETER	FROM (INPUT)	TO (OUTPUT)	V_{CC}	T_A = 25°C MIN	TYP	MAX	SN54HC241 MIN	MAX	SN74HC241 MIN	MAX	UNIT
t_{pd}	A	Y	2 V		50	165		245		210	ns
			4.5 V		16	33		49		42	
			6 V		14	28		42		35	
t_{en}	$\overline{G}$ or G	Y	2 V		100	200		300		250	ns
			4.5 V		20	40		60		50	
			6 V		17	34		51		43	
t_t		Y	2 V		45	210		315		265	ns
			4.5 V		17	42		63		53	
			6 V		13	36		53		45	

NOTE 1: For load circuit and voltage waveforms, see page 1-14.

TYPES SN54HCT240, SN54HCT241, SN74HCT240, SN74HCT241 OCTAL BUFFERS AND LINE DRIVERS CMOS LOGIC WITH 3-STATE OUTPUTS

D2804, MARCH 1984

- Inputs are TTL-Voltage Compatible
- 3-State Outputs Drive Bus Lines or Buffer Memory Address Registers
- High-Current Outputs Drive up to 15 LSTTL Loads
- Package Options Include Both Plastic and Ceramic Chip Carriers in Addition to Plastic and Ceramic DIPs
- Dependable Texas Instruments Quality and Reliability

SN54HCT' . . . J PACKAGE
SN74HCT' . . . J OR N PACKAGE
(TOP VIEW)

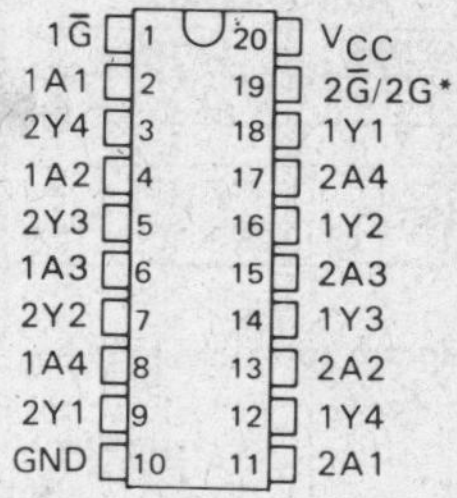

SN54HCT' . . . FH OR FK PACKAGE
SN74HCT' . . . FH OR FN PACKAGE
(TOP VIEW)

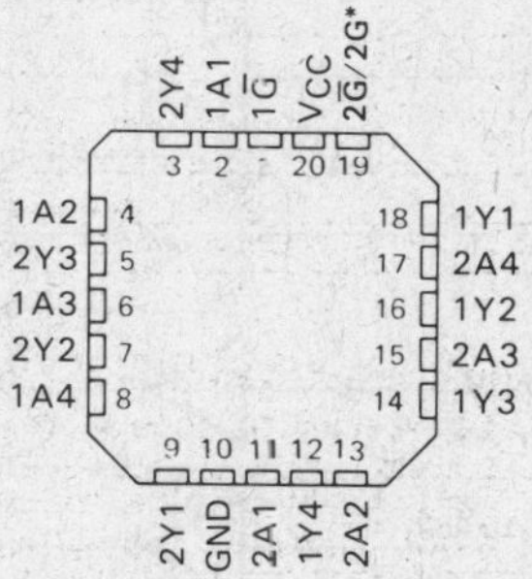

*2G̅ for 'HCT240, or 2G for 'HCT241

description

These octal buffers and line drivers are designed specifically to improve both the performance and density of three-state memory address drivers, clock drivers, and bus-oriented receivers and transmitters. The designer has a choice of selected combinations of inverting and noninverting outputs, symmetrical G̅ (active-low output control) inputs, and complementary G and G̅ inputs. These devices feature high fan-out.

The SN54HCT' family is characterized for operation over the full military temperature range of −55 °C to 125 °C. The SN74HCT' family is characterized for operation from −40 °C to 85 °C.

logic symbols

'HCT240

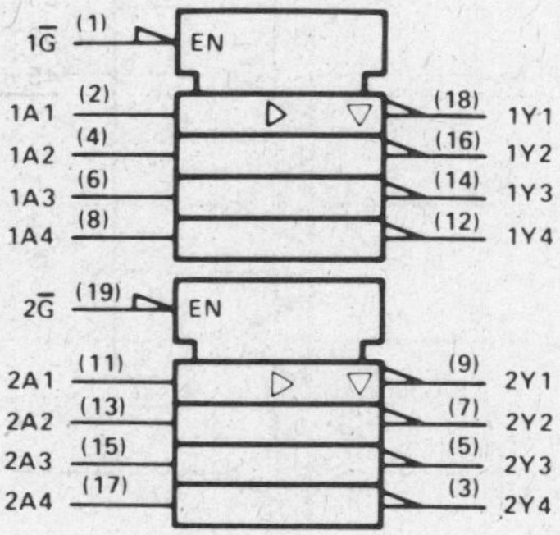

'HCT241

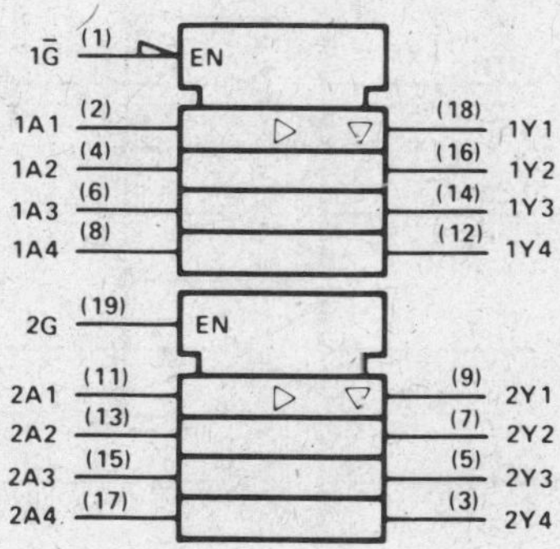

FUNCTION TABLES

'HCT240
(EACH BUFFER)

INPUTS		OUTPUT
$\overline{G}$	A	Y
L	H	L
L	L	H
H	X	Z

'HCT241
(EACH BUFFER IN FIRST SET)

INPUTS		OUTPUT
$1\overline{G}$	1A	1Y
L	H	H
L	L	L
H	X	Z

'HCT241
(EACH BUFFER IN SECOND SET)

INPUTS		OUTPUT
2G	2A	2Y
H	H	H
H	L	L
L	X	Z

logic diagram (positive logic)

'HCT240

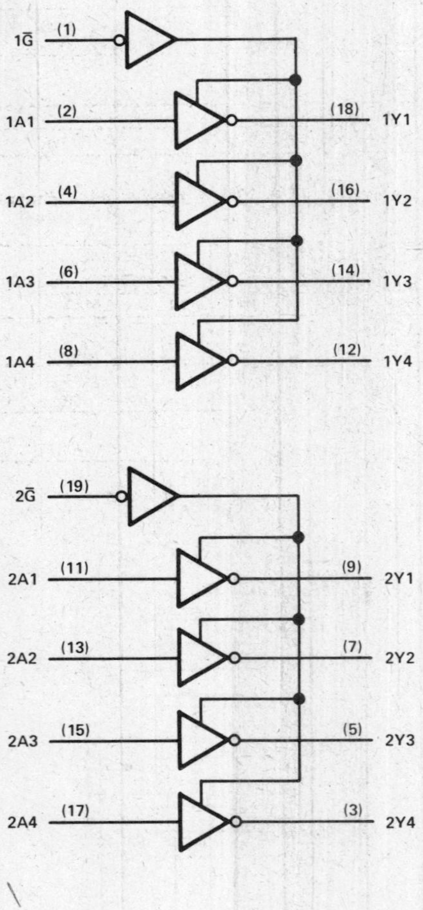

'HCT241

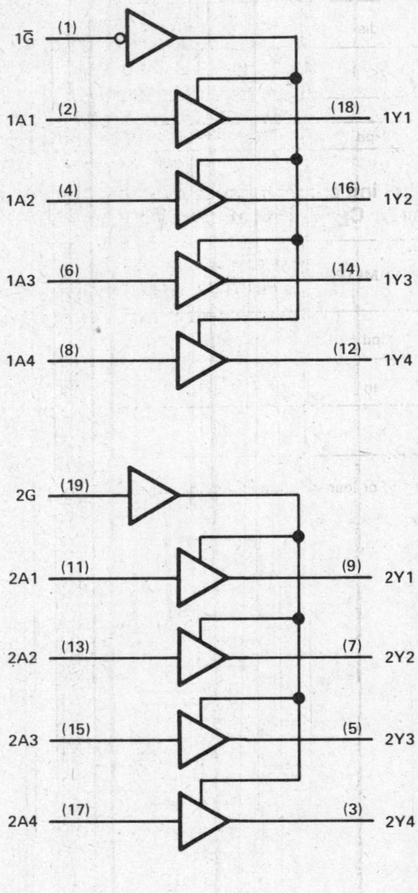

maximum ratings, recommended operating conditions, and electrical characteristics

See Table VII, page 2-14.

switching characteristics over recommended operating free-air temperature range (unless otherwise noted), C_L = 50 pF (see Note 1)

PARAMETER	FROM (INPUT)	TO (OUTPUT)	V_{CC}	T_A = 25°C MIN	TYP	MAX	SN54HCT240 SN54HCT241 MIN	MAX	SN74HCT240 SN74HCT241 MIN	MAX	UNIT
t_{pd}	A	Y	4.5 V		13	25		37		32	ns
			5.5 V		12	23		33		29	
t_{en}	G or $\overline{G}$	Y	4.5 V		21	35		53		44	ns
			5.5 V		19	32		48		40	
t_{dis}	G or $\overline{G}$	Y	4.5 V		19	35		53		44	ns
			5.5 V		18	32		48		40	
t_t		Y	4.5 V		8	12		18		15	ns
			5.5 V		7	11		16		14	

C_{pd}	Power dissipation capacitance per buffer	No load, T_A = 25°C	40 pF typ

switching characteristics over recommended operating free-air temperature range (unless otherwise noted), C_L = 150 pF (see Note 1)

PARAMETER	FROM (INPUT)	TO (OUTPUT)	V_{CC}	T_A = 25°C MIN	TYP	MAX	SN54HCT240 SN54HCT241 MIN	MAX	SN74HCT240 SN74HCT241 MIN	MAX	UNIT
t_{pd}	A	Y	4.5 V		20	42		63		53	ns
			5.5 V		19	38		56		48	
t_{en}	G or $\overline{G}$	Y	4.5 V		25	52		79		65	ns
			5.5 V		22	47		71		59	
t_t		Y	4.5 V		17	42		63		53	ns
			5.5 V		14	38		57		48	

Note 1: For load circuit and voltage waveforms, see page 1-14.

TYPES SN54HC242, SN54HC243 SN74HC242, SN74HC243 QUADRUPLE BUS TRANSCEIVERS WITH 3-STATE OUTPUTS

D2684, DECEMBER 1982—REVISED MARCH 1984

- **2-Way Asynchronous Communication Between Data Buses**
- **High-Current Outputs Can Drive up to 15 LSTTL Loads**
- **Package Options Include Both Plastic and Ceramic Chip Carriers in Addition to Plastic and Cermic DIPs**
- **Dependable Texas Instruments Quality and Reliability**

SN54HC242, SN54HC243 . . . J PACKAGE
SN74HC242, SN74HC243 . . . J OR N PACKAGE
(TOP VIEW)

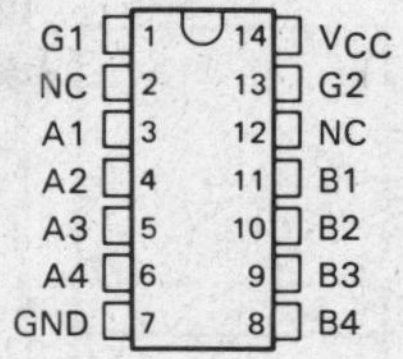

SN54HC242, SN54HC243 . . . FH OR FK PACKAGE
SN74HC242, SN74HC243 . . . FH OR FN PACKAGE
(TOP VIEW)

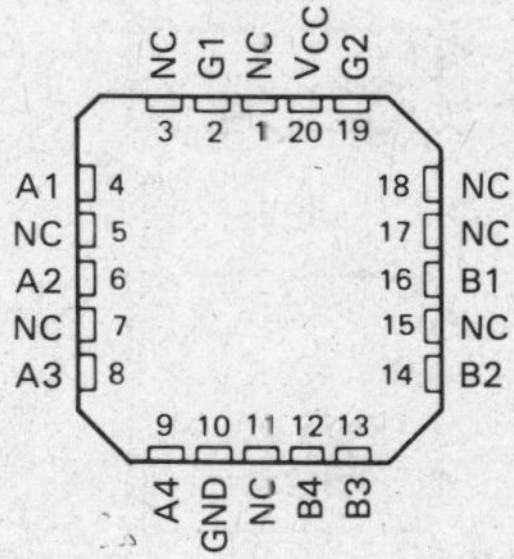

NC—No internal connection

description

These four-data line transceivers are designed for asynchronous two-way communications between data buses. The SN74HC' devices can be used to drive terminated lines down to 133 ohms.

These parts differ from their TTL counterparts (LS, ALS, and AS) in that these CMOS parts do not have a bus-latching mode in which both the outputs are simultaneously enabled. Instead of this latched mode, the buses are isolated, thus preventing potential bus conflicts if both buses are active. However, with the exception of the fourth line of the function table, their functional operation is identical to their TTL counterparts. The two enables have been renamed G1 and G2 since they work together to determine the direction of transmission rather than each enable controlling one direction independently of the other. Whenever G1 and G2 are at opposite logic levels with respect to each other, isolation between buses results.

The SN54HC' family is characterized for operation over the full military temperature range of −55°C to 125°C. The SN74HC' family is characterized for operation from −40°C to 85°C.

FUNCTION TABLE

INPUTS		'HC242	'HC243
G1	G2		
L	L	$\overline{A}$ to B	A to B
H	H	$\overline{B}$ to A	B to A
H	L	Isolation	Isolation
L	H	Isolation	Isolation

3

HCMOS DEVICES

logic symbol

'HC242

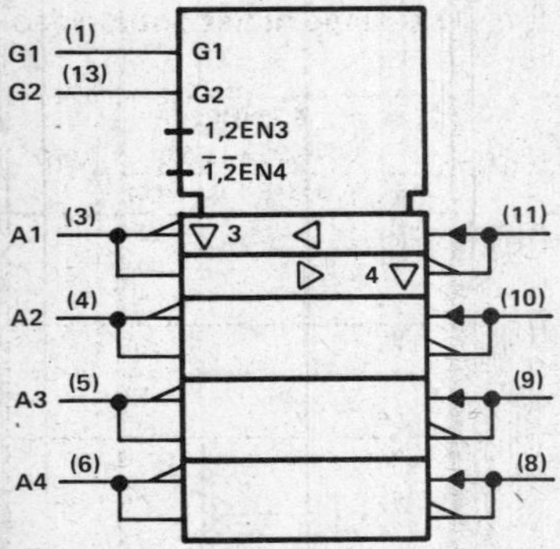

'HC243

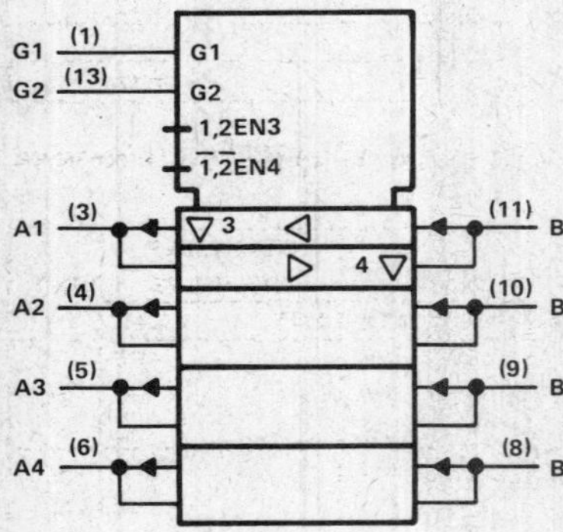

logic diagrams (positive logic)

'HC242

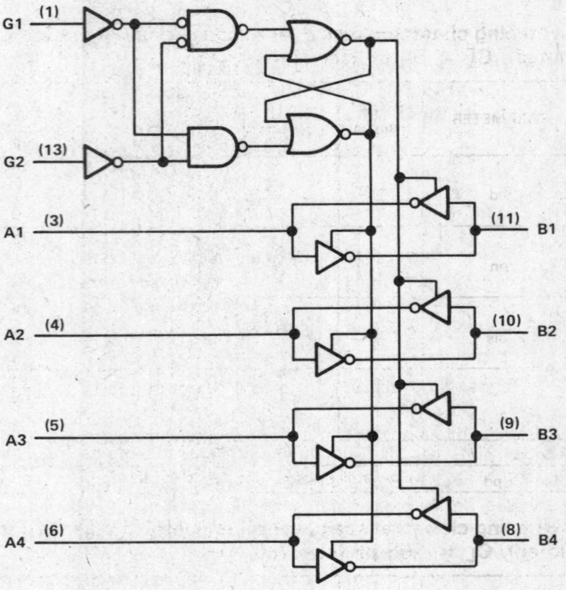

'HC243

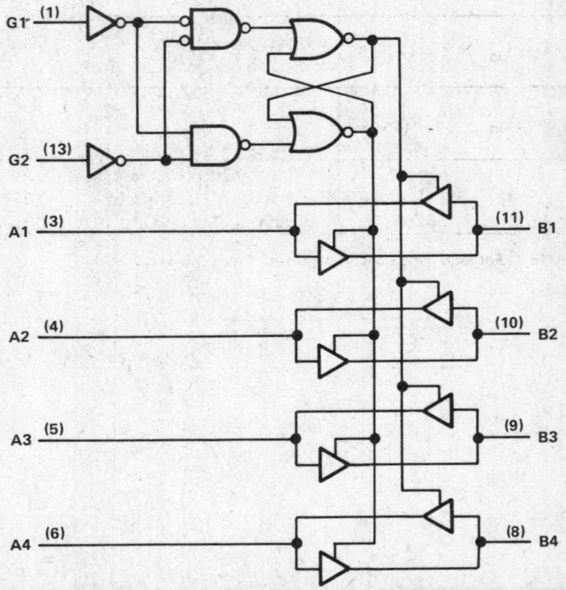

Pin numbers shown are for J and N packages.

3
HCMOS DEVICES

TEXAS INSTRUMENTS
POST OFFICE BOX 225012 • DALLAS, TEXAS 75265

maximum ratings, recommended operating conditions, and electrical characteristics

See Table III, page 2-8.

switching characteristics over recommended operating free-air temperature range (unless otherwise noted), C_L = 50 pF (see Note 1)

PARAMETER	FROM (INPUT)	TO (OUTPUT)	V_{CC}	T_A = 25°C			SN54HC242 SN54HC243		SN74HC242 SN74HC243		UNIT
				MIN	TYP	MAX	MIN	MAX	MIN	MAX	
			2 V		45	100		150		125	
t_{pd}	A or B	B or A	4.5 V		12	20		30		25	ns
			6 V		10	17		26		21	
			2 V		75	150		225		190	
t_{en}	G1 or G2	A or B	4.5 V		21	30		45		38	ns
			6 V		17	26		38		32	
			2 V		48	150		225		190	
t_{dis}	G1 or G2	A or B	4.5 V		23	30		45		38	ns
			6 V		20	26		38		32	
			2 V		28	60		90		75	
t_t		A or B	4.5 V		8	12		18		15	ns
			6 V		6	10		15		13	

C_{pd}	Power dissipation capacitance per transceiver	No load, T_A = 25°C	34 pF typ

switching characteristics over recommended operating free-air temperature range (unless otherwise noted), C_L = 150 pF (see Note 1)

PARAMETER	FROM (INPUT)	TO (OUTPUT)	V_{CC}	T_A = 25°C			SN54HC243 SN54HC243		SN74HC243 SN74HC243		UNIT
				MIN	TYP	MAX	MIN	MAX	MIN	MAX	
			2 V		63	150		225		190	
t_{pd}	A or B	B or A	4.5 V		17	30		45		38	ns
			6 V		14	26		38		32	
			2 V		100	200		300		250	
t_{en}	G1 or G2	A or B	4.5 V		26	40		60		50	ns
			6 V		21	34		51		43	
			2 V		45	210		315		265	
t_t		A or B	4.5 V		17	42		63		53	ns
			6 V		13	36		53		45	

NOTE 1: For load circuit and voltage waveforms, see page 1-14.

TYPES SN54HCT242, SN54HCT243 SN74HCT242, SN74HCT243 QUADRUPLE BUS TRANSCEIVERS WITH 3-STATE OUTPUTS

D2804, MARCH 1984

- **Inputs are TTL-Voltage Compatible**
- **2-Way Asynchronous Communication Between Data Buses**
- **High-Current Outputs Can Drive up to 15 LSTTL Loads**
- **Package Options Include Both Plastic and Ceramic Chip Carriers in Addition to Plastic and Ceramic DIPs**
- **Dependable Texas Instruments Quality and Reliability**

SN54HCT242, SN54HCT243 . . . J PACKAGE
SN74HCT242, SN74HCT243 . . . J OR N PACKAGE
(TOP VIEW)

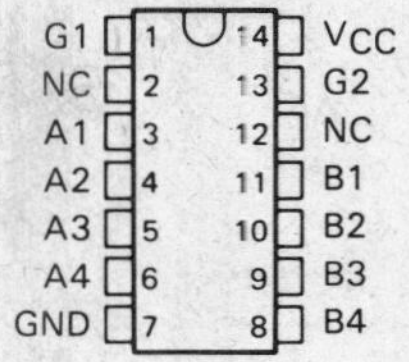

SN54HCT242, SN54HCT243 . . FH OR FK PACKAGE
SN74HCT242, SN74HCT243 . . FH OR FN PACKAGE
(TOP VIEW)

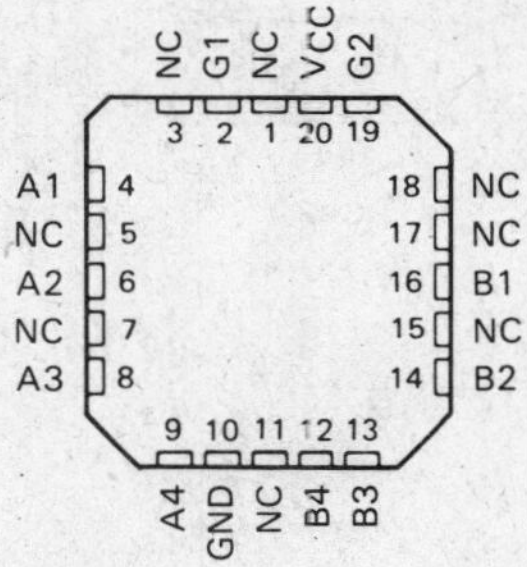

NC—No internal connection

description

These four-data line transceivers are designed for asynchronous two-way communications between data buses. The SN74HCT' devices can be used to drive terminated lines down to 133 ohms.

These parts differ from their TTL counterparts (LS, ALS, and AS) in that these CMOS parts do not have a bus-latching mode in which both the outputs are simultaneously enabled. Instead of this latched mode, the buses are isolated, thus preventing potential bus conflicts if both buses are active. However, with the exception of the fourth line of the function table, their functional operation is identical to their TTL counterparts. The two enables have been renamed G1 and G2 since they work together to determine the direction of transmission rather than each enable controlling one direction independently of the other. Whenever G1 and G2 are at opposite logic levels with respect to each other, isolation between buses results.

The SN54HCT' family is characterized for operation over the full military temperature range of −55 °C to 125 °C. The SN74HCT' family is characterized for operation from −40 °C to 85 °C.

FUNCTION TABLE

INPUTS		'HCT242	'HCT243
G1	G2		
L	L	$\overline{A}$ to B	A to B
H	H	$\overline{B}$ to A	B to A
H	L	Isolation	Isolation
L	H	Isolation	Isolation

TYPES SN54HCT242, SN54HCT243 SN74HCT242, SN74HCT243 QUADRUPLE BUS TRANSCEIVERS WITH 3-STATE OUTPUTS

logic symbol

'HCT242

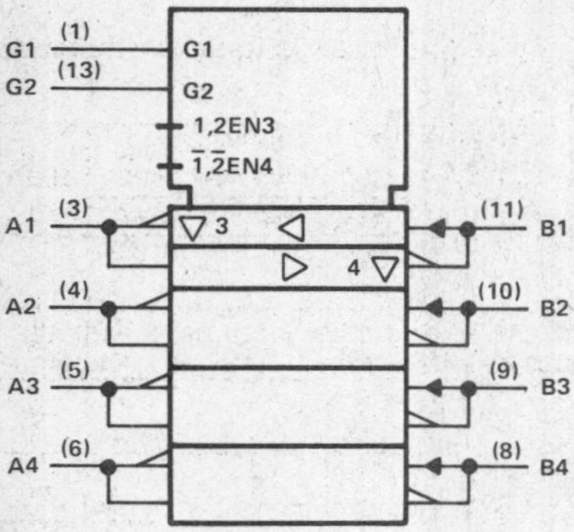

'HCT243

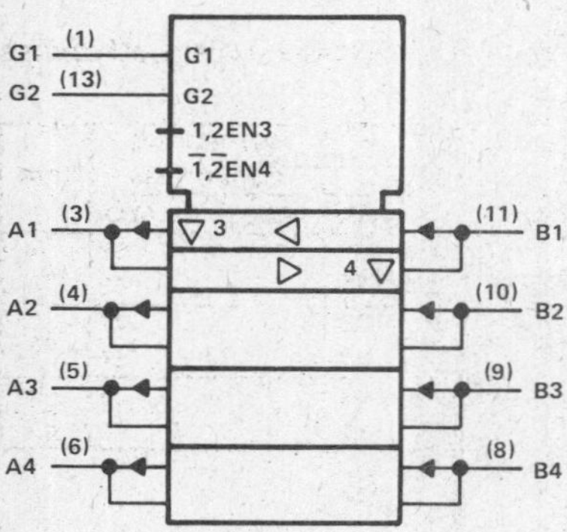

logic diagrams (positive logic)

'HCT242

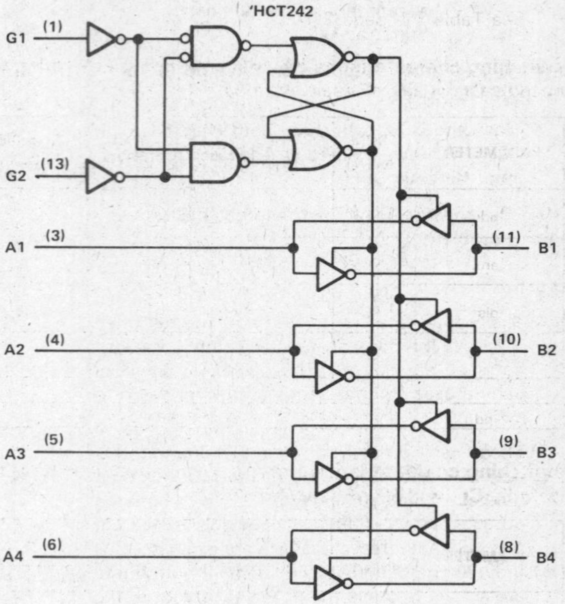

'HCT243

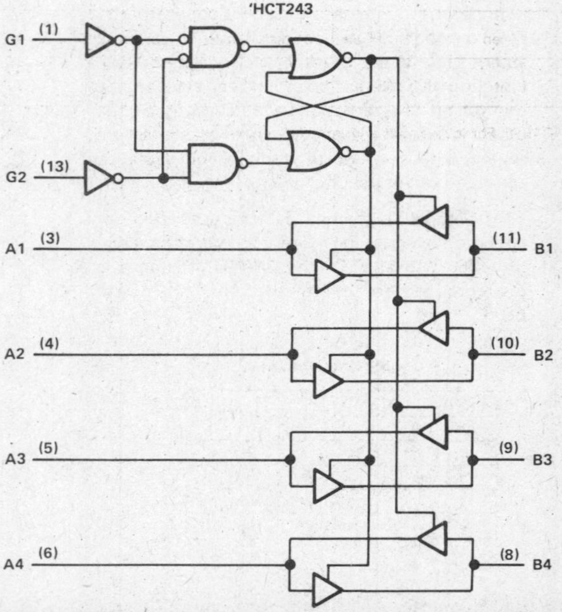

Pin numbers shown are for J and N packages.

TEXAS INSTRUMENTS
POST OFFICE BOX 225012 • DALLAS, TEXAS 75265

maximum ratings, recommended operating conditions, and electrical characteristics

See Table VII, page 2-14

switching characteristics over recommended operating free-air temperature range (unless otherwise noted), C_L = 50 pF (see Note 1)

PARAMETER	FROM (INPUT)	TO (OUTPUT)	V_{CC}	T_A = 25°C			SN54HCT242 SN54HCT243		SN74HCT242 SN74HCT243		UNIT
				MIN	TYP	MAX	MIN	MAX	MIN	MAX	
t_{pd}	A or B	B or A	4.5 V		15	30		45		38	ns
			5.5 V		13	27		41		34	
t_{en}	G1 or G2	A or B	4.5 V		21	40		60		50	ns
			5.5 V		19	36		54		45	
t_{dis}	G1 or G2	A or B	4.5 V		19	40		60		50	ns
			5.5 V		18	36		54		45	
t_t		A or B	4.5 V		8	12		18		15	ns
			5.5 V		7	11		16		14	

C_{pd}	Power dissipation capacitance per transceiver	No load, T_A = 25°C	40 pF typ

switching characteristics over recommended operating free-air temperature range (unless otherwise noted), C_L = 150 pF (see Note 1)

PARAMETER	FROM (INPUT)	TO (OUTPUT)	V_{CC}	T_A = 25°C			SN54HCT242 SN54HCT243		SN74HCT242 SN74HCT243		UNIT
				MIN	TYP	MAX	MIN	MAX	MIN	MAX	
t_{pd}	A or B	B or A	4.5 V		21	47		71		59	ns
			5.5 V		18	42		64		53	
t_{en}	G1 or G2	A or B	4.5 V		27	57		86		71	ns
			5.5 V		24	51		77		64	
t_t		A or B	4.5 V		17	42		63		53	ns
			5.5 V		14	38		57		48	

NOTE 1: For load circuit and voltage waveforms, see page 1-14.

3

HCMOS DEVICES

TYPES SN54HC244, SN74HC244
OCTAL BUFFERS AND LINE DRIVERS WITH 3-STATE OUTPUTS

D2684, DECEMBER 1982—REVISED MARCH 1984

- **3-State Outputs Drive Bus Lines or Buffer Memory Address Registers**
- **High-Current Outputs Can Drive up to 15 LSTTL Loads**
- **Package Options Include Both Plastic and Ceramic Chip Carriers in Addition to Plastic and Ceramic DIPs**
- **Dependable Texas Instruments Quality and Reliability**

SN54HC244 . . . J PACKAGE
SN74HC244 . . . J OR N PACKAGE
(TOP VIEW)

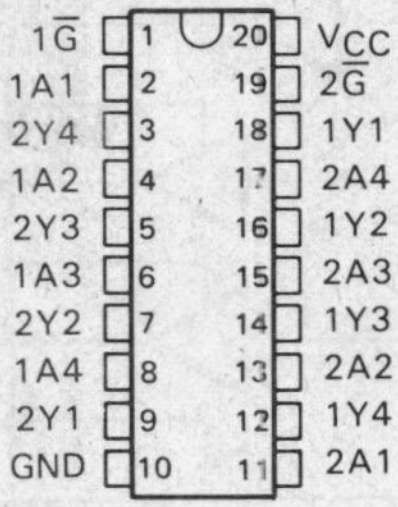

SN54HC244 . . . FH OR FK PACKAGE
SN74HC244 . . . FH OR FN PACKAGE
(TOP VIEW)

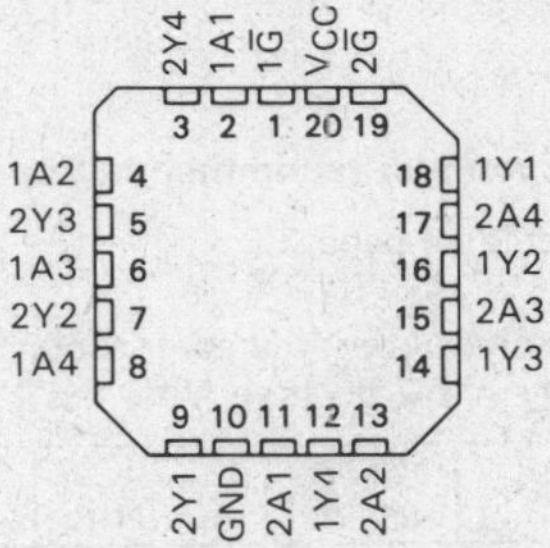

description

These octal buffers and line drivers are designed specifically to improve both the performance and density of the three-state memory address drivers, clock drivers, and bus-oriented receivers and transmitters. Taken together with the 'HC240 and 'HC241, these devices provide the choice of selected combinations of inverting outputs, symmetrical $\overline{G}$ (active-low input control) inputs and complementary G and $\overline{G}$ inputs.

The SN54HC244 is characterized for operation over the full military temperature range of −55 °C to 125 °C. The SN74HC244 is characterized for operation from −40 °C to 85 °C.

logic symbol

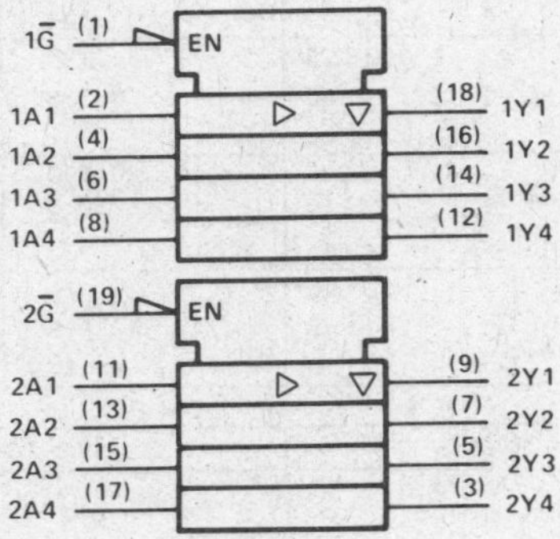

logic diagram (positive logic)

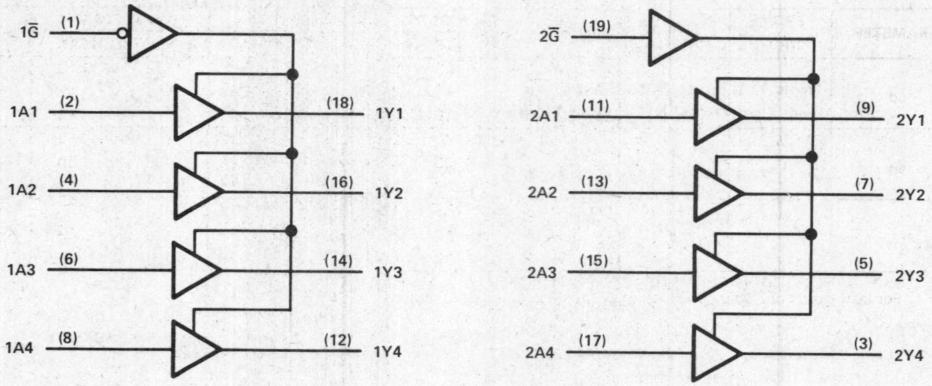

maximum ratings, recommended operating conditions, and electrical characteristics

See Table III, page 2-8.

switching characteristics over recommended operating free-air temperature range (unless otherwise noted), C_L = 50 pF (see Note 1)

PARAMETER	FROM (INPUT)	TO (OUTPUT)	V_{CC}	T_A = 25°C			SN54HC244		SN74HC244		UNIT
				MIN	TYP	MAX	MIN	MAX	MIN	MAX	
t_{pd}	A	Y	2 V		40	115		170		145	ns
			4.5 V		13	23		34		29	
			6 V		11	20		29		25	
t_{en}	$\overline{G}$	Y	2 V		75	150		225		190	ns
			4.5 V		15	30		45		38	
			6 V		13	26		38		32	
t_{dis}	$\overline{G}$	Y	2 V		75	150		225		190	ns
			4.5 V		15	30		45		38	
			6 V		13	26		38		32	
t_t		Y	2 V		28	60		90		75	ns
			4.5 V		8	12		18		15	
			6 V		6	10		15		13	

C_{pd}	Power dissipation capacitance per gate	No load, T_A = 25°C	35 pF typ

NOTE 1: For load circuit and voltage waveforms, see page 1-14.

switching characteristics over recommended operating free-air temperature range (unless otherwise noted), C_L = 150 pF (see Note 1)

PARAMETER	FROM (INPUT)	TO (OUTPUT)	V_{CC}	T_A = 25°C MIN	TYP	MAX	SN54HC244 MIN	TYP	MAX	SN74HC244 MIN	TYP	MAX	UNIT
t_{pd}	A	Y	2 V		56	165			245			210	ns
			4.5 V		18	33			49			42	
			6 V		15	28			42			35	
t_{en}	$\overline{G}$	Y	2 V		100	200			300			250	ns
			4.5 V		20	40			60			50	
			6 V		17	34			51			43	
t_t		Y	2 V		45	210			315			265	ns
			4.5 V		17	42			63			53	
			6 V		13	36			53			45	

NOTE 1: For load circuit and voltage waveforms, see page 1-14.

TYPES SN54HCT244, SN74HCT244
OCTAL BUFFERS AND LINE DRIVERS WITH 3-STATE OUTPUTS

D2804, MARCH 1984

- **Inputs are TTL-Voltage Compatible**
- **3-State Outputs Drive Bus Lines or Buffer Memory Address Registers**
- **High-Current Outputs Can Drive up to 15 LSTTL Loads**
- **Package Options Include Both Plastic and Ceramic Chip Carriers in Addition to Plastic and Ceramic DIPs**
- **Dependable Texas Instruments Quality and Reliability**

SN54HCT244 . . . J PACKAGE
SN74HCT244 . . . J OR N PACKAGE
(TOP VIEW)

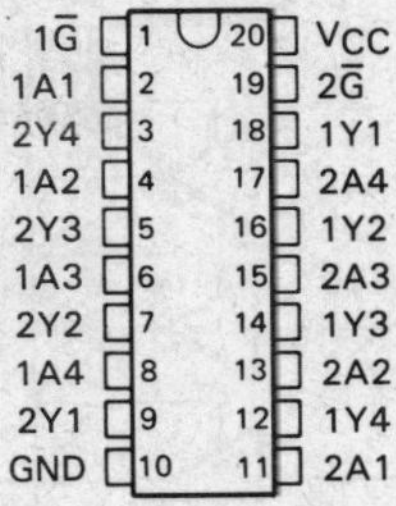

SN54HCT244 . . . FH OR FK PACKAGE
SN74HCT244 . . . FH OR FN PACKAGE
(TOP VIEW)

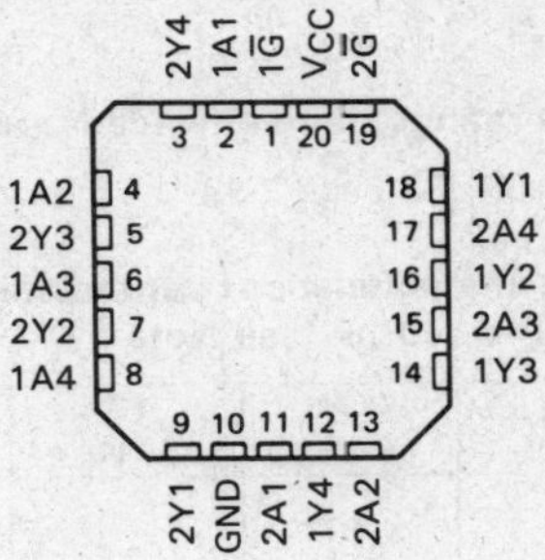

description

These octal buffers and line drivers are designed specifically to improve both the performance and density of three-state memory address drivers, clock drivers, and bus-oriented receivers and transmitters. Taken together with the 'HCT240 and 'HCT241, these devices provide the choice of selected combinations of inverting outputs, symmetrical $\overline{G}$ (active-low input control) inputs, and complementary G and $\overline{G}$ inputs.

The SN54HCT244 is characterized for operation over the full military temperature range of −55°C to 125°C. The SN74HCT244 is characterized for operation from −40°C to 85°C.

logic symbol

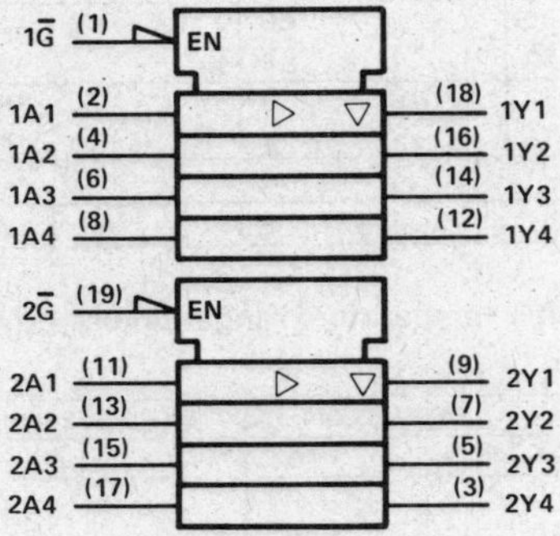

TEXAS INSTRUMENTS
POST OFFICE BOX 225012 • DALLAS, TEXAS 75265

logic diagram (positive logic)

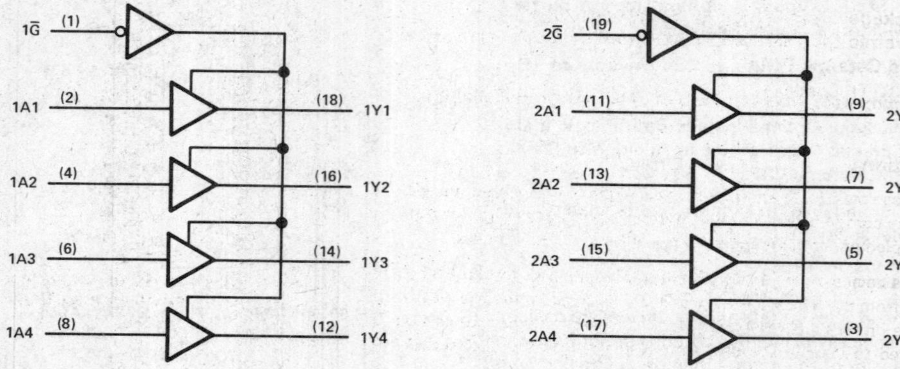

maximum ratings, recommended operating conditions, and electrical characteristics

See Table VII, page 2-14.

switching characteristics over recommended operating free-air temperature range (unless otherwise noted), C_L = 50 pF (see Note 1)

PARAMETER	FROM (INPUT)	TO (OUTPUT)	V_{CC}	T_A = 25°C MIN	TYP	MAX	SN54HCT244 MIN	MAX	SN74HCT244 MIN	MAX	UNIT
t_{pd}	A	Y	4.5 V		15	28		42		35	ns
			5.5 V		13	25		38		32	
t_{en}	G̅	Y	4.5 V		21	35		53		44	ns
			5.5 V		19	32		48		40	
t_{dis}	G̅	Y	4.5 V		19	35		53		44	ns
			5.5 V		18	32		48		40	
t_t		Y	4.5 V		8	12		18		15	ns
			5.5 V		7	11		16		14	

C_{pd}	Power dissipation capacitance per buffer	No load, T_A = 25°C	40 pF typ

switching characteristics over recommended operating free-air temperature range (unless otherwise noted), C_L = 150 pF (see Note 1)

PARAMETER	FROM (INPUT)	TO (OUTPUT)	V_{CC}	T_A = 25°C MIN	TYP	MAX	SN54HCT244 MIN	MAX	SN74HCT244 MIN	MAX	UNIT
t_{pd}	A	Y	4.5 V		21	45		68		56	ns
			5.5 V		18	40		61		51	
t_{en}	G̅	Y	4.5 V		25	52		79		65	ns
			5.5 V		22	47		71		59	
t_t		Y	4.5 V		17	42		63		53	ns
			5.5 V		14	38		57		48	

NOTE 1: For load circuit and voltage waveforms, see page 1-14.

TEXAS INSTRUMENTS
POST OFFICE BOX 225012 • DALLAS, TEXAS 75265

TYPES SN54HC245, SN74HC245 OCTAL BUS TRANSCEIVERS WITH 3-STATE OUTPUTS

D2684, DECEMBER 1982—REVISED MARCH 1984

- **High-Current 3-State Outputs Drive Bus Lines Directly or Up to 15 LSTTL Loads**
- **Package Options Include Both Plastic and Ceramic Chip Carriers in Addition to Plastic and Ceramic DIPs**
- **Dependable Texas Instruments Quality and Reliability**

SN54HC245 . . . J PACKAGE
SN74HC245 . . . J OR N PACKAGE
(TOP VIEW)

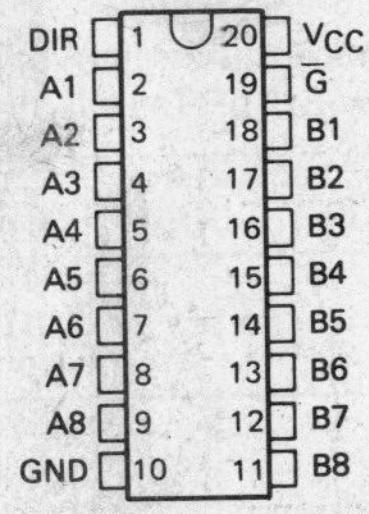

description

These octal bus transceivers are designed for synchronous two-way communication between data buses. The control function implementation minimizes external timing requirements.

The devices allow data transmission from the A bus to the B bus or from the B bus to the A bus depending upon the logic level at the direction control (DIR) input. The enable input ($\overline{G}$) can be used to disable the device so that the buses are effectively isolated.

The SN54HC245 is characterized for operation over the full military temperature range of −55°C to 125°C. The SN74HC245 is characterized for operation from −40°C to 85°C.

SN54HC245 . . . FH OR FK PACKAGE
SN74HC245 . . . FH OR FN PACKAGE
(TOP VIEW)

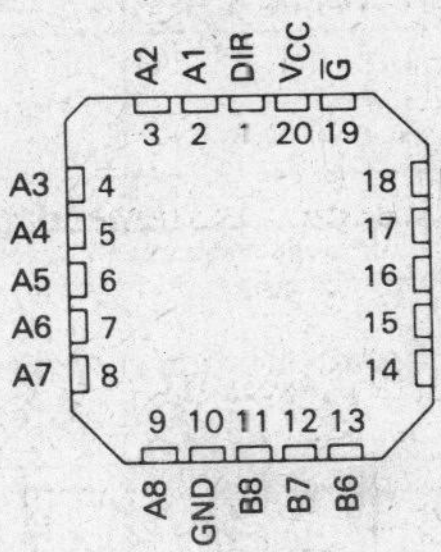

logic diagram (positive logic)

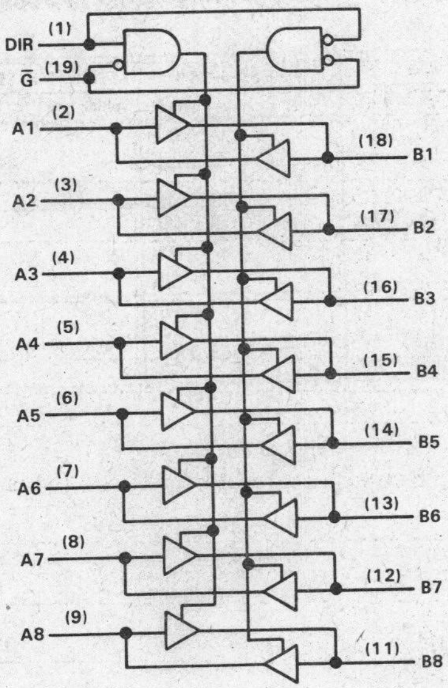

FUNCTION TABLE

CONTROL INPUTS		OPERATION
$\overline{G}$	DIR	
L	L	B data to A bus
L	H	A data to B bus
H	X	Isolation

logic symbol

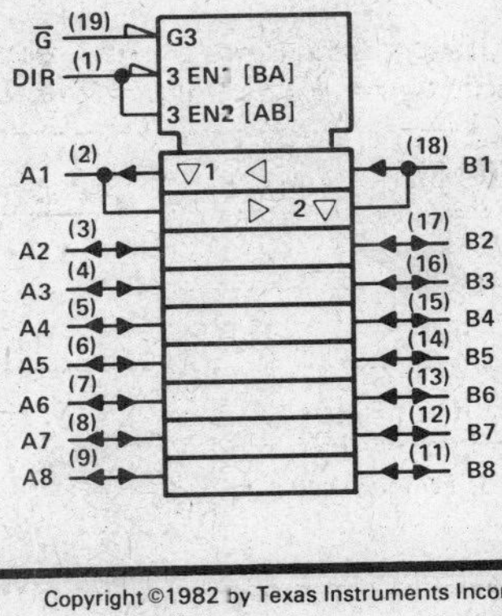

maximum ratings, recommended operating conditions, and electrical characteristics

See Table III, page 2-8.

switching characteristics over recommended operating free-air temperature range (unless otherwise noted), C_L = 50 pF (see Note 1)

PARAMETER	FROM (INPUT)	TO (OUTPUT)	V_{CC}	T_A = 25°C			SN54HC245		SN74HC245		UNIT
				MIN	TYP	MAX	MIN	MAX	MIN	MAX	
t_{pd}	A or B	B or A	2 V		40	105		160		130	ns
			4.5 V		15	21		32		26	
			6 V		12	18		27		22	
t_{en}	$\overline{G}$	A or B	2 V		125	230		340		290	ns
			4.5 V		23	46		68		58	
			6 V		20	39		58		49	
t_{dis}	$\overline{G}$	A or B	2 V		74	200		300		250	ns
			4.5 V		25	40		60		50	
			6 V		21	34		51		43	
t_t		A or B	2 V		20	60		90		75	ns
			4.5 V		8	12		18		15	
			6 V		6	10		15		13	

C_{pd}	Power dissipation capacitance per transceiver	No load, T_A = 25°C	40 pF typ

switching characteristics over recommended operating free-air temperature range (unless otherwise noted), C_L = 150 pF (see Note 1)

PARAMETER	FROM (INPUT)	TO (OUTPUT)	V_{CC}	T_A = 25°C			SN54HC245		SN74HC245		UNIT
				MIN	TYP	MAX	MIN	MAX	MIN	MAX	
t_{pd}	A or B	B or A	2 V		54	135		200		170	ns
			4.5 V		18	27		40		34	
			6 V		15	23		34		29	
t_{en}	$\overline{G}$	A or B	2 V		150	270		405		335	ns
			4.5 V		31	54		81		67	
			6 V		25	46		69		56	
t_t		A or B	2 V		45	210		315		265	ns
			4.5 V		17	42		63		53	
			6 V		13	36		53		45	

NOTE 1: For load circuit and voltage waveforms, see page 1-14.

3

HCMOS DEVICES

TYPES SN54HC251, SN74HC251 DATA SELECTORS/MULTIPLEXERS WITH 3-STATE OUTPUTS

D2684, DECEMBER 1982—REVISED MARCH 1984

- **3-State Version of 'HC151**
- **High-Current 3-State Outputs Interface Directly with System Bus or Can Drive up to 15 LSTTL Loads**
- **Performs Parallel-to-Serial Conversion**
- **Complementary Outputs Provide True and Inverted Data**
- **Package Options Include Both Plastic and Ceramic Chip Carriers in Addition to Plastic and Ceramic DIPs**
- **Dependable Texas Instruments Quality and Reliability**

SN54HC251 . . . J PACKAGE
SN74HC251 . . . J OR N PACKAGE
(TOP VIEW)

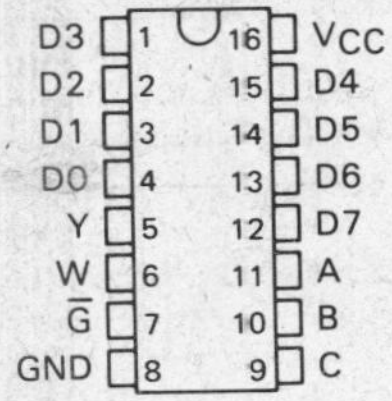

SN54HC251 . . . FH OR FK PACKAGE
SN74HC251 . . . FH OR FN PACKAGE
(TOP VIEW)

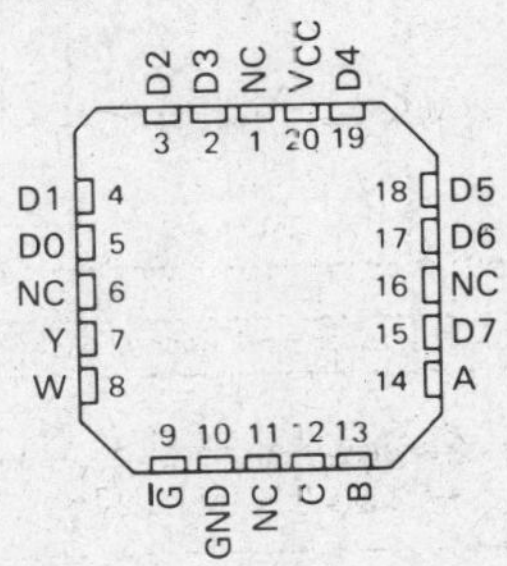

NC—No internal connection

description

These data selectors/multiplexers contain full binary decoding to select one-of-eight data sources and feature strobe-controlled complementary three-state outputs.

The three-state outputs can interface with and drive data lines of bus-organized systems. With all but one of the common outputs disabled (at a high-impedance state), the low-impedance of the single enabled output will drive the bus line to a high or low logic level. Both outputs are controlled by the strobe ($\overline{G}$). The outputs are disabled when $\overline{G}$ is high.

The SN54HC251 is characterized for operation over the full military temperature range of −55°C to 125°C. The SN74HC251 is characterized for operation from −40°C to 85°C.

FUNCTION TABLE

INPUTS				OUTPUTS	
SELECT			STROBE	Y	W
C	B	A	$\overline{G}$		
X	X	X	H	Z	Z
L	L	L	L	D0	$\overline{D0}$
L	L	H	L	D1	$\overline{D1}$
L	H	L	L	D2	$\overline{D2}$
L	H	H	L	D3	$\overline{D3}$
H	L	L	L	D4	$\overline{D4}$
H	L	H	L	D5	$\overline{D5}$
H	H	L	L	D6	$\overline{D6}$
H	H	H	L	D7	$\overline{D7}$

D0, D1 . . . D7 = the level of the respective D input

logic symbol

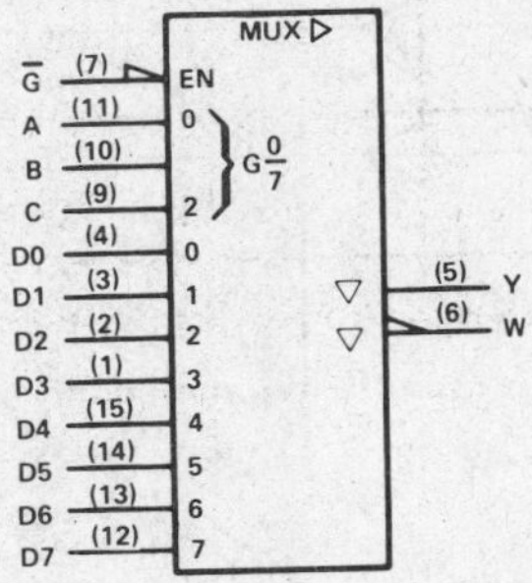

Pin numbers shown are for J and N package.

logic diagram (positive logic)

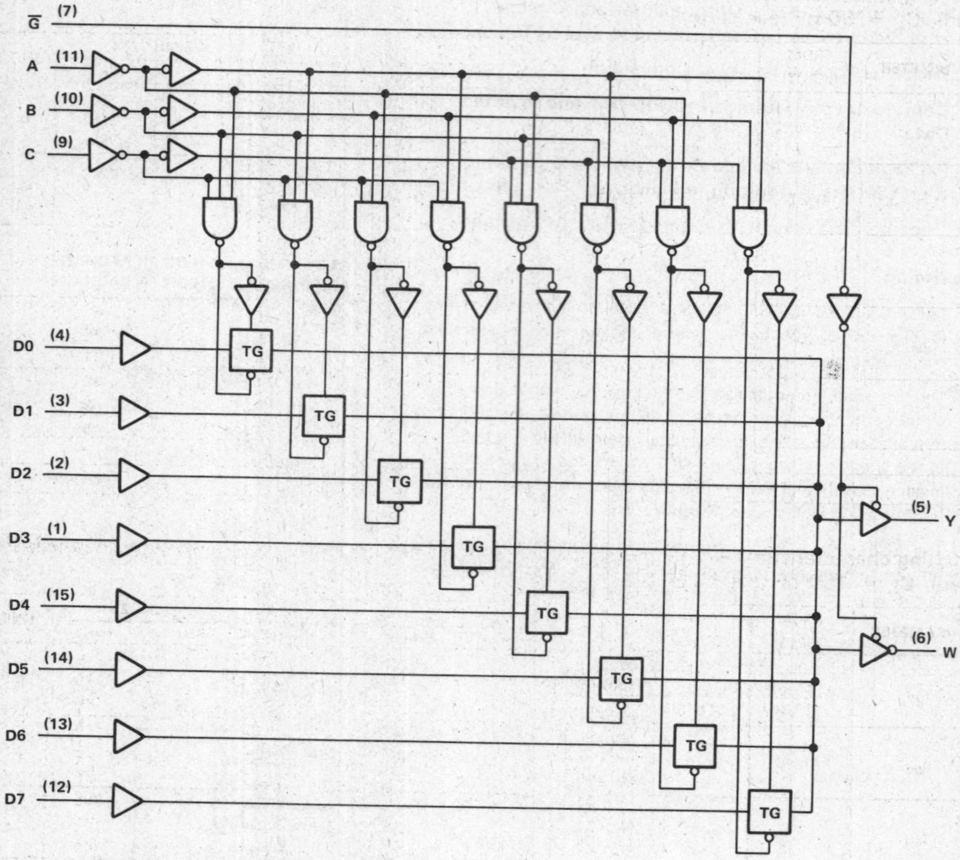

Pin numbers shown are for J and N packages.

maximum ratings, recommended operating conditions, and electrical characteristics

See Table III, page 2-8.

switching characteristics over recommended operating free-air temperature range (unless otherwise noted), C_L = 50 pF (see Note 1)

PARAMETER	FROM (INPUT)	TO (OUTPUT)	V_{CC}	T_A = 25°C MIN	T_A = 25°C TYP	T_A = 25°C MAX	SN54HC251 MIN	SN54HC251 MAX	SN74HC251 MIN	SN74HC251 MAX	UNIT
t_{pd}	A, B, or C	W or Y	2 V		58	205		300		256	ns
			4.5 V		21	41		60		51	
			6 V		19	35		51		44	
t_{pd}	Any D	W or Y	2 V		44	195		283		244	ns
			4.5 V		17	39		57		49	
			6 V		15	33		48		41	
t_{en}	$\overline{G}$	W or Y	2 V		30	145		210		181	ns
			4.5 V		10	29		42		36	
			6 V		9	25		36		31	
t_{dis}	$\overline{G}$	W or Y	2 V		25	195		283		244	ns
			4.5 V		15	39		57		49	
			6 V		14	33		48		41	
t_t			2 V		20	75		110		95	ns
			4.5 V		8	15		22		19	
			6 V		6	13		19		16	

C_{pd}	Power dissipation capacitance	No load, T_A = 25°C	70 pF typ

switching characteristics over recommended operating free-air temperature range (unless otherwise noted), C_L = 150 pF (see Note 1)

PARAMETER	FROM (INPUT)	TO (OUTPUT)	V_{CC}	T_A = 25°C MIN	T_A = 25°C TYP	T_A = 25°C MAX	SN54HC251 MIN	SN54HC251 MAX	SN74HC251 MIN	SN74HC251 MAX	UNIT
t_{pd}	A, B, or C	W or Y	2 V		72	300		450		375	ns
			4.5 V		25	60		90		75	
			6 V		22	52		77		65	
t_{pd}	Any D	W or Y	2 V		59	300		450		375	ns
			4.5 V		21	60		90		75	
			6 V		18	52		77		65	
t_{en}	$\overline{G}$	W or Y	2 V		50	230		340		335	ns
			4.5 V		17	46		68		57	
			6 V		15	40		58		50	
t_t			2 V		45	210		315		265	ns
			4.5 V		17	42		63		53	
			6 V		13	36		53		45	

NOTE 1: For load circuit and voltage waveforms, see page 1-14.

TYPES SN54HC253, SN74HC253 DUAL 4-LINE TO 1-LINE DATA SELECTORS/MULTIPLEXERS WITH 3-STATE OUTPUTS

D2684, DECEMBER 1982—REVISED MARCH 1984

- 3-State Versions of 'HC153
- High-Current Outputs Drive up to 15 LSTTL Loads
- Permits Multiplexing from N Lines to 1 Line
- Performs Parallel-to-Serial Conversion
- Package Options Include Both Plastic and Ceramic Chip Carriers in Addition to Plastic and Ceramic DIPs
- Dependable Texas Instruments Quality and Reliability

SN54HC253 . . . J PACKAGE
SN74HC253 . . . J OR N PACKAGE
(TOP VIEW)

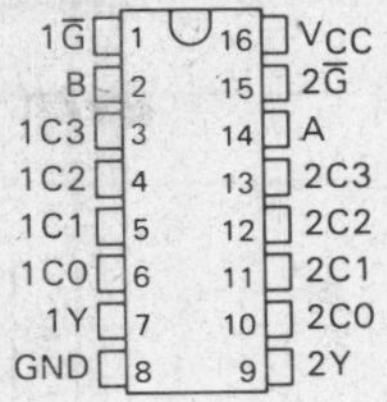

SN54HC253 . . . FH OR FK PACKAGE
SN74HC253 . . . FH OR FN PACKAGE
(TOP VIEW)

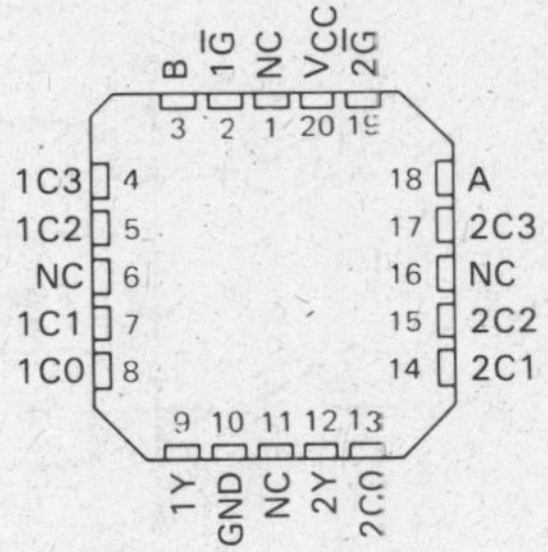

NC—No internal connection

description

Each of these data selectors/multiplexers contains inverters and drivers to supply full binary decoding data selection to the AND-OR gates. Separate output control inputs are provided for each of the two four-line sections.

The three-state outputs can interface with and drive data lines of bus-organized systems. With all but one of the common outputs disabled (at a high-impedance state) the low-impedance of the single enabled output will drive the bus line to a high or low logic level. Each output has its own strobe (G̅). The output is disabled when its strobe is high.

The SN54HC253 is characterized for operation over the full military temperature range of −55°C to 125°C. The SN74HC253 is characterized for operation from −40°C to 85°C.

logic symbol

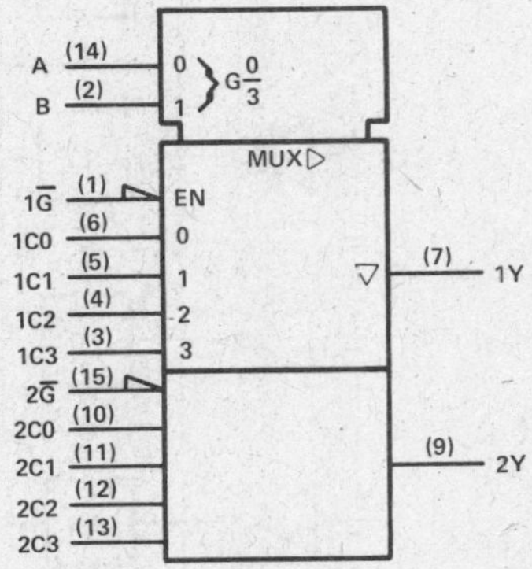

Pin numbers shown are for J and N packages

FUNCTION TABLE

SELECT INPUTS		DATA INPUTS				OUTPUT CONTROL	OUTPUT
B	A	C0	C1	C2	C3	G̅	Y
X	X	X	X	X	X	H	Z
L	L	L	X	X	X	L	L
L	L	H	X	X	X	L	H
L	H	X	L	X	X	L	L
L	H	X	H	X	X	L	H
H	L	X	X	L	X	L	L
H	L	X	X	H	X	L	H
H	H	X	X	X	L	L	L
H	H	X	X	X	H	L	H

Address inputs A and B are common to both sections.

logic diagram (positive logic)

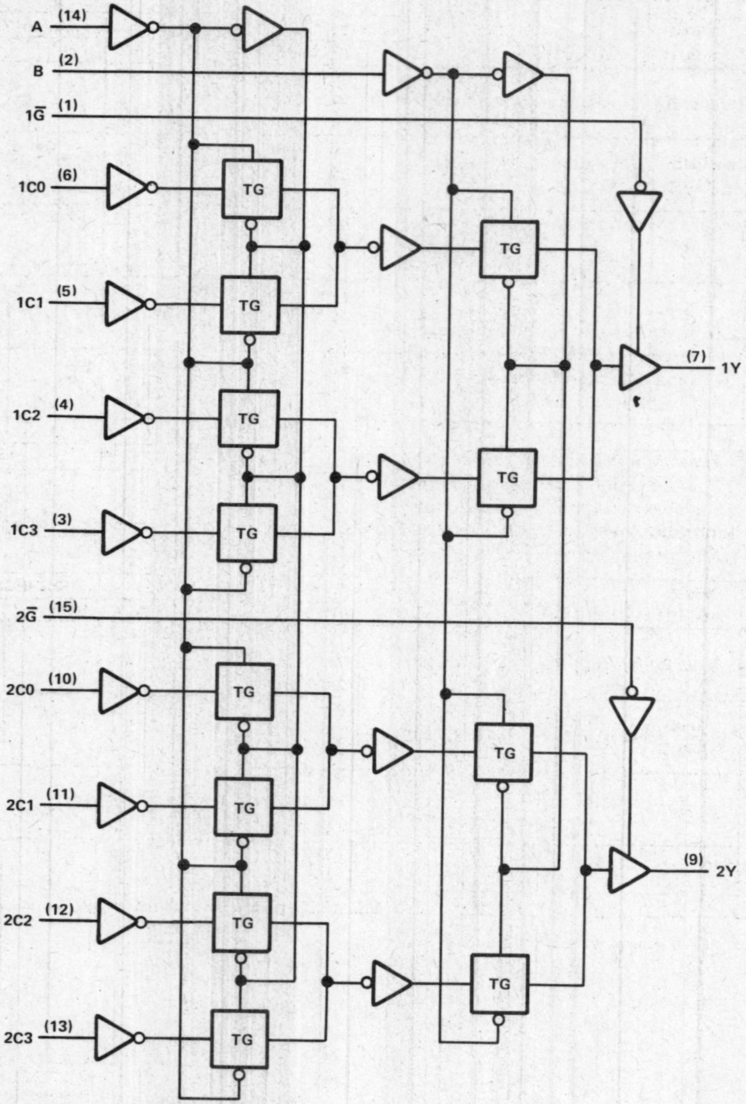

Pin numbers shown are for J and N packages

maximum ratings, recommended operating conditions, and electrical characteristics

See Table III, page 2-8.

switching characteristics over recommended operating free-air temperature range (unless otherwise noted), C_L = 50 pF (see Note 1)

PARAMETER	FROM (INPUT)	TO (OUTPUT)	V_{CC}	T_A = 25°C MIN	TYP	MAX	SN54HC253 MIN	MAX	SN74HC253 MIN	MAX	UNIT
t_{pd}	A or B	Any Y	2 V		62	150		225		190	ns
			4.5 V		19	30		45		38	
			6 V		16	26		38		32	
t_{pd}	Data (Any C)	Y	2 V		54	126		210		175	ns
			4.5 V		16	28		42		35	
			6 V		13	23		36		30	
t_{en}	$\overline{G}$	Y	2 V		28	100		150		125	ns
			4.5 V		11	20		30		25	
			6 V		9	17		26		21	
t_{dis}	$\overline{G}$	Y	2 V		21	135		203		170	ns
			4.5 V		14	30		45		38	
			6 V		12	25		38		31	
t_t		Y	2 V		28	60		90		75	ns
			4.5 V		8	12		18		15	
			6 V		6	10		15		13	

C_{pd}	Power dissipation capacitance per multiplexer	No load, T_A = 25°C	45 pF typ

switching characteristics over recommended operating free-air temperature range (unless otherwise noted), C_L = 150 pF (see Note 1)

PARAMETER	FROM (INPUT)	TO (OUTPUT)	V_{CC}	T_A = 25°C MIN	TYP	MAX	SN54HC253 MIN	MAX	SN74HC253 MIN	MAX	UNIT
t_{pd}	A or B	Any Y	2 V		76	235		355		295	ns
			4.5 V		23	47		71		59	
			6 V		20	41		60		51	
t_{pd}	Data (Any C)	Y	2 V		68	220		335		275	ns
			4.5 V		20	44		67		55	
			6 V		17	38		57		51	
t_{en}	$\overline{G}$	Y	2 V		44	185		280		230	ns
			4.5 V		16	37		56		46	
			6 V		14	32		48		40	
t_t		Y	2 V		45	210		315		265	ns
			4.5 V		17	42		63		53	
			6 V		13	36		53		45	

NOTE 1: For load circuit and voltage waveforms, see page 1-14.

TYPES SN54HC257, SN54HC258, SN74HC257, SN74HC258 QUAD 2-LINE TO 1-LINE DATA SELECTORS/MULTIPLEXERS WITH 3-STATE OUTPUTS

D2684, DECEMBER 1982—REVISED MARCH 1984

- High-Current 3-State Outputs Interface Directly with System Bus or Can Drive up to 15 LSTTL Loads
- Provides Bus Interface from Multiple Sources in High Performance Systems
- Package Options Include Both Plastic and Ceramic Chip Carriers in Addition to Plastic and Ceramic DIPs
- Dependable Texas Instruments Quality and Reliability

SN54HC257, SN54HC258 . . . J PACKAGE
SN74HC257, SN74HC258 . . . J OR N PACKAGE
(TOP VIEW)

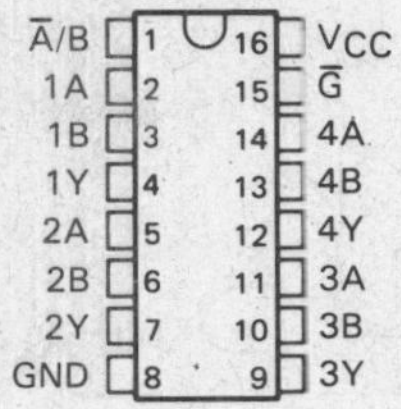

SN54HC257, SN54HC258 . . . FH OR FK PACKAGE
SN74HC257, SN74HC258 . . . FH OR FN PACKAGE
(TOP VIEW)

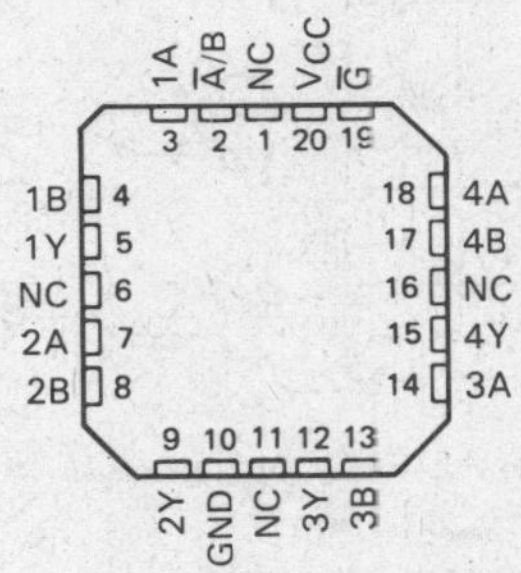

NC—No internal connection

description

These devices are designed to multiplex signals from four-bit data sources to four-output data lines in bus-organized systems. The 3-state outputs will not load the data lines when the output control pin ($\overline{G}$) is at a high-logic level.

The SN54HC257 and SN54HC258 are characterized for operation over the full military temperature range of −55°C to 125°C. The SN74HC257 and SN74HC258 are characterized for operation from −40°C to 85°C.

FUNCTION TABLE

INPUTS				OUTPUT Y	
OUTPUT CONTROL $\overline{G}$	SELECT $\overline{A}$/B	DATA A	DATA B	'HC257	'HC258
H	X	X	X	Z	Z
L	L	L	X	L	H
L	L	H	X	H	L
L	H	X	L	L	H
L	H	X	H	H	L

logic symbols

'HC257

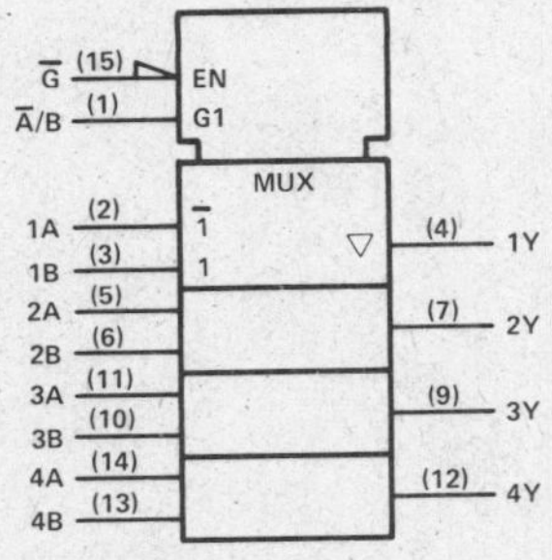

'HC258

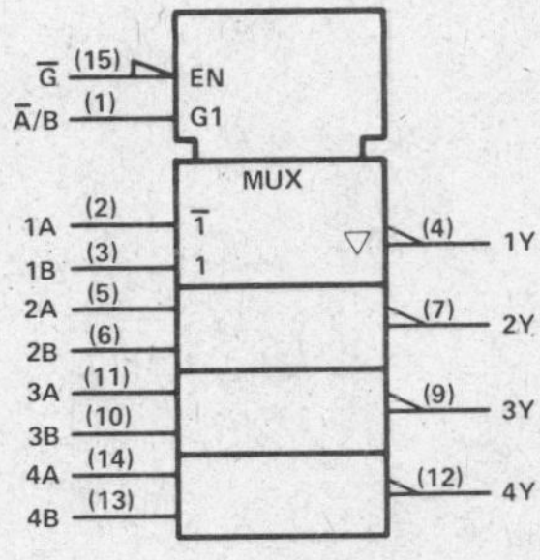

Pin numbers shown are for J and N packages.

logic diagrams (positive logic)

'HC257

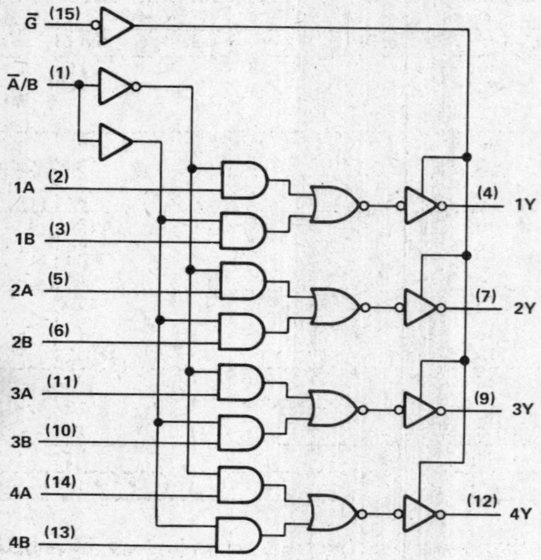

'HC258

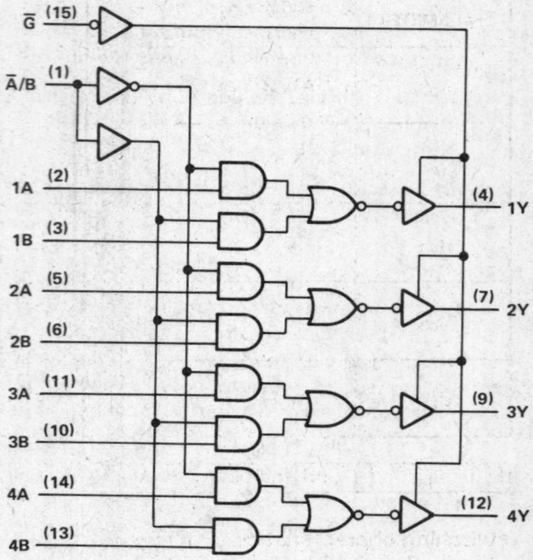

Pin numbers shown are for J and N packages.

maximum ratings, recommended operating conditions, and electrical characteristics

See Table III, page 2-8.

switching characteristics over recommended operating free-air temperature range (unless otherwise noted), C_L = 50 pF (see Note 1)

PARAMETER	FROM (INPUT)	TO (OUTPUT)	V_{CC}	T_A = 25°C MIN	TYP	MAX	SN54HC257 MIN	MAX	SN74HC257 MIN	MAX	UNIT
t_{pd}	A or B	Any Y	2 V		50	100		150		125	ns
			4.5 V		10	20		30		25	
			6 V		9	17		25		21	
t_{pd}	$\overline{A}$/B	Any Y	2 V		50	100		150		125	ns
			4.5 V		10	20		30		25	
			6 V		9	17		25		21	
t_{en}	$\overline{G}$	Any Y	2 V		75	150		225		190	ns
			4.5 V		15	30		45		38	
			6 V		13	26		38		32	
t_{dis}	$\overline{G}$	Any Y	2 V		75	150		225		190	ns
			4.5 V		15	30		45		38	
			6 V		13	26		38		32	
t_t		Any	2 V		28	60		90		75	ns
			4.5 V		8	12		18		15	
			6 V		6	10		15		13	

C_{pd}	Power dissipation capacitance per multiplexer	No load, T_A = 25°C	40 pF typ

switching characteristics over recommended operating free-air temperature range (unless otherwise noted), C_L = 150 pF (see Note 1)

PARAMETER	FROM (INPUT)	TO (OUTPUT)	V_{CC}	T_A = 25°C MIN	TYP	MAX	SN54HC257 MIN	MAX	SN74HC257 MIN	MAX	UNIT
t_{pd}	A or B	Any Y	2 V		75	150		245		190	ns
			4.5 V		15	30		45		38	
			6 V		13	26		38		32	
t_{pd}	$\overline{A}$/B	Any Y	2 V		75	150		245		190	ns
			4.5 V		15	30		45		38	
			6 V		13	26		38		32	
t_{en}	$\overline{G}$	Any Y	2 V		100	200		250		300	ns
			4.5 V		24	40		60		50	
			6 V		18	34		51		43	
t_t		Any	2 V		45	210		315		265	ns
			4.5 V		17	42		63		53	
			6 V		13	36		53		45	

NOTE 1: For load circuits and voltage waveforms, see page 1-14.

switching characteristics over recommended operating free-air temperature range (unless otherwise noted), C_L = 50 pF (see Note 1)

PARAMETER	FROM (INPUT)	TO (OUTPUT)	V_{CC}	T_A = 25°C MIN	TYP	MAX	SN54HC258 MIN	MAX	SN74HC258 MIN	MAX	UNIT
t_{pd}	A or B	Any Y	2 V		60	100		150		125	ns
			4.5 V		13	20		30		25	
			6 V		12	17		25		21	
t_{pd}	$\overline{A}$/B	Any Y	2 V		60	115		175		145	ns
			4.5 V		13	23		35		29	
			6 V		12	20		30		25	
t_{en}	$\overline{G}$	Any Y	2 V		70	150		225		190	ns
			4.5 V		15	30		45		38	
			6 V		13	26		38		32	
t_{dis}	$\overline{G}$	Any Y	2 V		75	150		225		190	ns
			4.5 V		15	30		45		38	
			6 V		13	26		38		32	
t_t		Any	2 V		28	60		90		75	ns
			4.5 V		8	12		18		15	
			6 V		6	10		15		13	

C_{pd}	Power dissipation capacitance per multiplexer	No load, T_A = 25°C	40 pF typ

switching characteristics over recommended operating free-air temperature range (unless otherwise noted), C_L = 150 pF (see Note 1)

PARAMETER	FROM (INPUT)	TO (OUTPUT)	V_{CC}	T_A = 25°C MIN	TYP	MAX	SN54HC258 MIN	MAX	SN74HC258 MIN	MAX	UNIT
t_{pd}	A or B	Any Y	2 V		95	150		245		190	ns
			4.5 V		23	30		45		38	
			6 V		21	26		38		32	
t_{pd}	$\overline{A}$/B	Any Y	2 V		95	165		240		210	ns
			4.5 V		23	33		48		42	
			6 V		21	28		41		36	
t_{en}	$\overline{G}$	Any Y	2 V		100	200		300		250	ns
			4.5 V		24	40		60		50	
			6 V		18	34		51		43	
t_t		Any	2 V		45	210		315		265	ns
			4.5 V		17	42		63		53	
			6 V		13	36		53		45	

NOTE 1: For load circuits and voltage waveforms, see page 1-14.

TYPES SN54HC259, SN74HC259
8-BIT ADDRESSABLE LATCHES

D2684, DECEMBER 1982—REVISED MARCH 1984

- **8-Bit Parallel-Out Storage Register Performs Serial-to-Parallel Conversion with Storage**
- **Asynchronous Parallel Clear**
- **Active-High Decoder**
- **Enable Input Simplifies Expansion**
- **Expandable for N-Bit Applications**
- **Four Distinct Functional Modes**
- **Package Options Include Both Plastic and Ceramic Chip Carriers in Addition to Plastic and Ceramic DIPs**
- **Dependable Texas Instruments Quality and Reliability**

SN54HC259 . . . J PACKAGE
SN74HC259 . . . J OR N PACKAGE
(TOP VIEW)

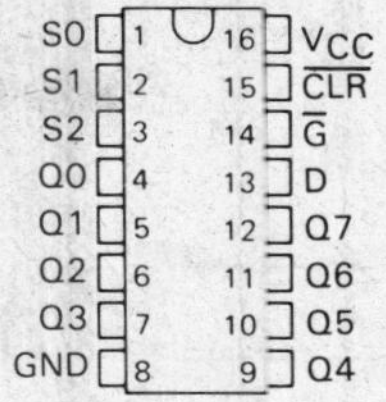

SN54HC259 . . . FH OR FK PACKAGE
SN74HC259 . . . FH OR FN PACKAGE
(TOP VIEW)

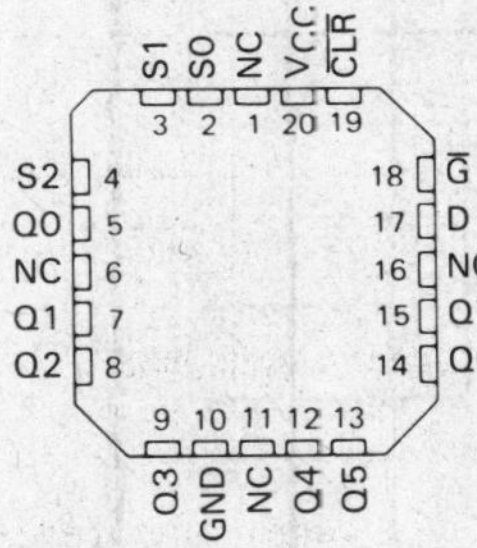

NC—No internal connection

description

These 8-bit addressable latches are designed for general purpose storage applications in digital systems. Specific uses include working registers, serial-holding registers, and active-high decoders or demultiplexers. They are multifunctional devices capable of storing single-line data in eight addressable latches, and being a 1-of-8 decoder or demultiplexer with active-high outputs.

Four distinct modes of operation are selectable by controlling the clear ($\overline{\mathrm{CLR}}$) and enable ($\overline{\mathrm{G}}$) inputs as enumerated in the function table. In the addressable-latch mode, data at the data-in terminal is written into the addressed latch. The addressed latch will follow the data input with all unaddressed latches remaining in their previous states. In the memory mode, all latches remain in their previous states and are unaffected by the data or address inputs. To eliminate the possibility of entering erroneous data in the latches, enable $\overline{\mathrm{G}}$ should be held high (inactive) while the address lines are changing. In the 1-of-8 decoding or demultiplexing mode, the addressed output will follow the level of the D input with all other outputs low. In the clear mode, all outputs are low and unaffected by the address and data inputs.

The SN54HC259 is characterized for operation over the full military temperature range of −55°C to 125°C. The SN74HC259 is characterized for operation from −40°C to 85°C.

FUNCTION TABLE

INPUTS		OUTPUT OF ADDRESSED LATCH	EACH OTHER OUTPUT	FUNCTION
$\overline{\mathrm{CLR}}$	$\overline{\mathrm{G}}$			
H	L	D	Q_{iO}	Addressable Latch
H	H	Q_{iO}	Q_{iO}	Memory
L	L	D	L	8-Line Demultiplexer
L	H	L	L	Clear

LATCH SELECTION TABLE

SELECT INPUTS			LATCH ADDRESSED
S2	S1	S0	
L	L	L	0
L	L	H	1
L	H	L	2
L	H	H	3
H	L	L	4
H	L	H	5
H	H	L	6
H	H	H	7

logic symbol

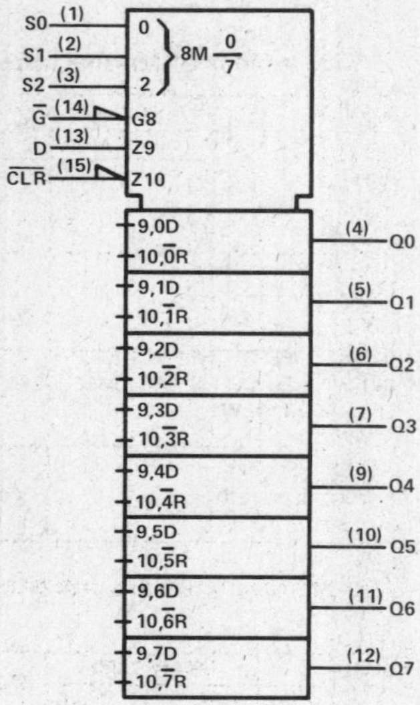

Pin numbers shown are for J and N packages.

logic diagram (positive logic)

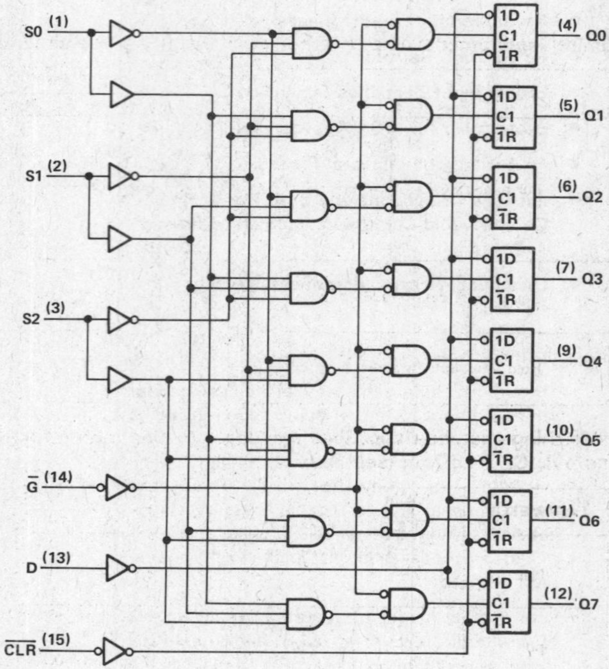

logic symbol and logic diagram, each internal latch (positive logic)

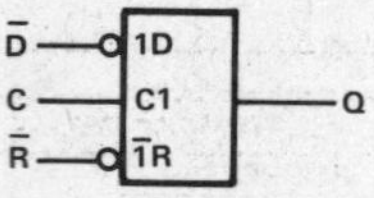

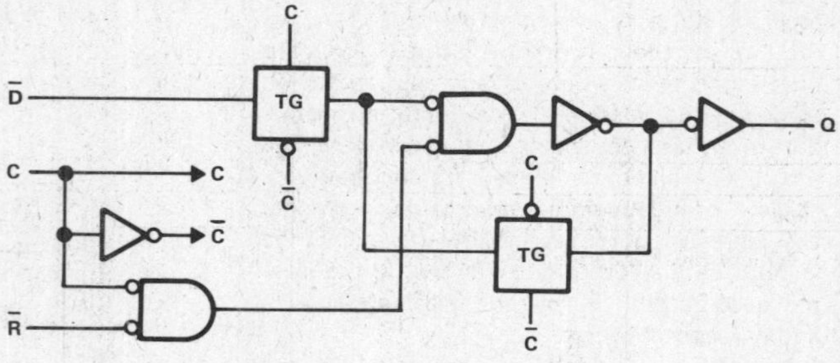

TEXAS INSTRUMENTS
POST OFFICE BOX 225012 • DALLAS, TEXAS 75265

maximum ratings, recommended operating conditions, and electrical characteristics

See Table IV, page 2-10.

timing requirements over recommended operating free-air temperature range (unless otherwise noted)

			V_{CC}	$T_A = 25°C$ MIN	$T_A = 25°C$ MAX	SN54HC259 MIN	SN54HC259 MAX	SN74HC259 MIN	SN74HC259 MAX	UNIT
t_w	Pulse duration	$\overline{CLR}$ low	2 V	80		120		100		ns
			4.5 V	16		24		20		
			6 V	14		20		17		
		$\overline{G}$ low	2 V	80		120		100		
			4.5 V	16		24		20		
			6 V	14		20		17		
t_{su}	Setup time, data or address before $\overline{G}$↑		2 V	75		115		95		ns
			4.5 V	15		23		19		
			6 V	13		20		16		
t_h	Hold time, data or address after $\overline{G}$↑		2 V	5		5		5		ns
			4.5 V	5		5		5		
			6 V	5		5		5		

switching characteristics over recommended operating free-air temperature range (unless otherwise noted), C_L = 50 pF (see Note 1)

PARAMETER	FROM (INPUT)	TO (OUTPUT)	V_{CC}	$T_A = 25°C$ MIN	$T_A = 25°C$ TYP	$T_A = 25°C$ MAX	SN54HC259 MIN	SN54HC259 MAX	SN74HC259 MIN	SN74HC259 MAX	UNIT
t_{PHL}	$\overline{CLR}$	Any Q	2 V		60	150		225		190	ns
			4.5 V		18	30		45		38	
			6 V		14	26		38		32	
t_{pd}	Data	Any Q	2 V		56	130		195		165	ns
			4.5 V		17	26		39		33	
			6 V		13	22		33		28	
t_{pd}	Address	Any Q	2 V		74	200		300		250	ns
			4.5 V		21	40		60		50	
			6 V		17	34		51		43	
t_{pd}	$\overline{G}$	Any Q	2 V		66	170		255		215	ns
			4.5 V		20	34		51		43	
			6 V		16	29		43		37	
t_t		Any	2 V		28	75		110		95	ns
			4.5 V		8	15		22		19	
			6 V		6	13		19		16	

C_{pd}	Power dissipation capacitance per latch	No load, $T_A = 25°C$	33 pF typ

NOTE 1: For load circuit and voltage waveforms, see page 1-14.

3

HCMOS DEVICES

TYPES SN54HC266, SN74HC266 QUADRUPLE 2-INPUT EXCLUSIVE-NOR GATES WITH OPEN-DRAIN OUTPUTS

D2684, DECEMBER 1982—REVISED MARCH 1984

- **Package Options Include Both Plastic and Ceramic Chip Carriers in Addition to Plastic and Ceramic DIPs**
- **Dependable Texas Instruments Quality and Reliability**

SN54HC266 . . . J PACKAGE
SN74HC266 . . . J OR N PACKAGE
(TOP VIEW)

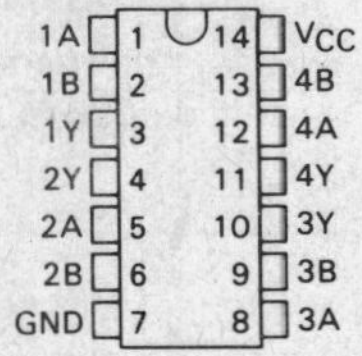

description

These devices are composed of four independent 2-input exclusive-NOR gates and feature open-drain outputs. They perform the Boolean functions: $Y = \overline{A \oplus B} = \bar{A}\bar{B} + AB$ in positive logic.

The SN54HC266 is characterized for operation over the full military temperature range of −55 °C to 125 °C. The SN74HC266 is characterized for operation from −40 °C to 85 °C.

SN54HC266 . . . FH OR FK PACKAGE
SN74HC266 . . . FH OR FN PACKAGE
(TOP VIEW)

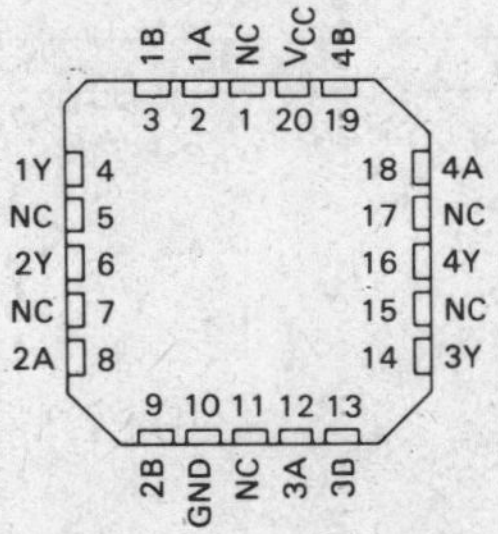

NC—No internal connection

logic symbol

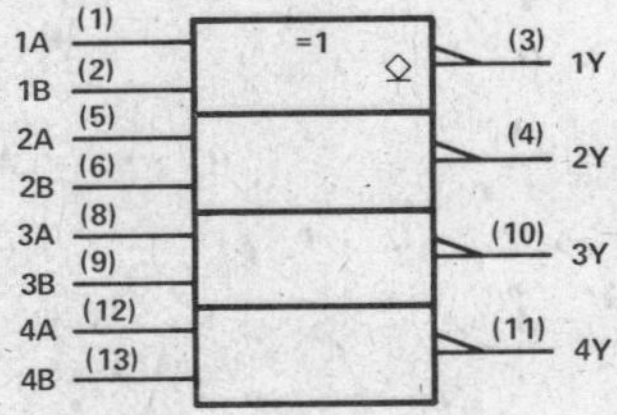

Pin numbers shown are for J and N packages.

FUNCTION TABLE

INPUTS		OUTPUT
A	B	Y
L	L	H
L	H	L
H	L	L
H	H	H

maximum ratings, recommended operating conditions, and electrical characteristics

See Table I, page 2-4.

switching characteristics over recommended operating free-air temperature range (unless otherwise noted), C_L = 50 pF (see Note 1)

PARAMETER	FROM (INPUT)	TO (OUTPUT)	V_{CC}	T_A = 25°C			SN54HC266		SN74HC266		UNIT
				MIN	TYP	MAX	MIN	MAX	MIN	MAX	
t_{PLH}	A or B	Y	2 V		60	125		190		155	ns
			4.5 V		13	25		38		31	
			6 V		10	23		32		26	
t_{PHL}	A or B	Y	2 V		60	100		150		125	ns
			4.5 V		13	20		30		25	
			6 V		10	17		25		21	
t_t		Y	2 V		28	75		110		95	ns
			4.5 V		8	15		22		19	
			6 V		6	13		19		16	

C_{pd}	Power dissipation capacitance per gate	No load, T_A = 25°C	35 pF typ

NOTE 1: For load circuit and voltage waveforms, see page 1-14.

TYPES SN54HC273, SN74HC273
OCTAL D-TYPE FLIP-FLOPS WITH CLEAR

D2684, DECEMBER 1982—REVISED MARCH 1984

- **Contains Eight Flip-Flops with Single-Rail Outputs**
- **Direct Clear Input**
- **Individual Data Input to Each Flip-Flop**
- **Applications Include:**
  - **Buffer/Storage Registers**
  - **Shift Registers**
  - **Pattern Generators**
- **Package Options Include Both Plastic and Ceramic Chip Carriers in Addition to Plastic and Ceramic DIPs**
- **Dependable Texas Instruments Quality and Reliability**

SN54HC273 . . . J PACKAGE
SN74HC273 . . . J OR N PACKAGE
(TOP VIEW)

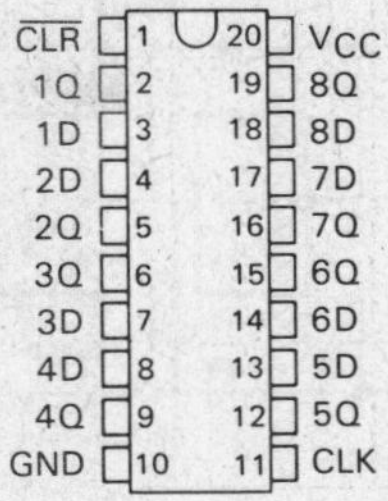

SN54HC273 . . . FH OR FK PACKAGE
SN74HC273 . . . FH OR FN PACKAGE
(TOP VIEW)

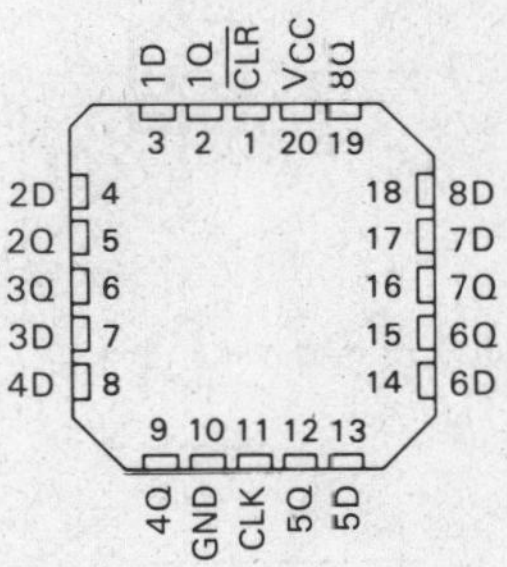

description

These circuits are positive-edge-triggered D-type flip-flops with a direct clear input.

Information at the D inputs meeting the setup time requirements is transferred to the Q outputs on the positive-going edge of the clock pulse. Clock triggering occurs at a particular voltage level and is not directly related to the transition time of the positive-going pulse. When the clock input is at either the high or low level, the D input signal has no effect at the output.

The SN54HC273 is characterized for operation over the full military temperature range of −55°C to 125°C. The SN74HC273 is characterized for operation from −40°C to 85°C.

FUNCTION TABLE
(EACH FLIP-FLOPS)

INPUTS			OUTPUT
CLEAR	CLOCK	D	Q
L	X	X	L
H	↑	H	H
H	↑	L	L
H	L	X	Q_0

logic symbol

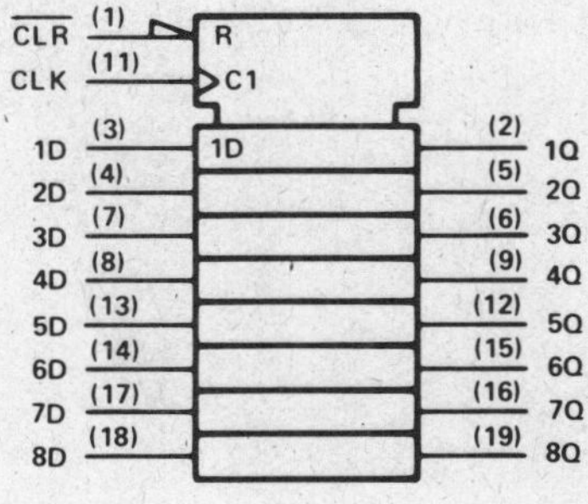

logic diagram, total device (positive logic)

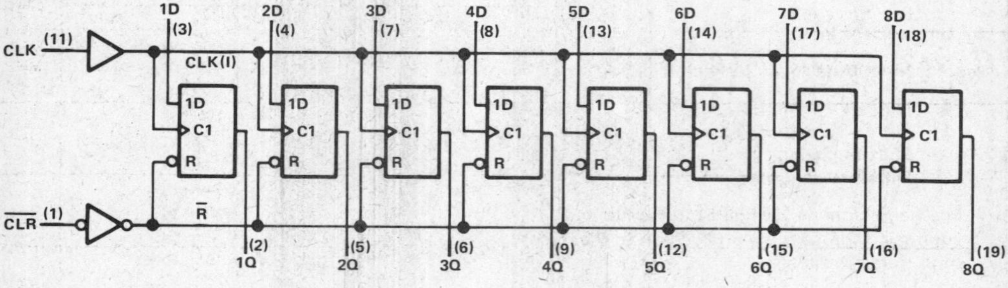

logic diagram each flip-flop (positive logic)

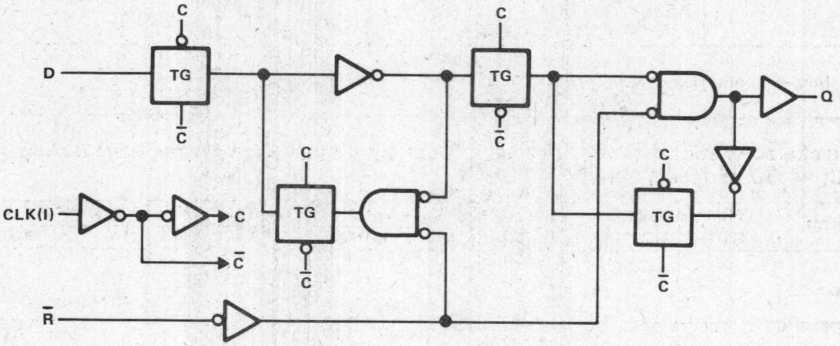

maximum ratings, recommended operating conditions, and electrical characteristics

See Table IV, page 2-10.

TEXAS INSTRUMENTS
POST OFFICE BOX 225012 • DALLAS, TEXAS 75265

timing requirements over recommended operating free-air temperature range (unless otherwise noted)

			V_{CC}	T_A = 25°C MIN	T_A = 25°C MAX	SN54HC273 MIN	SN54HC273 MAX	SN74HC273 MIN	SN74HC273 MAX	UNIT
f_{clock}	Clock frequency		2 V	0	5	0	4	0	4	MHz
			4.5 V	0	27	0	18	0	21	
			6 V	0	32	0	21	0	25	
t_w	Pulse duration	$\overline{CLR}$ low	2 V	80		120		100		ns
			4.5 V	16		24		20		
			6 V	14		20		17		
		CLK high or low	2 V	80		120		100		ns
			4.5 V	16		24		20		
			6 V	14		20		17		
t_{su}	Setup time before CLK↑	Data	2 V	100		150		125		ns
			4.5 V	20		30		25		
			6 V	17		25		21		
		$\overline{CLR}$ inactive	2 V	100		150		125		ns
			4.5 V	20		30		25		
			6 V	17		25		21		
t_h	Hold time, data after CLK↑		2 V	0		0		0		ns
			4.5 V	0		0		0		
			6 V	0		0		0		

switching characteristics over recommended operating free-air temperature range (unless otherwise noted), C_L = 50 pF (see Note 1)

PARAMETER	FROM (INPUT)	TO (OUTPUT)	V_{CC}	T_A = 25°C MIN	T_A = 25°C TYP	T_A = 25°C MAX	SN54HC273 MIN	SN54HC273 MAX	SN74HC273 MIN	SN74HC273 MAX	UNIT
f_{max}			2 V	5	11		4		4		MHz
			4.5 V	27	50		18		21		
			6 V	32	60		28		25		
t_{PHL}	$\overline{CLR}$	Any	2 V		55	160		240		200	ns
			4.5 V		15	32		48		40	
			6 V		12	27		41		34	
t_{pd}	CLK	Any	2 V		56	160		240		200	ns
			4.5 V		15	32		48		40	
			6 V		13	27		41		34	
t_t		Any	2 V		38	75		110		95	ns
			4.5 V		8	15		22		19	
			6 V		6	13		19		16	

C_{pd}	Power dissipation capacitance per flip-flop	No load, T_A = 25°C	35 pF typ

NOTE 1: For load circuit and voltage waveforms, see page 1-14.

3

HCMOS DEVICES

TYPES SN54HC280, SN74HC280
9-BIT ODD/EVEN PARITY GENERATORS/CHECKERS

D2684, DECEMBER 1982—REVISED MARCH 1984

- **Generates Either Odd or Even Parity for Nine Data Lines**
- **Cascadable for n-Bits**
- **Can Be Used to Upgrade Existing Systems Using MSI Parity Circuits**
- **Package Options Include Both Plastic and Ceramic Chip Carriers in Addition to Plastic and Ceramic DIPs**
- **Dependable Texas Instruments Quality and Reliability**

SN54HC280 . . . J PACKAGE
SN74HC280 . . . J OR N PACKAGE
(TOP VIEW)

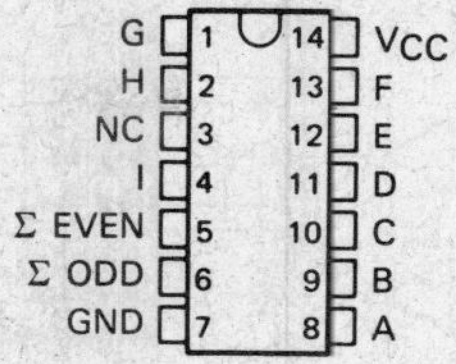

SN54HC280 . . . FH OR FK PACKAGE
SN74HC280 . . . FH OR FN PACKAGE
(TOP VIEW)

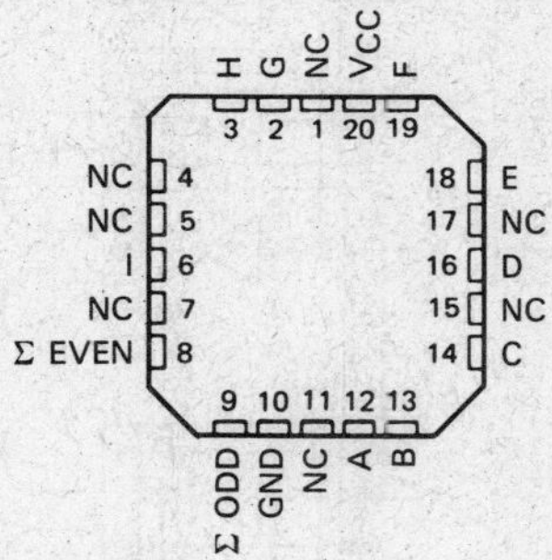

NC—No internal connection

description

These universal, monolithic, nine-bit parity generators/checkers feature odd and even outputs to facilitate operation of either odd or even parity application. The word-length capability is easily expanded by cascading.

The SN54HC280 is characterized for operation over the full military temperature range of −55°C to 125°C. The SN74HC280 is characterized for operation from −40°C to 85°C.

FUNCTION TABLE

NUMBER OF INPUTS A THRU I THAT ARE HIGH	OUTPUTS	
	Σ EVEN	Σ ODD
0, 2, 4, 6, 8	H	L
1, 3, 5, 7, 9	L	H

logic symbol

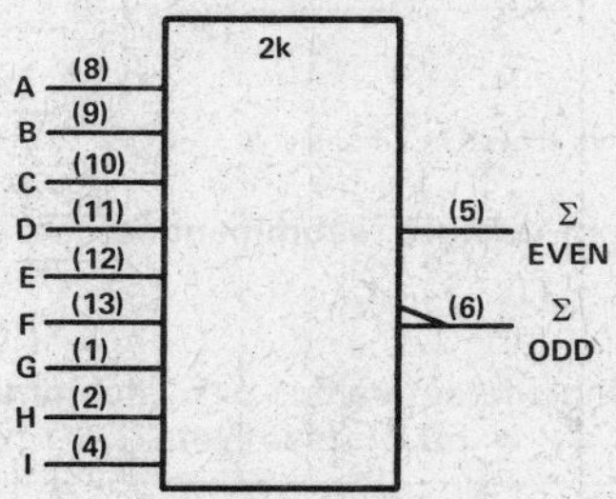

Pin numbers shown are for J and N packages.

logic diagram (positive logic)

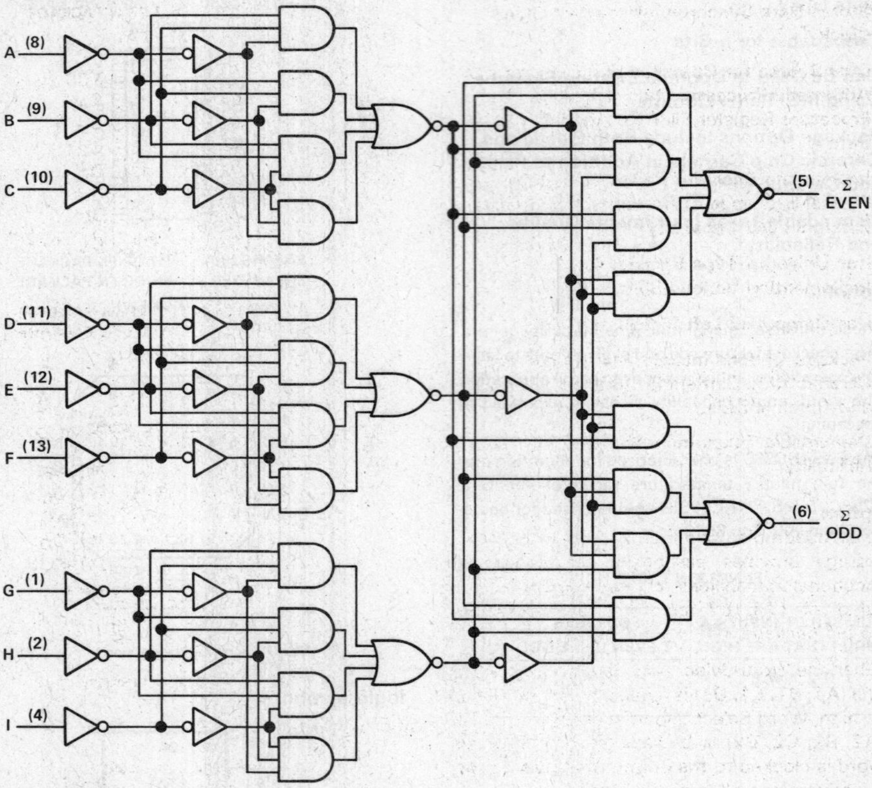

Pin numbers shown are for J and N packages.

maximum ratings, recommended operating conditions, and electrical characteristics

See Table IV, page 2-10.

switching characteristics over recommended operating free-air temperature range (unless otherwise noted), C_L = 50 pF (see Note 1)

PARAMETER	FROM (INPUT)	TO (OUTPUT)	V_{CC}	T_A = 25°C MIN	TYP	MAX	SN54HC280 MIN	MAX	SN74HC280 MIN	MAX	UNIT
t_{pd}	A thru I	Σ Even	2 V		103	205		305		260	ns
		or	4.5 V		21	41		61		52	
		Σ Odd	6 V		17	35		52		44	
t_t		Any	2 V		38	75		110		95	ns
			4.5 V		8	15		22		19	
			6 V		6	13		19		16	

C_{pd}	Power dissipation capacitance	No load, T_A = 25°C	60 pF typ

NOTE 1: For load circuits and voltage waveforms, see page 1-14.

TEXAS INSTRUMENTS
POST OFFICE BOX 225012 • DALLAS, TEXAS 75265

TYPES SN54HC298, SN74HC298
QUADRUPLE 2-INPUT MULTIPLEXER WITH STORAGE

D2804, MARCH 1984

- **Selects One of Two 4-Bit Data Sources and Stores Data Synchronously with System Clock**
- **Dual Source for Operands and Constants in Arithmetic Processor; Can Release Processor Register Files for Acquiring New Data**
- **Implements Separate Registers Capable of Parallel Exchange of Contents, yet Retains External Load Capability**
- **Has Universal-Type Register for Implementing Various Shift Patterns**
- **Has Compound Left-Right Capability**
- **Package Options Include Both Plastic and Ceramic Chip Carriers in Addition to Plastic and Ceramic DIPs**
- **Dependable Texas Instruments Quality and Reliability**

SN54HC298 . . . J PACKAGE
SN74HC298 . . . J OR N PACKAGE
(TOP VIEW)

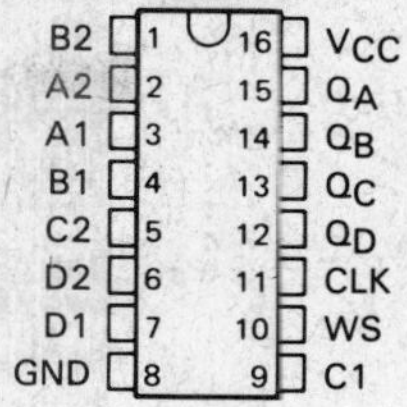

SN54HC298 . . . FH OR FK PACKAGE
SN74HC298 . . . FH OR FN PACKAGE
(TOP VIEW)

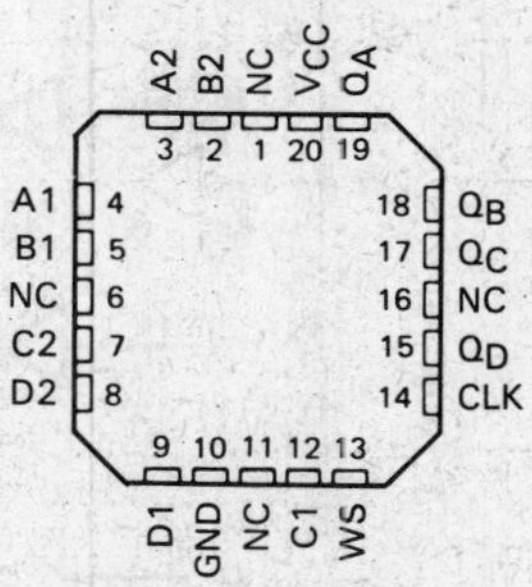

NC—No internal connection

description

This quadruple two-input multiplexer with storage provides essentially the equivalent functional capabilities of two separate MSI functions ('HC157 and 'HC175) in a single 16-pin package.

When the Word-Select (WS) input is low, word one (A1, B1, C1, D1) is applied to the flip-flops. A high Word-Select input causes word two (A2, B2, C2, D2) to be selected. The selected word is clocked to the output terminals on the negative-going edge of the clock pulse.

The SN54HC298 is characterized for operation over the full military temperature range of −55°C to 125°C. The SN74HC298 is characterized for operation from −40°C to 85°C.

logic symbol

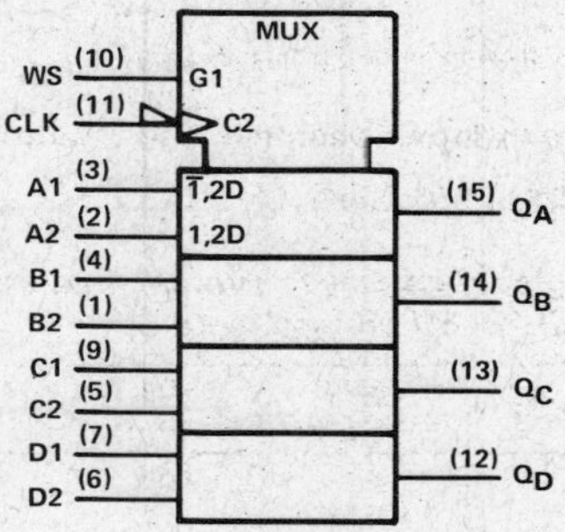

Pin numbers shown are for J and N packages.

logic diagram (positive logic)

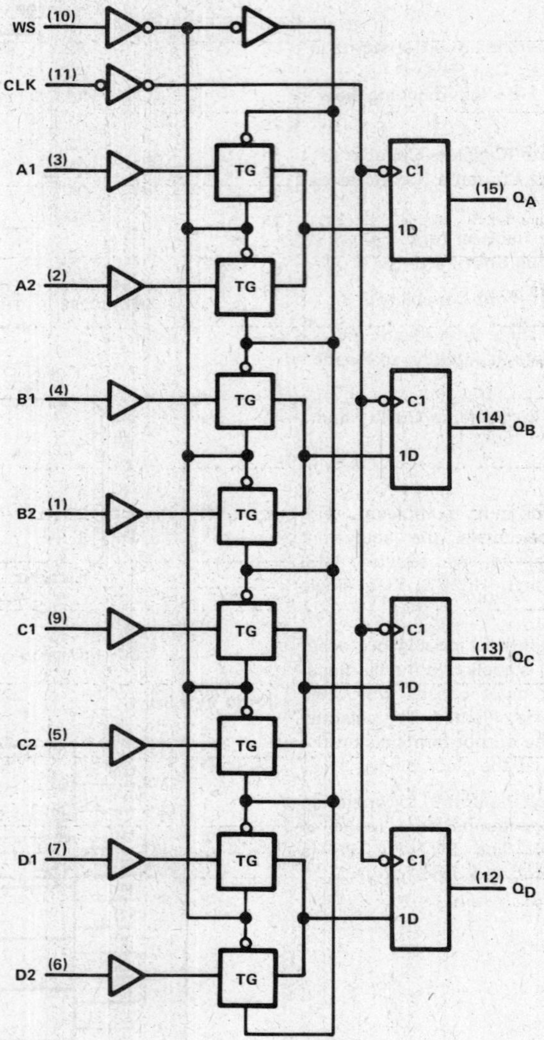

Pin numbers shown are for J and N packages.

absolute maximum ratings, recommended operating conditions, and electrical characteristics

See Table IV, page 2-10.

TEXAS INSTRUMENTS
POST OFFICE BOX 225012 • DALLAS, TEXAS 75265

timing requirements over recommended operating free-air temperature range (unless otherwise noted)

			V_{CC}	$T_A = 25°C$ MIN	$T_A = 25°C$ MAX	SN54HC298 MIN	SN54HC298 MAX	SN74HC298 MIN	SN74HC298 MAX	UNIT
f_{clock}	Clock frequency		2 V		6.5		4.3		5.5	MHz
			4.5 V		33		22		27	
			6 V		38		25		31	
t_w	Pulse duration, CLK high or low		2 V	75		115		95		ns
			4.5 V	15		23		19		
			6 V	13		20		16		
t_{su}	Setup time	Data before CLK↓	2 V	80		125		105		ns
			4.5 V	16		25		21		
			6 V	14		21		18		
		WS before CLK↓	2 V	80		125		105		
			4.5 V	16		25		21		
			6 V	14		21		18		
t_h	Hold time	Data after CLK↓	2 V	0		0		0		ns
			4.5 V	0		0		0		
			6 V	0		0		0		
		WS after CLK↓	2 V	0		0		0		
			4.5 V	0		0		0		
			6 V	0		0		0		

switching characteristics over recommended operating free-air temperature range (unless otherwise noted), C_L = 50 pF (see Note 1)

PARAMETER	FROM (INPUT)	TO (OUTPUT)	V_{CC}	$T_A = 25°C$ MIN	$T_A = 25°C$ TYP	$T_A = 25°C$ MAX	SN54HC298 MIN	SN54HC298 MAX	SN74HC298 MIN	SN74HC298 MAX	UNIT
f_{max}			2 V	6.5			4.3		5.5		MHz
			4.5 V	33			22		27		
			6 V	38			25		31		
t_{pd}	CLK	Any	2 V		46	125		190		155	ns
			4.5 V		15	25		38		31	
			6 V		12	21		32		26	
t_t		Any	2 V		38	75		110		95	ns
			4.5 V		8	15		22		19	
			6 V		6	13		19		16	

C_{pd}	Power dissipation capacitance per multiplexer	No load, $T_A = 25°C$	33 pF typ

NOTE 1: For load circuit and voltage waveforms, see page 1-14.

3 HCMOS DEVICES

TYPES SN54HC352, SN74HC352 DUAL 4-LINE TO 1-LINE DATA SELECTORS/MULTIPLEXERS

D2684, DECEMBER 1982—REVISED MARCH 1984

- **Inverting Versions of 'HC153**
- **High-Current Inverting Outputs Can Drive up to 15 LSTTL Loads**
- **Permits Multiplexing from n Lines to 1 Line**
- **Performs Parallel-to-Serial Conversion**
- **Strobe (Enable) Line Provided for Cascading (N Lines to n Lines)**
- **Package Options Include Both Plastic and Ceramic Chip Carriers in Addition to Plastic and Ceramic DIPs**
- **Dependable Texas Instruments Quality and Reliability**

SN54HC352 . . . J PACKAGE
SN74HC352 . . . J OR N PACKAGE
(TOP VIEW)

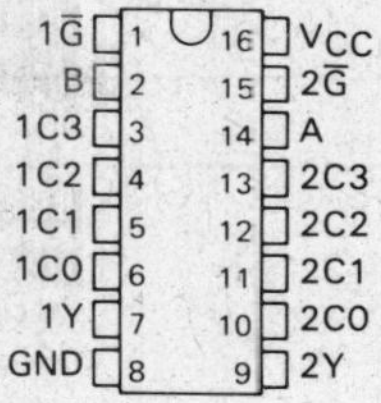

description

Separate output enable inputs (G̅) are provided for each of the two four-line sections of these data selectors/multiplexers.

The SN54HC352 is characterized for operation over the full military temperature range of −55°C to 125°C. The SN74HC352 is characterized for operation from −40°C to 85°C.

SN54HC352 . . . FH OR FK PACKAGE
SN74HC352 . . . FH OR FN PACKAGE
(TOP VIEW)

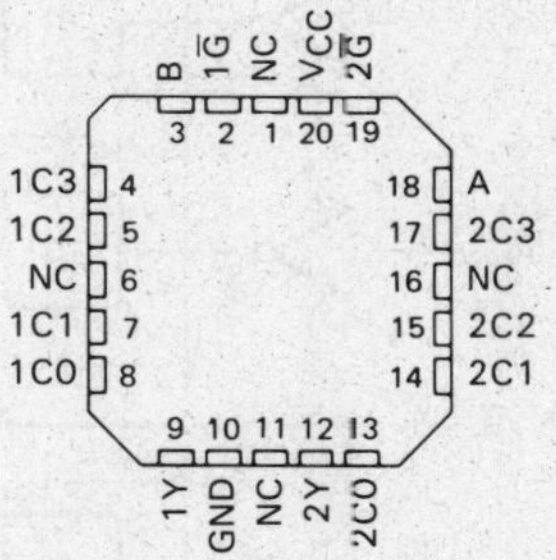

NC—No internal connection

FUNCTION TABLE

SELECT INPUTS		DATA INPUTS				OUTPUT ENABLE	OUTPUT
B	A	C0	C1	C2	C3	G̅	Y
X	X	X	X	X	X	H	H
L	L	L	X	X	X	L	H
L	L	H	X	X	X	L	L
L	H	X	L	X	X	L	H
L	H	X	H	X	X	L	L
H	L	X	X	L	X	L	H
H	L	X	X	H	X	L	L
H	H	X	X	X	L	L	H
H	H	X	X	X	H	L	L

Select inputs A and B are common to both sections.

logic symbol

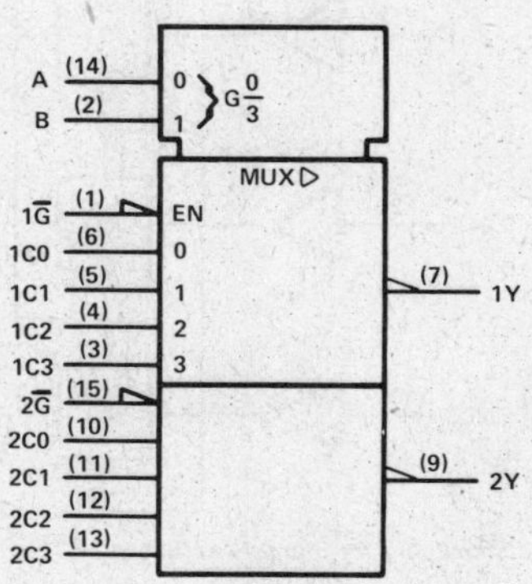

Pin numbers shown are for J and N packages.

logic diagram (positive logic)

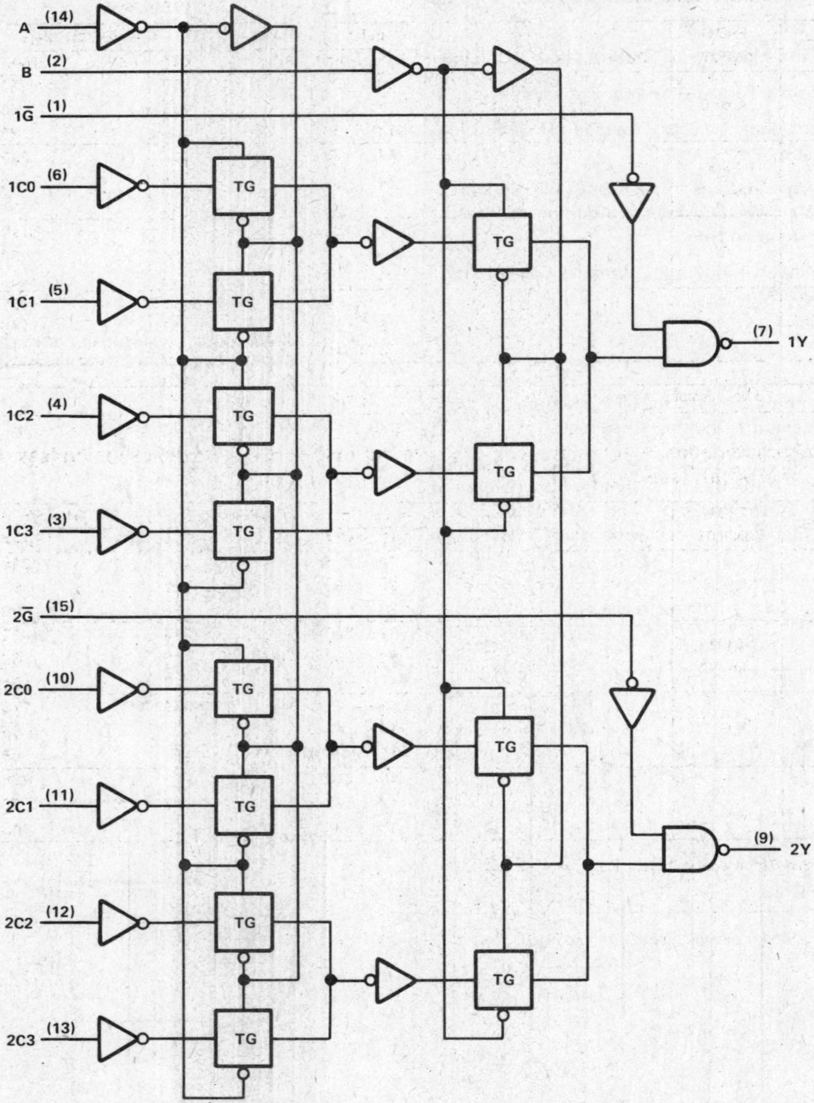

Pin numbers shown are for J and N packages.

maximum ratings, recommended operating conditions, and electrical characteristics

See Table III, page 2-8.

TEXAS INSTRUMENTS
POST OFFICE BOX 225012 • DALLAS, TEXAS 75265

switching characteristics over recommended operating free-air temperature range (unless otherwise noted), C_L = 50 pF (see Note 1)

PARAMETER	FROM (INPUT)	TO (OUTPUT)	V_{CC}	T_A = 25°C			SN54HC352		SN74HC352		UNIT
				MIN	TYP	MAX	MIN	MAX	MIN	MAX	
t_{pd}	A or B	Y	2 V		58	185		280		230	ns
			4.5 V		17	37		56		46	
			6 V		14	32		48		39	
t_{pd}	Data (Any C)	Y	2 V		47	175		265		220	ns
			4.5 V		14	35		53		44	
			6 V		12	30		45		37	
t_{pd}	$\overline{G}$	Y	2 V		27	135		205		170	ns
			4.5 V		10	27		41		34	
			6 V		8	23		35		29	
t_t		Y	2 V		20	60		90		75	ns
			4.5 V		8	12		18		15	
			6 V		6	10		15		13	

C_{pd}	Power dissipation capacitance per data selector	No load, T_A = 25°C	40 pF typ

switching characteristics over recommended operating free-air temperature range (unless otherwise noted), C_L = 150 pF (see Note 1)

PARAMETER	FROM (INPUT)	TO (OUTPUT)	V_{CC}	T_A = 25°C			SN54HC352		SN74HC352		UNIT
				MIN	TYP	MAX	MIN	MAX	MIN	MAX	
t_{pd}	A or B	Y	2 V		72	270		410		335	ns
			4.5 V		22	54		82		67	
			6 V		19	47		70		58	
t_{pd}	Data (Any C)	Y	2 V		62	260		395		325	ns
			4.5 V		19	52		79		63	
			6 V		16	45		67		56	
t_{pd}	$\overline{G}$	Y	2 V		43	220		335		275	ns
			4.5 V		14	44		67		55	
			6 V		12	38		57		48	
t_t		Y	2 V		45	210		315		265	ns
			4.5 V		17	42		63		53	
			6 V		13	36		53		45	

NOTE 1: For load circuit and voltage waveforms, see page 1-14.

TYPES SN54HC353, SN74HC353 DUAL 4-LINE TO 1-LINE DATA SELECTORS/MULTIPLEXERS WITH 3-STATE OUTPUTS

D2684, DECEMBER 1982—REVISED MARCH 1984

- Inverting Versions of 'HC253
- Permits Multiplexing from N Lines to 1 Line
- Performs Parallel-to-Serial Conversion
- High-Current Outputs Can Drive up to 15 LSTTL Loads
- Package Options Include Both Plastic and Ceramic Chip Carriers in Addition to Plastic and Ceramic DIPs
- Dependable Texas Instruments Quality and Reliability

SN54HC353 . . . J PACKAGE
SN74HC353 . . . J OR N PACKAGE
(TOP VIEW)

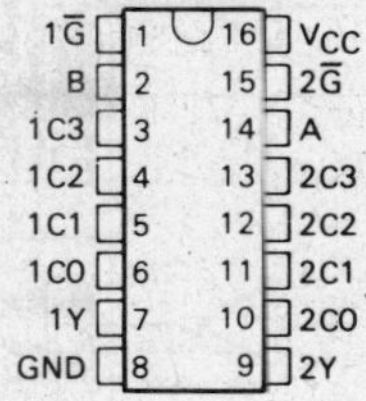

SN54HC353 . . . FH OR FK PACKAGE
SN74HC353 . . . FH OR FN PACKAGE
(TOP VIEW)

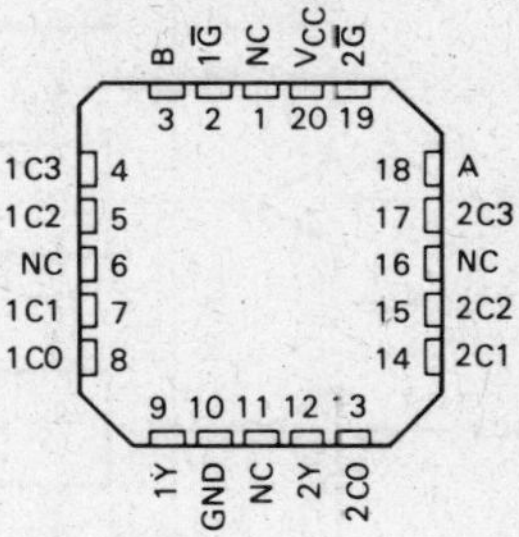

NC—No internal connection

description

Separate output enable inputs ($\overline{G}$) are provided for each of the two four-line sections of these data selectors/multiplexers.

The three-state outputs can interface with and drive data lines of bus-organized systems. With all but one of the common outputs disabled (at a high-impedance state), the low-impedance of the single enable output will drive the bus line to a high or low logic level. Each output has its own output enable ($\overline{G}$). The output is disabled when its output enable is high.

The SN54HC353 is characterized for operation over the full military temperature range of −55°C to 125°C. The SN74HC353 is characterized for operation from −40°C to 85°C.

FUNCTION TABLE

SELECT INPUTS		DATA INPUTS				OUTPUT ENABLE	OUTPUT
B	A	C0	C1	C2	C3	$\overline{G}$	Y
X	X	X	X	X	X	H	Z
L	L	L	X	X	X	L	H
L	L	H	X	X	X	L	L
L	H	X	L	X	X	L	H
L	H	X	H	X	X	L	L
H	L	X	X	L	X	L	H
H	L	X	X	H	X	L	L
H	H	X	X	X	L	L	H
H	H	X	X	X	H	L	L

Select inputs A and B are common to both sections.

logic symbol

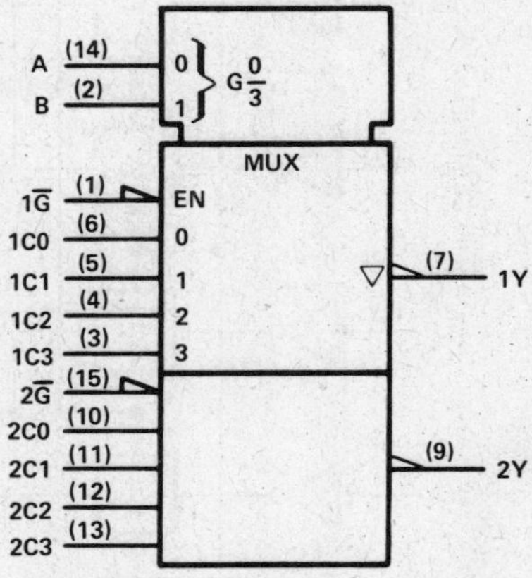

Pin numbers shown are for J and N packages.

logic diagram (positive logic)

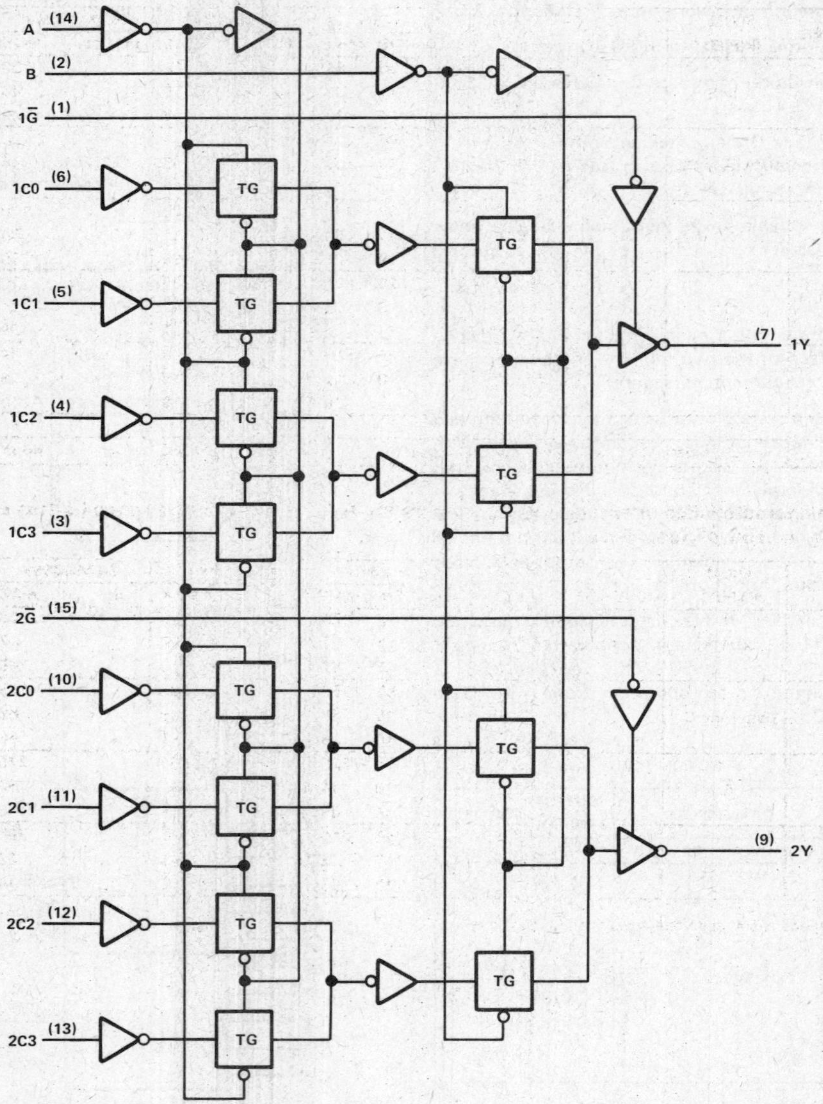

Pin numbers shown are for J and N packages.

maximum ratings, recommended operating conditions, and electrical characteristics

See Table III, page 2-8.

TEXAS INSTRUMENTS
POST OFFICE BOX 225012 • DALLAS, TEXAS 75265

switching characteristics over recommended operating free-air temperature range (unless otherwise noted), C_L = 50 pF (see Note 1)

PARAMETER	FROM (INPUT)	TO (OUTPUT)	V_{CC}	T_A = 25°C MIN	TYP	MAX	SN54HC353 MIN	MAX	SN74HC353 MIN	MAX	UNIT
t_{pd}	A or B	Y	2 V		60	185		280		230	ns
			4.5 V		17	37		56		46	
			6 V		14	32		48		39	
t_{pd}	Data (Any C)	Y	2 V		48	175		265		220	ns
			4.5 V		14	35		53		44	
			6 V		11	30		45		37	
t_{en}	$\overline{G}$	Y	2 V		37	135		205		170	ns
			4.5 V		11	27		41		34	
			6 V		9	23		35		29	
t_{dis}	$\overline{G}$	Y	2 V		22	135		205		170	ns
			4.5 V		13	27		41		34	
			6 V		11	23		35		29	
t_t		Any	2 V		20	60		90		75	ns
			4.5 V		8	12		18		15	
			6 V		6	10		15		13	

C_{pd}	Power dissipation capacitance per multiplexer	No load, T_A = 25°C	40 pF typ

switching characteristics over recommended operating free-air temperature range (unless otherwise noted), C_L = 150 pF (see Note 1)

PARAMETER	FROM (INPUT)	TO (OUTPUT)	V_{CC}	T_A = 25°C MIN	TYP	MAX	SN54HC353 MIN	MAX	SN74HC353 MIN	MAX	UNIT
t_{pd}	A or B	Y	2 V		75	270		410		335	ns
			4.5 V		21	54		82		67	
			6 V		18	47		70		58	
t_{pd}	Data (Any C)	Y	2 V		67	260		395		325	ns
			4.5 V		19	52		79		63	
			6 V		16	45		67		56	
t_{en}	$\overline{G}$	Y	2 V		54	220		335		275	ns
			4.5 V		16	44		67		55	
			6 V		14	38		57		48	
t_t		Any	2 V		45	210		315		265	ns
			4.5 V		17	42		63		53	
			6 V		13	36		53		45	

NOTE 1: For load circuit and voltage waveforms, see page 1-14.

3

HCMOS DEVICES

TYPES SN54HC365 THRU SN54HC368
SN74HC365 THRU SN74HC368
HEX BUS DRIVERS WITH 3-STATE OUTPUTS

D2684, DECEMBER 1982—REVISED MARCH 1984

- **High-Current 3-State Outputs Drive Bus Lines, Buffer Memory Address Registers, or up to 15 LSTTL Loads**
- **Choice of True or Inverting Outputs**
- **Package Options Include Both Plastic and Ceramic Chip Carriers in Addition to Plastic and Ceramic DIPs**
- **Dependable Texas Instruments Quality and Reliability**

'HC365, 'HC367	True Outputs
'HC366, 'HC368	Inverting Outputs

description

These Hex buffers and line drivers are designed specifically to improve both the performance and density of three-state memory address drivers, clock drivers, and bus-oriented receivers and transmitters. The designer has a choice of selected combinations of inverting and noninverting outputs, symmetrical G (active-low control) inputs.

The SN54HC' family is characterized for operation over the full military temperature range of −55°C to 125°C. The SN74HC' family is characterized for operation from −40°C to 85°C.

SN54HC365, SN54HC366 . . . J PACKAGE
SN74HC365, SN74HC365 . . . J OR N PACKAGE
(TOP VIEW)

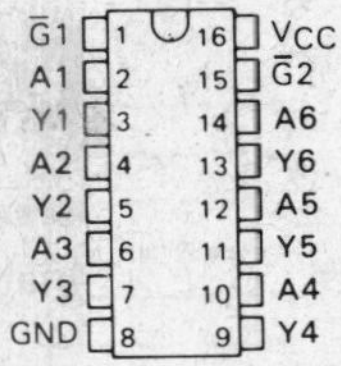

SN54HC365, SN54HC366 . . . FH OR FK PACKAGE
SN74HC365, SN74HC366 . . . FH OR FN PACKAGE
(TOP VIEW)

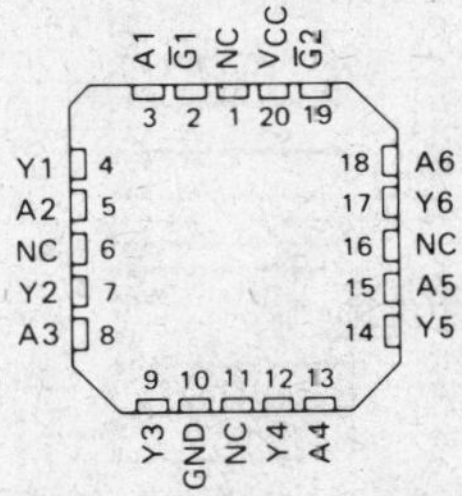

SN54HC367, SN54HC368 . . . J PACKAGE
SN74HC367, SN74HC368 . . . J OR N PACKAGE
(TOP VIEW)

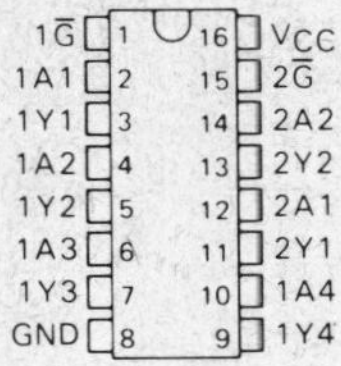

SN54HC367, SN54HC368 . . . FH OR FK PACKAGE
SN74HC367, SN74HC368 . . . FH OR FN PACKAGE
(TOP VIEW)

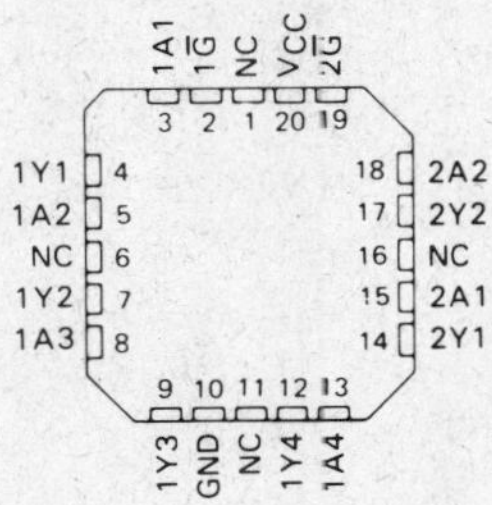

NC—No internal connection

maximum ratings, recommended operation conditions, and electrical characteristics

See Table III, page 2-8.

logic symbols

'HC365

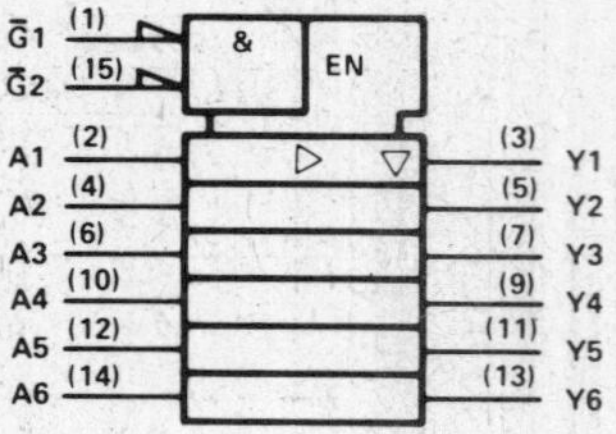

'HC366

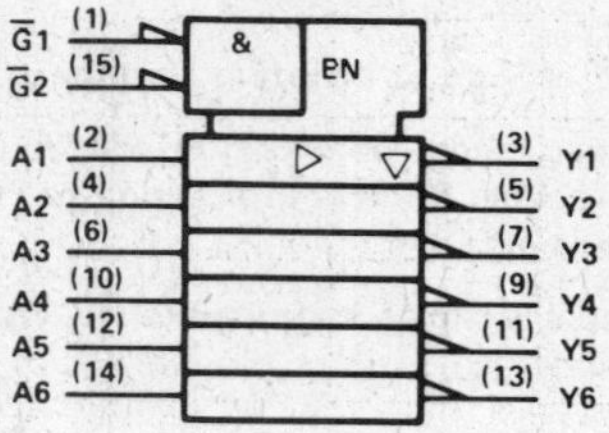

'HC367

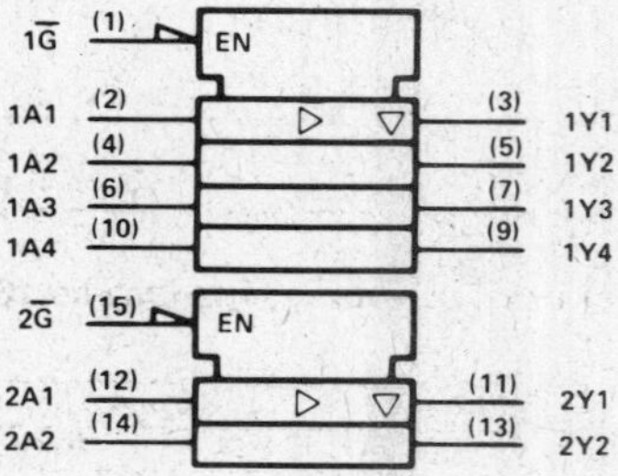

'HC368

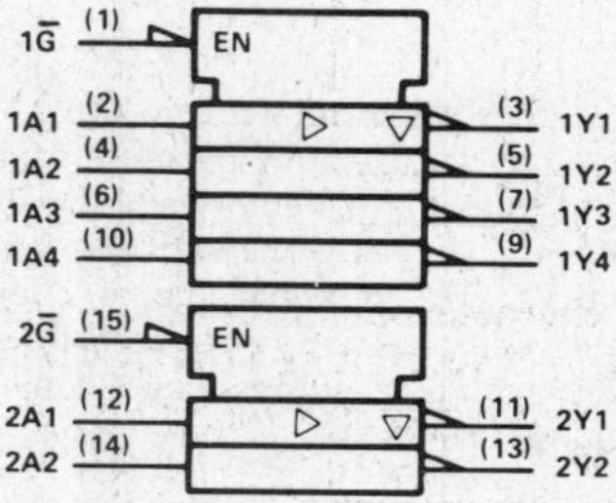

logic diagrams (positive logic)

'HC365

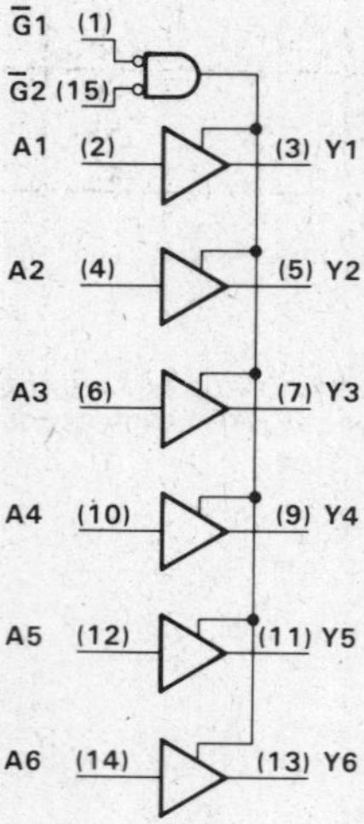

'HC366

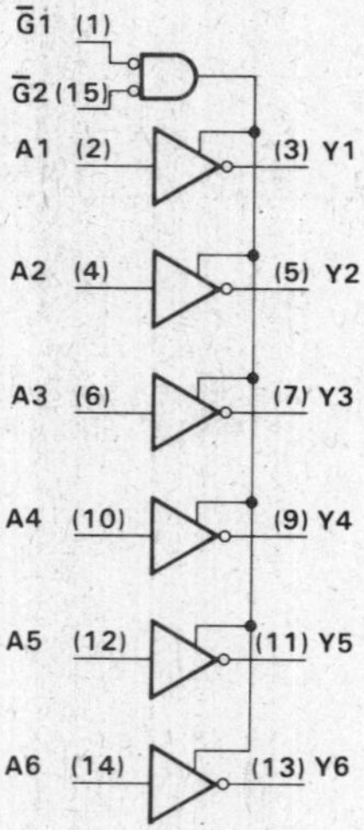

'HC367

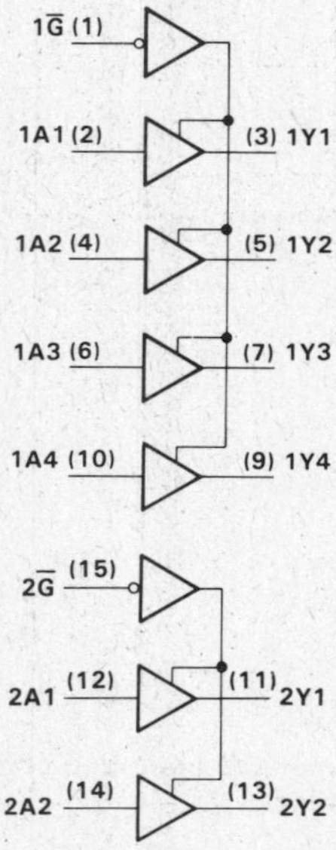

'HC368

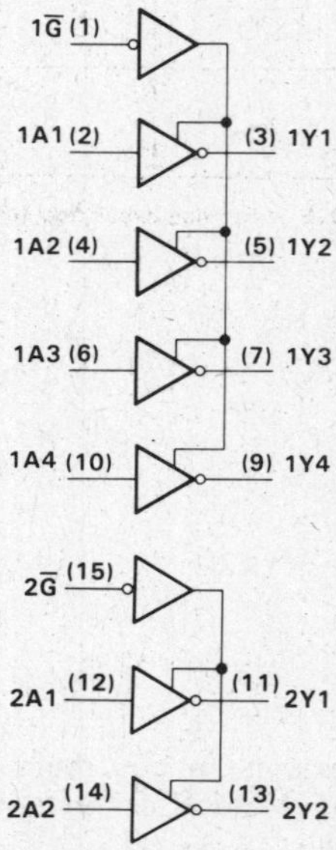

Pin numbers shown are for J and N packages.

switching characteristics over recommended operating free-air temperature range (unless otherwise noted), C_L = 50 pF (see Note 1)

PARAMETER	FROM (INPUT)	TO (OUTPUT)	V_{CC}	T_A = 25°C MIN	TYP	MAX	SN54HC′ MIN	MAX	SN74HC′ MIN	MAX	UNIT
t_{pd}	A	Y	2 V		50	95		145		120	ns
			4.5 V		12	19		29		24	
			6 V		10	16		25		20	
t_{en}	$\overline{G}$	Y	2 V		100	190		285		238	ns
			4.5 V		26	38		57		48	
			6 V		21	32		48		41	
t_{dis}	$\overline{G}$	Y	2 V		50	175		265		240	ns
			4.5 V		21	35		53		48	
			6 V		19	30		45		41	
t_t		Any	2 V		28	60		90		75	ns
			4.5 V		8	12		18		15	
			6 V		6	10		15		13	

C_{pd}	Power dissipation capacitance per driver	No load, T_A = 25°C	35 pF typ

switching characteristics over recommended operating free-air temperature range (unless otherwise noted), C_L = 150 pF (see Note 1)

PARAMETER	FROM (INPUT)	TO (OUTPUT)	V_{CC}	T_A = 25°C MIN	TYP	MAX	SN54HC′ MIN	MAX	SN74HC′ MIN	MAX	UNIT
t_{pd}	A	Y	2 V		70	120		180		150	ns
			4.5 V		17	24		36		30	
			6 V		14	20		31		25	
t_{en}	$\overline{G}$	Y	2 V		140	230		345		285	ns
			4.5 V		30	46		69		57	
			6 V		28	39		59		48	
t_t			2 V		45	210		315		265	ns
			4.5 V		17	42		63		53	
			6 V		13	36		53		45	

NOTE 1: For load circuit and voltage waveforms, see page 1-14.

3 HCMOS DEVICES

TYPES SN54HC373, SN74HC373 OCTAL D-TYPE TRANSPARENT LATCHES WITH 3-STATE OUTPUTS

D2684, DECEMBER 1982—REVISED MARCH 1984

- 8 High-Current Latches in a Single Package
- High-Current 3-State True Outputs Can Drive up to 15 LSTTL Loads
- Full Parallel Access for Loading
- Package Options Include Both Plastic and Ceramic Chip Carriers in Addition to Plastic and Ceramic DIPs
- Dependable Texas Instruments Quality and Reliability

SN54HC373 . . . J PACKAGE
SN74HC373 . . . J OR N PACKAGE
(TOP VIEW)

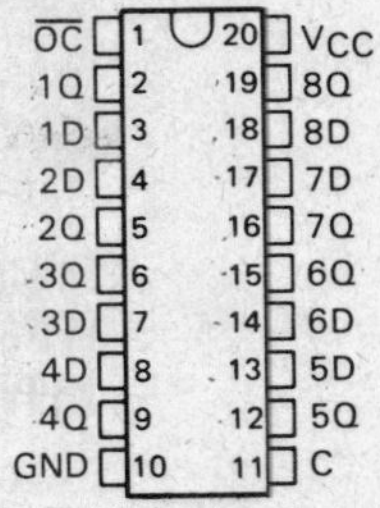

SN54HC373 . . . FH OR FK PACKAGE
SN74HC373 . . . FH OR FN PACKAGE
(TOP VIEW)

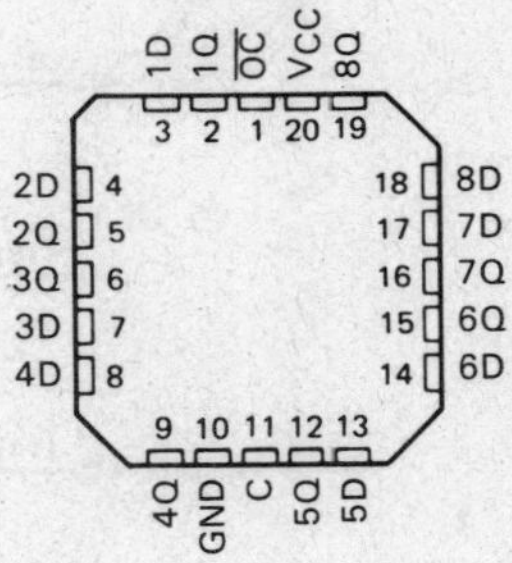

description

These 8-bit latches feature three-state outputs designed specifically for driving highly capacitive or relatively low-impedance loads. They are particularly suitable for implementing buffer registers, I/O ports, bidirectional bus drivers, and working registers.

The eight latches of the 'HC373 are transparent D-type latches. While the enable (C) is high the Q outputs will follow the data (D) inputs. When the enable is taken low, the Q outputs will be latched at the levels that were set up at the D inputs.

An output-control input ($\overline{OC}$) can be used to place the eight outputs in either a normal logic state (high or low logic levels) or a high-impedance state. In the high-impedance state the outputs neither load nor drive the bus lines significantly. The high-impedance third state and increased drive provide the capability to drive the bus lines in a bus-organized system without need for interface or pull-up components.

The output control $\overline{OC}$ does not affect the internal operations of the latches. Old data can be retained or new data can be entered while the outputs are off.

The SN54HC373 is characterized for operation over the full military temperature range of −55°C to 125°C. The SN74HC373 is characterized for operation from −40°C to 85°C.

logic symbol

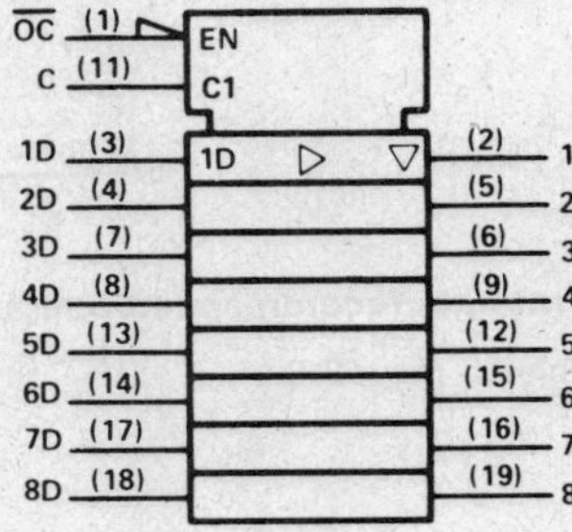

FUNCTION TABLE (EACH LATCH)

INPUTS			OUTPUT
$\overline{OC}$	ENABLE C	D	Q
L	H	H	H
L	H	L	L
L	L	X	Q_0
H	X	X	Z

logic diagram (positive logic)

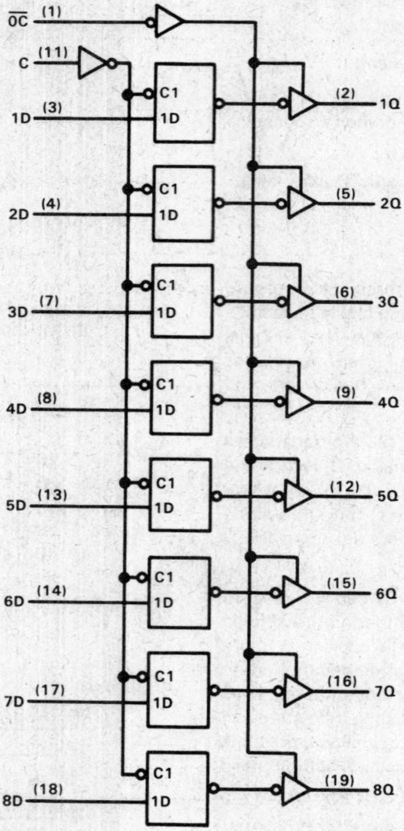

maximum ratings, recommended operating conditions, and electrical characteristics

See Table III, page 2-8.

TEXAS INSTRUMENTS
POST OFFICE BOX 225012 • DALLAS, TEXAS 75265

timing requirements over recommended operating free-air temperature range (unless otherwise noted)

		V_{CC}	T_A = 25°C MIN	T_A = 25°C MAX	SN54HC373 MIN	SN54HC373 MAX	SN74HC373 MIN	SN74HC373 MAX	UNIT
t_w	Pulse duration, enable C high	2 V	80		120		100		ns
		4.5 V	16		24		20		
		6 V	14		20		17		
t_{su}	Setup time, data before enable C↓	2 V	50		75		63		ns
		4.5 V	10		15		13		
		6 V	9		13		11		
t_h	Hold time, data after enable C↓	2 V	5		5		5		ns
		4.5 V	5		5		5		
		6 V	5		5		5		

switching characteristics over recommended operating free-air temperature range (unless otherwise noted), C_L = 50 pF (see Note 1)

PARAMETER	FROM (INPUT)	TO (OUTPUT)	V_{CC}	T_A = 25°C MIN	T_A = 25°C TYP	T_A = 25°C MAX	SN54HC373 MIN	SN54HC373 MAX	SN74HC373 MIN	SN74HC373 MAX	UNIT
t_{pd}	D	Q	2 V		58	150		225		190	ns
			4.5 V		15	30		45		38	
			6 V		13	26		38		32	
t_{pd}	C	Any Q	2 V		73	175		265		220	ns
			4.5 V		18	35		53		44	
			6 V		15	30		45		38	
t_{en}	$\overline{OC}$	Any Q	2 V		65	150		225		190	ns
			4.5 V		17	30		45		38	
			6 V		14	26		38		32	
t_{dis}	$\overline{OC}$	Any Q	2 V		50	150		225		190	ns
			4.5 V		15	30		45		38	
			6 V		13	26		38		32	
t_t		Any	2 V		28	60		90		75	ns
			4.5 V		8	12		18		15	
			6 V		6	10		15		13	

C_{pd}	Power dissipation capacitance per latch	No load, T_A = 25°C	100 pF typ

NOTE 1: For load circuit and voltage waveforms, see page 1-14.

3 HCMOS DEVICES

switching characteristics over recommended operating free-air temperature range (unless otherwise noted), C_L = 150 pF (see Note 1)

PARAMETER	FROM (INPUT)	TO (OUTPUT)	V_{CC}	T_A = 25°C MIN	TYP	MAX	SN54HC373 MIN	TYP	MAX	SN74HC373 MIN	TYP	MAX	UNIT
t_{pd}	D	Q	2 V		82	200			300			250	ns
			4.5 V		22	40			60			50	
			6 V		19	34			51			43	
t_{pd}	C	Any Q	2 V		100	225			335			285	ns
			4.5 V		24	45			67			57	
			6 V		20	38			57			48	
t_{en}	$\overline{OC}$	Any Q	2 V		90	200			300			250	ns
			4.5 V		23	40			60			50	
			6 V		19	34			51			43	
t_t		Any	2 V		45	210			315			265	ns
			4.5 V		17	42			63			53	
			6 V		13	36			53			45	

NOTE 1: For load circuit and voltage waveforms, see page 1-14.

D latch signal conventions

It is TI practice to name the outputs and other inputs of a D-type latch and to draw its logic symbol based on the assumption of true data (D) inputs. Then outputs that produce data in phase with the data inputs are called Q and those producing complementary data are called $\overline{Q}$. An input that causes a Q output to go high or a $\overline{Q}$ output to go low is called Preset; an input that causes a $\overline{Q}$ output to go high or a Q output to go low is called Clear. Bars are used over these pin names ($\overline{PRE}$ and $\overline{CLR}$) if they are active-low.

In some applications it may be advantageous to redesignate the data input $\overline{D}$. In that case all the other inputs and outputs should be renamed as shown below. Also shown are corresponding changes in the graphical symbol. Arbitrary pin numbers are shown in parentheses.

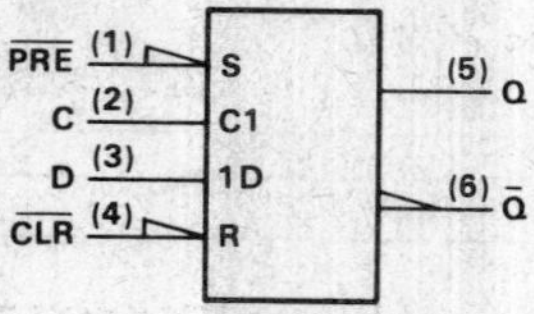

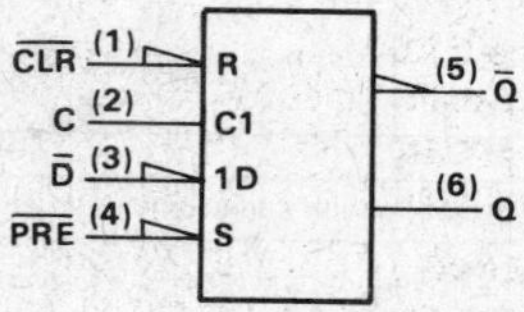

Notice that Q and $\overline{Q}$ exchange names, which causes Preset and Clear to do likewise. Also notice that the polarity indicators (⊵) on $\overline{PRE}$ and $\overline{CLR}$ remain since these inputs are still active-low, but that the presence or absence of the polarity indicator changes at $\overline{D}$, Q, and $\overline{Q}$. Of course pin 5 ($\overline{Q}$) is still in phase with the data input D, but now both are considered active low.

TYPES SN54HC374, SN74HC374 OCTAL D-TYPE EDGE-TRIGGERED FLIP-FLOPS WITH 3-STATE OUTPUTS

D2684, DECEMBER 1982—REVISED MARCH 1984

- 8 D-Type Flip-Flops in a Single Package
- High-Current 3-State True Outputs Can Drive up to 15 LSTTL Loads
- Full Parallel Access for Loading
- Package Options Include Both Plastic and Ceramic Chip Carriers in Addition to Plastic and Ceramic DIPs
- Dependable Texas Instruments Quality and Reliability

SN54HC374 . . . J PACKAGE
SN74HC374 . . . J OR N PACKAGE
(TOP VIEW)

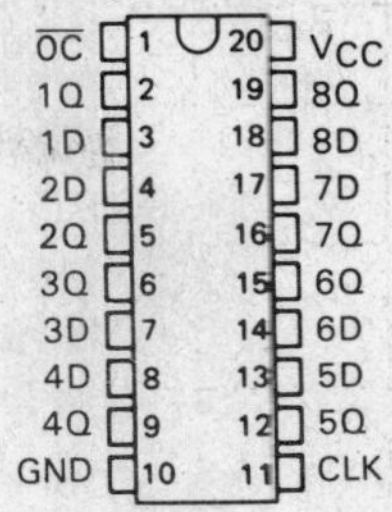

SN54HC374 . . . FH OR FK PACKAGE
SN74HC374 . . . FH OR FN PACKAGE
(TOP VIEW)

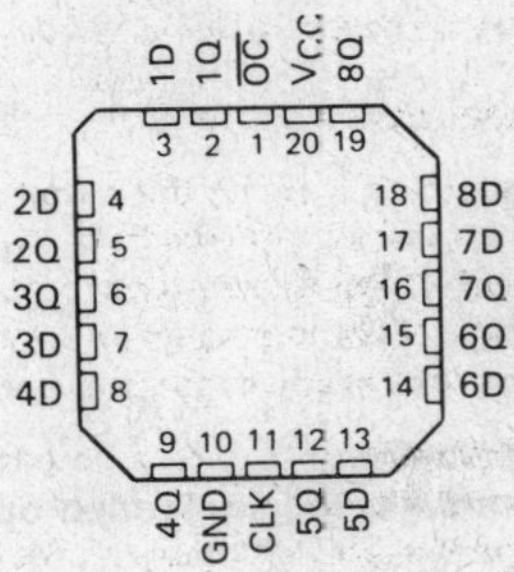

description

These 8-bit flip-flops feature three-state outputs designed specifically for driving highly capacitive or relatively low-impedance loads. They are particularly suitable for implementing buffer registers, I/O ports, bidirectional bus drivers, and working registers.

The eight flip-flops of the 'HC374 are edge-triggered D-type flip-flops. On the positive transition of the clock the Q outputs will be set to the logic levels that were set up at the D inputs.

An output-control input can be used to place the eight outputs in either a normal logic state (high or low logic levels) or a high-impedance state. In the high-impedance state the outputs neither load nor drive the bus lines significantly. The high-impedance third state and increased drive provide the capability to drive the bus lines in a bus-organized system without need for interface or pull-up components.

The output control ($\overline{OC}$) does not affect the internal operation of the flip-flops. Old data can be retained or new data can be entered while the outputs are in the high-impedance state.

The SN54HC374 is characterized for operation over the full military temperature range of −55°C to 125°C. The SN74HC374 is characterized for operation from −40°C to 85°C.

FUNCTION TABLE (EACH FLIP-FLOP)

INPUTS			OUTPUT
$\overline{OC}$	CLK	D	Q
L	↑	H	H
L	↑	L	L
L	L	X	Q_0
H	X	X	Z

logic symbol

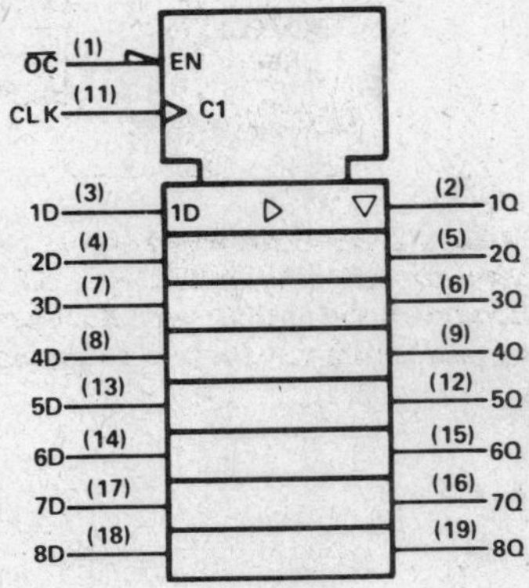

logic diagram (positive logic)

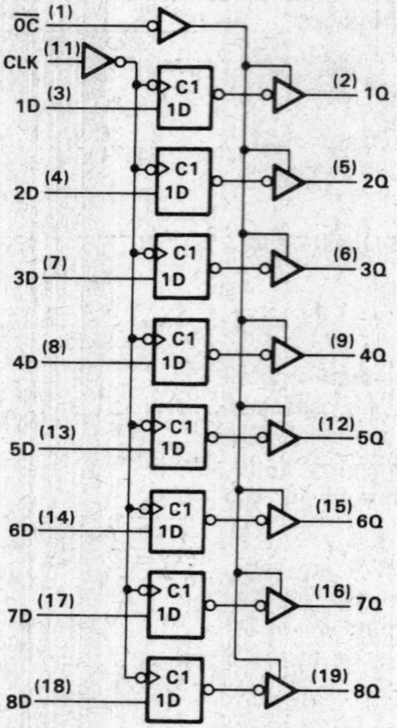

maximum ratings, recommended operating conditions, and electrical characteristics

See Table III, page 2-8.

timing requirements over recommended operating free-air temperature range (unless otherwise noted)

			V_{CC}	T_A = 25°C		SN54HC374		SN74HC374		UNIT
				MIN	MAX	MIN	MAX	MIN	MAX	
f_{clock}	Clock frequency		2 V	0	6	0	4	0	5	MHz
			4.5 V	0	30	0	20	0	24	
			6 V	0	35	0	24	0	28	
t_w	Pulse duration	CLK high or low	2 V	80		120		100		ns
			4.5 V	16		24		20		
			6 V	14		20		17		
t_{su}	Setup time, data before CLK↑		2 V	100		150		125		ns
			4.5 V	20		30		25		
			6 V	17		25		21		
t_h	Hold time, data after CLK↑		2 V	5		5		5		ns
			4.5 V	5		5		5		
			6 V	5		5		5		

switching characteristics over recommended operating free-air temperature range (unless otherwise noted), C_L = 50 pF (see Note 1)

PARAMETER	FROM (INPUT)	TO (OUTPUT)	V_{CC}	T_A = 25°C			SN54HC374		SN74HC374		UNIT
				MIN	TYP	MAX	MIN	MAX	MIN	MAX	
f_{max}			2 V	6	12		4		5		MHz
			4.5 V	30	60		20		24		
			6 V	35	70		24		28		
t_{pd}	CLK	Any	2 V		63	180		270		225	ns
			4.5 V		17	36		54		45	
			6 V		15	31		46		38	
t_{en}	$\overline{OC}$	Any	2 V		60	150		225		190	ns
			4.5 V		16	30		45		38	
			6 V		14	26		38		32	
t_{dis}	$\overline{OC}$	Any	2 V		36	150		225		190	ns
			4.5 V		17	30		45		38	
			6 V		16	26		38		32	
t_t		Any	2 V		28	60		90		75	ns
			4.5 V		8	12		18		15	
			6 V		6	10		15		13	

C_{pd}	Power dissipation capacitance per flip-flop	No load, T_A = 25°C	100 pF typ

NOTE 1: For load circuit and voltage waveforms, see page 1-14.

switching characteristics over recommended operating free-air temperature range (unless otherwise noted), C_L = 150 pF (see Note 1)

PARAMETER	FROM (INPUT)	TO (OUTPUT)	V_{CC}	T_A = 25°C			SN54HC374		SN74HC374		UNIT
				MIN	TYP	MAX	MIN	MAX	MIN	MAX	
f_{max}			2 V	6	12		4		5		MHz
			4.5 V	30	60		20		24		
			6 V	35	70		24		28		
t_{pd}	CLK	Any	2 V		80	230		345		290	ns
			4.5 V		22	46		69		58	
			6 V		19	39		58		49	
t_{en}	$\overline{OC}$	Any	2 V		70	200		300		250	ns
			4.5 V		25	40		60		50	
			6 V		22	34		51		43	
t_t		Any	2 V		45	210		315		265	ns
			4.5 V		17	42		63		53	
			6 V		13	36		53		45	

NOTE 1: For load circuit and voltage waveforms, see page 1-14.

D flip-flop signal conventions

It is TI practice to name the outputs and other inputs of a D-type flip-flop and to draw its logic symbol based on the assumption of true data (D) inputs. Then outputs that produce data in phase with the data inputs are called Q and those producing complementary data are called $\overline{Q}$. An input that causes a Q output to go high or a $\overline{Q}$ output to go low is called Preset; an input that causes a $\overline{Q}$ output to go high or a Q output to go low is called Clear. Bars are used over these pin names ($\overline{PRE}$ and $\overline{CLR}$) if they are active-low.

In some applications it may be advantageous to redesignate the data input $\overline{D}$. In that case all the other inputs and outputs should be renamed as shown below. Also shown are corresponding changes in the graphical symbol. Arbitrary pin numbers are shown in parentheses.

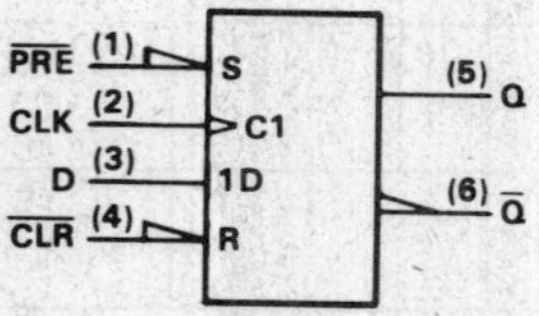

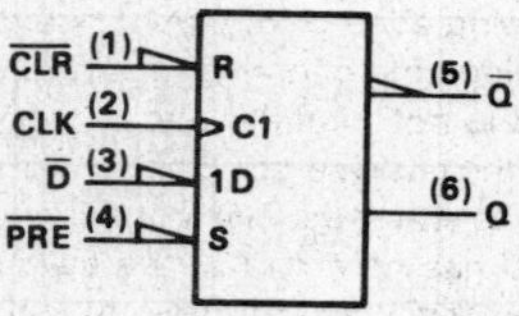

Notice that Q and $\overline{Q}$ exchange names, which causes Preset and Clear to do likewise. Also notice that the polarity indicators (⊳) on $\overline{PRE}$ and $\overline{CLR}$ remain since these inputs are still active-low, but that the presence or absence of the polarity indicator changes at $\overline{D}$, Q, and $\overline{Q}$. Of course pin 5 ($\overline{Q}$) is still in phase with the data input $\overline{D}$, but now both are considered active-low.

TYPES SN54HC377, SN54HC378, SN54HC379 SN74HC377, SN74HC378, SN74HC379 OCTAL, HEX, AND QUAD D-TYPE FLIP-FLOPS WITH CLOCK ENABLE

D2684, DECEMBER 1982—REVISED MARCH 1984

- **'HC377 and 'HC378 Contain Eight and Six Flip-Flops, Respectively, with Single-Rail Outputs**
- **'HC379 Contains Four Flip-Flops with Double-Rail Outputs**
- **Clock Enable Latched to Avoid False Clocking**
- **Applications Include:**
  - **Buffer/Storage Registers**
  - **Shift Registers**
  - **Pattern Generators**
- **Package Options Include Both Plastic and Ceramic Chip Carriers in Addition to Plastic and Ceramic DIPs**
- **Dependable Texas Instruments Quality and Reliability**

SN54HC377 . . . J PACKAGE
SN74HC377 . . . J OR N PACKAGE
(TOP VIEW)

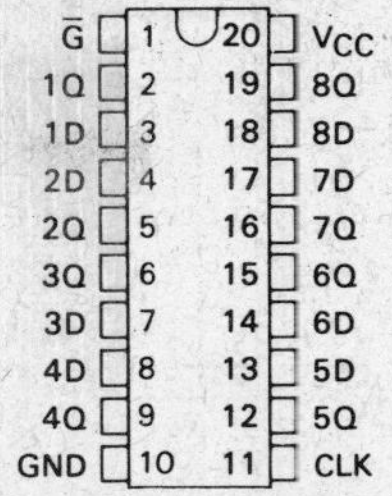

SN54HC377 . . . FH OR FK PACKAGE
SN74HC377 . . . FH OR FN PACKAGE
(TOP VIEW)

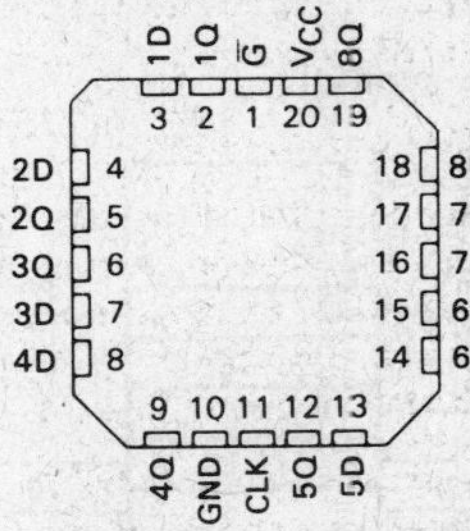

SN54HC378 . . . J PACKAGE
SN74HC378 . . . J OR N PACKAGE
(TOP VIEW)

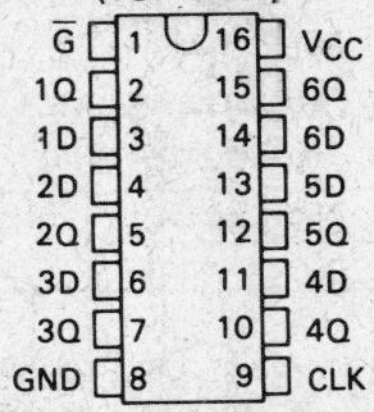

SN54HC378 . . . FH OR FK PACKAGE
SN74HC378 . . . FH OR FN PACKAGE
(TOP VIEW)

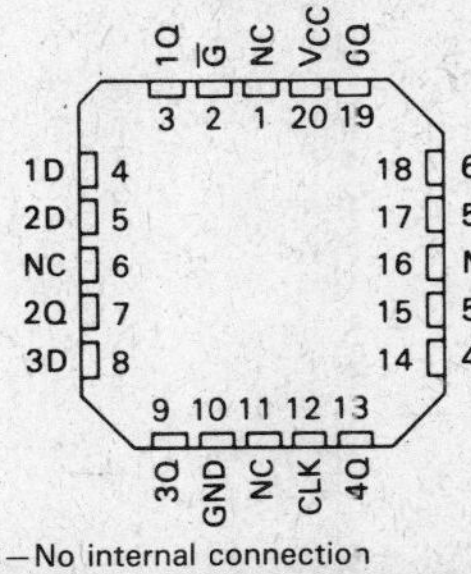

NC—No internal connection

description

These circuits are positive-edge-triggered D-type flip-flops with an enable input. The 'HC377, 'HC378, and 'HC379 devices are similar to 'HC273, 'HC174, and 'HC175 respectively, but feature a latched clock enable ($\overline{G}$) instead of a common clear.

Information at the D inputs meeting the setup time requirements is transferred to the Q outputs on the positive-going edge of the clock pulse if $\overline{G}$ is low. Clock triggering occurs at a particular voltage level and is not directly related to the transition time of the positive-going pulse. When the clock input is at either the high or low level, the D input signal has no effect at the output. The circuits are designed to prevent false clocking by transitions at the $\overline{G}$ input.

The SN54HC377, SN54HC378, and SN54HC379 are characterized for operation over the full military temperature range of −55°C to 125°C. The SN74HC377, SN74HC378, and SN74HC379 are characterized for operation from −40°C to 85°C.

SN54HC379 . . . J PACKAGE
SN74HC379 . . . J OR N PACKAGE
(TOP VIEW)

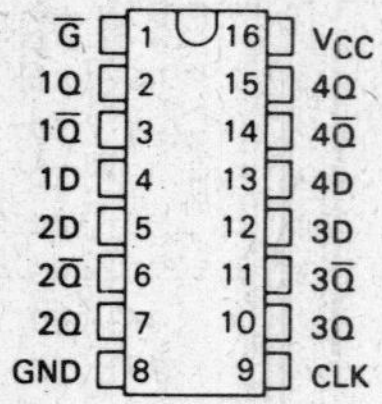

SN54HC379 . . . FH OR FK PACKAGE
SN74HC379 . . . FH OR FN PACKAGE
(TOP VIEW)

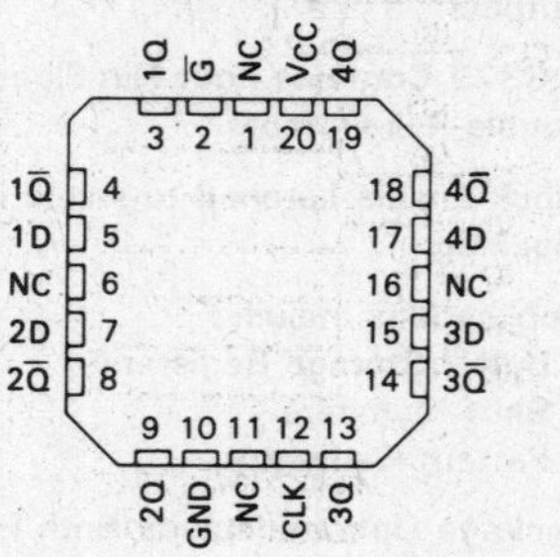

NC—No internal connection

'HC377 logic symbol

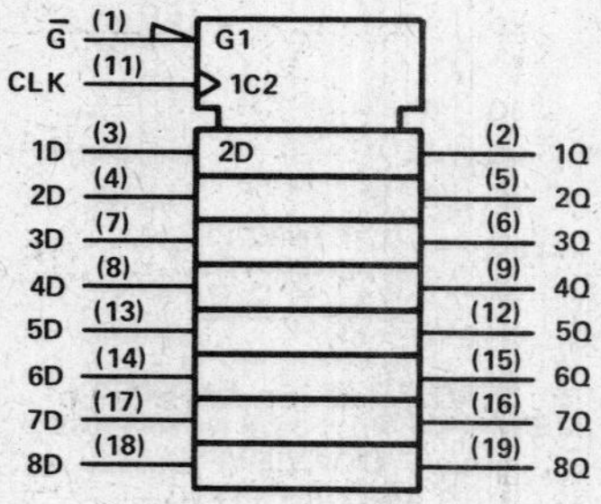

'HC377 logic diagram (positive logic)

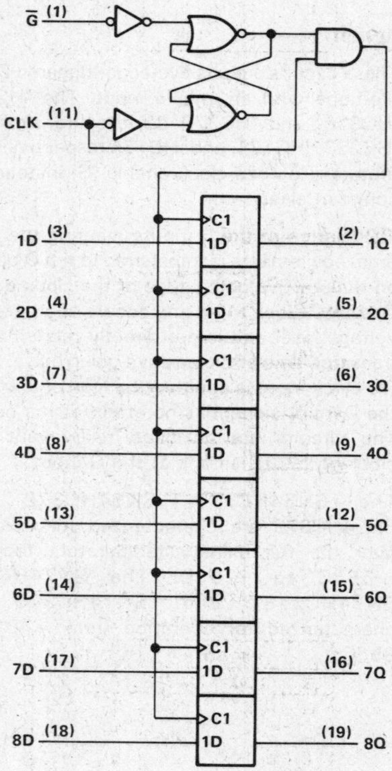

FUNCTION TABLE
(EACH FLIP-FLOP)

INPUTS			OUTPUT
G̅	CLOCK	DATA	Q
H	X	X	Q_0
L	↑	H	H
L	↑	L	L
X	L	X	Q_0

'HC378 logic symbol

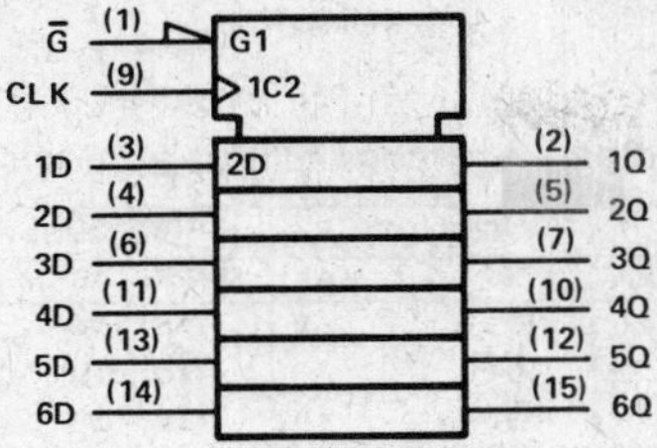

FUNCTION TABLE
(EACH FLIP-FLOP)

INPUTS			OUTPUT
$\overline{G}$	CLOCK	DATA	Q
H	X	X	Q_0
L	↑	H	H
L	↑	L	L
X	L	X	Q_0

'HC378 logic diagram (positive logic)

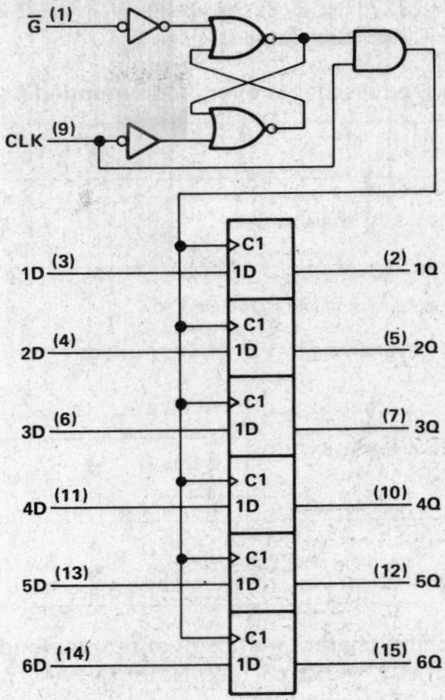

'HC379 logic symbol

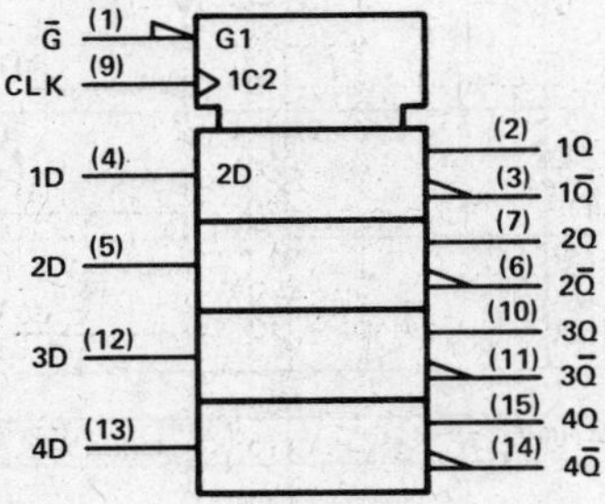

FUNCTION TABLE
(EACH FLIP-FLOP)

INPUTS			OUTPUTS	
$\overline{G}$	CLOCK	DATA	Q	$\overline{Q}$
H	X	X	Q_0	$\overline{Q}_0$
L	↑	H	H	L
L	↑	L	L	H
X	L	X	Q_0	$\overline{Q}_0$

'HC379 logic diagram (positive logic)

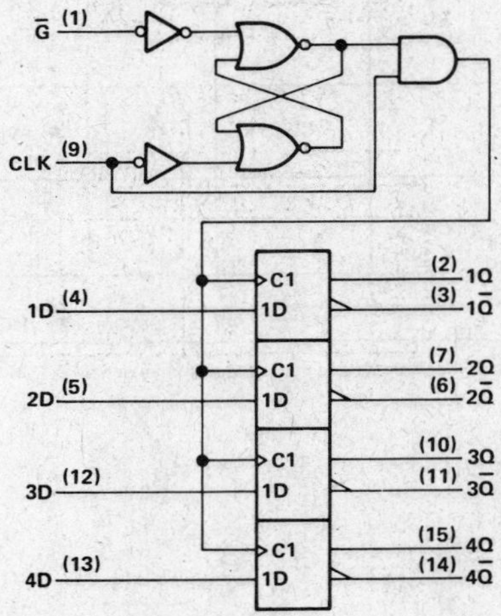

Pin numbers shown are for J and N packages.

maximum ratings, recommended operating conditions, and electrical characteristics

'HC377, 'HC378: See Table IV, page 2-6.
'HC379: See Table II, page 2-4.

timing requirements over recommended operating free-air temperature range (unless otherwise noted)

			V_{CC}	T_A = 25°C		SN54HC'		SN74HC'		UNIT
				MIN	MAX	MIN	MAX	MIN	MAX	
f_{clock}	Clock frequency		2 V	0	5	0	3	0	4	MHz
			4.5 V	0	25	0	16	0	20	
			6 V	0	29	0	19	0	23	
t_w	Pulse duration, CLK high or low		2 V	100		150		125		ns
			4.5 V	20		30		25		
			6 V	17		25		21		
t_{su}	Set up time before CLK↑	D	2 V	100		150		125		ns
			4.5 V	20		30		25		
			6 V	17		25		21		
		$\overline{G}$ high or low	2 V	100		150		125		ns
			4.5 V	20		30		25		
			6 V	17		25		21		
t_h	Hold time after CLK↑		2 V	0		0		0		ns
			4.5 V	0		0		0		
			6 V	0		0		0		

switching characteristics over recommended operating free-air temperature range (unless otherwise noted), C_L = 50 pF (see Note 1)

PARAMETER	FROM (INPUT)	TO (OUTPUT)	V_{CC}	T_A = 25°C			SN54HC'		SN74HC'		UNIT
				MIN	TYP	MAX	MIN	MAX	MIN	MAX	
f_{max}			2 V	5	11		3		4		MHz
			4.5 V	25	54		16		20		
			6 V	29	64		19		23		
t_{pd}	CLK	Any	2 V		56	160		240		200	ns
			4.5 V		15	32		48		40	
			6 V		12	27		41		34	
t_t		Any	2 V		38	75		110		95	ns
			4.5 V		8	15		22		19	
			6 V		6	13		19		16	

C_{pd}	Power dissipation capacitance	No load, T_A = 25°C	30 pF typ

NOTE 1: For load circuit and voltage waveforms, see page 1-14.

D2684, DECEMBER 1982—REVISED MARCH 1984

- **Package Options Include Both Plastic and Ceramic Chip Carriers in Addition to Plastic and Ceramic DIPs**
- **Dependable Texas Instruments Quality and Reliability**

SN54HC386 . . . J PACKAGE
SN74HC386 . . . J OR N PACKAGE
(TOP VIEW)

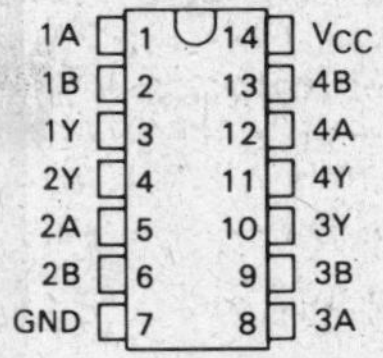

description

These devices contain four independent 2-input Exclusive-OR gates. They perform the Boolean functions $Y = A \oplus B = \bar{A}B + A\bar{B}$ in positive logic.

A common application is as a true/complement element. If one of the inputs is low, the other input will be reproduced in true form at the output. If one of the inputs is high, the signal on the other input will be reproduced inverted at the output.

The SN54HC386 is characterized for operation over the full military temperature range of −55°C to 125°C. The SN74HC386 is characterized for operation from −40°C to 85°C.

SN54HC386 . . . FH OR FK PACKAGE
SN74HC386 . . . FH OR FN PACKAGE
(TOP VIEW)

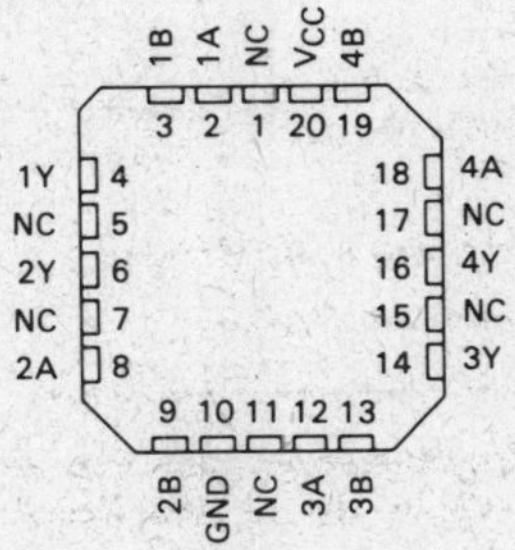

NC—No internal connection

logic symbol

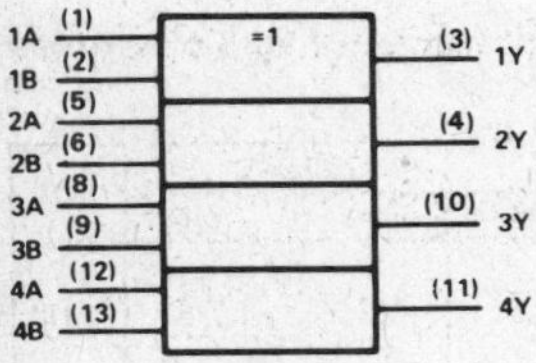

Pin numbers shown are for J and N packages.

FUNCTION TABLE
(each gate)

INPUTS		OUTPUT
A	B	Y
L	L	L
L	H	H
H	L	H
H	H	L

maximum ratings, recommended operating conditions, and electrical characteristics

See Table I, page 2-4.

switching characteristics over recommended operating free-air temperature range (unless otherwise noted), C_L = 50 pF (see Note 1)

PARAMETER	FROM (INPUT)	TO (OUTPUT)	V_{CC}	T_A = 25°C MIN	TYP	MAX	SN54HC386 MIN	MAX	SN74HC386 MIN	MAX	UNIT
t_{pd}	A or B	Y	2 V		40	100		150		125	ns
			4.5 V		12	20		30		25	
			6 V		10	17		25		21	
t_t		Y	2 V		28	75		110		95	ns
			4.5 V		8	15		22		19	
			6 V		6	13		19		16	

C_{pd}	Power dissipation capacitance per gate	No load, T_A = 25°C	35 pF typ

NOTE 1: For load circuit and voltage waveforms, see page 1-14.

D2684, DECEMBER 1982—REVISED MARCH 1984

- **'HC390 . . . Individual Clock for A and B Flip-Flops Provide Dual ÷2 and ÷5 Counters**
- **'HC393 . . . Dual 4-Bit Binary Counter with Individual Clocks**
- **All Have Direct Clear for Each 4-Bit Counter**
- **Dual 4-Bit Versions Can Significantly Improve System Densities by Reducing Counter Package Count by 50%**
- **Package Options Include Both Plastic and Ceramic Chip Carriers in Addition to Plastic and Ceramic DIPs**
- **Dependable Texas Instruments Quality and Reliability**

description

Each of these monolithic circuits contains eight flip-flops and additional gating to implement two individual four-bit counters in a single package. The 'HC390 incorporates dual divide-by-two and divide-by-five counters, which can be used to implement cycle lengths equal to any whole and/or cumulative multiples of 2 and/or 5 up to divide-by-100. When connected as a biquinary counter, the separate divide-by-two circuit can be used to provide symmetry (a square wave) at the final output stage. The 'HC393 comprises two independent four-bit binary counters each having a clear and a clock input. N-bit binary counters can be implemented with each package providing the capability of divide-by-256. The 'HC390 and 'HC393 have parallel outputs from each counter stage so that any submultiple of the input count frequency is available for system-timing signals.

The SN54HC390 and SN54HC393 are characterized for operation over the full military temperature range of −55°C to 125°C. The SN74HC390 and SN74HC393 are characterized for operation from −40°C to 85°C.

SN54HC390 . . . J PACKAGE
SN74HC390 . . . J OR N PACKAGE
(TOP VIEW)

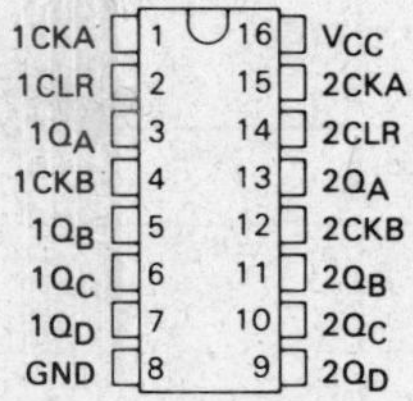

SN54HC390 . . . FH OR FK PACKAGE
SN74HC390 . . . FH OR FN PACKAGE
(TOP VIEW)

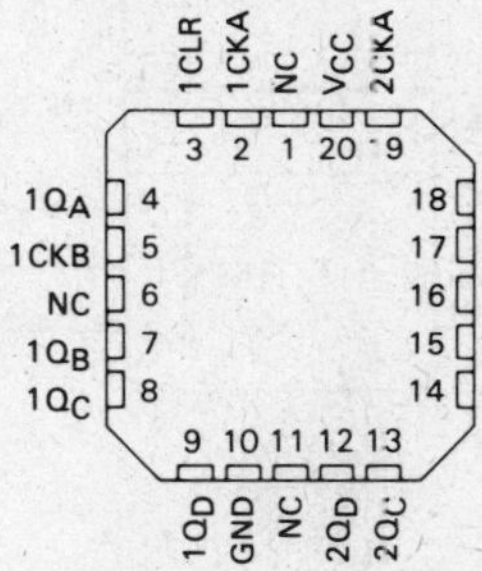

SN54HC393 . . . J PACKAGE
SN74HC393 . . . J OR N PACKAGE
(TOP VIEW)

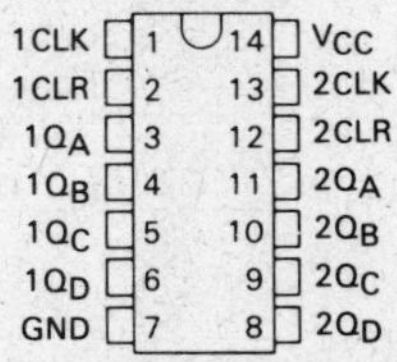

SN54HC393 . . . FH OR FK PACKAGE
SN74HC393 . . . FH OR FN PACKAGE
(TOP VIEW)

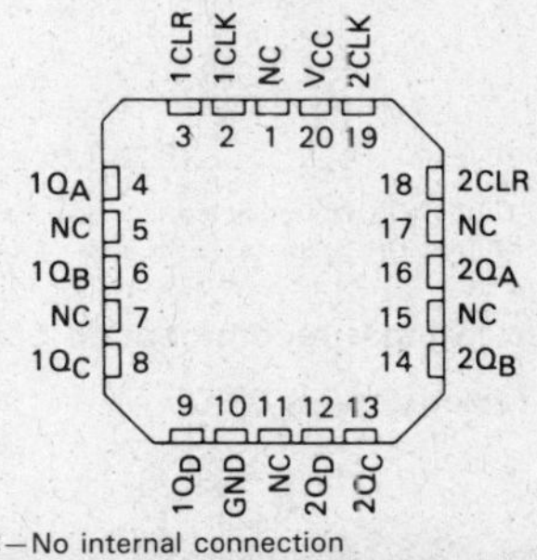

NC—No internal connection

logic symbol

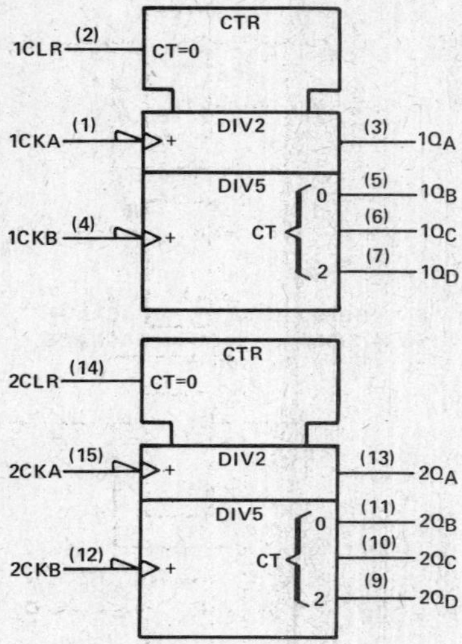

Pin numbers shown are for J and N packages.

logic diagram, each counter (positive logic)

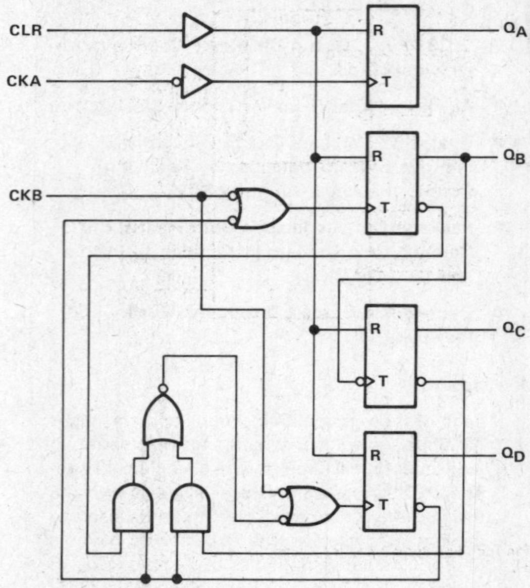

FUNCTION TABLES

BCD COUNT SEQUENCE (EACH COUNTER) (See Note A)

COUNT	OUTPUT			
	Q_D	Q_C	Q_B	Q_A
0	L	L	L	L
1	L	L	L	H
2	L	L	H	L
3	L	L	H	H
4	L	H	L	L
5	L	H	L	H
6	L	H	H	L
7	L	H	H	H
8	H	L	L	L
9	H	L	L	H

BIQUINARY (5-2) (EACH COUNTER) (See Note B)

COUNT	OUTPUT			
	Q_A	Q_D	Q_C	Q_B
0	L	L	L	L
1	L	L	L	H
2	L	L	H	L
3	L	L	H	H
4	L	H	L	L
5	H	L	L	L
6	H	L	L	H
7	H	L	H	L
8	H	L	H	H
9	H	H	L	L

NOTES: A. Output Q_A is connected to input CKB for BCD count.
B. Output Q_D is connected to input CKA for biquinary count.

maximum ratings, recommended operating conditions, and electrical characteristics

See Table IV, page 2-10.

TEXAS INSTRUMENTS
POST OFFICE BOX 225012 • DALLAS, TEXAS 75265

logic symbol

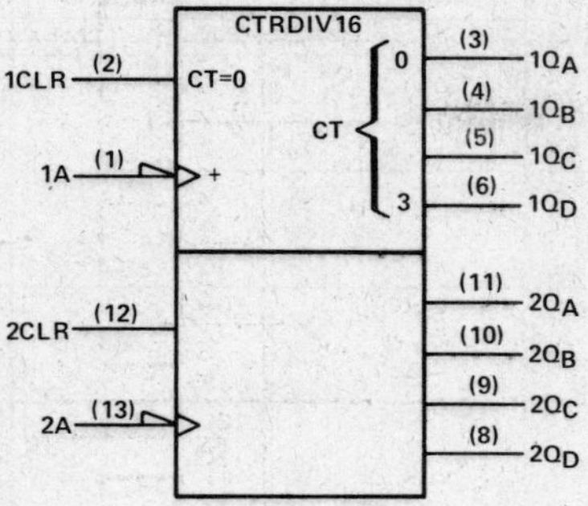

Pin numbers shown are for J and N packages.

logic diagram, each counter (positive logic)

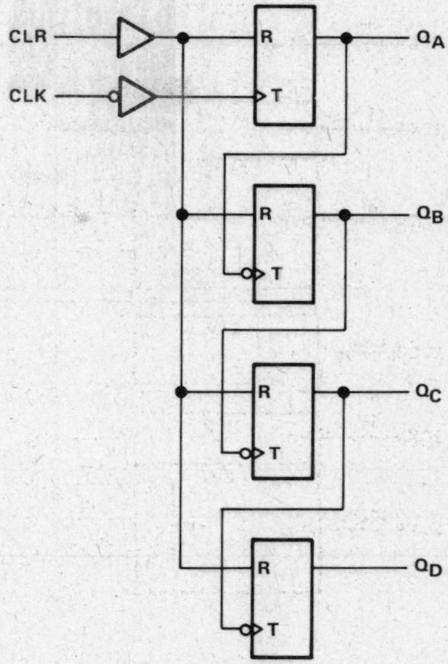

FUNCTION TABLE

COUNT SEQUENCE

(EACH COUNTER)

COUNT	OUTPUT			
	Q_D	Q_C	Q_B	Q_A
0	L	L	L	L
1	L	L	L	H
2	L	L	H	L
3	L	L	H	H
4	L	H	L	L
5	L	H	L	H
6	L	H	H	L
7	L	H	H	H
8	H	L	L	L
9	H	L	L	H
10	H	L	H	L
11	H	L	H	H
12	H	H	L	L
13	H	H	L	H
14	H	H	H	L
15	H	H	H	H

maximum ratings, recommended operating conditions, and electrical characteristics

See Table IV, page 2-10.

3

HCMOS DEVICES

timing requirements over recommended operating free-air temperature range (unless otherwise noted)

			V_{CC}	T_A = 25°C		SN54HC390		SN74HC390		UNIT
				MIN	MAX	MIN	MAX	MIN	MAX	
f_{clock}	Clock frequency	CKA	2 V	0	6	0	4.2	0	5	MHz
			4.5 V	0	31	0	21	0	25	
			6 V	0	36	0	25	0	28	
		CKB	2 V	0	6	0	4.2	0	5	
			4.5 V	0	31	0	21	0	25	
			6 V	0	36	0	25	0	28	
t_w	Pulse duration	CKA high or low	2 V	80		120		100		ns
			4.5 V	16		24		20		
			6 V	14		20		18		
		CKB high or low	2 V	80		120		100		
			4.5 V	16		24		20		
			6 V	14		20		18		
		CLR high	2 V	80		120		100		
			4.5 V	16		24		20		
			6 V	14		20		18		
t_{su}	Setup time, CLR inactive		2 V	25		25		25		ns
			4.5 V	5		5		5		
			6 V	5		5		5		

switching characteristics over recommended operating free-air temperature range (unless otherwise noted), C_L = 50 pF (see Note 1)

PARAMETER	FROM (INPUT)	TO (OUTPUT)	V_{CC}	T_A = 25°C MIN	T_A = 25°C TYP	T_A = 25°C MAX	SN54HC390 MIN	SN54HC390 MAX	SN74HC390 MIN	SN74HC390 MAX	UNIT
f_{max}	CKA	Q_A	2 V	6	10		4.2		5		MHz
			4.5 V	31	50		21		25		
			6 V	36	60		25		28		
	CKB	Q_B	2 V	6	10		4.2		5		
			4.5 V	31	50		21		25		
			6 V	36	60		25		28		
t_{pd}	CKA	Q_A	2 V		50	120		180		150	ns
			4.5 V		16	24		35		30	
			6 V		13	20		31		26	
t_{pd}	CKA	Q_C	2 V		100	290		430		365	ns
			4.5 V		35	58		87		72	
			6 V		30	50		74		62	
t_{pd}	CKB	Q_B	2 V		58	130		195		165	ns
			4.5 V		18	26		39		33	
			6 V		15	22		33		28	
t_{pd}	CKB	Q_C	2 V		83	185		280		230	ns
			4.5 V		26	37		55		46	
			6 V		21	32		48		40	
t_{pd}	CKB	Q_D	2 V		60	130		195		160	ns
			4.5 V		18	26		39		33	
			6 V		14	22		33		28	
t_{PHL}	CLR	Any	2 V		45	165		250		205	ns
			4.5 V		17	33		49		41	
			6 V		14	28		42		35	
t_t		Any	2 V		28	75		110		95	ns
			4.5 V		8	15		22		19	
			6 V		6	13		19		16	

C_{pd}	Power dissipation capacitance per counter	No load, T_A = 25°C	40 pF typ

NOTE 1: For load circuit and voltage waveforms, see page 1-14.

timing requirements over recommended operating free-air temperature range (unless otherwise noted)

			V_{CC}	$T_A = 25°C$ MIN	$T_A = 25°C$ MAX	SN54HC393 MIN	SN54HC393 MAX	SN74HC393 MIN	SN74HC393 MAX	UNIT
f_{clock}	Clock frequency	CLK	2 V	0	6	0	4.2	0	5	MHz
			4.5 V	0	31	0	21	0	25	
			6 V	0	36	0	25	0	28	
t_w	Pulse duration	CLK high or low	2 V	80		120		100		ns
			4.5 V	16		24		20		
			6 V	14		20		18		
		CLR high	2 V	80		120		100		
			4.5 V	16		24		20		
			6 V	14		20		18		
t_{su}	Setup time, CLR inactive		2 V	25		25		25		ns
			4.5 V	5		5		5		
			6 V	5		5		5		

3 HCMOS DEVICES

switching characteristics over recommended operating free-air temperature range (unless otherwise noted), C_L = 50 pF (see Note 1)

PARAMETER	FROM (INPUT)	TO (OUTPUT)	V_{CC}	$T_A = 25°C$ MIN	$T_A = 25°C$ TYP	$T_A = 25°C$ MAX	SN54HC393 MIN	SN54HC393 MAX	SN74HC393 MIN	SN74HC393 MAX	UNIT
f_{max}	CLK	Q_A	2 V	6	10		4.2		5		MHz
			4.5 V	31	50		21		25		
			6 V	36	60		25		28		
t_{pd}	CLK	Q_A	2 V		50	120		180		150	ns
			4.5 V		15	24		36		30	
			6 V		13	20		31		26	
t_{pd}	CLK	Q_D	2 V		100	290		430		360	ns
			4.5 V		32	58		87		72	
			6 V		24	50		74		62	
t_{PHL}	CLR	Any	2 V		45	165		250		205	ns
			4.5 V		17	33		49		41	
			6 V		14	28		42		35	
t_t		Any	2 V		28	75		110		95	ns
			4.5 V		8	15		22		19	
			6 V		6	13		19		16	

C_{pd}	Power dissipation capacitance per counter	No load, $T_A = 25°C$	40 pF typ

NOTE 1: For load circuit and voltage waveforms, see page 1-14.

TEXAS INSTRUMENTS
POST OFFICE BOX 225012 • DALLAS, TEXAS 75265

TYPES SN54HC490, SN74HC490 DUAL 4-BIT DECADE COUNTERS

D2684, DECEMBER 1982—REVISED MARCH 1984

- Individual Clock, Direct Clear, and Set-to-9 Inputs for Each Decade Counter
- Dual Counters Can Significantly Improve System Densities as Package Count Can be Reduced by 50%
- Package Options Include Both Plastic and Ceramic Chip Carriers in Addition to Plastic and Ceramic DIPs
- Dependable Texas Instruments Quality and Reliability

SN54HC490 . . . J PACKAGE
SN74HC490 . . . J OR N PACKAGE
(TOP VIEW)

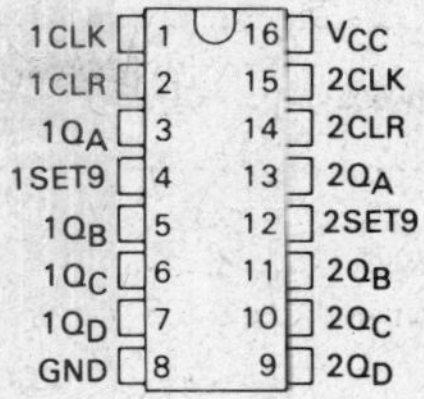

SN54HC490 . . . FH OR FK PACKAGE
SN74HC490 . . . FH OR FN PACKAGE
(TOP VIEW)

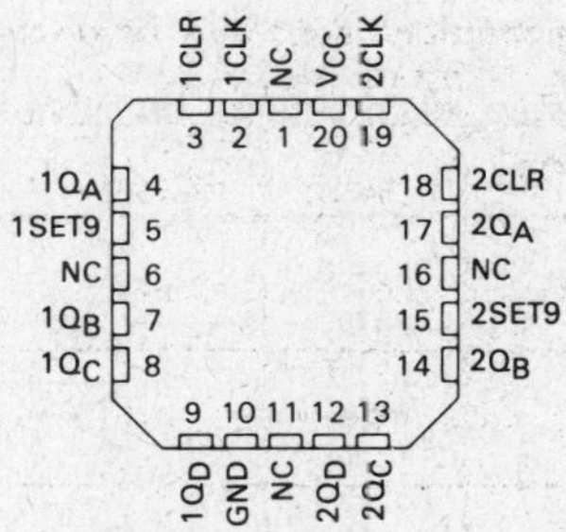

NC—No internal connection

description

Each of these monolithic circuits contains eight master-slave flip-flops and additional gating to implement two individual 4-bit decade counters in a single package. Each decade counter has individual clock, clear, and set-to-9 inputs. BCD count sequences of any length up to divide-by-100 may be implemented with a single 'HC490. The counters have parallel outputs from each counter stage so that submultiples of the input count frequency are available for system timing signals.

The SN54HC490 is characterized for operation over the full military temperature range of −55°C to 125°C. The SN74HC490 is characterized for operation from −40°C to 85°C.

CLEAR/SET-TO-9 FUNCTION TABLE (EACH COUNTER)

INPUTS		OUTPUTS			
CLEAR	SET-TO-9	Q_A	Q_B	Q_C	Q_D
H	L	L	L	L	L
L	H	H	L	L	H
L	L	COUNT			

BCD COUNT SEQUENCE (EACH COUNTER)

COUNT	OUTPUT			
	Q_D	Q_C	Q_B	Q_A
0	L	L	L	L
1	L	L	L	H
2	L	L	H	L
3	L	L	H	H
4	L	H	L	L
5	L	H	L	H
6	L	H	H	L
7	L	H	H	H
8	H	L	L	L
9	H	L	L	H

logic symbol

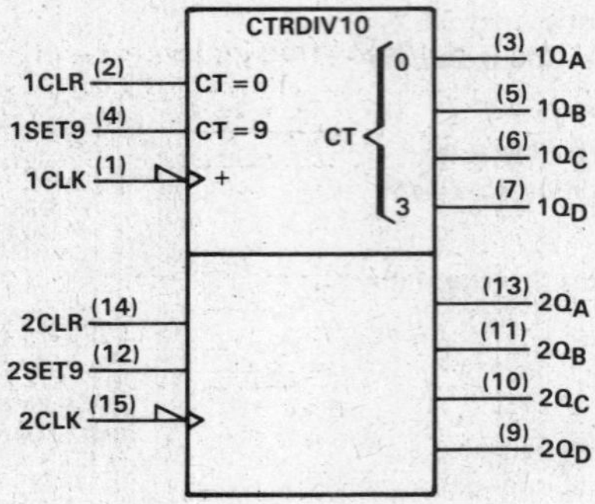

Pin numbers shown are for J and N packages.

logic diagram, each counter (positive logic)

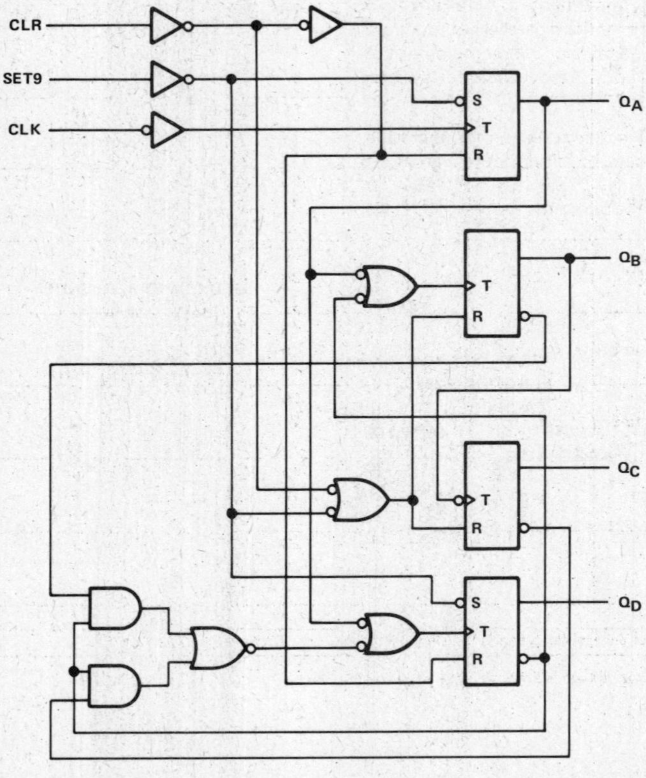

TEXAS INSTRUMENTS
POST OFFICE BOX 225012 • DALLAS, TEXAS 75265

maximum ratings, recommended operating conditions, and electrical characteristics

See Table IV, page 2-10.

timing requirements over recommended operating free-air temperature range (unless otherwise noted)

		V_{CC}	T_A = 25°C MIN	T_A = 25°C MAX	SN54HC490 MIN	SN54HC490 MAX	SN74HC490 MIN	SN74HC490 MAX	UNIT
f_{clock}	Clock frequency	2 V	0	6	0	4.2	0	5	MHz
		4.5 V	0	31	0	21	0	25	
		6 V	0	36	0	25	0	28	
t_w	Pulse duration, any input	2 V	80		120		100		ns
		4.5 V	16		24		20		
		6 V	14		20		18		
t_{su}	Setup time, CLR or set-to-9 inactive	2 V	25		25		25		ns
		4.5 V	5		5		5		
		6 V	5		5		5		

switching characteristics over recommended operating free-air temperature range (unless otherwise noted), C_L = 50 pF (see Note 1)

PARAMETER	FROM (INPUT)	TO (OUTPUT)	V_{CC}	T_A = 25°C MIN	T_A = 25°C TYP	T_A = 25°C MAX	SN54HC490 MIN	SN54HC490 MAX	SN74HC490 MIN	SN74HC490 MAX	UNIT
f_{max}			2 V	6			4.2		5		MHz
			4.5 V	31			21		25		
			6 V	36			25		28		
t_{pd}	CLK	Q_A	2 V		50	125		190		155	ns
			4.5 V		15	25		38		31	
			6 V		12	21		32		26	
	CLK	Q_B, Q_D	2 V		80	185		280		230	
			4.5 V		23	37		56		46	
			6 V		18	31		48		39	
	CLK	Q_C	2 V		100	235		355		295	
			4.5 V		30	47		71		59	
			6 V		23	40		60		50	
t_{PLH}	Set-to-9	Q_A, Q_D	2 V		60	185		280		230	ns
			4.5 V		19	37		56		46	
			6 V		16	31		48		39	
t_{PHL}	Set-to-9	Q_B, Q_C	2 V		54	140		210		175	ns
			4.5 V		18	28		42		35	
			6 V		16	24		36		30	
	Clear	Any	2 V		50	130		195		165	
			4.5 V		17	26		39		33	
			6 V		15	22		33		28	
t_t		Any	2 V		28	75		110		95	ns
			4.5 V		8	15		22		19	
			6 V		6	13		19		16	

C_{pd}	Power dissipation capacitance per counter	No load, T_A = 25°C	40 pF typ

NOTE 1: For load circuits and voltage waveforms, see page 1-14.

TYPES SN54HC563, SN74HC563
OCTAL D-TYPE TRANSPARENT LATCHES WITH 3-STATE OUTPUTS

D2684, DECEMBER 1982—REVISED MARCH 1984

- High-Current 3-State Outputs Drive Bus-Lines Directly or up to 15 LSTTL Loads
- Bus-Structured Pinout
- Package Options Include Both Plastic and Ceramic Chip Carriers in Addition to Plastic and Ceramic DIPs
- Dependable Texas Instruments Quality and Reliability

SN54HC563 . . . J PACKAGE
SN74HC563 . . . J OR N PACKAGE
(TOP VIEW)

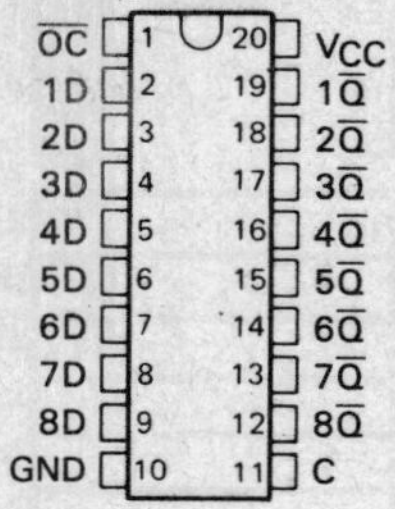

SN54HC563 . . . FH OR FK PACKAGE
SN74HC563 . . . FH OR FN PACKAGE
(TOP VIEW)

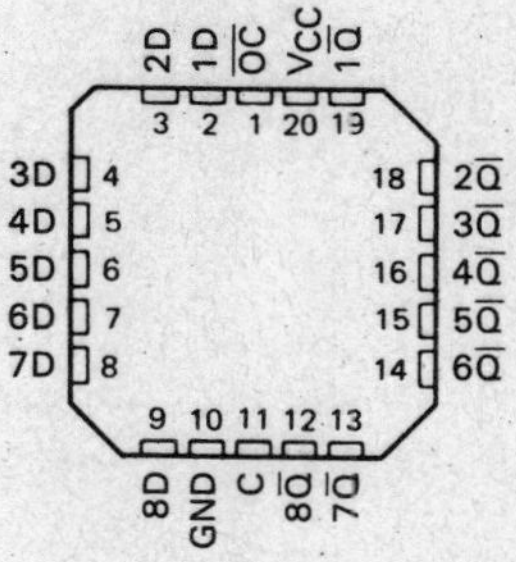

description

These 8-bit latches feature three-state outputs designed specifically for driving highly capacitive or relatively low-impedance loads. They are particularly suitable for implementing buffer registers, I/O ports, bidirectional bus drivers, and working registers.

The eight latches are transparent D-type latches. While the enable (C) is high the $\overline{Q}$ outputs will follow the complements of data (D) inputs. When the enable is taken low the outputs will be latched at the inverses of the levels that were set up at the D inputs.

An output-control input can be used to place the eight outputs in either a normal logic state (high or low logic levels) or a high-impedance state. In the high-impedance state the outputs neither load nor drive the bus lines significantly. The high-impedance state and increased high-logic level provide the capability to drive the bus lines in a bus-organized system without need for interface or pull-up components.

The output control ($\overline{OC}$) does not affect the internal operation of the latches. Old data can be retained or new data can be entered while the outputs are in the high-impedance state.

The SN54HC563 is characterized for operation over the full military temperature range of −55°C to 125°C. The SN74HC563 is characterized for operation from −40°C to 85°C.

FUNCTION TABLE
(Each Latch)

INPUTS			OUTPUT
$\overline{OC}$	ENABLE C	D	$\overline{Q}$
L	H	H	L
L	H	L	H
L	L	X	Q_0
H	X	X	Z

logic symbol

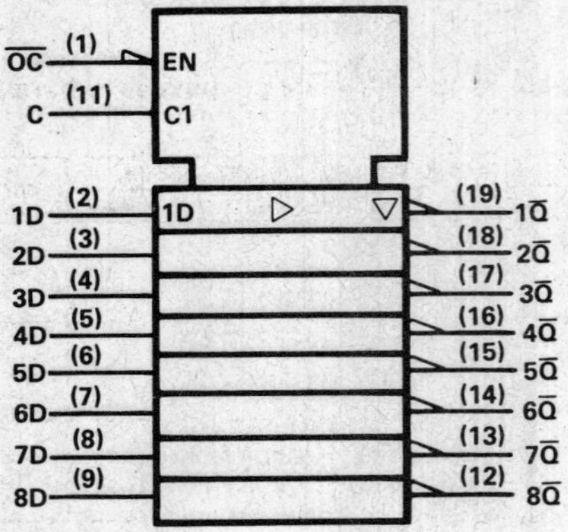

logic diagram (positive logic)

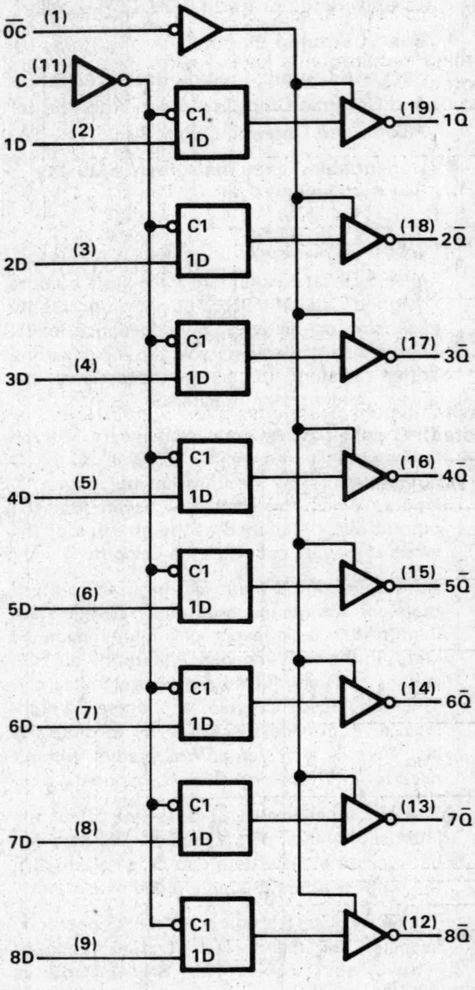

maximum ratings, recommended operating conditions, and electrical characteristics

See Table III, page 2-8.

timing requirements over recommended operating free-air temperature range (unless otherwise noted)

		V_{CC}	T_A = 25°C		SN54HC563		SN74HC563		UNIT
			MIN	MAX	MIN	MAX	MIN	MAX	
t_w	Pulse duration, enable C high	2 V	80		120		100		ns
		4.5 V	16		24		20		
		6 V	14		20		17		
t_{su}	Setup time, data before enable C↓	2 V	50		75		63		ns
		4.5 V	10		15		13		
		6 V	9		13		11		
t_h	Hold time, data after enable C↓	2 V	5		5		5		ns
		4.5 V	5		5		5		
		6 V	5		5		5		

switching characteristics over recommended operating free-air temperature range (unless otherwise noted), C_L = 50 pF (see Note 1)

PARAMETER	FROM (INPUT)	TO (OUTPUT)	V_{CC}	T_A = 25°C			SN54HC563		SN74HC563		UNIT
				MIN	TYP	MAX	MIN	MAX	MIN	MAX	
t_{pd}	D	$\overline{Q}$	2 V		77	175		265		220	ns
			4.5 V		26	35		53		44	
			6 V		23	30		45		38	
t_{pd}	C	Any	2 V		90	175		265		220	ns
			4.5 V		27	35		53		44	
			6 V		23	30		45		38	
t_{en}	$\overline{OC}$	Any	2 V		70	150		225		190	ns
			4.5 V		24	30		45		38	
			6 V		21	26		38		32	
t_{dis}	$\overline{OC}$	Any	2 V		47	150		225		190	ns
			4.5 V		23	30		45		38	
			6 V		21	26		38		32	
t_t		Any	2 V		28	60		90		75	ns
			4.5 V		8	12		18		15	
			6 V		6	10		15		13	

C_{pd}	Power dissipation capacitance per latch	No load, T_A = 25°C	50 pF typ

NOTE 1: For load circuits and voltage waveforms, see page 1-14.

switching characteristics over recommended operating free-air temperature range (unless otherwise noted), C_L = 150 pF (see Note 1)

PARAMETER	FROM (INPUT)	TO (OUTPUT)	V_{CC}	T_A = 25°C			SN54HC563		SN74HC563		UNIT
				MIN	TYP	MAX	MIN	MAX	MIN	MAX	
t_{pd}	D	$\overline{Q}$	2 V		95	200		300		250	ns
			4.5 V		33	40		60		50	
			6 V		29	34		51		43	
t_{pd}	C	Any	2 V		103	225		335		285	ns
			4.5 V		33	45		67		57	
			6 V		29	38		57		48	
t_{en}	$\overline{OC}$	Any	2 V		85	200		300		250	ns
			4.5 V		29	40		60		50	
			6 V		26	34		51		43	
t_t		Any	2 V		60	210		315		265	ns
			4.5 V		17	42		63		53	
			6 V		14	36		53		45	

NOTE 1: For load circuits and voltage waveforms, see page 1-14.

D latch signal conventions

It is TI practice to name the outputs and other inputs of a D-type latch and to draw its logic symbol based on the assumption of true data (D) inputs. Then outputs that produce data in phase with the data inputs are called Q and those producing complementary data are called $\overline{Q}$. An input that causes a Q output to go high or a $\overline{Q}$ output to go low is called Preset; an input that causes a $\overline{Q}$ output to go high or a Q output to go low is called Clear. Bars are used over these pin names ($\overline{PRE}$ and $\overline{CLR}$) if they are active-low.

In some applications it may be advantageous to redesignate the data input $\overline{D}$. In that case all the other inputs and outputs should be renamed as shown below. Also shown are corresponding changes in the graphical symbol. Arbitrary pin numbers are shown in parentheses.

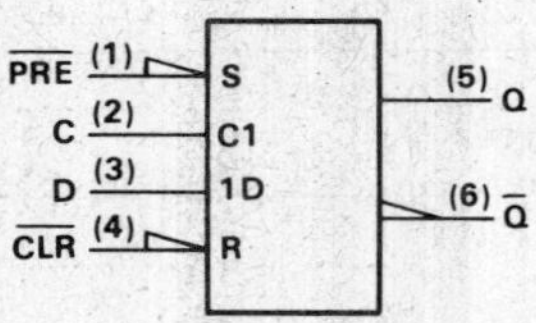

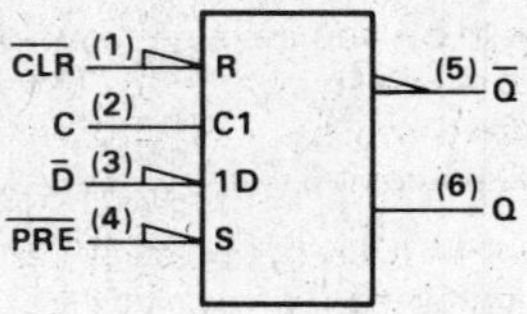

Notice that Q and $\overline{Q}$ exchange names, which causes Preset and Clear to do likewise. Also notice that the polarity indicators (⊳) on $\overline{PRE}$ and $\overline{CLR}$ remain since these inputs are still active-low, but that the presence or absence of the polarity indicator changes at $\overline{D}$, Q, and $\overline{Q}$. Of course pin 5 ($\overline{Q}$) is still in phase with the data input $\overline{D}$, but now both are considered active-low.

TYPES SN54HCT563, SN74HCT563
OCTAL D-TYPE TRANSPARENT LATCHES WITH 3-STATE OUTPUTS

D2804, MARCH 1984

- **Inputs are TTL-Voltage Compatible**
- **High-Current 3-State Outputs Drive Bus-Lines Directly or up to 15 LSTTL Loads**
- **Bus-Structured Pinout**
- **Package Options Include Both Plastic and Ceramic Chip Carriers in Addition to Plastic and Ceramic DIPs**
- **Dependable Texas Instruments Quality and Reliability**

SN54HCT563 . . . J PACKAGE
SN74HCT563 . . . J OR N PACKAGE
(TOP VIEW)

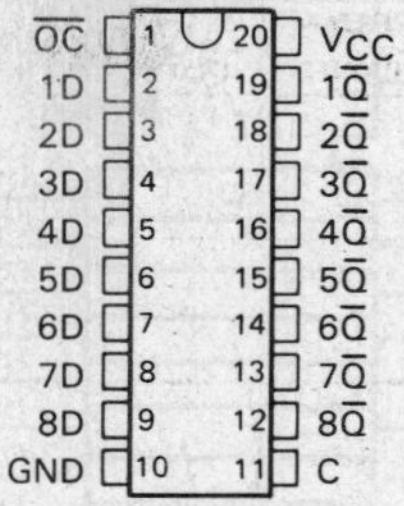

SN54HCT563 . . . FH OR FK PACKAGE
SN54HCT563 . . . FH OR FN PACKAGE
(TOP VIEW)

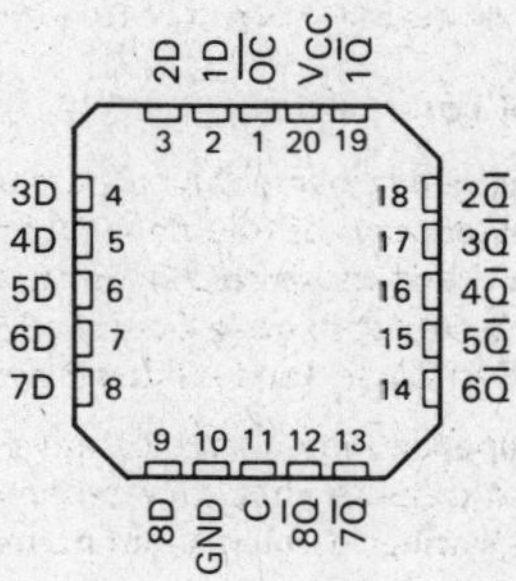

description

These 8-bit latches feature three-state outputs designed specifically for driving highly capacitive or relatively low-impedance loads. They are particularly suitable for implementing buffer registers, I/O ports, bidirectional bus drivers, and working registers.

The eight latches are transparent D-type latches. While the enable (C) is high the $\overline{Q}$ outputs will follow the complements of the data (D) inputs. When the enable is taken low the outputs will be latched at the inverses of the levels that were set up at the D inputs.

An output-control input can be used to place the eight outputs in either a normal logic state (high or low logic levels) or a high-impedance state. In the high-impedance state the outputs neither load nor drive the bus lines significantly. The high-impedance state and increased high-logic level provide the capability to drive the bus lines in a bus-organized system without need for interface or pull-up components.

The output control ($\overline{OC}$) does not affect the internal operation of the latches. Old data can be retained or new data can be entered while the outputs are in the high-impedance state.

The SN54HCT563 is characterized for operation over the full military temperature range of −55°C to 125°C. The SN74HCT563 is characterized for operation from −40°C to 85°C.

FUNCTION TABLE
(Each Latch)

INPUTS			OUTPUT
	ENABLE		$\overline{Q}$
$\overline{OC}$	C	D	
L	H	H	L
L	H	L	H
L	L	X	Q_0
H	X	X	Z

logic symbol

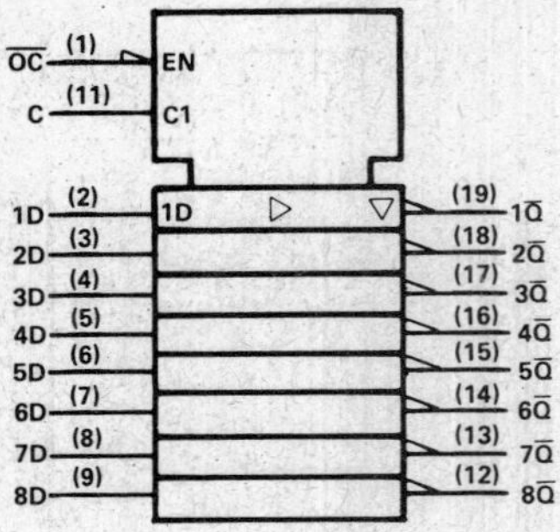

logic diagram (positive logic)

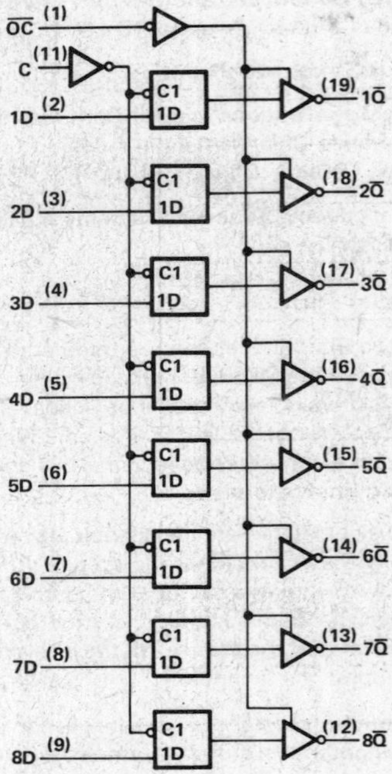

maximum ratings, recommended operating conditions, and electrical characteristics

See Table VII, page 2-14.

timing requirements over recommended operating free-air temperature range (unless otherwise noted)

		V_{CC}	T_A = 25°C		SN54HCT563		SN74HCT563		UNIT
			MIN	MAX	MIN	MAX	MIN	MAX	
t_w	Pulse duration, enable C high	4.5 V	20		30		25		ns
		5.5 V	17		27		23		
t_{su}	Setup time, data before enable C↓	4.5 V	10		15		13		ns
		5.5 V	9		14		12		
t_h	Hold time, data after enable C↓	4.5 V	5		5		5		ns
		5.5 V	5		5		5		

3

HCMOS DEVICES

TEXAS INSTRUMENTS
POST OFFICE BOX 225012 • DALLAS, TEXAS 75265

384

switching characteristics over recommended operating free-air temperature range (unless otherwise noted), C_L = 50 pF (see Note 1)

PARAMETER	FROM (INPUT)	TO (OUTPUT)	V_{CC}	T_A = 25°C MIN	T_A = 25°C TYP	T_A = 25°C MAX	SN54HCT563 MIN	SN54HCT563 MAX	SN74HCT563 MIN	SN74HCT563 MAX	UNIT
t_{pd}	D	$\overline{Q}$	4.5 V		28	35		53		44	ns
			5.5 V		24	32		48		40	
t_{pd}	C	Any	4.5 V		30	35		53		44	ns
			5.5 V		28	32		48		40	
t_{en}	$\overline{OC}$	Any	4.5 V		29	35		53		44	ns
			5.5 V		25	32		48		40	
t_{dis}	$\overline{OC}$	Any	4.5 V		25	35		53		44	ns
			5.5 V		24	32		48		40	
t_t		Any	4.5 V		10	12		18		15	ns
			5.5 V		9	11		16		14	

C_{pd}	Power dissipation capacitance per latch	No load, T_A = 25°C	50 pF typ

switching characteristics over recommended operating free-air temperature range (unless otherwise noted), C_L = 150 pF (see Note 1)

PARAMETER	FROM (INPUT)	TO (OUTPUT)	V_{CC}	T_A = 25°C MIN	T_A = 25°C TYP	T_A = 25°C MAX	SN54HCT563 MIN	SN54HCT563 MAX	SN74HCT563 MIN	SN74HCT563 MAX	UNIT
t_{pd}	D	$\overline{Q}$	4.5 V		36	52		79		65	ns
			5.5 V		32	47		71		59	
t_{pd}	C	Any	4.5 V		40	52		79		65	ns
			5.5 V		38	47		71		59	
t_{en}	$\overline{OC}$	Any	4.5 V		35	52		79		65	ns
			5.5 V		29	47		71		59	
t_t		Any	4.5 V		18	42		63		53	ns
			5.5 V		16	38		57		48	

NOTE 1: For load circuits and voltage waveforms, see page 1-14.

3
HCMOS DEVICES

D latch signal conventions

It is TI practice to name the outputs and other inputs of a D-type latch and to draw its logic symbol based on the assumption of true data (D) inputs. Then outputs that produce data in phase with the data inputs are called Q and those producing complementary data are called $\overline{Q}$. An input that causes a Q output to go high or a $\overline{Q}$ output to go low is called Preset; an input that causes a $\overline{Q}$ output to go high or a Q output to go low is called Clear. Bars are used over these pin names ($\overline{PRE}$ and $\overline{CLR}$) if they are active-low.

In some applications it may be advantageous to redesignate the data input $\overline{D}$. In that case all the other inputs and outputs should be renamed as shown below. Also shown are corresponding changes in the graphical symbol. Arbitrary pin numbers are shown in parentheses.

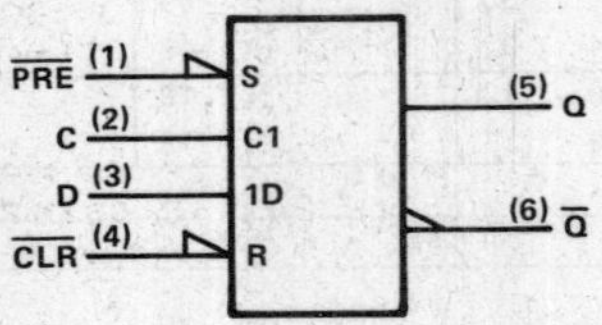

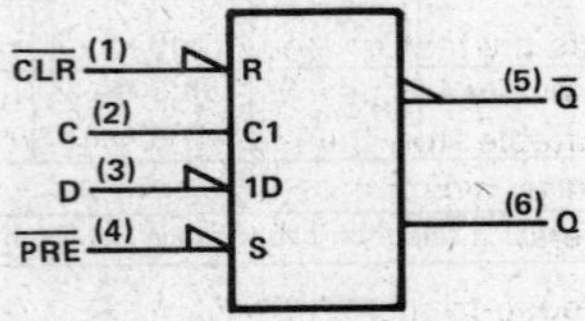

Notice that Q and $\overline{Q}$ exchange names, which causes Preset and Clear to do likewise. Also notice that the polarity indicators (⊵) on $\overline{PRE}$ and $\overline{CLR}$ remain since these inputs are still active-low, but that the presence or absence of the polarity indicator changes at $\overline{D}$, Q, and $\overline{Q}$. Of course pin 5 ($\overline{Q}$) is still in phase with the data input $\overline{D}$, but now both are considered active-low.

TYPES SN54HC564, SN74HC564
OCTAL D-TYPE EDGE-TRIGGERED FLIP-FLOPS WITH 3-STATE OUTPUTS

D2684, DECEMBER 1982—REVISED MARCH 1984

- **High-Current 3-State Inverting Outputs Drive Bus-Lines Directly or up to 15 LSTTL Loads**
- **Bus-Structured Pinout**
- **Package Options Include Both Plastic and Ceramic Chip Carriers in Addition to Plastic and Ceramic DIPs**
- **Dependable Texas Instruments Quality and Reliability**

SN54HC564 . . . J PACKAGE
SN74HC564 . . . J OR N PACKAGE
(TOP VIEW)

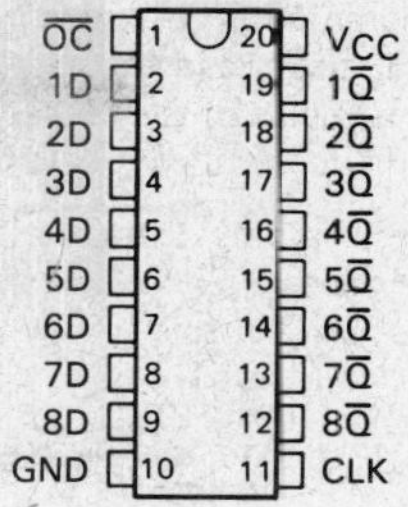

SN54HC564 . . . FH OR FK PACKAGE
SN74HC564 . . . FH OR FN PACKAGE
(TOP VIEW)

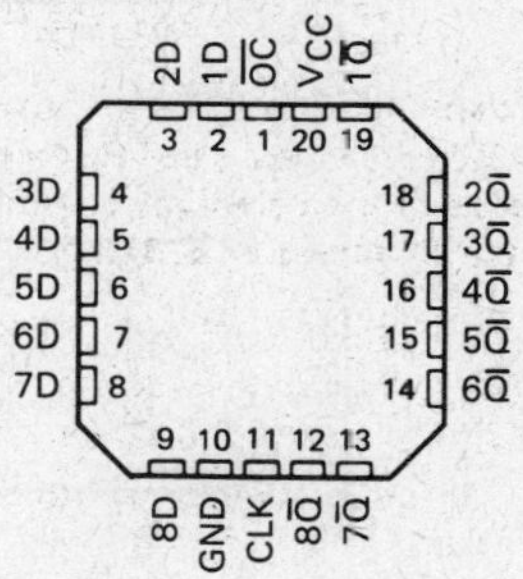

description

These 8-bit registers feature three-state outputs designed specifically for bus driving. They are particularly suitable for implementing buffer registers, I/O ports, bidirectional bus drivers, and working registers.

The eight-bit edge-triggered D-type flip-flops enter data on the low-to-high transition of the clock.

An output control does not affect the internal operation of the flip-flops. Old data can be retained or new data can be entered while the outputs are in the high-impedance state.

The SN54HC564 is characterized for operation over the full military temperature range of −55°C to 125°C. The SN74HC564 is characterized for operation from −40°C to 85°C.

FUNCTION TABLE
(EACH FLIP-FLOP)

INPUTS			OUTPUT
$\overline{OC}$	CLK	D	$\overline{Q}$
L	↑	H	L
L	↑	L	H
L	L	X	$\overline{Q}_0$
H	X	X	Z

logic symbol

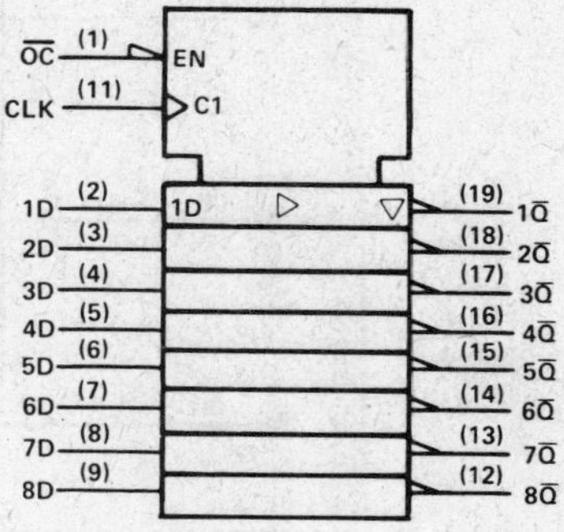

logic diagram (positive logic)

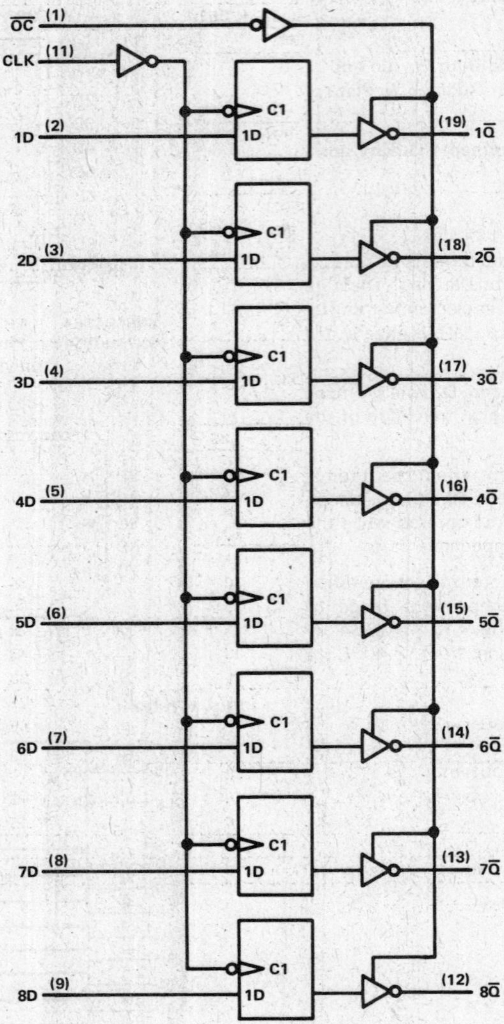

maximum ratings, recommended operating conditions, and electrical characteristics

See Table III, page 2-8.

TEXAS INSTRUMENTS
POST OFFICE BOX 225012 • DALLAS, TEXAS 75265

timing requirements over recommended operating free-air temperature range (unless otherwise noted)

		V_{CC}	$T_A = 25°C$		SN54HC564		SN74HC564		UNIT
			MIN	MAX	MIN	MAX	MIN	MAX	
f_{clock}	Clock frequency	2 V	0	6	0	4.2	0	5	MHz
		4.5 V	0	31	0	21	0	25	
		6 V	0	36	0	25	0	29	
t_w	Pulse duration, CLK high or low	2 V	80		120		100		ns
		4.5 V	16		24		20		
		6 V	14		20		17		
t_{su}	Setup time, data before CLK↑	2 V	100		150		125		ns
		4.5 V	20		30		25		
		6 V	17		26		21		
t_h	Hold time, data after CLK↑	2 V	5		5		5		ns
		4.5 V	5		5		5		
		6 V	5		5		5		

switching characteristics over recommended operating free-air temperature range (unless otherwise noted), $C_L = 50$ pF (see Note 1)

PARAMETER	FROM (INPUT)	TO (OUTPUT)	V_{CC}	$T_A = 25°C$			SN54HC564		SN74HC564		UNIT
				MIN	TYP	MAX	MIN	MAX	MIN	MAX	
f_{max}			2 V	6	11		4.2		5		MHz
			4.5 V	31	36		21		25		
			6 V	36	40		25		29		
t_{pd}	CLK	Any	2 V		54	180		270		225	ns
			4.5 V		18	36		54		45	
			6 V		15	31		46		38	
t_{en}	$\overline{OC}$	Any	2 V		45	150		225		190	ns
			4.5 V		15	30		45		38	
			6 V		13	26		38		32	
t_{dis}	$\overline{OC}$	Any	2 V		45	150		225		190	ns
			4.5 V		15	30		45		38	
			6 V		13	26		38		32	
t_t		Any	2 V		28	60		90		75	ns
			4.5 V		8	12		18		15	
			6 V		6	10		15		13	

C_{pd}	Power dissipation capacitance per flip-flop	No load, $T_A = 25°C$	100 pF typ

NOTE 1: For load circuit and voltage waveforms, see page 1-14.

switching characteristics over recommended operating free-air temperature range (unless otherwise noted), C_L = 150 pF (see Note 1)

PARAMETER	FROM (INPUT)	TO (OUTPUT)	V_{CC}	T_A = 25°C MIN	TYP	MAX	SN54HC564 MIN	MAX	SN74HC564 MIN	MAX	UNIT
t_{pd}	CLK	Any	2 V		75	230		345		290	ns
			4.5 V		24	46		69		58	
			6 V		21	34		58		49	
t_{en}	$\overline{OC}$	Any	2 V		57	200		300		250	ns
			4.5 V		19	40		60		50	
			6 V		17	34		51		43	
t_t		Any	2 V		60	210		315		265	ns
			4.5 V		17	42		63		53	
			6 V		14	36		53		45	

NOTE 1: For load circuit and voltage waveforms, see page 1-14.

D flip-flop signal conventions

It is TI practice to name the outputs and other inputs of a D-type flip-flop and to draw its logic symbol based on the assumption of true data (D) inputs. Then outputs that produce data in phase with the data inputs are called Q and those producing complementary data are called $\overline{Q}$. An input that causes a Q output to go high or a $\overline{Q}$ output to go low is called Preset; an input that causes a $\overline{Q}$ output to go high or a Q output to go low is called Clear. Bars are used over these pin names ($\overline{PRE}$ and $\overline{CLR}$) if they are active-low.

In some applications it may be advantageous to redesignate the data input $\overline{D}$. In that case all the other inputs and outputs should be renamed as shown below. Also shown are corresponding changes in the graphical symbol. Arbitrary pin numbers are shown in parentheses.

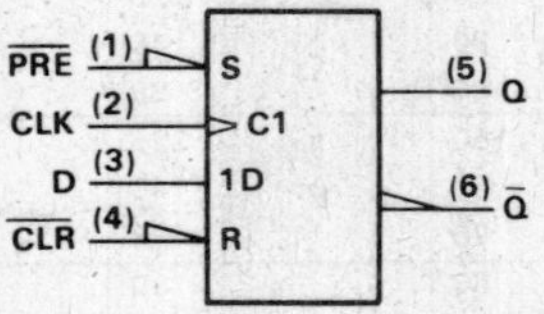

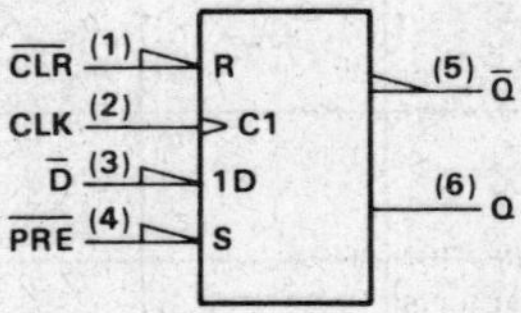

Notice that Q and Q exchange names, which causes Preset and Clear to do likewise. Also notice that the polarity indicators (⊵) on $\overline{PRE}$ and $\overline{CLR}$ remain since these inputs are still active-low, but that the presence or absence of the polarity indicator changes at $\overline{D}$, Q, and $\overline{Q}$. Of course pin 5 ($\overline{Q}$) is still in phase with the data input $\overline{D}$, but now both are considered active-low.

TEXAS INSTRUMENTS
POST OFFICE BOX 225012 • DALLAS, TEXAS 75265

TYPES SN54HCT564, SN74HCT564 OCTAL D-TYPE EDGE-TRIGGERED FLIP-FLOPS WITH 3-STATE OUTPUTS

D2804, MARCH 1984

- **Inputs are TTL-Voltage Compatible**
- **High-Current 3-State Inverting Outputs Drive Bus-Lines Directly or up to 15 LSTTL Loads**
- **Bus-Structured Pinout**
- **Package Options Include Both Plastic and Ceramic Chip Carriers in Addition to Plastic and Ceramic DIPs**
- **Dependable Texas Instruments Quality and Reliability**

SN54HCT564 . . . J PACKAGE
SN74HCT564 . . . J OR N PACKAGE
(TOP VIEW)

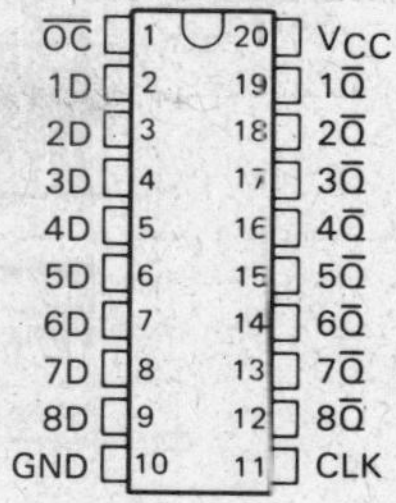

SN54HCT564 . . . FH OR FK PACKAGE
SN74HCT564 . . . FH OR FN PACKAGE
(TOP VIEW)

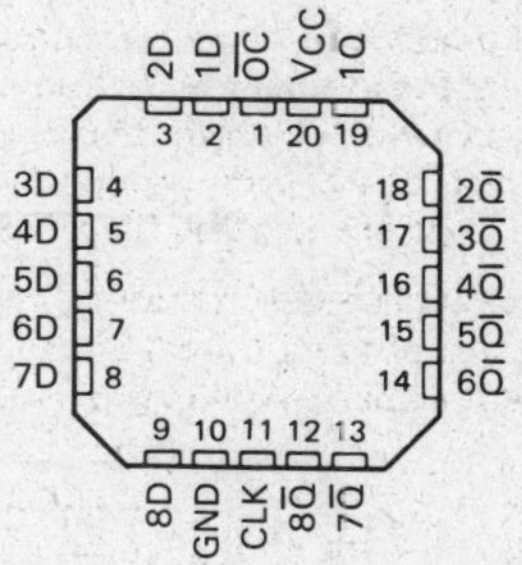

description

These 8-bit registers feature three-state outputs designed specifically for bus driving. They are particularly suitable for implementing buffer registers, I/O ports, bidirectional bus drivers, and working registers.

The eight-bit edge-triggered D-type flip-flops enter data on the low-to-high transition of the clock.

An output control does not affect the internal operation of the flip-flops. Old data can be retained or new data can be entered while the outputs are in the high-impedance state.

The SN54HCT564 is characterized for operation over the full military temperature range of −55 °C to 125 °C. The SN74HCT564 is characterized for operation from −40 °C to 85 °C.

FUNCTION TABLE (EACH FLIP-FLOP)

INPUTS			OUTPUT
$\overline{OC}$	CLK	D	$\overline{Q}$
L	↑	H	L
L	↑	L	H
L	L	X	$\overline{Q}_0$
H	X	X	Z

logic symbol

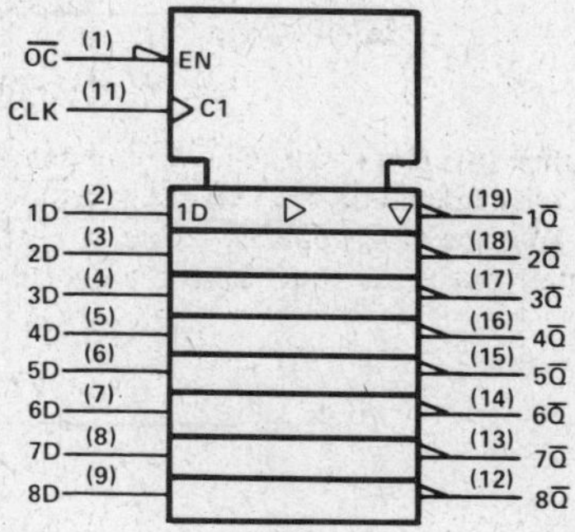

logic diagram (positive logic)

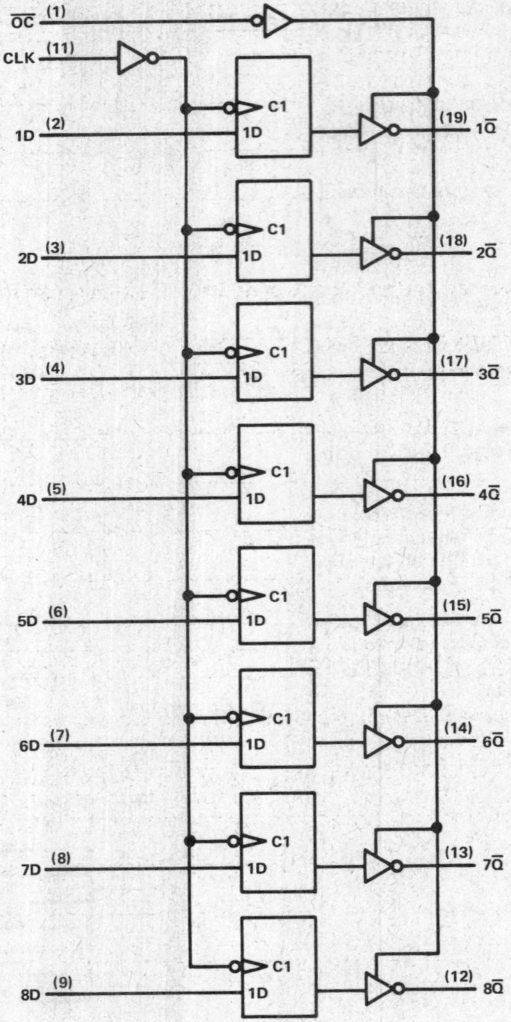

maximum ratings, recommended operating conditions, and electrical characteristics

See Table VII, page 2-14.

timing requirements over recommended operating free-air temperature range (unless otherwise noted)

		V_{CC}	T_A = 25°C MIN	T_A = 25°C MAX	SN54HCT564 MIN	SN54HCT564 MAX	SN74HCT564 MIN	SN74HCT564 MAX	UNIT
f_{clock}	Clock frequency	4.5 V	0	31		21		25	MHz
		5.5 V	0	36		23		28	
t_w	Pulse duration, CLK high or low	4.5 V	16		24		20		ns
		5.5 V	14		22		18		
t_{su}	Setup time, data before CLK↑	4.5 V	20		30		25		ns
		5.5 V	17		27		23		
t_h	Hold time, data after CLK↑	4.5 V	5		5		5		ns
		5.5 V	5		5		5		

switching characteristics over recommended operating free-air temperature range (unless otherwise noted), C_L = 50 pF (see Note 1)

PARAMETER	FROM (INPUT)	TO (OUTPUT)	V_{CC}	T_A = 25°C MIN	T_A = 25°C TYP	T_A = 25°C MAX	SN54HCT564 MIN	SN54HCT564 MAX	SN74HCT564 MIN	SN74HCT564 MAX	UNIT
f_{max}			4.5 V	31	36		21		25		MHz
			5.5 V	36	40		23		28		
t_{pd}	CLK	Any	4.5 V		18	36		54		45	ns
			5.5 V		16	32		48		41	
t_{en}	$\overline{OC}$	Any	4.5 V		14	30		45		38	ns
			5.5 V		10	27		41		34	
t_{dis}	$\overline{OC}$	Any	4.5 V		22	30		45		38	ns
			5.5 V		20	27		41		34	
t_t		Any	4.5 V		10	12		18		15	ns
			5.5 V		9	11		16		14	

C_{pd}	Power dissipation capacitance per flip-flop	No load, T_A = 25°C	93 pF typ

switching characteristics over recommended operating free-air temperature range (unless otherwise noted), C_L = 150 pF (see Note 1)

PARAMETER	FROM (INPUT)	TO (OUTPUT)	V_{CC}	T_A = 25°C MIN	T_A = 25°C TYP	T_A = 25°C MAX	SN54HCT564 MIN	SN54HCT564 MAX	SN74HCT564 MIN	SN74HCT564 MAX	UNIT
t_{pd}	CLK	Any	4.5 V		38	38		80		66	ns
			5.5 V		36	47		71		60	
t_{en}	$\overline{OC}$	Any	4.5 V		30	47		71		59	ns
			5.5 V		27	39		59		49	
t_t		Any	4.5 V		18	42		63		53	ns
			5.5 V		16	38		57		48	

NOTE 1: For load circuit and voltage waveforms, see page 1-14.

D flip-flop signal conventions

It is TI practice to name the outputs and other inputs of a D-type flip-flop and to draw its logic symbol based on the assumption of true data (D) inputs. Then outputs that produce data in phase with the data inputs are called Q and those producing complementary data are called $\overline{Q}$. An input that causes a Q output to go high or a $\overline{Q}$ output to go low is called Preset; an input that causes a $\overline{Q}$ output to go high or a Q output to go low is called Clear. Bars are used over these pin names ($\overline{PRE}$ and $\overline{CLR}$) if they are active-low.

In some applications it may be advantageous to redesignate the data input $\overline{D}$. In that case all the other inputs and outputs should be renamed as shown below. Also shown are corresponding changes in the graphical symbol. Arbitrary pin numbers are shown in parentheses.

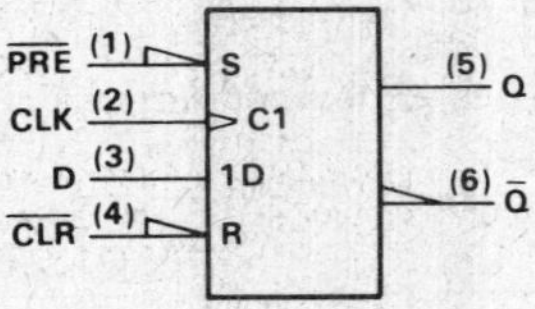

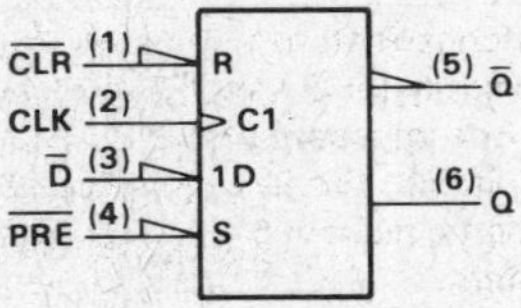

Notice that Q and $\overline{Q}$ exchange names, which causes Preset and Clear to do likewise. Also notice that the polarity indicators (⊵) on $\overline{PRE}$ and $\overline{CLR}$ remain since these inputs are still active-low, but that the presence or absence of the polarity indicator changes at $\overline{D}$, Q, and $\overline{Q}$. Of course pin 5 ($\overline{Q}$) is still in phase with the data input $\overline{D}$, but now both are considered active-low.

TYPES SN54HC573, SN74HC573
OCTAL D-TYPE TRANSPARENT LATCHES WITH 3-STATE OUTPUTS

D2684, DECEMBER 1982—REVISED MARCH 1984

- **High-Current 3-State Outputs Drive Bus-Lines Directly or up to 15 LSTTL Loads**
- **Bus-Structured Pinout**
- **Package Options Include Both Plastic and Ceramic Chip Carriers in Addition to Plastic and Ceramic DIPs**
- **Dependable Texas Instruments Quality and Reliability**

SN54HC573 . . . J PACKAGE
SN74HC573 . . . J OR N PACKAGE
(TOP VIEW)

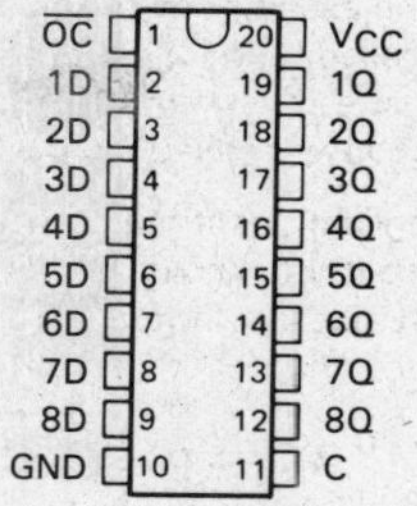

SN54HC573 . . . FH OR FK PACKAGE
SN74HC573 . . . FH OR FN PACKAGE
(TOP VIEW)

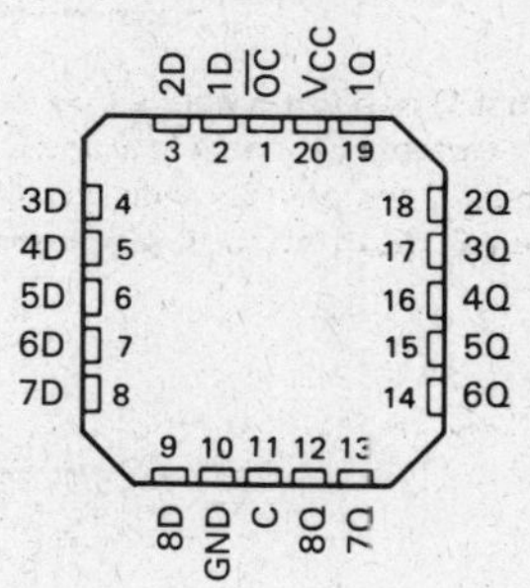

description

These 8-bit latches feature three-state outputs designed specifically for driving highly capacitive or relatively low-impedance loads. They are particularly suitable for implementing buffer registers, I/O ports, bidirectional bus drivers, and working registers.

The eight latches are transparent D-type latches. While the enable (C) is high the outputs (Q) will respond to the data (D) inputs. When the enable is taken low the outputs will be latched to retain the data that was set up.

An output-control input can be used to place the eight outputs in either a normal logic state (high or low logic levels) or a high-impedance state. In the high-impedance state the outputs neither load nor drive the bus lines significantly. The high-impedance state and increased drive provide the capability to drive the bus lines in a bus-organized system without need for interface or pull-up components.

The output control ($\overline{OC}$) does not affect the internal operation of the latches. Old data can be retained or new data can be entered while the outputs are in the high impedance state.

The SN54HC573 is characterized for operation over the full military temperature range of −55 °C to 125 °C. The SN74HC573 is characterized for operation from −40 °C to 85 °C.

FUNCTION TABLE
(EACH LATCH)

INPUTS			OUTPUT
ENABLE			Q
$\overline{OC}$	C	D	
L	H	H	H
L	H	L	L
L	L	X	Q_0
H	X	X	Z

logic symbol

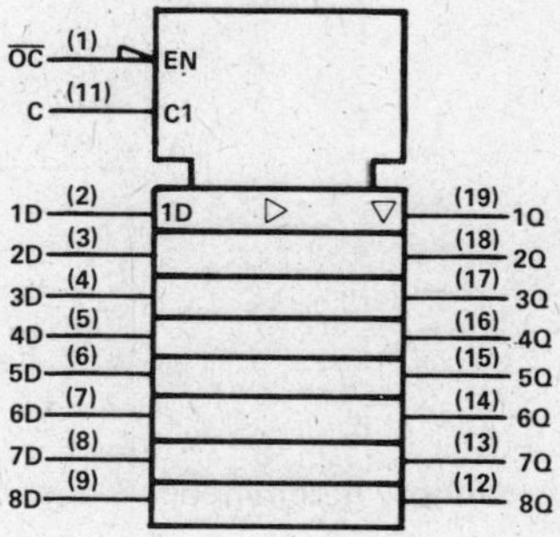

logic diagram (positive logic)

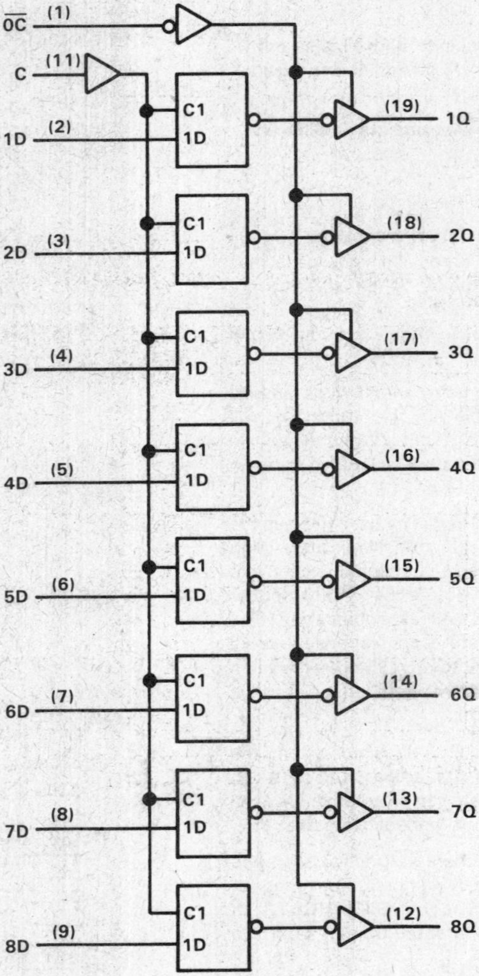

maximum ratings, recommended operating conditions, and electrical characteristics

See Table III, page 2-8.

TEXAS INSTRUMENTS
POST OFFICE BOX 225012 • DALLAS, TEXAS 75265

timing requirements over recommended operating free-air temperature range (unless otherwise noted)

		V_{CC}	T_A = 25°C		SN54HC573		SN74HC573		UNIT
			MIN	MAX	MIN	MAX	MIN	MAX	
t_w	Pulse duration, C high	2 V	80		120		100		ns
		4.5 V	16		24		20		
		6 V	14		20		17		
t_{su}	Setup time, data before enable C↓	2 V	50		75		63		ns
		4.5 V	10		15		13		
		6 V	9		13		11		
t_h	Hold time, data after enable C↓	2 V	5		5		5		ns
		4.5 V	5		5		5		
		6 V	5		5		5		

switching characteristics over recommended operating free-air temperature range (unless otherwise noted), C_L = 50 pF (see Note 1)

PARAMETER	FROM (INPUT)	TO (OUTPUT)	V_{CC}	T_A = 25°C			SN54HC573		SN74HC573		UNIT
				MIN	TYP	MAX	MIN	MAX	MIN	MAX	
t_{pd}	D	Q	2 V		77	175		265		220	ns
			4.5 V		26	35		53		44	
			6 V		23	30		45		38	
t_{pd}	C	Any	2 V		87	175		265		220	ns
			4.5 V		27	35		53		44	
			6 V		23	30		45		38	
t_{en}	$\overline{OC}$	Any	2 V		68	150		225		190	ns
			4.5 V		24	30		45		38	
			6 V		21	26		38		32	
t_{dis}	$\overline{OC}$	Any	2 V		47	150		225		190	ns
			4.5 V		23	30		45		38	
			6 V		21	26		38		32	
t_t		Any	2 V		28	60		90		75	ns
			4.5 V		8	12		18		15	
			6 V		6	10		15		13	

C_{pd}	Power dissipation capacitance per latch	No load, T_A = 25°C	50 pF typ

NOTE 1: For load circuit and voltage waveforms, see page 1-14.

3

HCMOS DEVICES

switching characteristics over recommended operating free-air temperature range (unless otherwise noted), C_L = 150 pF (see Note 1)

PARAMETER	FROM (INPUT)	TO (OUTPUT)	V_{CC}	T_A = 25°C MIN	TYP	MAX	SN54HC573 MIN	MAX	SN74HC573 MIN	MAX	UNIT
t_{pd}	D	Q	2 V		95	200		300		250	ns
			4.5 V		33	40		60		50	
			6 V		21	34		51		43	
t_{pd}	C	Any	2 V		103	225		335		285	ns
			4.5 V		33	45		67		57	
			6 V		29	38		57		48	
t_{en}	$\overline{OC}$	Any	2 V		85	200		300		250	ns
			4.5 V		29	40		60		50	
			6 V		26	34		51		43	
t_t		Any	2 V		60	210		315		265	ns
			4.5 V		17	42		63		53	
			6 V		14	36		53		45	

NOTE 1: For load circuit and voltage waveforms, see page 1-14.

D latch signal conventions

It is TI practice to name the outputs and other inputs of a D-type latch and to draw its logic symbol based on the assumption of true data (D) inputs. Then outputs that produce data in phase with the data inputs are called Q and those producing complementary data are called $\overline{Q}$. An input that causes a Q output to go high or a $\overline{Q}$ output to go low is called Preset; an input that causes a $\overline{Q}$ output to go high or a Q output to go low is called Clear. Bars are used over these pin names ($\overline{PRE}$ and $\overline{CLR}$) if they are active-low.

In some applications it may be advantageous to redesignate the data input $\overline{D}$. In that case all the other inputs and outputs should be renamed as shown below. Also shown are corresponding changes in the graphical symbol. Arbitrary pin numbers are shown in parentheses.

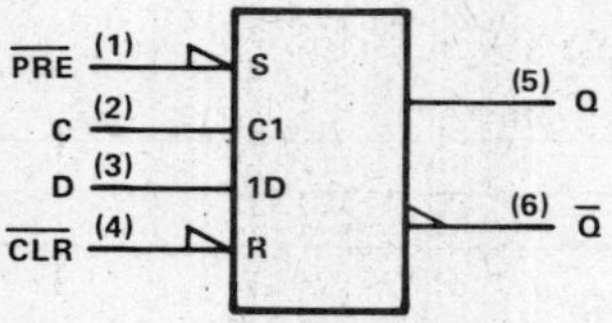

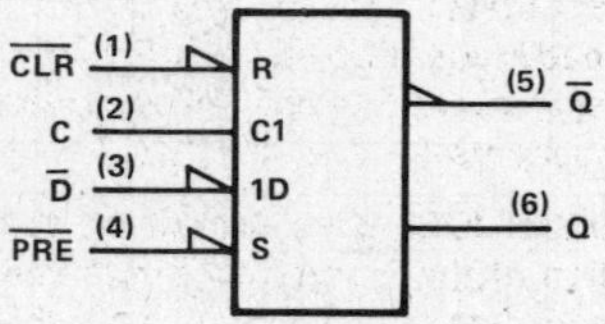

Notice that Q and $\overline{Q}$ exchange names, which causes Preset and Clear to do likewise. Also notice that the polarity indicators (⊵) on $\overline{PRE}$ and $\overline{CLR}$ remain since these inputs are active-low, but that the presence or absence of the polarity indicator changes at $\overline{D}$, Q and $\overline{Q}$. Of course pin 5 ($\overline{Q}$) is still in phase with the data input $\overline{D}$, but now both are considered active-low.

TEXAS INSTRUMENTS
POST OFFICE BOX 225012 • DALLAS, TEXAS 75265

TYPES SN54HCT573, SN74HCT573
OCTAL D-TYPE TRANSPARENT LATCHES WITH 3-STATE OUTPUTS

D2804, MARCH 1984

- Inputs are TTL-Voltage Compatible
- High-Current 3-State Outputs Drive Bus-Lines Directly or up to 15 LSTTL Loads
- Bus-Structured Pinout
- Package Options Include Both Plastic and Ceramic Chip Carriers in Addition to Plastic and Ceramic DIPs
- Dependable Texas Instruments Quality and Reliability

SN54HCT573 . . . J PACKAGE
SN74HCT573 . . . J OR N PACKAGE
(TOP VIEW)

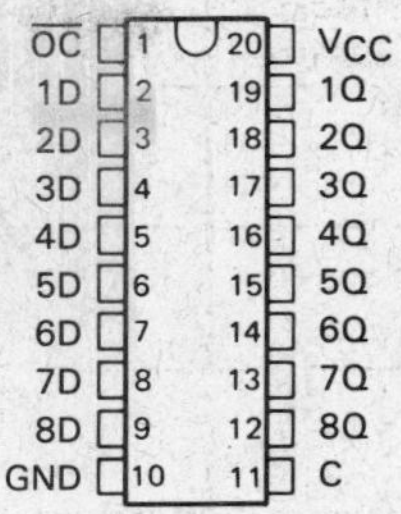

SN54HCT573 . . . FH OR FK PACKAGE
SN74HCT573 . . . FH OR FN PACKAGE
(TOP VIEW)

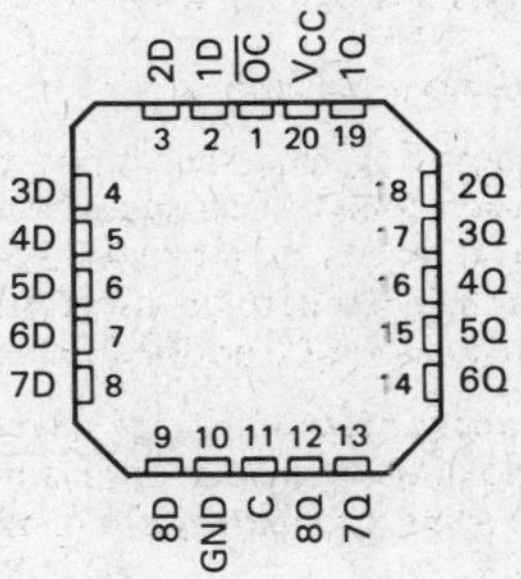

description

These 8-bit latches feature three-state outputs designed specifically for driving highly capacitive or relatively low-impedance loads. They are particularly suitable for implementing buffer registers, I/O ports, bidirectional bus drivers, and working registers.

The eight latches are transparent D-type latches. While the enable (C) is high the outputs (Q) will respond to the data (D) inputs. When the enable is taken low the outputs will be latched to retain the data that was set up.

An output-control input can be used to place the eight outputs in either a normal logic state (high or low logic levels) or a high-impedance state. In the high-impedance state the outputs neither load nor drive the bus lines significantly. The high-impedance state and increased drive provide the capability to drive the bus lines in a bus-organized system without need for interface or pull-up components.

The output control ($\overline{OC}$) does not affect the internal operation of the latches. Old data can be retained or new data can be entered while the outputs are in the high impedance state.

The SN54HCT573 is characterized for operation over the full military temperature range of −55°C to 125°C. The SN74HCT573 is characterized for operation from −40°C to 85°C.

logic symbol

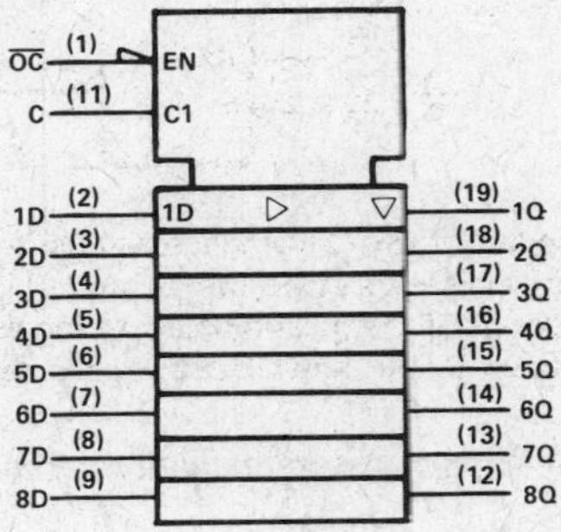

FUNCTION TABLE
(EACH LATCH)

INPUTS			OUTPUT
ENABLE $\overline{OC}$	C	D	Q
L	H	H	H
L	H	L	L
L	L	X	Q_0
H	X	X	Z

logic diagram (positive logic)

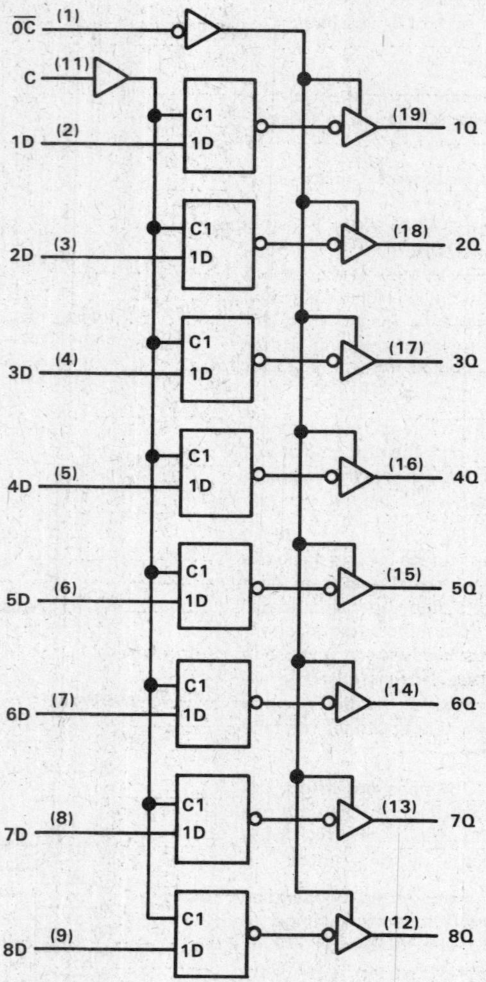

maximum ratings, recommended operating conditions, and electrical characteristics

See Table VII, page 2-14.

TEXAS INSTRUMENTS
POST OFFICE BOX 225012 • DALLAS, TEXAS 75265

timing requirements over recommended operating free-air temperature range (unless otherwise noted)

		V_{CC}	T_A = 25°C		SN54HCT573		SN74HCT573		UNIT
			MIN	MAX	MIN	MAX	MIN	MAX	
t_w	Pulse duration, C high	4.5 V	20		30		25		ns
		5.5 V	17		27		23		
t_{su}	Setup time, data before enable C↓	4.5 V	10		15		13		ns
		5.5 V	9		14		12		
t_h	Hold time, data after enable C↓	4.5 V	5		5		5		ns
		5.5 V	5		5		5		

switching characteristics over recommended operating free-air temperature range (unless otherwise noted), C_L = 50 pF (see Note 1)

PARAMETER	FROM (INPUT)	TO (OUTPUT)	V_{CC}	T_A = 25°C			SN54HCT573		SN74HCT573		UNIT
				MIN	TYP	MAX	MIN	MAX	MIN	MAX	
t_{pd}	D	Q	4.5 V		25	35		53		44	ns
			5.5 V		21	32		48		40	
t_{pd}	C	Any	4.5 V		28	35		53		44	ns
			5.5 V		25	32		48		40	
t_{en}	$\overline{OC}$	Any	4.5 V		26	35		53		44	ns
			5.5 V		23	32		48		40	
t_{dis}	$\overline{OC}$	Any	4.5 V		23	35		53		44	ns
			5.5 V		22	32		48		40	
t_t		Any	4.5 V		9	12		18		15	ns
			5.5 V		9	11		16		14	

C_{pd}	Power dissipation capacitance per latch	No load, T_A = 25°C	50 pF typ

switching characteristics over recommended operating free-air temperature range (unless otherwise noted), C_L = 150 pF (see Note 1)

PARAMETER	FROM (INPUT)	TO (OUTPUT)	V_{CC}	T_A = 25°C			SN54HCT573		SN74HCT573		UNIT
				MIN	TYP	MAX	MIN	MAX	MIN	MAX	
t_{pd}	D	Q	4.5 V		32	52		79		65	ns
			5.5 V		27	47		71		59	
t_{pd}	C	Any	4.5 V		38	52		79		65	ns
			5.5 V		36	47		71		59	
t_{en}	$\overline{OC}$	Any	4.5 V		33	52		79		65	ns
			5.5 V		28	47		71		59	
t_t		Any	4.5 V		18	42		63		53	ns
			5.5 V		16	38		57		48	

C_{pd}	Power dissipation capacitance per latch	No load, T_A = 25°C	50 pF typ

NOTE 1: For load circuit and voltage waveforms, see page 1-14.

D latch signal conventions

It is TI practice to name the outputs and other inputs of a D-type latch and to draw its logic symbol based on the assumption of true data (D) inputs. Then outputs that produce data in phase with the data inputs are called Q and those producing complementary data are called $\overline{Q}$. An input that causes a Q output to go high or a $\overline{Q}$ output to go low is called Preset; an input that causes a $\overline{Q}$ output to go high or a Q output to go low is called Clear. Bars are used over these pin names ($\overline{PRE}$ and $\overline{CLR}$) if they are active-low.

In some applications it may be advantageous to redesignate the data input $\overline{D}$. In that case all the other inputs and outputs should be renamed as shown below. Also shown are corresponding changes in the graphical symbol. Arbitrary pin numbers are shown in parentheses.

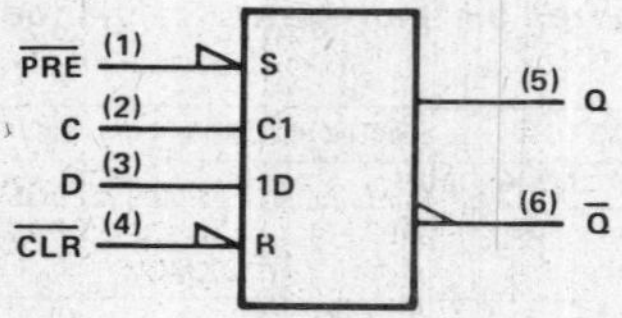

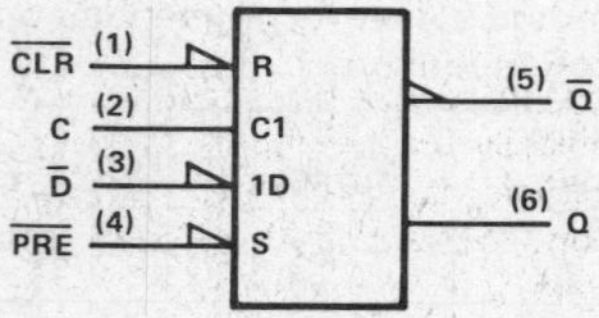

Notice that Q and $\overline{Q}$ exchange names, which causes Preset and Clear to do likewise. Also notice that the polarity indicators (⊵) on $\overline{PRE}$ and $\overline{CLR}$ remain since these inputs are still active-low, but that the presence or absence of the polarity indicator changes at $\overline{D}$, Q, and $\overline{Q}$. Of course pin 5 ($\overline{Q}$) is still in phase with the data input $\overline{D}$, but now both are considered active-low.

TYPES SN54HC574, SN74HC574
OCTAL D-TYPE EDGE-TRIGGERED FLIP-FLOPS WITH 3-STATE OUTPUTS

D2684, DECEMBER 1982—REVISED MARCH 1984

- **High-Current 3-State Noninverting Outputs Drive Bus-Lines Directly or up to 15 LSTTL Loads**
- **Bus-Structured Pinout**
- **Package Options Include Both Plastic and Ceramic Chip Carriers in Addition to Plastic and Ceramic DIPs**
- **Dependable Texas Instruments Quality and Reliability**

SN54HC574 . . . J PACKAGE
SN74HC574 . . . J OR N PACKAGE
(TOP VIEW)

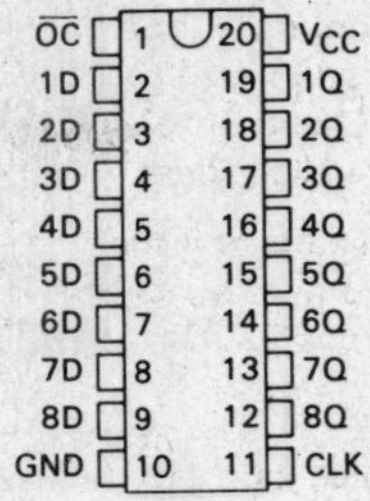

SN54HC574 . . . FH OR FK PACKAGE
SN74HC574 . . . FH OR FN PACKAGE
(TOP VIEW)

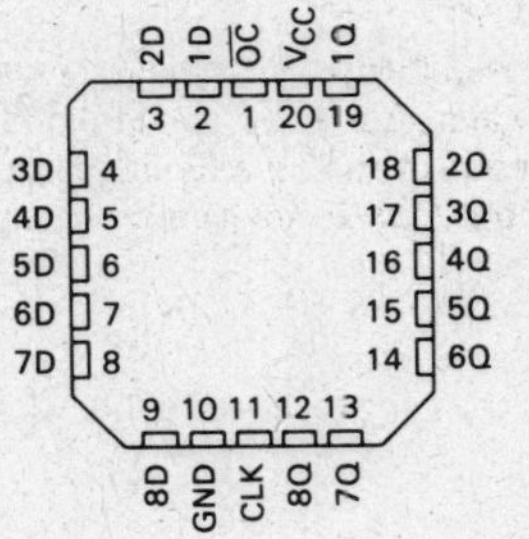

description

These 8-bit registers feature three-state outputs designed specifically for bus driving. They are particularly suitable for implementing buffer registers, I/O ports, bidirectional bus drivers, and working registers.

The eight edge-triggered D-type flip-flops enter data on the low-to-high transition of the clock.

The output-control does not affect the internal operation of the flip-flops. Old data can be retained or new data can be entered while the outputs are in the high-impedance state.

The SN54HC574 is characterized for operation over the full military temperature range of −55°C to 125°C. The SN74HC574 is characterized for operation from −40°C to 85°C.

FUNCTION TABLE
(EACH FLIP-FLOP)

INPUTS			OUTPUT
$\overline{OC}$	CLK	D	Q
L	↑	H	H
L	↑	L	L
L	L	X	Q_0
H	X	X	Z

logic symbol

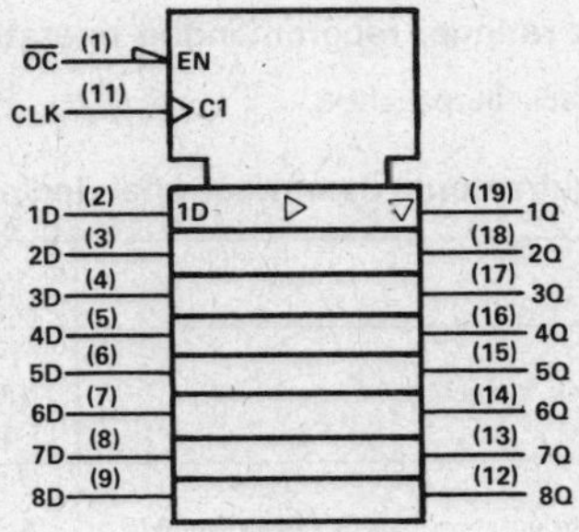

logic diagram (positive logic)

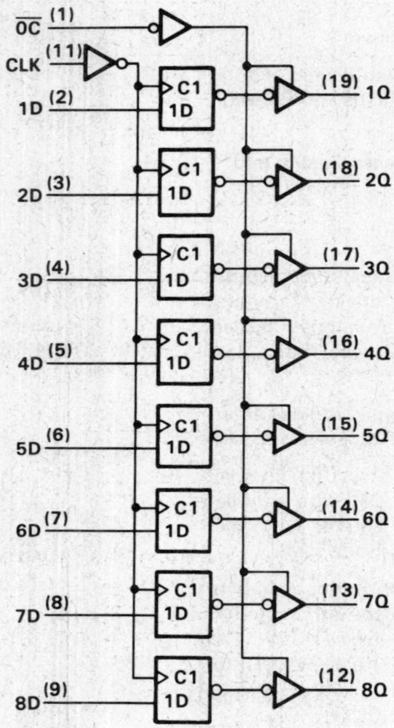

maximum ratings, recommended operating conditions, and electrical characteristics

See Table III, page 2-8.

timing requirements over recommended operating free-air temperature range (unless otherwise noted)

			V_{CC}	$T_A = 25°C$		SN54HC574		SN74HC574		UNIT
				MIN	MAX	MIN	MAX	MIN	MAX	
f_{clock}	Clock frequency		2 V	0	6	0	4	0	5	MHz
			4.5 V	0	30	0	20	0	24	
			6 V	0	36	0	24	0	28	
t_w	Pulse duration	CLK high or low	2 V	80		120		100		ns
			4.5 V	16		24		20		
			6 V	14		20		17		
t_{su}	Setup time, data before CLK↑		2 V	100		150		125		ns
			4.5 V	20		30		25		
			6 V	17		26		21		
t_h	Hold time, data after CLK↑		2 V	5		5		5		ns
			4.5 V	5		5		5		
			6 V	5		5		5		

switching characteristics over recommended operating free-air temperature range (unless otherwise noted), C_L = 50 pF (see Note 1)

PARAMETER	FROM (INPUT)	TO (OUTPUT)	V_{CC}	T_A = 25°C MIN	T_A = 25°C TYP	T_A = 25°C MAX	SN54HC574 MIN	SN54HC574 MAX	SN74HC574 MIN	SN74HC574 MAX	UNIT
f_{max}			2 V	6	11		4		5		MHz
			4.5 V	30	36		20		24		
			6 V	36	40		24		28		
t_{pd}	CLK	Any	2 V		90	180		270		225	ns
			4.5 V		28	36		54		45	
			6 V		24	31		46		38	
t_{en}	$\overline{OC}$	Any	2 V		77	150		225		190	ns
			4.5 V		26	30		45		38	
			6 V		23	26		38		32	
t_{dis}	$\overline{OC}$	Any	2 V		52	150		225		190	ns
			4.5 V		24	30		45		38	
			6 V		22	26		38		32	
t_t		Any	2 V		28	60		90		75	ns
			4.5 V		8	12		18		15	
			6 V		6	10		15		13	

C_{pd}	Power dissipation capacitance	No load, T_A = 25°C	100 pF typ

NOTE 1: For load circuit and voltage waveforms, see page 1-14.

switching characteristics over recommended operating free-air temperature range (unless otherwise noted), C_L = 150 pF (see Note 1)

PARAMETER	FROM (INPUT)	TO (OUTPUT)	V_{CC}	T_A = 25°C MIN	T_A = 25°C TYP	T_A = 25°C MAX	SN54HC574 MIN	SN54HC574 MAX	SN74HC574 MIN	SN74HC574 MAX	UNIT
f_{max}			2 V	6			4		5		MHz
			4.5 V	30			20		24		
			6 V	36			24		28		
t_{pd}	CLK	Any	2 V		105	265		400		330	ns
			4.5 V		36	53		80		66	
			6 V		31	46		68		57	
t_{en}	$\overline{OC}$	Any	2 V		95	235		355		295	ns
			4.5 V		32	47		71		59	
			6 V		28	41		60		51	
t_t		Any	2 V		60	210		315		265	ns
			4.5 V		17	42		63		53	
			6 V		14	36		53		45	

NOTE 1: For load circuit and voltage waveforms, see page 1-14.

D flip-flop signal conventions

It is TI practice to name the outputs and other inputs of a D-type flip-flop and to draw its logic symbol based on the assumption of true data (D) inputs. Then outputs that produce data in phase with the data inputs are called Q and those producing complementary data are called $\overline{Q}$. An input that causes a Q output to go high or a $\overline{Q}$ output to go low is called Preset; an input that causes a $\overline{Q}$ output to go high or a Q output to go low is called Clear. Bars are used over these pin names ($\overline{PRE}$ and $\overline{CLR}$) if they are active-low.

In some applications it may be advantageous to redesignate the data input $\overline{D}$. In that case all the other inputs and outputs should be renamed as shown below. Also shown are corresponding changes in the graphical symbol. Arbitrary pin numbers are shown in parentheses.

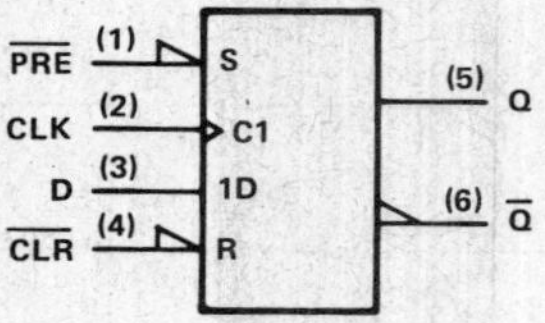

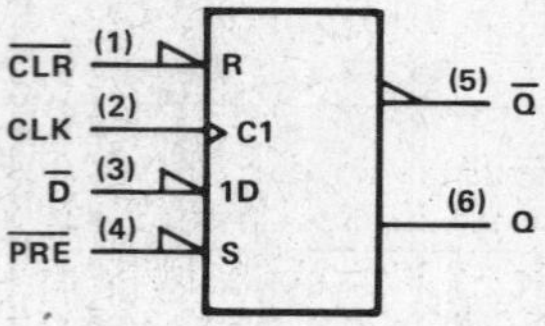

Notice that Q and $\overline{Q}$ exchange names, which causes Preset and Clear to do likewise. Also notice that the polarity indicators (◺) on $\overline{PRE}$ and $\overline{CLR}$ remain since these inputs are still active-low, but that the presence or absence of the polarity indicator changes at $\overline{D}$, Q, and $\overline{Q}$. Of course pin 5 ($\overline{Q}$) is still in phase with the data input $\overline{D}$, but now both are considered active-low.

TYPES SN54HCT574, SN74HCT574
OCTAL D-TYPE EDGE-TRIGGERED FLIP-FLOPS WITH 3-STATE OUTPUTS

D2804, MARCH 1984

- Inputs Are TTL-Voltage Compatible
- High-Current 3-State Noninverting Outputs Drive Bus-Lines Directly or up to 15 LSTTL Loads
- Bus-Structured Pinout
- Package Options Include Both Plastic and Ceramic Chip Carriers in Addition to Plastic and Ceramic DIPs
- Dependable Texas Instruments Quality and Reliability

SN54HCT574 . . . J PACKAGE
SN74HCT574 . . . J OR N PACKAGE
(TOP VIEW)

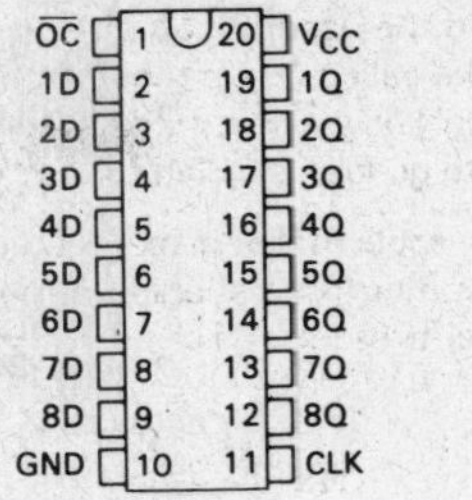

SN54HCT574 . . . FH OR FK PACKAGE
SN74HCT574 . . . FH OR FN PACKAGE
(TOP VIEW)

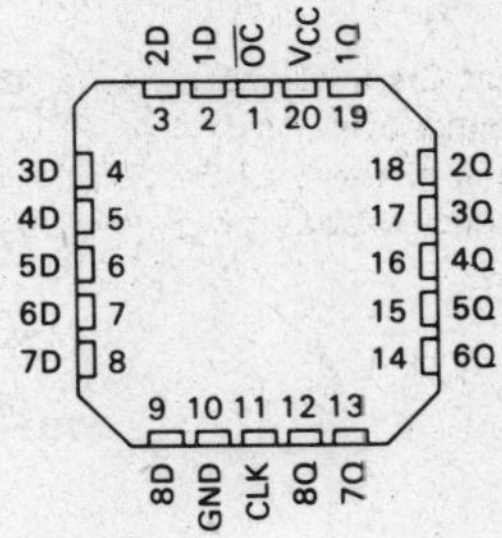

description

These 8-bit registers feature three-state outputs designed specifically for bus driving. They are particularly suitable for implementing buffer registers, I/O ports, bidirectional bus drivers, and working registers.

The eight edge-triggered D-type flip-flops enter data on the low-to-high transition of the clock.

The output-control does not affect the internal operation of the flip-flops. Old data can be retained or new data can be entered while the outputs are in the high-impedance state.

The SN54HCT574 is characterized for operation over the full military temperature range of −55°C to 125°C. The SN74HCT574 is characterized for operation from −40°C to 85°C.

FUNCTION TABLE
(EACH FLIP-FLOP)

INPUTS			OUTPUT
$\overline{OC}$	CLK	D	Q
L	↑	H	H
L	↑	L	L
L	L	X	Q_0
H	X	X	Z

logic symbol

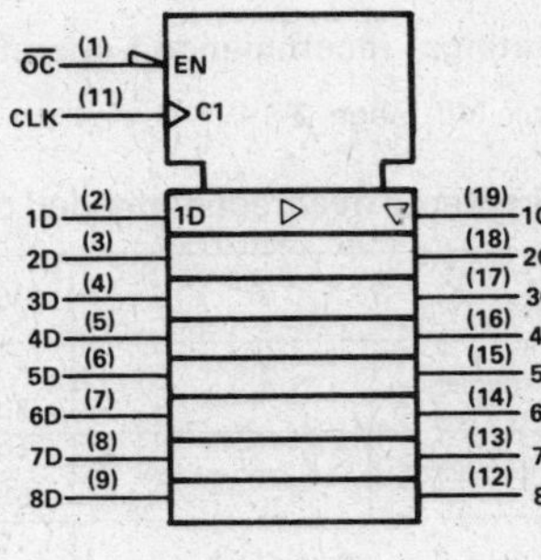

logic diagram (positive logic)

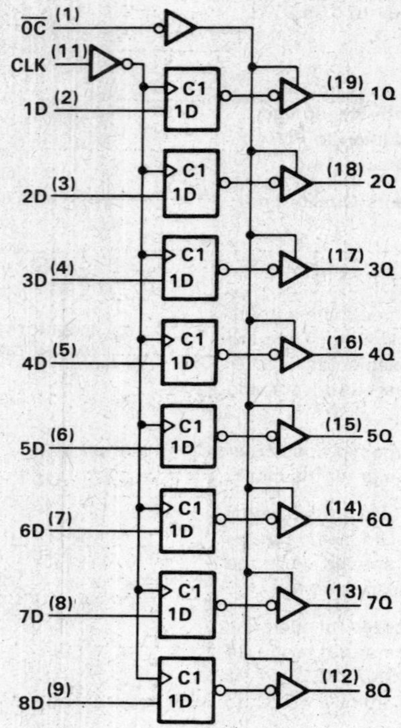

3 HCMOS DEVICES

maximum ratings, recommended operating conditions, and electrical characteristics

See Table VII, page 2-14.

timing requirements over recommended operating free-air temperature range (unless otherwise noted)

			V_{CC}	T_A = 25°C		SN54HCT574		SN74HCT574		UNIT
				MIN	MAX	MIN	MAX	MIN	MAX	
f_{clock}	Clock frequency		4.5 V	0	30	0	20	0	24	MHz
			5.5 V	0	33	0	22	0	27	
t_w	Pulse duration	CLK high or low	4.5 V	16		24		20		ns
			5.5 V	14		22		18		
t_{su}	Setup time, data before CLK↑		4.5 V	20		30		25		ns
			5.5 V	17		27		23		
t_h	Hold time, data after CLK↑		4.5 V	5		5		5		ns
			5.5 V	5		5		5		

TEXAS INSTRUMENTS
POST OFFICE BOX 225012 • DALLAS, TEXAS 75265

switching characteristics over recommended operating free-air temperature range (unless otherwise noted), C_L = 50 pF (see Note 1)

PARAMETER	FROM (INPUT)	TO (OUTPUT)	V_{CC}	T_A = 25°C MIN	TYP	MAX	SN54HCT574 MIN	MAX	SN74HCT574 MIN	MAX	UNIT
f_{max}			4.5 V	30	36		20		24		MHz
			5.5 V	33	40		22		27		
t_{pd}	CLK	Any	4.5 V		30	36		54		45	ns
			5.5 V		25	32		48		41	
t_{en}	$\overline{OC}$	Any	4.5 V		26	30		45		38	ns
			5.5 V		23	27		41		34	
t_{dis}	$\overline{OC}$	Any	4.5 V		23	30		45		38	ns
			5.5 V		22	27		41		34	
t_t		Any	4.5 V		10	12		18		15	ns
			5.5 V		9	11		16		14	

C_{pd}	Power dissipation capacitance per flip-flop	No load, T_A = 25°C	93 pF typ

switching characteristics over recommended operating free-air temperature range (unless otherwise noted) C_L = 150 pF (see Note 1)

PARAMETER	FROM (INPUT)	TO (OUTPUT)	V_{CC}	T_A = 25°C MIN	TYP	MAX	SN54HCT574 MIN	MAX	SN74HCT574 MIN	MAX	UNIT
f_{max}			4.5 V	30	36		20		24		MHz
			5.5 V	33	40		22		27		
t_{pd}	CLK	Any	4.5 V		40	53		80		66	ns
			5.5 V		35	47		71		60	
t_{en}	$\overline{OC}$	Any	4.5 V		34	47		71		59	ns
			5.5 V		29	39		94		78	
t_t		Any	4.5 V		18	42		63		53	ns
			5.5 V		16	38		57		48	

NOTE 1: For load circuit and voltage waveforms, see page 1-14.

3

HCMOS DEVICES

D flip-flop signal conventions

It is TI practice to name the outputs and other inputs of a D-type flip-flop and to draw its logic symbol based on the assumption of true data (D) inputs. Then outputs that produce data in phase with the data inputs are called Q and those producing complementary data are called $\overline{Q}$. An input that causes a Q output to go high or a $\overline{Q}$ output to go low is called Preset; an input that causes a $\overline{Q}$ output to go high or a Q output to go low is called Clear. Bars are used over these pin names ($\overline{PRE}$ and $\overline{CLR}$) if they are active-low.

In some applications it may be advantageous to redesignate the data input $\overline{D}$. In that case all the other inputs and outputs should be renamed as shown below. Also shown are corresponding changes in the graphical symbol. Arbitrary pin numbers are shown in parentheses.

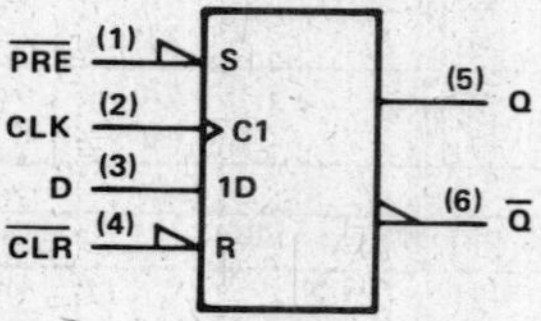

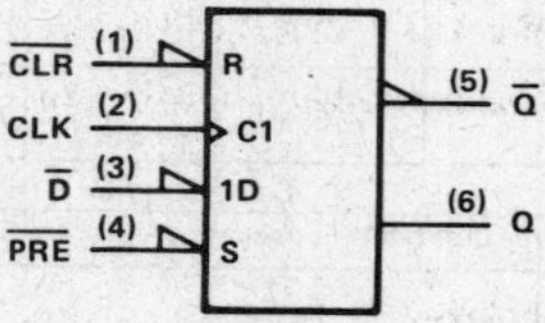

Notice that Q and $\overline{Q}$ exchange names, which causes Preset and Clear to do likewise. Also notice that the polarity indicators (▷) on $\overline{PRE}$ and $\overline{CLR}$ remain since these inputs are still active-low, but that the presence or absence of the polarity indicator changes at $\overline{D}$, Q, and $\overline{Q}$. Of course pin 5 ($\overline{Q}$) is still in phase with the data input $\overline{D}$, but now both are considered active-low.

TYPES SN54HC604, SN74HC604 OCTAL 2-INPUT MULTIPLEXED LATCHES WITH 3-STATE OUTPUTS

D2804, MARCH 1984

- High-Current 3-State Outputs Can Drive up to 15 LSTTL Loads
- 16 D-Type Registers, One for Each Data Input
- Multiplexer Selects Stored Data from Either A Bus or B Bus
- Application-Oriented for Maximum Speed
- Package Options Include Both Plastic and Ceramic Chip Carriers in Addition to Plastic and Ceramic DIPs
- Dependable Texas Instruments Quality and Reliability

SN54HC604 . . . J PACKAGE
SN74HC604 . . . J OR N PACKAGE
(TOP VIEW)

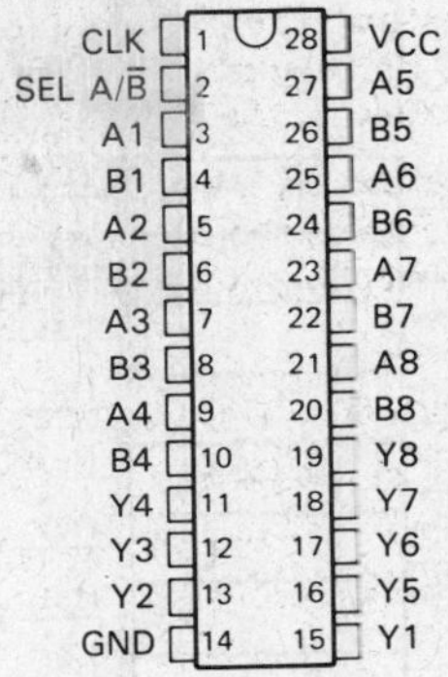

SN54HC604 . . . FH OR FK PACKAGE
SN74HC604 . . . FH OR FN PACKAGE
(TOP VIEW)

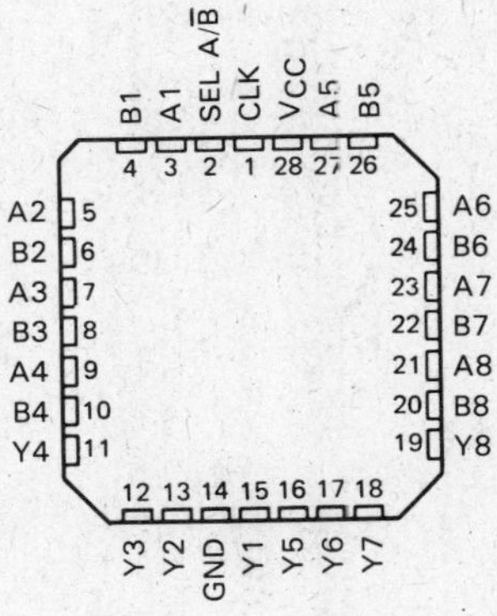

description

The 'HC604 multiplexed latch is ideal for storing data from two input buses, A and B, and for providing the output bus with stored data from either the A or B register.

The clock loads data on the positive-going (low-level to high-level) transition. The clock pin also controls the active and high-impedance states of the outputs. When the clock pin is low, the outputs are in the high-impedance or off state. When the clock pin is high, the outputs are enabled.

The device is optimized for high-speed operation.

These functions are ideal for interfacing from a 16-bit microprocessor to a 64K RAM board. The row and column addresses can be loaded as one word from the microprocessor and then multiplexed sequentially to the RAM during the time that RAS and CAS are active.

The SN54HC604 is characterized for operation over the full military range of −55 °C to 125 °C. The SN74HC604 is characterized for operation from −40 °C to 85 °C.

FUNCTION TABLE

INPUTS				OUTPUTS
A1–A8	B1–B8	A/B̄	CLOCK	Y1–Y8
A data	B data	L	↑	B data
A data	B data	H	↑	A data
X	X	X	L	Z
X	X	L	H	B register stored data
X	X	H	H	A register stored data

logic symbol

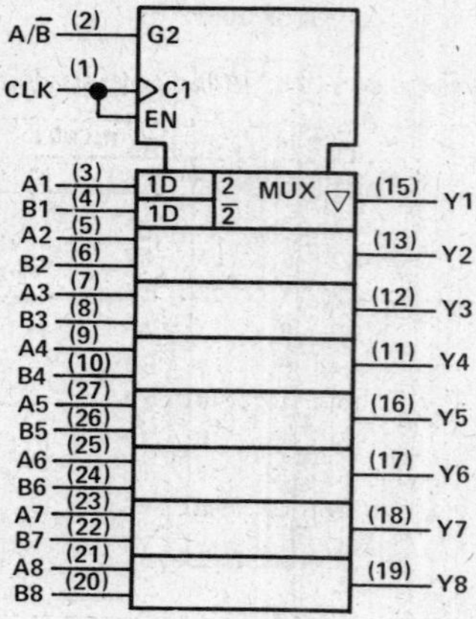

logic diagram (positive logic)

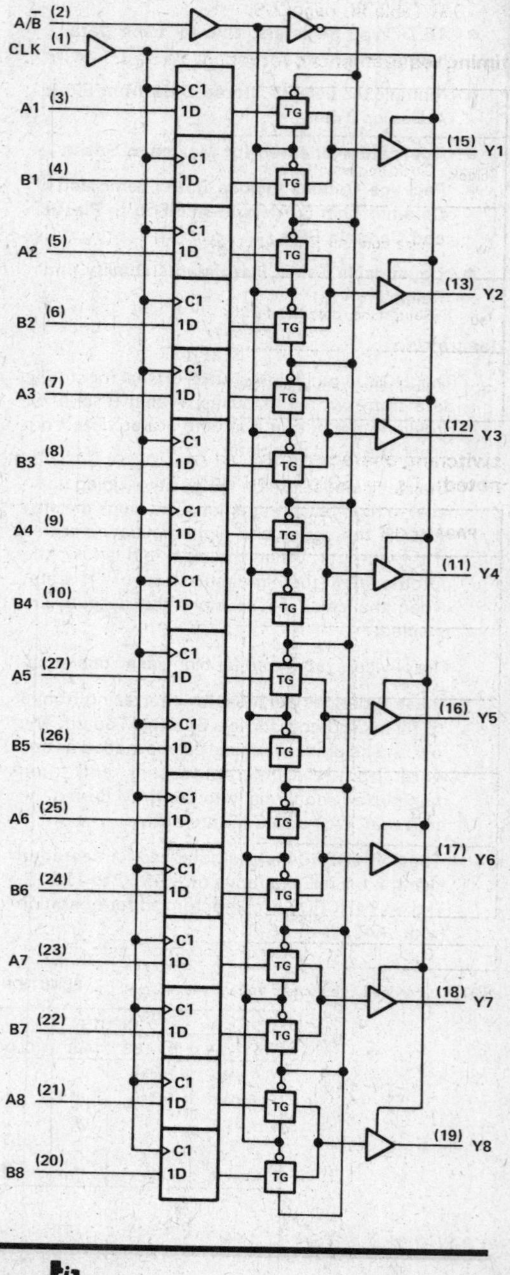

maximum ratings, recommended operating conditions, and electrical characteristics

See Table III, page 2-8.

timing requirements over recommended operating free-air temperature range (unless otherwise noted)

		V_{CC}	T_A = 25°C		SN54HC604		SN74HC604		UNIT
			MIN	MAX	MIN	MAX	MIN	MAX	
f_{clock}	Clock frequency	2 V	0	5	0	3.3	0	4	MHz
		4.5 V	0	25	0	17	0	20	
		6 V	0	29	0	20	0	24	
t_w	Pulse duration, CLK high or low	2 V	100		150		125		ns
		4.5 V	20		30		25		
		6 V	17		25		21		
t_{su}	Setup time, data before CLK↑	2 V	75		115		95		ns
		4.5 V	15		23		19		
		6 V	13		20		16		
t_h	Hold time, data after CLK↑	2 V	5		5		5		ns
		4.5 V	5		5		5		
		6 V	5		5		5		

switching characteristics over recommended operating free-air temperature range (unless otherwise noted), C_L = 50 pF (see Note 1)

PARAMETER	FROM (INPUT)	TO (OUTPUT)	V_{CC}	T_A = 25°C			SN54HC604		SN74HC604		UNIT
				MIN	TYP	MAX	MIN	MAX	MIN	MAX	
f_{max}			2 V	5			3.3		4		MHz
			4.5 V	25			17		20		
			6 V	29			20		24		
t_{pd}	A/$\overline{B}$	Y	2 V		92	170		255		215	ns
			4.5 V		23	34		51		43	
			6 V		17	29		43		37	
t_{en}	CLK	Y	2 V		96	195		295		245	ns
			4.5 V		25	39		59		49	
			6 V		19	33		50		42	
t_{dis}	CLK	Y	2 V		84	200		300		250	ns
			4.5 V		30	40		60		50	
			6 V		26	34		51		43	
t_t		Any	2 V		20	60		90		75	ns
			4.5 V		8	12		18		15	
			6 V		6	10		15		13	

C_{pd}	Power dissipation capacitance per latch	No load, T_A = 25°C	100 pF typ

NOTE 1: For load circuits and voltage waveforms, see page 1-14.

switching characteristics over recommended operating free-air temperature range (unless otherwise noted), C_L = 150 pF (see Note 1)

PARAMETER	FROM (INPUT)	TO (OUTPUT)	V_{CC}	T_A = 25°C			SN54HC604		SN74HC604		UNIT
				MIN	TYP	MAX	MIN	MAX	MIN	MAX	
t_{pd}	A/$\overline{B}$	Y	2 V		110	255		385		320	ns
			4.5 V		28	51		77		64	
			6 V		21	44		65		56	
t_{en}	CLK	Y	2 V		120	280		425		350	ns
			4.5 V		30	56		85		70	
			6 V		23	48		72		61	
t_t		Any	2 V		45	210		315		265	ns
			4.5 V		17	42		63		53	
			6 V		13	36		53		45	

NOTE 1: For load circuits and voltage waveforms, see page 1-14.

TEXAS INSTRUMENTS
POST OFFICE BOX 225012 • DALLAS, TEXAS 75265

TYPES SN54HC620, SN54HC623, SN74HC620, SN74HC623
OCTAL BUS TRANSCEIVERS WITH 3-STATE OUTPUTS

D2684, DECEMBER 1982—REVISED MARCH 1984

- **Lock Bus-Latch Capability**
- **Choice of True or Inverting Logic**
- **High-Current 3-State Outputs Can Drive up to 15 LSTTL Loads**
- **Package Options Include Both Plastic and Ceramic Chip Carriers in Addition to Plastic and Ceramic DIPs**
- **Dependable Texas Instruments Quality and Reliability**

DEVICE	LOGIC
'HC620	Inverting
'HC623	True

SN54HC' . . . J PACKAGE
SN74HC' . . . J OR N PACKAGE
(TOP VIEW)

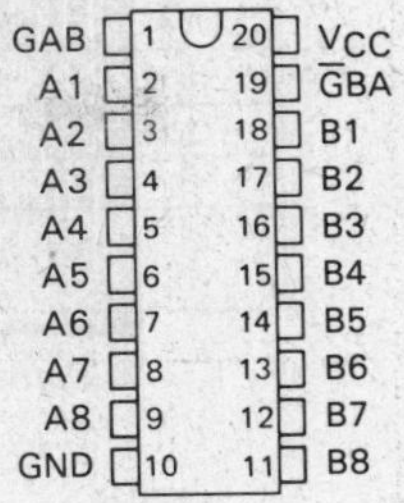

SN54HC' . . . FH OR FK PACKAGE
SN74HC' . . . FH OR FN PACKAGE
(TOP VIEW)

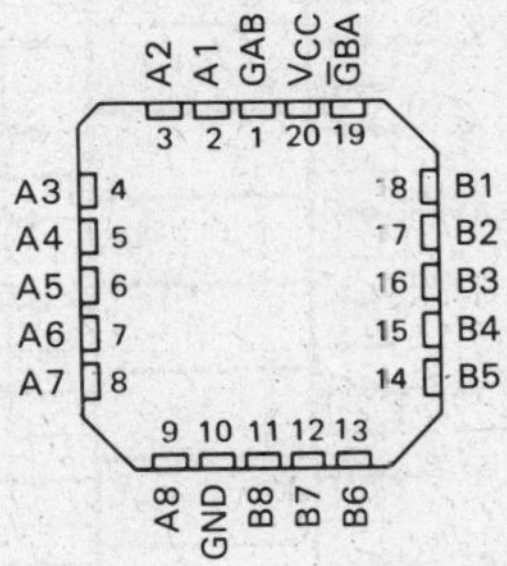

description

These octal bus transceivers are designed for asynchronous two-way communication between data buses. The control function implementation allows for maximum flexibility in timing.

These devices allow data transmission from the A bus to the B bus or from the B bus to the A bus depending upon the logic levels at the enable inputs ($\overline{G}$BA and GAB).

The enable inputs can be used to disable the device so that the buses are effectively isolated.

The dual-enable configuration gives these devices the capability to store data by simultaneous enabling of $\overline{G}$BA and GAB. Each output reinforces its input in this transceiver configuration. Thus when both control inputs are enabled and all other data sources to the two sets of bus lines are at high impedance, both sets of bus lines (16 in all) will remain at their last states. The 8-bit codes appearing on the two sets of buses will be identical for the 'HC623 or complementary for the 'HC620.

The SN54HC620 and SN54HC623 are characterized for operation over the full military temperature range of −55°C to 125°C. The SN74HC620 and SN74HC623 are characterized for operation from −40°C to 85°C.

FUNCTION TABLE

ENABLE INPUTS		OPERATION	
$\overline{G}BA$	GAB	'HC620	'HC623
L	L	$\overline{B}$ data to A bus	B data to A bus
H	H	$\overline{A}$ data to B bus	A data to B bus
H	L	Isolation	Isolation
L	H	$\overline{B}$ data to A bus, $\overline{A}$ data to B bus	B data to A bus, A data to B bus

logic symbols

'HC620

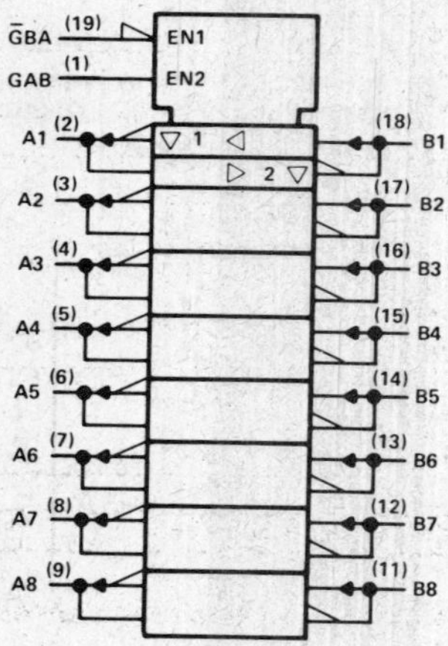

'HC623

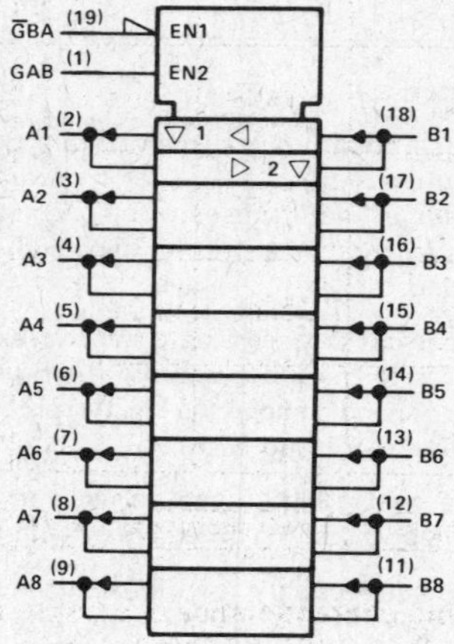

logic diagrams (positive logic)

'HC620

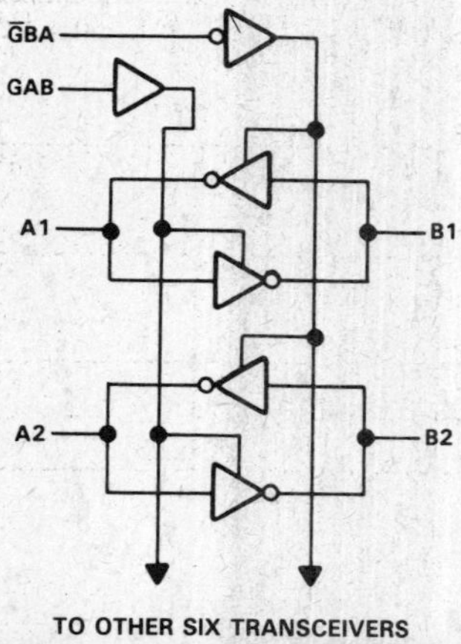

'HC623

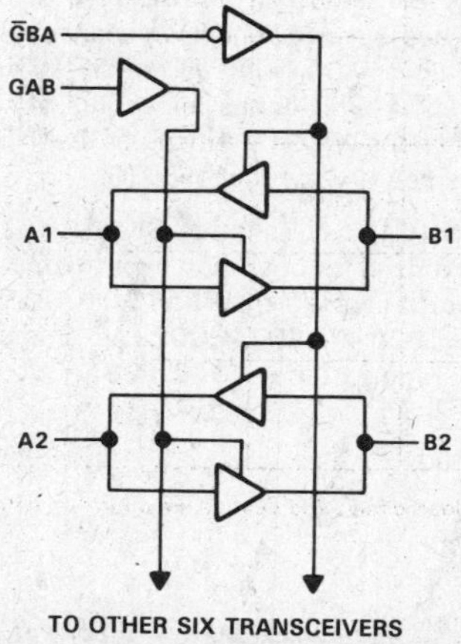

maximum ratings, recommended operating conditions, and electrical characteristics

See Table III, page 2-8.

switching characteristics over recommended operating free-air temperature range (unless otherwise noted), C_L = 50 pF (see Note 1)

PARAMETER	FROM (INPUT)	TO (OUTPUT)	V_{CC}	T_A = 25°C			SN54HC620 SN54HC623		SN74HC620 SN74HC623		UNIT
				MIN	TYP	MAX	MIN	MAX	MIN	MAX	
t_{pd}	A or B	B or A	2 V		29	105		160		130	ns
			4.5 V		10	21		32		26	
			6 V		8	18		27		22	
t_{en}	$\overline{G}$BA	A	2 V		112	210		315		265	ns
			4.5 V		27	42		63		53	
			6 V		20	36		54		45	
t_{dis}	$\overline{G}$BA	A	2 V		40	150		225		190	ns
			4.5 V		18	30		45		38	
			6 V		16	26		38		32	
t_{en}	GAB	B	2 V		112	210		315		265	ns
			4.5 V		27	42		63		53	
			6 V		20	36		54		45	
t_{dis}	GAB	B	2 V		40	150		225		190	ns
			4.5 V		18	30		45		38	
			6 V		16	26		38		32	
t_t		A or B	2 V		20	60		90		75	ns
			4.5 V		8	12		18		15	
			6 V		6	10		15		13	

C_{pd}	Power dissipation capacitance per transceiver	No load, T_A = 25°C	40 pF typ

switching characteristics over recommended operating free-air temperature range (unless otherwise noted), C_L = 150 pF (see Note 1)

PARAMETER	FROM (INPUT)	TO (OUTPUT)	V_{CC}	T_A = 25°C			SN54HC620 SN54HC623		SN74HC620 SN74HC623		UNIT
				MIN	TYP	MAX	MIN	MAX	MIN	MAX	
t_{pd}	A or B	B or A	2 V		44	135		200		170	ns
			4.5 V		14	27		40		34	
			6 V		11	23		34		29	
t_{en}	$\overline{G}$BA	A	2 V		130	270		405		335	ns
			4.5 V		31	54		81		67	
			6 V		23	46		69		56	
t_{en}	GAB	B	2 V		130	270		405		335	ns
			4.5 V		31	54		81		67	
			6 V		23	46		69		56	
t_t		A or B	2 V		45	210		315		265	ns
			4.5 V		17	42		63		53	
			6 V		13	36		53		45	

NOTE 1: For load circuit and voltage waveforms, see page 1-14.

TYPES SN54HC646, SN54HC648, SN74HC646, SN74HC648 OCTAL BUS TRANSCEIVERS AND REGISTERS WITH 3-STATE OUTPUTS

D2684, DECEMBER 1982—REVISED MARCH 1984

- **Independent Registers for A and B Buses**
- **Multiplexed Real-Time and Stored Data**
- **Choice of True or Inverting Data Paths**
- **High-Current 3-State Outputs Can Drive up to 15 LSTTL Loads**
- **Package Options Include Both Plastic and Ceramic Chip Carriers in Addition to Plastic and Ceramic DIPs**
- **Dependable Texas Instruments Quality and Reliability**

SN54HC' . . . JT PACKAGE
SN74HC' . . . JT OR NT PACKAGE
(TOP VIEW)

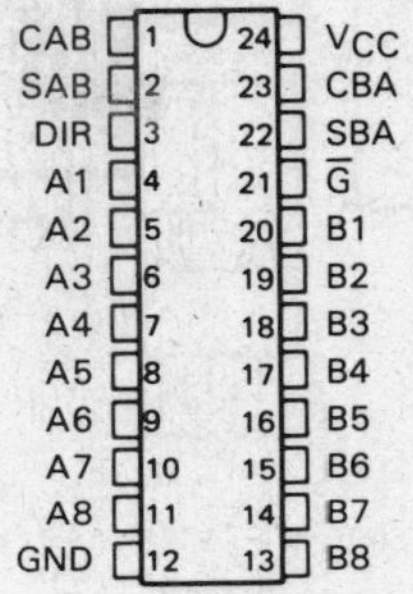

SN54HC' . . . FH OR FK PACKAGE
SN74HC' . . . FH OR FN PACKAGE
(TOP VIEW)

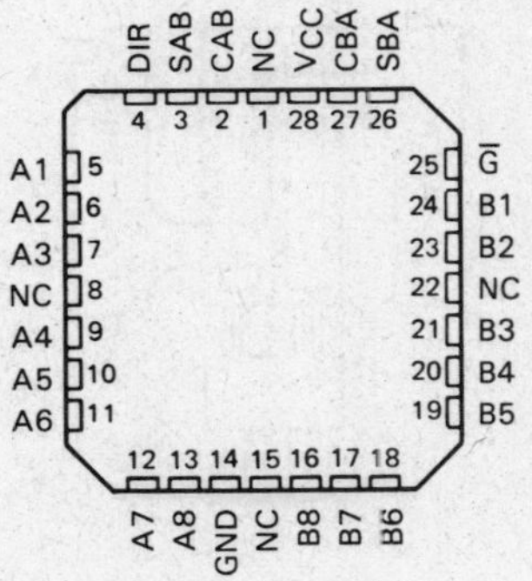

NC—No internal connection

description

These devices consist of bus transceiver circuits with 3-state outputs, D-type flip-flops, and control circuitry arranged for multiplexed transmission of data directly from the input bus or from the internal registers. Data on the A or B bus will be clocked into the registers on the low-to-high transition of the appropriate clock pin (CAB or CBA). The examples on the following page demonstrate the four fundamental bus-management functions that can be performed with the 'HC646 or 'HC648.

Enable ($\overline{G}$) and direction (DIR) pins are provided to control the transceiver functions. In the transceiver mode, data present at the high-impedance port may be stored in either register or in both. The select controls (SAB and SBA) can multiplex stored and real-time (transparent mode) data. The direction control determines which bus will receive data when enable $\overline{G}$ is active (low). In the isolation mode (enable $\overline{G}$ high), A data may be stored in one register and/or B data may be stored in the other register.

When an output function is disabled, the input function is still enabled and may be used to store and transmit data. Only one of the two buses, A or B, may be driven at a time.

The SN54HC' family is characterized for operation over the full military temperature range of −55°C to 125°C. The SN74HC' family is characterized for operation from −40°C to 85°C.

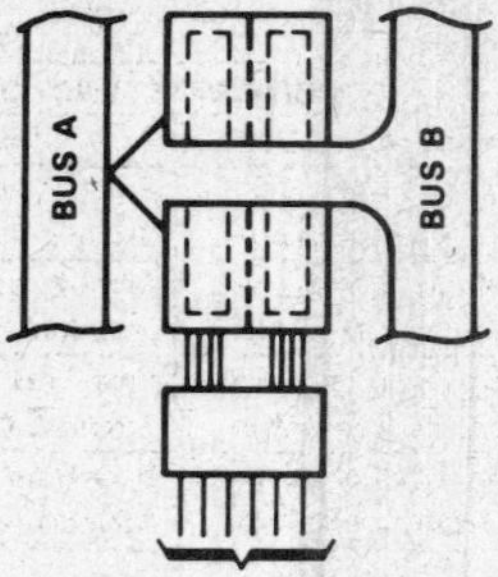

(21) $\overline{G}$	(3) DIR	(1) CAB	(23) CBA	(2) SAB	(22) SBA
L	L	X	X	X	L

**REAL-TIME TRANSFER
BUS B TO BUS A**

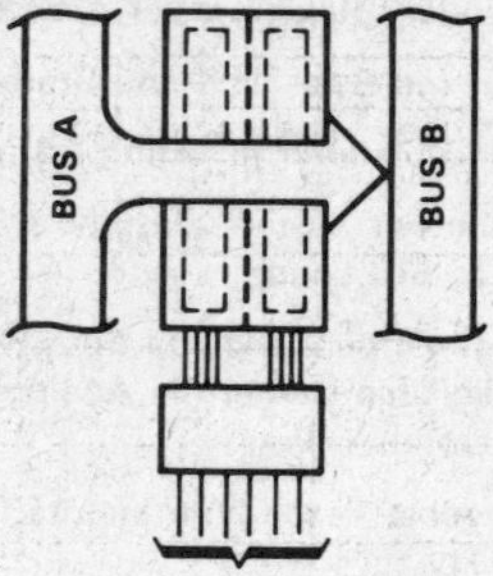

(21) $\overline{G}$	(3) DIR	(1) CAB	(23) CBA	(2) SAB	(22) SBA
L	H	X	X	L	X

**REAL-TIME TRANSFER
BUS A TO BUS B**

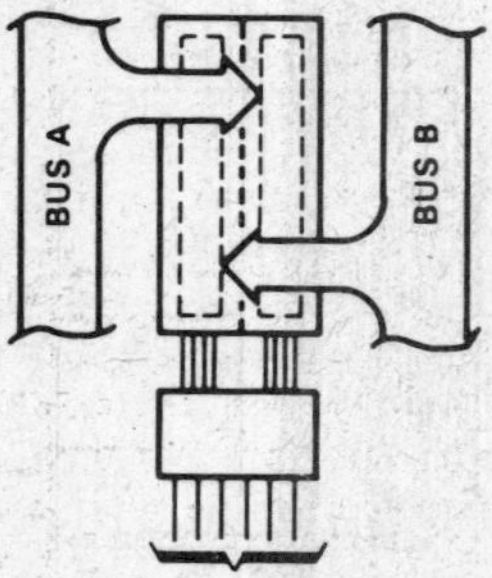

(21) $\overline{G}$	(3) DIR	(1) CAB	(23) CBA	(2) SAB	(22) SBA
X	X	↑	X	X	X
X	X	X	↑	X	X
H	X	↑	↑	X	X

**STORAGE FROM
A, B, OR A AND B**

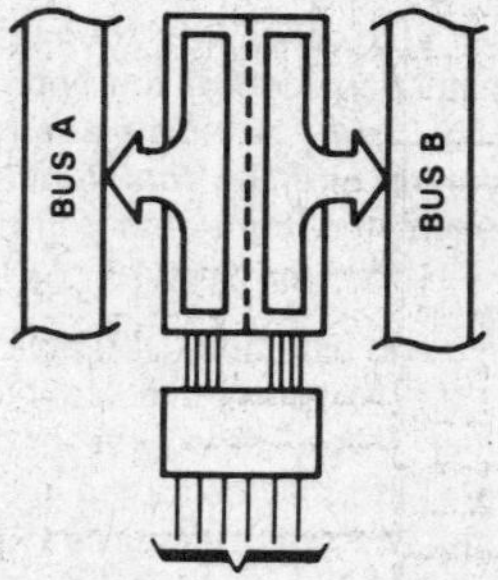

(21) $\overline{G}$	(3) DIR	(1) CAB	(23) CBA	(2) SAB	(22) SBA
L	L	X	H or L	X	H
L	H	H or L	X	H	X

**TRANSFER STORED DATA
TO A OR B**

Pin numbers shown are for JT and NT packages.

FUNCTION TABLE

INPUTS						DATA I/O†		OPERATION OR FUNCTION	
$\overline{G}$	DIR	CAB	CBA	SAB	SBA	A1 THRU A8	B1 THRU B8	'HC646	'HC648
X	X	↑	X	X	X	Input	Not specified	Store A, B unspecified	Store A, B unspecified
X	X	X	↑	X	X	Not specified	Input	Store B, A unspecified	Store B, A unspecified
H	X	↑	↑	X	X	Input	Input	Store A and B Data	Store A and B Data
H	X	H or L	H or L	X	X			Isolation, hold storage	Isolation, hold storage
L	L	X	X	X	L	Output	Input	Real-Time B Data to A Bus	Real-Time $\overline{B}$ Data to A Bus
L	L	X	H or L	X	H			Stored B Data to A Bus	Stored $\overline{B}$ Data to A Bus
L	H	X	X	L	X	Input	Output	Real-Time A Data to B Bus	Real-Time $\overline{A}$ Data to B Bus
L	H	H or L	X	H	X			Stored A Data to B Bus	Stored $\overline{A}$ Data to B Bus

† The data output functions may be enabled or disabled by various signals at the $\overline{G}$ and DIR inputs. Data input functions are always enabled, i.e., data at the bus pins will be stored on every low-to-high transition on the clock inputs.

logic symbols

'HC646

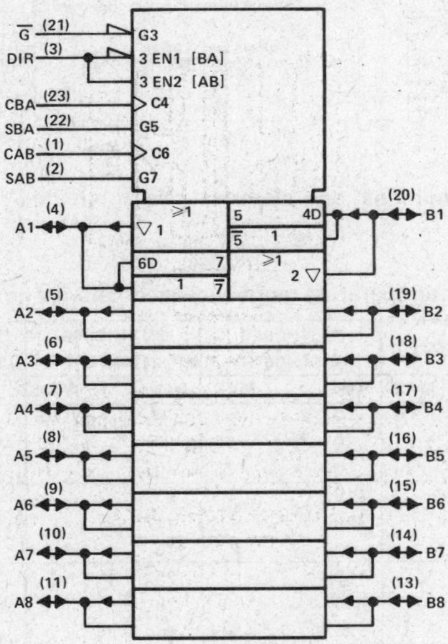

'HC648

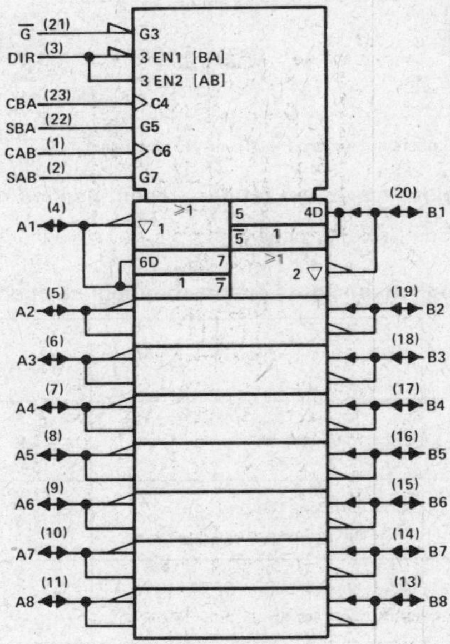

Pin numbers shown are for JT and NT packages.

logic diagram (positive logic)

'HC646

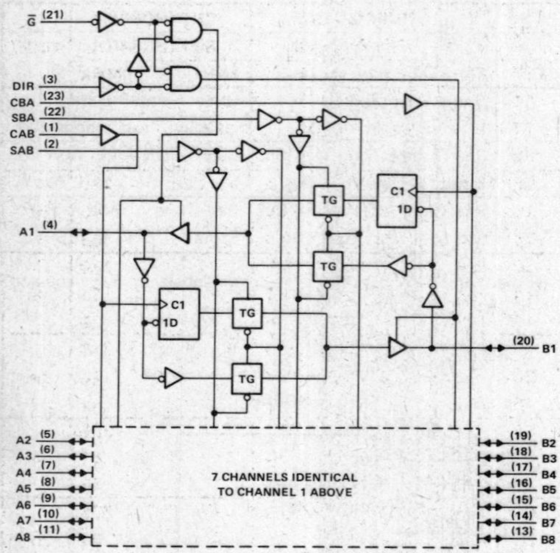

'HC648

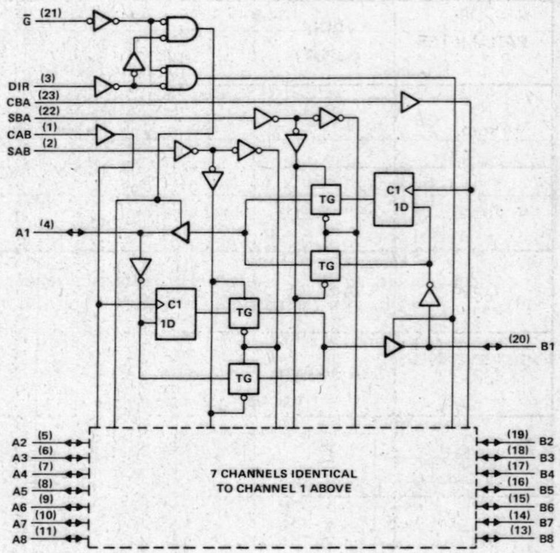

Pin numbers shown are for JT and NT packages.

absolute maximum ratings, recommended operating conditions, and electrical characteristics

See Table III, page 2-8.

timing requirements over recommended operating free-air temperature range (unless otherwise noted)

		V_{CC}	T_A = 25°C		SN54HC646 SN54HC648		SN74HC646 SN74HC648		UNIT
			MIN	MAX	MIN	MAX	MIN	MAX	
f_{clock}	Clock frequency	2 V	0	6	0	4.3	0	5.5	MHz
		4.5 V	0	31	0	22	0	27	
		6 V	0	36	0	25	0	31	
t_w	Pulse duration, CBA or CAB high or low	2 V	80		115		95		ns
		4.5 V	16		23		19		
		6 V	14		20		16		
t_{su}	Setup time, A before CAB↑ or B before CBA↑	2 V	100		150		125		ns
		4.5 V	20		30		25		
		6 V	17		26		21		
t_h	Hold time, A after CAB↑ or B after CBA↑	2 V	5		5		5		ns
		4.5 V	5		5		5		
		6 V	5		5		5		

switching characteristics over recommended operating free-air temperature range (unless otherwise noted), C_L = 50 pF (see Note 1)

PARAMETER	FROM (INPUT)	TO (OUTPUT)	V_{CC}	T_A = 25°C MIN	T_A = 25°C TYP	T_A = 25°C MAX	SN54HC646 SN54HC648 MIN	SN54HC646 SN54HC648 MAX	SN74HC646 SN74HC648 MIN	SN74HC646 SN74HC648 MAX	UNIT
f_{max}			2 V	6	11		4.4		5.5		MHz
			4.5 V	31	54		22		27		
			6 V	36	64		25		31		
t_{pd}	CBA or CAB	A or B	2 V		65	180		270		225	ns
			4.5 V		18	36		54		45	
			6 V		14	31		46		38	
t_{pd}	A or B	B or A	2 V		50	135		205		170	ns
			4.5 V		14	27		41		34	
			6 V		11	23		35		29	
t_{pd}	SBA or SAB†	A or B	2 V		70	190		285		240	ns
			4.5 V		20	38		57		48	
			6 V		16	32		48		41	
t_{en}	$\overline{G}$	A or B	2 V		85	245		370		305	ns
			4.5 V		25	49		74		61	
			6 V		20	42		63		52	
t_{dis}	$\overline{G}$	A or B	2 V		85	245		370		305	ns
			4.5 V		25	49		74		61	
			6 V		20	42		63		52	
t_{en}	DIR	A or B	2 V		80	245		370		305	ns
			4.5 V		25	49		74		61	
			6 V		20	42		63		52	
t_{dis}	DIR	A or B	2 V		80	245		370		305	ns
			4.5 V		25	49		74		61	
			6 V		20	42		63		52	
t_t		Any	2 V		28	60		90		75	ns
			4.5 V		8	12		18		15	
			6 V		6	10		15		13	

C_{pd}	Power dissipation capacitance	No load, T_A = 25°C	50 pF typ

NOTE 1: For load circuit and voltage waveforms, see page 1-14.

†These parameters are measured with the internal output state of the storage register opposite to that of the bus input.

switching characteristics over recommended operating free-air temperature range (unless otherwise noted), C_L = 150 pF (see Note 1)

PARAMETER	FROM (INPUT)	TO (OUTPUT)	V_{CC}	T_A = 25°C MIN	TYP	MAX	SN54HC646 SN54HC648 MIN	MAX	SN74HC648 SN74HC648 MIN	MAX	UNIT
t_{pd}	CBA or CAB	A or B	2 V		90	265		400		330	ns
			4.5 V		24	53		80		66	
			6 V		20	46		68		57	
t_{pd}	A or B	B or A	2 V		70	220		335		280	ns
			4.5 V		20	44		67		56	
			6 V		15	38		57		49	
t_{pd}	SBA or SAB†	A or B	2 V		80	275		415		345	ns
			4.5 V		24	55		83		69	
			6 V		20	47		70		60	
t_{en}	$\overline{G}$	A or B	2 V		113	330		500		410	ns
			4.5 V		33	66		100		82	
			6 V		27	57		85		71	
t_{en}	DIR	A or B	2 V		113	330		500		410	ns
			4.5 V		33	66		100		82	
			6 V		27	57		85		71	
t_t		Any	2 V		45	210		315		265	ns
			4.5 V		17	42		63		53	
			6 V		13	36		53		43	

NOTE 1: For load circuit and voltage waveforms, see page 1-14.
† These parameters are measured with the internal output state of the storage register opposite to that of the bus input.

TYPES SN54HCT646, SN54HCT648, SN74HCT646, SN74HCT648 OCTAL BUS TRANSCEIVERS AND REGISTERS WITH 3-STATE OUTPUTS

D2804, MARCH 1984

- **Inputs are TTL-Voltage Compatible**
- **Independent Registers for A and B Buses**
- **Multiplexed Real-Time and Stored Data**
- **Choice of True or Inverting Data Paths**
- **High-Current 3-State Outputs Can Drive up to 15 LSTTL Loads**
- **Package Options Include Both Plastic and Ceramic Chip Carriers in Addition to Plastic and Ceramic DIPs**
- **Dependable Texas Instruments Quality and Reliability**

SN54HCT' . . . JT PACKAGE
SN74HCT' . . . JT OR NT PACKAGE
(TOP VIEW)

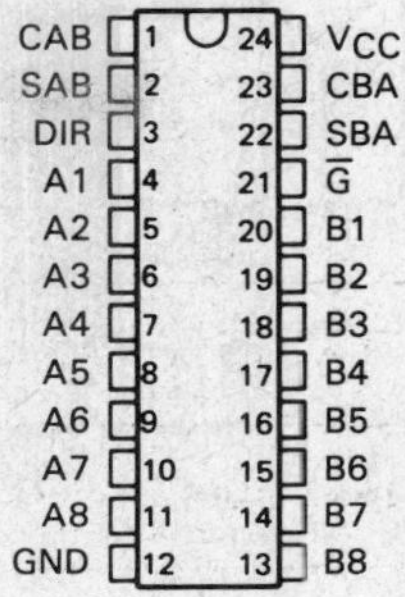

SN54HCT' . . . FH OR FK PACKAGE
SN74HCT' . . . FH OR FN PACKAGE
(TOP VIEW)

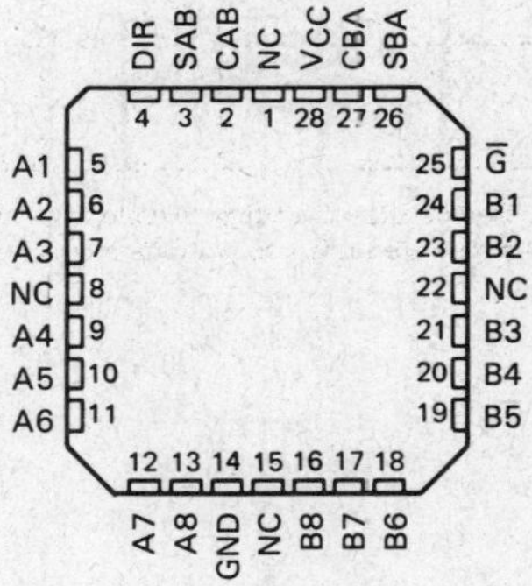

NC—No internal connection

description

These devices consist of bus transceiver circuits with 3-state outputs, D-type flip-flops, and control circuitry arranged for multiplexed transmission of data directly from the input bus or from the internal registers. Data on the A or B bus will be clocked into the registers on the low-to-high transition of the appropriate clock pin (CAB or CBA). The examples on the following page demonstrate the four fundamental bus-management functions that can be performed with the 'HCT646 or 'HCT648.

Enable ($\overline{G}$) and direction (DIR) pins are provided to control the transceiver functions. In the transceiver mode, data present at the high-impedance port may be stored in either register or in both. The select controls (SAB and SBA) can multiplex stored and real-time (transparent mode) data. The direction control determines which bus will receive data when enable $\overline{G}$ is active (low). In the isolation mode (enable $\overline{G}$ high), A data may be stored in one register and/or B data may be stored in the other register.

When an output function is disabled, the input function is still enabled and may be used to store and transmit data. Only one of the two buses, A or B, may be driven at a time.

The SN54HCT' family is characterized for operation over the full military temperature range of −55°C to 125°C. The SN74HCT' family is characterized for operation from −40°C to 85°C.

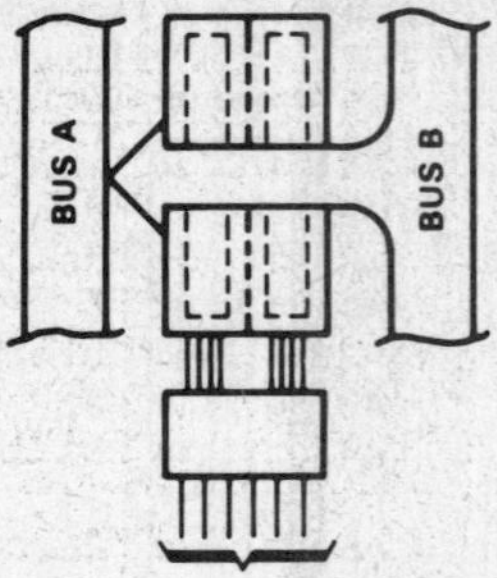

(21) $\overline{G}$	(3) DIR	(1) CAB	(23) CBA	(2) SAB	(22) SBA
L	L	X	X	X	L

REAL-TIME TRANSFER BUS B TO BUS A

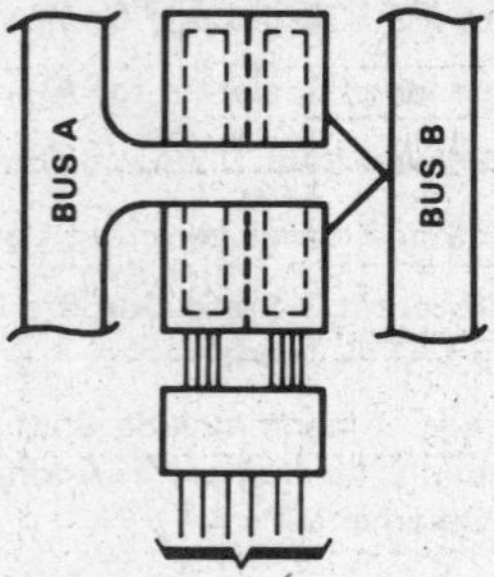

(21) $\overline{G}$	(3) DIR	(1) CAB	(23) CBA	(2) SAB	(22) SBA
L	H	X	X	L	X

REAL-TIME TRANSFER BUS A TO BUS B

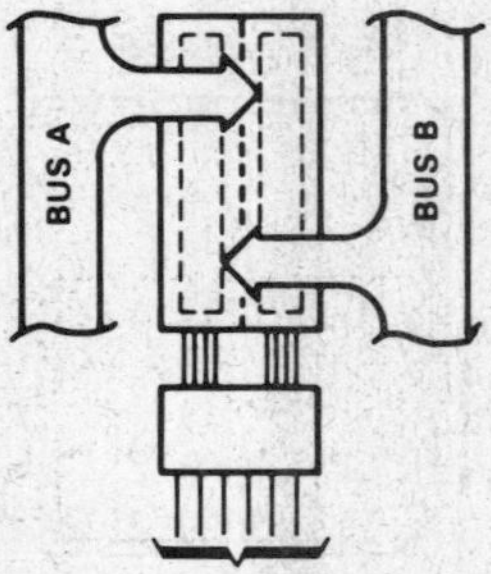

(21) $\overline{G}$	(3) DIR	(1) CAB	(23) CBA	(2) SAB	(22) SBA
X	X	↑	X	X	X
X	X	X	↑	X	X
H	X	↑	↑	X	X

STORAGE FROM A, B, OR A AND B

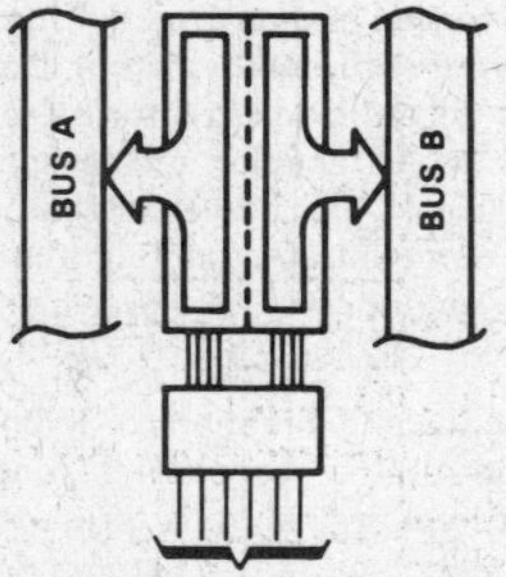

(21) $\overline{G}$	(3) DIR	(1) CAB	(23) CBA	(2) SAB	(22) SBA
L	L	X	H or L	X	H
L	H	H or L	X	H	X

TRANSFER STORED DATA TO A OR B

Pin numbers shown are for JT and NT packages.

FUNCTION TABLE

INPUTS						DATA I/O†		OPERATION OR FUNCTION	
$\overline{G}$	DIR	CAB	CBA	SAB	SBA	A1 THRU A8	B1 THRU B8	'HCT646	'HCT648
X	X	↑	X	X	X	Input	Not specified	Store A, B unspecified	Store A, B unspecified
X	X	X	↑	X	X	Not specified	Input	Store B, A unspecified	Store B, A unspecified
H	X	↑	↑	X	X	Input	Input	Store A and B Data	Store A and B Data
H	X	H or L	H or L	X	X			Isolation, hold storage	Isolation, hold storage
L	L	X	X	X	L	Output	Input	Real-Time B Data to A Bus	Real-Time $\overline{B}$ Data to A Bus
L	L	X	H or L	X	H			Stored B Data to A Bus	Stored $\overline{B}$ Data to A Bus
L	H	X	X	L	X	Input	Output	Real-Time A Data to B Bus	Real-Time $\overline{A}$ Data to B Bus
L	H	H or L	X	H	X			Stored A Data to B Bus	Stored $\overline{A}$ Data to B Bus

† The data output functions may be enabled or disabled by various signals at the $\overline{G}$ and DIR inputs. Data input functions are always enabled, i.e., data at the bus pins will be stored on every low-to-high transition on the clock inputs.

logic symbols

'HCT646

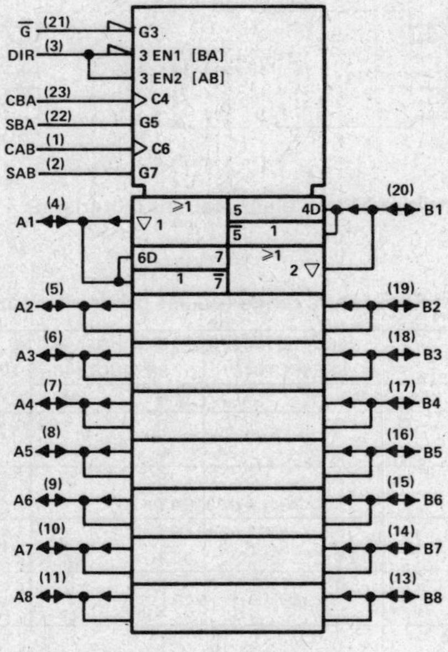

'HCT648

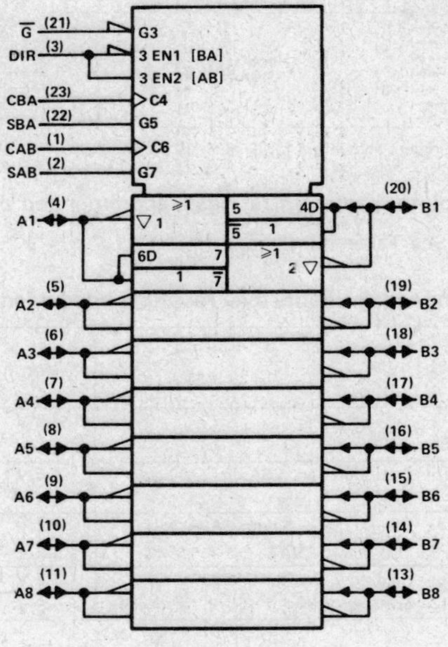

Pin numbers shown are for JT and NT packages.

logic diagram (positive logic)

'HCT646

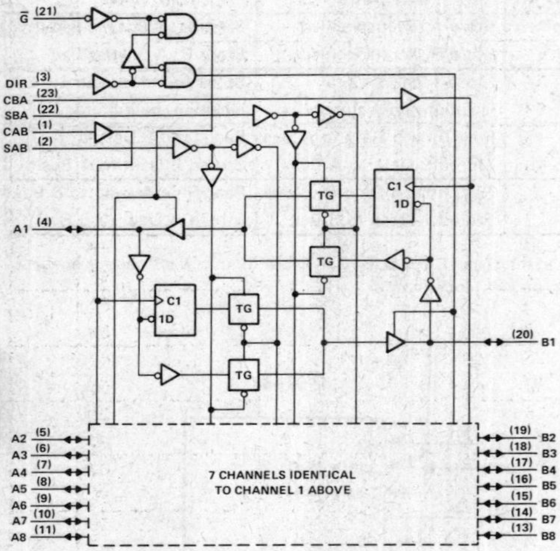

'HCT648

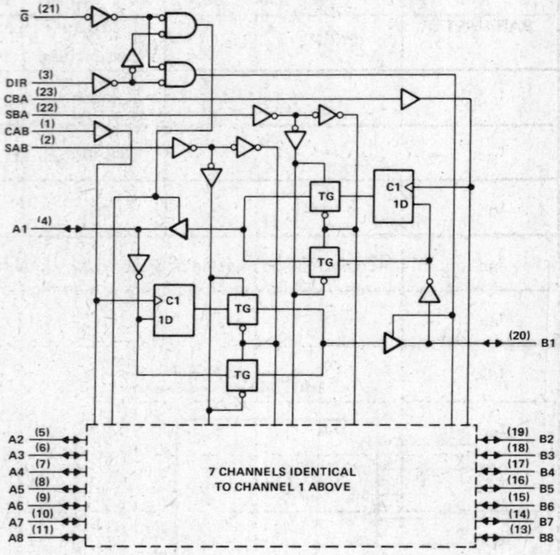

Pin numbers shown are for JT and NT packages.

absolute maximum ratings, recommended operating conditions, and electrical characteristics

See Table VII, page 2-14.

timing requirements over recommended operating free-air temperature range (unless otherwise noted)

		V_{CC}	$T_A = 25°C$		SN54HCT646 SN54HCT648		SN74HCT646 SN74HCT648		UNIT
			MIN	MAX	MIN	MAX	MIN	MAX	
f_{clock}	Clock frequency	4.5 V	0	31	0	22	0	27	MHz
		5.5 V	0	36	0	24	0	29	
t_w	Pulse duration, CBA or CAB high or low	4.5 V	16		23		19		ns
		5.5 V	14		21		17		
t_{su}	Setup time, A before CAB↑ or B before CBA↑	4.5 V	20		30		25		ns
		5.5 V	18		27		23		
t_h	Hold time, A after CAB↑ or B after CBA↑	4.5 V	5		5		5		ns
		5.5 V	5		5		5		

TEXAS INSTRUMENTS
POST OFFICE BOX 225012 • DALLAS, TEXAS 75265

switching characteristics over recommended operating free-air temperature range (unless otherwise noted), C_L = 50 pF (see Note 1)

PARAMETER	FROM (INPUT)	TO (OUTPUT)	V_{CC}	T_A = 25°C			SN54HCT646 SN54HCT648		SN74HCT646 SN74HCT648		UNIT
				MIN	TYP	MAX	MIN	MAX	MIN	MAX	
f_{max}			4.5 V	31	54		22		27		MHz
			5.5 V	36	64		24		29		
t_{pd}	CBA or CAB	A or B	4.5 V		18	36		54		45	ns
			5.5 V		16	32		49		41	
t_{pd}	A or B	B or A	4.5 V		14	27		41		34	ns
			5.5 V		12	24		37		31	
t_{pd}	SBA or SAB†	A or B	4.5 V		20	38		57		48	ns
			5.5 V		17	34		51		43	
t_{en}	$\overline{G}$	A or B	4.5 V		25	49		74		61	ns
			5.5 V		22	44		67		55	
t_{dis}	$\overline{G}$	A or B	4.5 V		25	49		74		61	ns
			5.5 V		22	44		67		55	
t_{en}	DIR	A or B	4.5 V		25	49		74		61	ns
			5.5 V		22	44		67		55	
t_{dis}	DIR	A or B	4.5 V		25	49		74		61	ns
			5.5 V		22	44		67		55	
t_t		Any	4.5 V		9	12		18		15	ns
			5.5 V		7	11		16		14	

C_{pd}	Power dissipation capacitance	No load, T_A = 25°C	50 pF typ

switching characteristics over recommended operating free-air temperature range (unless otherwise noted), C_L = 150 pF (see Note 1)

PARAMETER	FROM (INPUT)	TO (OUTPUT)	V_{CC}	T_A = 25°C			SN54HCT646 SN54HCT648		SN74HCT646 SN74HCT648		UNIT
				MIN	TYP	MAX	MIN	MAX	MIN	MAX	
t_{pd}	CBA or CAB	A or B	4.5 V		24	53		80		66	ns
			5.5 V		22	47		52		60	
t_{pd}	A or B	B or A	4.5 V		22	44		67		55	ns
			5.5 V		20	39		60		50	
t_{pd}	SBA or SAB†	A or B	4.5 V		26	55		83		69	ns
			5.5 V		24	49		74		62	
t_{en}	$\overline{G}$	A or B	4.5 V		33	66		100		87	ns
			5.5 V		22	59		90		74	
t_{en}	DIR	A or B	4.5 V		33	66		100		87	ns
			5.5 V		22	59		90		74	
t_t		Any	4.5 V		17	42		63		53	ns
			5.5 V		14	38		57		48	

NOTE 1: For load circuit and voltage waveforms, see page 1-14.
† These parameters are measured with the internal output state of the storage register opposite to that of the bus input.

TYPES SN54HC651, SN54HC652, SN74HC651, SN74HC652 OCTAL BUS TRANSCEIVERS AND REGISTERS WITH 3-STATE OUTPUTS

D2684, DECEMBER 1982—REVISED MARCH 1984

- **Bus Transceivers and Registers**
- **Independent Registers and Enables for A and B Buses**
- **High-Current 3-State Outputs Can Drive up to 15 LSTTL Loads**
- **Multiplexed Real-Time and Stored Data**
- **Choice of True and Inverting Data Paths**
- **Package Options Include Both Plastic and Ceramic Chip Carriers in Addition to Plastic and Ceramic DIPs**
- **Dependable Texas Instruments Quality and Reliability**

SN54HC651, SN54HC652 . . . JT PACKAGE
SN74HC651, SN74HC652 . . . JT OR NT PACKAGE
(TOP VIEW)

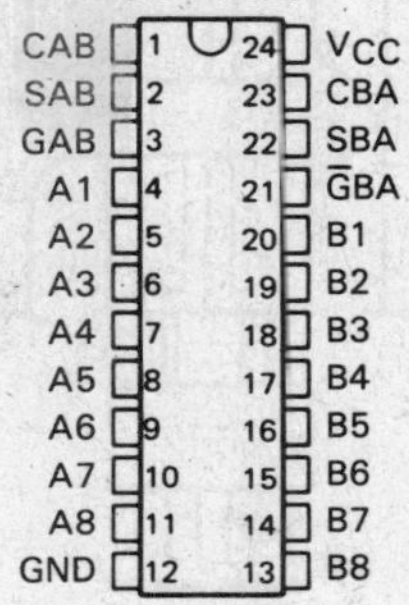

SN54HC651, SN54HC652 . . . FH OR FK PACKAGE
SN74HC651, SN74HC652 . . . FH OR FN PACKAGE
(TOP VIEW)

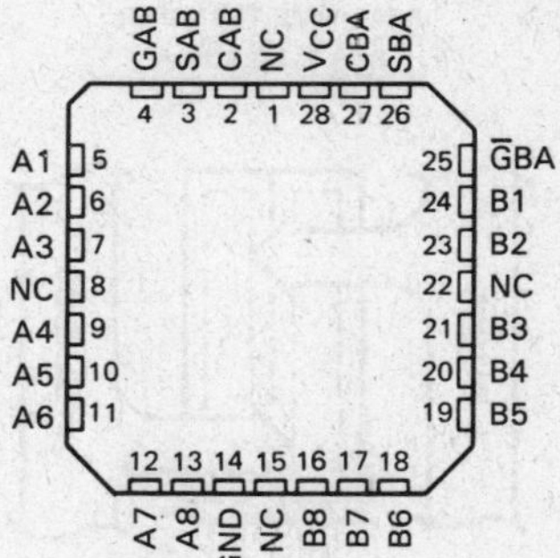

NC—No internal connection

description

These devices consist of bus transceiver circuits, D-type flip-flops, and control circuitry arranged for multiplexed transmission of data directly from the data bus or from the internal storage registers. Enable GAB and $\overline{G}$BA are provided to control the transceiver functions. SAB and SBA control pins are provided to select whether real-time or stored data is transferred. A low input level selects real-time data, and a high selects stored data. The examples on the following page demonstrate the four fundamental bus-management functions that can be performed with the 'HC651 and 'HC652.

Data on the A or B data bus, or both, can be stored in the internal D flip-flops by low-to-high transitions at the appropriate clock pins (CAB or CBA) regardless of the select or enable control pins. When SAB and SBA are in the real-time transfer mode, it is also possible to store data without using the internal D-type flip-flops by simultaneously enabling GAB and $\overline{G}$BA. In this configuration each output reinforces its input. Thus, when all other data sources to the two sets of bus lines are at high impedance, each set of bus lines will remain at its last state.

The SN54HC651 and SN54HC652 are characterized for operation over the full military temperature range of −55 °C to 125 °C. The SN74HC651 and SN74HC652 are characterized for operation from −40 °C to 85 °C.

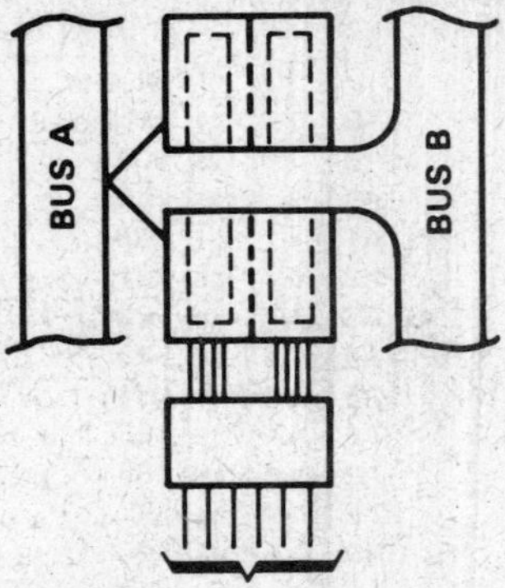

(3) GAB	(21) $\overline{G}$BA	(1) CAB	(23) CBA	(2) SAB	(22) SBA
L	L	X	X	X	L

REAL-TIME TRANSFER BUS B TO BUS A

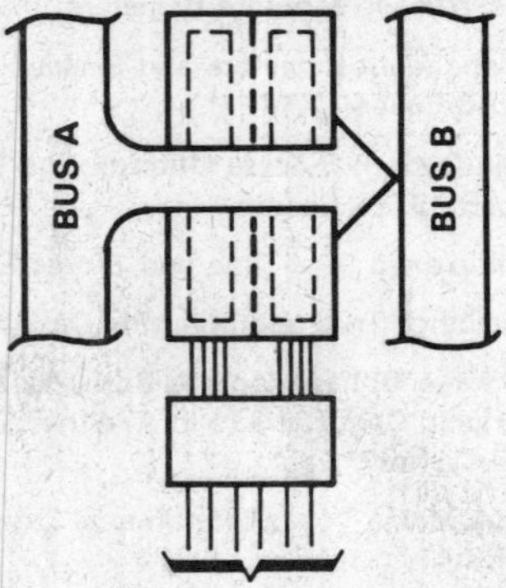

(3) GAB	(21) $\overline{G}$BA	(1) CAB	(23) CBA	(2) SAB	(22) SBA
H	H	X	X	L	X

REAL-TIME TRANSFER BUS A TO BUS B

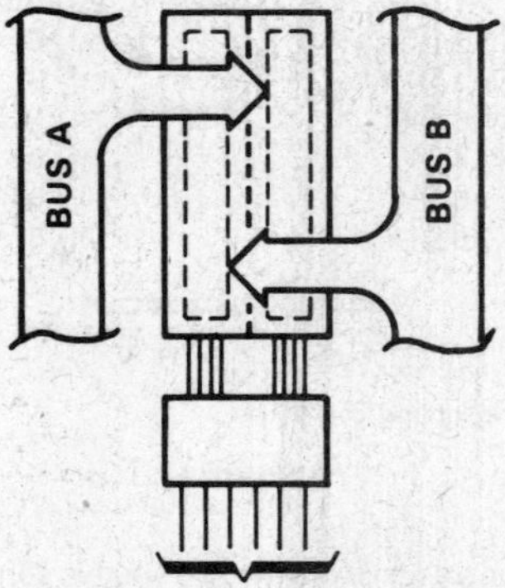

(3) GAB	(21) $\overline{G}$BA	(1) CAB	(23) CBA	(2) SAB	(22) SBA
X	H	↑	X	X	X
L	X	X	↑	X	X
L	H	↑	↑	X	X

STORAGE FROM A AND/OR B

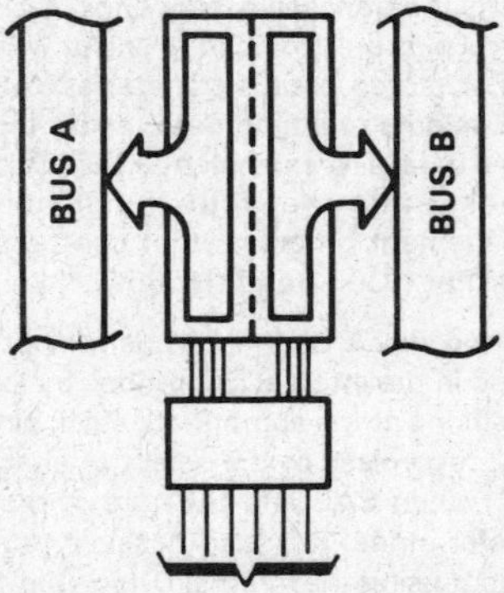

(3) GAB	(21) $\overline{G}$BA	(1) CAB	(23) CBA	(2) SAB	(22) SBA
H	L	H or L	H or L	H	H

TRANSFER STORED DATA TO A AND/OR B

Pin numbers shown are for JT and NT packages.

FUNCTION TABLE

INPUTS						DATA I/O†		OPERATION OR FUNCTION	
GAB	$\overline{G}$BA	CAB	CBA	SAB	SBA	A1 THRU A8	B1 THRU B8	'HC651	'HC652
L	H	H or L	H or L	X	X	Input	Input	Isolation	Isolation
L	H	↑	↑	X	X			Store A and B Data	Store A and B Data
X	H	↑	H or L	X	X	Input	Not specified	Store A, Hold B	Store A, Hold B
H	H	↑	↑	X	X	Input	Output	Store A in both registers	Store A in both registers
L	X	H or L	↑	X	X	Not specified	Input	Hold A, Store B	Hold A, Store B
L	L	↑	↑	X	X	Output	Input	Store B in both registers	Store B in both registers
L	L	X	X	X	L	Output	Input	Real-Time $\overline{B}$ Data to A Bus	Real-Time B Data to A Bus
L	L	X	H or L	X	H			Stored $\overline{B}$ Data to A Bus	Stored B Data to A Bus
H	H	X	X	L	X	Input	Output	Real-Time $\overline{A}$ Data to B Bus	Real-Time A Data to B Bus
H	H	H or L	X	H	X			Stored $\overline{A}$ Data to B Bus	Stored A Data to B Bus
H	L	H or L	H or L	H	H	Output	Output	Stored $\overline{A}$ Data to B Bus and Stored $\overline{B}$ Data to A Bus	Stored A Data to B Bus and Stored B Data to A Bus

† The data output functions may be enabled or disabled by various signals at the GAB and $\overline{G}$BA inputs. Data input functions are always enabled, i.e., data at the bus pins will be stored on every low-to-high transition on the clock inputs.

logic symbols

'HC651

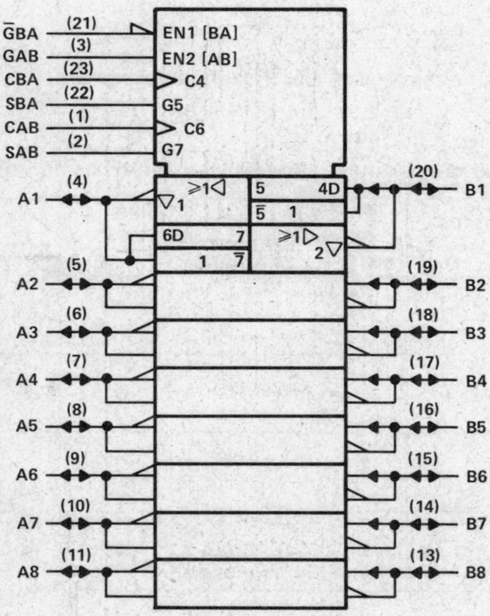

'HC652

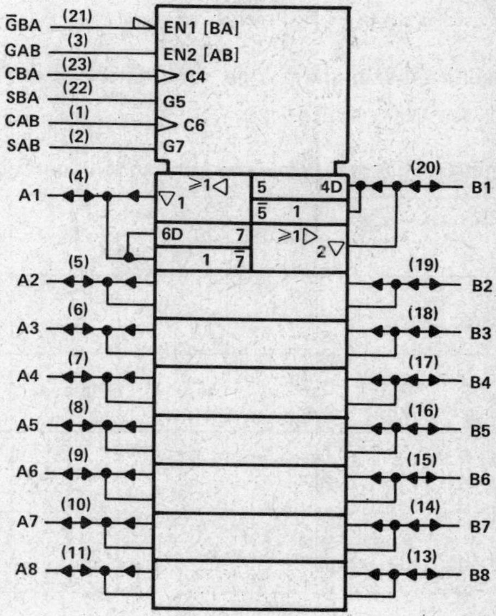

Pin numbers shown are for JT and NT packages.

logic diagram (positive logic)

'HC651

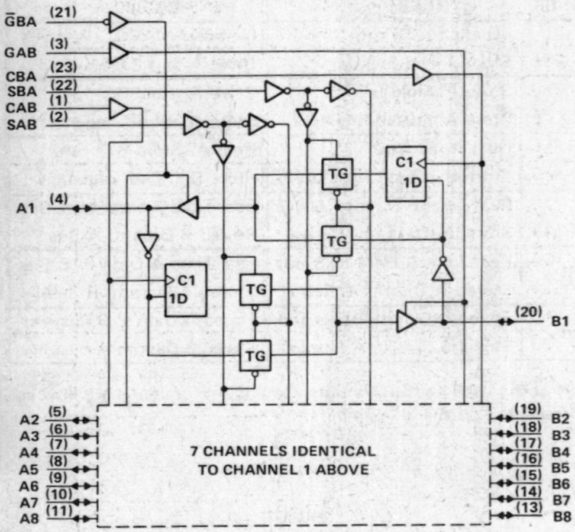

'HC652

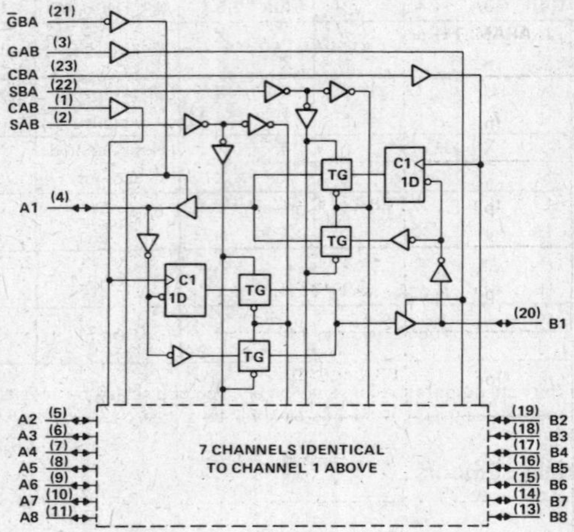

Pin numbers shown are for JT and NT packages.

absolute maximum ratings, recommended operating conditions, and electrical characteristics

See Table III, page 2-8.

timing requirements over recommended operating free-air temperature range (unless otherwise noted)

		V_{CC}	T_A = 25°C		SN54HC651 SN54HC652		SN74HC651 SN74HC652		UNIT
			MIN	MAX	MIN	MAX	MIN	MAX	
f_{clock}	Clock frequency	2 V	0	6	0	4.3	0	5.5	MHz
		4.5 V	0	31	0	22	0	27	
		6 V	0	36	0	25	0	31	
t_w	Pulse duration, CBA or CAB high or low	2 V	80		115		95		ns
		4.5 V	16		23		19		
		6 V	14		20		16		
t_{su}	Setup time, A before CAB↑ or B before CBA↑	2 V	100		150		125		ns
		4.5 V	20		30		25		
		6 V	17		26		21		
t_h	Hold time, A after CAB↑ or B after CBA↑	2 V	5		5		5		ns
		4.5 V	5		5		5		
		6 V	5		5		5		

switching characteristics over recommended operating free-air temperature range (unless otherwise noted), C_L = 50 pF (see Note 1)

PARAMETER	FROM (INPUT)	TO (OUTPUT)	V_{CC}	T_A = 25°C			SN54HC651 SN54HC652		SN74HC652 SN74HC652		UNIT
				MIN	TYP	MAX	MIN	MAX	MIN	MAX	
			2 V	6	10		4.3		5.5		
f_{max}			4.5 V	31	40		22		27		MHz
			6 V	36	45		25		31		
			2 V		65	180		270		225	
t_{pd}	CBA or CAB	A or B	4.5 V		18	36		54		45	ns
			6 V		14	31		46		38	
			2 V		50	135		205		170	
t_{pd}	A or B	B or A	4.5 V		14	27		41		34	ns
			6 V		11	23		35		29	
			2 V		70	190		285		240	
t_{pd}	SBA or SAB†	A or B	4.5 V		20	38		57		48	ns
			6 V		16	32		48		41	
			2 V		85	245		370		305	
t_{en}	$\overline{G}$BA or GAB	A or B	4.5 V		25	49		74		61	ns
			6 V		20	42		63		52	
			2 V		50	245		370		305	
t_{dis}	$\overline{G}$BA or GAB	A or B	4.5 V		23	49		74		61	ns
			6 V		20	42		63		52	
			2 V		28	60		90		75	
t_t		Any	4.5 V		8	12		18		15	ns
			6 V		6	10		15		13	

C_{pd}	Power dissipation capacitance	No load, T_A = 25°C	50 pF typ

switching characteristics over recommended operating free-air temperature range (unless otherwise noted), C_L = 150 pF (see Note 1)

PARAMETER	FROM (INPUT)	TO (OUTPUT)	V_{CC}	T_A = 25°C			SN54HC651 SN54HC652		SN74HC651 SN74HC652		UNIT
				MIN	TYP	MAX	MIN	MAX	MIN	MAX	
			2 V		90	265		400		330	
t_{pd}	CBA or CAB	A or B	4.5 V		24	53		80		66	ns
			6 V		18	46		68		57	
			2 V		70	220		335		275	
t_{pd}	A or B	B or A	4.5 V		20	44		70		55	ns
			6 V		15	38		57		48	
			2 V		80	275		415		345	
t_{pd}	SBA or SAB†	A or B	4.5 V		24	55		83		69	ns
			6 V		20	47		70		60	
			2 V		100	330		500		410	
t_{en}	$\overline{G}$BA or GAB	A or B	4.5 V		33	66		100		82	ns
			6 V		27	57		85		71	
			2 V		45	210		315		265	
t_t		Any	4.5 V		17	42		63		53	ns
			6 V		13	36		53		43	

NOTE 1: For load circuit and voltage waveforms, see page 1-14.

†These parameters are measured with the internal output state of the storage register opposite to that of the bus input.

TYPES SN54HCT651, SN54HCT652, SN74HCT651, SN74HCT652 OCTAL BUS TRANSCEIVERS AND REGISTERS WITH 3-STATE OUTPUTS

D2804, MARCH 1984

- Inputs are TTL-Voltage Compatible
- Bus Transceivers and Registers
- Independent Registers and Enables for A and B Buses
- High-Current 3-State Outputs Can Drive up to 15 LSTTL Loads
- Multiplexed Real-Time and Stored Data
- Choice of True and Inverting Data Paths
- Package Options Include Both Plastic and Ceramic Chip Carriers in Addition to Plastic and Ceramic DIPs
- Dependable Texas Instruments Quality and Reliability

SN54HCT651, SN54HCT652 . . . JT PACKAGE
SN74HCT651, SN74HCT652 . . . JT OR NT PACKAGE
(TOP VIEW)

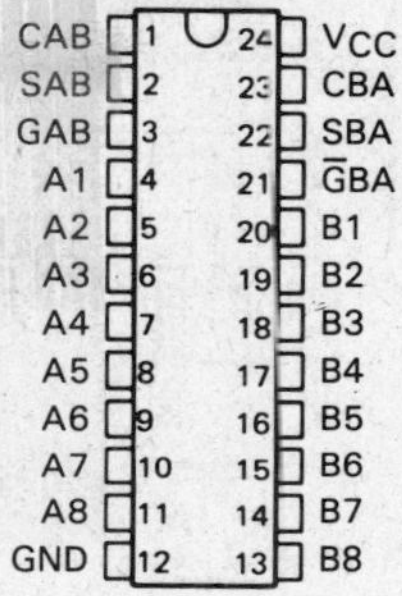

SN54HCT651, SN54HCT652 . . . FH OR FK PACKAGE
SN74HCT651, SN74HCT652 . . . FH OR FN PACKAGE
(TOP VIEW)

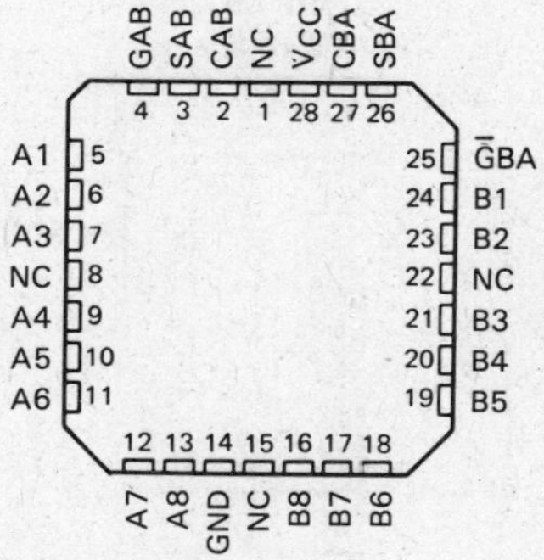

NC—No internal connection

description

These devices consist of bus transceiver circuits, D-type flip-flops, and control circuitry arranged for multiplexed transmission of data directly from the data bus or from the internal storage registers. Enable GAB and $\overline{G}$BA are provided to control the transceiver functions. SAB and SBA control pins are provided to select whether real-time or stored data is transferred. A low input level selects real-time data, and a high selects stored data. The examples on the following page demonstrate the four fundamental bus-management functions that can be performed with the 'HCT651 and 'HCT652.

Data on the A or B data bus, or both, can be stored in the internal D flip-flops by low-to-high transitions at the appropriate clock pins (CAB or CBA) regardless of the select or enable control pins. When SAB and SBA are in the real-time transfer mode, it is also possible to store data without using the internal D-type flip-flops by simultaneously enabling GAB and $\overline{G}$BA. In this configuration each output reinforces its input. Thus, when all other data sources to the two sets of bus lines are at high impedance, each set of bus lines will remain at its last state.

The SN54HCT651 and SN54HCT652 are characterized for operation over the full military temperature range of −55 °C to 125 °C. The SN74HCT651 and SN74HCT652 are characterized for operation from −40 °C to 85 °C.

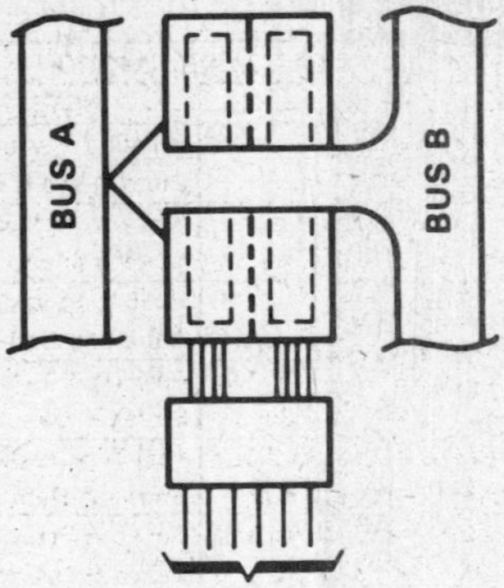

(3) GAB	(21) $\overline{G}$BA	(1) CAB	(23) CBA	(2) SAB	(22) SBA
L	L	X	X	X	L

REAL-TIME TRANSFER BUS B TO BUS A

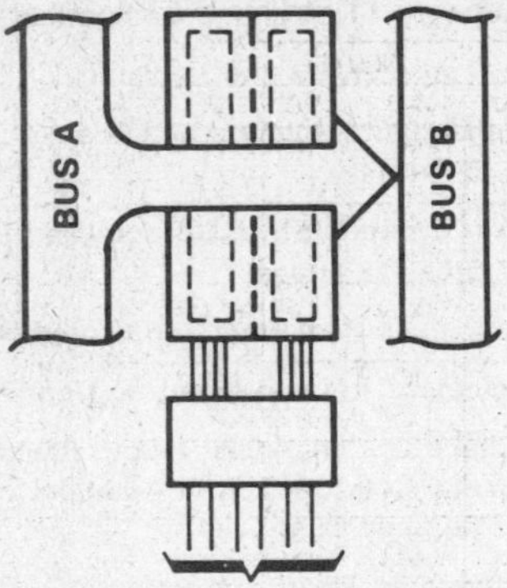

(3) GAB	(21) $\overline{G}$BA	(1) CAB	(23) CBA	(2) SAB	(22) SBA
H	H	X	X	L	X

REAL-TIME TRANSFER BUS A TO BUS B

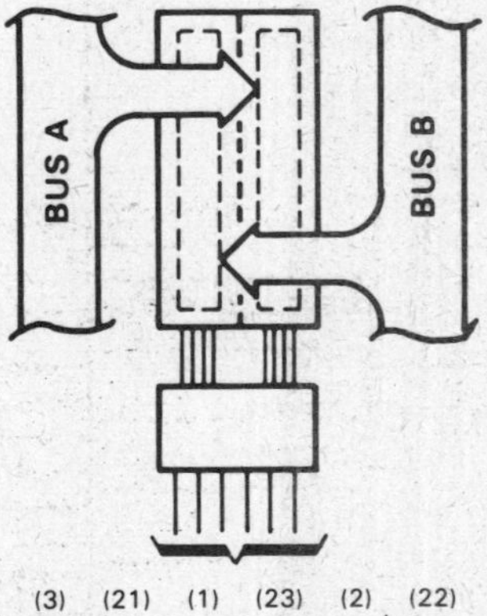

(3) GAB	(21) $\overline{G}$BA	(1) CAB	(23) CBA	(2) SAB	(22) SBA
X	H	↑	X	X	X
L	X	X	↑	X	X
L	H	↑	↑	X	X

STORAGE FROM A AND/OR B

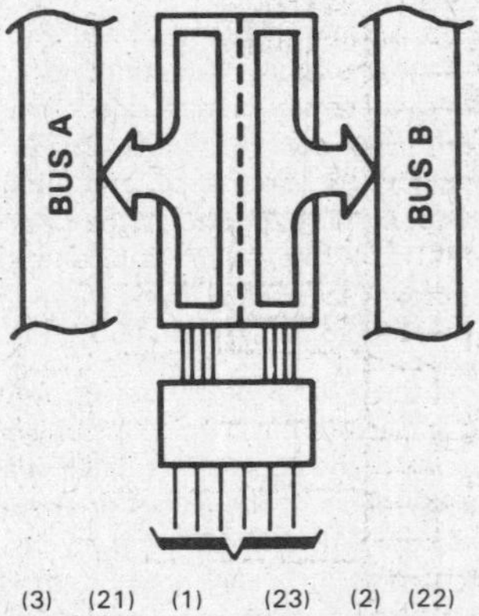

(3) GAB	(21) $\overline{G}$BA	(1) CAB	(23) CBA	(2) SAB	(22) SBA
H	L	H or L	H or L	H	H

TRANSFER STORED DATA TO A AND/OR B

Pin numbers shown are for JT and NT packages.

3 HCMOS DEVICES

TEXAS INSTRUMENTS
POST OFFICE BOX 225012 • DALLAS, TEXAS 75265

FUNCTION TABLE

INPUTS						DATA I/O†		OPERATION OR FUNCTION	
GAB	$\overline{G}$BA	CAB	CBA	SAB	SBA	A1 THRU A8	B1 THRU B8	'HCT651	'HCT652
L	H	H or L	H or L	X	X	Input	Input	Isolation	Isolation
L	H	↑	↑	X	X			Store A and B Data	Store A and B Data
X	H	↑	H or L	X	X	Input	Not specified	Store A, Hold B	Store A, Hold B
H	H	↑	↑	X	X	Input	Output	Store A in both registers	Store A in both registers
L	X	H or L	↑	X	X	Not specified	Input	Hold A, Store B	Hold A, Store B
L	L	↑	↑	X	X	Output	Input	Store B in both registers	Store B in both registers
L	L	X	X	X	L	Output	Input	Real-Time $\overline{B}$ Data to A Bus	Real-Time B Data to A Bus
L	L	X	H or L	X	H			Stored $\overline{B}$ Data to A Bus	Stored B Data to A Bus
H	H	X	X	L	X	Input	Output	Real-Time $\overline{A}$ Data to B Bus	Real-Time A Data to B Bus
H	H	H or L	X	H	X			Stored $\overline{A}$ Data to B Bus	Stored A Data to B Bus
H	L	H or L	H or L	H	H	Output	Output	Stored $\overline{A}$ Data to B Bus and Stored $\overline{B}$ Data to A Bus	Stored A Data to B Bus and Stored B Data to A Bus

† The data output functions may be enabled or disabled by various signals at the GAB and $\overline{G}$BA inputs. Data input functions are always enabled, i.e., data at the bus pins will be stored on every low-to-high transition on the clock inputs.

3

HCMOS DEVICES

logic symbols

'HCT651

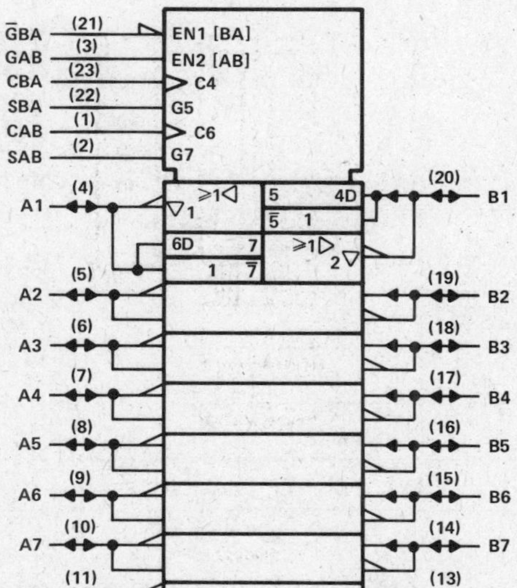

'HCT652

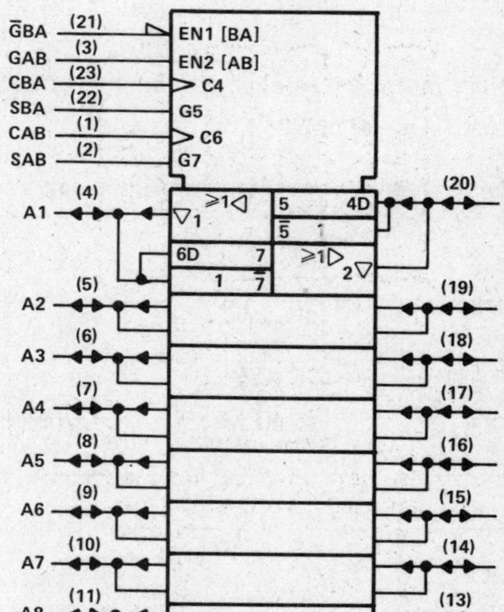

Pin numbers shown are for JT and NT packages.

logic diagram (positive logic)

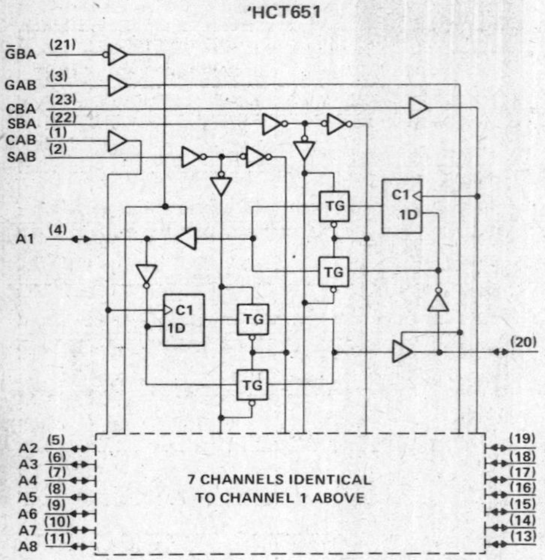

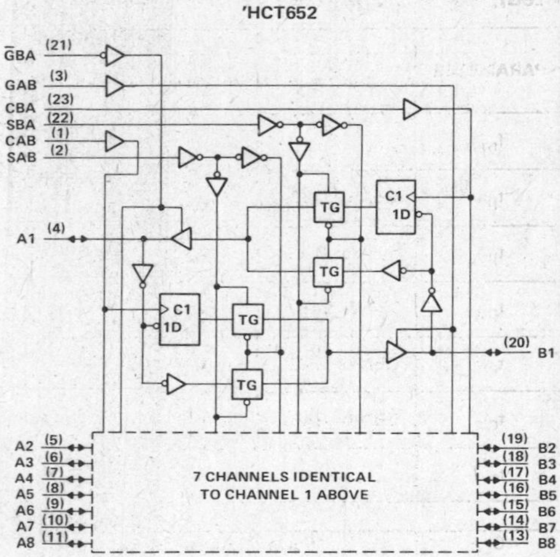

Pin numbers shown are for JT and NT packages.

absolute maximum ratings, recommended operating conditions, and electrical characteristics

See Table VII, page 2-14.

timing requirements over recommended operating free-air temperature range (unless otherwise noted)

		V_{CC}	T_A = 25°C MIN	T_A = 25°C MAX	SN54HCT651 SN54HCT652 MIN	SN54HCT651 SN54HCT652 MAX	SN74HCT651 SN74HCT652 MIN	SN74HCT651 SN74HCT652 MAX	UNIT
f_{clock}	Clock frequency	4.5 V	0	25	0	17	0	20	MHz
		5.5 V	0	28	0	19	0	22	
t_w	Pulse duration, CBA or CAB high or low	4.5 V	20		30		25		ns
		5.5 V	18		27		23		
t_{su}	Setup time, A before CAB↑ or B before CBA↑	4.5 V	15		23		19		ns
		5.5 V	14		21		17		
t_h	Hold time, A after CAB↑ or B after CBA↑	4.5 V	5		5		5		ns
		5.5 V	5		5		5		

switching characteristics over recommended operating free-air temperature range (unless otherwise noted), C_L = 50 pF (see Note 1)

PARAMETER	FROM (INPUT)	TO (OUTPUT)	V_{CC}	T_A = 25°C			SN54HCT651 SN54HCT652		SN74HCT652 SN74HCT652		UNIT
				MIN	TYP	MAX	MIN	MAX	MIN	MAX	
f_{max}			4.5 V	25	35		17		20		MHz
			5.5 V	28	40		19		22		
t_{pd}	CBA or CAB	A or B	4.5 V		18	36		54		45	ns
			5.5 V		16	32		49		41	
t_{pd}	A or B	B or A	4.5 V		14	27		41		34	ns
			5.5 V		12	24		37		31	
t_{pd}	SBA or SAB†	A or B	4.5 V		20	38		57		48	ns
			5.5 V		17	34		51		43	
t_{en}	$\overline{G}$BA or GAB	A or B	4.5 V		25	49		74		61	ns
			5.5 V		22	44		67		55	
t_{dis}	$\overline{G}$BA or GAB	A or B	4.5 V		25	49		74		61	ns
			5.5 V		22	44		67		55	
t_t		Any	4.5 V		9	12		18		15	ns
			5.5 V		7	11		16		14	

C_{pd}	Power dissipation capacitance	No load, T_A = 25°C	50 pF typ

switching characteristics over recommended operating free-air temperature range (unless otherwise noted), C_L = 150 pF (see Note 1)

PARAMETER	FROM (INPUT)	TO (OUTPUT)	V_{CC}	T_A = 25°C			SN54HCT651 SN54HCT652		SN74HCT651 SN74HCT652		UNIT
				MIN	TYP	MAX	MIN	MAX	MIN	MAX	
t_{pd}	CBA or CAB	A or B	4.5 V		24	53		80		66	ns
			5.5 V		22	47		72		60	
t_{pd}	A or B	B or A	4.5 V		22	44		70		55	ns
			5.5 V		20	39		60		50	
t_{pd}	SBA or SAB†	A or B	4.5 V		26	55		83		69	ns
			5.5 V		24	49		74		62	
t_{en}	$\overline{G}$BA or GAB	A or B	4.5 V		33	66		100		82	ns
			5.5 V		30	59		90		74	
t_t		Any	4.5 V		17	42		63		53	ns
			5.5 V		14	38		57		48	

NOTE 1: For load circuit and voltage waveforms, see page 1-14.
†These parameters are measured with the internal output state of the storage register opposite to that of the bьs input.

TYPES SN54HC658, SN54HC659, HC74HC658, SN74HC659 OCTAL BUS TRANSCEIVERS WITH PARITY

D2839, MARCH 1984

- Bus Transceivers with Inverting Outputs ('HC658) or True Outputs ('HC659)
- Generates a Parity Bit for A Bus and B Bus
- Easily Cascadable
- Internal Active Pull-Ups and Pull-Downs
- High-Current 3-State Outputs Can Drive up to 15 LSTTL Loads
- Package Options Include Both Plastic and Ceramic Chip Carriers in Addition to Plastic and Ceramic DIPs
- Dependable Texas Instruments Quality and Reliability

SN54HC658, SN54HC659 . . . JT PACKAGE
SN74HC658, SN74HC659 . . . JT OR NT PACKAGE
(TOP VIEW)

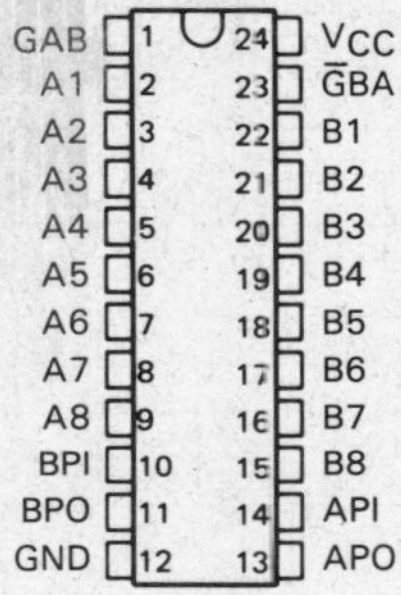

SN54HC658, SN54HC659 . . . FH OR FK PACKAGE
SN74HC658, SN74HC659 . . . FH OR FN PACKAGE
(TOP VIEW)

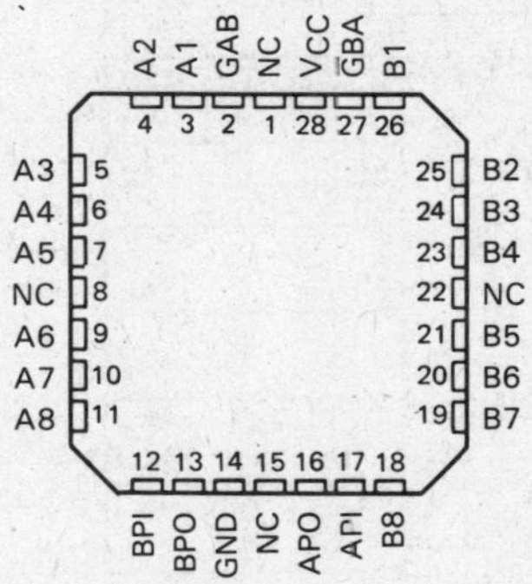

NC—No internal connection

description

These octal bus transceivers are designed for asynchronous, bidirectional communication between data buses. The devices transmit data from the A Bus to the B Bus or from the B Bus to the A Bus, depending on the level at the direction control inputs, GAB and $\overline{G}$BA. These devices also generate parity outputs, APO and BPO, which reflect the number of high levels at the A Bus and B Bus, respectively, taking into account the parity inputs API and BPI.

The bidirectional I/O ports feature active circuitry on the input stage that, when the output shared by that pin is disabled, will maintain the input in the last state taken by the output. This state will be maintained until changed by activity on the bus. The advantage of this arrangement is that when all outputs on the bus are disabled, the inputs will be prevented from floating, resulting in minimum power dissipation and minimum susceptibility to noise. This eliminates any need for external pull-up or pull-down resistors. The parity inputs API and BPI have similar circuits. For further information, see Typical Application Data.

The SN54HC658 and SN54HC659 are characterized for operation over the full military temperature range of −55°C to 125°C. The SN74HC658 and SN75HC659 are characterized for operation from −40°C to 85°C.

FUNCTION TABLE

CONTROL INPUTS		NUMBER OF HIGH INPUTS ON A BUS AND API	NUMBER OF HIGH INPUTS ON B BUS AND BPI	OUTPUTS		OPERATION	
$\overline{\text{GBA}}$	GAB			APO	BPO	'HC658	'HC659
L	L	X	0, 2, 4, 6, 8	Z	H	$\overline{\text{B}}$ Data to A Bus	B Data to A Bus
		X	1, 3, 5, 7, 9	Z	L		
H	H	0, 2, 4, 6, 8	X	H	Z	$\overline{\text{A}}$ Data to B Bus	A Data to B Bus
		1, 3, 5, 7, 9	X	L	Z		
H	L	X	X	Z	Z	Isolation	Isolation
L	H	X	0, 2, 4, 6, 8		H	$\overline{\text{B}}$ Data to A Bus, $\overline{\text{A}}$ Data to B Bus	B Data to A Bus, A Data to B Bus
		X	1, 3, 5, 7, 9		L		
		0, 2, 4, 6, 8	X	H			
		1, 3, 5, 7, 9	X	L			

logic symbols

'HC658

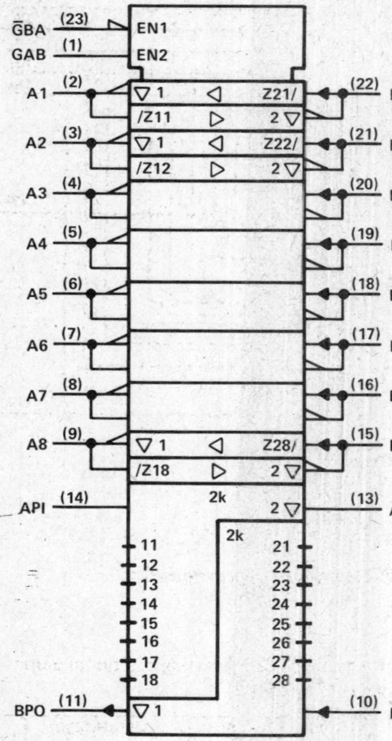

'HC659

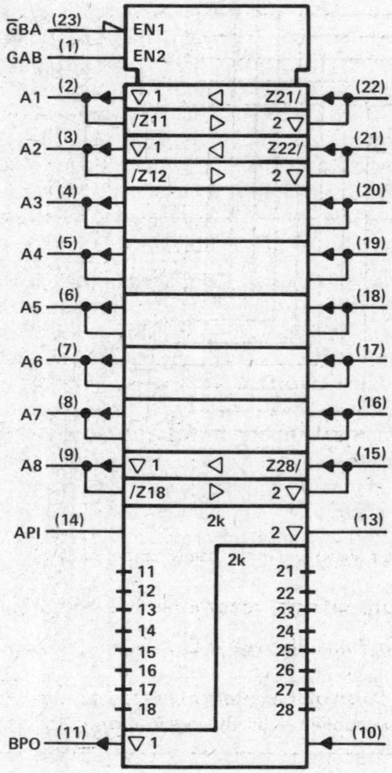

Pin numbers shown are for JT and NT packages.

logic diagram (positive logic)

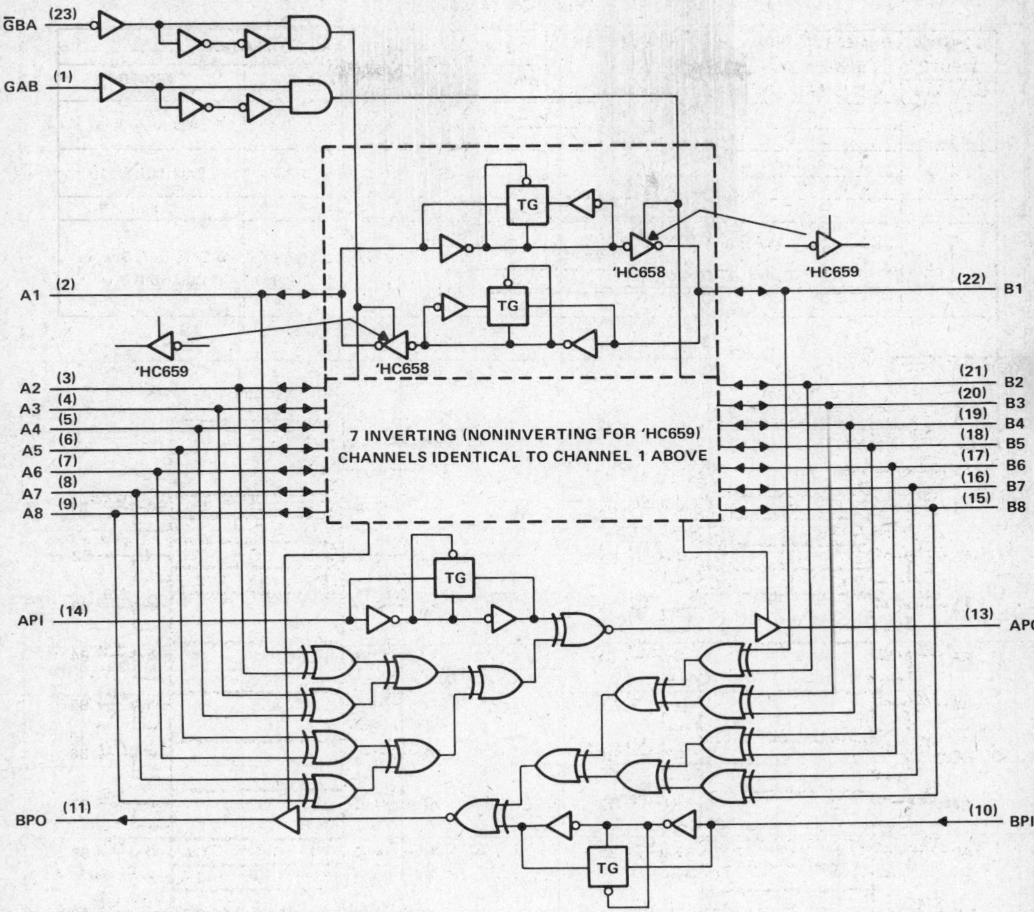

Pin numbers shown are for JT and NT packages.

maximum ratings, recommended operating conditions, and electrical characteristics

See Table III, page 2-8.

Because of the nature of the transceiver I/O ports and the parity inputs, the following additional parameter also applies. It is the peak current as the input changes from 0 V to V_{CC}.

PARAMETER	TEST CONDITION	V_{CC}	$T_A = 25°C$			SN54HC658 SN54HC659		SN74HC658 SN74HC659		UNIT
			MIN	TYP	MAX	MIN	MAX	MIN	MAX	
I_{IM}	$V_I = 0$ to V_{CC}	6 V			±400		±520		±520	μA

'HC658 switching characteristics over recommended operating free-air temperature range (unless otherwise noted), C_L = 50 pF (see Note 1)

PARAMETER	FROM (INPUT)	TO (OUTPUT)	V_{CC}	T_A = 25°C MIN	TYP	MAX	SN54HC658 MIN	MAX	SN74HC658 MIN	MAX	UNIT
t_{pd}	A or B	B or A	2 V		75	150		225		190	ns
			4.5 V		15	30		45		38	
			6 V		13	26		38		32	
t_{pd}	A or B	APO or BPO	2 V		115	230		345		290	ns
			4.5 V		23	46		69		58	
			6 V		20	39		59		49	
t_{pd}	API or BPI	APO or BPO	2 V		77	155		235		195	ns
			4.5 V		15	31		47		39	
			6 V		13	26		40		33	
t_{en}	GAB or $\overline{G}$BA	APO or BPO	2 V		117	235		355		295	ns
			4.5 V		23	47		71		59	
			6 V		20	40		60		50	
t_{dis}	GAB or $\overline{G}$BA	APO or BPO	2 V		117	235		355		295	ns
			4.5 V		23	47		71		59	
			6 V		20	40		60		50	
t_t		Any	2 V		28	60		90		75	ns
			4.5 V		8	12		18		15	
			6 V		6	10		15		13	

C_{pd}	Power dissipation capacitance	No load, T_A = 25°C	56 pF typ

'HC658 switching characteristics over recommended operating free-air temperature range (unless otherwise noted), C_L = 150 pF (see Note 1)

PARAMETER	FROM (INPUT)	TO (OUTPUT)	V_{CC}	T_A = 25°C MIN	TYP	MAX	SN54HC658 MIN	MAX	SN74HC658 MIN	MAX	UNIT
t_{pd}	A or B	B or A	2 V		117	235		355		295	ns
			4.5 V		23	47		71		59	
			6 V		20	41		60		51	
t_{pd}	A or B	APO or BPO	2 V		157	315		475		395	ns
			4.5 V		31	63		95		79	
			6 V		27	54		81		68	
t_{pd}	API or BPI	APO or BPO	2 V		120	240		365		300	ns
			4.5 V		24	48		73		60	
			6 V		20	41		62		52	
t_{en}	GAB or $\overline{G}$BA	APO or BPO	2 V		160	320		485		400	ns
			4.5 V		32	64		97		80	
			6 V		27	55		82		69	
t_t		Any	2 V		37	210		315		265	ns
			4.5 V		12	42		63		53	
			6 V		10	36		53		45	

NOTE 1: For load circuit and voltage waveforms, see page 1-14.

'HC659 switching characteristics over recommended operating free-air temperature range (unless otherwise noted), C_L = 50 pF (see Note 1)

PARAMETER	FROM (INPUT)	TO (OUTPUT)	V_{CC}	T_A = 25°C MIN	TYP	MAX	SN54HC659 MIN	MAX	SN74HC659 MIN	MAX	UNIT
t_{pd}	A or B	B or A	2 V		70	140		210		175	ns
			4.5 V		14	28		42		35	
			6 V		12	24		36		30	
t_{pd}	A or B	APO or BPO	2 V		115	230		345		290	ns
			4.5 V		23	46		69		58	
			6 V		20	39		59		49	
t_{pd}	API or BPI	APO or BPO	2 V		77	155		235		195	ns
			4.5 V		15	31		47		39	
			6 V		13	26		40		33	
t_{en}	GAB or $\overline{G}$BA	APO or BPO	2 V		117	235		355		295	ns
			4.5 V		23	47		71		59	
			6 V		20	40		60		50	
t_{dis}	GAB or $\overline{G}$BA	APO or BPO	2 V		117	235		355		295	ns
			4.5 V		23	47		71		59	
			6 V		20	40		60		50	
t_t		Any	2 V		28	60		90		75	ns
			4.5 V		8	12		18		15	
			6 V		6	10		15		13	

C_{pd}	Power dissipation capacitance	No load, T_A = 25°C	56 pF typ

'HC659 switching characteristics over recommended operating free-air temperature range (unless otherwise noted), C_L = 150 pF (see Note 1)

PARAMETER	FROM (INPUT)	TO (OUTPUT)	V_{CC}	T_A = 25°C MIN	TYP	MAX	SN54HC659 MIN	MAX	SN74HC659 MIN	MAX	UNIT
t_{pd}	A or B	B or A	2 V		117	225		340		280	ns
			4.5 V		23	45		68		56	
			6 V		20	39		58		49	
t_{pd}	A or B	APO or BPO	2 V		157	315		475		395	ns
			4.5 V		31	63		95		79	
			6 V		27	54		81		68	
t_{pd}	API or BPI	APO or BPO	2 V		120	240		365		300	ns
			4.5 V		24	48		73		60	
			6 V		20	41		62		52	
t_{en}	GAB or $\overline{G}$BA	APO or BPO	2 V		160	320		485		400	ns
			4.5 V		32	64		97		80	
			6 V		27	55		82		69	
t_t		Any	2 V		37	210		315		265	ns
			4.5 V		12	42		63		53	
			6 V		10	36		53		45	

NOTE 1: For load circuit and voltage waveforms, see page 1-14.

TYPICAL APPLICATION DATA

The unique structure used on the I/O ports and the parity inputs of these devices deserves some special consideration (see Figure 1). Only the input structure is shown. The conventional 3-state output structure associated with each I/O port has been omitted to facilitate understanding

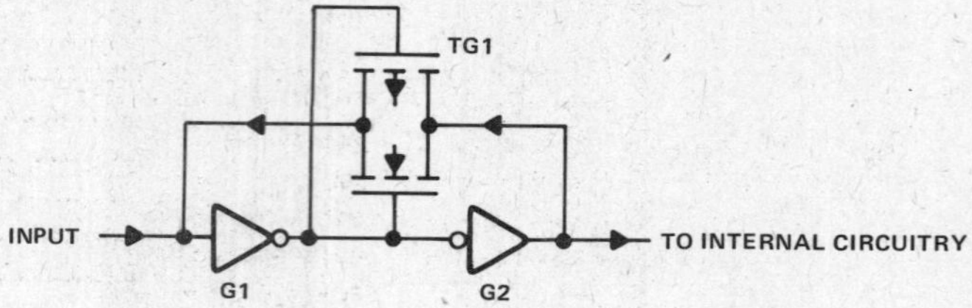

FIGURE 1. INPUT STRUCTURE

The two inverters (G1 and G2) have a transmission gate (TG1) connected in a feedback loop around them. This transmission gate is connected in an unusual fashion, that is, with the gates of both transistors connected to the output of G1. Thus, with the output of G1 at either a high or a low level, one or the other of the transistors will be turned on allowing feedback of the output of G2 to the input of G1. The effect of TG1 is that the input level will be maintained at whatever level existed prior to the bus being disabled or the level currently existing on the bus will be reinforced.

To understand the operation of this input, assume that initially the input is at a low logic level. As the input voltage is raised, TG1 sinks current to attempt to maintain the low level. However, TG1 consists of small geometry transistors and appears resistive as current flows thus allowing the input voltage to rise toward the threshold voltage of G1. When the threshold voltage is reached, G1 changes state causing G2 to change state. G2 then attempts to maintain a high level on the input through TG1. A similar operation occurs when the input voltage is decreased toward the threshold voltge of G1. G2 sources current through TG1 until the threshold is reached.

This characteristic of the input stage has some implications for the input current levels. With the input held at either V_{CC} or GND, there is no voltage across TG1 and negligible input current. However, as the input voltage is raised from GND or lowered from V_{CC}, the input current rises as the voltage across TG1 increases. The input current continues to rise until it reaches a maximum just as the threshold voltage of G1 is reached. This maximum corresponds to parameter I_{IM} shown in the electrical characteristics table. Because G1 consists of small geometry transistors, I_{IM} has a value much lower than the output drive capability of a conventional 'HC output stage. Also for this reason, the input configuration has negligible effect when the I/O port is used as an output.

This configuration provides for minimum power dissipation when the bus is inactive (all outputs on the bus in the high-impedance state) and minimum susceptibility to noise on the bus during this time. The increase in input current may go unnoticed as it only occurs during transitions on the bus. Care must be taken when measuring input currents (e.g., at incoming inspection) to ensure that the input voltage is set to the correct value.

The use of these devices for interfacing to 8-, 16-, and 24-bit-wide memory arrays with parity is illustrated in Figures 2, 3, and 4.

TYPICAL APPLICATION DATA

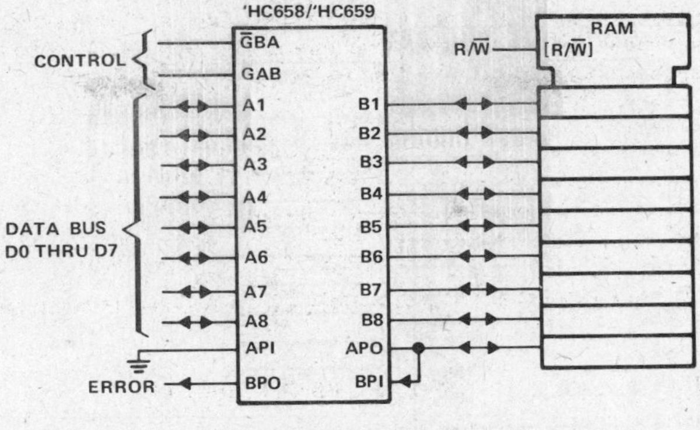

FIGURE 2. 8-BIT-WIDE MEMORY ARRAY WITH PARITY

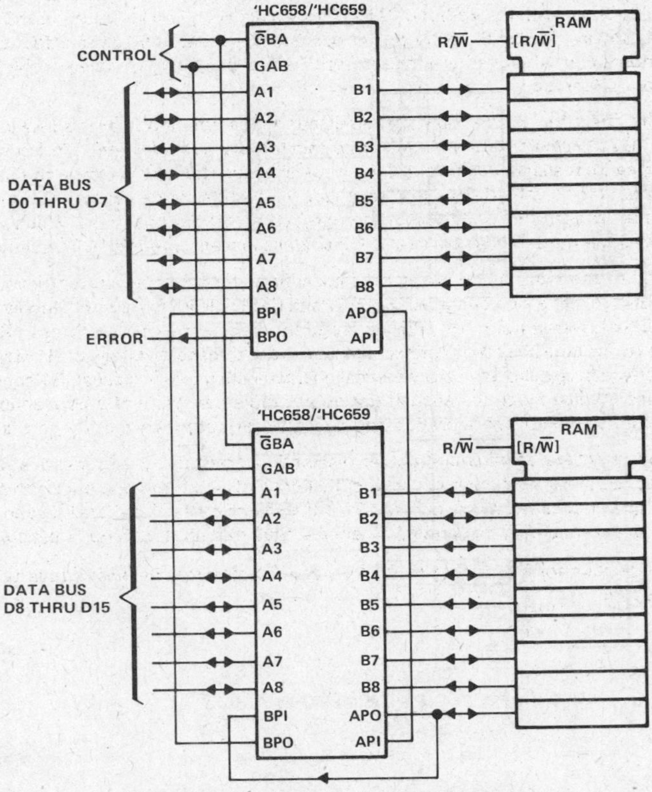

FIGURE 3. 16-BIT-WIDE MEMORY ARRAY WITH PARITY

TYPICAL APPLICATION DATA

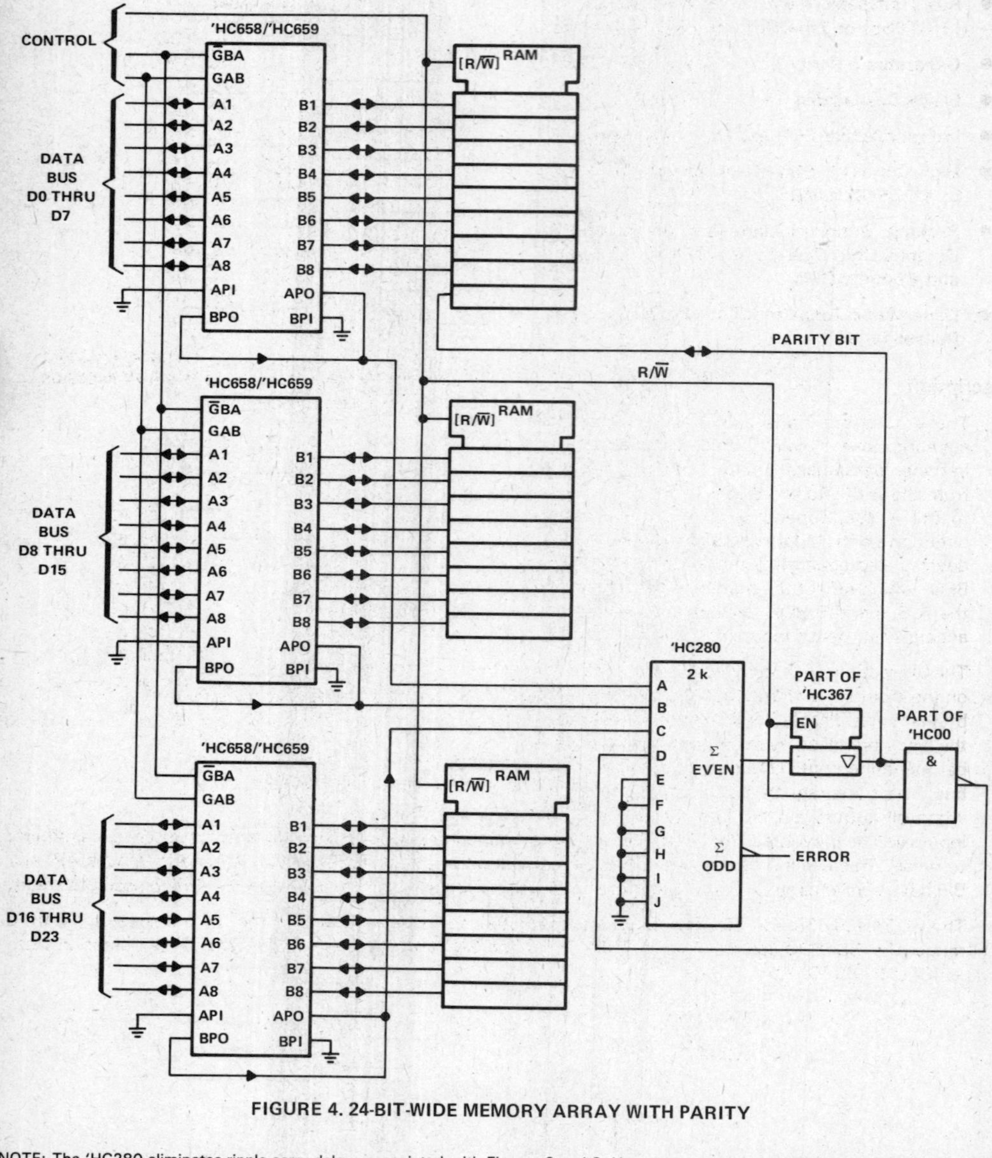

FIGURE 4. 24-BIT-WIDE MEMORY ARRAY WITH PARITY

NOTE: The 'HC280 eliminates ripple carry delays associated with Figures 2 and 3. However, in those two cases the delays are probably too small to be of concern.

TEXAS INSTRUMENTS
POST OFFICE BOX 225012 • DALLAS, TEXAS 75265

TYPES SN54HCT658, SN54HCT659, SN74HCT658, SN74HCT659 OCTAL BUS TRANSCEIVERS WITH PARITY

D2839, MARCH 1984

- Inputs are TTL-Voltage Compatible
- Bus Transceivers with Inverting Outputs ('HCT658) or True Outputs ('HCT659)
- Generates a Parity Bit for A Bus and B Bus
- Easily Cascadable
- Internal Active Pull-Ups and Pull-Downs
- High-Current 3-State Outputs Can Drive up to 15 LSTTL Loads
- Package Options Include Both Plastic and Ceramic Chip Carriers in Addition to Plastic and Ceramic DIPs
- Dependable Texas Instruments Quality and Reliability

SN54HCT658, SN54HCT659 . . . JT PACKAGE
SN74HCT658, SN74HCT659 . . . JT OR NT PACKAGE
(TOP VIEW)

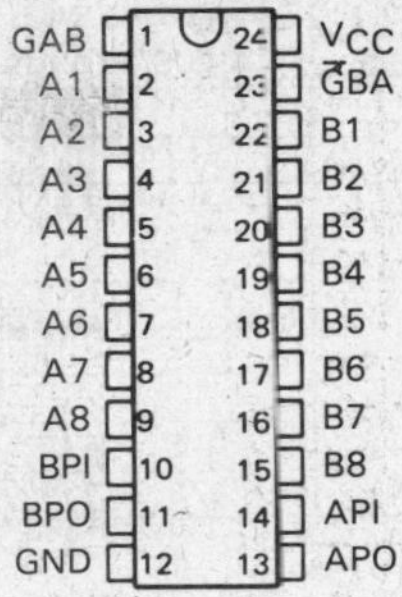

SN54HCT658, SN54HCT659 . . . FH OR FK PACKAGE
SN74HCT658, SN74HCT659 . . . FH OR FN PACKAGE
(TOP VIEW)

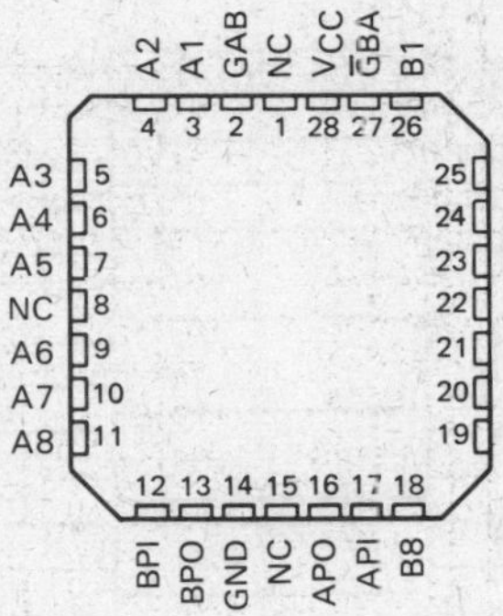

NC—No internal connection.

description

These octal bus transceivers are designed for asynchronous, bidirectional communication between data buses. The devices transmit data from the A Bus to the B Bus, or from the B Bus to the A Bus, depending on the levels at the direction control inputs, GAB and ḠBA. These devices also generate parity outputs, APO and BPO, which reflect the number of high levels at the A Bus and B Bus, respectively, taking into account the parity inputs API and BPI.

The bidirectional I/O ports feature active circuits on the input stage that, when the output shared by that pin is disabled, will maintain the input in the last state taken by the output. This state will be maintained until changed by activity on the bus. The advantage of this arrangement is that when all outputs on the bus are disabled, the inputs will be prevented from floating, resulting in minimum power dissipation and minimum susceptibility to noise. This eliminates any need for external pull-up or pull-down resistors. The parity inputs API and BPI have similar circuits. For further information, see Typical Application Data in 'HC658 series data sheet.

The SN54HCT658 and SN54HCT659 are characterized for operation over the full military temperature range of −55 °C to 125 °C. The SN74HCT658 and SN74HCT659 are characterized for operation from −40 °C to 85 °C.

FUNCTION TABLE

CONTROL INPUTS		NUMBER OF HIGH INPUTS ON A BUS AND API	NUMBER OF HIGH INPUTS ON B BUS AND BPI	OUTPUTS		OPERATION	
$\overline{GBA}$	GAB			APO	BPO	'HCT658	'HCT659
L	L	X	0, 2, 4, 6, 8	Z	H	$\overline{B}$ Data to A Bus	B Data to A Bus
		X	1,3,5,7,9	Z	L		
H	H	0, 2,4,6,8	X	H	Z	$\overline{A}$ Data to B Bus	A Data to B Bus
		1,3,5,7,9	X	L	Z		
H	L	X	X	Z	Z	Isolation	Isolation
L	H	X	0, 2,4,6,8		H	$\overline{B}$ Data to A Bus, $\overline{A}$ Data to B Bus	B Data to A Bus, A Data to B Bus
		X	1,3,5,7,9		L		
		0, 2, 4, 6, 8	X	H			
		1, 3, 5, 7, 9	X	L			

logic symbols

'HCT658

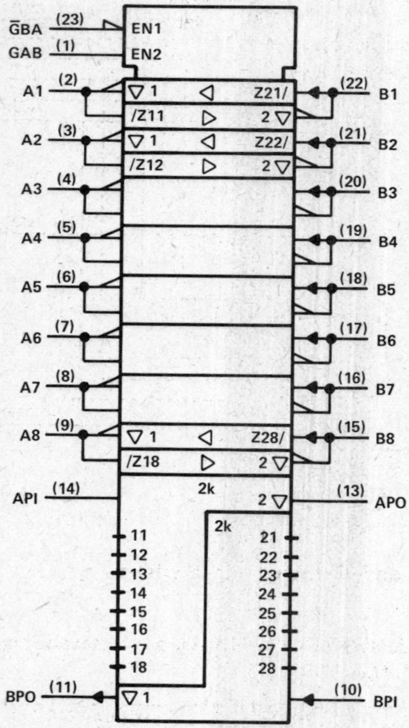

'HCT659

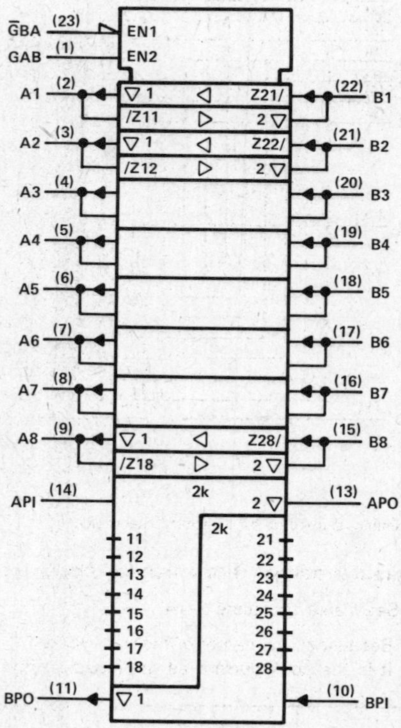

Pin numbers shown are for JT and NT packages.

logic diagram (positive logic)

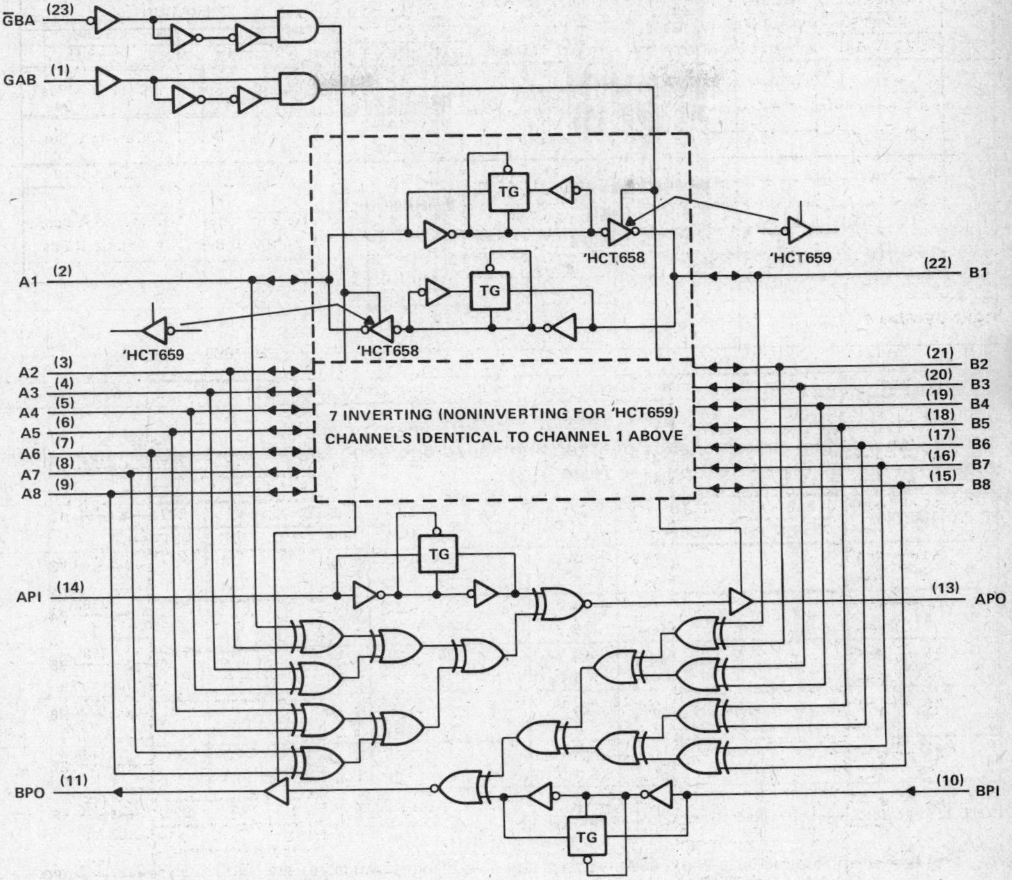

Pin numbers shown are for JT and NT packages.

maximum ratings, recommended operating conditions, and electrical characteristics

See Table VII, page 2-14.

Because of the nature of the transceiver I/O ports and the parity inputs, the following parameter also applies. It is the peak current as the input changes from 0 V to V_{CC}.

PARAMETER	TEST CONDITION	V_{CC}	$T_A = 25°C$			SN54HCT658 SN54HCT659		SN74HCT658 SN74HCT659		UNIT
			MIN	TYP	MAX	MIN	MAX	MIN	MAX	
I_{IM}	V_I = 0 to V_{CC}	5.5 V			±400		±520		±520	μA

'HCT658 switching characteristics over recommended operating free-air temperature range (unless otherwise noted), C_L = 50 pF (see Note 1)

PARAMETER	FROM (INPUT)	TO (OUTPUT)	V_{CC}	T_A = 25°C MIN	TYP	MAX	SN54HCT658 MIN	MAX	SN74HCT658 MIN	MAX	UNIT
t_{pd}	A or B	B or A	4.5 V		15	30		45		38	ns
			5.5 V		13	27		41		34	
t_{pd}	A or B	APO or BPO	4.5 V		23	46		69		58	ns
			5.5 V		20	41		62		52	
t_{pd}	API or BPI	APO or BPO	4.5 V		15	31		47		39	ns
			5.5 V		14	28		42		35	
t_{en}	GAB or $\overline{G}$BA	APO or BPO	4.5 V		24	47		71		59	ns
			5.5 V		21	42		64		53	
t_{dis}	GAB or $\overline{G}$BA	APO or BPO	4.5 V		24	47		71		59	ns
			5.5 V		21	42		64		53	
t_t		Any	4.5 V		8	12		18		15	ns
			5.5 V		7	11		16		14	

C_{pd}	Power dissipation capacitance	No load, T_A = 25°C	62 pF typ

'HCT658 switching characteristics over recommended operating free-air temperature range (unless otherwise noted), C_L = 150 pF (see Note 1)

PARAMETER	FROM (INPUT)	TO (OUTPUT)	V_{CC}	T_A = 25°C MIN	TYP	MAX	SN54HCT658 MIN	MAX	SN74HCT658 MIN	MAX	UNIT
t_{pd}	A or B	B or A	4.5 V		23	47		71		59	ns
			5.5 V		21	42		64		53	
t_{pd}	A or B	APO or BPO	4.5 V		31	63		95		79	ns
			5.5 V		28	56		85		71	
t_{pd}	API or BPI	APO or BPO	4.5 V		24	48		73		60	ns
			5.5 V		21	43		65		54	
t_{en}	GAB or $\overline{G}$BA	APO or BPO	4.5 V		32	64		97		80	ns
			5.5 V		28	57		87		72	
t_t		Any	4.5 V		17	42		63		53	ns
			5.5 V		14	38		57		48	

NOTE 1: For load circuit and voltage waveforms, see page 1-14.

For typical application data and a description of the unique input structure, see the 'HC658 series data sheet.

'HCT659 switching characteristics over recommended operating free-air temperature range (unless otherwise noted), C_L = 50 pF (see Note 1)

PARAMETER	FROM (INPUT)	TO (OUTPUT)	V_{CC}	T_A = 25°C MIN	TYP	MAX	SN54HCT659 MIN	MAX	SN74HCT659 MIN	MAX	UNIT
t_{pd}	A or B	B or A	4.5 V		14	28		42		35	ns
			5.5 V		12	25		61		50	
t_{pd}	A or B	APO or BPO	4.5 V		23	46		69		58	ns
			5.5 V		20	41		62		52	
t_{pd}	API or BPI	APO or BPO	4.5 V		15	31		47		39	ns
			5.5 V		14	28		42		35	
t_{en}	GAB or $\overline{G}$BA	APO or BPO	4.5 V		24	47		71		59	ns
			5.5 V		21	42		64		53	
t_{dis}	GAB or $\overline{G}$BA	APO or BPO	4.5 V		24	47		71		59	ns
			5.5 V		21	42		64		53	
t_t		Any	4.5 V		8	12		18		15	ns
			5.5 V		7	11		16		14	

C_{pd}	Power dissipation capacitance	No load, T_A = 25°C	62 pF typ

'HCT659 switching characteristics over recommended operating free-air temperature range (unless otherwise noted), C_L = 150 pF (see Note 1)

PARAMETER	FROM (INPUT)	TO (OUTPUT)	V_{CC}	T_A = 25°C MIN	TYP	MAX	SN54HCT659 MIN	MAX	SN74HCT659 MIN	MAX	UNIT
t_{pd}	A or B	B or A	4.5 V		23	45		68		56	ns
			5.5 V		20	40		84		67	
t_{pd}	A or B	APO or BPO	4.5 V		32	63		95		79	ns
			5.5 V		28	56		85		71	
t_{pd}	API or BPI	APO or BPO	4.5 V		24	48		73		60	ns
			5.5 V		21	43		65		54	
t_{en}	GAB or $\overline{G}$BA	APO or BPO	4.5 V		32	64		97		80	ns
			5.5 V		29	57		87		72	
t_t		Any	4.5 V		21	42		63		53	ns
			5.5 V		19	38		57		48	

NOTE 1: For load circuit and voltage waveforms, see page 1-14.

For typical application data and a description of the unique input structure, see the 'HC658 series data sheet.

TYPES SN54HC664, SN54HC665, SN74HC664, SN74HC665 OCTAL BUS TRANSCEIVERS WITH PARITY

D2839, MARCH 1984

- **Bus Transceivers with Inverting Outputs ('HC664) or True Outputs ('HC665)**
- **Generates a Parity Bit for A Bus and B Bus**
- **Easily Cascadable**
- **Internal Active Pull-Ups and Pull-Downs**
- **High-Current 3-State Outputs Can Drive up to 15 LSTTL Loads**
- **Package Options Include Both Plastic and Ceramic Chip Carriers in Addition to Plastic and Ceramic DIPs**
- **Dependable Texas Instruments Quality and Reliability**

SN54HC664, SN54HC665 . . . JT PACKAGE
SN74HC664, SN74HC665 . . . JT OR NT PACKAGE
(TOP VIEW)

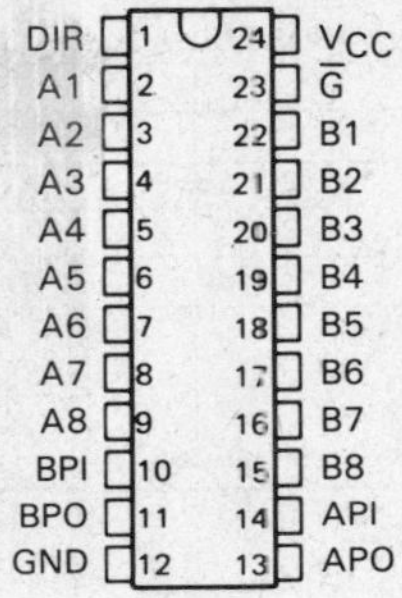

SN54HC664, SN54HC665 . . FH OR FK PACKAGE
SN74HC664, SN74HC665 . . . FH OR FN PACKAGE
(TOP VIEW)

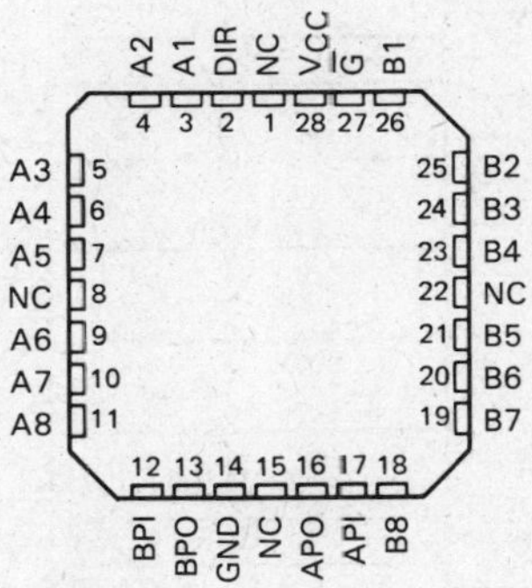

NC—No internal connection

description

These octal bus transceivers are designed for asynchronous, bidirectional communication between data buses. The devices transmit data from the A Bus to the B Bus or from the B Bus to the A Bus, depending on the level at the direction control input, DIR. The enable input, $\overline{G}$, can be used to disable the device so that the buses are isolated. These devices will also generate parity outputs, APO and BPO, which reflect the number of high levels at the A Bus and B Bus, respectively, taking into account the parity inputs API and BPI.

The bidirectional I/O ports feature active circuitry on the input stage that, when the output shared by that pin is disabled, will maintain the input in the last state taken by the output. This state will be maintained until changed by the activity on the bus. The advantage of this arrangement is that when all outputs on the bus are disabled, the inputs will be prevented from floating, resulting in minimum power dissipation and minimum susceptibility to noise. This eliminates any need for external pull-up or pull-down resistors. The parity inputs API and BPI have similar circuitry. For further information, see the Typical Application Data.

The SN54HC664 and SN54HC665 are characterized for operation over the full military temperature range of −55 °C to 125 °C. The SN74HC664 and SN74HC665 are characterized for operation from −40 °C to 85 °C.

FUNCTION TABLE

CONTROL INPUTS		NUMBER OF HIGH INPUTS ON A BUS AND API	NUMBER OF HIGH INPUTS ON B BUS AND BPI	OUTPUTS		OPERATION	
$\overline{G}$	DIR			APO	BPO	'HC664	'HC665
L	L	X	0, 2, 4, 6, 8	Z	H	$\overline{B}$ Data to A Bus	B Data to A Bus
		X	1, 3, 5, 7, 9	Z	L		
L	H	0, 2, 4, 6, 8	X	H	Z	$\overline{A}$ Data to B Bus	A Data to B Bus
		1, 3, 5, 7, 9	X	L	Z		
H	X	X	X	Z	Z	Isolation	Isolation

logic symbols

'HC664

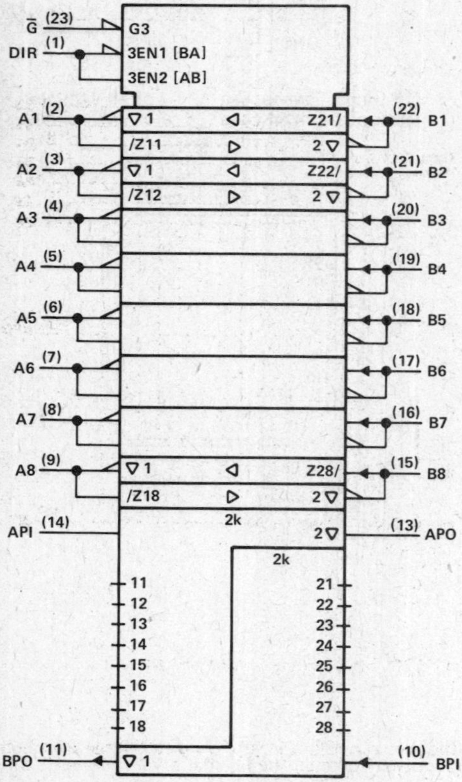

'HC665

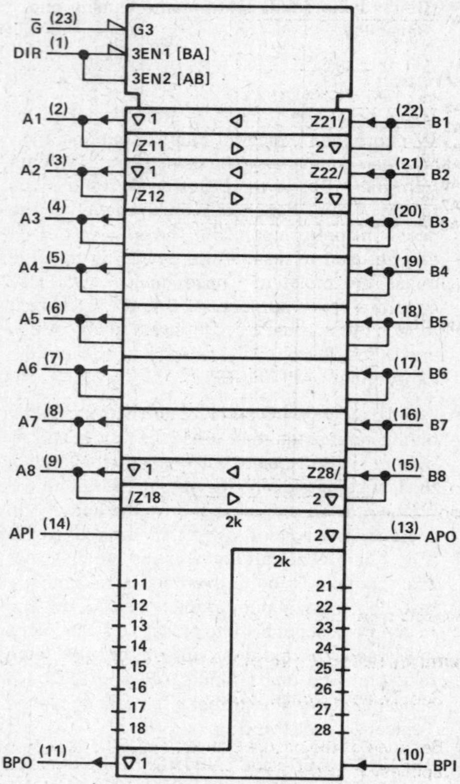

Pin numbers shown are for JT and NT packages.

TEXAS INSTRUMENTS
POST OFFICE BOX 225012 • DALLAS, TEXAS 75265

logic diagram (positive logic)

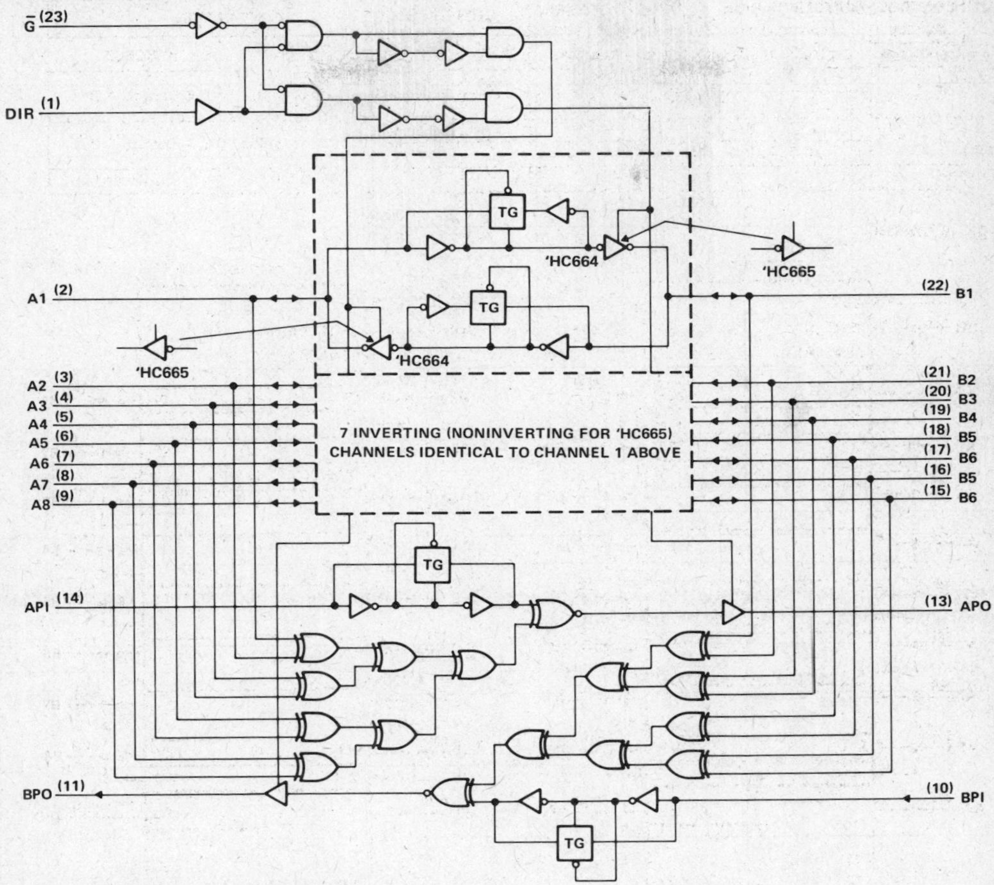

Pin numbers shown are for JT and NT packages.

maximum ratings, recommended operating conditions, and electrical characteristics

See Table III, page 2-8.

Because of the nature of the transceiver I/O ports and the parity inputs, the following additional parameter applies. It is the peak current as the input changes from 0 V to V_{CC}.

PARAMETER	TEST CONDITION	V_{CC}	T_A = 25°C			SN54HC664 SN54HC665		SN74HC664 SN74HC665		UNIT
			MIN	TYP	MAX	MIN	MAX	MIN	MAX	
I_{IM}	V_I = 0 to V_{CC}	6 V			±400		±520		±520	μA

'HC664 switching characteristics over recommended operating free-air temperature range (unless otherwise noted), C_L = 50 pF (see Note 1)

PARAMETER	FROM (INPUT)	TO (OUTPUT)	V_{CC}	T_A = 25°C MIN	TYP	MAX	SN54HC664 MIN	MAX	SN74HC664 MIN	MAX	UNIT
t_{pd}	A or B	B or A	2 V		75	150		225		190	ns
			4.5 V		15	30		45		38	
			6 V		13	26		38		32	
t_{pd}	A or B	APO or BPO	2 V		115	230		345		290	ns
			4.5 V		23	46		69		58	
			6 V		20	39		59		49	
t_{pd}	API or BPI	APO or BPO	2 V		77	155		235		195	ns
			4.5 V		15	31		47		39	
			6 V		13	26		40		33	
t_{en}	$\overline{G}$ or DIR	A or B	2 V		125	255		385		320	ns
			4.5 V		25	51		77		64	
			6 V		22	43		65		54	
t_{dis}	$\overline{G}$ or DIR	A or B	2 V		125	255		385		320	ns
			4.5 V		25	51		77		64	
			6 V		22	43		65		54	
t_t		Any	2 V		28	60		90		75	ns
			4.5 V		8	12		18		15	
			6 V		6	10		15		13	

C_{pd}	Power dissipation capacitance	No load, T_A = 25°C	56 pF typ

'HC664 switching characteristics over recommended operating free-air temperature range (unless otherwise noted), C_L = 150 pF (see Note 1)

PARAMETER	FROM (INPUT)	TO (OUTPUT)	V_{CC}	T_A = 25°C MIN	TYP	MAX	SN54HC664 MIN	MAX	SN74HC664 MIN	MAX	UNIT
t_{pd}	A or B	B or A	2 V		116	235		355		295	ns
			4.5 V		23	47		71		59	
			6 V		20	41		60		51	
t_{pd}	A or B	APO or BPO	2 V		157	315		475		395	ns
			4.5 V		31	63		95		79	
			6 V		27	54		81		68	
t_{pd}	API or BPI	APO or BPO	2 V		120	240		365		300	ns
			4.5 V		24	48		73		60	
			6 V		20	41		62		52	
t_{en}	$\overline{G}$ or DIR	A or B	2 V		170	340		515		425	ns
			4.5 V		34	68		103		85	
			6 V		29	58		87		73	
t_t		Any	2 V		37	210		315		265	ns
			4.5 V		12	42		63		53	
			6 V		10	36		53		45	

NOTE 1: For load circuit and voltage waveforms, see page 1-14.

TEXAS INSTRUMENTS
POST OFFICE BOX 225012 • DALLAS, TEXAS 75265

'HC665 switching characteristics over recommended operating free-air temperature range (unless otherwise noted), C_L = 50 pF (see Note 1)

PARAMETER	FROM (INPUT)	TO (OUTPUT)	V_{CC}	T_A = 25°C MIN	T_A = 25°C TYP	T_A = 25°C MAX	SN54HC665 MIN	SN54HC665 MAX	SN74HC665 MIN	SN74HC665 MAX	UNIT
t_{pd}	A or B	B or A	2 V		70	140		210		175	ns
			4.5 V		14	28		42		35	
			6 V		12	24		36		30	
t_{pd}	A or B	APO or BPO	2 V		115	230		345		290	ns
			4.5 V		23	46		69		58	
			6 V		20	39		59		49	
t_{pd}	API or BPI	APO or BPO	2 V		77	155		235		195	ns
			4.5 V		15	31		47		39	
			6 V		13	26		40		33	
t_{en}	$\overline{G}$ or DIR	A or B	2 V		125	255		385		320	ns
			4.5 V		25	51		77		64	
			6 V		22	43		65		54	
t_{dis}	$\overline{G}$ or DIR	A or B	2 V		125	255		385		320	ns
			4.5 V		25	51		77		64	
			6 V		22	43		65		54	
t_t		Any	2 V		28	60		90		75	ns
			4.5 V		8	12		18		15	
			6 V		6	10		15		13	

C_{pd}	Power dissipation capacitance	No load, T_A = 25°C	56 pF typ

'HC665 switching characteristics over recommended operating free-air temperature range (unless otherwise noted), C_L = 150 pF (see Note 1)

PARAMETER	FROM (INPUT)	TO (OUTPUT)	V_{CC}	T_A = 25°C MIN	T_A = 25°C TYP	T_A = 25°C MAX	SN54HC665 MIN	SN54HC665 MAX	SN74HC665 MIN	SN74HC665 MAX	UNIT
t_{pd}	A or B	B or A	2 V		112	225		340		280	ns
			4.5 V		22	45		68		56	
			6 V		20	39		58		49	
t_{pd}	A or B	APO or BPO	2 V		157	315		475		395	ns
			4.5 V		31	63		95		79	
			6 V		27	54		81		68	
t_{pd}	API or BPI	APO or BPO	2 V		120	240		365		300	ns
			4.5 V		24	48		73		60	
			6 V		20	41		62		52	
t_{en}	$\overline{G}$ or DIR	A or B	2 V		170	340		515		425	ns
			4.5 V		34	68		103		85	
			6 V		29	58		87		73	
t_t		Any	2 V		37	210		315		265	ns
			4.5 V		12	42		63		53	
			6 V		10	36		53		45	

NOTE 1: For load circuit and voltage waveforms, see page 1-14.

TYPICAL APPLICATION DATA

The unique structure used on the I/O ports and the parity inputs of these devices deserves some special consideration (see Figure 1). Only the input structure is shown. The conventional 3-state output structure associated with each I/O port has been omitted to facilitate understanding.

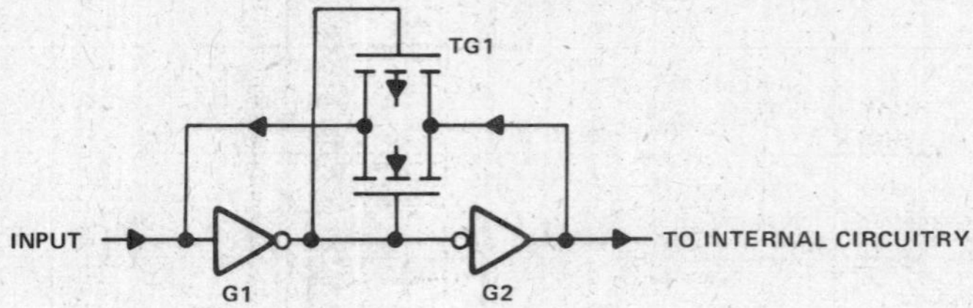

FIGURE 1. INPUT STRUCTURE

The two inverters (G1 and G2) have a transmission gate (TG1) connected in a feedback loop around them. This transmission gate is connected in an unusual fashion, that is, with the gates of both transistors connected to the output of G1. Thus, with the output of G1 at either a high or a low level, one or the other of the transistors will be turned on allowing feedback of the output of G2 to the input of G1. The effect of TG1 is that the input level will be maintained at whatever level existed prior to the bus being disabled or the level currently existing on the bus will be reinforced.

To understand the operation of this input, assume that initially the input is at a low logic level. As the input voltage is raised, TG1 sinks current to attempt to maintain the low level. However, TG1 consists of small geometry transistors and appears resistive as current flows thus allowing the input voltage to rise toward the threshold voltage of G1. When the threshold voltage is reached, G1 changes state causing G2 to change state. G2 then attempts to maintain a high level on the input through TG1. A similar operation occurs when the input voltage is decreased toward the threshold voltage of G1. G2 sources current through TG1 until the threshold is reached.

This characteristic of the input stage has some implications for the input current levels. With the input held at either V_{CC} or GND, there is no voltage across TG1 and negligible input current. However, as the input voltage is raised from GND or lowered from V_{CC}, the input current rises as the voltage across TG1 increases. The input current continues to rise until it reaches a maximum just as the threshold voltage of G1 is reached. This maximum corresponds to parameter I_{IM} shown in the electrical characteristics table. Because G1 consists of small geometry transistors, I_{IM} has a value much lower than the output drive capability of a conventional 'HC output stage. Also for this reason, the input configuration has negligible effect when the I/O port is used as an output.

This configuration provides for minimum power dissipation when the bus is inactive (all outputs on the bus in the high-impedance state) and minimum susceptibility to noise on the bus during this time. The increase in input current may go unnoticed as it only occurs during transitions on the bus. Care must be taken when measuring input currents (e.g., at incoming inspection) to ensure that the input voltage is set to the correct value.

The use of these devices for interfacing to 8-, 16-, 24-bit-wide memory arrays with parity is illustrated in Figures 2, 3, and 4.

TYPICAL APPLICATION DATA

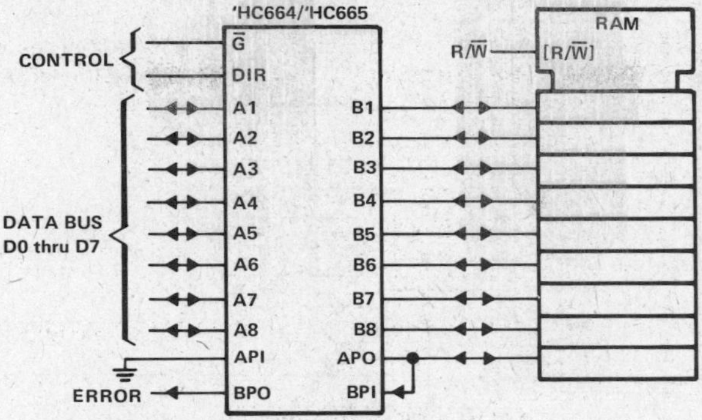

FIGURE 2. 8-BIT-WIDE MEMORY ARRAY WITH PARITY

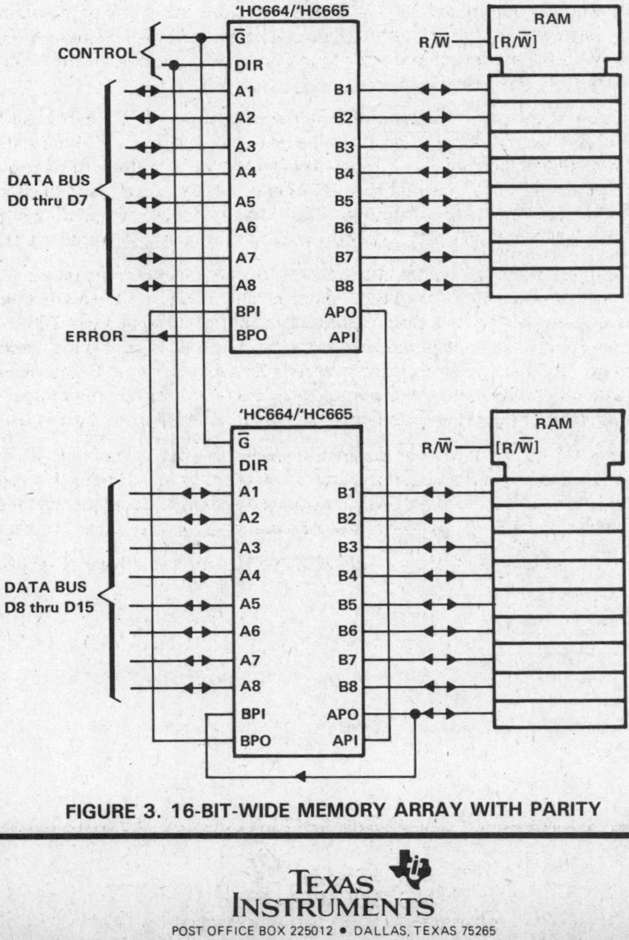

FIGURE 3. 16-BIT-WIDE MEMORY ARRAY WITH PARITY

3

HCMOS DEVICES

TYPICAL APPLICATION DATA

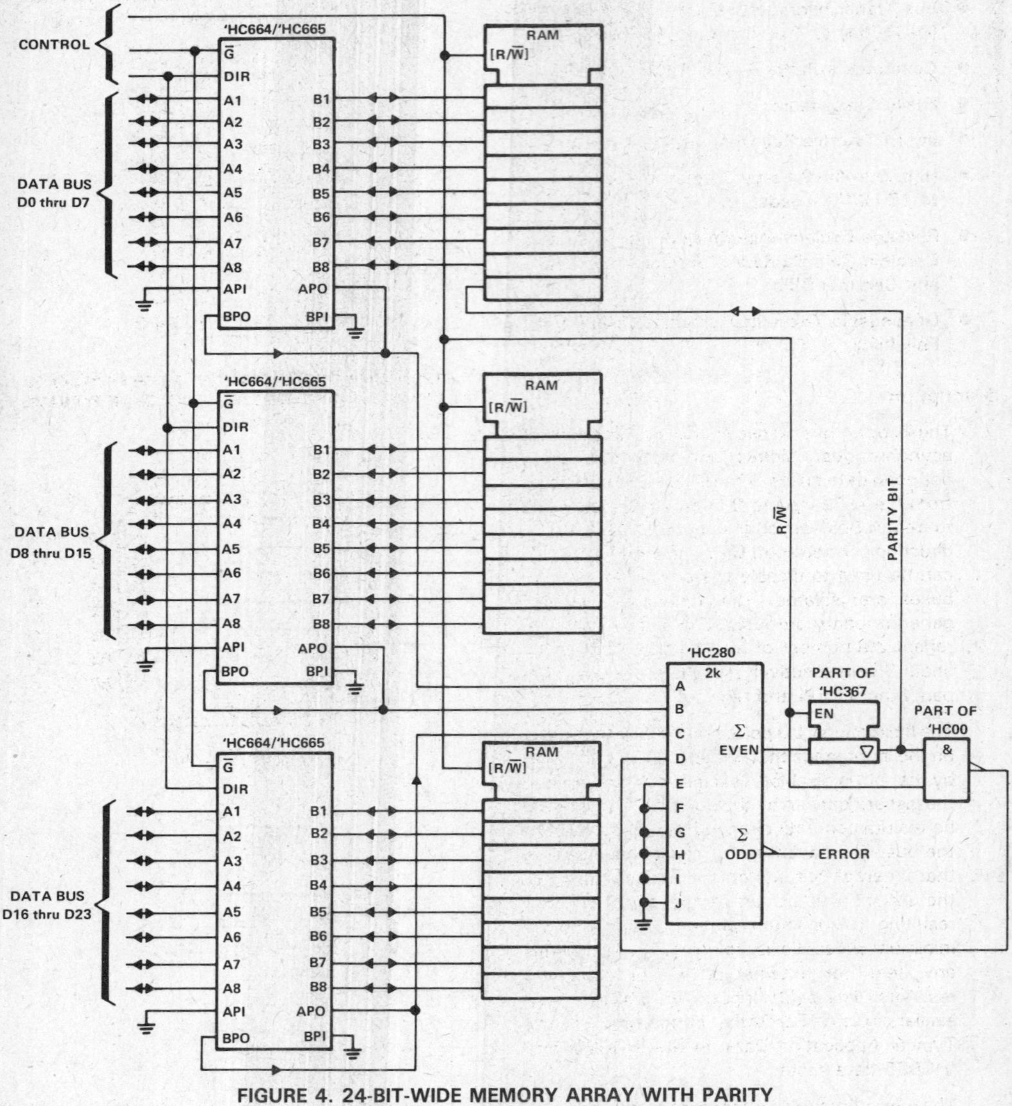

FIGURE 4. 24-BIT-WIDE MEMORY ARRAY WITH PARITY

NOTE: The 'HC280 eliminates ripple carry delays associated with Figures 2 and 3. However, in those two cases the delays are probably too small to be of concern.

TYPES SN54HCT664, SN54HCT665, SN74HCT664, SN74HCT665 OCTAL BUS TRANSCEIVERS WITH PARITY

D2839, MARCH 1984

- Inputs Are TTL-Voltage Compatible
- Bus Transceivers with Inverting Outputs ('HCT664) or True Outputs ('HCT665)
- Generates a Parity Bit for A Bus and B Bus
- Easily Cascadable
- Internal Active Pull-Ups and Pull-Downs
- High-Current 3-State Outputs Can Drive up to 15 LSTTL Loads
- Package Options Include Both Plastic and Ceramic Chip Carriers in Addition to Plastic and Ceramic DIPs
- Dependable Texas Instruments Quality and Reliability

SN54HCT664, SN54HCT665 . . . JT PACKAGE
SN74HCT664, SN74HCT665 . . . JT OR NT PACKAGE
(TOP VIEW)

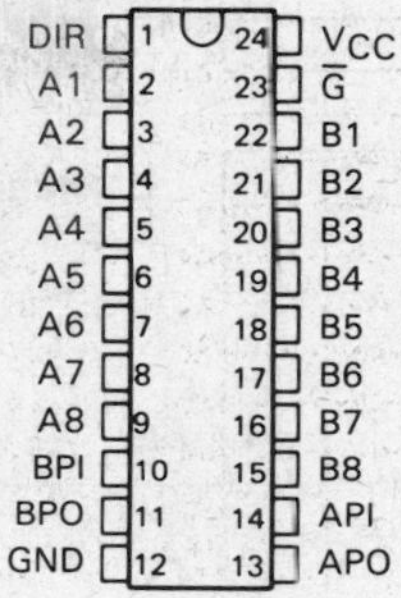

SN54HCT664, SN54HCT665 . . . FH OR FK PACKAGE
SN74HCT664, SN74HCT665 . . . FH OR FN PACKAGE
(TOP VIEW)

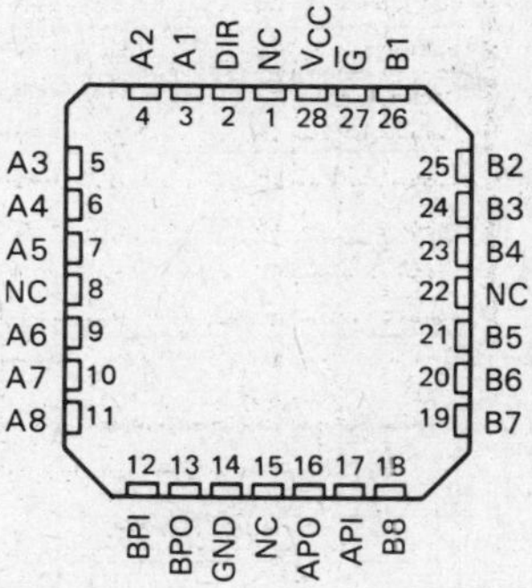

NC—No internal connection

description

These octal bus transceivers are designed for asynchronous, bidirectional communication between data buses. The devices transmit data from the A Bus to the B Bus or from the B Bus to the A Bus, depending on the level at the direction control input, DIR. The enable input, $\overline{G}$, can be used to disable the device so that the buses are isolated. These devices will also generate parity outputs, APO and BPO, which reflect the number of high levels at the A Bus and B Bus, respectively, taking into account the parity inputs API and BPI.

The bidirectional I/O ports feature active circuitry on the input stage that, when the output shared by that pin is disabled, will maintain the input in the last state taken by the output. This state will be maintained until changed by the activity on the bus. The advantage of this arrangement is that when all outputs on the bus are disabled, the inputs will be prevented from floating, resulting in minimum power dissipation and minimum susceptibility to noise. This eliminates any need for external pull-up or pull-down resistors. The parity inputs API and BPI have similar circuitry. For further information, see the Typical Application Data on the 'HC664, and 'HC665 data sheet.

The input threshold voltages on these devices are adjusted to be TTL compatible, allowing direct interface to TTL levels on the bus or to memories with TTL output voltage levels.

The SN54HCT664 and SN54HCT665 are characterized for operation over the full military temperature range of −55°C to 125°C. The SN74HCT664 and SN74HCT665 are characterized for operation from −40°C to 85°C.

FUNCTION TABLE

CONTROL INPUTS		NUMBER OF HIGH INPUTS ON B BUS AND BPI	NUMBER OF HIGH INPUTS ON B BUS AND BPI	OUTPUTS		OPERATION	
$\overline{G}$	DIR			APO	BPO	'HCT664	'HCT665
L	L	X	0, 2, 4, 6, 8	Z	H	$\overline{B}$ Data to A Bus	B Data to A Bus
		X	1, 3, 5, 7, 9	Z	L		
L	H	0, 2, 4, 6, 8	X	H	Z	$\overline{A}$ Data to B Bus	A Data to B Bus
		1, 3, 5, 7, 9	X	L	Z		
H	X	X	X	Z	Z	Isolation	Isolation

logic symbols

'HCT664

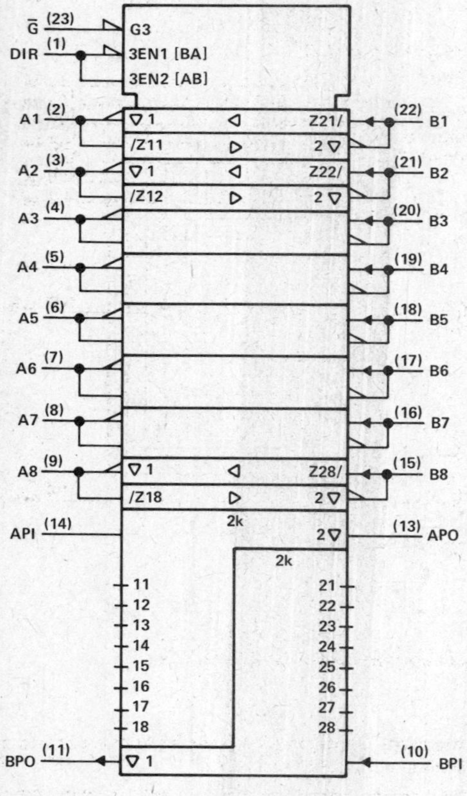

'HCT665

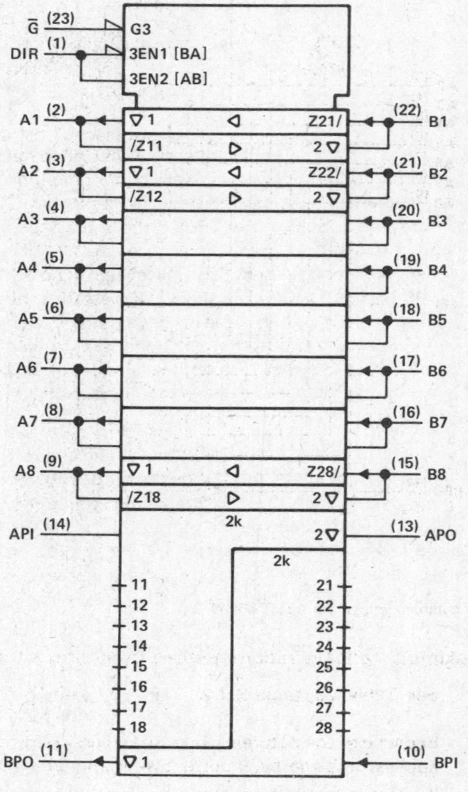

Pin numbers shown are for JT and NT packages.

logic diagram (positive logic)

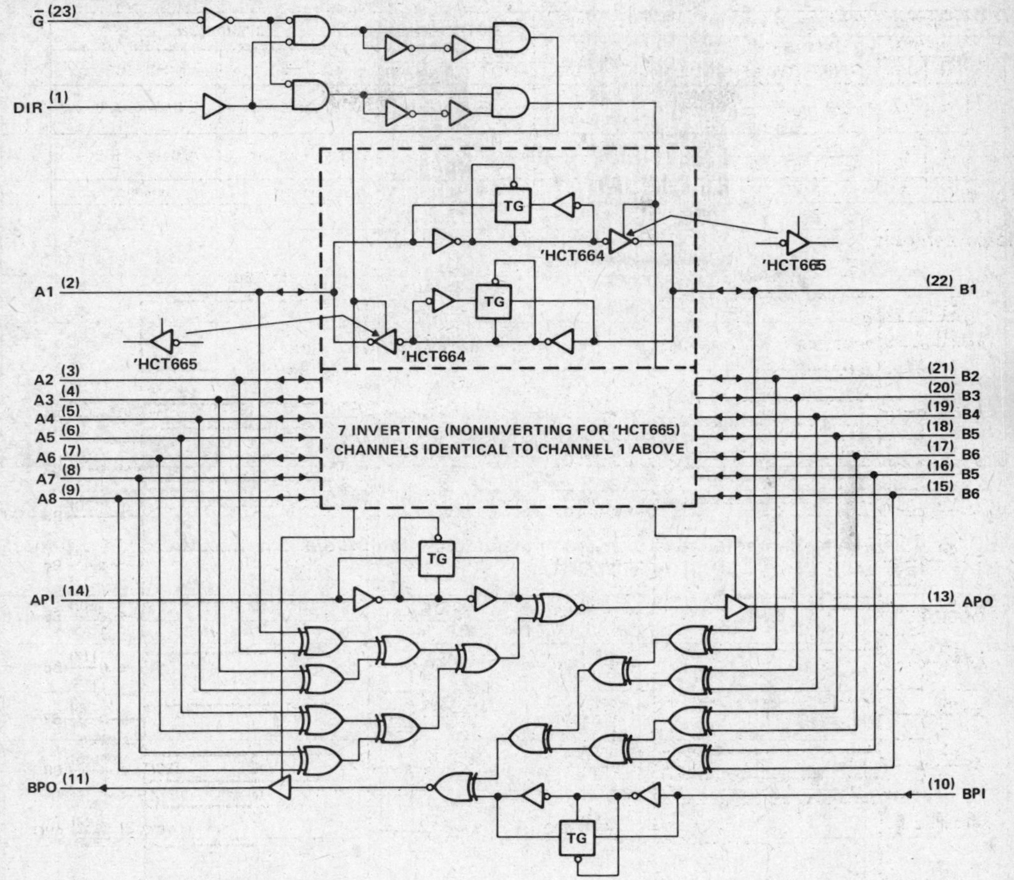

Pin numbers shown are for JT and NT packages.

maximum ratings, recommended operating conditions, and electrical characteristics

See Table VII, page 2-14.

Because of the nature of the transceiver I/O ports and the parity inputs, the following additional parameter applies. It is the peak current as the input changes from 0 V to V_{CC}.

PARAMETER	TEST CONDITION	V_{CC}	T_A = 25°C			SN54HCT664 SN54HCT665		SN74HCT664 SN74HCT665		UNIT
			MIN	TYP	MAX	MIN	MAX	MIN	MAX	
I_{IM}	V_I = 0 to V_{CC}	6 V			±400		±520		±520	μA

'HCT664 switching characteristics over recommended operating free-air temperature range (unless otherwise noted), C_L = 50 pF (see Note 1)

PARAMETER	FROM (INPUT)	TO (OUTPUT)	V_{CC}	T_A = 25°C			SN54HCT664		SN74HCT664		UNIT
				MIN	TYP	MAX	MIN	MAX	MIN	MAX	
t_{pd}	A or B	B or A	4.5 V		15	30		45		38	ns
			5.5 V		13	27		41		34	
t_{pd}	A or B	APO or BPO	4.5 V		23	46		69		58	ns
			5.5 V		20	41		62		52	
t_{pd}	API or BPI	APO or BPO	4.5 V		15	31		47		39	ns
			5.5 V		14	28		42		35	
t_{en}	$\overline{G}$	A or B	4.5 V		25	51		77		64	ns
			5.5 V		23	46		69		58	
t_{dis}	$\overline{G}$	Aor B	4.5 V		25	51		77		64	ns
			5.5 V		23	46		69		58	
t_{en}	DIR	A or B	4.5 V		25	51		77		64	ns
			5.5 V		23	46		69		58	
t_{dis}	DIR	A or B	4.5 V		25	51		77		64	ns
			5.5 V		23	46		69		58	
t_t		Any	4.5 V		8	12		18		15	ns
			5.5 V		7	11		16		14	

C_{pd}	Power dissipation capacitance	No load, T_A = 25°C	62 pF typ

'HCT664 switching characteristics over recommended operating free-air temperature range (unless otherwise noted), C_L = 150 pF (see Note 1)

PARAMETER	FROM (INPUT)	TO (OUTPUT)	V_{CC}	T_A = 25°C			SN54HCT664		SN74HCT664		UNIT
				MIN	TYP	MAX	MIN	MAX	MIN	MAX	
t_{pd}	A or B	B or A	4.5 V		23	47		71		59	ns
			5.5 V		21	42		64		53	
t_{pd}	A or B	APO or BPO	4.5 V		31	63		95		79	ns
			5.5 V		28	56		85		71	
t_{pd}	API or BPI	APO or BPO	4.5 V		24	48		73		60	ns
			5.5 V		21	43		65		54	
t_{en}	$\overline{G}$	A or B	4.5 V		34	68		103		85	ns
			5.5 V		30	61		92		77	
t_{en}	DIR	Aor B	4.5 V		34	68		103		85	ns
			5.5 V		30	61		92		77	
t_t		Any	4.5 V		17	42		63		53	ns
			5.5 V		14	38		57		48	

NOTE 1: For load circuit and voltage waveforms, see page 1-14.

For typical application data and a description of the unique input structure, see the 'HC664 series data sheet.

TEXAS INSTRUMENTS
POST OFFICE BOX 225012 • DALLAS, TEXAS 75265

'HCT665 switching characteristics over recommended operating free-air temperature range (unless otherwise noted), C_L = 50 pF (see Note 1)

PARAMETER	FROM (INPUT)	TO (OUTPUT)	V_{CC}	T_A = 25°C MIN	TYP	MAX	SN54HCT665 MIN	MAX	SN74HCT665 MIN	MAX	UNIT
t_{pd}	A or B	B or A	4.5 V		14	28		61		50	ns
			5.5 V		12	25		42		35	
t_{pd}	A or B	APO or BPO	4.5 V		23	46		69		58	ns
			5.5 V		20	41		62		52	
t_{pd}	API or BPI	APO or BPO	4.5 V		15	31		47		39	ns
			5.5 V		14	28		42		35	
t_{en}	$\overline{G}$	A or B	4.5 V		25	51		77		64	ns
			5.5 V		23	46		69		58	
t_{dis}	$\overline{G}$	Aor B	4.5 V		25	51		77		64	ns
			5.5 V		23	46		69		58	
t_{en}	DIR	A or B	4.5 V		25	51		77		64	ns
			5.5 V		23	46		69		58	
t_{dis}	DIR	A or B	4.5 V		25	51		77		64	ns
			5.5 V		23	46		69		58	
t_t		Any	4.5 V		8	12		18		15	ns
			5.5 V		7	11		16		14	

C_{pd}	Power dissipation capacitance	No load, T_A = 25°C	62 pF typ

'HCT665 switching characteristics over recommended operating free-air temperature range (unless otherwise noted), C_L = 150 pF (see Note 1)

PARAMETER	FROM (INPUT)	TO (OUTPUT)	V_{CC}	T_A = 25°C MIN	TYP	MAX	SN54HCT665 MIN	MAX	SN74HCT665 MIN	MAX	UNIT
t_{pd}	A or B	B or A	4.5 V		22	45		84		69	ns
			5.5 V		20	40		68		56	
t_{pd}	A or B	APO or BPO	4.5 V		31	63		95		79	ns
			5.5 V		28	56		85		71	
t_{pd}	API or BPI	APO or BPO	4.5 V		24	48		73		60	ns
			5.5 V		21	43		65		54	
t_{en}	$\overline{G}$	A or B	4.5 V		34	68		103		85	ns
			5.5 V		30	61		92		77	
t_{en}	DIR	Aor B	4.5 V		34	68		103		85	ns
			5.5 V		30	61		92		77	
t_t		Any	4.5 V		17	42		63		53	ns
			5.5 V		14	38		57		48	

NOTE 1: For load circuit and voltage waveforms, see page 1-14.

For typical application data and a description of the unique input structure, see the 'HC664 series data sheet.

TYPES SN54HC804, SN74HC804 HEX 2-INPUT NAND DRIVERS

D2804, MARCH 1984

- High-Current Outputs Can Drive up to 15 LSTTL Loads
- Package Options Include Both Plastic and Ceramic Chip Carriers in Addition to Plastic and Ceramic DIPs
- Dependable Texas Instruments Quality and Reliability

SN54HC804 . . . J PACKAGE
SN74HC804 . . . J OR N PACKAGE
(TOP VIEW)

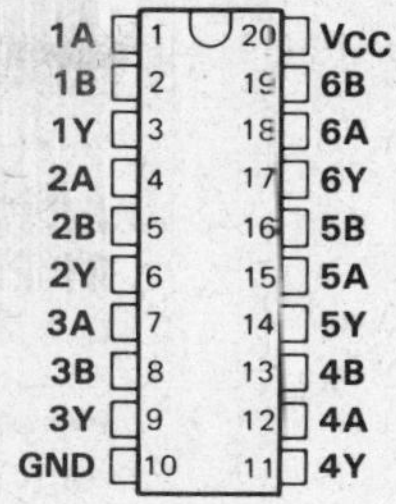

SN54HC804 . . . FH OR FK PACKAGE
SN74HC804 . . . FH OR FN PACKAGE
(TOP VIEW)

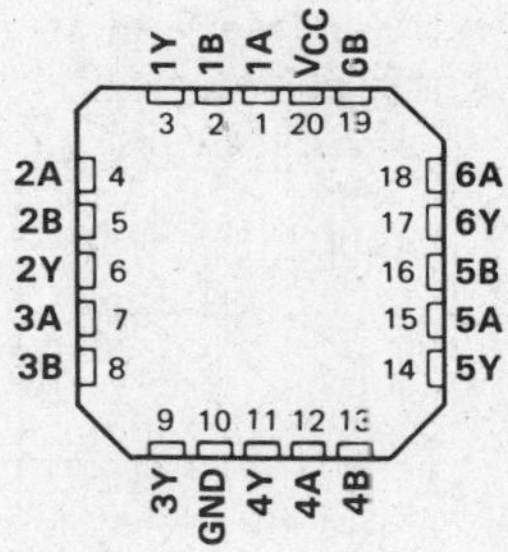

description

These devices contain six independent 2-input NAND drivers. They perform the Boolean functions $Y = \overline{A \cdot B}$ or $Y = \overline{A} + \overline{B}$ in positive logic.

The SN54HC804 is characterized for operation over the full military temperature range of −55°C to 125°C. The SN74HC804 is characterized for operation from −40°C to 85°C.

FUNCTION TABLE (each driver)

INPUTS		OUTPUT
A	B	Y
H	H	L
L	X	H
X	L	H

logic symbol

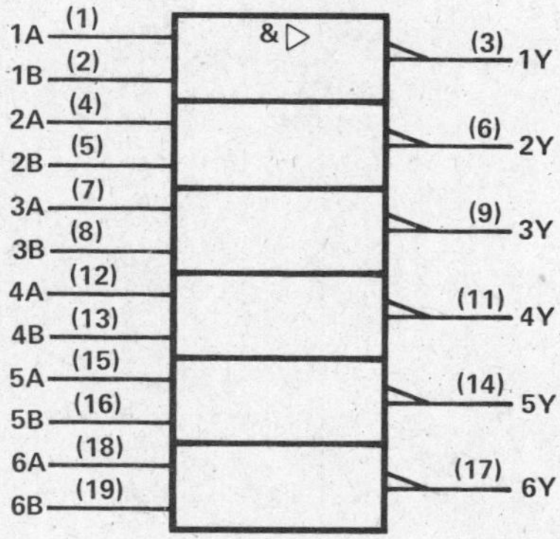

maximum ratings, recommended operating conditions, and electrical characteristics

See Table III, page 2-8.

switching characteristics over recommended operating free-air temperature range (unless otherwise noted), C_L = 50 pF (see Note 1)

PARAMETER	FROM (INPUT)	TO (OUTPUT)	V_{CC}	T_A = 25°C MIN	T_A = 25°C TYP	T_A = 25°C MAX	SN54HC804 MIN	SN54HC804 MAX	SN74HC804 MIN	SN74HC804 MAX	UNIT
t_{pd}	A or B	Y	2 V		40	100		150		125	ns
			4.5 V		12	20		30		25	
			6 V		10	17		26		22	
t_t		Any	2 V		28	60		90		75	ns
			4.5 V		8	12		18		15	
			6 V		6	10		15		13	

C_{pd}	Power dissipation capacitance per gate	No load, T_A = 25°C	40 pF typ

switching characteristics over recommended operating free-air temperature range (unless otherwise noted), C_L = 150 pF (see Note 1)

PARAMETER	FROM (INPUT)	TO (OUTPUT)	V_{CC}	T_A = 25°C MIN	T_A = 25°C TYP	T_A = 25°C MAX	SN54HC804 MIN	SN54HC804 MAX	SN74HC804 MIN	SN74HC804 MAX	UNIT
t_{pd}	A or B	Y	2 V		60	185		280		230	ns
			4.5 V		20	37		56		46	
			6 V		16	32		48		41	
t_t		Any	2 V		45	210		315		265	ns
			4.5 V		17	42		63		53	
			6 V		13	36		53		45	

NOTE 1: For load circuit and voltage waveforms, see page 1-14.

TYPES SN54HC805, SN74HC805 HEX 2-INPUT NOR DRIVERS

D2804, MARCH 1984

- **High-Current Outputs Can Drive up to 15 LSTTL Loads**
- **Package Options Include Both Plastic and Ceramic Chip Carriers in Addition to Plastic and Ceramic DIPs**
- **Dependable Texas Instruments Quality and Reliability**

SN54HC805 . . . J PACKAGE
SN74HC805 . . . J OR N PACKAGE
(TOP VIEW)

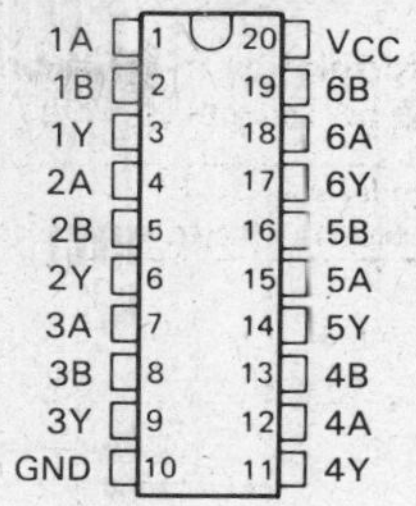

SN54HC805 . . . FH OR FK PACKAGE
SN74HC805 . . . FH OR FN PACKAGE
(TOP VIEW)

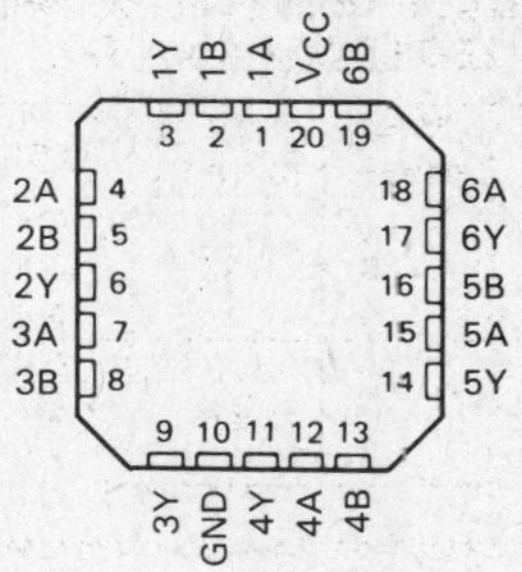

description

These devices contain six independent 2-input NOR drivers. They perform the Boolean functions $Y = \overline{A+B}$ or $Y = \overline{A} \cdot \overline{B}$ in positive logic.

The SN54HC805 is characterized for operation over the full military temperature range of −55°C to 125°C. The SN74HC805 is characterized for operation from −40°C to 85°C.

FUNCTION TABLE (each driver)

INPUTS		OUTPUT
A	B	Y
H	X	L
X	H	L
L	L	H

logic symbol

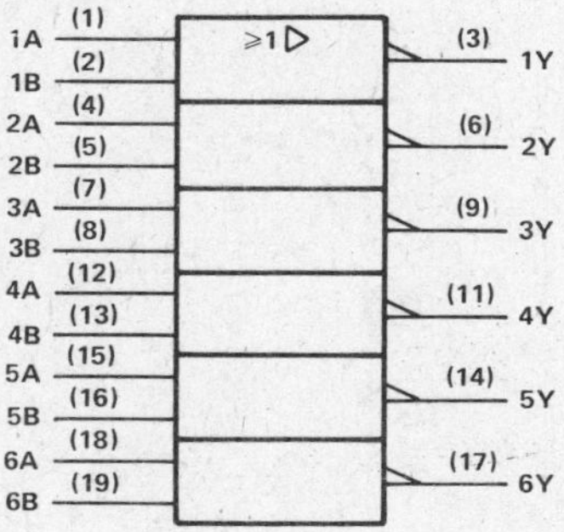

maximum ratings, recommended operating conditions, and electrical characteristics

See Table III, page 2-8.

switching characteristics over recommended operating free-air temperature range (unless otherwise noted), C_L = 50 pF (see Note 1)

PARAMETER	FROM (INPUT)	TO (OUTPUT)	V_{CC}	T_A = 25°C			SN54HC805		SN74HC805		UNIT
				MIN	TYP	MAX	MIN	MAX	MIN	MAX	
t_{pd}	A or B	Y	2 V		31	95		145		120	ns
			4.5 V		10	19		29		24	
			6 V		8	16		25		20	
t_t		Any	2 V		28	60		90		75	ns
			4.5 V		8	12		18		15	
			6 V		6	10		15		13	

C_{pd}	Power dissipation capacitance per gate	No load, T_A = 25°C	40 pF typ

switching characteristics over recommended operating free-air temperature range (unless otherwise noted), C_L = 150 pF (see Note 1)

PARAMETER	FROM (INPUT)	TO (OUTPUT)	V_{CC}	T_A = 25°C			SN54HC805		SN74HC805		UNIT
				MIN	TYP	MAX	MIN	MAX	MIN	MAX	
t_{pd}	A or B	Y	2 V		44	180		275		225	ns
			4.5 V		14	36		55		45	
			6 V		11	31		47		39	
t_t		Any	2 V		45	210		315		265	ns
			4.5 V		17	42		63		53	
			6 V		13	36		53		45	

NOTE 1: For load circuit and voltage waveforms, see page 1-14.

TEXAS INSTRUMENTS
POST OFFICE BOX 225012 • DALLAS, TEXAS 75265

TYPES SN54HC808, SN74HC808 HEX 2-INPUT AND DRIVERS

D2804, MARCH 1984

- High-Current Outputs Can Drive up to 15 LSTTL Loads
- Package Options Include Both Plastic and Ceramic Chip Carriers in Addition to Plastic and Ceramic DIPs
- Dependable Texas Instruments Quality and Reliability

description

These devices contain six independent 2-input AND drivers. They perform the Boolean functions $Y = A \cdot B$ or $Y = \overline{\overline{A} + \overline{B}}$ in positive logic.

The SN54HC808 is characterized for operation over the full military temperature range of −55°C to 125°C. The SN74HC808 is characterized for operation from −40°C to 85°C.

FUNCTION TABLE (each driver)

INPUTS		OUTPUT
A	B	Y
H	H	H
L	X	L
X	L	L

logic symbol

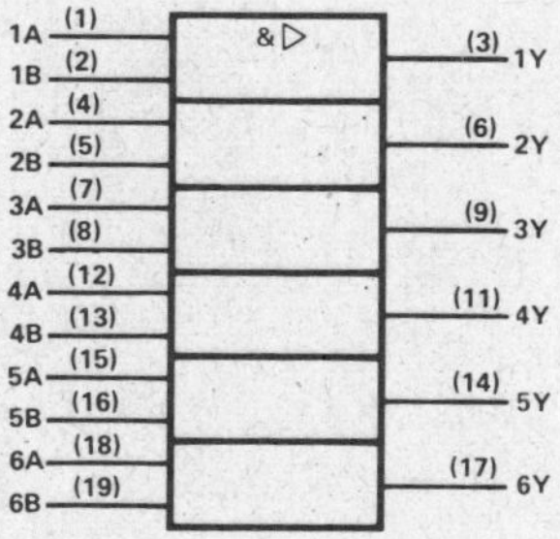

SN54HC808 . . . J PACKAGE
SN74HC808 . . . J OR N PACKAGE
(TOP VIEW)

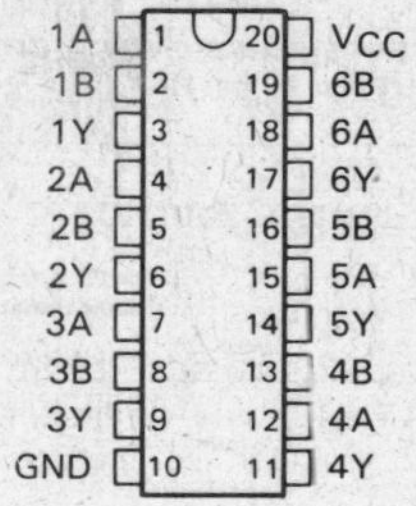

SN54HC808 . . . FH OR FK PACKAGE
SN74HC808 . . . FH OR FN PACKAGE
(TOP VIEW)

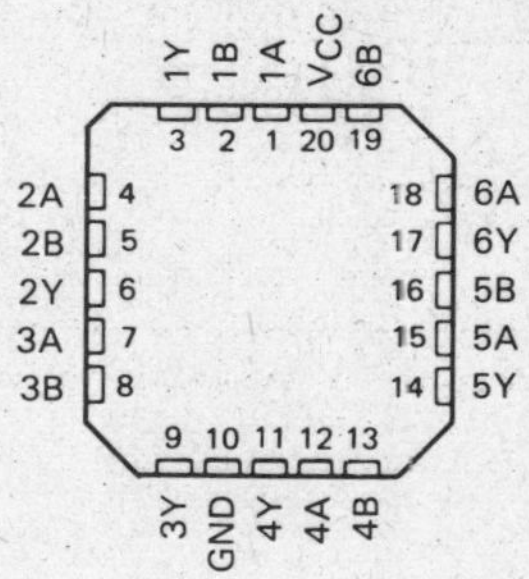

absolute maximum ratings, recommended operating conditions, and electrical characteristics

See Table III, page 2-8.

switching characteristics over recommended operating free-air temperature range (unless otherwise noted), C_L = 50 pF (see Note 1)

PARAMETER	FROM (INPUT)	TO (OUTPUT)	V_{CC}	T_A = 25°C			SN54HC808		SN74HC808		UNIT
				MIN	TYP	MAX	MIN	MAX	MIN	MAX	
t_{pd}	A or B	Y	2 V		50	100		150		125	ns
			4.5 V		10	20		30		25	
			6 V		8	17		25		21	
t_t		Y	2 V		28	60		90		75	ns
			4.5 V		8	12		18		15	
			6 V		6	10		15		13	

C_{pd}	Power dissipation capacitance per gate	No load, T_A = 25°C	20 pF typ

NOTE 1: For load circuit and voltage waveforms, see page 1-14.

TEXAS INSTRUMENTS
POST OFFICE BOX 225012 • DALLAS, TEXAS 75265

TYPES SN54HC832, SN74HC832 HEX 2-INPUT OR DRIVERS

D2804, MARCH 1984

- High-Current Outputs Can Drive up to 15 LSTTL Loads
- Package Options Include Both Plastic and Ceramic Chip Carriers in Addition to Plastic and Ceramic DIPs
- Dependable Texas Instruments Quality and reliability

SN54HC832 . . . J PACKAGE
SN74HC832 . . . J OR N PACKAGE
(TOP VIEW)

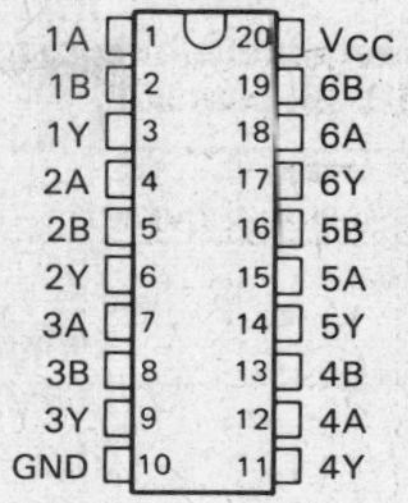

SN54HC832 . . . FH OR FK PACKAGE
SN74HC832 . . . FH OR FN PACKAGE
(TOP VIEW)

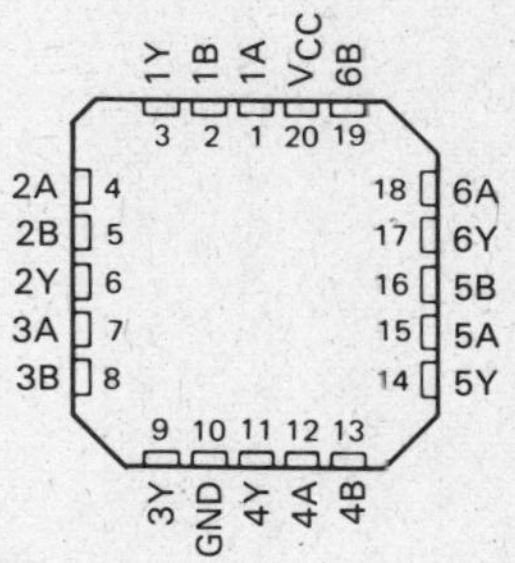

description

These devices contain six independent 2-input OR drivers. They perform the Boolean functions $Y = A+B$ or $Y = \overline{\overline{A} \cdot \overline{B}}$ in positive logic.

The SN54HC832 is characterized for operation over the full military temperature range of −55°C to 125°C. The SN74HC832 is characterized for operation from −40°C to 85°C.

FUNCTION TABLE (each driver)

INPUTS		OUTPUT
A	B	Y
H	X	H
X	H	H
L	L	L

logic symbol

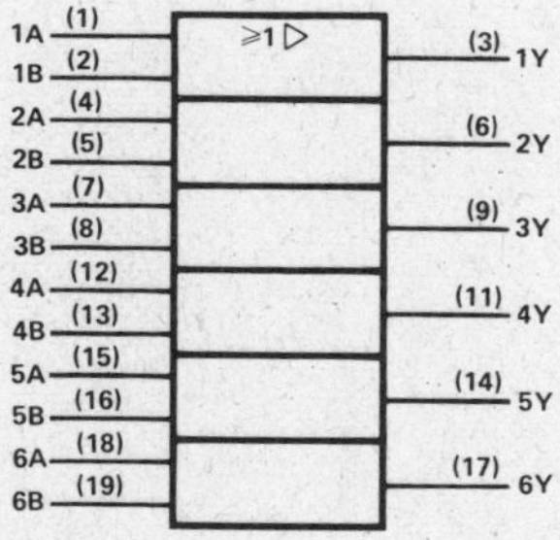

absolute maximum ratings, recommended operating conditions, and electrical characteristics

See Table III, page 2-8.

switching characteristics over recommended operating free-air temperature range (unless otherwise noted), C_L = 50 pF (see Note 1)

PARAMETER	FROM (INPUT)	TO (OUTPUT)	V_{CC}	T_A = 25°C			SN54HC832		SN74HC832		UNIT
				MIN	TYP	MAX	MIN	MAX	MIN	MAX	
t_{pd}	A or B	Y	2 V		50	100		150		125	ns
			4.5 V		10	20		30		25	
			6 V		8	17		25		21	
t_t		Y	2 V		28	60		90		75	ns
			4.5 V		8	12		18		15	
			6 V		6	10		15		13	

C_{pd}	Power dissipation capacitance per gate	No load, T_A = 25°C	20 pF typ

NOTE 1: For load circuit and voltage waveforms, see page 1-14.

TEXAS INSTRUMENTS
POST OFFICE BOX 225012 • DALLAS, TEXAS 75265

TYPES SN54HC4002, SN74HC4002
DUAL 4-INPUT POSITIVE-NOR GATES

D2684, DECEMBER 1982—REVISED MARCH 1984

- **Package Options Include Both Plastic and Ceramic Chip Carriers in Addition to Plastic and Ceramic DIPs**
- **Dependable Texas Instruments Quality and Reliability**

SN54HC4002 . . . J PACKAGE
SN74HC4002 . . . J OR N PACKAGE
(TOP VIEW)

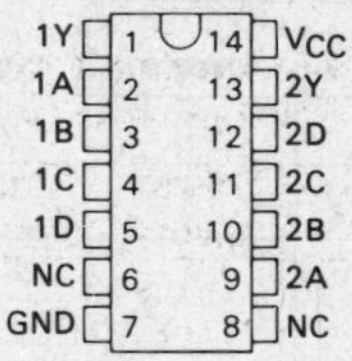

SN54HC4002 . . . FH OR FK PACKAGE
SN74HC4002 . . . FH OR FN PACKAGE
(TOP VIEW)

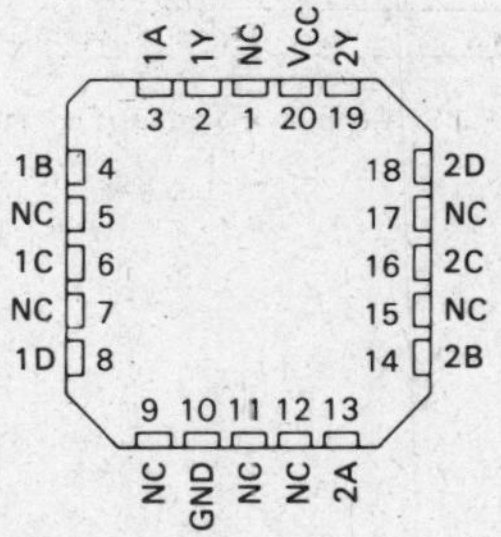

NC—No internal connection

description

These devices contain two independent 4-input positive NOR gates. They perform the Boolean functions:

$$Y = \overline{A + B + C + D} \text{ or } Y = \overline{A} \cdot \overline{B} \cdot \overline{C} \cdot \overline{D}$$

in positive logic.

The SN54HC4002 is characterized for operation over the full military temperature range of −55°C to 125°C. The SN74HC4002 is characterized for operation from −40°C to 85°C.

FUNCTION TABLE

INPUTS				OUTPUT
A	B	C	D	Y
H	X	X	X	L
X	H	X	X	L
X	X	H	X	L
X	X	X	H	L
L	L	L	L	H

logic symbol

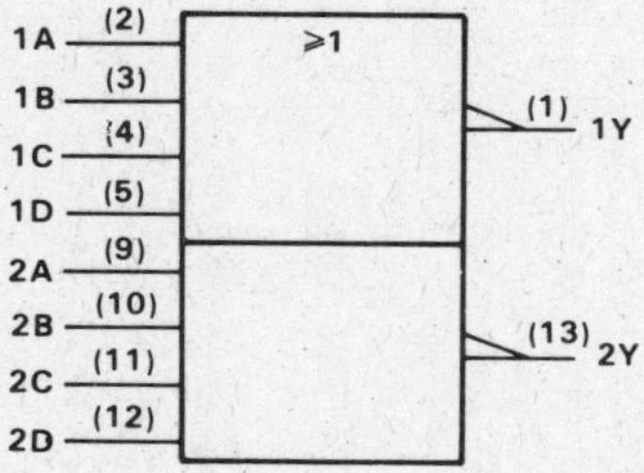

Pin numbers shown are for J and N packages.

maximum ratings, recommended operating conditions, and electrical characteristics

See Table I, page 2-4.

switching characteristics over recommended operating free-air temperature range (unless otherwise noted), C_L = 50 pF (see Note 1)

PARAMETER	FROM (INPUT)	TO (OUTPUT)	V_{CC}	T_A = 25°C MIN	TYP	MAX	SN54HC4002 MIN	MAX	SN74HC4002 MIN	MAX	UNIT
t_{pd}	A thru D	Y	2 V		44	110		165		140	ns
			4.5 V		12	22		33		28	
			6 V		11	19		28		24	
t_t		Y	2 V		38	75		110		95	ns
			4.5 V		8	15		22		19	
			6 V		6	13		19		16	

C_{pd}	Power dissipation capacitance per gate	No load, T_A = 25°C	25 pF typ

NOTE 1: For load circuit and voltage waveforms, see page 1-14.

TYPES SN54HC4020, SN74HC4020 ASYNCHRONOUS 14-BIT BINARY COUNTERS

D2684, DECEMBER 1982—REVISED MARCH 1984

- **Package Options Include Both Plastic and Ceramic Chip Carriers in Addition to Plastic and Ceramic DIPs**
- **Dependable Texas Instruments Quality and Reliability**

description

These devices are 14-stage binary ripple-carry counters that advance on the negative-going edge of the clock pulse. The counters are reset to zero (all outputs low) independently of the clock when CLR goes high.

The SN54HC4020 is characterized for operation over the full military temperature range of −55°C to 125°C. The SN74HC4020 is characterized for operation from −40°C to 85°C.

SN54HC4020 . . . J PACKAGE
SN74HC4020 . . . J OR N PACKAGE
(TOP VIEW)

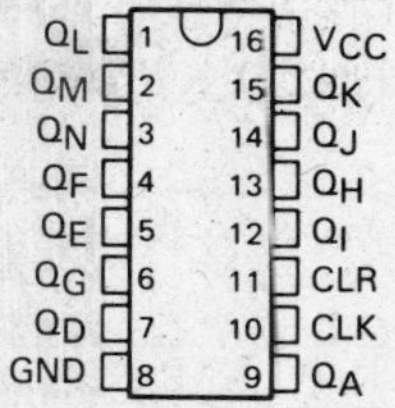

SN54HC4020 . . . FH OR FK PACKAGE
SN74HC4020 . . . FH OR FN PACKAGE
(TOP VIEW)

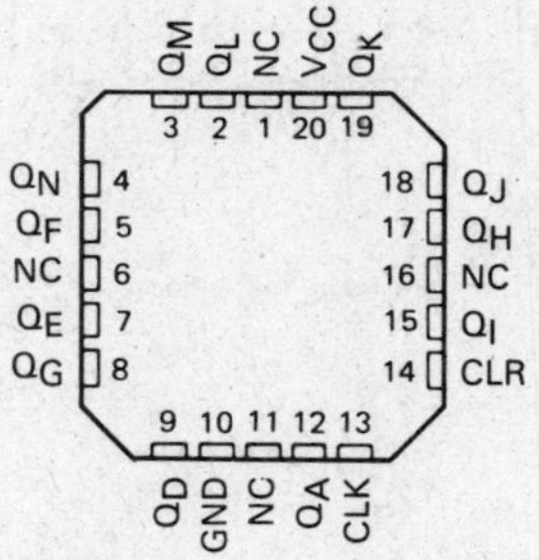

NC—No internal connection

logic symbol

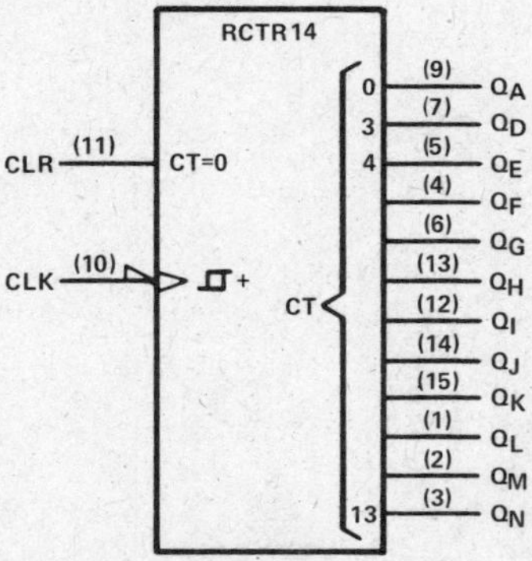

Pin numbers shown are for J and N packages.

logic diagram (positive logic)

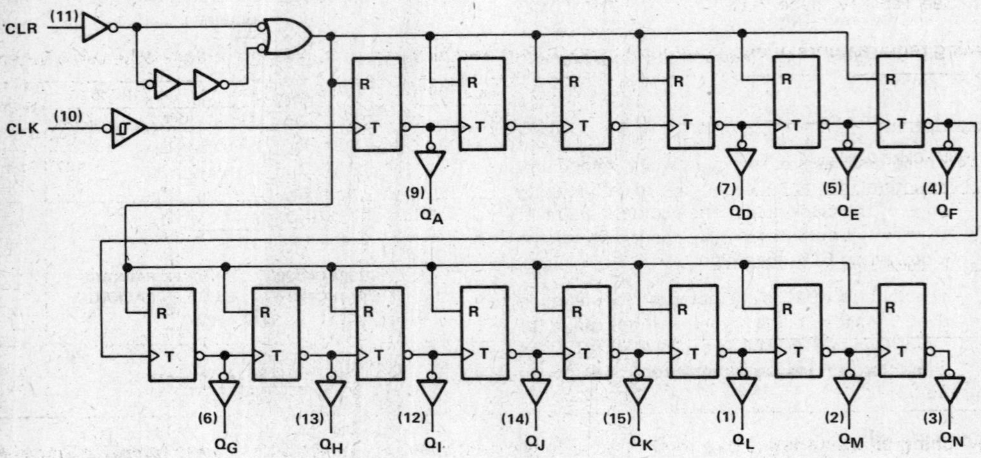

Pin numbers shown are for J and N packages.

TEXAS INSTRUMENTS
POST OFFICE BOX 225012 • DALLAS, TEXAS 75265

maximum ratings, recommended operating conditions, and electrical characteristics

See Table IV, page 2-10.

timing requirements over recommended operating free-air temperature range (unless otherwise noted)

			V_{CC}	T_A = 25 °C		SN54HC4020		SN74HC4020		UNIT
				MIN	MAX	MIN	MAX	MIN	MAX	
f_{clock}	Clock frequency		2 V	0	5.5	0	3.7	0	4.3	MHz
			4.5 V	0	28	0	19	0	22	
			6 V	0	33	0	22	0	25	
t_w	Pulse duration	CLK high or low	2 V	90		135		115		ns
			4.5 V	18		27		23		
			6 V	15		23		20		
		CLR high	2 V	70		105		90		ns
			4.5 V	14		21		18		
			6 V	12		18		25		
t_{su}	Setup time, CLR inactive before CLK↓		2 V	60		90		75		ns
			4.5 V	12		18		15		
			6 V	10		15		13		

switching characteristics over recommended operating free-air temperature range (unless otherwise noted), C_L = 50 pF (see Note 1)

PARAMETER	FROM (INPUT)	TO (OUTPUT)	V_{CC}	T_A = 25 °C			SN54HC4020		SN74HC4020		UNIT
				MIN	TYP	MAX	MIN	MAX	MIN	MAX	
f_{max}			2 V	5.5	10		3.7		4.3		MHz
			4.5 V	28	45		19		22		
			6 V	33	53		22		25		
t_{pd}	CLK	Q_A	2 V		62	150		225		190	ns
			4.5 V		16	30		45		38	
			6 V		12	26		38		32	
t_{PHL}	CLR	Any	2 V		63	140		210		175	ns
			4.5 V		17	28		42		35	
			6 V		13	24		36		30	
t_t		Any	2 V		28	75		110		95	ns
			4.5 V		8	15		22		19	
			6 V		6	13		19		16	

C_{pd}	Power dissipation capacitance	No load, T_A = 25 °C	88 pF typ

NOTE 1: For load circuit and voltage waveforms, see page 1-14.

TYPES SN54HC4024, SN74HC4024 ASYNCHRONOUS 7-BIT BINARY COUNTERS

D2804, MARCH 1984

- **Package Options Include Both Plastic and Ceramic Chip Carriers in Addition to Plastic and Ceramic DIPs**
- **Dependable Texas Instruments Quality and Reliability**

SN54HC4024 . . . J PACKAGE
SN74HC4024 . . . J OR N PACKAGE
(TOP VIEW)

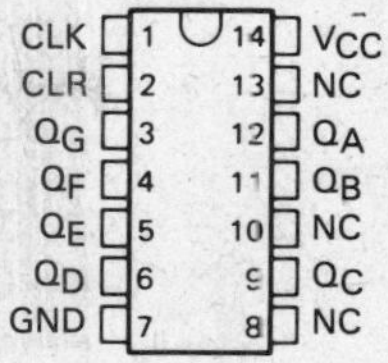

SN54HC4024 . . . FH OR FK PACKAGE
SN74HC4024 . . . FH OR FN PACKAGE
(TOP VIEW)

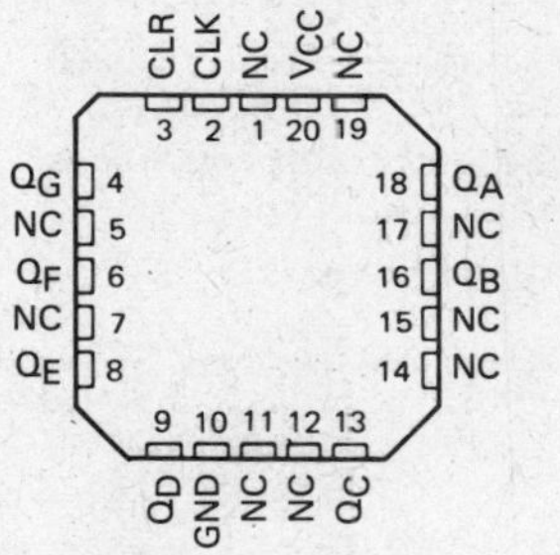

NC—No internal connection.

description

The 'HC4024 is an asynchronous 7-stage binary counter designed with an input pulse-shaping circuit. The outputs of all stages are available externally. A high clear signal asynchronously clears the counter and resets all outputs low. The count is advanced on the high-to-low transition of the clock pulse. Applications include time-delay circuits, counter controls, and frequency-dividing circuits.

The SN54HC4024 is characterized for operation over the full military temperature range of −55°C to 125°C. The SN74HC4024 is characterized for operation from −40°C to 85°C.

logic symbol

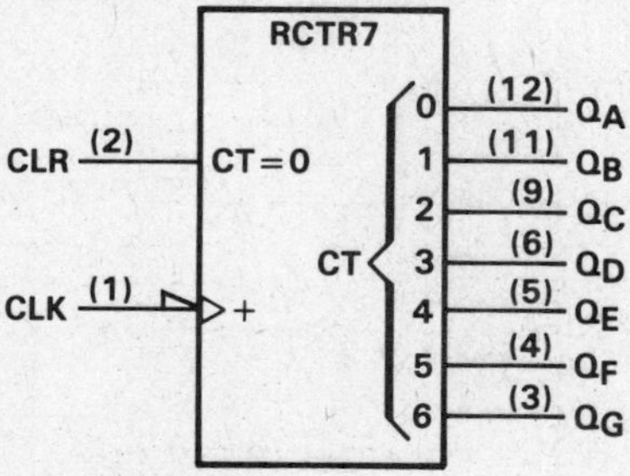

Pin numbers shown are for J and N packages.

3 HCMOS DEVICES

logic diagram (positive logic)

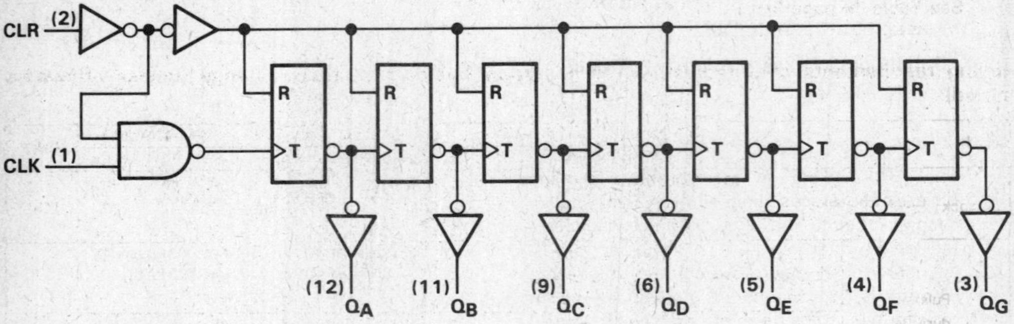

Pin numbers shown are for J and N packages.

typical clear and count sequences

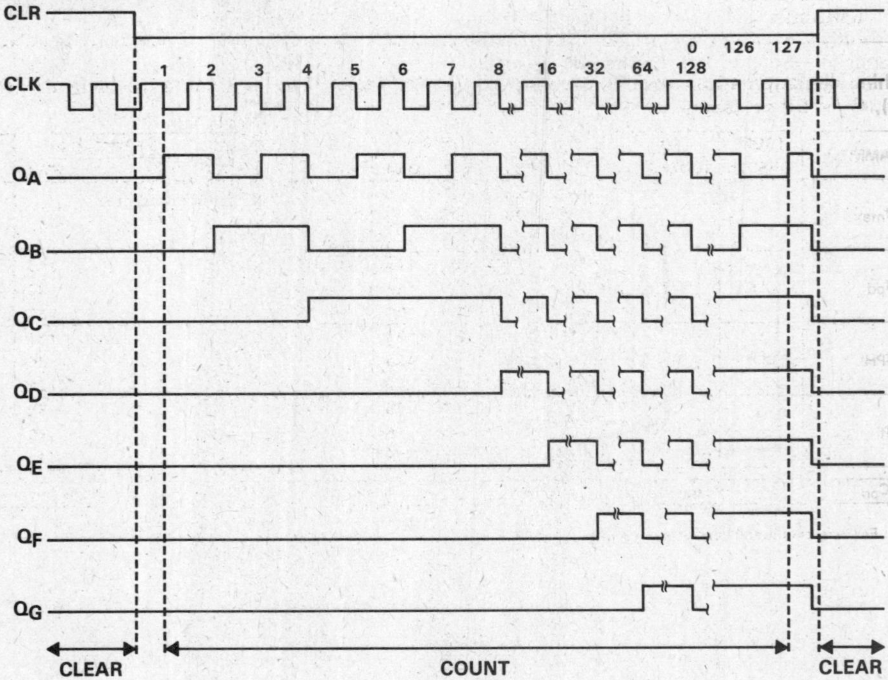

maximum ratings, recommended operating conditions, and electrical characteristics

See Table III, page 2-8.

timing requirements over recommended operating free-air temperature range (unless otherwise noted)

		V_{CC}	T_A = 25°C		SN54HC4024		SN74HC4024		UNIT
			MIN	MAX	MIN	MAX	MIN	MAX	
f_{clock} Clock frequency		2 V	0	5.5	0	3.7	0	4.3	MHz
		4.5 V	0	28	0	19	0	22	
		6 V	0	33	0	22	0	25	
t_w Pulse duration	CLK high or low	2 V	90		135		115		ns
		4.5 V	18		27		23		
		6 V	15		23		20		
	CLR high	2 V	80		120		100		
		4.5 V	16		24		20		
		6 V	14		20		17		
t_{su} Setup time, CLR low before CLK↓		2 V	80		120		100		ns
		4.5 V	16		24		20		
		6 V	14		20		17		

switching characteristics over recommended operating free-air temperature range (unless otherwise noted), C_L = 50 pF (see Note 1)

PARAMETER	FROM (INPUT)	TO (OUTPUT)	V_{CC}	T_A = 25°C			SN54HC4024		SN74HC4024		UNIT
				MIN	TYP	MAX	MIN	MAX	MIN	MAX	
f_{max}			2 V	5.5	10		3.7		4.3		MHz
			4.5 V	28	50		19		22		
			6 V	33	60		22		26		
t_{pd}	CLK	Q_A	2 V		56	120		180		150	ns
			4.5 V		16	24		36		30	
			6 V		12	20		31		26	
t_{PHL}	CLR	Any	2 V		61	130		195		165	ns
			4.5 V		17	26		39		33	
			6 V		13	22		33		28	
t_t			2 V		28	75		110		95	ns
			4.5 V		8	15		22		19	
			6 V		6	13		19		16	

C_{pd}	Power dissipation capacitance	No load, T_A = 25°C	40 pF typ

NOTE 1: For load circuit and voltage waveforms, see page 1-14.

TYPES SN54HC4040, SN74HC4040 ASYNCHRONOUS 12-BIT BINARY COUNTERS

D2684, DECEMBER 1982—REVISED MARCH 1984

- **Package Options Include Both Plastic and Ceramic Chip Carriers in Addition to Plastic and Ceramic DIPs**
- **Dependable Texas Instruments Quality and Reliability**

SN54HC4040 . . . J PACKAGE
SN74HC4040 . . . J OR N PACKAGE
(TOP VIEW)

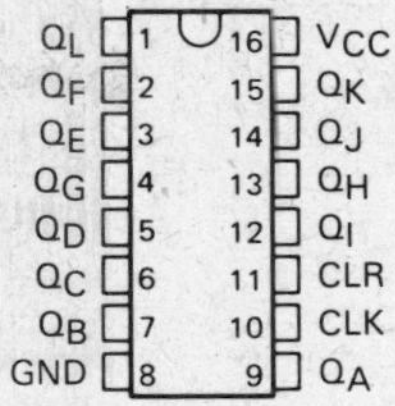

description

This device is an asynchronous 12-state binary counter with the outputs of all stages available externally. A high level at CLR asynchronously clears the counter and resets all outputs low. The count is advanced on a high-to-low transition at CLK. Applications include time delay circuits, counter controls, and frequency-dividing circuits.

The SN54HC4040 is characterized for operation over the full military temperature range of −55 °C to 125 °C. The SN74HC4040 is characterized for operation from −40 °C to 85 °C.

SN54HC4040 . . . FH OR FK PACKAGE
SN74HC4040 . . . FH OR FN PACKAGE
(TOP VIEW)

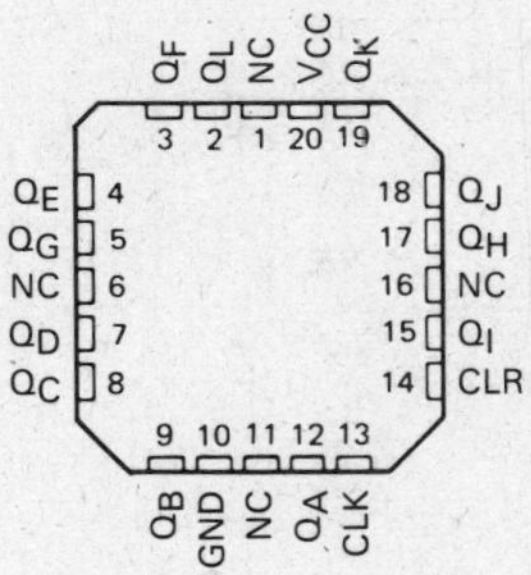

NC—No internal connection

logic symbol

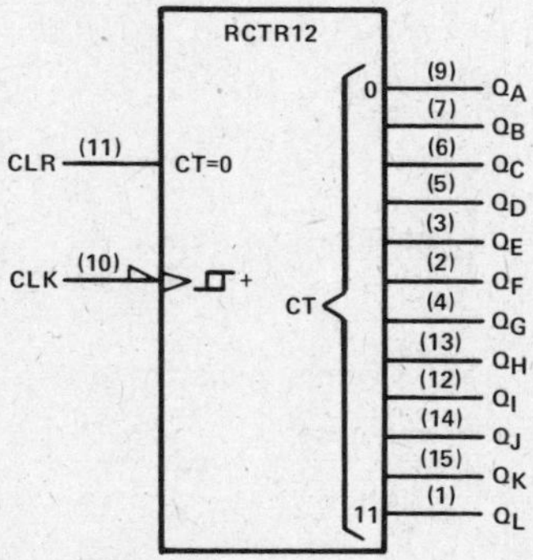

Pin numbers shown are for J and N packages.

logic diagram (positive logic)

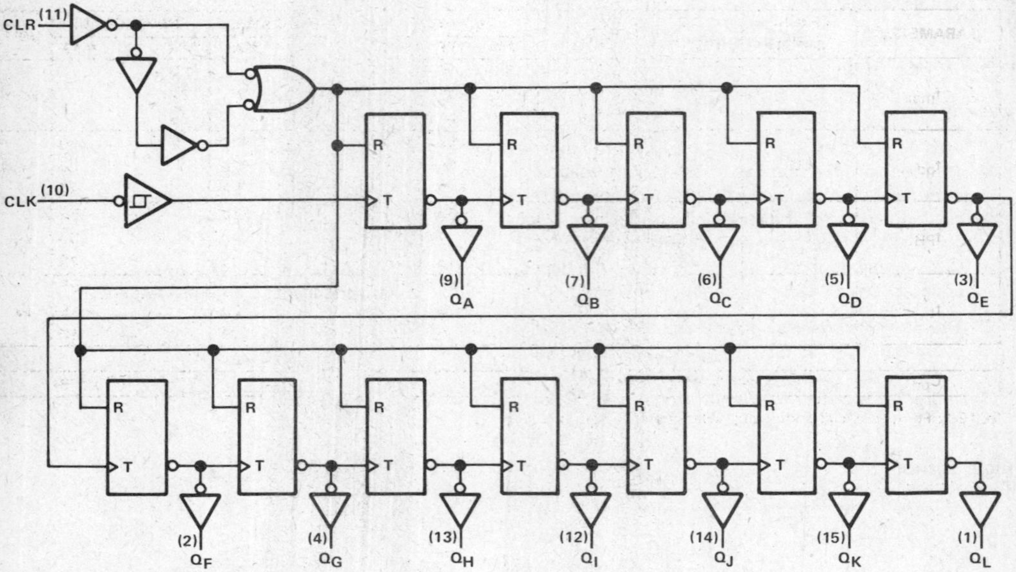

Pin numbers shown are for J and N packages.

3 HCMOS DEVICES

maximum ratings, recommended operating conditions, and electrical characteristics

See Table IV, page 2-10.

timing requirements over recommended operating free-air temperature range (unless otherwise noted)

			V_{CC}	T_A = 25°C MIN	T_A = 25°C MAX	SN54HC4040 MIN	SN54HC4040 MAX	SN74HC4040 MIN	SN74HC4040 MAX	UNIT
f_{clock}	Clock frequency		2 V	0	5.5	0	3.7	0	4.3	MHz
			4.5 V	0	28	0	19	0	22	
			6 V	0	33	0	22	0	25	
t_w	Pulse duration	CLK high or low	2 V	90		135		115		ns
			4.5 V	18		27		23		
			6 V	15		23		20		
		CLR high	2 V	70		105		90		ns
			4.5 V	14		21		18		
			6 V	12		18		15		
t_{su}	Setup time, CLR inactive before CLK↓		2 V	60		90		75		ns
			4.5 V	12		18		15		
			6 V	10		15		13		

switching characteristics over recommended operating free-air temperature range (unless otherwise noted), C_L = 50 pF (see Note 1)

PARAMETER	FROM (INPUT)	TO (OUTPUT)	V_{CC}	T_A = 25°C			SN54HC4040		SN74HC4040		UNIT
				MIN	TYP	MAX	MIN	MAX	MIN	MAX	
f_{max}			2 V	5.5	10		3.7		4.3		MHz
			4.5 V	28	45		19		22		
			6 V	33	53		22		25		
t_{pd}	CLK	Q_A	2 V		62	150		225		190	ns
			4.5 V		16	30		45		38	
			6 V		12	26		38		32	
t_{PHL}	CLR	Any	2 V		63	140		210		175	ns
			4.5 V		17	28		42		35	
			6 V		13	24		36		30	
t_t		Any	2 V		28	75		110		95	ns
			4.5 V		8	15		22		19	
			6 V		6	30		19		16	

C_{pd}	Power dissipation capacitance	No load, T_A = 25°C	88 pF typ

NOTE 1: For load circuit and voltage waveforms, see page 1-14.

3

HCMOS DEVICES

TYPES SN54HC4060, SN74HC4060 ASYNCHRONOUS 14-STAGE BINARY COUNTERS AND OSCILLATORS

D2684, DECEMBER 1982—REVISED MARCH 1984

- Allows Design of Either RC or Crystal Oscillator Circuits
- Package Options Include Both Plastic and Ceramic Chip Carriers in Addition to Plastic and Ceramic DIPs
- Dependable Texas Instruments Quality and Reliability

SN54HC4060 . . . J PACKAGE
SN74HC4060 . . . J OR N PACKAGE
(TOP VIEW)

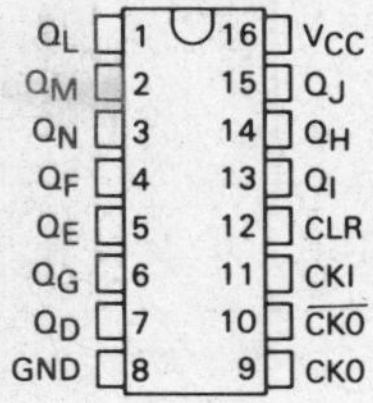

description

The 'HC4060 consists of an oscillator section and 14 ripple-carry binary counter stages. The oscillator configuration allows design of either RC or crystal oscillator circuits. A high-to-low transition on the clock input increments the counter. A high level at CLR disables the oscillator ($\overline{CKO}$ goes high and CKO goes low) and resets the counter to zero (all Q outputs low).

The SN54HC4060 is characterized for operation over the full military temperature range of −55 °C to 125 °C. The SN74HC4060 is characterized for operation from −40 °C to 85 °C.

SN54HC4060 . . . FH OR FK PACKAGE
SN74HC4060 . . . FH OR FN PACKAGE
(TOP VIEW)

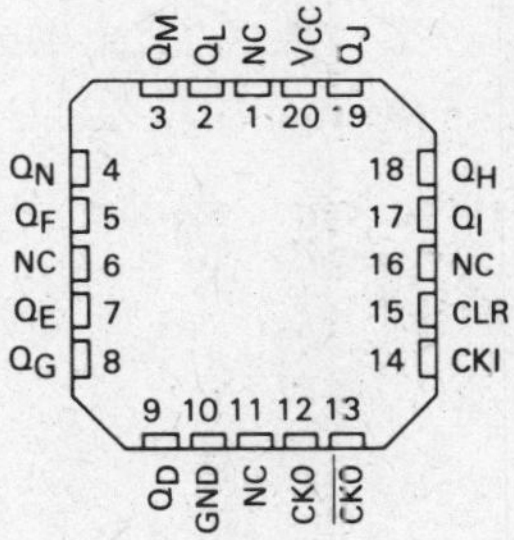

NC—No internal connection

logic symbol

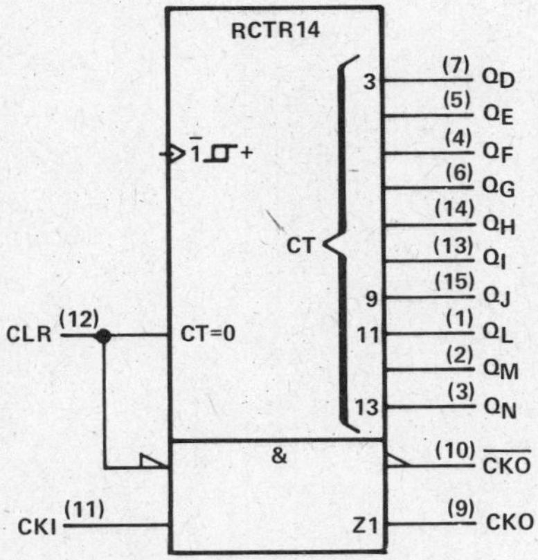

Pin numbers shown are for J and N packages.

logic diagram (positive logic)

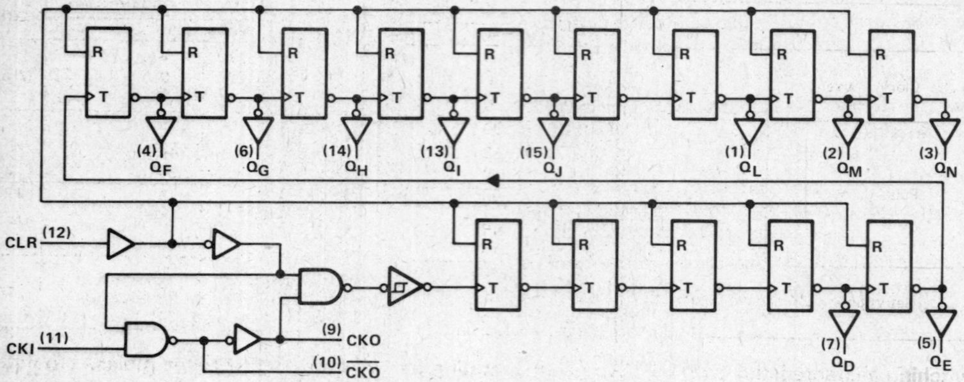

Pin numbers shown are for J and N packages.

maximum ratings, recommended operating conditions, and electrical characteristics

See Table IV, page 2-10.

TEXAS INSTRUMENTS
POST OFFICE BOX 225012 • DALLAS, TEXAS 75265

timing requirements over recommended operating free-air temperature range (unless otherwise noted)

			V_{CC}	T_A = 25°C		SN54HC4060		SN74HC4060		UNIT
				MIN	MAX	MIN	MAX	MIN	MAX	
f_{clock}	Clock frequency		2 V	0	5.5	0	3.7	0	4.3	MHz
			4.5 V	0	28	0	19	0	22	
			6 V	0	33	0	22	0	25	
t_w	Pulse duration	CKI high or low	2 V	90		135		115		ns
			4.5 V	18		27		23		
			6 V	15		23		20		
		CLR high	2 V	90		135		115		ns
			4.5 V	18		27		23		
			6 V	15		23		20		
t_{su}	Setup time, CLR inactive before CKI↓		2 V	160		240		200		ns
			4.5 V	32		48		40		
			6 V	27		41		34		

switching characteristics over recommended operating free-air temperature range (unless otherwise noted), C_L = 50 pF (see Note 1)

PARAMETER	FROM (INPUT)	TO (OUTPUT)	V_{CC}	T_A = 25°C			SN54HC4060		SN74HC4060		UNIT
				MIN	TYP	MAX	MIN	MAX	MIN	MAX	
f_{max}			2 V	5.5	10		3.7		4.3		MHz
			4.5 V	28	45		19		22		
			6 V	33	53		22		25		
t_{pd}	CKI	Q_D	2 V		240	490		735		615	ns
			4.5 V		58	98		147		123	
			6 V		42	83		125		105	
t_{PHL}	CLR	Any Q	2 V		66	140		210		175	ns
			4.5 V		18	28		42		35	
			6 V		14	24		36		30	
t_t		Any	2 V		28	75		110		95	ns
			4.5 V		8	15		22		19	
			6 V		6	13		19		16	

C_{pd}	Power dissipation capacitance	No load, T_A = 25°C	88 pF typ

NOTE 1: For load circuit and voltage waveforms, see page 1-14.

TYPES SN54HC4061, SN74HC4061 ASYNCHRONOUS 14-STAGE BINARY COUNTERS AND OSCILLATORS

D2804, MARCH 1984

- Allows Design of Either RC or Crystal Oscillator Circuits
- Package Options Include Both Plastic and Ceramic Chip Carriers in Addition to Plastic and Ceramic DIPs
- Dependable Texas Instruments Quality and Reliability

SN54HC4061 . . . J PACKAGE
SN74HC4061 . . . J OR N PACKAGE
(TOP VIEW)

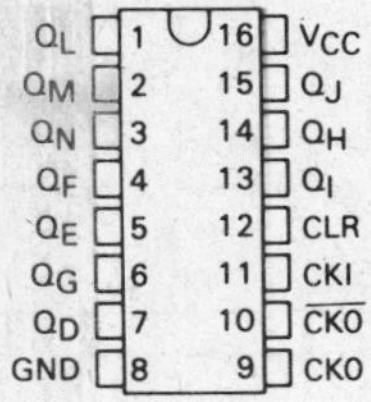

SN54HC4061 . . . FH OR FK PACKAGE
SN74HC4061 . . . FH OR FN PACKAGE
(TOP VIEW)

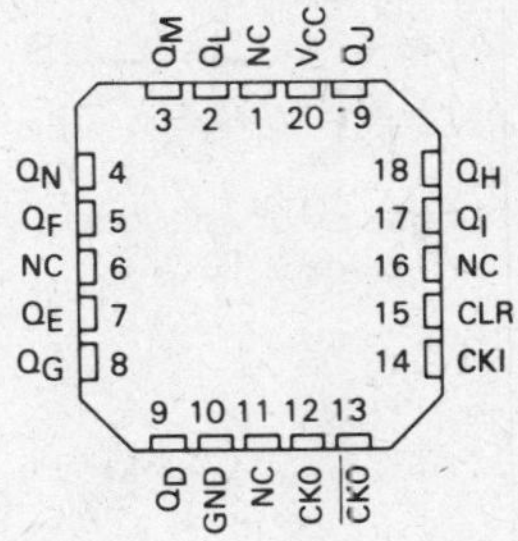

NC—No internal connection

description

The 'HC4061 consists of an oscillator section and 14 ripple-carry binary counter stages. The oscillator configuration allows design of either RC or crystal oscillator circuits. A high-to-low transition on the clock input increments the counter. A high level at CLR resets the counter to zero (all Q outputs low) but has no effect on the oscillator.

The SN54HC4061 is characterized for operation over the full military temperature range of −55°C to 125°C. The SN74HC4061 is characterized for operation from −40°C to 85°C.

logic symbol

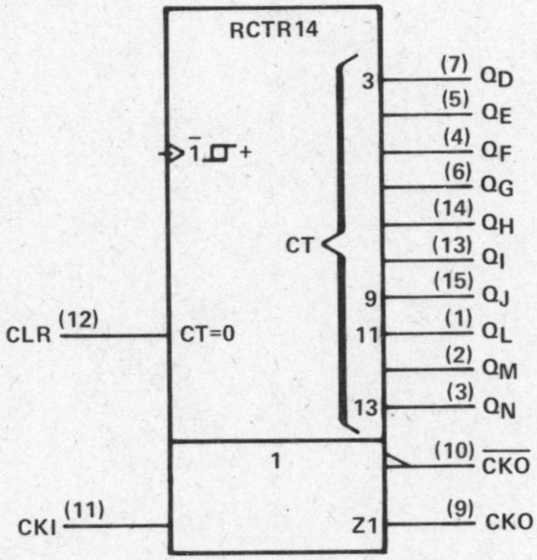

Pin numbers shown are for J and N packages.

logic diagram (positive logic)

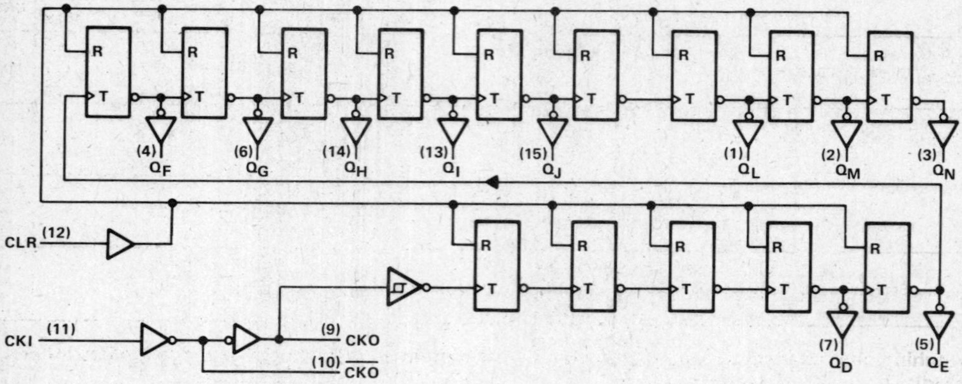

Pin numbers shown are for J and N packages.

maximum ratings, recommended operating conditions, and electrical characteristics

See Table IV, page 2-10.

timing requirements over recommended operating free-air temperature range (unless otherwise noted)

			V_{CC}	$T_A = 25°C$ MIN	$T_A = 25°C$ MAX	SN54HC4061 MIN	SN54HC4061 MAX	SN74HC4061 MIN	SN74HC4061 MAX	UNIT
f_{clock}	Clock frequency		2 V		5.5		3.7		4.3	MHz
			4.5 V		28		19		22	
			6 V		33		22		25	
t_w	Pulse duration	CKI high or low	2 V	90		135		115		ns
			4.5 V	18		27		23		
			6 V	15		23		20		
		CLR high	2 V	90		135		115		ns
			4.5 V	18		27		23		
			6 V	15		23		20		
t_{su}	Setup time, CLR inactive before CKI↓		2 V	160		240		200		ns
			4.5 V	32		48		40		
			6 V	27		41		34		

switching characteristics over recommended operating free-air temperature range (unless otherwise noted), C_L = 50 pF (see Note 1)

PARAMETER	FROM (INPUT)	TO (OUTPUT)	V_{CC}	$T_A = 25°C$ MIN	$T_A = 25°C$ TYP	$T_A = 25°C$ MAX	SN54HC4061 MIN	SN54HC4061 MAX	SN74HC4061 MIN	SN74HC4061 MAX	UNIT
f_{max}			2 V	5.5	10		3.7		4.3		MHz
			4.5 V	28	45		19		22		
			6 V	33	53		22		25		
t_{pd}	CKI	Q_D	2 V		240	490		735		615	ns
			4.5 V		58	98		147		123	
			6 V		42	83		125		105	
t_{PHL}	CLR	Any Q	2 V		66	140		210		175	ns
			4.5 V		18	28		42		35	
			6 V		14	24		36		30	
t_t		Any	2 V		28	75		110		95	ns
			4.5 V		8	15		22		19	
			6 V		6	13		19		16	

C_{pd}	Power dissipation capacitance	No load, T_A = 25°C	88 pF typ

NOTE 1: For load circuit and voltage waveforms, see page 1-14.

D2684, DECEMBER 1982—REVISED MARCH 1984

- **Package Options Include Both Plastic and Ceramic Chip Carriers in Addition to Plastic and Ceramic DIPs**
- **Dependable Texas Instruments Quality and Reliability**

SN54HC4075 . . . J PACKAGE
SN74HC4075 . . . J OR N PACKAGE
(TOP VIEW)

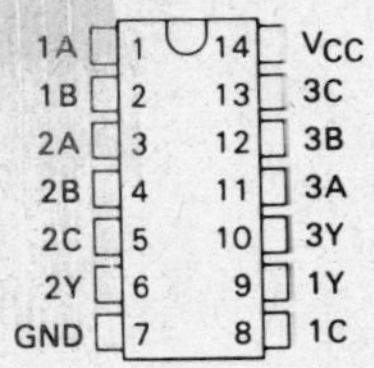

description

These devices contain three independent 3-input OR gates and perform the Boolean functions $Y = A + B + C$ or $Y = \overline{\overline{A} \cdot \overline{B} \cdot \overline{C}}$ in positive logic.

The SN54HC4075 is characterized for operation over the full military temperature range of −55°C to 125°C. The SN74HC4075 is characterized for operation from −40°C to 85°C.

SN54HC4075 . . . FH OR FK PACKAGE
SN74HC4075 . . . FH OR FN PACKAGE
(TOP VIEW)

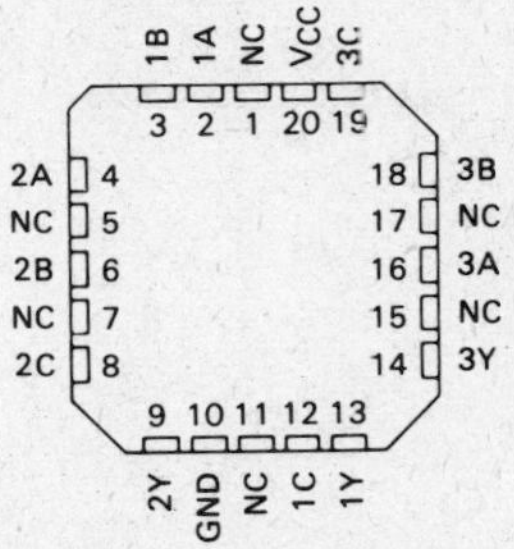

NC—No internal connection

logic symbol

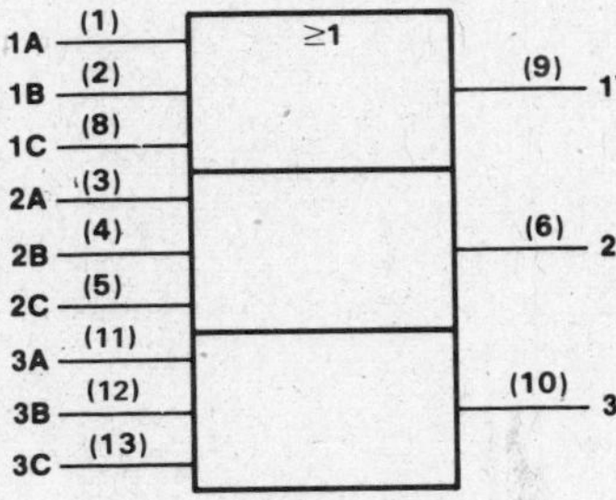

Pin numbers shown are for J and N packages.

FUNCTION TABLE

INPUTS			OUTPUT
A	B	C	Y
H	X	X	H
X	H	X	H
X	X	H	H
L	L	L	L

maximum ratings, recommended operating conditions, and electrical characteristics

See Table I, page 2-4.

switching characteristics over recommended operating free-air temperature range (unless otherwise noted), C_L = 50 pF (see Note 1)

PARAMETER	FROM (INPUT)	TO (OUTPUT)	V_{CC}	T_A = 25°C MIN	TYP	MAX	SN54HC4075 MIN	MAX	SN74HC4075 MIN	MAX	UNIT
t_{pd}	A, B, or C	Y	2 V		38	100		150		125	ns
			4.5 V		11	20		30		25	
			6 V		9	17		25		21	
t_t		Y	2 V		38	75		110		95	ns
			4.5 V		8	15		22		19	
			6 V		6	13		19		16	

C_{pd}	Power dissipation capacitance per gate	No load, T_A = 25°C	26 pF	typ

NOTE 1: For load circuit and voltage waveforms, see page 1-14.

TYPES SN54HC4078A, SN74HC4078A 8-INPUT OR/NOR GATE

D2804, MARCH 1984

- **Package Options Include Both Plastic and Ceramic Chip Carriers in Addition to Plastic and Ceramic DIPs**
- **Dependable Texas Instruments Quality and Reliability**

SN54HC4078A . . . J PACKAGE
SN74HC4078A . . . J OR N PACKAGE
(TOP VIEW)

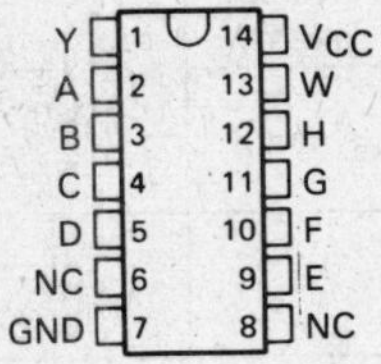

SN54HC4078A . . . FH OR FK PACKAGE
SN74HC4078A . . . FH OR FN PAKCAGE
(TOP VIEW)

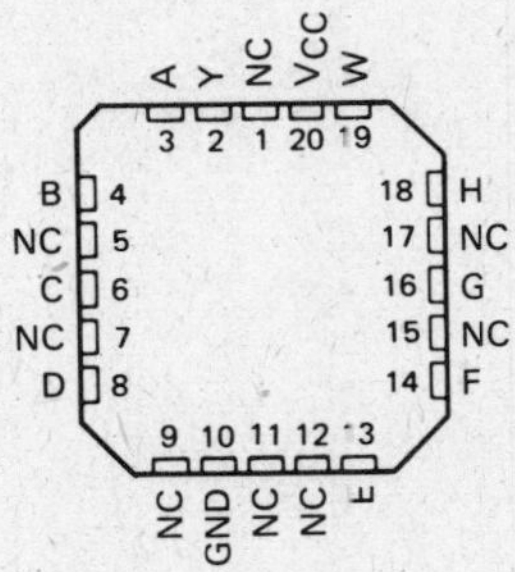

NC—No internal connection

description

These devices contain a single 8-input OR/NOR gate and perform the following Boolean functions in positive logic:

$$W = \overline{A + B + C + D + E + F + G + H}$$

or

$$W = \overline{A} \cdot \overline{B} \cdot \overline{C} \cdot \overline{D} \cdot \overline{E} \cdot \overline{F} \cdot \overline{G} \cdot \overline{H}$$

and

$$Y = A + B + C + D + E + F + G + H$$

or

$$Y = \overline{\overline{A} \cdot \overline{B} \cdot \overline{C} \cdot \overline{D} \cdot \overline{E} \cdot \overline{F} \cdot \overline{G} \cdot \overline{H}}$$

The SN54HC4078A is characterized for operation over the full military temperature range of −55°C to 125°C. The SN74HC4078A is characterized for operation from −40°C to 85°C.

FUNCTION TABLE

INPUTS A THRU H	OUTPUTS W	OUTPUTS Y
One or more inputs H	L	H
All inputs L	H	L

logic symbol

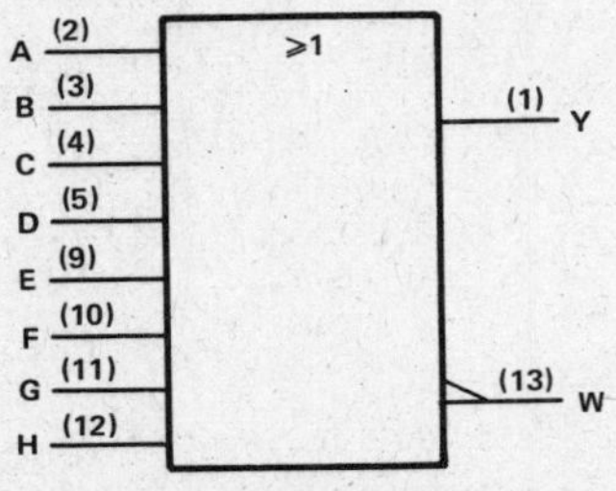

Pin numbers shown are for J and N packages.

maximum ratings, recommended operating conditions, and electrical characteristics

See Table I, page 2-4.

switching characteristics over recommended operating free-air temperature range (unless otherwise noted), C_L = 50 pF (see Note 1)

PARAMETER	FROM (INPUT)	TO (OUTPUT)	V_{CC}	T_A = 25°C MIN	T_A = 25°C TYP	T_A = 25°C MAX	SN54HC4078A MIN	SN54HC4078A MAX	SN74HC4078A MIN	SN74HC4078A MAX	UNIT
t_{pd}	A thru H	Y/$\overline{Y}$	2 V		40	130		195		165	ns
			4.5 V		12	26		39		33	
			6 V		10	22		33		28	
t_t		Y/$\overline{Y}$	2 V		38	75		110		95	ns
			4.5 V		8	15		22		19	
			6 V		6	13		19		16	

C_{pd}	Power dissipation capacitance per gate	No load, T_A = 25°C	25 pF typ

NOTE 1: For load circuit and voltage waveforms, see page 1-14.

D2684, DECEMBER 1982—REVISED MARCH 1984

- 8-Bit Parallel-Out Storage Register Performs Serial-to-Parallel Conversion with Storage
- Asynchronous Parallel Clear
- Active-High Decoder
- Enable Input Simplifies Expansion
- Expandable for N-Bit Applications
- Four Distinct Functional Modes
- Package Options Include Both Plastic and Ceramic Chip Carriers in Addition to Plastic and Ceramic DIPs
- Dependable Texas Instruments Quality and Reliability

SN54HC4724 . . . J PACKAGE
SN74HC4724 . . . J OR N PACKAGE
(TOP VIEW)

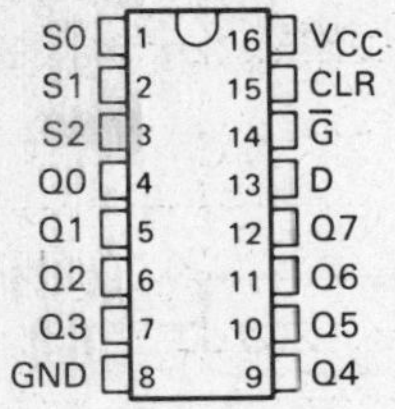

SN54HC4724 . . . FH OR FK PACKAGE
SN74HC4724 . . . FH OR FN PACKAGE
(TOP VIEW)

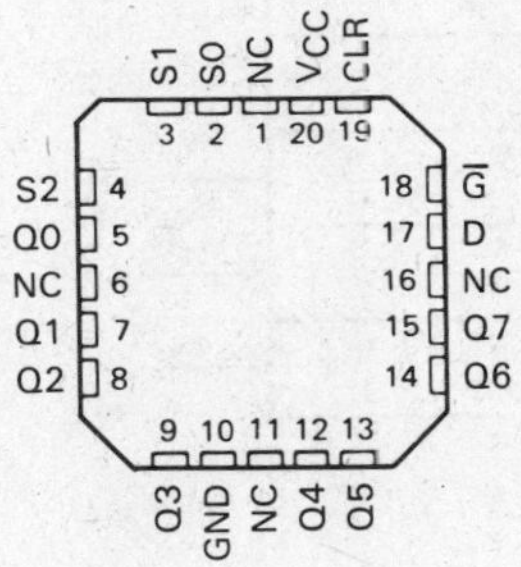

NC—No internal connection

description

These 8-bit addressable latches are designed for general purpose storage applications in digital systems. Specific uses include working registers, serial-holding registers, and active-high decoders or demultiplexers. They are multifunctional devices capable of storing single-line data in eight addressable latches, and being a 1-of-8 decoder or demultiplexer with active-high outputs.

Four distinct modes of operation are selectable by controlling the clear (CLR) and enable ($\overline{G}$) inputs as enumerated in the function table. In the addressable-latch mode, data at the data-in terminal is written into the addressed latch. The addressed latch will follow the data input with all unaddressed latches remaining in their previous states. In the memory mode, all latches remain in their previous states and are unaffected by the data or address inputs. To eliminate the possibility of entering erroneous data in the latches, enable $\overline{G}$ should be held high (inactive) while the address lines are changing. In the 1-of-8 decoding or demultiplexing mode, the addressed output will follow the level of the D input with all other outputs low. In the clear mode, all outputs are low and unaffected by the address and data inputs.

The SN54HC4724 is characterized for operation over the full military temperature range of −55°C to 125°C. The SN74HC4724 is characterized for operation from −40°C to 85°C.

FUNCTION TABLE

INPUTS		OUTPUT OF ADDRESSED LATCH	EACH OTHER OUTPUT	FUNCTION
CLR	$\overline{G}$			
L	L	D	Q_{iO}	Addressable Latch
L	H	Q_{iO}	Q_{iO}	Memory
H	L	D	L	8-Line Demultiplexer
H	H	L	L	Clear

LATCH SELECTION TABLE

SELECT INPUTS			LATCH ADDRESSED
S2	S1	S0	
L	L	L	0
L	L	H	1
L	H	L	2
L	H	H	3
H	L	L	4
H	L	H	5
H	H	L	6
H	H	H	7

logic symbol

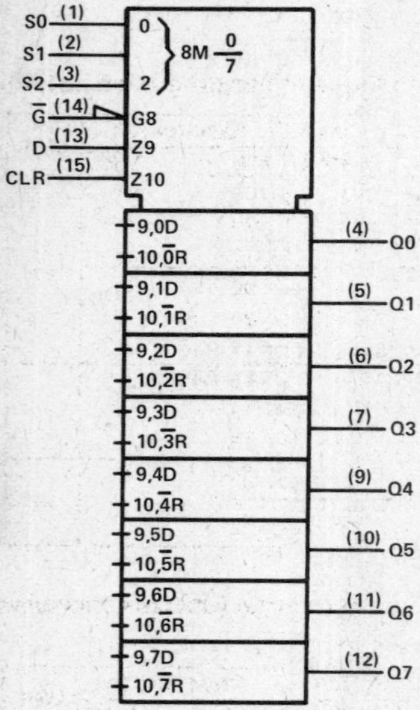

logic diagram (positive logic)

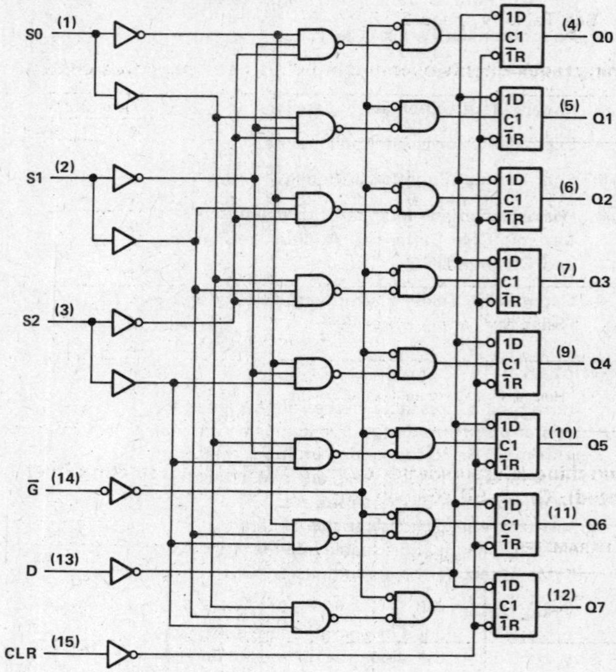

Pin numbers shown are for J and N packages.

logic symbol and logic diagram, each internal latch (positive logic)

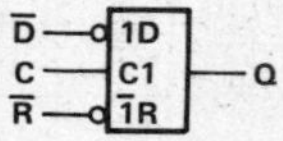

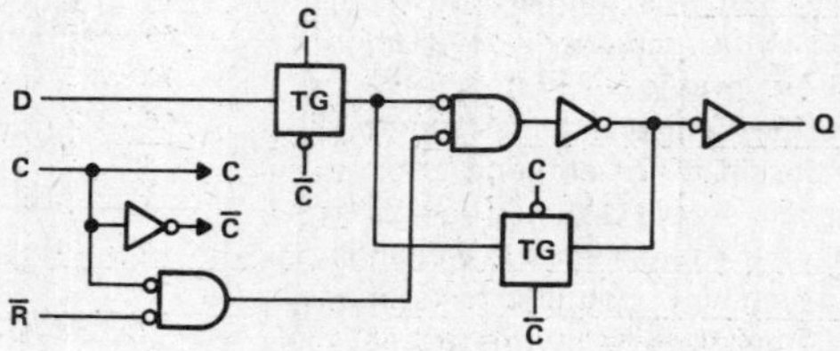

TEXAS INSTRUMENTS
POST OFFICE BOX 225012 • DALLAS, TEXAS 75265

maximum ratings, recommended operating conditions, and electrical characteristics

See Table IV, page 2-10.

timing requirements over recommended operating free-air temperature range (unless otherwise noted)

			V_{CC}	T_A = 25°C		SN54HC4724		SN74HC4724		UNIT
				MIN	MAX	MIN	MAX	MIN	MAX	
t_w	Pulse duration	CLR high	2 V	80		120		100		ns
			4.5 V	16		24		20		
			6 V	14		20		17		
		$\overline{G}$ low	2 V	80		120		100		
			4.5 V	16		24		20		
			6 V	14		20		17		
t_{su}	Setup time, data or address before $\overline{G}$↑		2 V	75		115		95		ns
			4.5 V	15		23		19		
			6 V	13		20		16		
t_h	Hold time, data or address after $\overline{G}$↑		2 V	5		5		5		ns
			4.5 V	5		5		5		
			6 V	5		5		5		

switching characteristics over recommended operating free-air temperature range (unless otherwise noted), C_L = 50 pF (see Note 1)

PARAMETER	FROM (INPUT)	TO (OUTPUT)	V_{CC}	T_A = 25°C			SN54HC4724		SN74HC4724		UNIT
				MIN	TYP	MAX	MIN	MAX	MIN	MAX	
t_{PHL}	CLR	Any Q	2 V		60	150		225		190	ns
			4.5 V		18	30		45		38	
			6 V		14	26		38		32	
t_{pd}	Data	Any Q	2 V		56	130		195		165	ns
			4.5 V		17	26		39		33	
			6 V		13	22		33		28	
t_{pd}	Address	Any Q	2 V		74	200		300		250	ns
			4.5 V		21	40		60		50	
			6 V		17	34		51		43	
t_{pd}	$\overline{G}$	Any Q	2 V		66	170		255		215	ns
			4.5 V		20	34		51		43	
			6 V		16	29		43		37	
t_t		Any	2 V		28	75		110		95	ns
			4.5 V		8	15		22		19	
			6 V		6	13		19		16	

C_{pd}	Power dissipation capacitance per latch	No load, T_A = 25°C	33 pF typ

NOTE 1: For load circuit and voltage waveforms, see page 1-14.

3 HCMOS DEVICES

TYPES SN54HC7266, SN74HC7266 QUADRUPLE 2-INPUT EXCLUSIVE-NOR GATES

D2804, MARCH 1984

- Package Options Include Both Plastic and Ceramic Chip Carriers in Addition to Plastic and Ceramic DIPs
- Dependable Texas Instruments Quality and Reliability

SN54HC7266 . . . J PACKAGE
SN74HC7266 . . . J OR N PACKAGE
(TOP VIEW)

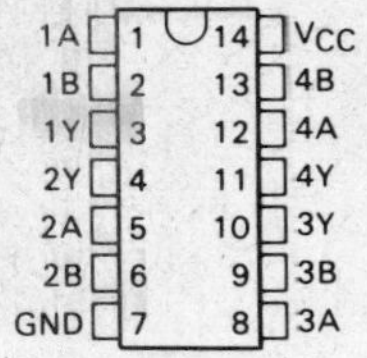

description

These devices are composed of four independent 2-input exclusive-NOR gates. 'HC7266 devices are totem-pole-output versions of the 'HC266. They perform the Boolean functions $Y = \overline{A \oplus B} = \overline{A}\overline{B} + AB$ in positive logic.

The SN54HC7266 is characterized for operation over the full military temperature range of −55 °C to 125 °C. The SN74HC7266 is characterized for operation from −40 °C to 85 °C.

SN54HC7266 . . . FH OR FK PACKAGE
SN74HC7266 . . . FH OR FN PACKAGE
(TOP VIEW)

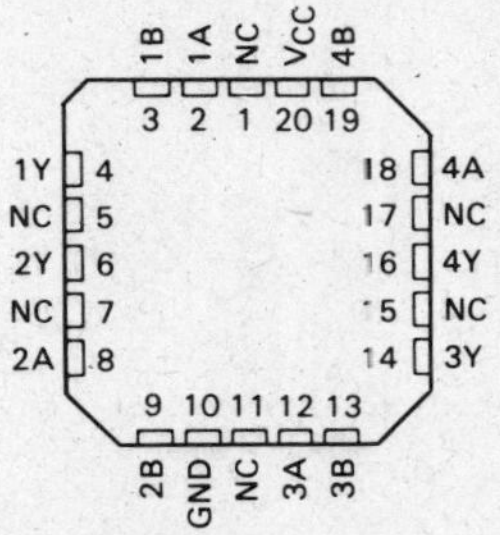

NC—No internal connection

logic symbol

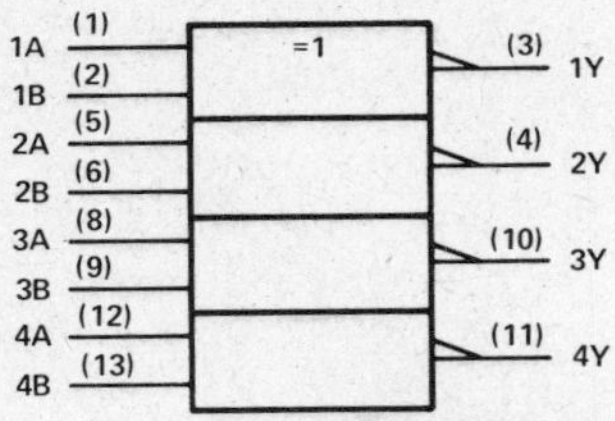

Pin numbers shown are for J and N packages.

FUNCTION TABLE

INPUTS		OUTPUT
A	B	Y
L	L	H
L	H	L
H	L	L
H	H	H

maximum ratings, recommended operating conditions, and electrical characteristics

See Table I, page 2-4.

switching characteristics over recommended operating free-air temperature range (unless otherwise noted), C_L = 50 pF (see Note 1)

PARAMETER	FROM (INPUT)	TO (OUTPUT)	V_{CC}	T_A = 25 °C MIN	TYP	MAX	SN54HC7266 MIN	MAX	SN74HC7266 MIN	MAX	UNIT
t_{pd}	A or B	Y	2 V		40	100		150		125	ns
			4.5 V		12	20		30		25	
			6 V		10	17		25		21	
t_t		Y	2 V		28	75		110		95	ns
			4.5 V		8	15		22		19	
			6 V		6	13		19		16	

C_{pd}	Power dissipation capacitance per gate	No load, T_A = 25 °C	35 pF typ

NOTE 1: For load circuit and voltage waveforms, see page 1-14.

3

HCMOS DEVICES

ADVANCE INFORMATION

This section contains information on new products in the sampling or preproduction stage. Characteristic data and other specifications are subject to change without notice.

D2684, DECEMBER 1982—REVISED MARCH 1984

- **Package Options Include Both Plastic and Ceramic Chip Carriers in Addition to Plastic and Ceramic DIPs**
- **Dependable Texas Instruments Quality and Reliability**

SN54HC73 . . . J PACKAGE
SN74HC73 . . . J OR N PACKAGE
(TOP VIEW)

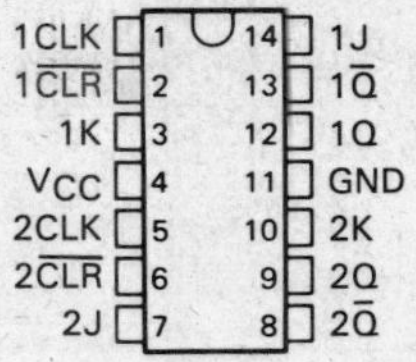

For functionally and electrically identical parts in chip carrier packages, see SN54HC107 and SN74HC107.

description

These devices contain two independent J-K negative-edge-triggered flip-flops. A low level at the Clear input resets the outputs regardless of the other inputs. When Clear is inactive (high), data at the J and K inputs meeting the setup time requirements are transferred to the outputs on the negative-going edge of the clock pulse. Following the hold time interval, data at the J and K inputs may be changed without affecting the levels at the outputs. These flip-flops can perform as toggle flip-flops by tying J and K high.

The SN54HC73 is characterized for operation over the full military temperature range of −55°C to 125°C. The SN74HC73 is characterized for operation from −40°C to 85°C.

FUNCTION TABLE
(EACH FLIP-FLOP)

INPUTS				OUTPUTS	
$\overline{CLR}$	CLK	J	K	Q	$\overline{Q}$
L	X	X	X	L	H
H	↓	L	L	Q_0	$\overline{Q}_0$
H	↓	H	L	H	L
H	↓	L	H	L	H
H	↓	H	H	TOGGLE	
H	H	X	X	Q_0	$\overline{Q}_0$

logic symbol

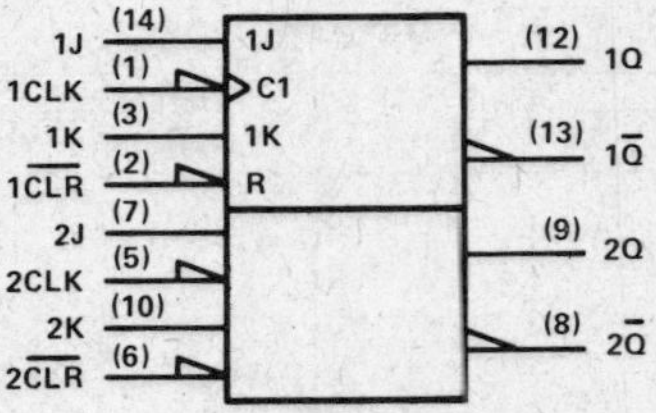

Pin numbers shown are for J and N packages.

logic diagram, each flip-flop (positive logic)

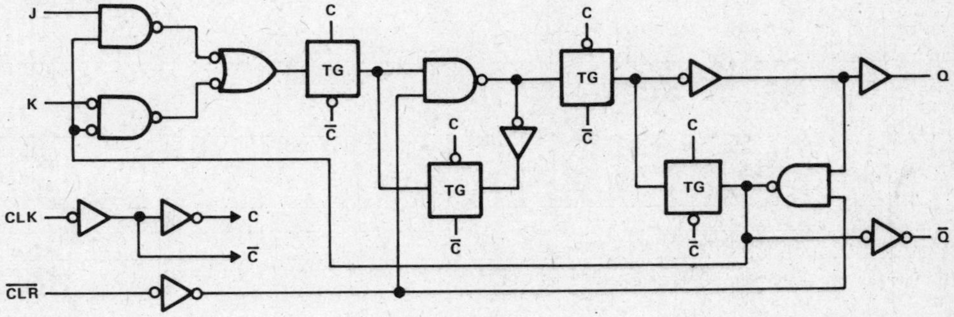

absolute maximum ratings, recommended operating conditions, and electrical characteristics

See Table II, page 2-6.

timing requirements over recommended operating free-air temperature range (unless otherwise noted)

			V_{CC}	T_A = 25°C		SN54HC73		SN74HC73		UNIT
				MIN	MAX	MIN	MAX	MIN	MAX	
f_{clock}	Clock frequency		2 V	0	5					MHz
			4.5 V	0	25					
			6 V	0	29					
t_w	Pulse duration	CLK high or low	2 V	100						ns
			4.5 V	20						
			6 V	17						
		$\overline{CLR}$ low	2 V	100						ns
			4.5 V	20						
			6 V	17						
t_{su}	Setup time, $\overline{CLR}$ inactive or data before CLK↓		2 V	100						ns
			4.5 V	20						
			6 V	17						
t_h	Hold time, data after CLK↓		2 V	0						ns
			4.5 V	0						
			6 V	0						

switching characteristics over recommended operating free-air temperature range (unless otherwise noted), C_L = 50 pF (see Note 1)

PARAMETER	FROM (INPUT)	TO (OUTPUT)	V_{CC}	T_A = 25°C			SN54HC73		SN74HC73		UNIT
				MIN	TYP	MAX	MIN	MAX	MIN	MAX	
f_{max}			2 V	5	9						MHz
			4.5 V	25	50						
			6 V	29	60						
t_{PHL}	$\overline{CLR}$	Q	2 V		42						ns
			4.5 V		16						
			6 V		14						
t_{PLH}	$\overline{CLR}$	$\overline{Q}$	2 V		35						ns
			4.5 V		11						
			6 V		9						
t_{pd}	CLK	Q	2 V		40						ns
			4.5 V		13						
			6 V		11						
t_{pd}	CLK	$\overline{Q}$	2 V		42						ns
			4.5 V		16						
			6 V		14						
t_t		Any	2 V		38						ns
			4.5 V		8						
			6 V		6						

C_{pd}	Power dissipation capacitance per flip-flop	No load, T_A = 25°C	30 pF typ

NOTE 1: For load circuit and voltage waveforms, see page 1-14.

TEXAS INSTRUMENTS
POST OFFICE BOX 225012 • DALLAS, TEXAS 75265

TYPES SN54HC78, SN74HC78 DUAL J-K NEGATIVE-EDGE-TRIGGERED FLIP-FLOPS WITH PRESET, COMMON CLEAR, AND COMMON CLOCK

D2684, DECEMBER 1982—REVISED MARCH 1984

- **Package Options Include Both Plastic and Ceramic Chip Carriers in Addition to Plastic and Ceramic DIPs**
- **Dependable Texas Instruments Quality and Reliability**

SN54HC78 . . . J PACKAGE
SN74HC78 . . . J OR N PACKAGE
(TOP VIEW)

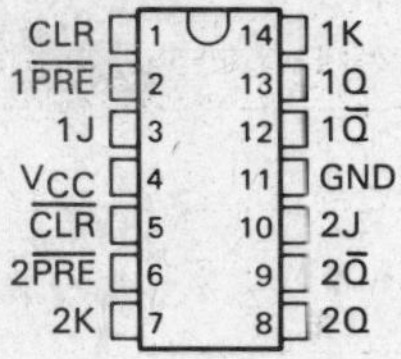

For functionally and electrically identical parts in chip carrier packages, see SN54HC114 and SN74HC114.

description

These devices contain two independent J-K negative-edge-triggered flip-flops. A low level at the Preset or Clear inputs sets or resets the outputs regardless of the levels of the other inputs. When the Preset and Clear are inactive (high), data at the J and K inputs meeting the setup time requirements are transferred to the outputs on the negative-going edge of the clock pulse. Clock triggering occurs at a voltage level and is not directly related to the rise time of the clock pulse. Following the hold time interval, data at the J and K inputs may be changed without affecting the levels at the outputs. These versatile flip-flops can perform as toggle flip-flops by tying J and K high.

FUNCTION TABLE

INPUTS					OUTPUTS	
$\overline{PRE}$	$\overline{CLR}$	CLK	J	K	Q	$\overline{Q}$
L	H	X	X	X	H	L
H	L	X	X	X	L	H
L	L	X	X	X	H*	H*
H	H	↓	L	L	Q_0	$\overline{Q}_0$
H	H	↓	H	L	H	L
H	H	↓	L	H	L	H
H	H	↓	H	H	TOGGLE	
H	H	H	X	X	Q_0	$\overline{Q}_0$

* This configuration is nonstable; that is, it will not persist when either Preset or Clear returns to its inactive (high) level.

logic symbol

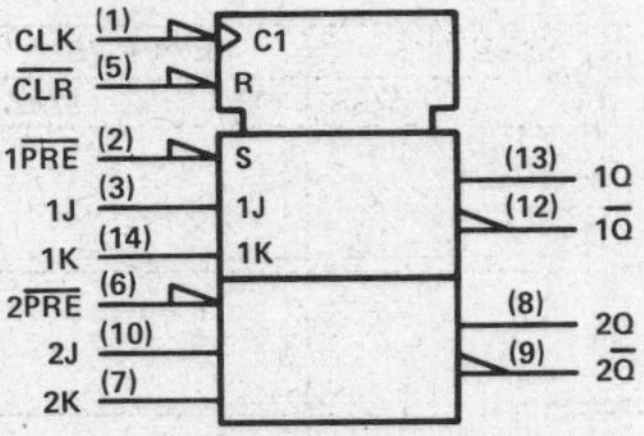

Pin numbers shown are for J and N packages.

logic diagram, each flip-flop (positive logic)

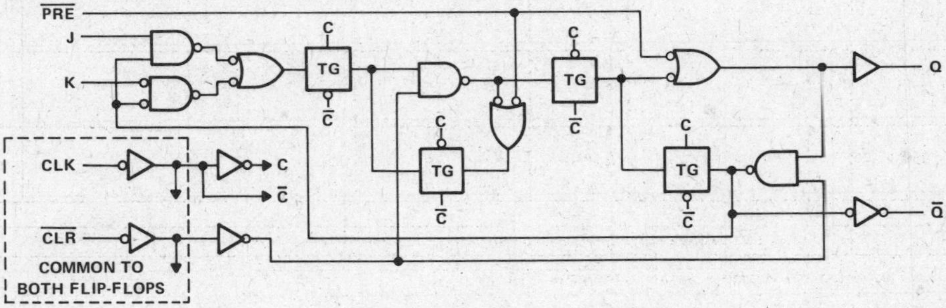

maximum ratings, recommended operating conditions, and electrical characteristics

See Table II, page 2-6.

TEXAS INSTRUMENTS
POST OFFICE BOX 225012 • DALLAS, TEXAS 75265

timing requirements over recommended operating free-air temperature range (unless otherwise noted)

			V_{CC}	$T_A = 25°C$		SN54HC78		SN74HC78		UNIT
				MIN	MAX	MIN	MAX	MIN	MAX	
f_{clock}	Clock frequency		2 V	0	5					MHz
			4.5 V	0	25					
			6 V	0	29					
t_w	Pulse duration	$\overline{CLR}$ or $\overline{PRE}$ low	2 V	100						ns
			4.5 V	20						
			6 V	17						
		CLK high or low	2 V	100						ns
			4.5 V	20						
			6 V	17						
t_{su}	Setup time before CLK↓	$\overline{CLR}$ or $\overline{PRE}$ inactive or data	2 V	100						ns
			4.5 V	20						
			6 V	17						
t_h	Hold time, data after CLK↓		2 V	0						ns
			4.5 V	0						
			6 V	0						

switching characteristics over recommended operating free-air temperature range (unless otherwise noted), $C_L = 50$ pF (see Note 1)

PARAMETER	FROM (INPUT)	TO (OUTPUT)	V_{CC}	$T_A = 25°C$			SN54HC78		SN74HC78		UNIT
				MIN	TYP	MAX	MIN	MAX	MIN	MAX	
f_{max}			2 V	5	9						ns
			4.5 V	25	50						
			6 V	29	60						
t_{PHL}	$\overline{CLR}$	Q	2 V		52						ns
			4.5 V		17						
			6 V		14						
t_{PLH}	$\overline{CLR}$	$\overline{Q}$	2 V		40						ns
			4.5 V		13						
			6 V		11						
t_{PHL}	$\overline{PRE}$	$\overline{Q}$	2 V		52						ns
			4.5 V		17						
			6 V		14						
t_{PLH}	$\overline{PRE}$	Q	2 V		36						ns
			4.5 V		12						
			6 V		10						
t_{pd}	CLK	Q	2 V		52						ns
			4.5 V		17						
			6 V		14						
t_{pd}	CLK	$\overline{Q}$	2 V		48						ns
			4.5 V		16						
			6 V		14						
t_t			2 V		38						ns
			4.5 V		8						
			6 V		6						

C_{pd}	Power dissipation capacitance per flip-flop	No load, $T_A = 25°C$	30 pF typ

NOTE 1: For load circuit and voltage waveforms, see page 1-14.

4

ADVANCE INFORMATION

TYPES SN54HC85, SN74HC85
4-BIT MAGNITUDE COMPARATORS

D2684, DECEMBER 1982—REVISED MARCH 1984

- **Package Options Include Both Plastic and Ceramic Chip Carriers in Addition to Plastic and Ceramic DIPs**
- **Dependable Texas Instruments Quality and Reliability**

SN54HC85 . . . J PACKAGE
SN74HC85 . . . J OR N PACKAGE
(TOP VIEW)

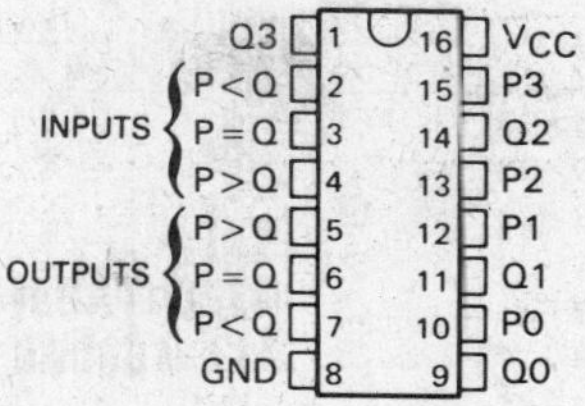

SN54HC85 . . . FH OR FK PACKAGE
SN74HC85 . . . FH OR FN PACKAGE
(TOP VIEW)

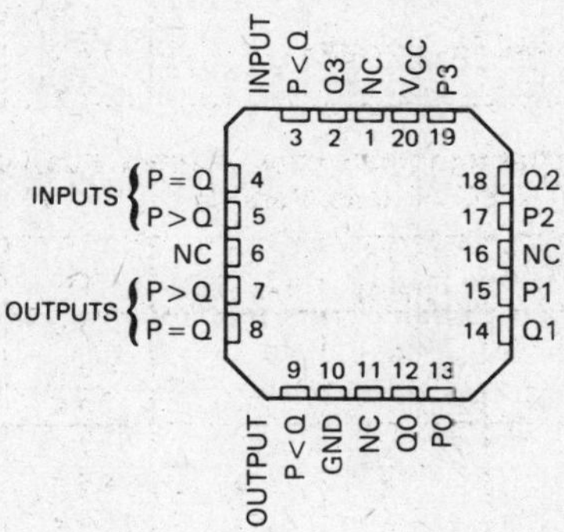

NC—No internal connection

description

These four-bit magnitude comparators perform comparison of straight binary and straight BCD (8-4-2-1) codes. Three fully decoded decisions about two 4-bit words (P, Q) are made and are externally available at three outputs. These devices are fully expandable to any number of bits without external gates. Words of greater length may be compared by connecting comparators in cascade. The P>Q, P<Q, and P=Q outputs of a stage handling less-significant bits are connected to the corresponding P>Q, P<Q, and P=Q inputs of the next stage handling more-significant bits. The stage handling the least-significant bits must have a high-level voltage applied to the P=Q input. The cascading path of the 'HC85 is implemented with only a two-gate-level delay to reduce overall comparison times for long words.

The SN54HC85 is characterized for operation over the full military temperature range of −55°C to 125°C. The SN74HC85 is characterized for operation from −40°C to 85°C.

logic symbol

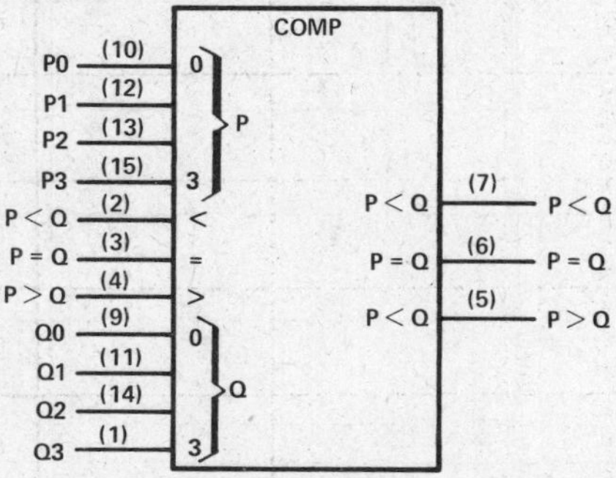

Pin numbers shown are for J and N packages.

FUNCTION TABLE

COMPARING INPUTS				CASCADING INPUTS			OUTPUTS		
P3, Q3	P2, Q2	P1, Q1	P0, Q0	P>Q	P<Q	P=Q	P>Q	P<Q	P=Q
P3>Q3	X	X	X	X	X	X	H	L	L
P3<Q3	X	X	X	X	X	X	L	H	L
P3=Q3	P2>Q2	X	X	X	X	X	H	L	L
P3=Q3	P2<Q2	X	X	X	X	X	L	H	L
P3=Q3	P2=Q2	P1>Q1	X	X	X	X	H	L	L
P3=Q3	P2=Q2	P1<Q1	X	X	X	X	L	H	L
P3=Q3	P2=Q2	P1=Q1	P0>Q0	X	X	X	H	L	L
P3=Q3	P2=Q2	P1=Q1	P0<Q0	X	X	X	L	H	L
P3=Q3	P2=Q2	P1=Q1	P0=Q0	H	L	L	H	L	L
P3=Q3	P2=Q2	P1=Q1	P0=Q0	L	H	L	L	H	L
P3=Q3	P2=Q2	P1=Q1	P0=Q0	X	X	H	L	L	H
P3=Q3	P2=Q2	P1=Q1	P0=Q0	H	H	L	L	L	L
P3=Q3	P2=Q2	P1=Q1	P0=Q0	L	L	L	H	H	L

maximum ratings, recommended operating conditions, and electrical characteristics

See Table IV, page 2-10.

switching characteristics over recommended operating free-air temperature range (unless otherwise noted), C_L = 50 pF (see Note 1)

PARAMETER	FROM (INPUT)	TO (OUTPUT)	V_{CC}	T_A = 25°C MIN	TYP	MAX	SN54HC85 MIN	MAX	SN74HC85 MIN	MAX	UNIT
t_{pd}	Any P or Q	P>Q or P<Q	2 V		80	230		345		290	ns
			4.5 V		26	46		69		58	
			6 V		22	39		59		49	
t_{pd}	Any P or Q	P=Q	2 V		66	200		300		250	ns
			4.5 V		22	40		60		50	
			6 V		19	34		51		43	
t_{pd}	P<Q or P=Q	P>Q	2 V		63	175		260		220	ns
			4.5 V		21	35		52		44	
			6 V		18	30		44		37	
t_{pd}	P>Q or P=Q	P<Q	2 V		72	175		260		220	ns
			4.5 V		24	35		52		44	
			6 V		20	30		44		37	
t_{pd}	P=Q	P=Q	2 V		51	145		215		185	ns
			4.5 V		17	29		43		37	
			6 V		14	25		37		31	
t_t		Any	2 V		38	75		110		95	ns
			4.5 V		8	15		22		19	
			6 V		6	13		19		16	

C_{pd}	Power dissipation capacitance	No load, T_A = 25°C	80 pF typ

NOTE 1: For load circuit and voltage waveforms, see page 1-14.

TYPES SN54HC154, SN74HC154
4-LINE TO 16-LINE DECODERS/DEMULTIPLEXERS

D2684, DECEMBER 1982—REVISED MARCH 1984

- Decodes 4 Binary-Coded Inputs into One of 16 Mutually Exclusive Outputs
- Performs the Demultiplexing Function by Distributing Data From One Input to Any One of 16 Outputs
- Package Options Include Both Plastic and Ceramic Chip Carriers in Addition to Plastic and Ceramic DIPs
- Dependable Texas Instruments Quality and Reliability

SN54HC154 . . . JT PACKAGE
SN74HC154 . . . JT OR NT PACKAGE
(TOP VIEW)

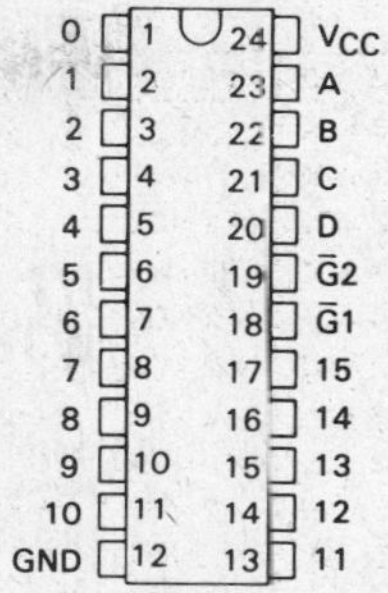

description

Each of these monolithic, 4-line to 16-line decoders decodes four binary-coded inputs into one of sixteen mutually exclusive outputs when both the strobe inputs, G̅1 and G̅2, are low. The demultiplexing function is performed by using the 4 input lines to address the output line, passing data from one of the strobe inputs with the other strobe input low. When either strobe input is high, all outputs are high. These demultiplexers are ideally suited for implementing high-performance memory decoders.

The SN54HC154 is characterized for operation over the full military temperature range of −55°C to 125°C. The SN74HC154 is characterized for operation from −40°C to 85°C.

SN54HC154 . . . FH OR FK PACKAGE
SN74HC154 . . . FH OR FN PACKAGE
(TOP VIEW)

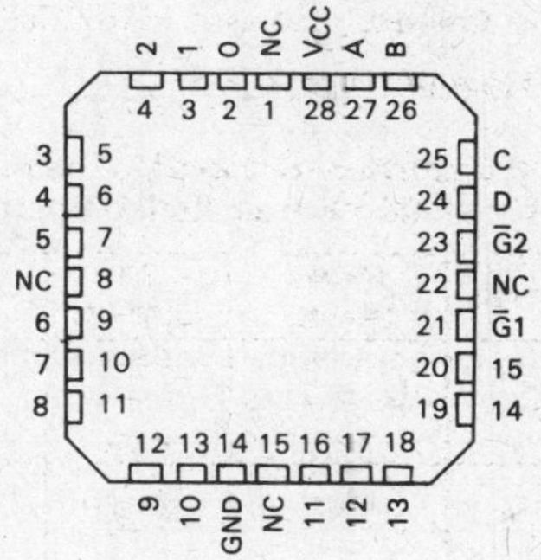

NC—No internal connection

logic symbols (alternatives)

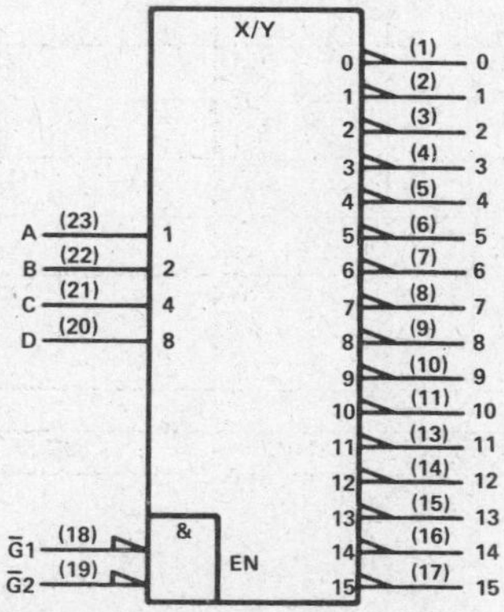

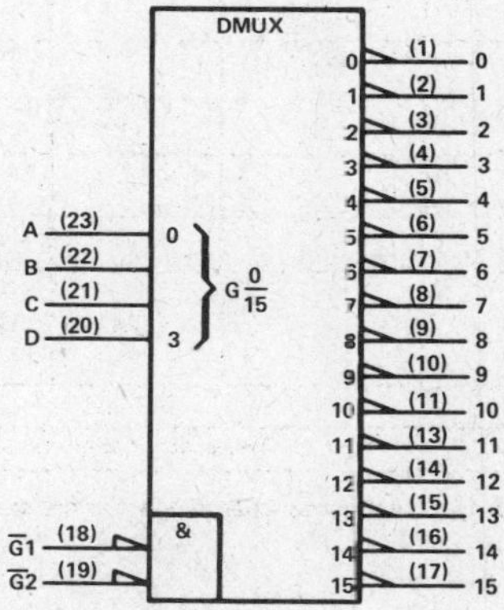

Pin numbers shown are for JT and NT packages.

4

ADVANCE INFORMATION

FUNCTION TABLE

INPUTS						OUTPUTS															
$\overline{G1}$	$\overline{G2}$	D	C	B	A	0	1	2	3	4	5	6	7	8	9	10	11	12	13	14	15
L	L	L	L	L	L	L	H	H	H	H	H	H	H	H	H	H	H	H	H	H	H
L	L	L	L	L	H	H	L	H	H	H	H	H	H	H	H	H	H	H	H	H	H
L	L	L	L	H	L	H	H	L	H	H	H	H	H	H	H	H	H	H	H	H	H
L	L	L	L	H	H	H	H	H	L	H	H	H	H	H	H	H	H	H	H	H	H
L	L	L	H	L	L	H	H	H	H	L	H	H	H	H	H	H	H	H	H	H	H
L	L	L	H	L	H	H	H	H	H	H	L	H	H	H	H	H	H	H	H	H	H
L	L	L	H	H	L	H	H	H	H	H	H	L	H	H	H	H	H	H	H	H	H
L	L	L	H	H	H	H	H	H	H	H	H	H	L	H	H	H	H	H	H	H	H
L	L	H	L	L	L	H	H	H	H	H	H	H	H	L	H	H	H	H	H	H	H
L	L	H	L	L	H	H	H	H	H	H	H	H	H	H	L	H	H	H	H	H	H
L	L	H	L	H	L	H	H	H	H	H	H	H	H	H	H	L	H	H	H	H	H
L	L	H	L	H	H	H	H	H	H	H	H	H	H	H	H	H	L	H	H	H	H
L	L	H	H	L	L	H	H	H	H	H	H	H	H	H	H	H	H	L	H	H	H
L	L	H	H	L	H	H	H	H	H	H	H	H	H	H	H	H	H	H	L	H	H
L	L	H	H	H	L	H	H	H	H	H	H	H	H	H	H	H	H	H	H	L	H
L	L	H	H	H	H	H	H	H	H	H	H	H	H	H	H	H	H	H	H	H	L
L	H	X	X	X	X	H	H	H	H	H	H	H	H	H	H	H	H	H	H	H	H
H	L	X	X	X	X	H	H	H	H	H	H	H	H	H	H	H	H	H	H	H	H
H	H	X	X	X	X	H	H	H	H	H	H	H	H	H	H	H	H	H	H	H	H

H = high level, L = low level, X = irrelevant

4

ADVANCE INFORMATION

maximum ratings, recommended operating conditions, and electrical characteristics

See Table IV, page 2-10.

switching characteristics over recommended operating free-air temperature range (unless otherwise noted), C_L = 50 pF (see Note 1)

PARAMETER	FROM (INPUT)	TO (OUTPUT)	V_{CC}	T_A = 25°C MIN	TYP	MAX	SN54HC154 MIN	MAX	SN74HC154 MIN	MAX	UNIT
t_{pd}	A, B, C, or D	Any	2 V		72	180		270		225	ns
			4.5 V		24	36		54		45	
			6 V		20	31		46		38	
t_{pd}	$\overline{G1}$ or $\overline{G2}$	Any	2 V		72	180		270		225	ns
			4.5 V		24	36		54		45	
			6 V		20	31		46		38	
t_t		Any	2 V		28	75		110		95	ns
			4.5 V		8	15		22		19	
			6 V		6	13		19		16	

NOTE 1: For load circuit and voltage waveforms, see page 1-14.

TYPES SN54HC189, SN54HC219, SN74HC189, SN74HC219 64-BIT RANDOM-ACCESS MEMORIES WITH 3-STATE OUTPUTS

D2684, DECEMBER 1982—REVISED MARCH 1984

- **Organized as 16 Words of Four Bits Each**
- **Choice of Noninverted or Inverted Outputs**
- **High-Current 3-State Inverting Outputs Can Drive up to 15 LSTTL Loads**
- **Package Options Include Both Plastic and Ceramic Chip Carriers in Addition to Plastic and Ceramic DIPs**
- **Dependable Texas Instruments Quality and Reliability**

SN54HC189 . . . J PACKAGE
SN74HC189 . . . J OR N PACKABE
(TOP VIEW)

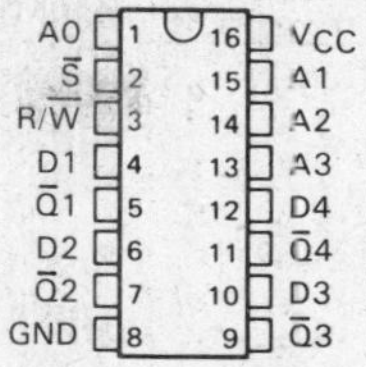

SN54HC189 . . . FH OR FK PACKAGE
SN74HC189 . . . FH OR FN PACKAGE
(TOP VIEW)

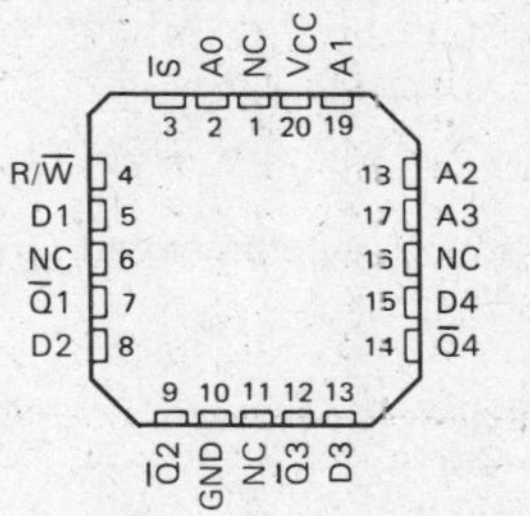

description

Information to be stored in the memory is written into the selected address location when the chip-select ($\overline{S}$) and the write-enable (R/$\overline{W}$) inputs are low. While the write-enable input is low, the memory outputs are off (Hi-Z). When a number of outputs are bus-connected, this off state neither loads nor drives the data bus; however, it permits the bus line to be driven by the other active outputs or a passive pull-up.

Information stored in the memory (see function table for input/output phase relationship) is available at the outputs when the write-enable input is high and the chip-select input is low. When the chip-select input is high, the outputs will be off.

The SN54HC189 and SN54HC219 are characterized for operation over the full military temperature range of −55°C to 125°C. The SN74HC189 and SN74HC219 are characterized for operation from −40°C to 85°C.

FUNCTION TABLE

FUNCTION	INPUTS		OUTPUTS	
	CHIP SELECT	WRITE ENABLE	'HC189	'HC219
Write	L	L	Z	Z
Read	L	H	Complement of data entered	Data entered
Inhibit	H	X	Z	Z

SN54HC219 . . . J PACKAGE
SN74HC219 . . . J OR N PACKAGE
(TOP VIEW)

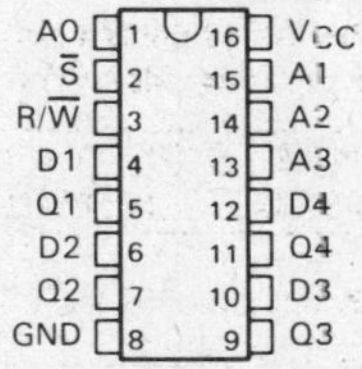

SN54HC219 . . . FH OR FK PACKAGE
SN74HC219 . . . FH OR FN PACKAGE
(TOP VIEW)

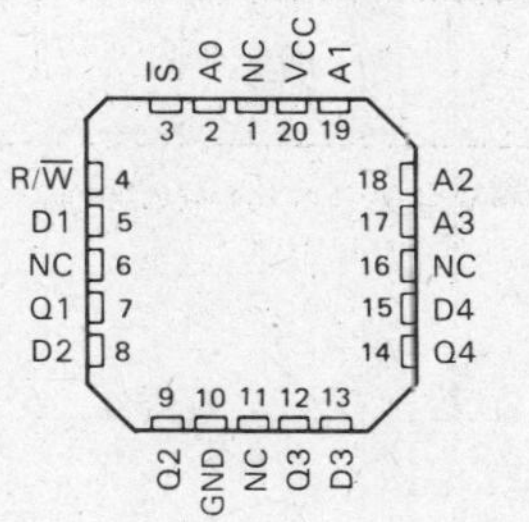

NC—No internal connection

ADVANCE INFORMATION
This document contains information on a new product. Specifications are subject to change without notice.

logic symbols

'HC189

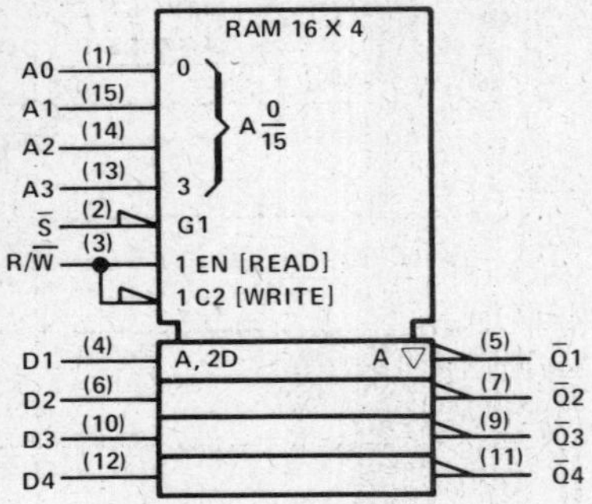

'HC219

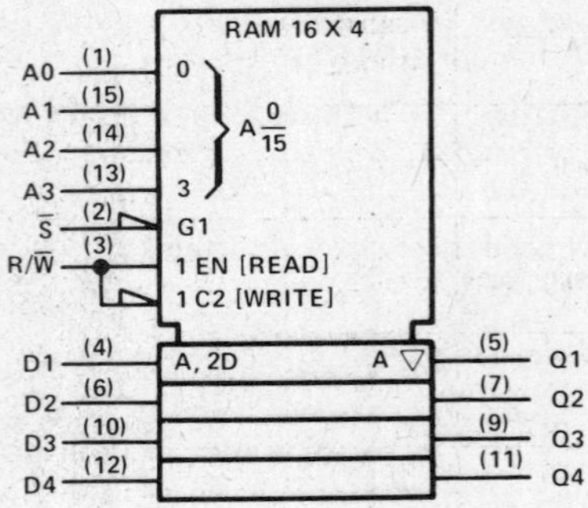

Pin numbers shown are for J and N packages.

maximum ratings, recommended operating conditions, and electrical characteristics

See Table VII, page 2-14.

timing requirements over recommended operating free-air temperature range (unless otherwise noted)

			V_{CC}	$T_A = 25°C$		SN54HC189 SN54HC219		SN74HC189 SN74HC219		UNIT
				MIN	MAX	MIN	MAX	MIN	MAX	
t_w	Pulse duration, R/W̄ low		2 V	275		400		350		ns
			4.5 V	55		80		70		
			6 V	47		68		60		
t_{su}	Setup time	Address before R/W̄↓	2 V	0		0		0		ns
			4.5 V	0		0		0		
			6 V	0		0		0		
		Data before R/W̄↑	2 V	275		400		350		
			4.5 V	55		80		70		
			6 V	47		68		60		
		Chip-select before R/W̄↑	2 V	275		400		350		
			4.5 V	55		80		70		
			6 V	47		68		60		
t_h	Hold time	Address after R/W̄↑	2 V	0		0		0		ns
			4.5 V	0		0		0		
			6 V	0		0		0		
		Data after R/W̄↑	2 V	0		0		0		
			4.5 V	0		0		0		
			6 V	0		0		0		
		Chip-select after R/W̄↑	2 V	0		0		0		
			4.5 V	0		0		0		
			6 V	0		0		0		

switching characteristics over recommended operating free-air temperature range (unless otherwise noted) C_L = 50 pF (see Note 1)

PARAMETER	FROM (FROM)	TO (OUTPUT)	V_{CC}	T_A = 25°C			SN54HC189 SN54HC219		SN74HC189 SN74HC219		UNIT
				MIN	TYP	MAX	MIN	MAX	MIN	MAX	
$t_{a(ad)}$	A	Any	2 V		81						ns
			4.5 V		27						
			6 V		23						
$t_{a(S)}$	$\overline{S}$	Any	2 V		81						ns
			4.5 V		27						
			6 V		23						
t_{en}	R/$\overline{W}$	Any	2 V		50						ns
			4.5 V		16						
			6 V		14						
t_{dis}	$\overline{S}$	Any	2 V		25						ns
			4.5 V		8						
			6 V		7						
	R/$\overline{W}$	Any	2 V		35						
			4.5 V		11						
			6 V		10						
t_t		Any	2 V		28						ns
			4.5 V		8						
			6 V		6						

C_{pd}	Power dissipation capacitance	No load, T_A = 25°C	55 pF typ

NOTE 1: For load circuit and voltage waveforms, see page 1-14.

TYPES SN54HCT189, SN54HCT219, SN74HCT189, SN74HCT219 64-BIT RANDOM-ACCESS MEMORIES WITH 3-STATE OUTPUTS

D2804, MARCH 1984

- **Inputs are TTL-Voltage Compatible**
- **Organized as 16 Words of Four Bits Each**
- **Choice of Noninverted or Inverted Outputs**
- **High-Current 3-State Inverting Outputs Can Drive up to 15 LSTTL Loads**
- **Package Options Include Both Plastic and Ceramic Chip Carriers in Addition to Plastic and Ceramic DIPs**
- **Dependable Texas Instruments Quality and Reliability**

SN54HCT189 . . . J PACKAGE
SN74HCT189 . . . J OR N PACKABE
(TOP VIEW)

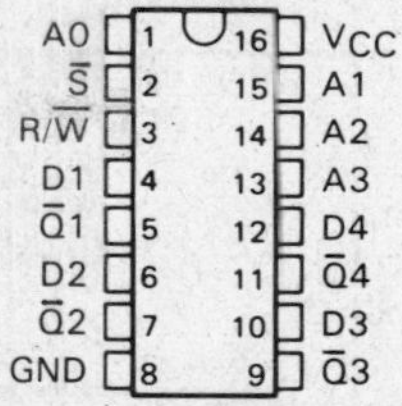

SN54HCT189 . . . FH OR FK PACKAGE
SN74HCT189 . . . FH OR FN PACKAGE
(TOP VIEW)

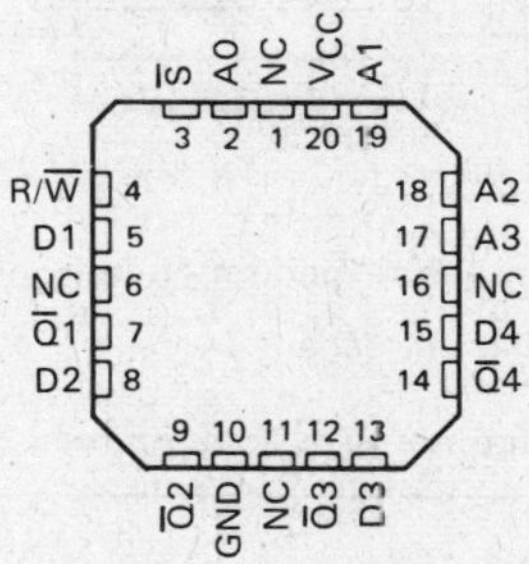

description

Information to be stored in the memory is written into the selected address location when the chip-select ($\overline{S}$) and the write-enable (R/$\overline{W}$) inputs are low. While the write-enable input is low, the memory outputs are off (Hi-Z). When a number of outputs are bus-connected, this off state neither loads nor drives the data bus; however, it permits the bus line to be driven by the other active outputs or a passive pull-up.

Information stored in the memory (see function table for input/output phase relationship) is available at the outputs when the write-enable input is high and the chip-select input is low. When the chip-select input is high, the outputs will be off.

The SN54HCT189 and SN54HCT219 are characterized for operation over the full military temperature range of −55°C to 125°C. The SN74HCT189 and SN74HCT219 are characterized for operation from −40°C to 85°C.

SN54HCT219 . . . J PACKAGE
SN74HCT219 . . . J OR N PACKAGE
(TOP VIEW)

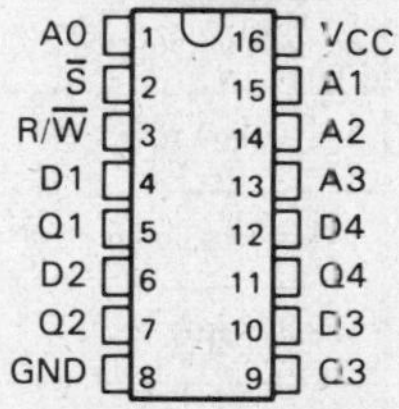

SN54HCT219 . . . FH OR FK PACKAGE
SN74HCT219 . . . FH OR FN PACKAGE
(TOP VIEW)

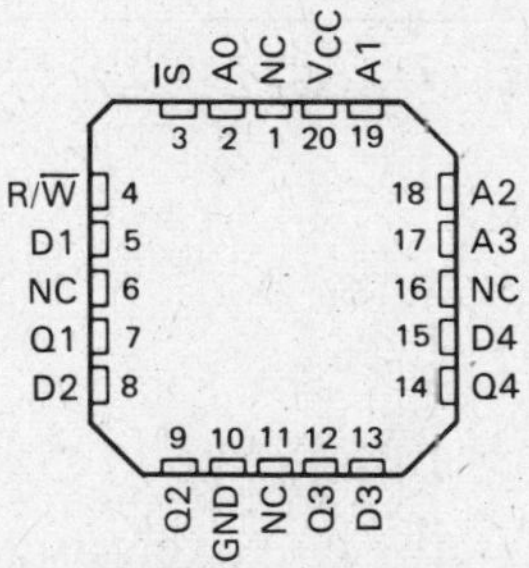

NC—No internal connection

FUNCTION TABLE

FUNCTION	INPUTS		OUTPUTS	
	CHIP SELECT	WRITE ENABLE	'HCT189	'HCT219
Write	L	L	Z	Z
Read	L	H	Complement of data entered	Data entered
Inhibit	H	X	Z	Z

TEXAS INSTRUMENTS
POST OFFICE BOX 225012 • DALLAS, TEXAS 75265

logic symbols

'HCT189

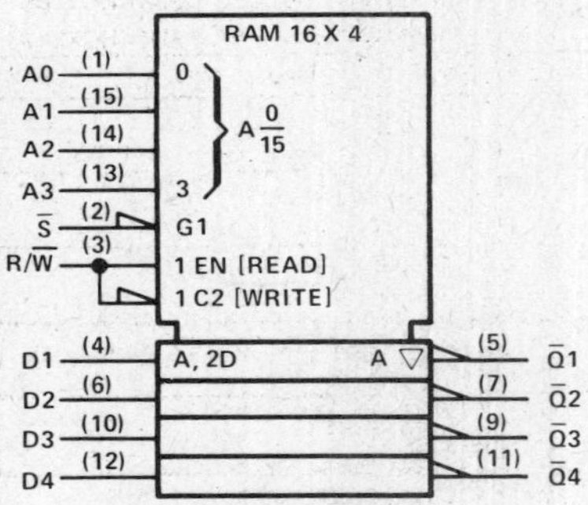

'HCT219

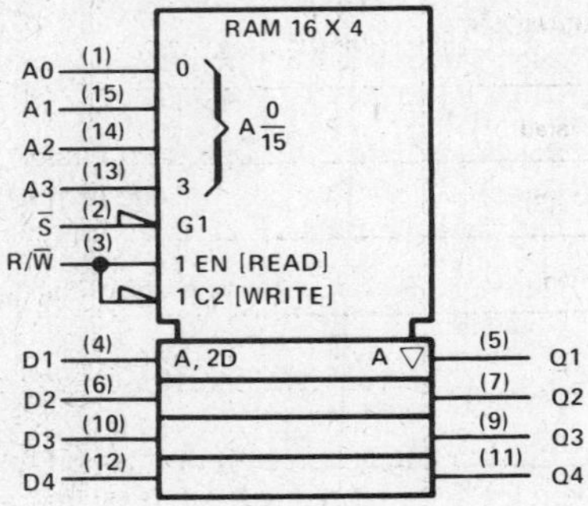

Pin numbers shown are for J and N packages.

maximum ratings, recommended operating conditions, and electrical characteristics

See Table VII, page 2-14.

timing requirements over recommended operating free-air temperature range (unless otherwise noted)

			V_{CC}	T_A = 25°C MIN	T_A = 25°C MAX	SN54HCT189 SN54HCT219 MIN	SN54HCT189 SN54HCT219 MAX	SN74HCT189 SN74HCT219 MIN	SN74HCT189 SN74HCT219 MAX	UNIT
t_w	Pulse duration, R/$\overline{W}$ low		4.5 V	55		80		70		ns
			5.5 V	50		75		63		
t_{su}	Setup time	Address before R/$\overline{W}$↓	4.5 V	0		0		0		ns
			5.5 V	0		0		0		
		Data before R/$\overline{W}$↑	4.5 V	55		80		70		ns
			5.5 V	50		75		63		
		Chip-select before R/$\overline{W}$↑	4.5 V	55		80		70		ns
			5.5 V	50		75		63		
t_h	Hold time	Address after R/$\overline{W}$↑	4.5 V	0		0		0		ns
			5.5 V	0		0		0		
		Data after R/$\overline{W}$↑	4.5 V	0		0		0		ns
			5.5 V	0		0		0		
		Chip-select after R/$\overline{W}$↑	4.5 V	0		0		0		ns
			5.5 V	0		0		0		

switching characteristics over recommended operating free-air temperature range (unless otherwise noted), C_L = 50 pF (see Note 1)

PARAMETER	FROM (INPUT)	TO (OUTPUT)	V_{CC}	T_A = 25°C MIN	TYP	MAX	SN54HCT189 SN54HCT219 MIN	MAX	SN74HCT189 SN74HCT219 MIN	MAX	UNIT
$t_{a(ad)}$	A	Any	4.5 V		27						ns
			5.5 V		23						
$t_{a(S)}$	$\overline{S}$	Any	4.5 V		27						ns
			5.5 V		23						
t_{en}	R/$\overline{W}$	Any	4.5 V		16						ns
			5.5 V		14						
t_{dis}	$\overline{S}$	Any	4.5 V		8						ns
			5.5 V		7						
	R/$\overline{W}$	Any	4.5 V		11						
			5.5 V		10						
t_t		Any	4.5 V		12						ns
			5.5 V		11						

C_{pd}	Power dissipation capacitance	No load, T_A = 25°C	55 pF typ

NOTE 1: For load circuit and voltage waveforms, see page 1-14.

4

ADVANCE INFORMATION

TYPES SN54HCT245, SN74HCT245
OCTAL BUS TRANSCEIVERS WITH 3-STATE OUTPUTS

D2804, MARCH 1984

- Inputs Are TTL-Voltage Compatible
- High-Current 3-State Outputs Drive Bus Lines Directly or Up to 15 LSTTL Loads
- Package Options Include Both Plastic and Ceramic Chip Carriers in Addition to Plastic and Ceramic DIPs
- Dependable Texas Instruments Quality and Reliability

SN54HCT245 . . . J PACKAGE
SN74HCT245 . . . J OR N PACKAGE
(TOP VIEW)

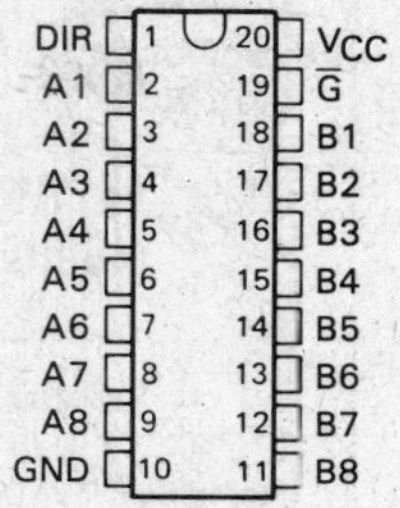

SN54HCT245 . . . FH OR FK PACKAGE
SN74HCT245 . . . FH OR FN PACKAGE
(TOP VIEW)

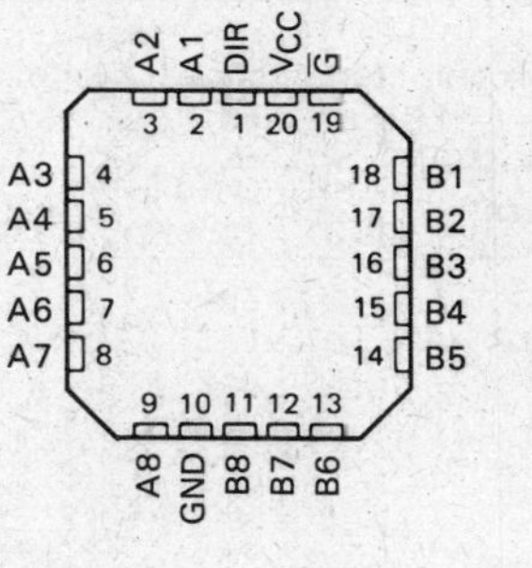

description

These octal bus transceivers are designed for asynchronous two-way communication between data buses. The control function implementation minimizes external timing requirements.

The devices allow data transmission from the A bus to the B bus or from the B bus to the A bus depending upon the logic level at the direction control (DIR) input. The enable input ($\overline{G}$) can be used to disable the device so that the buses are effectively isolated.

The SN54HCT245 is characterized for operation over the full military temperature range of −55°C to 125°C. The SN74HCT245 is characterized for operation from −40°C to 85°C.

FUNCTION TABLE

CONTROL INPUTS		OPERATION
$\overline{G}$	DIR	
L	L	B data to A bus
L	H	A data to B bus
H	X	Isolation

logic symbol

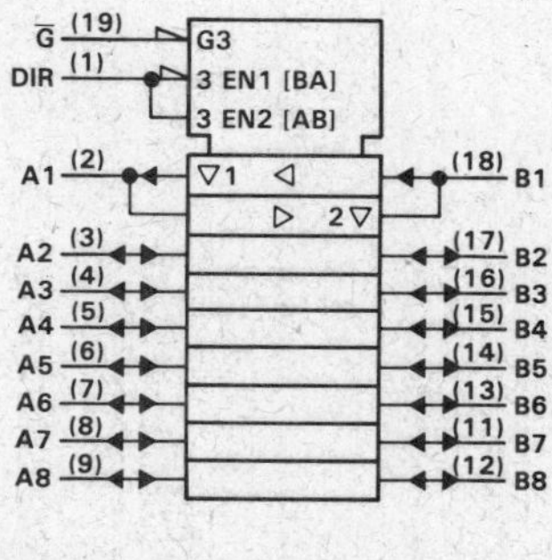

logic diagram

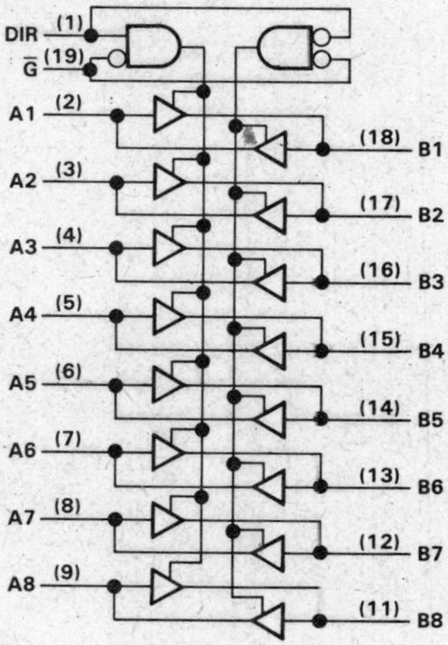

ADVANCE INFORMATION
This document contains information on a new product. Specifications are subject to change without notice.

maximum ratings, recommended operating conditions, and electrical characteristics

See Table VII, page 2-14.

switching characteristics over recommended operating free-air temperature range (unless otherwise noted), C_L = 50 pF (see Note 1)

PARAMETER	FROM (INPUT)	TO (OUTPUT)	V_{CC}	T_A = 25°C MIN	TYP	MAX	SN54HCT245 MIN	MAX	SN74HCT245 MIN	MAX	UNIT
t_{pd}	A or B	B or A	4.5 V		16	22		33		28	ns
			5.5 V		14	20		30		25	
t_{en}	$\overline{G}$	A or B	4.5 V		25	46		69		58	ns
			5.5 V		22	41		62		52	
t_{dis}	$\overline{G}$	A or B	4.5 V		26	40		60		50	ns
			5.5 V		23	36		54		45	
t_t		A or B	4.5 V		9	12		18		15	ns
			5.5 V		8	11		16		14	

C_{pd}	Power dissipation capacitance per transceiver	No load, T_A = 25°C	40 pF typ

switching characteristics over recommended operating free-air temperature range (unless otherwise noted), C_L = 150 pF (see Note 1)

PARAMETER	FROM (INPUT)	TO (OUTPUT)	V_{CC}	T_A = 25°C MIN	TYP	MAX	SN54HCT245 MIN	MAX	SN74HCT245 MIN	MAX	UNIT
t_{pd}	A or B	B or A	4.5 V		20	30		45		38	ns
			5.5 V		18	27		41		34	
t_{en}	$\overline{G}$	A or B	4.5 V		36	59		89		74	ns
			5.5 V		30	63		80		67	
t_t		A or B	4.5 V		17	42		63		53	ns
			5.5 V		14	38		57		48	

NOTE 1: For load circuit and voltage waveforms, see page 1-14.

TYPES SN54HC283, SN74HC283 4-BIT BINARY FULL ADDERS WITH FAST CARRY

D2804, MARCH 1984

- **Full-Carry Look-Ahead Across the Four Bits**
- **Systems Achieve Partial Look-Ahead Performance with the Economy of Ripple Carry**
- **Supply Voltage and Ground on Corner Pins to Simplify P-C Board Layout**
- **Package Options Include Both Plastic and Ceramic Chip Carriers in Addition to Plastic and Ceramic DIPs**
- **Dependable Texas Instruments Quality and Reliability**

SN54HC283 . . . J PACKAGE
SN74HC283 . . . J OR N PACKAGE
(TOP VIEW)

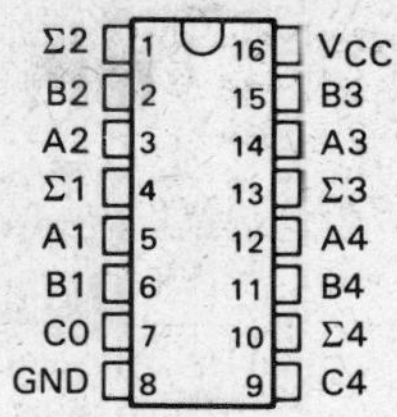

SN54HC283 . . . FH OR FK PACKAGE
SN74HC283 . . . FH OR FN PACKAGE
(TOP VIEW)

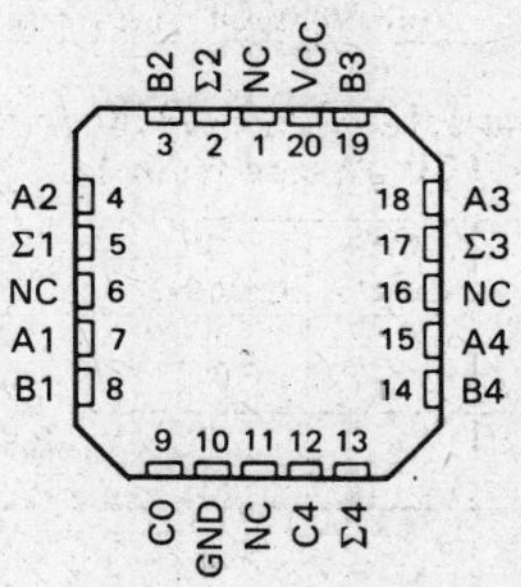

NC—No internal connection

description

These improved full adders perform the addition of two 4-bit binary words. The sum (Σ) outputs are provided for each bit and the resultant carry (C4) is obtained from the fourth bit.

These adders feature full internal look-ahead across all four bits generating the carry term. This capability provides the system designer with partial look-ahead performance at the economy and reduced package count of a ripple-carry implementation.

The adder logic, including the carry, is implemented in its true form. End around carry can be accomplished without the need for logic or level inversion.

The SN54HC283 is characterized for operation over the full military temperature range of −55°C to 125°C. The SN74HC283 is characterized for operation from −40°C to 85°C.

TEXAS INSTRUMENTS
POST OFFICE BOX 225012 • DALLAS, TEXAS 75265

FUNCTION TABLE

INPUT				OUTPUT					
				WHEN C0 = L / WHEN C2 = L			WHEN C0 = H / WHEN C2 = H		
A1 / A3	B1 / B3	A2 / A4	B2 / B4	Σ1 / Σ3	Σ2 / Σ4	C2 / C4	Σ1 / Σ3	Σ2 / Σ4	C2 / C4
L	L	L	L	L	L	L	H	L	L
H	L	L	L	H	L	L	L	H	L
L	H	L	L	H	L	L	L	H	L
H	H	L	L	L	H	L	H	H	L
L	L	H	L	L	H	L	H	H	L
H	L	H	L	H	H	L	L	L	H
L	H	H	L	H	H	L	L	L	H
H	H	H	L	L	L	H	H	L	H
L	L	L	H	L	H	L	H	H	L
H	L	L	H	H	H	L	L	L	H
L	H	L	H	H	H	L	L	L	H
H	H	L	H	L	L	H	H	L	H
L	L	H	H	L	L	H	H	L	H
H	L	H	H	H	L	H	L	H	H
L	H	H	H	H	L	H	L	H	H
H	H	H	H	L	H	H	H	H	H

H = high level, L = low level

NOTE: Input conditions at A1, B1, A2, B2, and C0 are used to determine outputs Σ1 and Σ2 and the value of the internal carry C2. The values at C2, A3, B3, A4, and B4 are then used to determine outputs Σ3, Σ4, and C4.

logic symbol

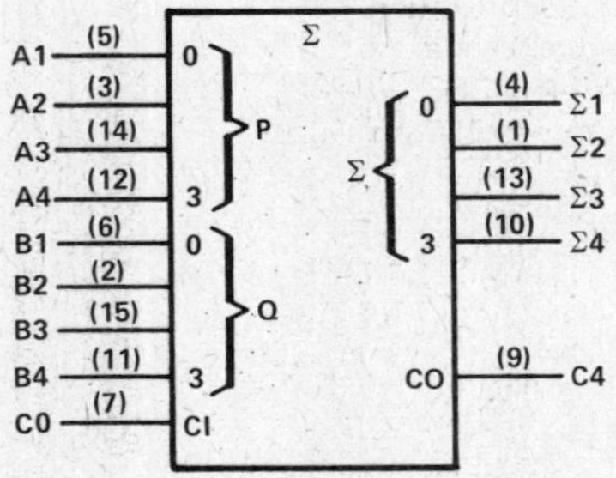

Pin numbers shown are for J and N packages.

logic diagram (positive logic)

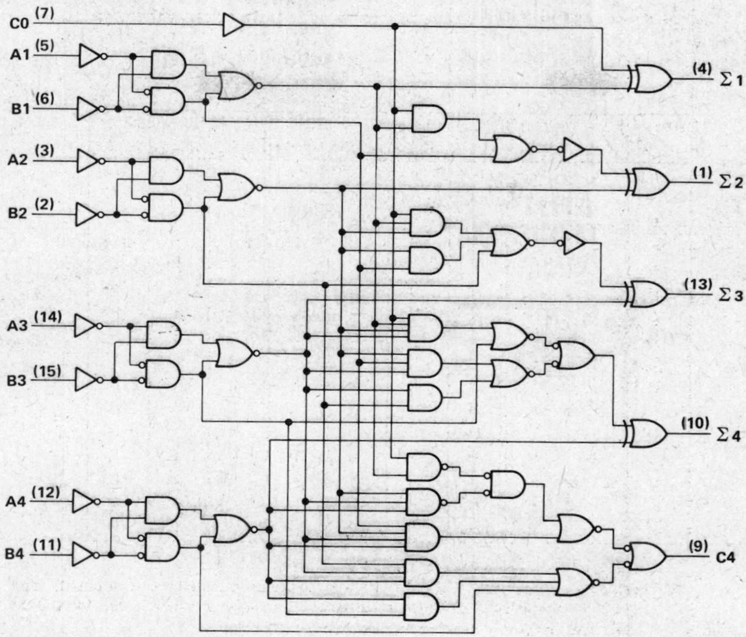

Pin numbers shown are for J and N packages.

maximum ratings, recommended operating conditions, and electrical characteristics

See Table IV, page 2-10.

switching characteristics over recommended operating free-air temperature range (unless otherwise noted), C_L = 50 pF (see Note 1)

PARAMETER	FROM (INPUT)	TO (OUTPUT)	V_{CC}	T_A = 25°C MIN	T_A = 25°C TYP	T_A = 25°C MAX	SN54HC283 MIN	SN54HC283 MAX	SN74HC283 MIN	SN74HC283 MAX	UNIT
t_{pd}	C0	Any Σ	2 V		70	150		225		190	ns
			4.5 V		23	30		45		38	
			6 V		20	26		38		32	
t_{pd}	A1 or B1	Σ1	2 V		70	150		225		190	ns
			4.5 V		23	30		45		38	
			6 V		20	26		38		32	
t_{pd}	C0	C4	2 V		55	125		190		155	ns
			4.5 V		17	25		38		31	
			6 V		14	21		32		26	
t_{pd}	A1 or B1	C4	2 V		55	130		195		165	ns
			4.5 V		17	26		39		33	
			6 V		14	22		33		28	
t_t		Any	2 V		28	75		110		95	ns
			4.5 V		8	15		22		19	
			6 V		6	13		19		16	

C_{pd}	Power dissipation capacitance	No load, T_A = 25°C	75 pF typ

NOTE 1: For load circuit and voltage waveforms, see page 1-14.

4

ADVANCE INFORMATION

TEXAS INSTRUMENTS
POST OFFICE BOX 225012 • DALLAS, TEXAS 75265

TYPES SN54HC299, SN74HC299
8-BIT UNIVERSAL SHIFT/STORAGE REGISTERS WITH DIRECT CLEAR AND 3-STATE OUTPUTS

D2684, DECEMBER 1982—REVISED MARCH 1984

- **Multiplexed I/O Ports Provide Improved Bit Density**
- **Four Modes of Operation: Hold (Store), Shift Right, Shift Left, and Load Data**
- **High Current 3-State Outputs Drive Bus Lines Directly or up to 15 LSTTL Loads**
- **Can Be Cascaded for N-Bit Word Lengths**
- **Package Options Include Both Plastic and Ceramic Chip Carriers in Addition to Plastic and Ceramic DIPs**
- **Dependable Texas Instruments Quality and Reliability**

SN54HC299 . . . J PACKAGE
SN74HC299 . . . J OR N PACKAGE
(TOP VIEW)

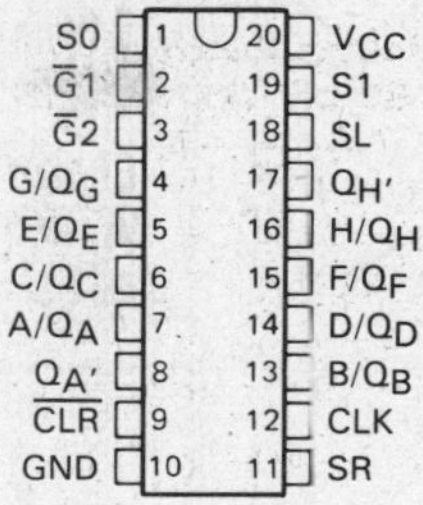

SN54HC299 . . . FH OR FK PACKAGE
SN74HC299 . . . FH OR FN PACKAGE
(TOP VIEW)

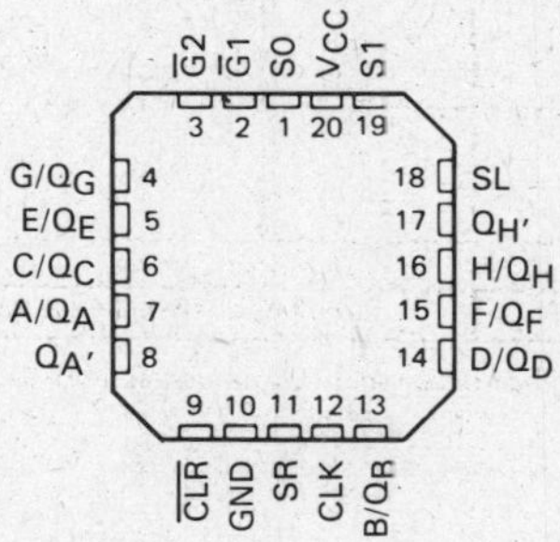

description

These eight-bit universal registers feature multiplexed I/O ports to achieve full eight-bit handling in a single 20-pin package. 'HC299 applications are as stacked or push-down registers, buffer storage, and accumulator registers.

Two function-select inputs and two output control inputs can be used to choose the modes of operation listed in the function table.

Synchronous parallel loading is accomplished by taking both function-select lines, S0 and S1, high. This places the three-state outputs in a high-impedance state. Which permits data that is applied on the I/O ports to be clocked into the register. Reading out of this register can be accomplished while the outputs enabled in any mode. A direct overriding input is provided to clear the register whether the outputs are enabled or off. Taking either of the output controls, $\overline{G}$1 or $\overline{G}$2, high disables the outputs but this has no effect on shifting or storage of data.

The SN54HC299 is characterized for operation over the full military temperature range of −55°C to 125°C. The SN74HC299 is characterized for operation from −40°C to 85°C.

logic symbol

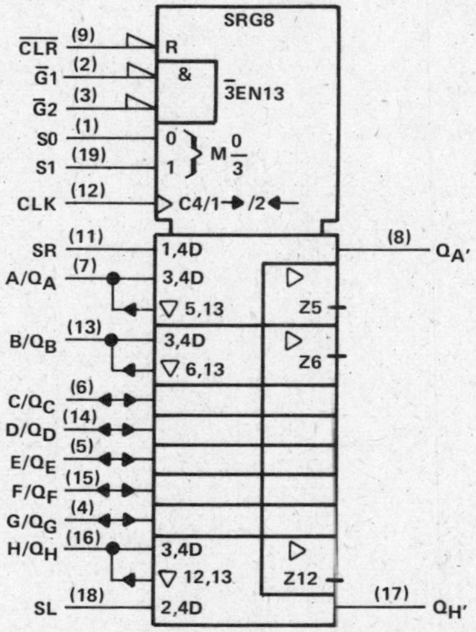

logic diagram (positive logic)

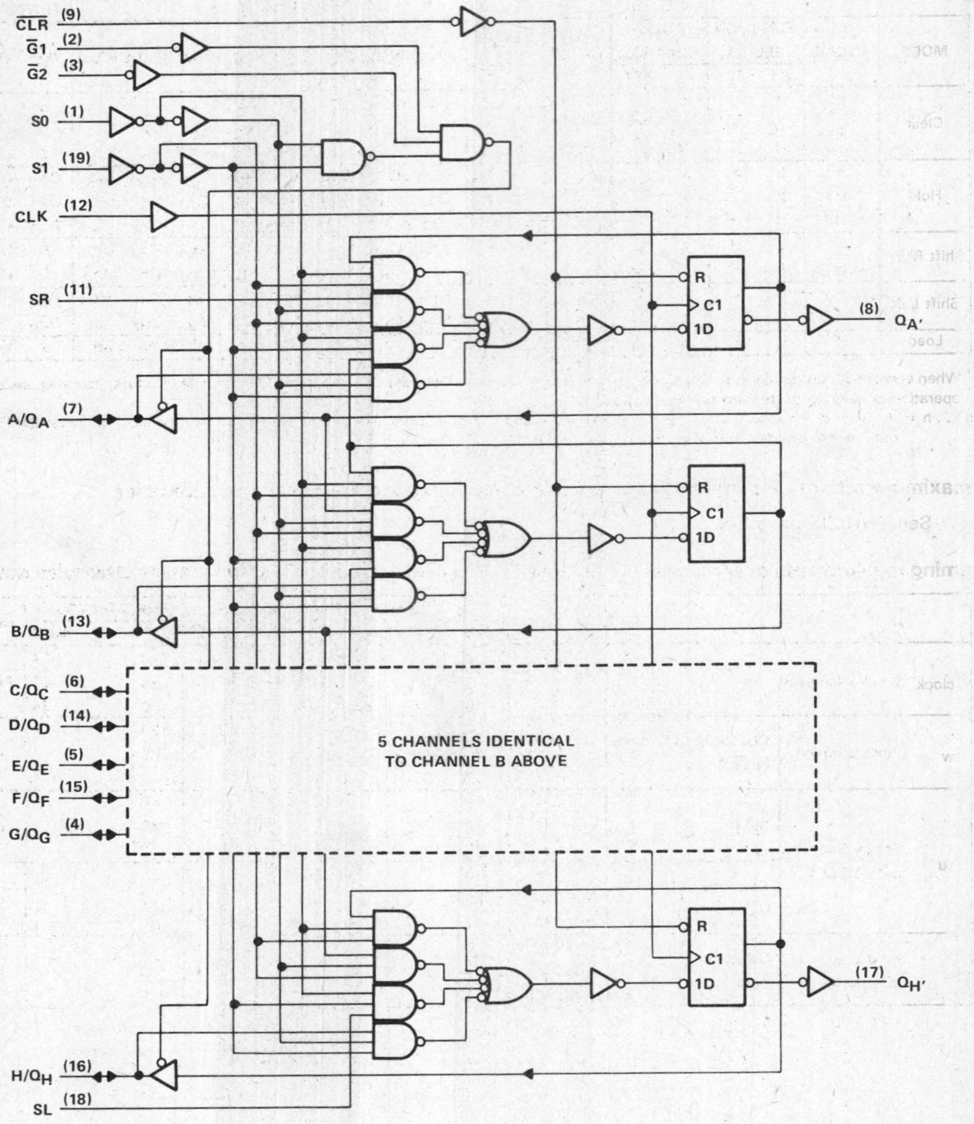

4

ADVANCE INFORMATION

FUNCTION TABLE

MODE	INPUTS								INPUTS/OUTPUTS								OUTPUTS	
	CLEAR	FUNCTION SELECT S1	FUNCTION SELECT S0	OUTPUT CONTROL $\overline{G1}$†	OUTPUT CONTROL $\overline{G2}$†	CLOCK	SERIAL SL	SERIAL SR	A/Q_A	B/Q_B	C/Q_C	D/Q_D	E/Q_E	F/Q_F	G/Q_G	H/Q_H	$Q_{A'}$	$Q_{H'}$
Clear	L	X	L	L	L	X	X	X	L	L	L	L	L	L	L	L	L	L
	L	L	X	L	L	X	X	X	L	L	L	L	L	L	L	L	L	L
	L	H	H	X	X	X	X	X	X	X	X	X	X	X	X	X	L	L
Hold	H	L	X	L	L	L	X	X	Q_{A0}	Q_{B0}	Q_{C0}	Q_{D0}	Q_{E0}	Q_{F0}	Q_{G0}	Q_{H0}	Q_{A0}	Q_{H0}
	H	X	L	L	L	L	X	X	Q_{A0}	Q_{B0}	Q_{C0}	Q_{D0}	Q_{E0}	Q_{F0}	Q_{G0}	Q_{H0}	Q_{A0}	Q_{H0}
	H	H	H	L	L	L	X	X	X	X	X	X	X	X	X	X	Q_{A0}	Q_{H0}
Shift Right	H	L	H	L	L	↑	X	H	H	Q_{An}	Q_{Bn}	Q_{Cn}	Q_{Dn}	Q_{En}	Q_{Fn}	Q_{Gn}	H	Q_{Gn}
	H	L	H	L	L	↑	X	L	L	Q_{An}	Q_{Bn}	Q_{Cn}	Q_{Dn}	Q_{En}	Q_{Fn}	Q_{Gn}	L	Q_{Gn}
Shift Left	H	H	L	L	L	↑	H	X	Q_{Bn}	Q_{Cn}	Q_{Dn}	Q_{En}	Q_{Fn}	Q_{Gn}	Q_{Hn}	H	Q_{Bn}	H
	H	H	L	L	L	↑	L	X	Q_{Bn}	Q_{Cn}	Q_{Dn}	Q_{En}	Q_{Fn}	Q_{Gn}	Q_{Hn}	L	Q_{Bn}	L
Load	H	H	H	X	X	↑	X	X	a	b	c	d	e	f	g	h	a	h

† When one or both output controls are high the eight input/output terminals are disabled to the high-impedance state; however, sequential operation or clearing of the register is not affected.

a . . . h = the level of the steady-state input at inputs A through H, respectively. These data are loaded into the flip-flops while the flip-flop outputs are isolated from the input/output terminals.

maximum ratings, recommended operating conditions, and electrical characteristics

See Table III, page 2-8.

timing requirements over recommended operating free-air temperature range (unless otherwise noted)

			V_{CC}	$T_A = 25°C$ MIN	$T_A = 25°C$ MAX	SN54HC299 MIN	SN54HC299 MAX	SN74HC299 MIN	SN74HC299 MAX	UNIT
f_{clock}	Clock frequency		2 V	0	5	0	3.3	0	4	MHz
			4.5 V	0	25	0	17	0	20	
			6 V	0	29	0	19	0	24	
t_w	Pulse duration	CLK high, CLK low, or $\overline{CLR}$ low	2 V	100		150		125		ns
			4.5 V	20		30		25		
			6 V	17		26		21		
t_{su}	Setup time before CLK↑	Select	2 V	150		225		190		ns
			4.5 V	30		45		38		
			6 V	25		38		32		
		Data or $\overline{CLR}$ inactive	2 V	100		150		125		ns
			4.5 V	20		30		25		
			6 V	17		26		21		
t_h	Hold time after CLK↑	Select or data	2 V	0		0		0		ns
			4.5 V	0		0		0		
			6 V	0		0		0		

switching characteristics over recommended operating free-air temperature range (unless otherwise noted), C_L = 50 pF (see Note 1)

PARAMETER	FROM (INPUT)	TO (OUTPUT)	V_{CC}	T_A = 25°C MIN	TYP	MAX	SN54HC299 MIN	MAX	SN74HC299 MIN	MAX	UNIT
f_{max}			2 V	5			3.3		4		MHz
			4.5 V	25			17		20		
			6 V	29			19		24		
t_{pd}	CLK	$Q_{A'}$ or $Q_{H'}$	2 V		35						ns
			4.5 V		11						
			6 V		9						
		Q_A thru Q_H	2 V		42						ns
			4.5 V		14						
			6 V		12						
t_{en}	$\overline{G}1$ or $\overline{G}2$	Q_A thru Q_H	2 V		50						ns
			4.5 V		15						
			6 V		12						
	S0 or S1		2 V		50						ns
			4.5 V		15						
			6 V		12						
t_{dis}	$\overline{G}1$ or $\overline{G}2$	Q_A thru Q_H	2 V		60						ns
			4.5 V		20						
			6 V		17						
	S0 or S1		2 V		60						ns
			4.5 V		20						
			6 V		17						
t_{PHL}	$\overline{CLR}$	$Q_{A'}$ or $Q_{H'}$	2 V		40						ns
			4.5 V		13						
			6 V		11						
		Q_A thru Q_H	2 V		55						ns
			4.5 V		16						
			6 V		14						
t_t		Y	2 V		38						ns
			4.5 V		8						
			6 V		6						

C_{pd}	Power dissipation capacitance	No load, T_A = 25°C	100 pF typ

NOTE 1: For load circuit and voltage waveforms, see page 1-14.

4 ADVANCE INFORMATION

switching characteristics over recommended operating free-air temperature range (unless otherwise noted), C_L = 150 pF (see Note 1)

PARAMETER	FROM (INPUT)	TO (OUTPUT)	V_{CC}	T_A = 25°C MIN	TYP	MAX	SN54HC299 MIN	MAX	SN74HC299 MIN	MAX	UNIT
t_{pd}	CLK	$Q_{A'}$ or $Q_{H'}$	2 V		50						ns
			4.5 V		17						
			6 V		12						
		Q_A thru Q_H	2 V		55						ns
			4.5 V		20						
			6 V		18						
t_{en}	$\overline{G}1$ or $\overline{G}2$	Q_A thru Q_H	2 V		90						ns
			4.5 V		30						
			6 V		24						
	S0 or S1		2 V		90						ns
			4.5 V		30						
			6 V		24						
t_{dis}	$\overline{G}1$ or $\overline{G}2$	Q_A thru Q_H	2 V		90						ns
			4.5 V		30						
			6 V		24						
	S0 or S1		2 V		90						ns
			4.5 V		30						
			6 V		24						
t_{PHL}	$\overline{CLR}$	$Q_{A'}$ or $Q_{H'}$	2 V		60						ns
			4.5 V		26						
			6 V		22						
		Q_A thru Q_H	2 V		70						ns
			4.5 V		28						
			6 V		24						

NOTE 1: For load circuit and voltage waveforms, see page 1-14.

4

ADVANCE INFORMATION

4

ADVANCE INFORMATION

TYPES SN54HC322, SN74HC322 8-BIT SHIFT REGISTERS WITH SIGN EXTEND AND 3-STATE OUTPUTS

D2804, MARCH 1984

- **Multiplexed Inputs/Outputs Provide Improved Bit Density**
- **High-Current 3-State Outputs Can Drive Up to 15 LSTTL Loads**
- **Sign Extend Function**
- **Direct Overriding Clear**
- **Package Options Include Both Plastic and Ceramic Chip Carriers in Addition to Plastic and Ceramic DIPs**
- **Dependable Texas Instruments Quality and Reliability**

SN54HC322 . . . J PACKAGE
SN74HC322 . . . J OR N PACKAGE
(TOP VIEW)

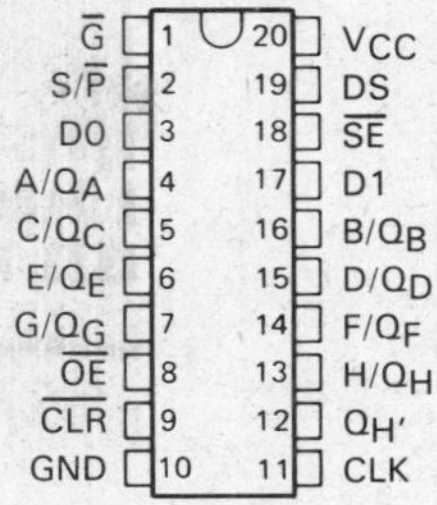

SN54HC322 . . . FH OR FK PACKAGE
SN74HC322 . . . FH OR FN PACKAGE
(TOP VIEW)

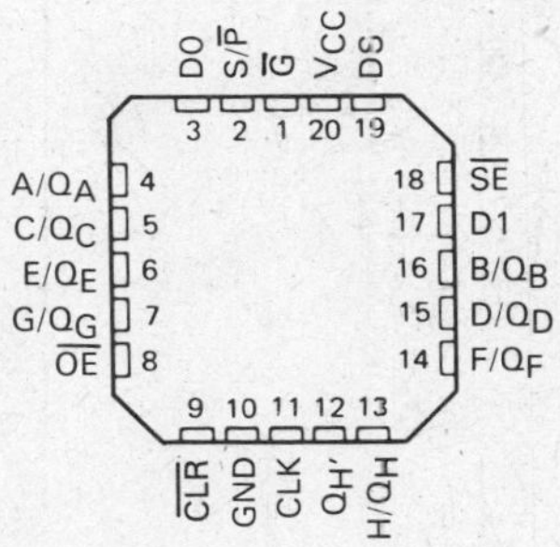

NC – No internal connection

description

These eight-bit registers feature multiplexed input/output data ports to achieve full eight-bit handling in a single 20-pin package. Serial data may be entered into the shift-register through either the D0 or the D1 input as selected by the data select input DS. A serial output $Q_{H'}$ is also provided to facilitate expansion. Synchronous parallel loading is accomplished by taking both the register enable $\overline{G}$ and the $S/\overline{P}$ input low. This places the three-state input/output ports in the data input mode. Data are entered on the low-to-high transition of the clock. The sign extend function repeats the sign in the Q_A flip-flop during shifting if the sign extend input $\overline{SE}$ is low. A direct overriding clear input clears the internal registers when taken low whether the outputs are enabled or off. The output enable does not interfere with synchronous operation of the register.

The SN54HC322 is characterized for operation over the full military range of −55°C to 125°C. The SN74HC322 is characterized for operation from −40°C to 85°C.

FUNCTION TABLE

OPERATION	INPUTS							INPUTS/OUTPUTS				OUTPUT
	$\overline{CLR}$	$\overline{G}$	$S/\overline{P}$	$\overline{SE}$	DS	$\overline{OE}$	CLK	A/Q_A	B/Q_B	C/Q_C . . .	H/Q_H	$Q_{H'}$
Clear	L	H	X	X	X	L	X	L	L	L	L	L
	L	X	H	X	X	L	X	L	L	L	L	L
Hold	H	H	X	X	X	L	X	Q_{A0}	Q_{B0}	Q_{C0}	Q_{H0}	Q_{H0}
Shift Right	H	L	H	H	L	L	↑	D0	Q_{An}	Q_{Bn}	Q_{Gn}	Q_{Gn}
	H	L	H	H	H	L	↑	D1	Q_{An}	Q_{Bn}	Q_{Gn}	Q_{Gn}
Sign Extend	H	L	H	L	X	L	↑	Q_{An}	Q_{An}	Q_{Bn}	Q_{Gn}	Q_{Gn}
Load	H	L	L	X	X	X	↑	a	b	c	h	h

When the output enable is high, the eight input/output terminals are disabled to the high-impedance state; however, sequential operation or clearing of the register is not affected. If both the register enable input and the $S/\overline{P}$ are low while the clear input is low, the register is cleared while the eight input/output terminals are disabled to the high-impedance state.

H = high level (steady state)
L = low level (steady state)
X = irrelevant (any input, including transitions)
↑ = transition from low to high level
Q_{A0} . . . Q_{H0} = the level of Q_A through Q_H, respectively, before the indicated steady-state conditions were established
Q_{An} . . . Q_{Hn} = the level of Q_A through Q_H, respectively, before the most recent ↑ transition of the clock.
D0, D1 = the level of steady-state inputs D0 and D1 respectively
a . . . h = the level of steady-state inputs at inputs A through H respectively

logic symbol

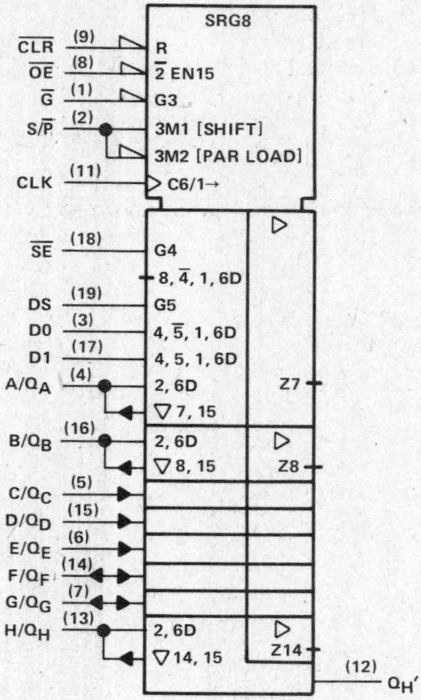

TEXAS INSTRUMENTS
POST OFFICE BOX 225012 • DALLAS, TEXAS 75265

logic diagram (positive logic)

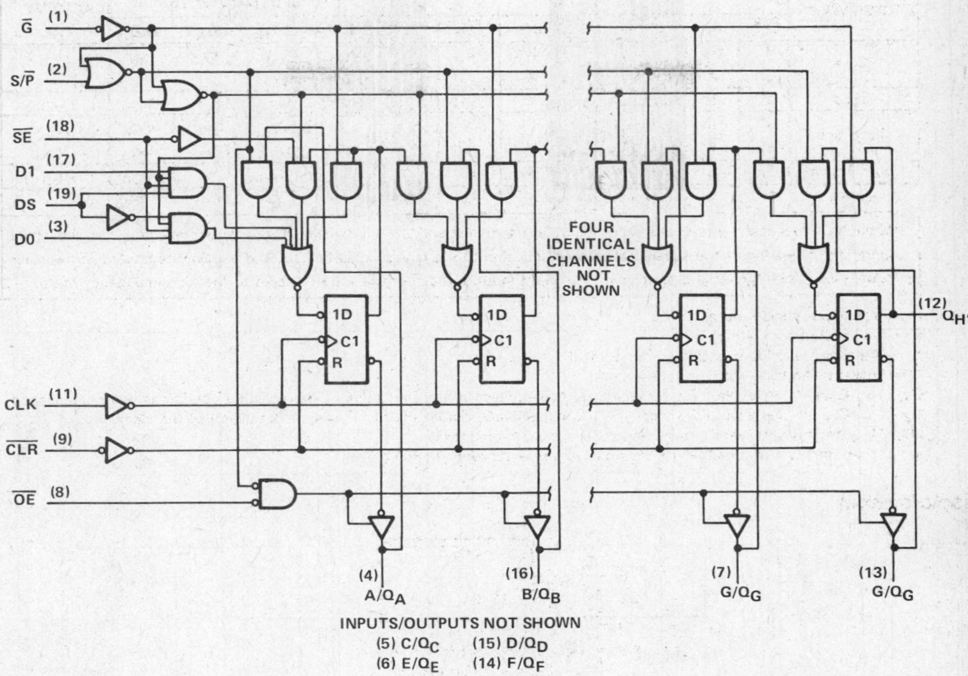

maximum ratings, recommended operating conditions, and electrical characteristics

See Table III, page 2-8.

timing requirements over recommended operating free-air temperature range (unless otherwise noted)

			V_{CC}	T_A = 25°C MIN	T_A = 25°C MAX	SN54HC322 MIN	SN54HC322 MAX	SN74HC322 MIN	SN74HC322 MAX	UNIT
f_{clock}	Clock frequency		2 V	0	3.3	0	2.2	0	2.7	MHz
			4.5 V	0	17	0	11	0	13	
			6 V	0	20	0	13	0	16	
t_w	Pulse duration	CLK high	2 V	200		300		250		ns
			4.5 V	40		60		50		
			6 V	34		51		43		
		CLK low	2 V	100		150		125		
			4.5 V	20		30		25		
			6 V	17		26		21		
		$\overline{CLR}$ low	2 V	150		225		190		
			4.5 V	30		45		38		
			6 V	26		38		32		
t_{su}	Setup time before CLK↑	DS	2 V	100		150		125		ns
			4.5 V	20		30		20		
			6 V	17		26		21		
		Data inputs	2 V	150		225		190		
			4.5 V	30		45		38		
			6 V	26		38		32		
		$\overline{CLR}$ inactive state	2 V	150		225		190		
			4.5 V	30		45		38		
			6 V	26		38		32		
t_h	Hold time after CLK↑	DS	2 V	75		115		95		ns
			4.5 V	15		23		19		
			6 V	13		20		16		
		Data inputs	2 V	0		0		0		
			4.5 V	0		0		0		
			6 V	0		0		0		

switching characteristics over recommended operating free-air temperature range (unless otherwise noted), C_L = 50 pF (see Note 1)

PARAMETER	FROM (INPUT)	TO (OUTPUT)	V_{CC}	T_A = 25°C MIN	TYP	MAX	SN54HC322 MIN	MAX	SN74HC322 MIN	MAX	UNIT
f_{max}			2 V	3.3	6		2.2		2.7		MHz
			4.5 V	17	35		11		13		
			6 V	20	40		13		16		
t_{pd}	Clock	$Q_{H'}$	2 V		95						ns
			4.5 V		31						
			6 V		26						
t_{PHL}	Clear	$Q_{H'}$	2 V		95						ns
			4.5 V		32						
			6 V		27						
t_{pd}	Clock	Q_A thru Q_H	2 V		80						ns
			4.5 V		26						
			6 V		22						
t_{PHL}	Clear	Q_A thru Q_H	2 V		80						ns
			4.5 V		26						
			6 V		22						
t_{en}	Output enable	Q_A thru Q_H	2 V		55						ns
			4.5 V		18						
			6 V		15						
t_{dis}	Output enable	Q_A thru Q_H	2 V		55						ns
			4.5 V		18						
			6 V		15						
t_t			2 V		38						ns
			4.5 V		8						
			6 V		6						

NOTE 1: For load circuit and voltage waveforms, see page 1-14.

TYPES SN54HC323, SN74HC323
8-BIT UNIVERSAL SHIFT/STORAGE REGISTERS
WITH SYNCHRONOUS CLEAR AND 3-STATE OUTPUTS

D2684, DECEMBER 1982—REVISED MARCH 1984

- **Multiplexed I/O Ports Provide Improved Bit Density**
- **Four Modes of Operation: Hold (Store), Shift Right, Shift Left, and Load Data**
- **High Current 3-State Outputs Drive Bus Lines Directly or up to 15 LSTTL Loads**
- **Can Be Cascaded for N-Bit Word Lengths**
- **Package Options Include Both Plastic and Ceramic Chip Carriers in Addition to Plastic and Ceramic DIPs**
- **Dependable Texas Instruments Quality and Reliability**

SN54HC323 . . . J PACKAGE
SN74HC323 . . . J OR N PACKAGE
(TOP VIEW)

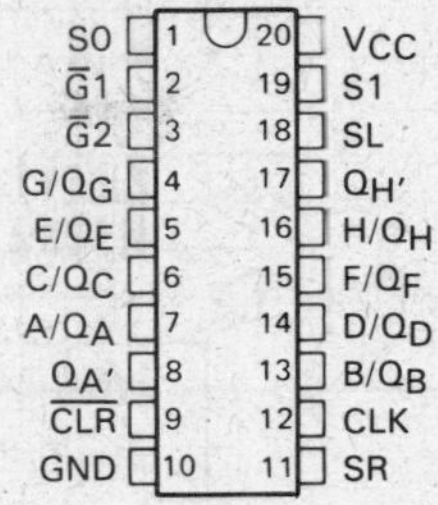

SN54HC323 . . . FH OR FK PACKAGE
SN74HC323 . . . FH OR FN PACKAGE
(TOP VIEW)

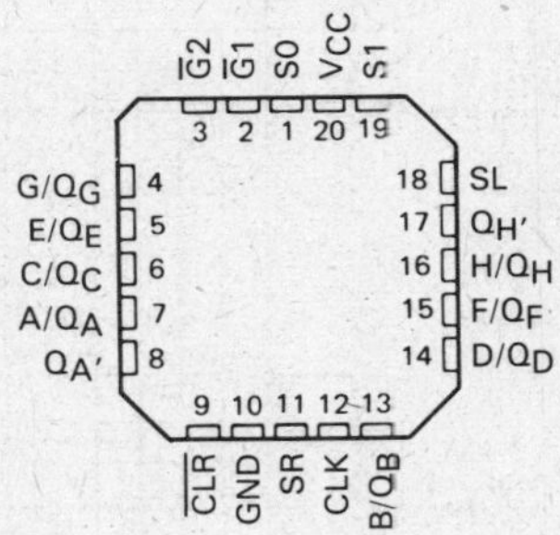

description

These eight-bit universal registers feature multiplexed I/O ports to achieve full eight-bit handling in a single 20-pin package. 'HC323 applications are as stacked or push-down registers, buffer storage, and accumulator registers.

Two function-select inputs and two output control inputs can be used to choose the modes of operation listed in the function table.

Synchronous parallel loading is accomplished by taking both function-select lines, S0 and S1, high. This places the three-state outputs in a high-impedance state, which permits data that is applied on the I/O ports to be clocked into the register. Reading out of this register can be accomplished while the outputs are enabled in any mode. The clear function is synchronous, and a low level at the clear input clears the register on the next low-to-high transition of the clock.

The SN54HC323 is characterized for operation over the full military temperature range of −55°C to 125°C. The SN74HC323 is characterized for operation from −40°C to 85°C.

logic symbol

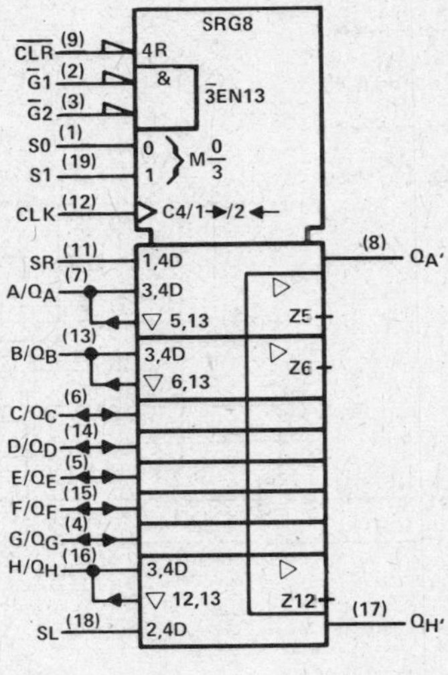

4

ADVANCE INFORMATION

ADVANCE INFORMATION
This document contains information on a new product. Specifications are subject to change without notice.

TEXAS INSTRUMENTS
POST OFFICE BOX 225012 • DALLAS, TEXAS 75265

logic diagram (positive logic)

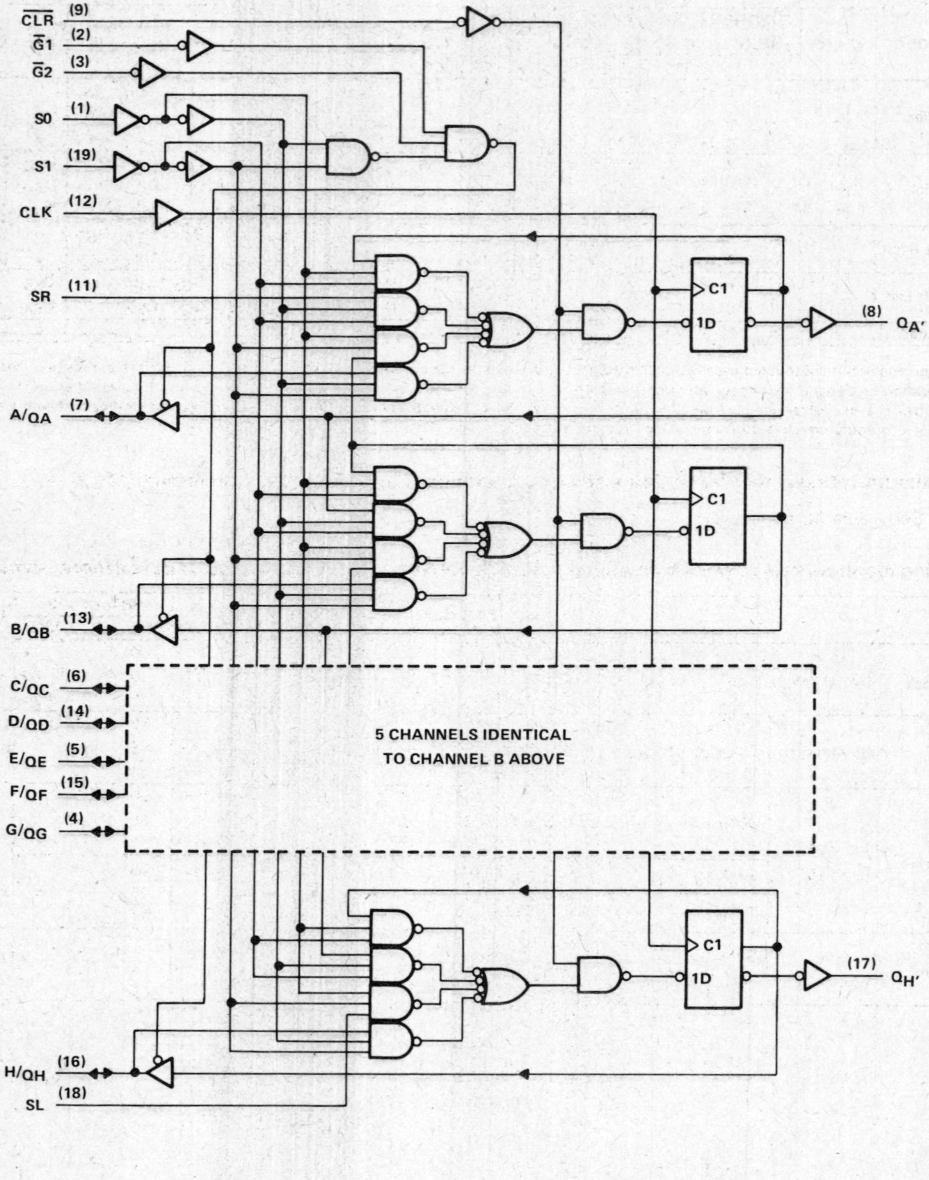

FUNCTION TABLE

MODE	INPUTS								INPUTS/OUTPUTS								OUTPUTS	
	CLEAR	FUNCTION SELECT		OUTPUT CONTROL		CLOCK	SERIAL		A/Q_A	B/Q_B	C/Q_C	D/Q_D	E/Q_E	F/Q_F	G/Q_G	H/Q_H	$Q_{A'}$	$Q_{H'}$
		S1	S0	$\overline{G}1$†	$\overline{G}2$†		SL	SR										
Clear	L	X	L	L	L	↑	X	X	L	L	L	L	L	L	L	L	L	L
	L	L	X	L	L	↑	X	X	L	L	L	L	L	L	L	L	L	L
	L	H	H	X	X	↑	X	X	X	X	X	X	X	X	X	X	L	L
Hold	H	L	X	L	L	L	X	X	Q_{A0}	Q_{B0}	Q_{C0}	Q_{D0}	Q_{E0}	Q_{F0}	Q_{G0}	Q_{H0}	Q_{A0}	Q_{H0}
	H	X	L	L	L	L	X	X	Q_{A0}	Q_{B0}	Q_{C0}	Q_{D0}	Q_{E0}	Q_{F0}	Q_{G0}	Q_{H0}	Q_{A0}	Q_{H0}
	H	H	H	L	L	L	X	X	X	X	X	X	X	X	X	X	Q_{A0}	Q_{H0}
Shift Right	H	L	H	L	L	↑	X	H	H	Q_{An}	Q_{Bn}	Q_{Cn}	Q_{Dn}	Q_{En}	Q_{Fn}	Q_{Gn}	H	Q_{Gn}
	H	L	H	L	L	↑	X	L	L	Q_{An}	Q_{Bn}	Q_{Cn}	Q_{Dn}	Q_{En}	Q_{Fn}	Q_{Gn}	L	Q_{Gn}
Shift Left	H	H	L	L	L	↑	H	X	Q_{Bn}	Q_{Cn}	Q_{Dn}	Q_{En}	Q_{Fn}	Q_{Gn}	Q_{Hn}	H	Q_{Bn}	H
	H	H	L	L	L	↑	L	X	Q_{Bn}	Q_{Cn}	Q_{Dn}	Q_{En}	Q_{Fn}	Q_{Gn}	Q_{Hn}	L	Q_{Bn}	L
Load	H	H	H	X	X	↑	X	X	a	b	c	d	e	f	g	h	a	h

† When one or both output controls are high the eight input/output terminals are disabled to the high-impedance state; however, sequential operation or clearing of the register is not affected.

a . . . h = the level of the steady-state input at inputs A through H, respectively. These data are loaded into the flip-flops while the flip-flop outputs are isolated from the input/output terminals.

maximum ratings, recommended operating conditions, and electrical characteristics

See Table III, page 2-8.

timing requirements over recommended operating free-air temperature range (unless otherwise noted)

			V_{CC}	T_A = 25°C MIN	T_A = 25°C MAX	SN54HC323 MIN	SN54HC323 MAX	SN74HC323 MIN	SN74HC323 MAX	UNIT
f_{clock}	Clock frequency		2 V	0	5	0	3.3	0	4	MHz
			4.5 V	0	25	0	17	0	20	
			6 V	0	29	0	19	0	24	
t_w	Pulse duration	CLK high or low	2 V	100		150		125		ns
			4.5 V	20		30		25		
			6 V	17		26		21		
t_{su}	Setup time before CLK↑	S0 or S1	2 V	150		225		190		ns
			4.5 V	30		45		38		
			6 V	26		38		32		
		$\overline{CLR}$ low or data	2 V	100		150		125		
			4.5 V	20		30		25		
			6 V	17		26		21		
t_h	Hold time after CLK↑	S0 or S1, data, or $\overline{CLR}$ low	2 V	0		0		0		ns
			4.5 V	0		0		0		
			6 V	0		0		0		

switching characteristics over recommended operating free-air temperature range (unless otherwise noted), C_L = 50 pF (see Note 1)

PARAMETER	FROM (INPUT)	TO (OUTPUT)	V_{CC}	T_A = 25°C MIN	TYP	MAX	SN54HC323 MIN	MAX	SN74HC323 MIN	MAX	UNIT
f_{max}			2 V	5			3.3		4		MHz
			4.5 V	25			17		20		
			6 V	29			19		24		
t_{pd}	CLK	$Q_{A'}$ or $Q_{H'}$	2 V		36						ns
			4.5 V		12						
			6 V		10						
		Q_A thru Q_H	2 V		50						
			4.5 V		14						
			6 V		12						
t_{en}	$\overline{G}1$, $\overline{G}2$	Q_A thru Q_H	2 V		50						ns
			4.5 V		15						
			6 V		12						
	S0 or S1		2 V		50						
			4.5 V		15						
			6 V		12						
t_{dis}	$\overline{G}1$, $\overline{G}2$	Q_A thru Q_H	2 V		60						ns
			4.5 V		20						
			6 V		17						
	S0 or S1		2 V		60						
			4.5 V		20						
			6 V		17						
t_t		Any	2 V		28						ns
			4.5 V		8						
			6 V		6						

C_{pd}	Power dissipation capacitance	No load, T_A = 25°C	100 pF typ

NOTE 1: For load circuit and voltage waveforms, see page 1-14.

4

ADVANCE INFORMATION

switching characteristics over recommended operating free-air temperature range (unless otherwise noted), C_L = 150 pF (see Note 1)

PARAMETER	FROM (INPUT)	TO (OUTPUT)	V_{CC}	T_A = 25°C MIN	TYP	MAX	SN54HC323 MIN	MAX	SN74HC323 MIN	MAX	UNIT
t_{pd}	CLK	$Q_{A'}$ or $Q_{H'}$	2 V		50						ns
			4.5 V		17						
			6 V		15						
		Q_A thru Q_H	2 V		50						
			4.5 V		17						
			6 V		15						
t_{en}	$\overline{G}1$, $\overline{G}2$	Q_A thru Q_H	2 V		90						ns
			4.5 V		30						
			6 V		24						
	S0 or S1		2 V		90						
			4.5 V		30						
			6 V		24						
t_{dis}	$\overline{G}1$, $\overline{G}2$	Q_A thru Q_H	2 V		90						ns
			4.5 V		30						
			6 V		24						
	S0 or S1		2 V		90						
			4.5 V		30						
			6 V		24						
t_t		Any	2 V		45						ns
			4.5 V		17						
			6 V		13						

NOTE 1: For load circuit and voltage waveforms, see page 1-14.

4

ADVANCE INFORMATION

4

ADVANCE INFORMATION

TYPES SN54HC354, SN74HC354 8-LINE TO 1-LINE DATA SELECTORS/MULTIPLEXERS/ TRANSPARENT REGISTERS WITH 3-STATE OUTPUTS

D2684, DECEMBER 1982—REVISED MARCH 1984

- Transparent Latches on Data Select Inputs
- Transparent Data Registers
- High-Current 3-State Outputs Can Drive up to 15 LSTTL Loads
- Complementary Outputs
- Package Options Include Both Plastic and Ceramic Chip Carriers in Addition to Plastic and Ceramic DIPs
- Dependable Texas Instruments Quality and Reliability

SN54HC354 . . . J PACKAGE
SN74HC354 . . . J OR N PACKAGE
(TOP VIEW)

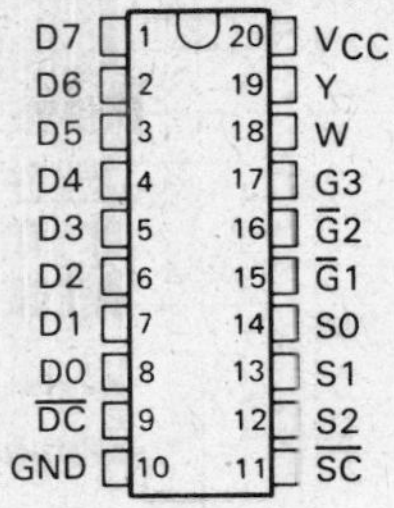

SN54HC354 . . . FH OR FK PACKAGE
SN74HC354 . . . FH OR FN PACKAGE
(TOP VIEW)

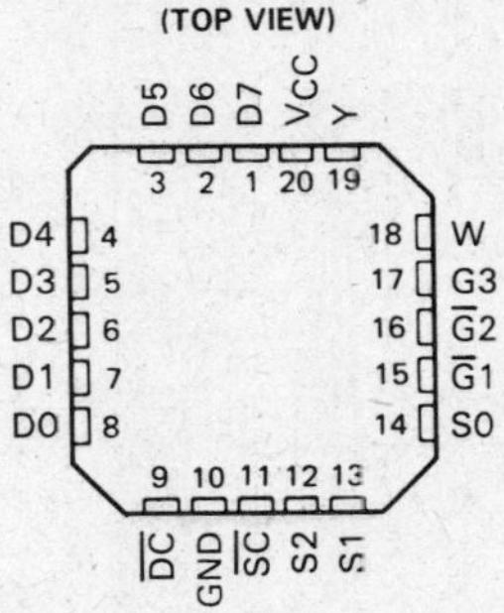

description

These monolithic data selectors/multiplexers contain full on-chip binary decoding to select one of eight data sources. The data-select is stored in transparent latches that are enabled by a low level on pin 11, $\overline{SC}$. A similar enable for data is obtained by a low level on pin 9, $\overline{DC}$.

The SN54HC354 is characterized for operation over the full military temperature range of −55°C to 125°C. The SN74HC354 is characterized for operation from −40°C to 85°C.

logic symbol

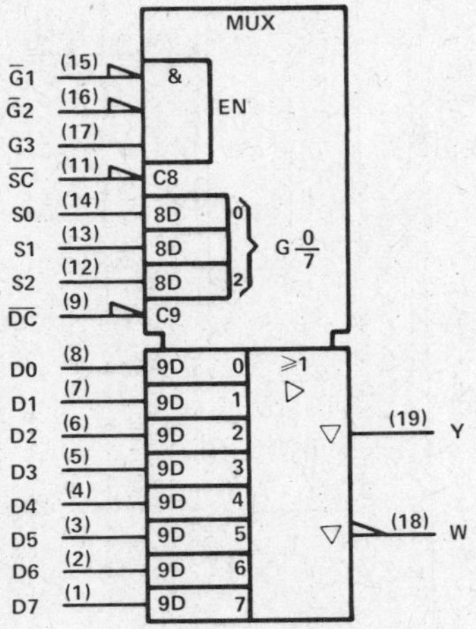

4

ADVANCE INFORMATION

logic diagram (positive logic)

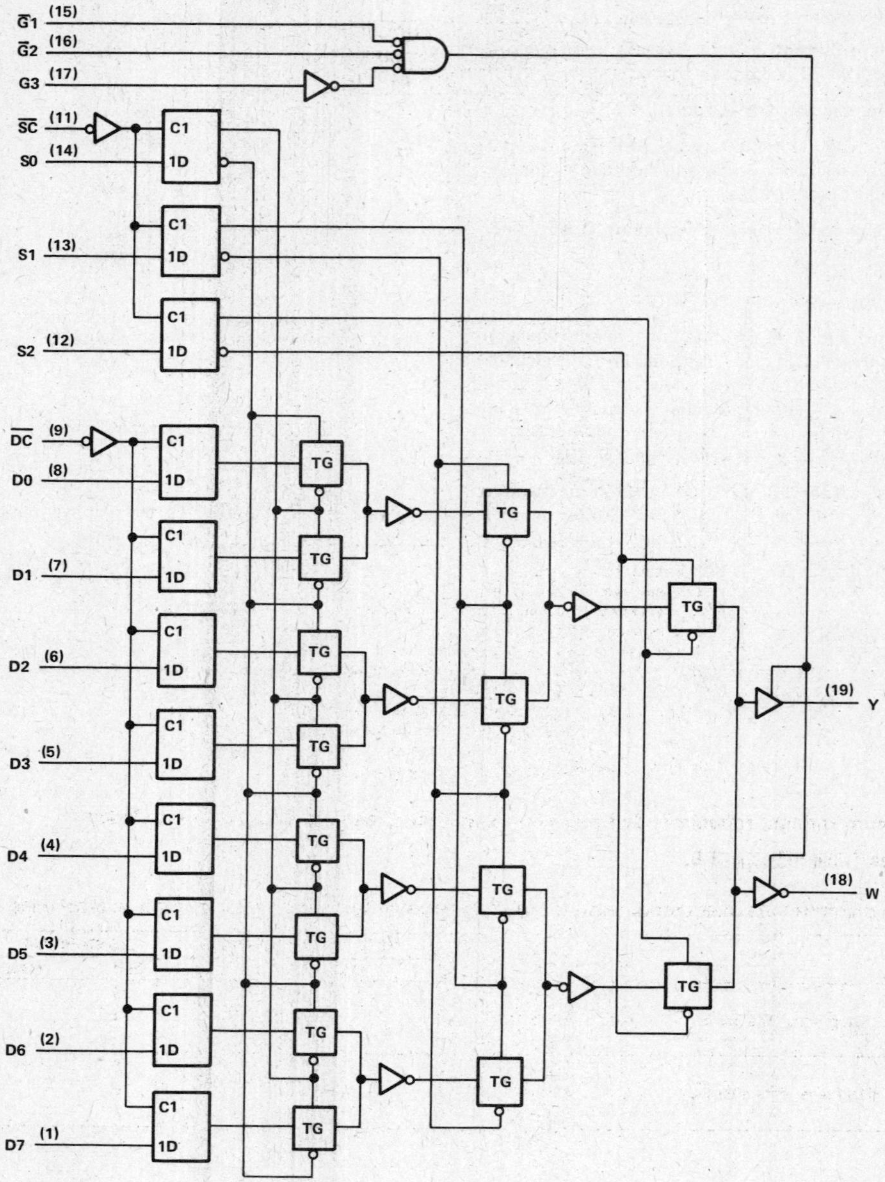

FUNCTION TABLE

INPUTS							OUTPUTS	
SELECT†			DATA CONTROL	OUTPUT ENABLES				
S2	S1	S0	$\overline{DC}$	$\overline{G1}$	$\overline{G2}$	G3	W	Y
X	X	X	X	H	X	X	Z	Z
X	X	X	X	X	H	X	Z	Z
X	X	X	X	X	X	L	Z	Z
L	L	L	L	L	L	H	$\overline{D0}$	D0
L	L	L	H	L	L	H	$\overline{D0}_n$	$D0_n$
L	L	H	L	L	L	H	$\overline{D1}$	D1
L	L	H	H	L	L	H	$\overline{D1}_n$	$D1_n$
L	H	L	L	L	L	H	$\overline{D2}$	D2
L	H	L	H	L	L	H	$\overline{D2}_n$	$D2_n$
L	H	H	L	L	L	H	$\overline{D3}$	D3
L	H	H	H	L	L	H	$\overline{D3}_n$	$D3_n$
H	L	L	L	L	L	H	$\overline{D4}$	D4
H	L	L	H	L	L	H	$\overline{D4}_n$	$D4_n$
H	L	H	L	L	L	H	$\overline{D5}$	D5
H	L	H	H	L	L	H	$\overline{D5}_n$	$D5_n$
H	H	L	L	L	L	H	$\overline{D6}$	D6
H	H	L	H	L	L	H	$\overline{D6}_n$	$D6_n$
H	H	H	L	L	L	H	$\overline{D7}$	D7
H	H	H	H	L	L	H	$\overline{D7}_n$	$D7_n$

H = high level (steady state)
L = low level (steady state)
X = irrelevant (any input, including transitions)
Z = high-impedance state (off state)
↑ = transition from low to high level
D0 . . . D7 = the level of stead-state inputs at inputs D0 through D7, respectively
$D0_n$ . . . $D7_n$ = the level of steady state inputs at inputs D0 through D7, respectively, before the most recent low-to-high transition of data control

†This column shows the input address setup with $\overline{SC}$ low.

maximum ratings, recommended operating conditions, and electrical characteristics

See Table III, page 2-8.

timing requirements over recommended operating free-air temperature range (unless otherwise noted)

		V_{CC}	T_A = 25°C		SN54HC354		SN74HC354		UNIT
			MIN	MAX	MIN	MAX	MIN	MAX	
t_{su}	Setup time, data before $\overline{DC}$↑	2 V	75		115		95		ns
		4.5 V	15		23		19		
		6 V	13		20		16		
t_h	Hold time, data after $\overline{DC}$↑	2 V	5		5		5		ns
		4.5 V	5		5		5		
		6 V	5		5		5		

4

ADVANCE INFORMATION

switching characteristics over recommended operating free-air temperature range (unless otherwise noted), C_L = 50 pF (see Note 1)

PARAMETER	FROM (INPUT)	TO (OUTPUT)	V_{CC}	T_A = 25°C MIN	TYP	MAX	SN54HC354 MIN	MAX	SN74HC354 MIN	MAX	UNIT
t_{pd}	D0–D07	Y	2 V		90						ns
			4.5 V		29						
			6 V		25						
		W	2 V		85						ns
			4.5 V		28						
			6 V		24						
t_{pd}	$\overline{DC}$	Y	2 V		85						ns
			4.5 V		28						
			6 V		24						
		W	2 V		85						ns
			4.5 V		28						
			6 V		24						
t_{pd}	S0, S1, S2	Y	2 V		100						ns
			4.5 V		32						
			6 V		27						
		W	2 V		90						ns
			4.5 V		30						
			6 V		26						
t_{pd}	$\overline{SC}$	Y	2 V		85						ns
			4.5 V		28						
			6 V		24						
		W	2 V		80						ns
			4.5 V		26						
			6 V		22						
t_{en}	$\overline{G}1$, $\overline{G}2$	Y	2 V		75						ns
			4.5 V		24						
			6 V		20						
		W	2 V		75						ns
			4.5 V		24						
			6 V		20						
t_{dis}	$\overline{G}1$, $\overline{G}2$	Y	2 V		45						ns
			4.5 V		15						
			6 V		13						
		W	2 V		45						ns
			4.5 V		15						
			6 V		13						
t_{en}	G3	Y	2 V		75						ns
			4.5 V		24						
			6 V		20						
		W	2 V		75						ns
			4.5 V		24						
			6 V		20						
t_{dis}	G3	Y	2 V		50						ns
			4.5 V		17						
			6 V		15						
		W	2 V		50						ns
			4.5 V		17						
			6 V		15						

C_{pd}	Power dissipation capacitance	No load, T_A = 25°C	40 pF typ

NOTE 1: For load circuit and voltage waveforms, see page 1-14.

4 ADVANCE INFORMATION

TEXAS INSTRUMENTS
POST OFFICE BOX 225012 • DALLAS, TEXAS 75265

TYPES SN54HC356, SN74HC356
8-LINE TO 1-LINE DATA SELECTORS/MULTIPLEXERS/ EDGE-TRIGGERED REGISTERS WITH 3-STATE OUTPUTS

D2684, DECEMBER 1982—REVISED MARCH 1984

- Transparent Latches on Data Select Inputs
- Edge-Triggered Data Registers
- High-Current 3-State Outputs Can Drive up to 15 LSTTL Loads
- Complementary Outputs
- Package Options Include Both Plastic and Ceramic Chip Carriers in Addition to Plastic and Ceramic DIPs
- Dependable Texas Instruments Quality and Reliability

SN54HC356 . . . J PACKAGE
SN74HC356 . . . J OR N PACKAGE
(TOP VIEW)

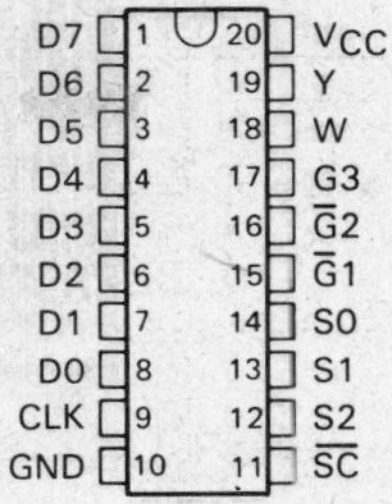

SN54HC356 . . . FH OR FK PACKAGE
SN74HC356 . . . FH OR FN PACKAGE
(TOP VIEW)

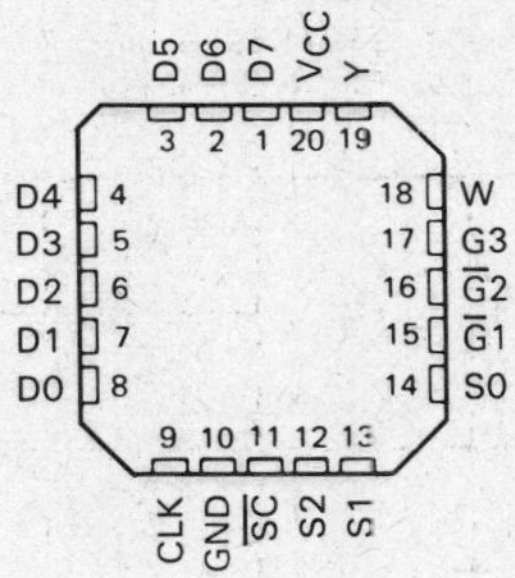

description

These monolithic data selectors/multiplexers contain full on-chip binary decoding to select one of eight data sources. The data-select address is stored in transparent latches that are enabled by a low level on pin 11, S̅C̅. The edge-triggered data registers are clocked by a low-to-high transition on pin 9, CLK. Both true and complementary outputs are available.

The SN54HC356 is characterized for operation over the full military temperature range of −55°C to 125°C. The SN74HC356 is characterized for operation from −40°C to 85°C.

logic symbol

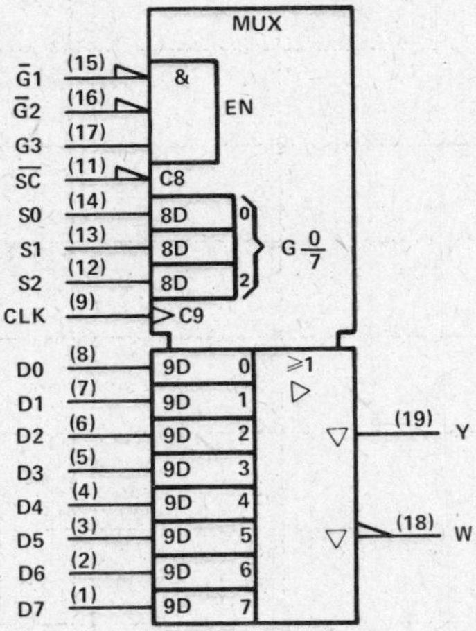

ADVANCE INFORMATION
This document contains information on a new product. Specifications are subject to change without notice.

TEXAS INSTRUMENTS
POST OFFICE BOX 225012 • DALLAS, TEXAS 75265

logic diagram (positive logic)

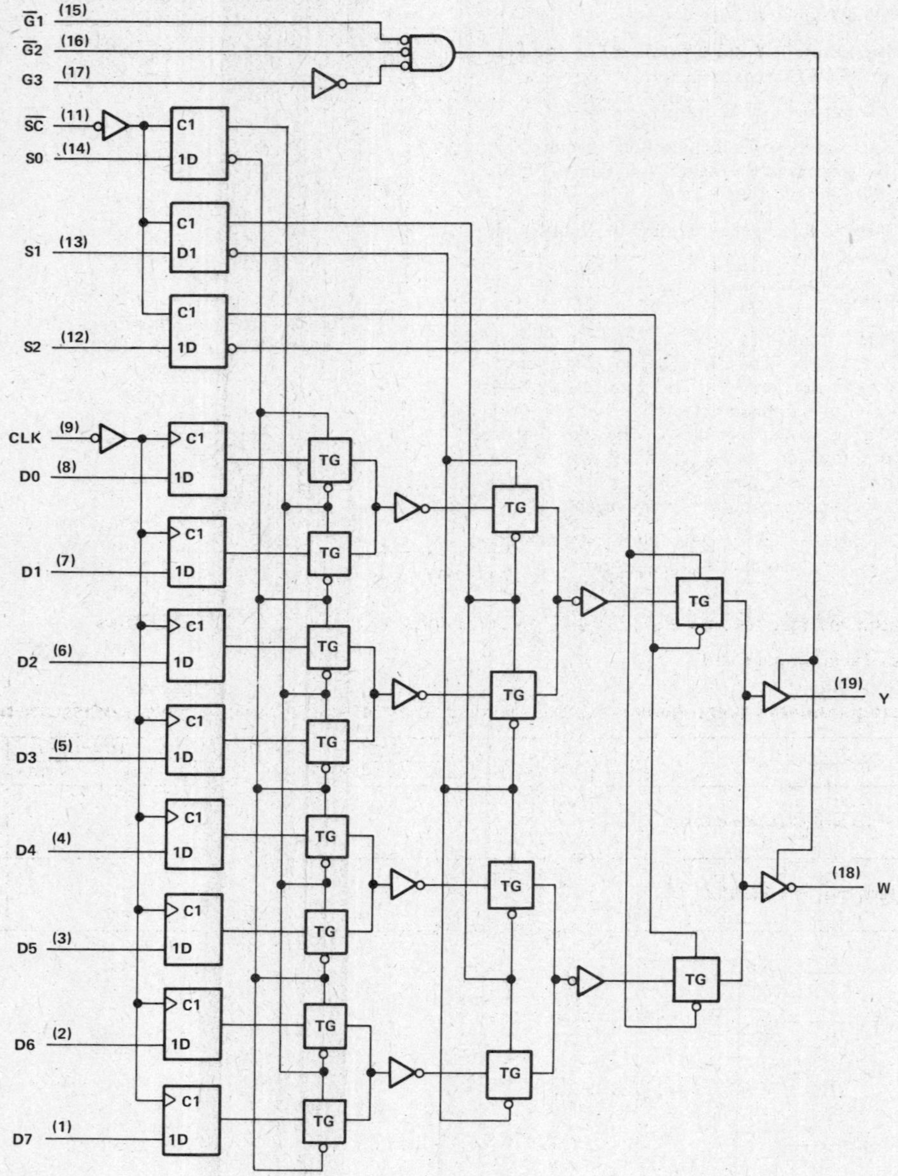

FUNCTION TABLE

INPUTS							OUTPUTS	
SELECT†			CLOCK	OUTPUT ENABLES				
S2	S1	S0		$\overline{G1}$	$\overline{G2}$	G3	W	Y
X	X	X	X	H	X	X	Z	Z
X	X	X	X	X	H	X	Z	Z
X	X	X	X	X	X	L	Z	Z
L	L	L	↑	L	L	H	$\overline{D0}$	D0
L	L	L	H or L	L	L	H	$\overline{D0}_n$	$D0_n$
L	L	H	↑	L	L	H	$\overline{D1}$	D1
L	L	H	H or L	L	L	H	$\overline{D1}_n$	$D1_n$
L	H	L	↑	L	L	H	$\overline{D2}$	D2
L	H	L	H or L	L	L	H	$\overline{D2}_n$	$D2_n$
L	H	H	↑	L	L	H	$\overline{D3}$	D3
L	H	H	H or L	L	L	H	$\overline{D3}_n$	$D3_n$
H	L	L	↑	L	L	H	$\overline{D4}$	D4
H	L	L	H or L	L	L	H	$\overline{D4}_n$	$D4_n$
H	L	H	↑	L	L	H	$\overline{D5}$	D5
H	L	H	H or L	L	L	H	$\overline{D5}_n$	$D5_n$
H	H	L	↑	L	L	H	$\overline{D6}$	D6
H	H	L	H or L	L	L	H	$\overline{D6}_n$	$D6_n$
H	H	H	↑	L	L	H	$\overline{D7}$	D7
H	H	H	H or L	L	L	H	$\overline{D7}_n$	$D7_n$

†This column shows the input address setup with $\overline{SC}$ low.

maximum ratings, recommended operating conditions, and electrical characteristics

See Table III, page 2-8.

timing requirements over recommended operating free-air temperature range (unless otherwise noted)

		V_{CC}	T_A = 25°C		SN54HC356		SN74HC356		UNIT
			MIN	MAX	MIN	MAX	MIN	MAX	
t_{su}	Setup time, data before CLK↑	2 V	75		115		95		ns
		4.5 V	15		23		19		
		6 V	13		20		16		
t_h	Hold time, data after CLK↑	2 V	5		5		5		ns
		4.5 V	5		5		5		
		6 V	5		5		5		

ADVANCE INFORMATION

4

switching characteristics over recommended operating free-air temperature range (unless otherwise noted), C_L = 50 pF (see Note 1)

PARAMETER	FROM (INPUT)	TO (OUTPUT)	V_{CC}	T_A = 25°C MIN	T_A = 25°C TYP	T_A = 25°C MAX	SN54HC356 MIN	SN54HC356 MAX	SN74HC356 MIN	SN74HC356 MAX	UNIT
t_{pd}	CLK	Y	2 V		95						ns
			4.5 V		31						
			6 V		26						
		W	2 V		85						ns
			4.5 V		28						
			6 V		24						
t_{pd}	S0, S1, S2	Y	2 V		100						ns
			4.5 V		32						
			6 V		27						
		W	2 V		90						ns
			4.5 V		30						
			6 V		26						
t_{pd}	$\overline{SC}$,	Y	2 V		80						ns
			4.5 V		26						
			6 V		22						
		W	2 V		80						ns
			4.5 V		26						
			6 V		22						
t_{en}	$\overline{G}1$, $\overline{G}2$	Y	2 V		75						ns
			4.5 V		24						
			6 V		20						
		W	2 V		75						ns
			4.5 V		24						
			6 V		20						
t_{dis}	$\overline{G}1$, $\overline{G}2$	Y	2 V		45						ns
			4.5 V		15						
			6 V		13						
		W	2 V		45						ns
			4.5 V		15						
			6 V		13						
t_{en}	G3	Y	2 V		75						ns
			4.5 V		24						
			6 V		20						
		W	2 V		75						ns
			4.5 V		24						
			6 V		20						
t_{dis}	G3	Y	2 V		50						ns
			4.5 V		17						
			6 V		15						
		W	2 V		50						ns
			4.5 V		17						
			6 V		15						

C_{pd}	Power dissipation capacitance	No load, T_A = 25°C	40 pF typ

NOTE 1: For load circuit and voltage waveforms, see page 1-14.

4 ADVANCE INFORMATION

TEXAS INSTRUMENTS
POST OFFICE BOX 225012 • DALLAS, TEXAS 75265

TYPES SN54HCT373, SN74HCT373
OCTAL D-TYPE TRANSPARENT LATCHES WITH 3-STATE OUTPUTS

D2804, MARCH 1984

- Inputs are TTL-Voltage Compatible
- 8 High-Current Latches in a Single Package
- High-Current 3-State True Outputs Can Drive up to 15 LSTTL Loads
- Full Parallel Access for Loading
- Package Options Include Both Plastic and Ceramic Chip Carriers in Addition to Plastic and Ceramic DIPs
- Dependable Texas Instruments Quality and Reliability

SN54HCT373 . . . J PACKAGE
SN74HCT373 . . . J OR N PACKAGE
(TOP VIEW)

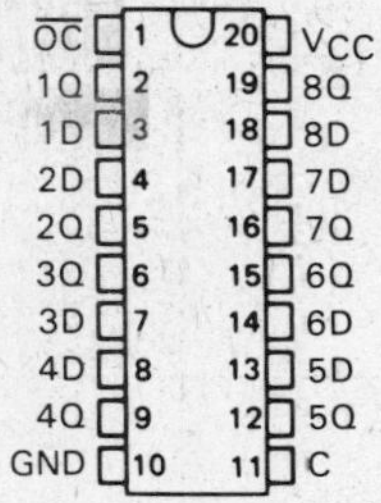

SN54HCT373 . . . FH OR FK PACKAGE
SN74HCT373 . . . FH OR FN PACKAGE
(TOP VIEW)

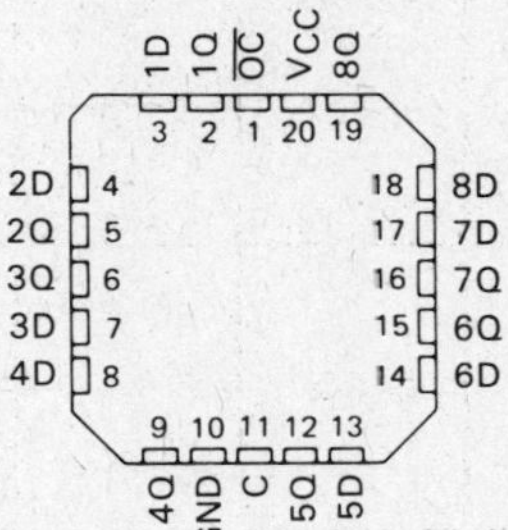

description

These 8-bit latches feature three-state outputs designed specifically for driving highly capacitive or relatively low-impedance loads. They are particularly suitable for implementing buffer registers, I/O ports, bidirectional bus drivers, and working registers.

The eight latches of the 'HCT373 are transparent D-type latches. While the enable (C) is high the Q outputs will follow the data (D) inputs. When the enable is taken low, the Q outputs will be latched at the levels that were set up at the D inputs.

An output-control input ($\overline{OC}$) can be used to place the eight outputs in either a normal logic state (high or low logic levels) or a high-impedance state. In the high-impedance state the outputs neither load nor drive the bus lines significantly. The high-impedance third state and increased drive provide the capability to drive the bus lines in a bus-organized system without need for interface or pull-up components.

The output control $\overline{OC}$ does not affect the internal operations of the latches. Old data can be retained or new data can be entered while the outputs are off.

The SN54HCT373 is characterized for operation over the full military temperature range of −55°C to 125°C. The SN74HCT373 is characterized for operation from −40°C to 85°C.

logic symbol

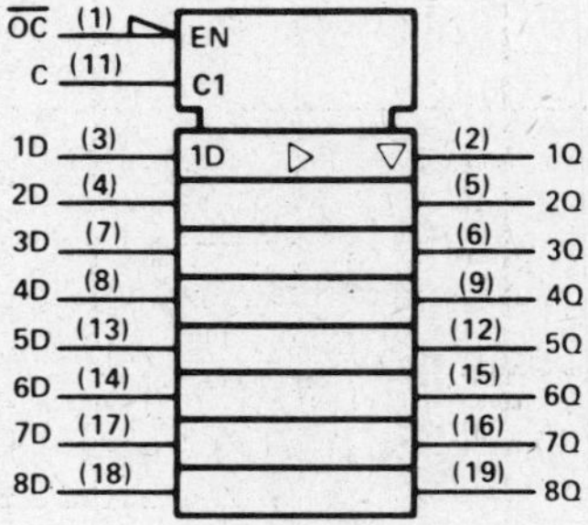

FUNCTION TABLE (EACH LATCH)

INPUTS			OUTPUT
$\overline{OC}$	ENABLE C	D	Q
L	H	H	H
L	H	L	L
L	L	X	Q_0
H	X	X	Z

TEXAS INSTRUMENTS
POST OFFICE BOX 225012 • DALLAS, TEXAS 75265

logic diagram (positive logic)

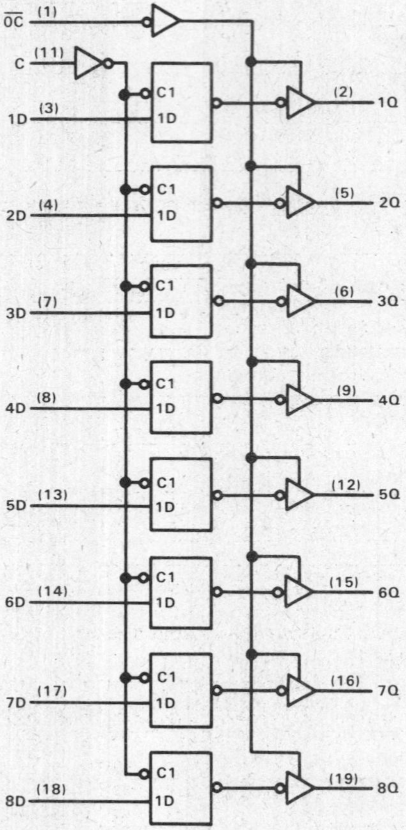

maximum ratings, recommended operating conditions, and electrical characteristics

See Table VII, page 2-14.

TEXAS INSTRUMENTS
POST OFFICE BOX 225012 • DALLAS, TEXAS 75265

timing requirements over recommended operating free-air temperature range (unless otherwise noted)

		V_{CC}	T_A = 25°C MIN	T_A = 25°C MAX	SN54HCT373 MIN	SN54HCT373 MAX	SN74HCT373 MIN	SN74HCT373 MAX	UNIT
t_w	Pulse duration, enable C high	4.5 V	20		30		25		ns
		5.5 V	17		27		23		
t_{su}	Setup time, data before enable C↓	4.5 V	10		15		13		ns
		5.5 V	9		14		12		
t_h	Hold time, data after enable C↓	4.5 V	5		5		5		ns
		5.5 V	5		5		5		

switching characteristics over recommended operating free-air temperature range (unless otherwise noted), C_L = 50 pF (see Note 1)

PARAMETER	FROM (INPUT)	TO (OUTPUT)	V_{CC}	T_A = 25°C MIN	T_A = 25°C TYP	T_A = 25°C MAX	SN54HCT373 MIN	SN54HCT373 MAX	SN74HCT373 MIN	SN74HCT373 MAX	UNIT
t_{pd}	D	Q	4.5 V		25	35		53		44	ns
			5.5 V		21	32		48		40	
t_{pd}	C	Any Q	4.5 V		28	35		53		44	ns
			5.5 V		25	32		48		40	
t_{en}	$\overline{OC}$	Any Q	4.5 V		26	35		53		44	ns
			5.5 V		23	32		48		40	
t_{dis}	$\overline{OC}$	Any Q	4.5 V		23	35		53		44	ns
			5.5 V		22	32		48		40	
t_t		Any	4.5 V		10	12		18		15	ns
			5.5 V		9	11		16		14	

C_{pd}	Power dissipation capacitance per latch	No load, T_A = 25°C	50 pF typ

switching characteristics over recommended operating free-air temperature range (unless otherwise noted), C_L = 150 pF (see Note 1)

PARAMETER	FROM (INPUT)	TO (OUTPUT)	V_{CC}	T_A = 25°C MIN	T_A = 25°C TYP	T_A = 25°C MAX	SN54HCT373 MIN	SN54HCT373 TYP	SN54HCT373 MAX	SN74HCT373 MIN	SN74HCT373 TYP	SN74HCT373 MAX	UNIT
t_{pd}	D	Q	4.5 V		32	52			79			65	ns
			5.5 V		27	47			71			59	
t_{pd}	C	Any Q	4.5 V		38	52			79			65	ns
			5.5 V		36	47			71			59	
t_{en}	$\overline{OC}$	Any Q	4.5 V		33	52			79			65	ns
			5.5 V		28	47			71			59	
t_t		Any	4.5 V		18	42			63			53	ns
			5.5 V		16	38			57			48	

NOTE 1: For load circuit and voltage waveforms, see page 1-14.

4

ADVANCE INFORMATION

D latch signal conventions

It is TI practice to name the outputs and other inputs of a D-type latch and to draw its logic symbol based on the assumption of true data (D) inputs. Then outputs that produce data in phase with the data inputs are called Q and those producing complementary data are called $\overline{Q}$. An input that causes a Q output to go high or a $\overline{Q}$ output to go low is called Preset; an input that causes a $\overline{Q}$ output to go high or a Q output to go low is called Clear. Bars are used over these pin names ($\overline{PRE}$ and $\overline{CLR}$) if they are active-low.

In some applications it may be advantageous to redesignate the data input D. In that case all the other inputs and outputs should be renamed as shown below. Also shown are corresponding changes in the graphical symbol. Arbitrary pin numbers are shown in parentheses.

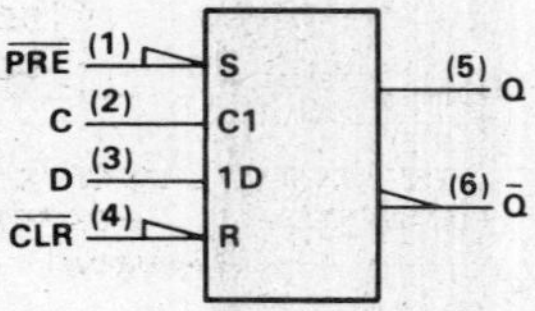

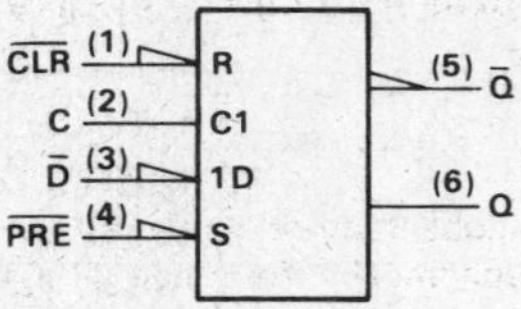

Notice that Q and $\overline{Q}$ exchange names, which causes Preset and Clear to do likewise. Also notice that the polarity indicators (⊵) on $\overline{PRE}$ and $\overline{CLR}$ remain since these inputs are still active-low, but that the presence or absence of the polarity indicator changes at $\overline{D}$, Q, and $\overline{Q}$. Of course pin 5 ($\overline{Q}$) is still in phase with the data input D, but now both are considered active low.

TEXAS INSTRUMENTS
POST OFFICE BOX 225012 • DALLAS, TEXAS 75265

TYPES SN54HCT374, SN74HCT374 OCTAL D-TYPE EDGE-TRIGGERED FLIP-FLOPS WITH 3-STATE OUTPUTS

D2804, MARCH 1984

- Inputs Are TTL-Voltage Compatible
- 8 D-Type Flip-Flops in a Single Package
- High-Current 3-State True Outputs Can Drive Up to 15 LSTTL Loads
- Full Parallel Access for Loading
- Package Options Include Both Plastic and Ceramic Chip Carriers in Addition to Plastic and Ceramic DIPs
- Dependable Texas Instruments Quality and Reliability

SN54HCT374 . . . J PACKAGE
SN74HCT374 . . . J OR N PACKAGE
(TOP VIEW)

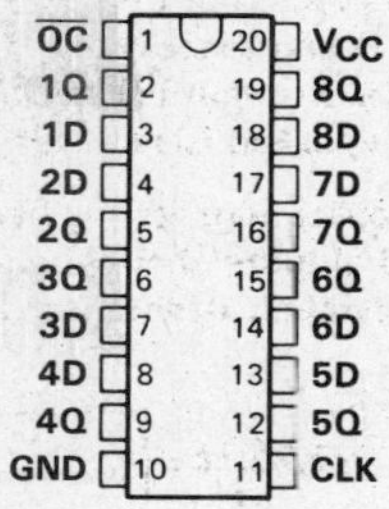

SN54HCT374 . . . FH OR FK PACKAGE
SN74HCT374 . . . FH OR FN PACKAGE
(TOP VIEW)

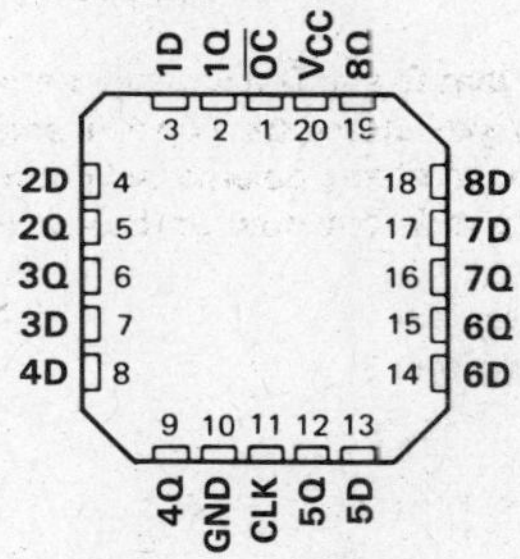

description

These 8-bit flip-flops feature three-state outputs designed specifically for driving highly capacitive or relatively low-impedance loads. They are particularly suitable for implementing buffer registers, I/O ports, bidirectional bus drivers, and working registers.

The eight flip-flops of the 'HCT374 are edge-triggered D-type flip-flops. On the positive transition of the clock the Q outputs will be set to the logic levels that were set up at the D inputs.

An output-control input can be used to place the eight outputs in either a normal logic state (high or low logic levels) or a high-impedance state. In the high-impedance state the outputs neither load nor drive the bus lines significantly. The high-impedance third state and increased drive provide the capability to drive the bus lines in a bus-organized system without need for interface or pull-up components.

The output control ($\overline{OC}$) does not affect the internal operation of the flip-flops. Old data can be retained or new data can be entered while the outputs are in the high-impedance state.

The SN54HCT374 is characterized for operation over the full military temperature range of −55°C to 125°C. The SN74HCT374 is characterized for operation from −40°C to 85°C.

FUNCTION TABLE (EACH FLIP-FLOP)

INPUTS			OUTPUT
$\overline{OC}$	CLK	D	Q
L	↑	H	H
L	↑	L	L
L	L	X	Q_0
H	X	X	Z

logic symbol

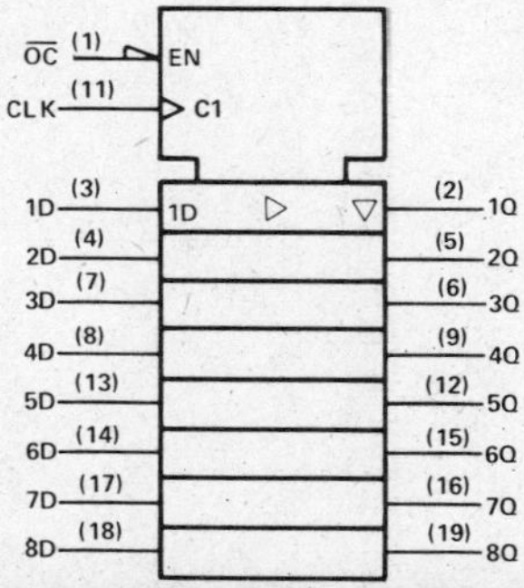

logic diagram (positive logic)

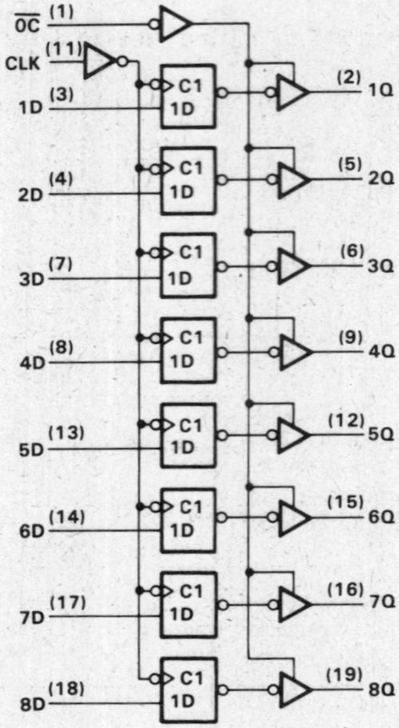

TEXAS INSTRUMENTS
POST OFFICE BOX 225012 • DALLAS, TEXAS 75265

maximum ratings, recommended operating conditions, and electrical characteristics

See Table VII, page 2-14.

timing requirements over recommended operating free-air temperature range (unless otherwise noted)

		V_{CC}	T_A = 25°C		SN54HCT374		SN74HCT374		UNIT
			MIN	MAX	MIN	MAX	MIN	MAX	
f_{clock}	Clock frequency	4.5 V	0	31	0	21	0	25	MHz
		5.5 V	0	36	0	23	0	28	
t_w	Pulse duration, CLK high or low	4.5 V	16		24		20		ns
		5.5 V	14		22		18		
t_{su}	Setup time, data before CLK ↑	4.5 V	20		30		25		ns
		5.5 V	17		27		23		
t_h	Hold time, data after CLK ↑	4.5 V	5		5		5		ns
		5.5 V	5		5		5		

switching characteristics over recommended operating free-air temperature range (unless otherwise noted), C_L = 50 pF (see Note 1)

PARAMETER	FROM (INPUT)	TO (OUTPUT)	V_{CC}	T_A = 25°C			SN54HCT374		SN74HCT374		UNIT
				MIN	TYP	MAX	MIN	MAX	MIN	MAX	
f_{max}			4.5 V	31	36		21		25		MHz
			5.5 V	36	40		23		28		
t_{pd}	CLK	Any	4.5 V		30			54		45	ns
			5.5 V		25			48		41	
t_{en}	$\overline{OC}$	Any	4.5 V		26			45		38	ns
			5.5 V		23			41		34	
t_{dis}	$\overline{OC}$	Any	4.5 V		23			45		38	ns
			5.5 V		22			41		34	
t_t		Any	4.5 V		10			18		15	ns
			5.5 V		9			16		14	

C_{pd}	Power dissipation capacitance	No load, T_A = 25°C	85 pF typ

switching characteristics over recommended operating free-air temperature range (unless otherwise noted), C_L = 150 pF (see Note 1)

PARAMETER	FROM (INPUT)	TO (OUTPUT)	V_{CC}	T_A = 25°C			SN54HCT374		SN74HCT374		UNIT
				MIN	TYP	MAX	MIN	MAX	MIN	MAX	
t_{pd}	CLK	Any	4.5 V		40	53		80		66	ns
			5.5 V		35	47		71		60	
t_{en}	$\overline{OC}$	Any	4.5 V		34	47		71		59	ns
			5.5 V		29	39		59		49	
t_t		Any	4.5 V		18	42		63		53	ns
			5.5 V		16	38		57		48	

NOTE 1: For load circuit and voltage waveforms, see page 1-14.

4

ADVANCE INFORMATION

D flip-flop signal conventions

It is TI practice to name the outputs and other inputs of a D-type flip-flop and to draw its logic symbol based on the assumption of true data (D) inputs. Then outputs that produce data in phase with the data inputs are called Q and those producing complementary data are called $\overline{Q}$. An input that causes a Q output to go high or a $\overline{Q}$ output to go low is called Preset; an input that causes a $\overline{Q}$ output to go high or a Q output to go low is called Clear. Bars are used over these pin names ($\overline{PRE}$ and $\overline{CLR}$) if they are active-low.

In some applications it may be advantageous to redesignate the data input $\overline{D}$. In that case all the other inputs and outputs should be renamed as shown below. Also shown are corresponding changes in the graphical symbol. Arbitrary pin numbers are shown in parentheses.

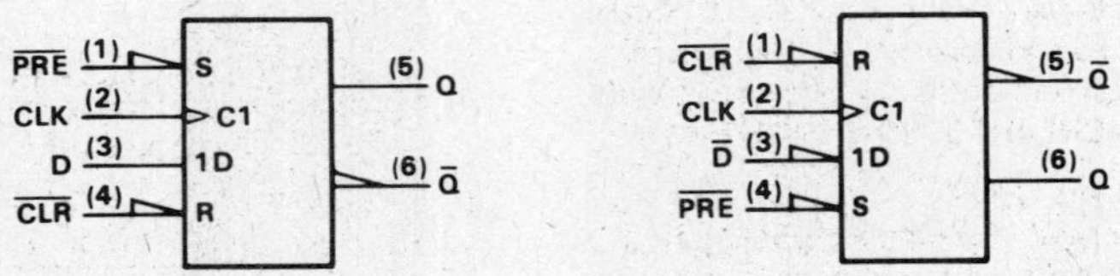

Notice that Q and $\overline{Q}$ exchange names, which causes Preset and Clear to do likewise. Also notice that the polarity indicators (◺) on $\overline{PRE}$ and $\overline{CLR}$ remain since these inputs are still active-low, but that the presence or absence of the polarity indicator changes at $\overline{D}$, Q, and $\overline{Q}$. Of course pin 5 ($\overline{Q}$) is still in phase with the data input $\overline{D}$, but now both are considered active-low.

D2804, MARCH 1984

- Supply Voltage and Ground on Corner Pins to Simplify PC-Board Layout
- Package Options Include Both Plastic and Ceramic Chip Carriers in Addition to Plastic and Ceramic DIPs
- Dependable Texas Instruments Quality and Reliability

description

The SN54HC375 and SN74HC375 bistable latches are electrically and functionally identical to the SN54HC75 and SN74HC75, respectively. Only the arrangement of the terminals has been changed in the SN54HC375 and SN74HC375.

These latches are ideally suited for use as temporary storage for binary information between processing units and input/output or indicator units. Information present at a data (D) input is transferred to the Q output when the enable (C) is high and the Q output will follow the data input as long as the enable remains high. When the enable goes low, the information (that was present at the data input at the time the transition occurred) is retained at the Q output until the enable goes high.

The SN54HC375 is characterized for operation over the full military temperature range of −55°C to 125°C. The SN74HC375 is characterized for operation from −40°C to 85°C.

SN54HC375 . . . J PACKAGE
SN74HC375 . . . J OR N PACKAGE
(TOP VIEW)

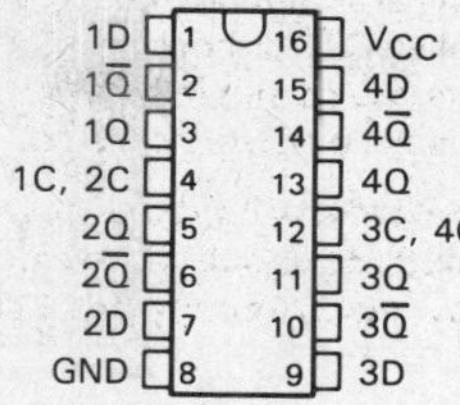

SN54HC375 . . . FH OR FK PACKAGE
SN74HC375 . . . FH OR FN PACKAGE
(TOP VIEW)

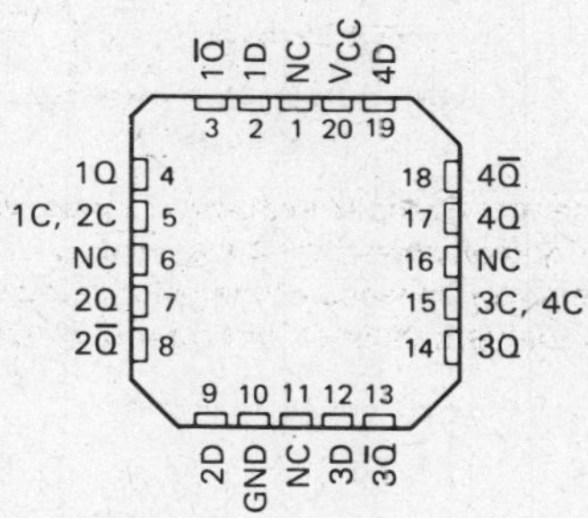

NC – No internal connection

logic symbol

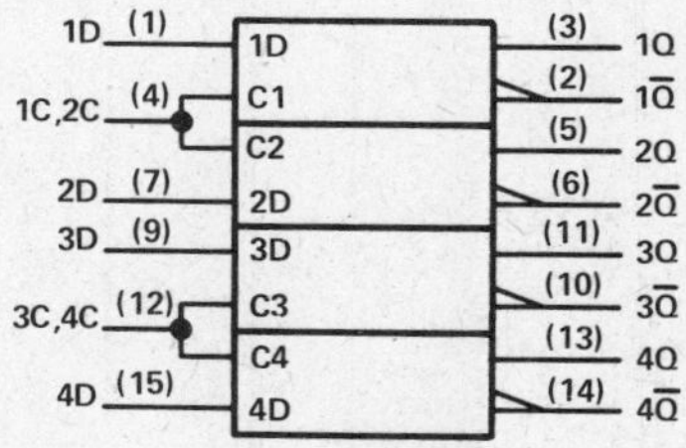

Pin numbers shown are for J and N packages.

FUNCTION TABLE
(EACH LATCH)

INPUTS		OUTPUTS	
D	C	Q	$\overline{Q}$
L	H	L	H
H	H	H	L
X	L	Q_0	$\overline{Q}_0$

logic diagram (positive logic)

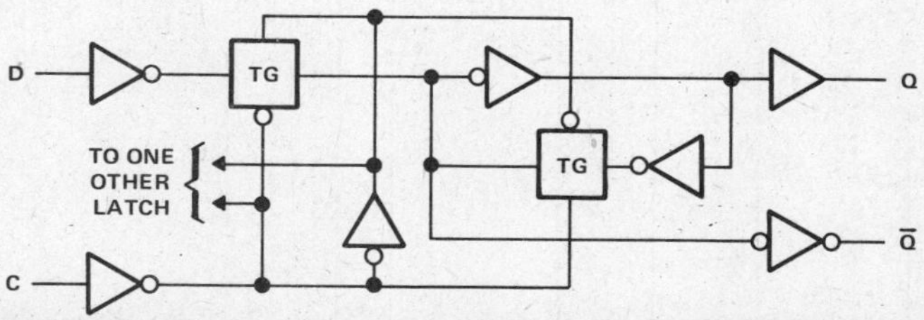

absolute maximum ratings, recommended operating conditions, electrical characteristics

See Table II, page 2-6.

timing requirements over recommended operating free-air temperature range (unless otherwise noted)

		V_{CC}	T_A = 25°C		SN54HC375		SN74HC375		UNIT
			MIN	MAX	MIN	MAX	MIN	MAX	
t_w	Pulse duration, C high	2 V	80		120		100		ns
		4.5 V	16		24		20		
		6 V	14		20		17		
t_{su}	Setup time, data before C↓	2 V	100		150		125		ns
		4.5 V	20		30		25		
		6 V	17		26		21		
t_h	Hold time, data after C↓	2 V	5		5		5		ns
		4.5 V	5		5		5		
		6 V	5		5		5		

switching characteristics over recommended operating free-air temperature range (unless otherwise noted), C_L = 50 pF (see Note 1)

PARAMETER	FROM (INPUT)	TO (OUTPUT)	V_{CC}	T_A = 25°C			SN54HC375		SN74HC375		UNIT
				MIN	TYP	MAX	MIN	MAX	MIN	MAX	
t_{pd}	D	Q or $\overline{Q}$	2 V		40	120		180		150	ns
			4.5 V		14	24		36		30	
			6 V		11	20		31		26	
t_{pd}	C	Q or $\overline{Q}$	2 V		42	130		195		165	ns
			4.5 V		15	26		39		33	
			6 V		12	22		33		28	
t_t		Any	2 V		38	75		110		95	ns
			4.5 V		8	15		22		19	
			6 V		6	13		19		16	

NOTE 1: For load circuit and voltage waveforms, see page 1-14.

TYPES SN54HC533, SN74HC533 OCTAL D-TYPE TRANSPARENT LATCHES WITH 3-STATE OUTPUTS

D2684, DECEMBER 1982—REVISED MARCH 1984

- 8 Latches in a Single Package
- High-Current 3-State Inverting Outputs Can Drive up to 15 LSTTL Loads
- Full Parallel Access for Loading
- Package Options Include Both Plastic and Ceramic Chip Carriers in Addition to Plastic and Ceramic DIPs
- Dependable Texas Instruments Quality and Reliability

SN54HC533 . . . J PACKAGE
SN74HC533 . . . J OR N PACKAGE
(TOP VIEW)

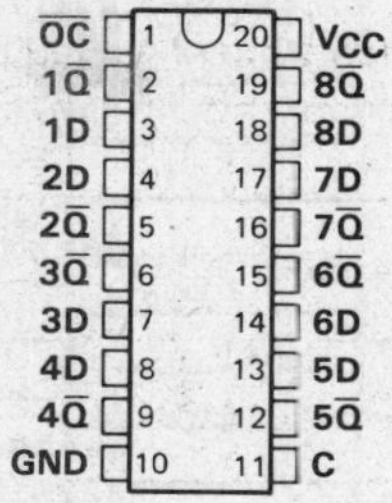

SN54HC533 . . . FH OR FK PACKAGE
SN74HC533 . . . FH OR FN PACKAGE
(TOP VIEW)

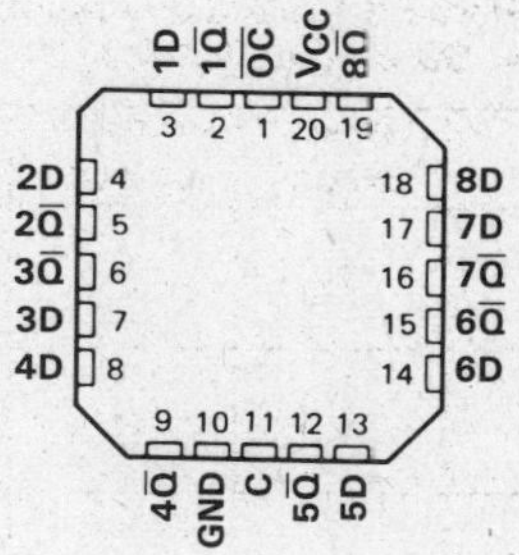

FUNCTION TABLE (EACH LATCH)

INPUTS			OUTPUT
$\overline{OC}$	ENABLE C	D	$\overline{Q}$
L	H	H	L
L	H	L	H
L	L	X	$\overline{Q}_0$
H	X	X	Z

description

These 8-bit latches feature three-state outputs designed specifically for driving highly capacitive or relatively low-impedance loads. They are particularly suitable for implementing buffer registers, I/O ports, bidirectional bus drivers, and working registers.

The eight latches of the 'HC533 are transparent D-type latches. While the enable (C) is high, the $\overline{Q}$ outputs will follow the complements of the D inputs. When the enable is taken low, the $\overline{Q}$ outputs will be latched at the inverses of the levels that were set up at the D inputs. The 'HC533 is functionally equivalent to the 'HC373 except for having inverted outputs.

An output-control ($\overline{OC}$) input can be used to place the eight outputs in either a normal logic state (high or low logic levels) or a high-impedance state. In the high-impedance state the outputs neither load nor drive the bus lines significantly. The high-impedance third state and increased drive provide the capability to drive the bus lines in a bus-organized system without need for interface or pull-up components.

The output control does not affect the internal operation of the latches. Old data can be retained or new data can be entered while the outputs are off.

The SN54HC533 is characterized for operation over the full military temperature range of −55°C to 125°C. The SN74HC533 is characterized for operation from −40°C to 85°C.

logic symbol

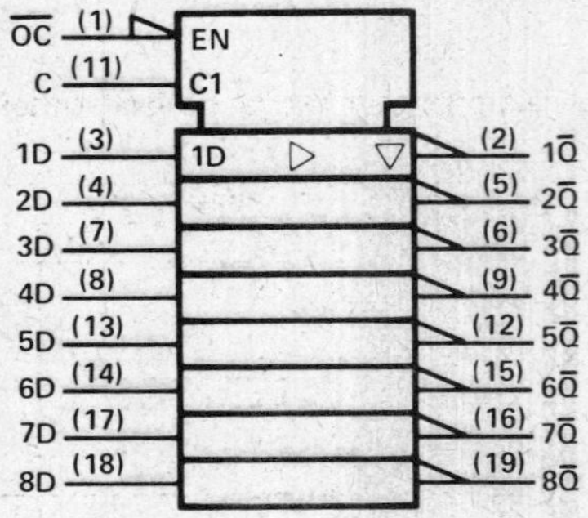

logic diagram (positive logic)

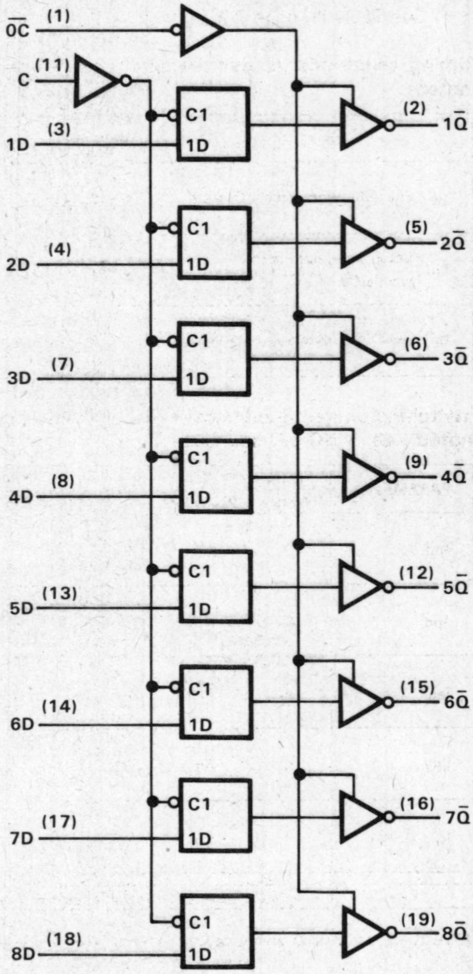

maximum ratings, recommended operating conditions, and electrical characteristics

See Table III, page 2-8.

timing requirements over recommended operating free-air temperature range (unless otherwise noted)

		V_{CC}	T_A = 25°C		SN54HC533		SN74HC533		UNIT
			MIN	MAX	MIN	MAX	MIN	MAX	
t_w	Pulse duration, enable C high	2 V	80		120		100		ns
		4.5 V	16		24		20		
		6 V	14		20		17		
t_{su}	Setup time, data before enable C↓	2 V	50		75		63		ns
		4.5 V	10		15		13		
		6 V	9		13		11		
t_h	Hold time, data after enable C↓	2 V	5		5		5		ns
		4.5 V	5		5		5		
		6 V	5		5		5		

switching characteristics over recommended operating free-air temperature range (unless otherwise noted), C_L = 50 pF (see Note 1)

PARAMETER	FROM (INPUT)	TO (OUTPUT)	V_{CC}	T_A = 25°C			SN54HC533		SN74HC533		UNIT
				MIN	TYP	MAX	MIN	MAX	MIN	MAX	
t_{pd}	D	$\overline{Q}$	2 V		77	175		265		220	ns
			4.5 V		26	35		53		44	
			6 V		23	30		45		38	
t_{pd}	C	Any	2 V		87	175		265		220	ns
			4.5 V		27	35		53		44	
			6 V		23	30		45		38	
t_{en}	$\overline{OC}$	Any	2 V		68	150		225		190	ns
			4.5 V		24	30		45		38	
			6 V		21	26		38		32	
t_{dis}	$\overline{OC}$	Any	2 V		47	150		225		190	ns
			4.5 V		23	30		45		38	
			6 V		21	26		38		32	
t_t		Any	2 V		28	60		90		75	ns
			4.5 V		8	12		18		15	
			6 V		6	10		15		13	

C_{pd}	Power dissipation capacitance per latch	No load, T_A = 25°C	50 pF typ

NOTE 1: For load circuits and voltage waveforms, see page 1-14.

4

ADVANCE INFORMATION

switching characteristics over recommended operating free-air temperature range (unless otherwise noted), C_L = 150 pF (see Note 1)

PARAMETER	FROM (INPUT)	TO (OUTPUT)	V_{CC}	T_A = 25°C MIN	TYP	MAX	SN54HC533 MIN	MAX	SN74HC533 MIN	MAX	UNIT
t_{pd}	D	$\overline{Q}$	2 V		95	200		300		250	ns
			4.5 V		33	40		60		50	
			6 V		21	34		51		43	
t_{pd}	C	Any	2 V		103	225		335		285	ns
			4.5 V		33	45		67		57	
			6 V		29	38		37		48	
t_{en}	$\overline{OC}$	Any	2 V		85	200		300		250	ns
			4.5 V		29	40		60		50	
			6 V		26	34		31		43	
t_t		Any	2 V		60	210		315		265	ns
			4.5 V		17	42		63		53	
			6 V		14	36		33		45	

NOTE 1: For load circuits and voltage waveforms, see page 1-14.

D latch signal conventions

It is TI practice to name the outputs and other inputs of a D-type latch and to draw its logic symbol based on the assumption of true data (D) inputs. Then outputs that produce data in phase with the data inputs are called Q and those producing complementary data are called $\overline{Q}$. An input that causes a Q output to go high or a $\overline{Q}$ output to go low is called Preset; an input that causes a $\overline{Q}$ output to go high or a Q output to go low is called Clear. Bars are used over these pin names ($\overline{PRE}$ and $\overline{CLR}$) if they are active-low.

In some applications it may be advantageous to redesignate the data input $\overline{D}$. In that case all the other inputs and outputs should be renamed as shown below. Also shown are corresponding changes in the graphical symbol. Arbitrary pin numbers are shown in parentheses.

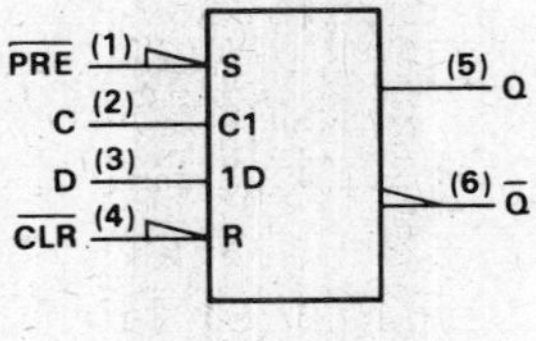

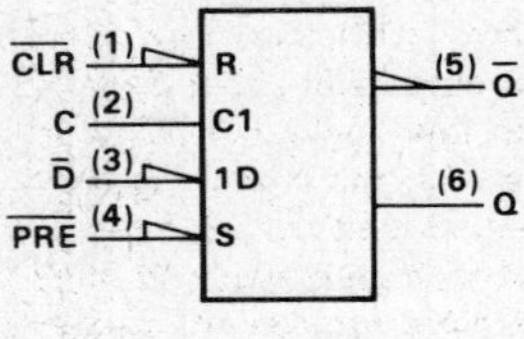

Notice that Q and $\overline{Q}$ exchange names, which causes Preset and Clear to do likewise. Also notice that the polarity indicators (⊳) on $\overline{PRE}$ and $\overline{CLR}$ remain since these inputs are still active-low, but that the presence or absence of the polarity indicator changes at $\overline{D}$, Q, and $\overline{Q}$. Of course pin 5 ($\overline{Q}$) is still in phase with the data input $\overline{D}$, but now both are considered active-low.

TYPES SN54HCT533, SN74HCT533 OCTAL D-TYPE TRANSPARENT LATCHES WITH 3-STATE OUTPUTS

D2804, MARCH 1984

- **Inputs Are TTL-Voltage Compatible**
- **8 Latches in a Single Package**
- **High-Current 3-State Inverting Outputs Can Drive up to 15 LSTTL Loads**
- **Full Parallel Access for Loading**
- **Package Options Include Both Plastic and Ceramic Chip Carriers in Addition to Plastic and Ceramic DIPs**
- **Dependable Texas Instruments Quality and Reliability**

SN54HCT533 . . . J PACKAGE
SN74HCT533 . . . J OR N PACKAGE
(TOP VIEW)

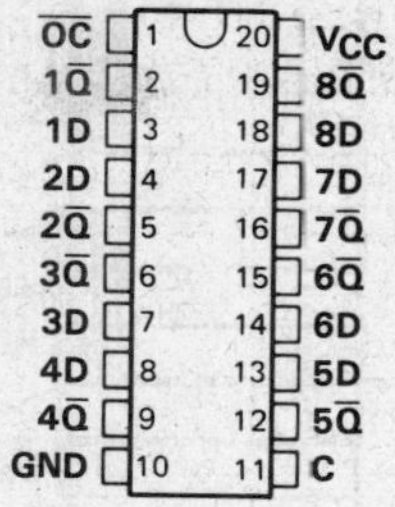

SN54HCT533 . . . FH OR FK PACKAGE
SN74HCT533 . . . FH OR FN PACKAGE
(TOP VIEW)

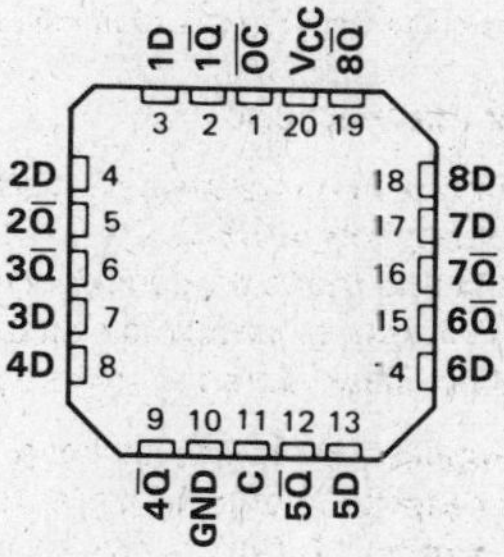

description

These 8-bit latches feature three-state outputs designed specifically for driving highly capacitive or relatively low-impedance loads. They are particularly suitable for implementing buffer registers, I/O ports, bidirectional bus drivers, and working registers.

The eight latches of the 'HCT533 are transparent D-type latches. While the enable (C) is high, the $\overline{Q}$ outputs will follow the complements of the D inputs. When the enable is taken low, the $\overline{Q}$ outputs will be latched at the inverses of the levels that were set up at the D inputs. The 'HCT533 is functionally equivalent to the 'HCT373 except for having inverted outputs.

An output-control ($\overline{OC}$) input can be used to place the eight outputs in either a normal logic state (high or low logic levels) or a high-impedance state. In the high-impedance state the outputs neither load nor drive the bus lines significantly. The high-impedance third state and increased drive provide the capability to drive the bus lines in a bus-organized system without need for interface or pull-up components.

The output control does not affect the internal operation of the latches. Old data can be retained or new data can be entered while the outputs are off.

The SN54HCT533 is characterized for operation over the full military temperature range of −55°C to 125°C. The SN74HCT533 is characterized for operation from −40°C to 85°C.

FUNCTION TABLE (EACH LATCH)

INPUTS			OUTPUT
$\overline{OC}$	ENABLE C	D	$\overline{Q}$
L	H	H	L
L	H	L	H
L	L	X	$\overline{Q}_0$
H	X	X	Z

4

ADVANCE INFORMATION

ADVANCE INFORMATION
This document contains information on a new product. Specifications are subject to change without notice.

TEXAS INSTRUMENTS
POST OFFICE BOX 225012 • DALLAS, TEXAS 75265

logic symbol

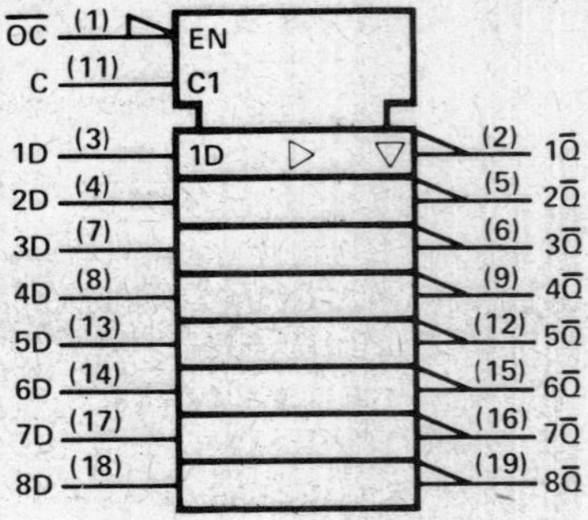

logic diagram (positive logic)

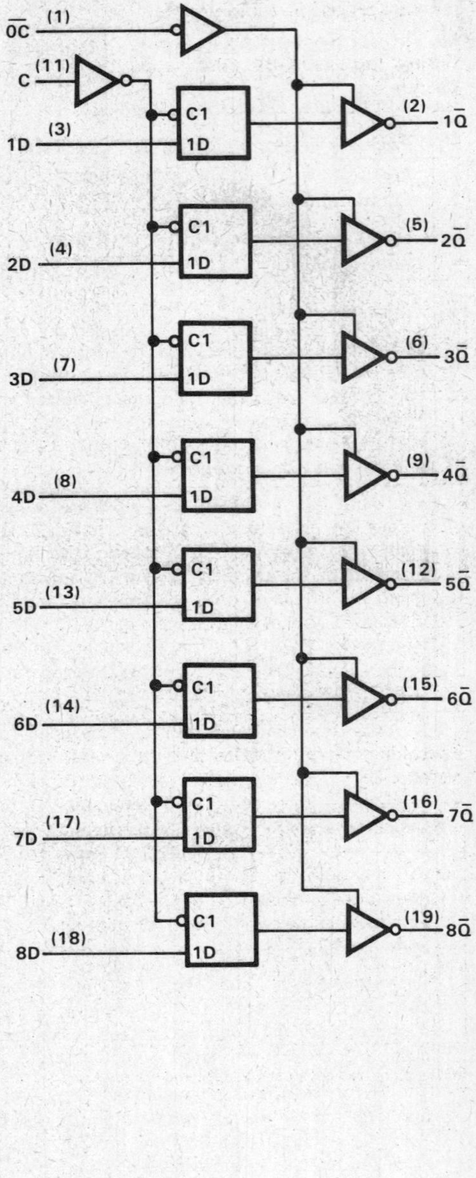

maximum ratings, recommended operating conditions, and electrical characteristics

See Table VII, page 2-14.

timing requirements over recommended operating free-air temperature range (unless otherwise noted)

		V_{CC}	T_A = 25°C		SN54HCT533		SN74HCT533		UNIT
			MIN	MAX	MIN	MAX	MIN	MAX	
t_w	Pulse duration, enable C high	4.5 V	20		30		25		ns
		5.5 V	17		27		23		
t_{su}	Setup time, data before enable C↓	4.5 V	10		15		13		ns
		5.5 V	9		14		12		
t_h	Hold time, data after enable C↓	4.5 V	5		5		5		ns
		5.5 V	5		5		5		

switching characteristics over recommended operating free-air temperature range (unless otherwise noted), C_L = 50 pF (see Note 1)

PARAMETER	FROM (INPUT)	TO (OUTPUT)	V_{CC}	T_A = 25°C			SN54HCT533		SN74HCT533		UNIT
				MIN	TYP	MAX	MIN	MAX	MIN	MAX	
t_{pd}	D	Q	4.5 V		38	35		53		44	ns
			5.5 V		24	32		48		40	
t_{pd}	C	Any	4.5 V		30	35		53		44	ns
			5.5 V		28	32		48		40	
t_{en}	$\overline{OC}$	Any	4.5 V		29	35		53		44	ns
			5.5 V		25	32		48		40	
t_{dis}	$\overline{OC}$	Any	4.5 V		25	35		53		44	ns
			5.5 V		24	32		48		40	
t_t		Any	4.5 V		10	12		18		15	ns
			5.5 V		9	11		16		14	

C_{pd}	Power dissipation capacitance per latch	No load, T_A = 25°C	50 pF typ

switching characteristics over recommended operating free-air temperature range (unless otherwise noted), C_L = 150 pF (see Note 1)

PARAMETER	FROM (INPUT)	TO (OUTPUT)	V_{CC}	T_A = 25°C			SN54HCT533		SN74HCT533		UNIT
				MIN	TYP	MAX	MIN	MAX	MIN	MAX	
t_{pd}	D	Q	4.5 V		36	52		79		65	ns
			5.5 V		32	47		71		59	
t_{pd}	C	Any	4.5 V		40	52		79		65	ns
			5.5 V		38	47		71		59	
t_{en}	$\overline{OC}$	Any	4.5 V		35	52		79		65	ns
			5.5 V		29	47		71		59	
t_t		Any	4.5 V		18	42		63		53	ns
			5.5 V		16	38		57		48	

NOTE 1: For load circuit and voltage waveforms, see page 1-14.

D latch signal conventions

It is TI practice to name the outputs and other inputs of a D-type latch and to draw its logic symbol based on the assumption of true data (D) inputs. Then outputs that produce data in phase with the data inputs are called Q and those producing complementary data are called $\overline{Q}$. An input that causes a Q output to go high or a $\overline{Q}$ output to go low is called Preset; an input that causes a $\overline{Q}$ output to go high or a Q output to go low is called Clear. Bars are used over these pin names ($\overline{PRE}$ and $\overline{CLR}$) if they are active-low.

In some applications it may be advantageous to redesignate the data input $\overline{D}$. In that case all the other inputs and outputs should be renamed as shown below. Also shown are corresponding changes in the graphical symbol. Arbitrary pin numbers are shown in parentheses.

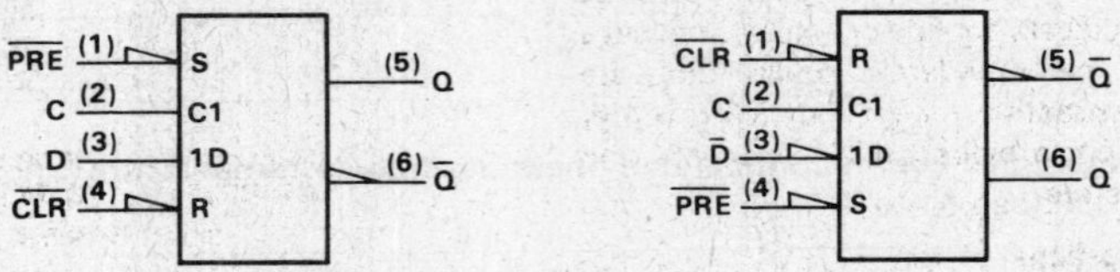

Notice that Q and $\overline{Q}$ exchange names, which causes Preset and Clear to do likewise. Also notice that the polarity indicators (⊳) on $\overline{PRE}$ and $\overline{CLR}$ remain since these inputs are still active-low, but that the presence or absence of the polarity indicator changes at $\overline{D}$, Q, and $\overline{Q}$. Of course pin 5 ($\overline{Q}$) is still in phase with the data input $\overline{D}$, but now both are considered active-low.

4

ADVANCE INFORMATION

TYPES SN54HC534, SN74HC534 OCTAL D-TYPE EDGE-TRIGGERED FLIP-FLOPS WITH 3-STATE OUTPUTS

D2684, DECEMBER 1982—REVISED MARCH 1984

- **High-Current 3-State Inverting Outputs Can Drive up to 15 LSTTL Loads**
- **Package Options Include Both Plastic and Ceramic Chip Carriers in Addition to Plastic and Ceramic DIPs**
- **Dependable Texas Instruments Quality and Reliability**

SN54HC534 . . . J PACKAGE
SN74HC534 . . . J OR N PACKAGE
(TOP VIEW)

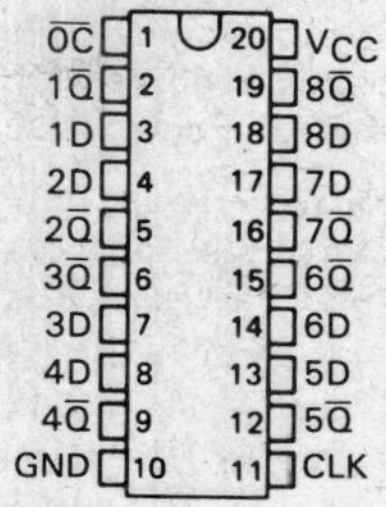

SN54HC534 . . . FH OR FK PACKAGE
SN74HC534 . . . FH OR FN PACKAGE
(TOP VIEW)

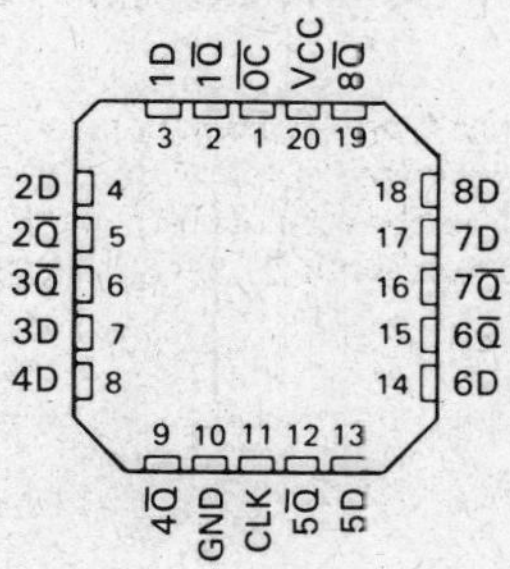

description

These 8-bit flip-flops feature three-state outputs designed specifically for driving highly capacitive or relatively low-impedance loads. They are particularly attractive for implementing buffer registers, I/O ports, bidirectional bus drivers, and working registers.

The eight flip-flops of the 'HC534 are edge-triggered D-type flip-flops. On the positive transition of the clock, the $\overline{Q}$ outputs will be set to the complement of the logic states that were set up at the D inputs. The 'HC534 is functionally equivalent to the 'HC374 except for having inverted outputs.

An output-control input can be used to place the eight outputs in either a normal logic state (high or low logic levels) or a high-impedance state. In the high-impedance state the outputs neither load nor drive the bus lines significantly. The high-impedance third state and increased drive provide the capability to drive the bus lines in a bus-organized system without need for interface or pull-up components.

The output control does not affect the internal operation of the flip-flops. Old data can be retained or new data can be entered while the outputs are off.

The SN54HC534 is characterized for operation over the full military temperature range of −55°C to 125°C. The SN74HC534 is characterized for operation from −40°C to 85°C.

FUNCTION TABLE (EACH FLIP-FLOP)

INPUTS			OUTPUT
$\overline{OC}$	CLK	D	$\overline{Q}$
L	↑	H	L
L	↑	L	H
L	L	X	$\overline{Q}_0$
H	X	X	Z

logic symbol

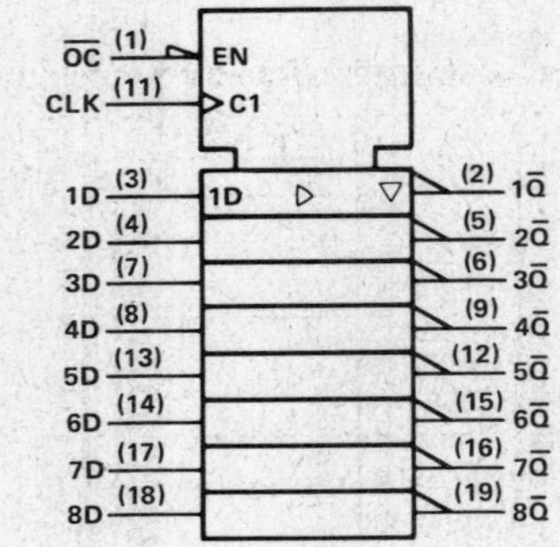

logic diagram (positive logic)

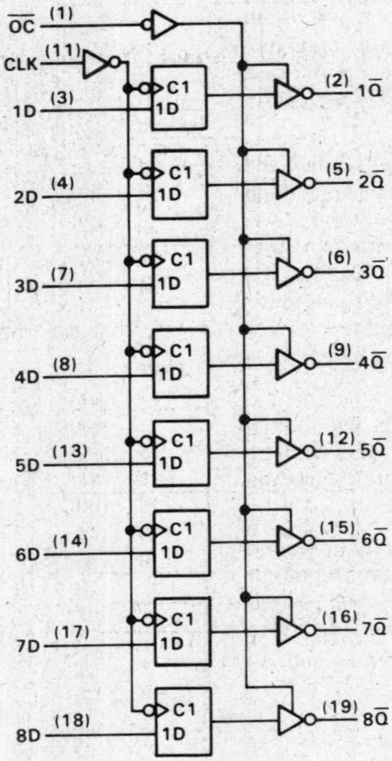

TEXAS INSTRUMENTS
POST OFFICE BOX 225012 • DALLAS, TEXAS 75265

maximum ratings, recommended operating conditions, and electrical characteristics

See Table III, page 2-8.

timing requirements over recommended operating free-air temperature range (unless otherwise noted)

			V_{CC}	T_A = 25°C		SN54HC534		SN74HC534		UNIT
				MIN	MAX	MIN	MAX	MIN	MAX	
f_{clock}	Clock frequency		2 V	0	6	0	4.2	0	5	MHz
			4.5 V	0	31	0	21	0	25	
			6 V	0	36	0	25	0	29	
t_w	Pulse duration	CLK high or low	2 V	80		120		100		ns
			4.5 V	16		24		20		
			6 V	14		20		17		
t_{su}	Setup time, data before CLK↑		2 V	100		150		125		ns
			4.5 V	20		30		25		
			6 V	17		26		21		
t_h	Hold time, data after CLK↑		2 V	5		5		5		ns
			4.5 V	5		5		5		
			6 V	5		5		5		

switching characteristics over recommended operating free-air temperature range (unless otherwise noted), C_L = 50 pF (see Note 1)

PARAMETER	FROM (INPUT)	TO (OUTPUT)	V_{CC}	T_A = 25°C			SN54HC534		SN74HC534		UNIT
				MIN	TYP	MAX	MIN	MAX	MIN	MAX	
f_{max}			2 V	6	11		4.2		5		MHz
			4.5 V	31	36		21		25		
			6 V	36	40		25		29		
t_{pd}	CLK	Any	2 V		88	180		270		225	ns
			4.5 V		28	36		54		45	
			6 V		24	31		46		38	
t_{en}	$\overline{OC}$	Any	2 V		77	150		225		190	ns
			4.5 V		26	30		45		38	
			6 V		23	26		38		32	
t_{dis}	$\overline{OC}$	Any	2 V		51	150		225		190	ns
			4.5 V		25	30		45		38	
			6 V		23	26		38		32	
t_t		Any	2 V		28	60		90		75	ns
			4.5 V		8	12		18		15	
			6 V		6	10		15		13	

C_{pd}	Power dissipation per flip-flop	No load, T_A = 25°C	100 pF typ

NOTE 1: For load circuit and voltage waveforms, see page 1-14.

switching characteristics over recommended operating free-air temperature range (unless otherwise noted), C_L = 150 pF (see Note 1)

PARAMETER	FROM (INPUT)	TO (OUTPUT)	V_{CC}	T_A = 25°C			SN54HC534		SN74HC534		UNIT
				MIN	TYP	MAX	MIN	MAX	MIN	MAX	
t_{pd}	CLK	Any	2 V		105	230		345		290	ns
			4.5 V		35	46		69		58	
			6 V		31	39		58		49	
t_{en}	$\overline{OC}$	Any	2 V		95	200		300		250	ns
			4.5 V		32	40		60		50	
			6 V		29	34		51		43	
t_{dis}	$\overline{OC}$	Any	2 V		60	210		315		265	ns
			4.5 V		17	42		63		53	
			6 V		14	36		53		45	

NOTE 1: For load circuit and voltage waveforms, see page 1-14.

D flip-flop signal conventions

It is TI practice to name the outputs and other inputs of a D-type flip-flop and to draw its logic symbol based on the assumption of true data (D) inputs. Then outputs that produce data in phase with the data inputs are called Q and those producing complementary data are called $\overline{Q}$. An input that causes a Q output to go high or a $\overline{Q}$ output to go low is called Preset; an input that causes a $\overline{Q}$ output to go high or a Q output to go low is called Clear. Bars are used over these pin names ($\overline{PRE}$ and $\overline{CLR}$) if they are active-low.

In some applications it may be advantageous to redesignate the data input $\overline{D}$. In that case all the other inputs and outputs should be renamed as shown below. Also shown are corresponding changes in the graphical symbol. Arbitrary pin numbers are shown in parentheses.

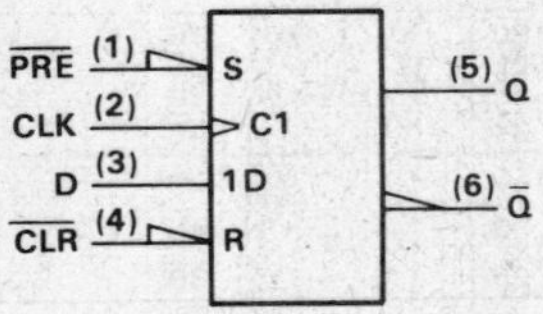

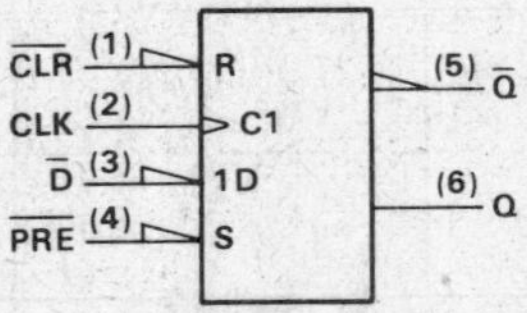

Notice that Q and $\overline{Q}$ exchange names, which causes Preset and Clear to do likewise. Also notice that the polarity indicators (⊵) on $\overline{PRE}$ and $\overline{CLR}$ remain since these inputs are still active-low, but that the presence or absence of the polarity indicator changes at $\overline{D}$, Q and $\overline{Q}$. Of course pin 5 ($\overline{Q}$) is still in phase with the data input $\overline{D}$, but now both are considered active-low.

TYPES SN54HC540, SN54HC541 SN74HC540, SN74HC541 OCTAL BUFFERS AND LINE DRIVERS WITH 3-STATE OUTPUTS

D2804, MARCH 1984

- High-Current 3-State Outputs Drive Bus Lines Directly or up to 15 LSTTL Loads
- Data Flow-Thru Pinout (All Inputs on Opposite Side from Outputs)
- Package Options Include Both Plastic and Ceramic Chip Carriers in Addition to Plastic and Ceramic DIPs
- Dependable Texas Instruments Quality and Reliability

SN54HC540, SN54HC541 . . . J PACKAGE
SN74HC540, SN74HC541 . . . J OR N PACKAGE
(TOP VIEW)

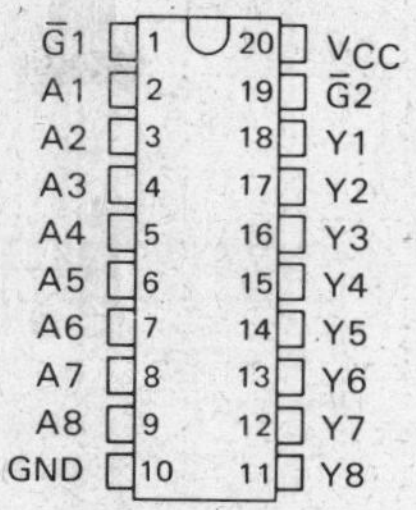

SN54HC540, SN54HC541 . . . FH OR FK PACKAGE
SN74HC540, SN74HC541 . . . FH OR FN PACKAGE
(TOP VIEW)

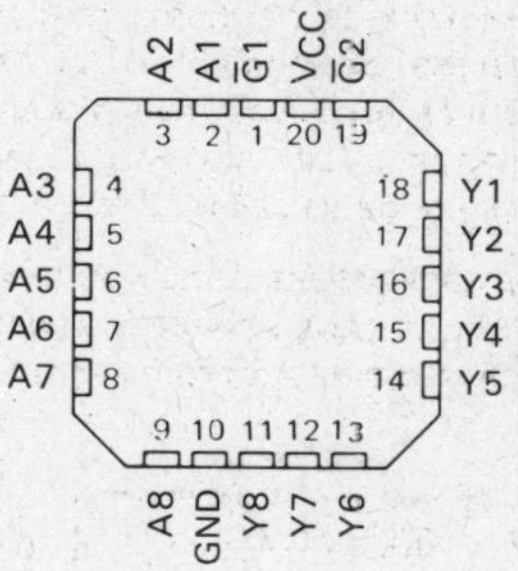

description

These octal buffers and line drivers are designed to have the performance of the popular SN54HC240/SN74HC240 series and, at the same time, offer a pinout with inputs and outputs on opposite sides of the package. This arrangement greatly enhances printed circuit board layout.

The three-state control gate is a 2-input NOR such that if either G1 or G2 is high, all eight outputs are in the high-impedance state.

The 'HC540 provides inverted data and the 'HC541 provides true data at the outputs.

The SN54HC540 and SN54HC541 are characterized for operation over the full military temperature range of −55°C to 125°C. The SN7HC540 and SN74HC541 are characterized for operation from −40°C to 85°C.

ADVANCE INFORMATION
This document contains information on a new product. Specifications are subject to change without notice.

logic symbols

'HC540

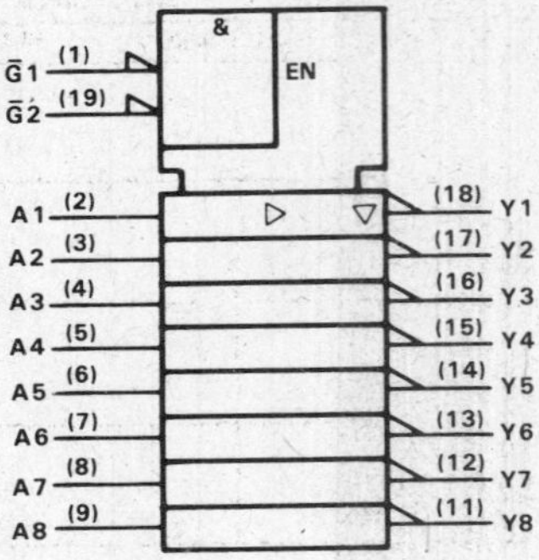

'HC541

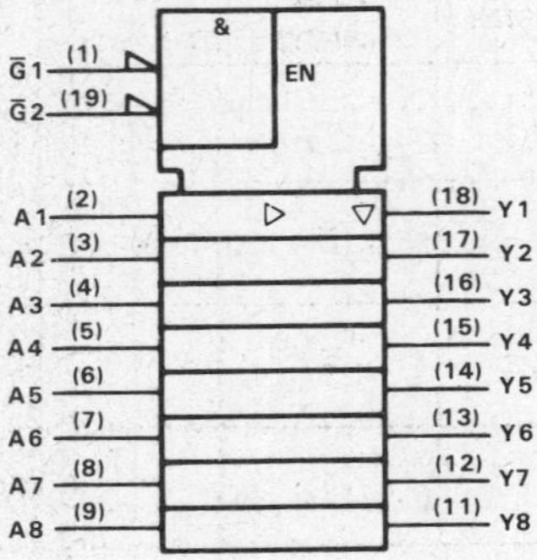

logic diagrams (positive logic)

'HC540

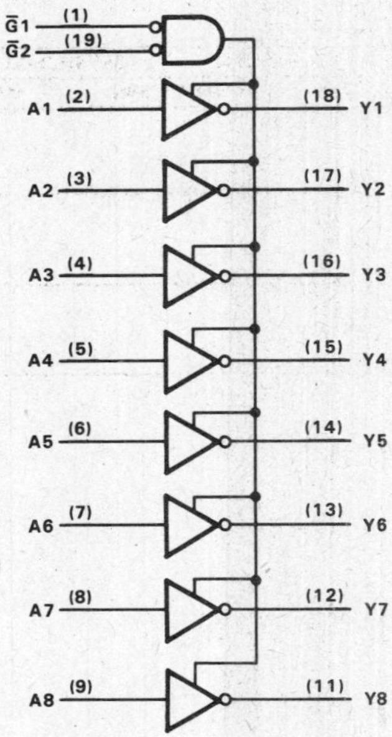

'HC541

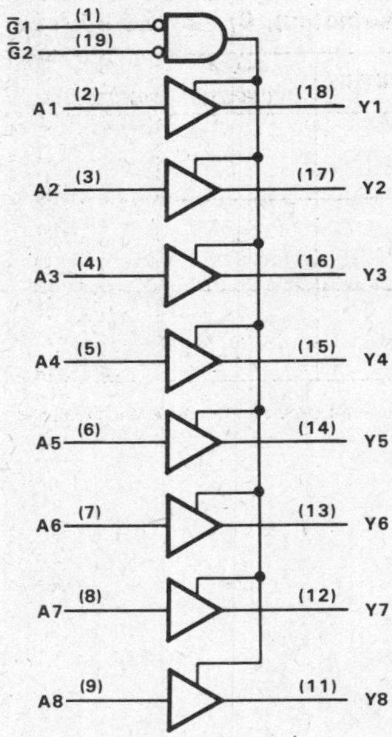

maximum ratings, recommended operating conditions, and electrical characteristics

See Table III, page 2-8.

'HC540 switching characteristics over recommended operating free-air temperature range (unless otherwise noted), C_L = 50 pF (see Note 1)

PARAMETER	FROM (INPUT)	TO (OUTPUT)	V_{CC}	T_A = 25°C MIN	TYP	MAX	SN54HC540 MIN	MAX	SN74HC540 MIN	MAX	UNIT
t_{pd}	A	Y	2 V		50						ns
			4.5 V		10						
			6 V		9						
t_{en}	$\overline{G}$	Y	2 V		75						ns
			4.5 V		15						
			6 V		13						
t_{dis}	$\overline{G}$	Y	2 V		40						ns
			4.5 V		18						
			6 V		17						
t_t		Y	2 V		28						ns
			4.5 V		8						
			6 V		6						

C_{pd}	Power dissipation capacitance	No load, T_A = 25°C	35 pF typ

'HC540 switching characteristics over recommended operating free-air temperature range (unless otherwise noted), C_L = 150 pF (see Note 1)

PARAMETER	FROM (INPUT)	TO (OUTPUT)	V_{CC}	T_A = 25°C MIN	TYP	MAX	SN54HC540 MIN	MAX	SN74HC540 MIN	MAX	UNIT
t_{pd}	A	Y	2 V		75						ns
			4.5 V		15						
			6 V		13						
t_{en}	$\overline{G}$	Y	2 V		100						ns
			4.5 V		20						
			6 V		17						
t_t		Y	2 V		45						ns
			4.5 V		17						
			6 V		13						

NOTE 1: For load circuit and voltage waveforms, see page 1-14.

4

ADVANCE INFORMATION

'HC541 switching characteristics over recommended operating free-air temperature range (unless otherwise noted), C_L = 50 pF (see Note 1)

PARAMETER	FROM (INPUT)	TO (OUTPUT)	V_{CC}	T_A = 25 °C MIN	TYP	MAX	SN54HC541 MIN	MAX	SN74HC541 MIN	MAX	UNIT
t_{pd}	A	Y	2 V		55						ns
			4.5 V		12						
			6 V		11						
t_{en}	$\overline{G}$	Y	2 V		80						ns
			4.5 V		17						
			6 V		15						
t_{dis}	$\overline{G}$	Y	2 V		40						ns
			4.5 V		18						
			6 V		17						
t_t		Y	2 V		28						ns
			4.5 V		8						
			6 V		6						

C_{pd}	Power dissipation capacitance	No load, T_A = 25 °C	35 pF typ

'HC541 switching characteristics over recommended operating free-air temperature range (unless otherwise noted), C_L = 150 pF (see Note 1)

PARAMETER	FROM (INPUT)	TO (OUTPUT)	V_{CC}	T_A = 25 °C MIN	TYP	MAX	SN54HC541 MIN	MAX	SN74HC541 MIN	MAX	UNIT
t_{pd}	A	Y	2 V		75						ns
			4.5 V		16						
			6 V		14						
t_{en}	$\overline{G}$	Y	2 V		100						ns
			4.5 V		20						
			6 V		17						
t_t		Y	2 V		45						ns
			4.5 V		17						
			6 V		13						

NOTE 1: For load circuit and voltage waveforms, see page 1-14.

TYPES SN54HCT540, SN54HCT541 SN74HCT540, SN74HCT541 OCTAL BUFFERS AND LINE DRIVERS WITH 3-STATE OUTPUTS

D2804, MARCH 1984

- Inputs are TTL-Voltage Compatible
- High-Current 3-State Outputs Interface Directly with System Bus or Can Drive up to 15 LSTTL Loads
- Data Flow-Thru Pinout (All Inputs on Opposite Side from Outputs)
- Package Options Include Both Plastic and Ceramic Chip Carriers in Addition to Plastic and Ceramic DIPs
- Dependable Texas Instruments Quality and Reliability

SN54HCT540, SN54HCT541 . . . J PACKAGE
SN74HCT540, SN74HCT541 . . . J OR N PACKAGE
(TOP VIEW)

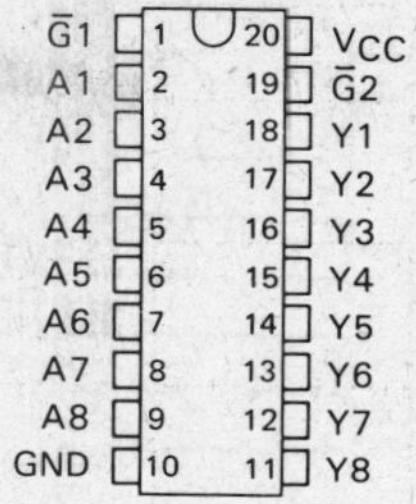

SN54HCT540, SN54HCT541 . . . FH OR FK PACKAGE
SN74HCT540, SN74HCT541 . . . FH OR FN PACKAGE
(TOP VIEW)

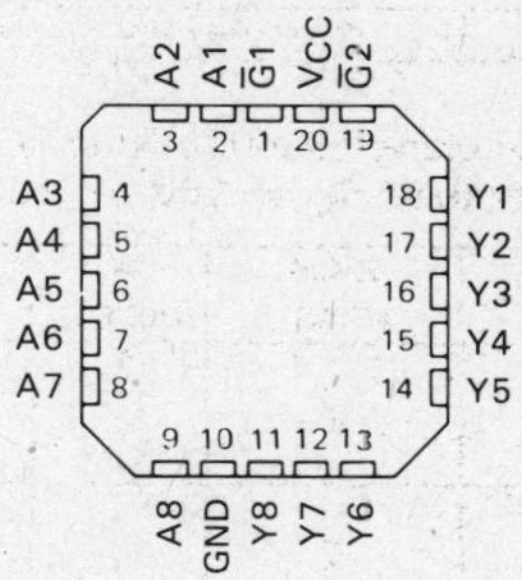

description

These octal buffers and line drivers are designed to have the performance of the popular SN54HCT240/SN74HCT240 series and, at the same time, offer a pinout with inputs and outputs on opposite sides of the package. This arrangement greatly enhances printed circuit board layout.

The three-state control gate is a 2-input NOR such that if either $\overline{G}1$ or $\overline{G}2$ is high, all eight outputs are in the high-impedance state.

The 'HCT540 provides inverted data and the 'HCT541 provides true data at the outputs.

The SN54HCT540 and SN54HCT541 are characterized for operation over the full military temperature range of −55°C to 125°C. The SN74HCT540 and SN74HCT541 are characterized for operation from −40°C to 85°C.

logic symbols

'HCT540

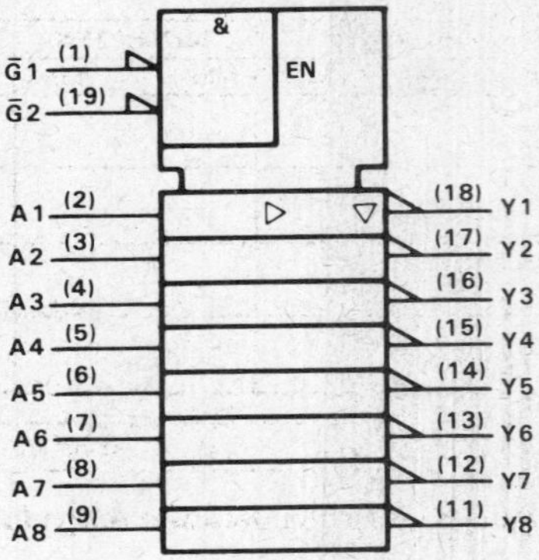

'HCT541

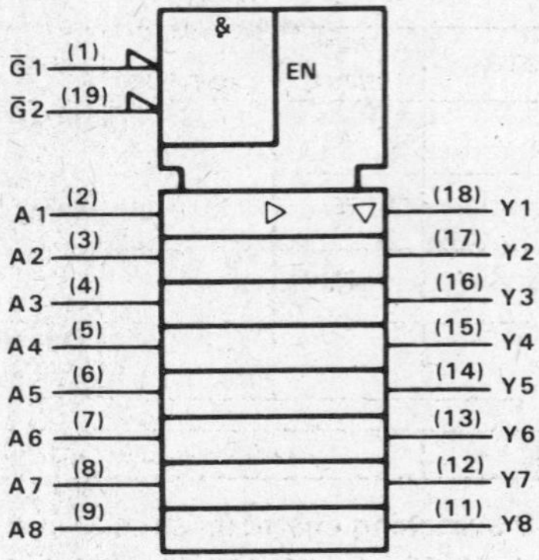

logic diagrams (positive logic)

'HCT540

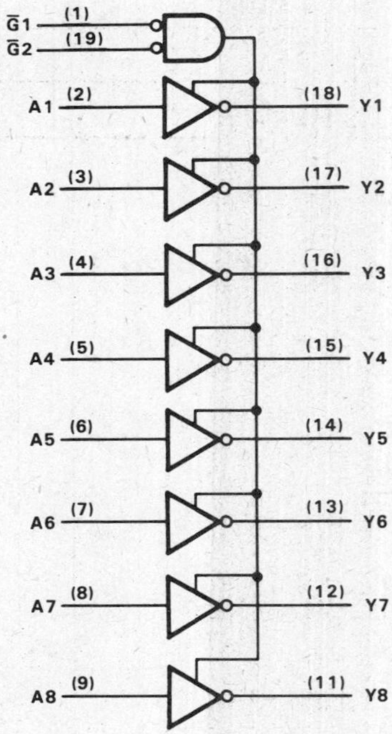

'HCT541

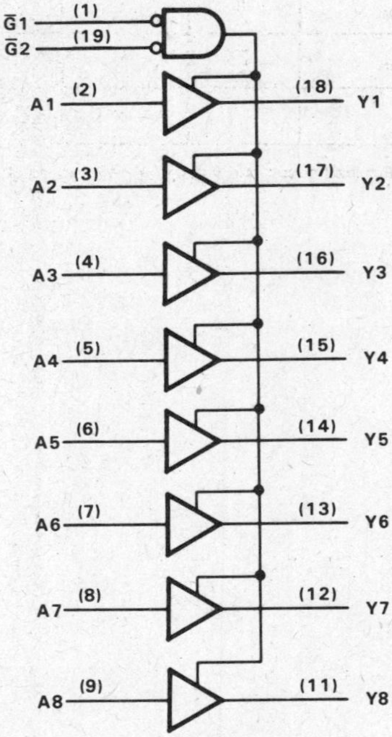

maximum ratings, recommended operating conditions, and electrical characteristics

See Table VII, page 2-14.

'HCT540 switching characteristics over recommended operating free-air temperature range (unless otherwise noted), C_L = 50 pF (see Note 1)

PARAMETER	FROM (INPUT)	TO (OUTPUT)	V_{CC}	T_A = 25°C MIN	T_A = 25°C TYP	T_A = 25°C MAX	SN54HCT540 MIN	SN54HCT540 MAX	SN74HCT540 MIN	SN74HCT540 MAX	UNIT
t_{pd}	A	Y	4.5 V		13						ns
			5.5 V		12						
t_{en}	$\overline{G}$	Y	4.5 V		21						ns
			5.5 V		19						
t_{dis}	$\overline{G}$	Y	4.5 V		19						ns
			5.5 V		18						
t_t		Y	4.5 V		8						ns
			5.5 V		7						

C_{pd}	Power dissipation capacitance	No load, T_A = 25°C	35 pF typ

'HCT540 switching characteristics over recommended operating free-air temperature range (unless otherwise noted), C_L = 150 pF (see Note 1)

PARAMETER	FROM (INPUT)	TO (OUTPUT)	V_{CC}	T_A = 25°C MIN	T_A = 25°C TYP	T_A = 25°C MAX	SN54HCT540 MIN	SN54HCT540 MAX	SN74HCT540 MIN	SN74HCT540 MAX	UNIT
t_{pd}	A	Y	4.5 V		20						ns
			5.5 V		19						
t_{en}	$\overline{G}$	Y	4.5 V		26						ns
			5.5 V		25						
t_t		Y	4.5 V		17						ns
			5.5 V		14						

NOTE 1: For load circuit and voltage waveforms, see page 1-14.

4

ADVANCE INFORMATION

'HCT541 switching characteristics over recommended operating free-air temperature range (unless otherwise noted), C_L = 50 pF (see Note 1)

PARAMETER	FROM (INPUT)	TO (OUTPUT)	V_{CC}	T_A = 25°C MIN	T_A = 25°C TYP	T_A = 25°C MAX	SN54HCT541 MIN	SN54HCT541 MAX	SN74HCT541 MIN	SN74HCT541 MAX	UNIT
t_{pd}	A	Y	4.5 V		13						ns
			5.5 V		12						
t_{en}	$\overline{G}$	Y	4.5 V		21						ns
			5.5 V		19						
t_{dis}	$\overline{G}$	Y	4.5 V		19						ns
			5.5 V		18						
t_t		Y	4.5 V		8						ns
			5.5 V		7						

C_{pd}	Power dissipation capacitance	No load, T_A = 25°C	35 pF typ

'HCT541 switching characteristics over recommended operating free-air temperature range (unless otherwise noted), C_L = 150 pF (see Note 1)

PARAMETER	FROM (INPUT)	TO (OUTPUT)	V_{CC}	T_A = 25°C MIN	T_A = 25°C TYP	T_A = 25°C MAX	SN54HCT541 MIN	SN54HCT541 MAX	SN74HCT541 MIN	SN74HCT541 MAX	UNIT
t_{pd}	A	Y	4.5 V		20						ns
			5.5 V		19						
t_{en}	$\overline{G}$	Y	4.5 V		26						ns
			5.5 V		25						
t_t		Y	4.5 V		17						ns
			5.5 V		14						

NOTE 1: For load circuit and voltage waveforms, see page 1-14.

4

ADVANCE INFORMATION

TYPES SN54HC589, SN74HC589 8-BIT SHIFT REGISTERS WITH INPUT LATCHES AND 3-STATE OUTPUTS

D2804, MARCH 1984

- 8-Bit Parallel Storage Inputs
- Parallel 3-State I/O Storage Register Inputs, Shift Register Output
- High-Current 3-State Output Can Drive Up to 15 LSTTL Loads
- Package Options Include Both Plastic and Ceramic Chip Carriers in Addition to Plastic and Ceramic DIPs
- Dependable Texas Instruments Quality and Reliability

SN54HC589 . . . J PACKAGE
SN74HC589 . . . J OR N PACKAGE
(TOP VIEW)

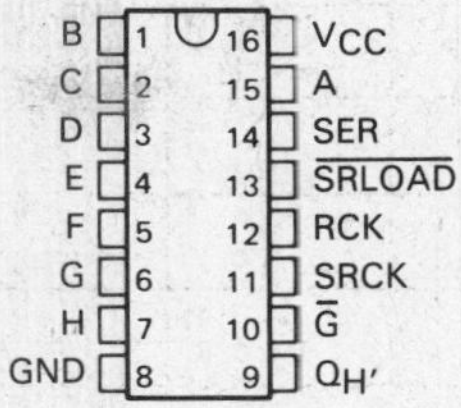

SN54HC589 . . . FH OR FK PACKAGE
SN74HC589 . . . FH OR FN PACKAGE
(TOP VIEW)

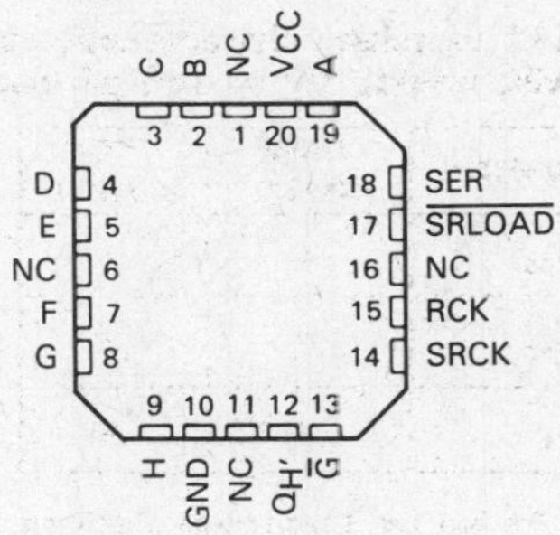

NC—No internal connection

description

The 'HC589 is similar to 'HC598 but has a three-state output whose control input replaces the direct clear input for the shift register. Like 'HC598, 'HC589 consists of an 8-bit storage register feeding a parallel-in, serial-out 8-bit shift register. Parallel loading of the storage register takes place on the positive-going edge of the RCK signal. If $\overline{\text{SRLOAD}}$ is low, data from the storage register is loaded into the shift register on the positive edge of the SRCK signal. If $\overline{\text{SRLOAD}}$ is high, data in the storage register is shifted one bit with new data entering serially at SER.

If the output enable $\overline{G}$ is high, the output is in the high-impedance state, but this does not affect loading, transfer of data from storage, or shifting.

The SN54HC589 is characterized for operation over the full military temperature range of −55°C to 125°C. The SN74HC589 is characterized for operation from −40°C to 85°C.

logic symbol

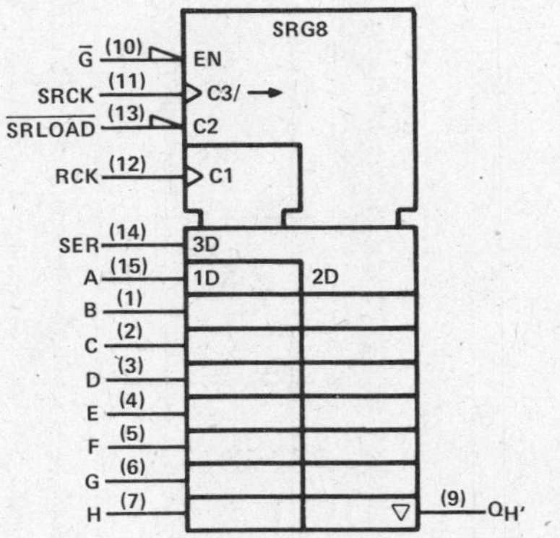

Pin numbers shown are for J and N packages.

logic diagram (positive logic)

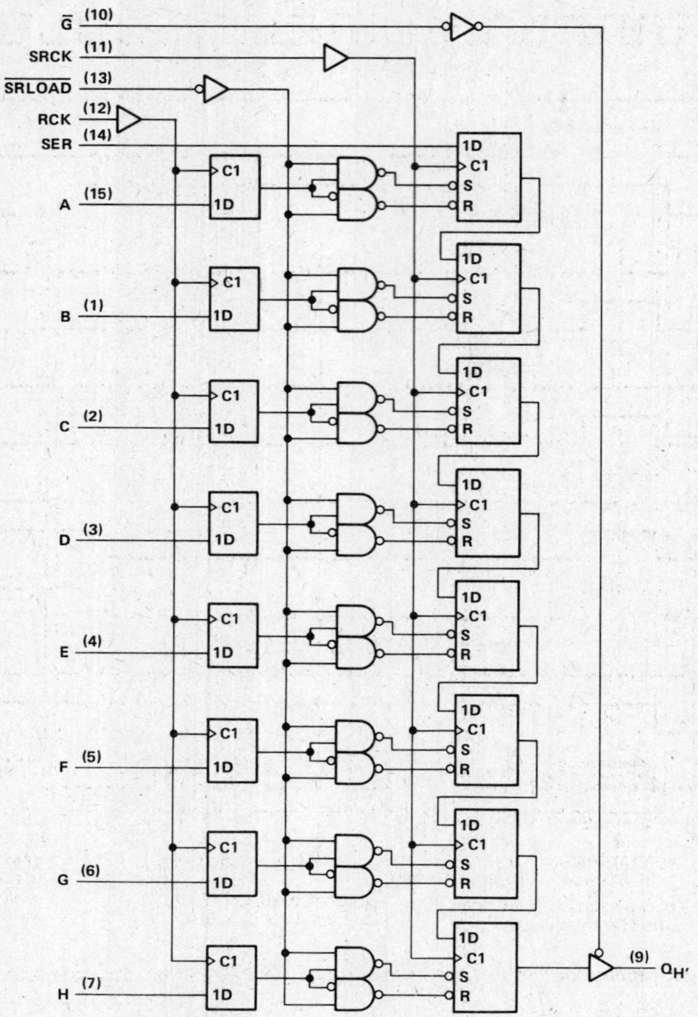

Pin numbers shown are for J and N packages.

TEXAS INSTRUMENTS
POST OFFICE BOX 225012 • DALLAS, TEXAS 75265

typical load and shift sequence

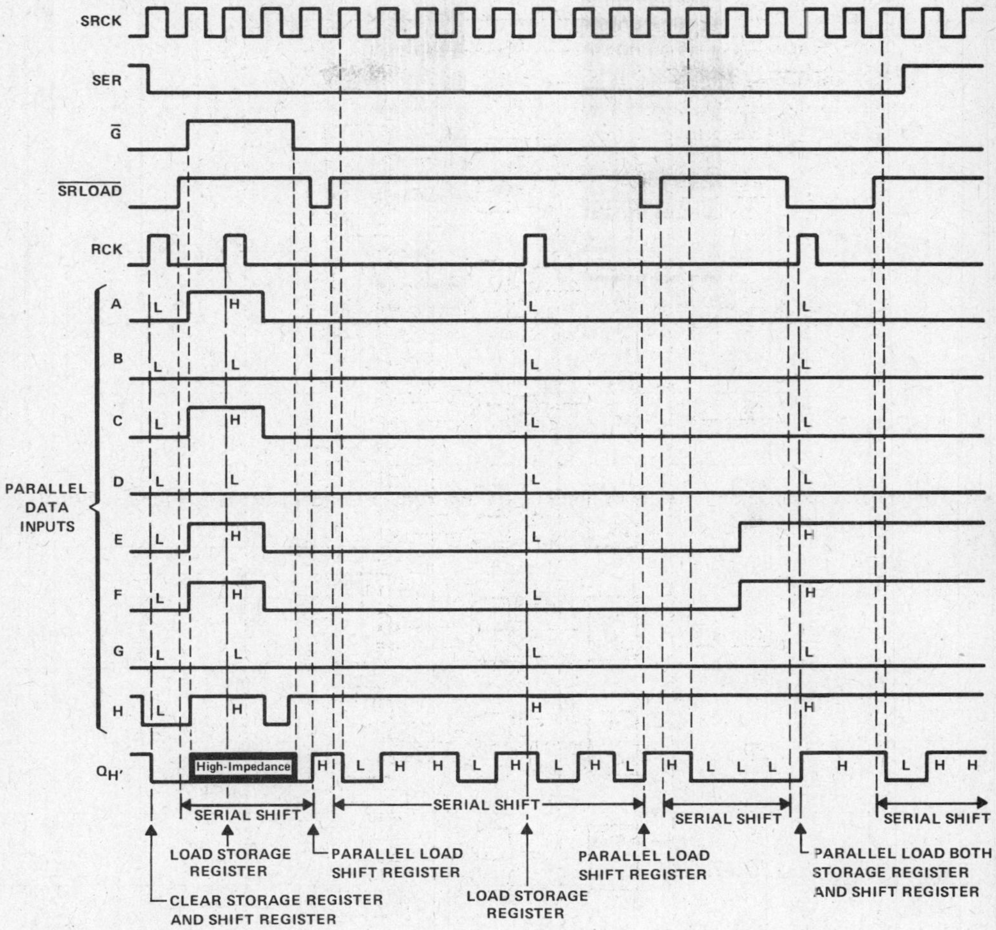

maximum ratings, recommended operating conditions, and electrical characteristics

See Table III, page 2-8.

timing requirements over recommended operating free-air temperature range (unless otherwise noted)

			V_{CC}	T_A = 25°C MIN	T_A = 25°C MAX	SN54HC589 MIN	SN54HC589 MAX	SN74HC589 MIN	SN74HC589 MAX	UNIT
f_{clock}	Shift clock frequency		2 V	0	6	0	5	0	4.2	MHz
			4.5 V	0	31	0	25	0	21	
			6 V	0	36	0	29	0	25	
t_w	Pulse duration, RCK or SRCK high or low, or $\overline{SRLOAD}$ low		2 V	80		100		120		ns
			4.5 V	16		20		24		
			6 V	14		17		20		
t_{su}	Setup time	A thru H before RCK↑	2 V	100		126		149		ns
			4.5 V	20		25		30		
			6 V	17		21		25		
		SER data or $\overline{SRLOAD}$ before SRCK↑	2 V	100		126		149		
			4.5 V	20		25		30		
			6 V	17		21		25		
t_h	Hold time	A thru H after RCK↑	2 V	25		32		37		ns
			4.5 V	5		6		7		
			6 V	4		5		6		
		SER data or $\overline{SRLOAD}$ after SRCK↑	2 V	5		5		5		
			4.5 V	5		5		5		
			6 V	5		5		5		

switching characteristics over recommended operating free-air temperature range (unless otherwise noted), C_L = 50 pF (see Note 1)

PARAMETER	FROM (INPUT)	TO (OUTPUT)	V_{CC}	T_A = 25°C MIN	T_A = 25°C TYP	T_A = 25°C MAX	SN54CH589 MIN	SN54CH589 MAX	SN74HC589 MIN	SN74HC589 MAX	UNIT
f_{max}	SRCK		2 V	6			5		4.2		MHz
			4.5 V	31			25		21		
			6 V	36			29		25		
t_{pd}	RCK	$Q_{H'}$	2 V		72	210		265		313	ns
			4.5 V		21	42		53		63	
			6 V		18	36		45		53	
t_{pd}	SRCK	$Q_{H'}$	2 V		88	175		221		261	ns
			4.5 V		18	35		44		52	
			6 V		15	30		37		44	
t_{pd}	$\overline{SRLOAD}$	$Q_{H'}$	2 V		88	175		221		261	ns
			4.5 V		18	35		44		52	
			6 V		15	30		37		44	
t_{en}	$\overline{G}$	$Q_{H'}$	2 V		75	150		189		224	ns
			4.5 V		15	30		38		45	
			6 V		13	26		32		38	
t_{dis}	$\overline{G}$	$Q_{H'}$	2 V		75	150		189		224	ns
			4.5 V		15	30		38		45	
			6 V		13	26		32		38	
t_t		Any	2 V		28	60		75		90	ns
			4.5 V		8	12		15		18	
			6 V		6	10		13		15	

NOTE 1: For load circuit and voltage waveforms, see page 1-14.

switching characteristics over recommended operating free-air temperature range (unless otherwise noted), C_L = 150 pF (see Note 1)

PARAMETER	FROM (INPUT)	TO (OUTPUT)	V_{CC}	T_A = 25°C MIN	T_A = 25°C MAX	SN54HC589 MIN	SN54HC589 MAX	SN74HC589 MIN	SN74HC589 MAX	UNIT
t_{pd}	RCK	$Q_{H'}$	2 V		260		328		387	ns
			4.5 V		52		66		77	
			6 V		44		56		66	
t_{pd}	SRCK	$Q_{H'}$	2 V		225		284		335	ns
			4.5 V		45		57		67	
			6 V		38		48		57	
t_{pd}	$\overline{\text{SRLOAD}}$	$Q_{H'}$	2 V		225		284		335	ns
			4.5 V		45		57		67	
			6 V		38		48		57	
t_{en}	$\overline{\text{G}}$	$Q_{H'}$	2 V		200		252		298	ns
			4.5 V		40		50		60	
			6 V		34		43		51	

NOTE 1: For load circuit and voltage waveforms, see page 1-14.

4

ADVANCE INFORMATION

TYPES SN54HC590, SN74HC590 8-BIT BINARY COUNTERS WITH 3-STATE OUTPUT REGISTERS

D2684, DECEMBER 1982—REVISED MARCH 1984

- **8-Bit Counter with Register**
- **High-Current 3-State Parallel Register Outputs Can Drive up to 15 LSTTL Loads**
- **Counter Has Direct Clear**
- **Package Options Include Both Plastic and Ceramic Chip Carriers in Addition to Plastic and Ceramic DIPs**
- **Dependable Texas Instruments Quality and Reliability**

SN54HC590 . . . J PACKAGE
SN74HC590 . . . J OR N PACKAGE
(TOP VIEW)

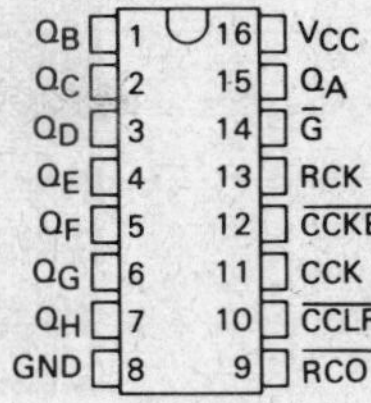

SN54HC590 . . . FH OR FK PACKAGE
SN74HC590 . . . FH OR FN PACKAGE
(TOP VIEW)

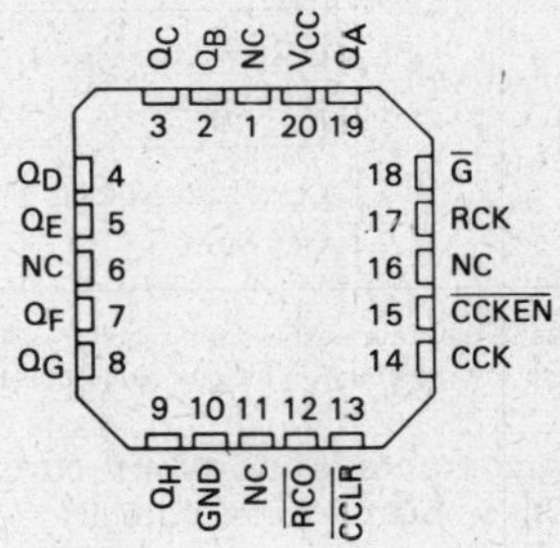

NC—No internal connection

description

These devices each contain an 8-bit binary counter that feeds an 8-bit storage register. The storage register has parallel outputs. Separate clocks are provided for both the binary counter and storage register. The binary counter features a direct clear input $\overline{\text{CCLR}}$ and a count enable input $\overline{\text{CCKEN}}$. For cascading a ripple carry output $\overline{\text{RCO}}$ is provided. Expansion is easily accomplished by tying $\overline{\text{RCO}}$ of the first stage to $\overline{\text{CCKEN}}$ of the second stage, etc.

Both the counter and register clocks are positive-edge triggered. If the user wishes to connect both clocks together, the counter state will always be one count ahead of the register. Internal circuitry prevents clocking from the clock enable.

The SN54HC590 is characterized for operation over the full military temperature range of −55°C to 125°C. The SN74HC590 is characterized for operation from −40°C to 85°C.

logic symbol

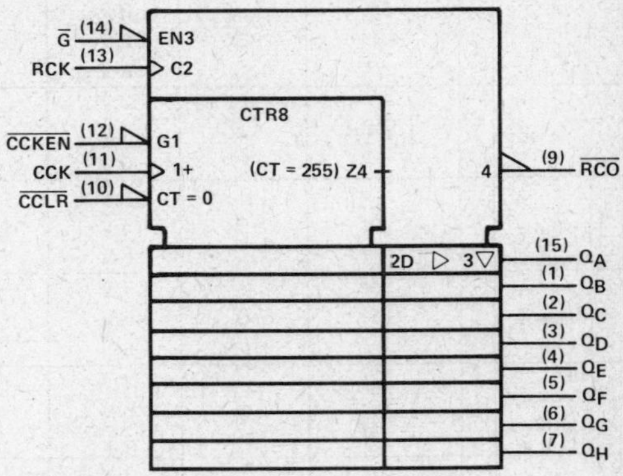

Pin numbers shown are for J and N packages.

maximum ratings, recommended operating conditions, and electrical characteristics

See Table III, page 2-8.

ADVANCE INFORMATION
This document contains information on a new product. Specifications are subject to change without notice.

timing requirements over recommended operating free-air temperature range (unless otherwise noted)

			V_{CC}	T_A = 25°C		SN54HC590		SN74HC590		UNIT
				MIN	MAX	MIN	MAX	MIN	MAX	
f_{clock}	Clock frequency, CCK or RCK		2 V	0	4	0	2.6	0	3.2	MHz
			4.5 V	0	20	0	13	0	16	
			6 V	0	24	0	16	0	19	
t_w	Pulse duration	CCK or RCK high or low	2 V	125		190		155		ns
			4.5 V	25		38		31		
			6 V	21		32		26		
		$\overline{\text{CCLR}}$ low	2 V	100		150		125		ns
			4.5 V	20		30		25		
			6 V	17		26		21		
t_{su}	Setup time	$\overline{\text{CCKEN}}$ low before CCK↑	2 V	100		150		125		ns
			4.5 V	20		30		25		
			6 V	17		26		21		
		$\overline{\text{CCLR}}$ high (inactive) before CCK↑	2 V	100		150		125		ns
			4.5 V	20		30		25		
			6 V	17		26		21		
		CLK↑ before RCK↑ (see Note 1)	2 V	200		300		250		ns
			4.5 V	40		60		50		
			6 V	34		51		43		

NOTE 1: This setup time ensures the register will see stable data from the counter outputs. The clocks may be tied together in which case the register will be one clock pulse behind the counter.

switching characteristics over recommended operating free-air temperature range (unless otherwise noted), C_L = 50 pF (see Note 2)

PARAMETER	FROM (INPUT)	TO (OUTPUT)	V_{CC}	T_A = 25°C			SN54HC590		SN74HC590		UNIT
				MIN	TYP	MAX	MIN	MAX	MIN	MAX	
f_{max}	CCK or RCK		2 V	4	8		2.6		3.2		MHz
			4.5 V	20	35		13		16		
			6 V	24	40		16		19		
t_{pd}	CCK↑	$\overline{\text{RCO}}$	2 V		75						ns
			4.5 V		25						
			6 V		23						
t_{PHL}	$\overline{\text{CCLR}}$↓	$\overline{\text{RCO}}$	2 V		90						ns
			4.5 V		30						
			6 V		27						
t_{pd}	$\overline{\text{RCK}}$↑	Q	2 V		75						ns
			4.5 V		25						
			6 V		23						
t_{en}	$\overline{\text{G}}$↓	Q	2 V		60						ns
			4.5 V		20						
			6 V		17						
t_{dis}	$\overline{\text{G}}$↑	Q	2 V		60						ns
			4.5 V		20						
			6 V		17						
t_t			2 V		28						ns
			4.5 V		8						
			6 V		6						

NOTE 2: For load circuit and voltage waveforms, see page 1-14.

D2804, MARCH 1984

- Inputs Are TTL-Voltage Compatible
- Lock Bus-Latch Capability
- Choice of True or Inverting Logic
- High-Current 3-State Outputs Can Drive Up to 15 LSTTL Loads
- Package Options Include Both Plastic and Ceramic Chip Carriers in Addition to Plastic and Ceramic DIPs
- Dependable Texas Instruments Quality and Reliability

DEVICE	LOGIC
'HCT620	Inverting
'HCT623	True

SN54HCT' . . . J PACKAGE
SN74HCT' . . . J OR N PACKAGE
(TOP VIEW)

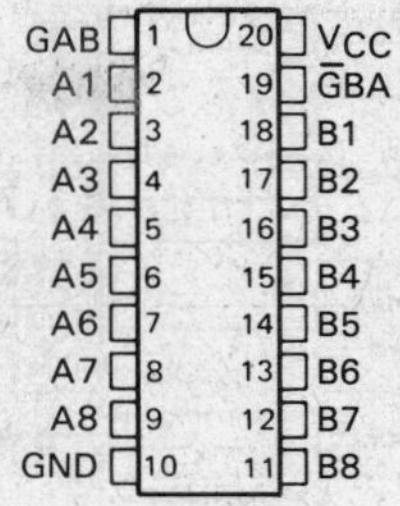

SN54HCT' . . . FH OR FK PACKAGE
SN74HCT' . . . FH OR FN PACKAGE
(TOP VIEW)

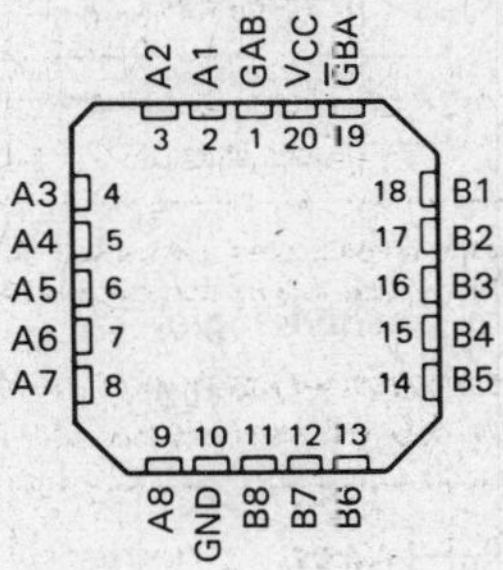

description

These octal bus transceivers are designed for asynchronous two-way communication between data buses. The control function implementation allows for maximum flexibility in timing.

These devices allow data transmission from the A bus to the B bus or from the B bus to the A bus depending upon the logic levels at the enable inputs ($\overline{G}$BA and GAB).

The enable inputs can be used to disable the device so that the buses are effectively isolated.

The dual-enable configuration gives these devices the capability to store data by simultaneous enabling of GBA and GAB. Each output reinforces its input in this transceiver configuration. Thus, when both control inputs are enabled and all other data sources to the two sets of bus lines are at high impedance, both sets of bus lines (16 in all) will remain at their last states. The 8-bit codes appearing on the two sets of buses will be identical for the 'HCT623 or complementary for the 'HCT620.

The SN54HCT620 and SN54HCT623 are characterized for operation over the full military temperature range of −55°C to 125°C. The SN74HCT620 and SN74HCT623 are characterized for operation from −40°C to 85°C.

FUNCTION TABLE

ENABLE INPUTS		OPERATION	
$\overline{G}$BA	GAB	'HCT620	'HCT623
L	L	$\overline{B}$ data to A bus	B data to A bus
H	H	$\overline{A}$ data to B bus	A data to B bus
H	L	Isolation	Isolation
L	H	$\overline{B}$ data to A bus, $\overline{A}$ data to B bus	B data to A bus, A data to B bus

ADVANCE INFORMATION
This document contains information on a new product. Specifications are subject to change without notice.

TEXAS INSTRUMENTS
POST OFFICE BOX 225012 • DALLAS, TEXAS 75265

logic symbol

'HCT620

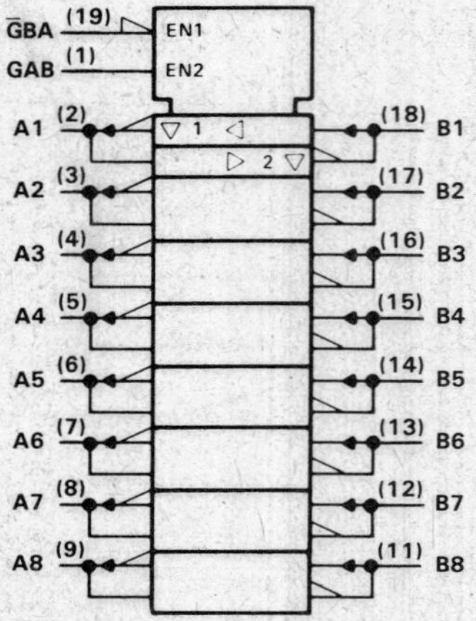

'HCT623

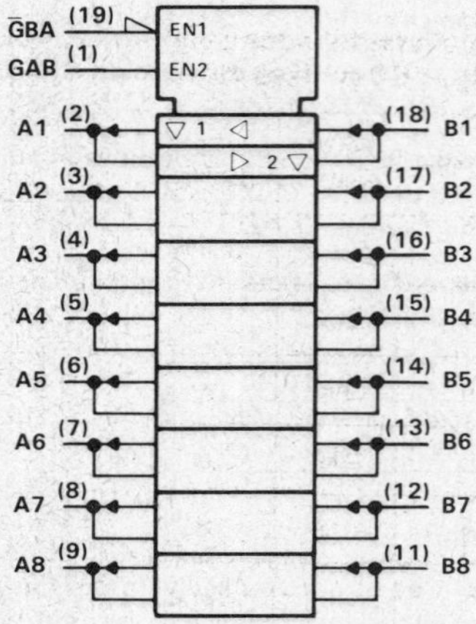

logic diagram (positive logic)

'HCT620

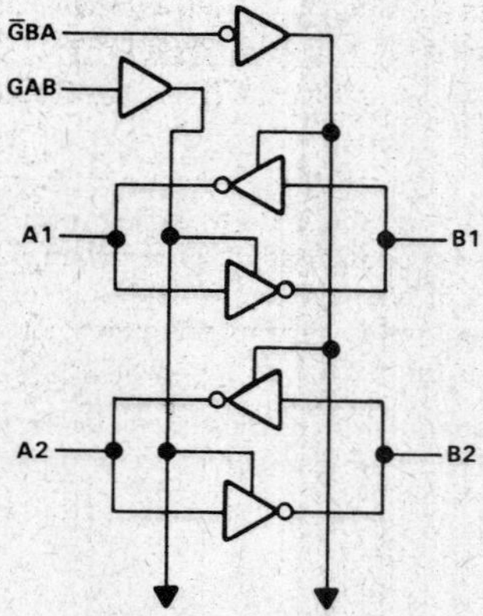

TO OTHER SIX TRANSCEIVERS

'HCT623

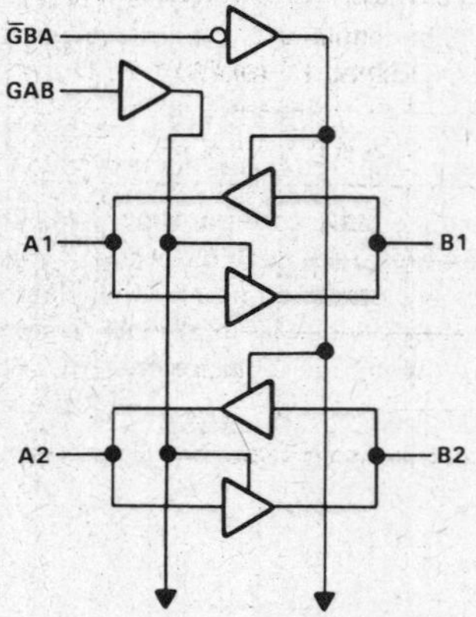

TO OTHER SIX TRANSCEIVERS

maximum ratings, recommended operating conditions, and electrical characteristics

See Table VII, page 2-14.

switching characteristics over recommended operating free-air temperature range (unless otherwise noted), C_L = 50 pF (see Note 1)

PARAMETER	FROM (INPUT)	TO (OUTPUT)	V_{CC}	T_A = 25°C MIN	TYP	MAX	SN54HCT620 SN54HCT623 MIN	MAX	SN74HCT620 SN74HCT623 MIN	MAX	UNIT
t_{pd}	A or B	B or A	4.5 V		13	21		32		26	ns
			5.5 V		11	18		27		23	
t_{en}	$\overline{G}$BA	A	4.5 V		30	42		63		53	ns
			5.5 V		23	38		57		48	
t_{dis}	$\overline{G}$BA	A	4.5 V		18	30		45		38	ns
			5.5 V		16	28		42		35	
t_{en}	GAB	B	4.5 V		30	42		63		53	ns
			5.5 V		23	38		57		48	
t_{dis}	GAB	B	4.5 V		18	30		45		38	ns
			5.5 V		16	28		42		35	
t_t		A or B	4.5 V		9	12		18		15	ns
			5.5 V		8	11		16		14	

C_{pd}	Power dissipation capacitance per transceiver	No load, T_A = 25°C	40 pF typ

switching characteristics over recommended operating free-air temperature range (unless otherwise noted), C_L = 150 pF (see Note 1)

PARAMETER	FROM (INPUT)	TO (OUTPUT)	V_{CC}	T_A = 25°C MIN	TYP	MAX	SN54HCT620 SN54HCT623 MIN	MAX	SN74HCT620 SN74HCT623 MIN	MAX	UNIT
t_{pd}	A or B	B or A	4.5 V		18	38		58		47	ns
			5.5 V		11	34		52		42	
t_{en}	$\overline{G}$BA	A	4.5 V		36	59		89		74	ns
			5.5 V		30	53		80		67	
t_{en}	GAB	B	4.5 V		36	59		89		74	ns
			5.5 V		30	53		80		67	
t_t		A or B	4.5 V		17	42		63		53	ns
			5.5 V		14	38		57		48	

NOTE 1: For load circuit and voltage waveforms, see page 1-14.

4

ADVANCE INFORMATION

TYPES SN54HC640, SN54HC643, SN54HC645 SN74HC640, SN74HC643, SN74HC645 OCTAL BUS TRANSCEIVERS WITH 3-STATE OUTPUTS

D2684, DECEMBER 1982—REVISED MARCH 1984

- Choice of True or Inverting Logic
- High-Current 3-State Outputs Can Drive up to 15 LSTTL Loads
- Package Options Include Both Plastic and Ceramic Chip Carriers in Addition to Plastic and Ceramic DIPs
- Dependable Texas Instruments Quality and Reliability

DEVICE	LOGIC
'HC640	Inverting
'HC643	True and Inverting
'HC645	True

SN54HC' . . . J PACKAGE
SN74HC' . . . J OR N PACKAGE
(TOP VIEW)

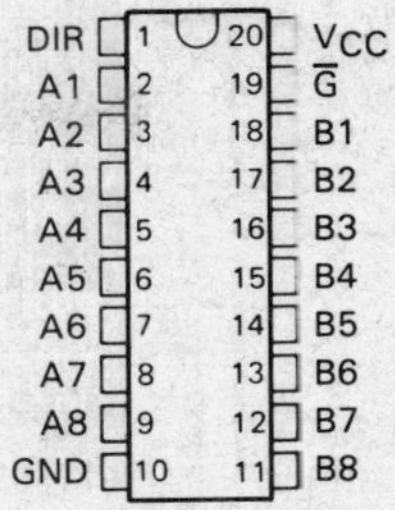

SN54HC' . . . FH OR FK PACKAGE
SN74HC' . . . FH OR FN PACKAGE
(TOP VIEW)

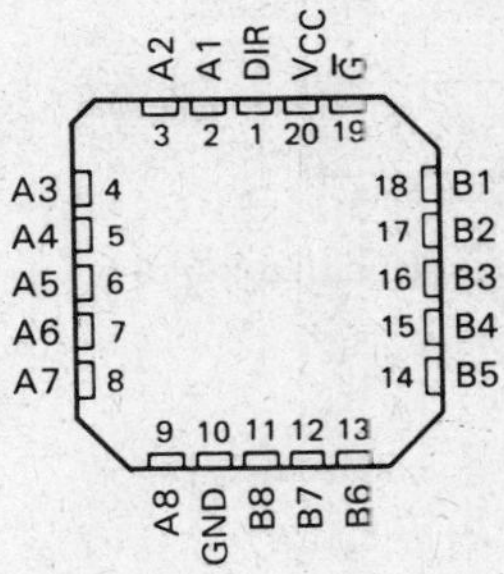

description

These octal bus transceivers are designed for asynchronous two-way communication between data buses. The devices transmit data from the A bus to the B bus or from the B bus to the A bus depending upon the level at the direction control (DIR) input. The enable input ($\overline{G}$) can be used to disable the device so the buses are effectively isolated.

The SN54HC640, SN54HC643, and SN54HC645 are characterized for operation over the full military temperature range of −55°C to 125°C. The SN74HC640, SN74HC643, and SN74HC645 are characterized for operation from −40°C to 85°C.

FUNCTION TABLE

CONTROL INPUTS		OPERATION		
$\overline{G}$	DIR	'HC640	'HC645	'HC643
L	L	$\overline{B}$ data to A bus	B data to A bus	B data to A bus
L	H	$\overline{A}$ data to B bus	A data to B bus	$\overline{A}$ data to B bus
H	X	Isolation	Isolation	Isolation

4

ADVANCE INFORMATION

logic symbols

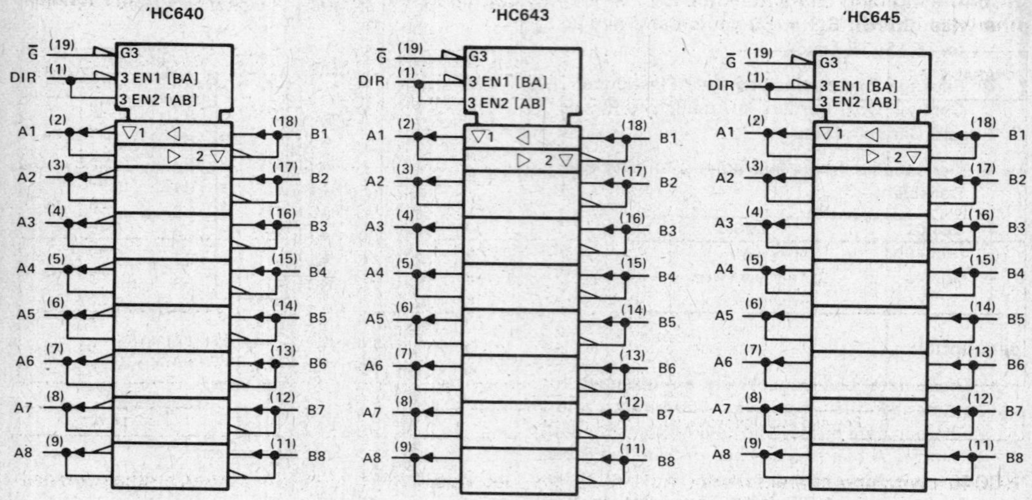

4

logic diagrams (positive logic)

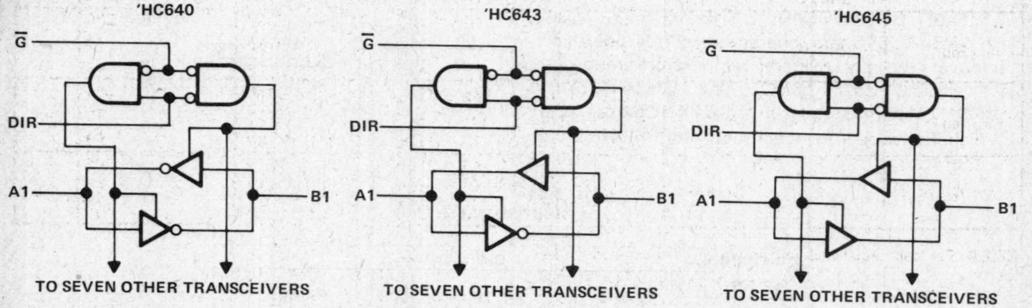

maximum ratings, recommended operating conditions, and electrical characteristics

See Table III, page 2-8.

TEXAS INSTRUMENTS
POST OFFICE BOX 225012 • DALLAS, TEXAS 75265

'HC640 switching characteristics over recommended operating free-air temperature range (unless otherwise noted), C_L = 50 pF (see Note 1)

PARAMETER	FROM (INPUT)	TO (OUTPUT)	V_{CC}	T_A = 25°C			SN54HC640		SN74HC640		UNIT
				MIN	TYP	MAX	MIN	MAX	MIN	MAX	
t_{pd}	A or B	B or A	2 V		29	105		160		130	ns
			4.5 V		10	21		32		26	
			6 V		8	18		27		22	
t_{en}	$\overline{G}$	A or B	2 V		109	230		340		290	ns
			4.5 V		27	46		68		58	
			6 V		20	39		58		49	
t_{dis}	$\overline{G}$	A or B	2 V		40	150		225		190	ns
			4.5 V		18	30		45		38	
			6 V		16	26		38		32	
t_t		A or B	2 V		20	60		90		75	ns
			4.5 V		8	12		18		15	
			6 V		6	10		15		13	

C_{pd}	Power dissipation capacitance per transceiver	No load, T_A = 25°C	40 pF typ

'HC640 switching characteristics over recommended operating free-air temperature range (unless otherwise noted), C_L = 150 pF (see Note 1)

PARAMETER	FROM (INPUT)	TO (OUTPUT)	V_{CC}	T_A = 25°C			SN54HC640		SN74HC640		UNIT
				MIN	TYP	MAX	MIN	MAX	MIN	MAX	
t_{pd}	A or B	B or A	2 V		44	190		290		235	ns
			4.5 V		14	38		58		47	
			6 V		11	33		49		41	
t_{en}	$\overline{G}$	A or B	2 V		124	315		470		395	ns
			4.5 V		31	63		94		79	
			6 V		23	54		88		68	
t_t		A or B	2 V		45	210		315		265	ns
			4.5 V		17	42		63		53	
			6 V		13	36		53		45	

NOTE 1: For load circuit and voltage waveforms, see page 1-14.

'HC643 switching characteristics over recommended operating free-air temperature range (unless otherwise noted), C_L = 50 pF (see Note 1)

PARAMETER	FROM (INPUT)	TO (OUTPUT)	V_{CC}	T_A = 25°C MIN	T_A = 25°C TYP	T_A = 25°C MAX	SN54HC643 MIN	SN54HC643 MAX	SN74HC643 MIN	SN74HC643 MAX	UNIT
t_{pd}	A or B	B or A	2 V		29	110		165		140	ns
			4.5 V		10	22		33		28	
			6 V		8	19		28		24	
t_{en}	$\overline{G}$	A or B	2 V		109	230		340		290	ns
			4.5 V		27	46		68		58	
			6 V		20	39		58		49	
t_{dis}	$\overline{G}$	A or B	2 V		40	150		225		190	ns
			4.5 V		18	30		45		38	
			6 V		16	26		38		32	
t_t		A or B	2 V		20	60		90		75	ns
			4.5 V		8	12		18		15	
			6 V		6	10		15		13	

C_{pd}	Power dissipation capacitance per transceiver	No load, T_A = 25°C	40 pF typ

'HC643 switching characteristics over recommended operating free-air temperature range (unless otherwise noted), C_L = 150 pF (see Note 1)

PARAMETER	FROM (INPUT)	TO (OUTPUT)	V_{CC}	T_A = 25°C MIN	T_A = 25°C TYP	T_A = 25°C MAX	SN54HC643 MIN	SN54HC643 MAX	SN74HC643 MIN	SN74HC643 MAX	UNIT
t_{pd}	A or B	B or A	2 V		44	195		295		245	ns
			4.5 V		14	39		59		49	
			6 V		11	34		50		43	
t_{en}	$\overline{G}$	A or B	2 V		124	315		470		395	ns
			4.5 V		31	63		94		79	
			6 V		23	54		80		68	
t_t		A or B	2 V		45	210		315		265	ns
			4.5 V		17	42		63		53	
			6 V		13	36		53		45	

NOTE 1: For load circuit and voltage waveforms, see page 1-14.

TEXAS INSTRUMENTS
POST OFFICE BOX 225012 • DALLAS, TEXAS 75265

'HC645 switching characteristics over recommended operating free-air temperature range (unless otherwise noted), C_L = 50 pF (see Note 1)

PARAMETER	FROM (INPUT)	TO (OUTPUT)	V_{CC}	T_A = 25°C MIN	TYP	MAX	SN54HC645 MIN	MAX	SN74HC645 MIN	MAX	UNIT
t_{pd}	A or B	B or A	2 V		40	105		160		130	ns
			4.5 V		15	21		32		26	
			6 V		12	18		27		22	
t_{en}	$\overline{G}$	A or B	2 V		125	230		340		290	ns
			4.5 V		23	46		68		58	
			6 V		20	39		58		49	
t_{dis}	$\overline{G}$	A or B	2 V		74	200		300		250	ns
			4.5 V		25	40		60		50	
			6 V		21	34		51		43	
t_t		A or B	2 V		20	60		90		75	ns
			4.5 V		8	12		18		15	
			6 V		6	10		15		13	

C_{pd}	Power dissipation capacitance per transceiver	No load, T_A = 25°C	40 pF typ

'HC645 switching characteristics over recommended operating free-air temperature range (unless otherwise noted), C_L = 150 pF (see Note 1)

PARAMETER	FROM (INPUT)	TO (OUTPUT)	V_{CC}	T_A = 25°C MIN	TYP	MAX	SN54HC645 MIN	MAX	SN74HC645 MIN	MAX	UNIT
t_{pd}	A or B	B or A	2 V		54	150		225		190	ns
			4.5 V		18	30		45		38	
			6 V		15	26		38		33	
t_{en}	$\overline{G}$	A or B	2 V		150	315		470		395	ns
			4.5 V		31	63		94		79	
			6 V		25	54		80		68	
t_t		A or B	2 V		45	210		315		265	ns
			4.5 V		17	42		63		53	
			6 V		13	36		53		45	

NOTE 1: For load circuit and voltage waveforms, see page 1-14.

TYPES SN54HCT640, SN54HCT643, SN54HCT645 SN74HCT640, SN74HCT643, SN74HCT645 OCTAL BUS TRANSCEIVERS WITH 3-STATE OUTPUTS

D2804, MARCH 1984

- **Inputs are TTL-Voltage Compatible**
- **Choice of True or Inverting Logic**
- **High-Current 3-State Outputs Can Drive up to 15 LSTTL Loads**
- **Package Options Include Both Plastic and Ceramic Chip Carriers in Addition to Plastic and Ceramic DIPs**
- **Dependable Texas Instruments Quality and Reliability**

DEVICE	LOGIC
'HCT640	Inverting
'HCT643	True and Inverting
'HCT645	True

SN54HCT' . . . J PACKAGE
SN74HCT' . . . J OR N PACKAGE
(TOP VIEW)

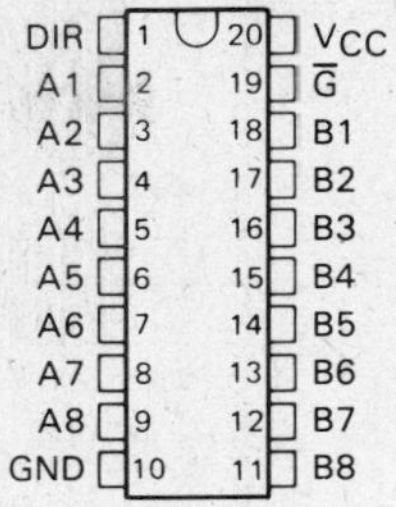

SN54HCT' . . . FH OR FK PACKAGE
SN74HCT' . . . FH OR FN PACKAGE
(TOP VIEW)

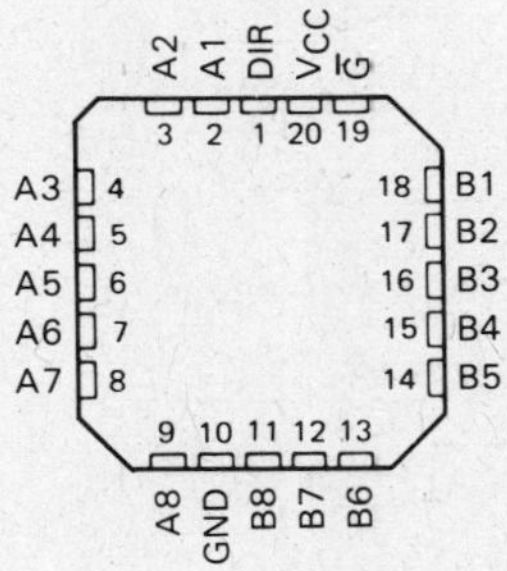

description

These octal bus transceivers are designed for asynchronous two-way communication between data buses. The devices transmit data from the A bus to the B bus or from the B bus to the A bus depending upon the level at the direction control (DIR) input. The enable input ($\overline{G}$) can be used to disable the device so the buses are effectively isolated.

The SN54HCT640, SN54HCT643, and SN54HCT645 are characterized for operation over the full military temperature range of −55°C to 125°C. The SN74HCT640, SN74HCT643 and SN74HCT645 are characterized for operation from −40°C to 85°C.

FUNCTION TABLE

CONTROL INPUTS		OPERATION		
$\overline{G}$	DIR	'HCT640	'HCT645	'HCT643
L	L	$\overline{B}$ data to A bus	B data to A bus	B data to A bus
L	H	$\overline{A}$ data to B bus	A data to B bus	$\overline{A}$ data to B bus
H	X	Isolation	Isolation	Isolation

4 ADVANCE INFORMATION

logic symbols

'HCT640

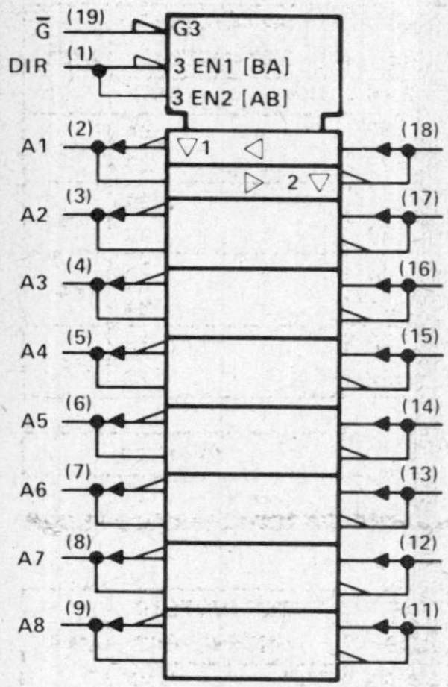

'HCT643

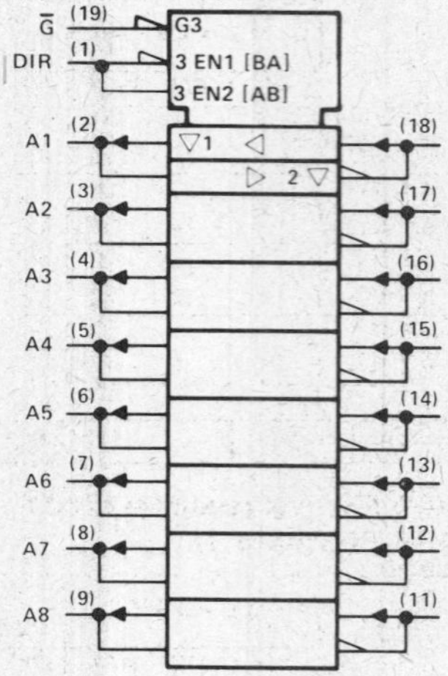

'HCT645

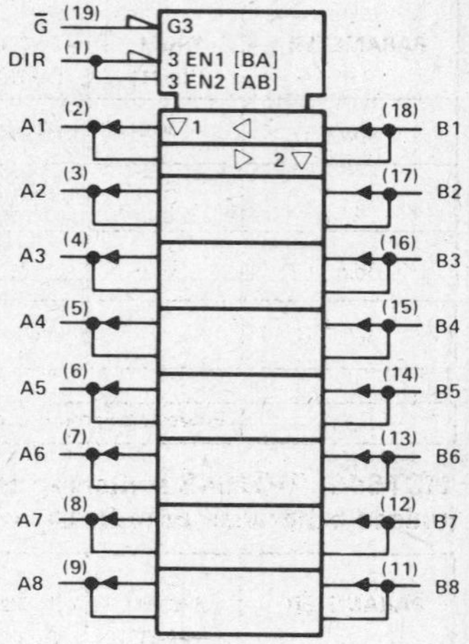

logic diagrams (positive logic)

'HCT640

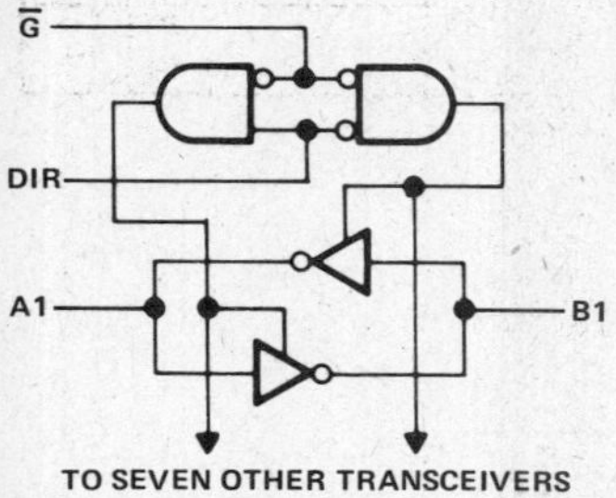

'HCT643

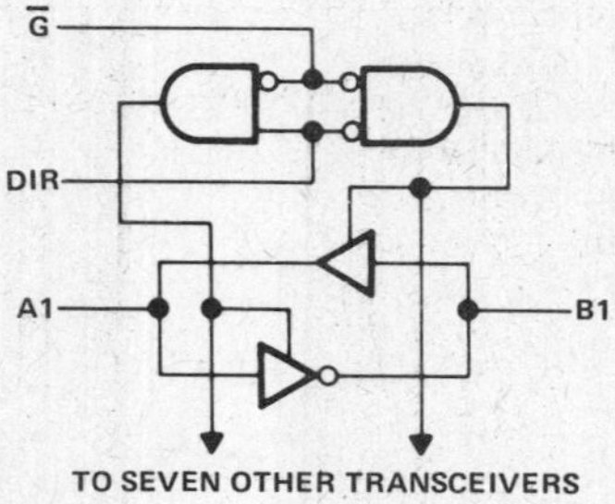

'HCT645

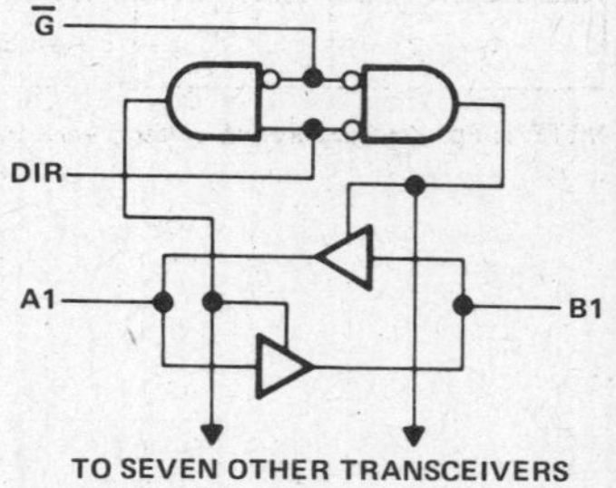

maximum ratings, recommended operating conditions, and electrical characteristics

See Table VII, page 2-14.

TEXAS INSTRUMENTS
POST OFFICE BOX 225012 • DALLAS, TEXAS 75265

'HCT640, 'HCT643 switching characteristics over recommended operating free-air temperature range (unless otherwise noted), C_L = 50 pF (see Note 1)

PARAMETER	FROM (INPUT)	TO (OUTPUT)	V_{CC}	T_A = 25°C MIN	T_A = 25°C TYP	T_A = 25°C MAX	SN54HCT640 SN54HCT643 MIN	SN54HCT640 SN54HCT643 MAX	SN74HCT640 SN74HCT643 MIN	SN74HCT640 SN74HCT643 MAX	UNIT
t_{pd}	A or B	B or A	4.5 V		14	21		32		26	ns
			5.5 V		12	18		27		23	
t_{en}	$\overline{G}$	A or B	4.5 V		27	35		53		44	ns
			5.5 V		24	32		47		39	
t_{dis}	$\overline{G}$	A or B	4.5 V		20	30		45		38	ns
			5.5 V		18	26		41		34	
t_t		A or B	4.5 V		9	12		18		15	ns
			5.5 V		8	11		16		14	

C_{pd}	Power dissipation capacitance per transceiver	No load, T_A = 25°C	40 pF typ

'HCT640, 'HCT643 switching characteristics over recommended operating free-air temperature range (unless otherwise noted), C_L = 150 pF (see Note 1)

PARAMETER	FROM (INPUT)	TO (OUTPUT)	V_{CC}	T_A = 25°C MIN	T_A = 25°C TYP	T_A = 25°C MAX	SN54HCT640 SN54HCT643 MIN	SN54HCT640 SN54HCT643 MAX	SN74HCT640 SN74HCT643 MIN	SN74HCT640 SN74HCT643 MAX	UNIT
t_{pd}	A or B	B or A	4.5 V		17	27		41		34	ns
			5.5 V		15	24		37		30	
t_{en}	$\overline{G}$	A or B	4.5 V		31	45		68		56	ns
			5.5 V		28	41		61		51	
t_t		A or B	4.5 V		17	42		63		53	ns
			5.5 V		14	38		57		48	

NOTE 1: For load circuit and voltage waveforms, see page 1-14.

4

ADVANCE INFORMATION

'HCT645 switching characteristics over recommended operating free-air temperature range (unless otherwise noted), C_L = 50 pF (see Note 1)

PARAMETER	FROM (INPUT)	TO (OUTPUT)	V_{CC}	T_A = 25°C MIN	TYP	MAX	SN54HCT645 MIN	MAX	SN74HCT645 MIN	MAX	UNIT
t_{pd}	A or B	B or A	4.5 V		16	22		33		28	ns
			5.5 V		14	20		30		25	
t_{en}	$\overline{G}$	A or B	4.5 V		25	46		69		58	ns
			5.5 V		22	41		62		52	
t_{dis}	$\overline{G}$	A or B	4.5 V		26	40		60		50	ns
			5.5 V		23	36		54		45	
t_t		A or B	4.5 V		9	12		18		15	ns
			5.5 V		8	11		16		14	

C_{pd}	Power dissipation capacitance per transceiver	No load, T_A = 25°C	40 pF typ

'HCT645 switching characteristics over recommended operating free-air temperature range (unless otherwise noted), C_L = 150 pF (see Note 1)

PARAMETER	FROM (INPUT)	TO (OUTPUT)	V_{CC}	T_A = 25°C MIN	TYP	MAX	SN54HCT645 MIN	MAX	SN74HCT645 MIN	MAX	UNIT
t_{pd}	A or B	B or A	4.5 V		20	30		45		38	ns
			5.5 V		18	27		41		34	
t_{en}	$\overline{G}$	A or B	4.5 V		36	59		89		74	ns
			5.5 V		30	63		80		67	
t_t		A or B	4.5 V		17	42		63		53	ns
			5.5 V		14	38		57		48	

NOTE 1: For load circuit and voltage waveforms, see page 1-14.

4

ADVANCE INFORMATION

D2804, MARCH 1984

'HC673

- **16-Bit Serial-In, Serial-Out Shift Register with 16-Bit Parallel-Out Storage Register**
- **Performs Serial-to-Parallel Conversion**

'HC674

- **16-Bit Parallel-In, Serial-Out Shift Register**
- **Performs Parallel-to-Serial Conversion**
- **Package Options Include Both Plastic and Ceramic Chip Carriers in Addition to Plastic and Ceramic DIPs**
- **Dependable Texas Instruments Quality and Reliability**

SN54HC673 . . . JT PACKAGE
SN74HC673 . . . JT OR NT PACKAGE
(TOP VIEW)

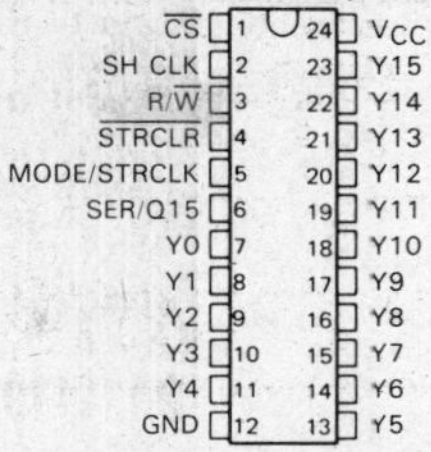

SN54HC673 . . . FH OR FK PACKAGE
SN74HC673 . . . FH OR FN PACKAGE
(TOP VIEW)

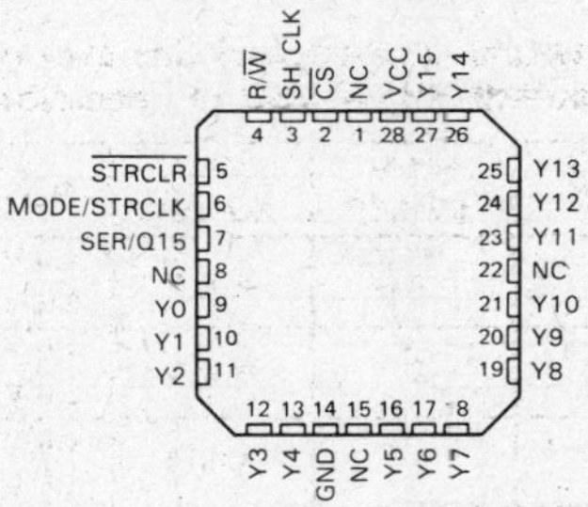

SN54HC674 . . . JT PACKAGE
SN74HC674 . . . JT OR NT PACKAGE
(TOP VIEW)

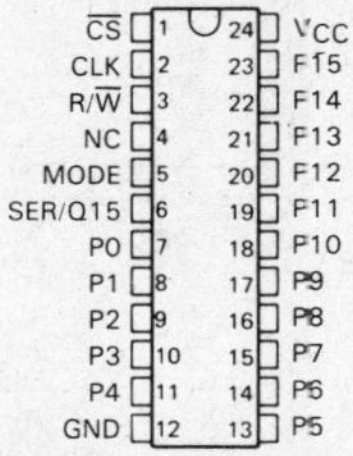

SN54HC674 . . . FH OR FK PACKAGE
SN74HC674 . . . FH OR FN PACKAGE
(TOP VIEW)

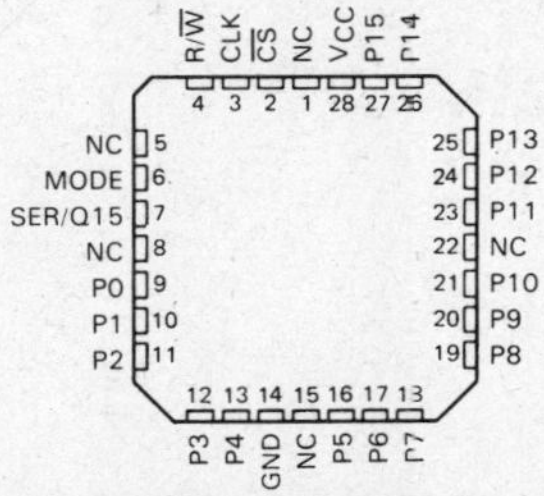

NC—No internal connection

description

SN54HC673, SN74HC673

The 'HC673 is a 16-bit shift register and a 16-bit storage register in a single 24-pin package. A three-state input/output (SER/Q15) port to the shift register allows serial entry and/or reading of data. The storage register is connected in a parallel data loop with the shift register and may be asynchronously cleared by taking the store-clear input low. The storage register may be parallel loaded with shift-register data to provide shift-register status via the parallel outputs. The shift register can be parallel loaded with the storage-register data upon command.

A high logic level at the chip-select (CS) input disables both the shift-register clock and the storage-register clock and places SER/Q15 in the high-impedance state. The store-clear function is not disabled by the chip select.

Caution must be exercised to prevent false clocking of either the shift register or the storage register via the chip-select input. The shift clock should be low during the low-to-high transition of chip select and the store clock should be low during the high-to-low transition of chip select.

The SN54HC673 is characterized for operation over the full military temperature range of −55°C to 125°C. The SN74HC673 is characterized for operation from −40°C to 85°C.

TEXAS INSTRUMENTS
POST OFFICE BOX 225012 • DALLAS, TEXAS 75265

SN54HC674, SN74HC674

The 'HC674 is a 16-bit parallel-in, serial-out shift register. A three-state input/output (SER/Q15) port provides access for entering serial data or reading the shift-register word in a recirculating loop.

The device has four basic modes of operation:

1) Hold (do nothing)
2) Write (serially via input/output)
3) Read (serially)
4) Load (parallel via data inputs)

Low-to-high-level changes at the chip select input should be made only when the clock input is low to prevent false clocking.

The SN54HC674 is characterized for operation over the full military temperature range of −55 °C to 125 °C. The SN74HC674 is characterized for operation from −40 °C to 85 °C.

'HC673
FUNCTION TABLE

INPUTS					SER/ Q15	SHIFT REGISTER FUNCTIONS				STORAGE REGISTER FUNCTIONS	
$\overline{CS}$	R/$\overline{W}$	SH CLK	$\overline{STRCLR}$	MODE/ STRCLK		SHIFT	READ FROM SERIAL OUTPUT	WRITE INTO SERIAL INPUT	PARALLEL LOAD	CLEAR	LOAD
H	X	X	X	X	Z	NO	NO	NO	NO		NO
X	X	X	L	X						YES	
L	L	↓	X	X	Z	YES	NO	YES	NO		
L	H	X	X	X	Q15		YES	NO			NO
L	H	↓	X	L	Q14n	YES	YES	NO	NO		NO
L	H	↓	L	H	L	NO	YES		YES	YES	NO
L	H	↓	H	H	Y15n	NO	YES		YES	NO	NO
L	L	X	H	↑	Z		NO		NO	NO	YES

'HC674 FUNCTION TABLE

INPUTS				SER/ Q15	OPERATION
$\overline{CS}$	R/$\overline{W}$	MODE	CLK		
H	X	X	X	Z	Do nothing
L	L	X	↓	Z	Shift and write (serial load)
L	H	L	↓	Q14n	Shift and read
L	H	H	↓	P15	Parallel load

H = high level (steady state)
L = low level (steady state)
↑ = transition from low to high level
↓ = transition from high to low level
X = irrelevant (any input including transitions)
Z = high impedance, input mode
Q14n = content of 14th bit of the shift register before the most recent ↓ transition of the clock
Q15 = present content of 15th bit of the shift register
Y15n = content of the 15th bit of the storage register before the most recent ↓ transition of the clock
P15 = level of input P15

4

ADVANCE INFORMATION

TEXAS INSTRUMENTS
POST OFFICE BOX 225012 • DALLAS, TEXAS 75265

logic symbols

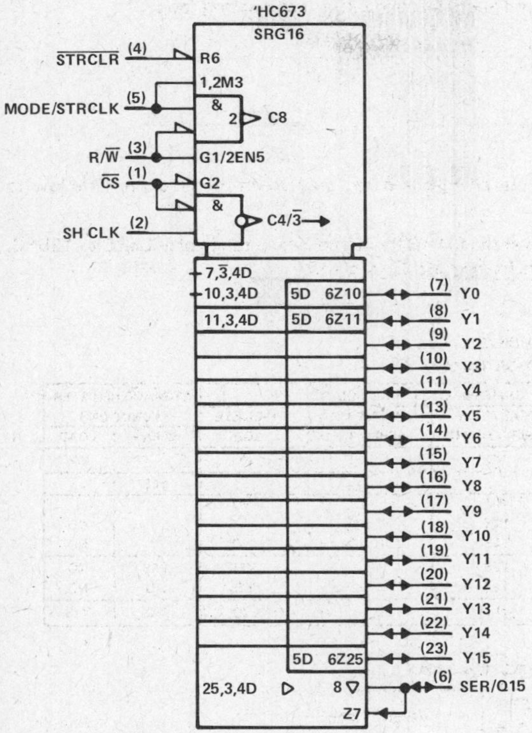

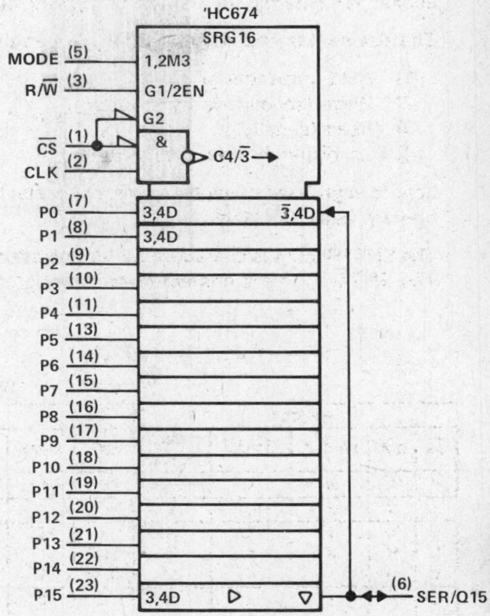

Pin numbers shown are for JT and NT packages.

4

ADVANCE INFORMATION

functional block diagrams (positive logic)

'HC673

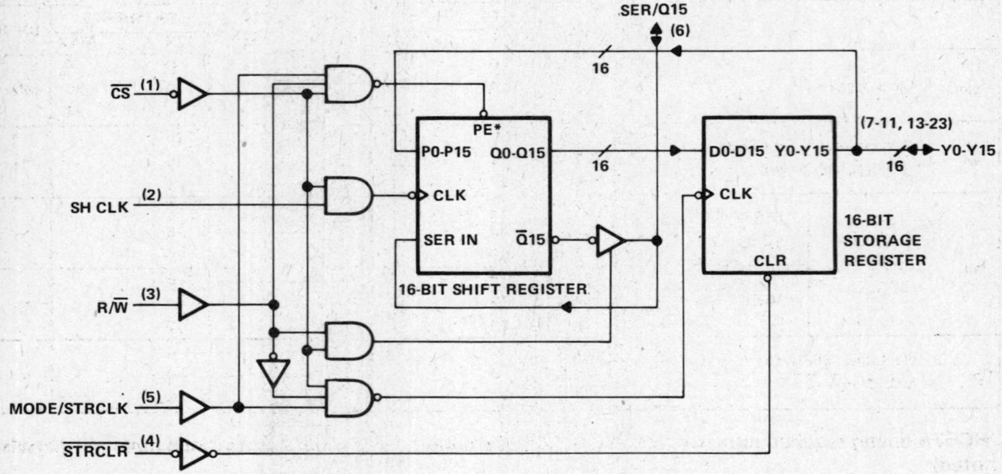

'HC674

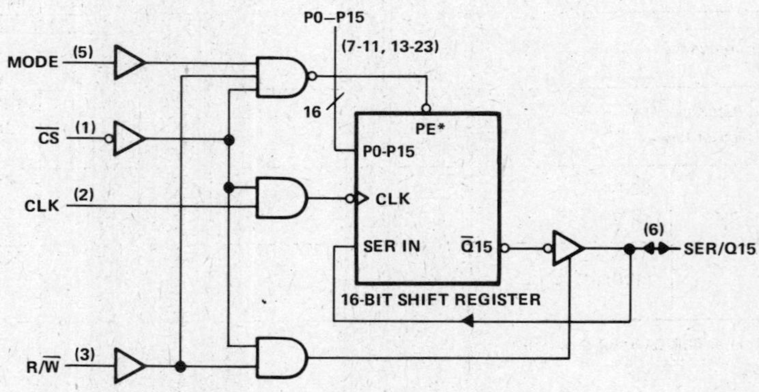

*When PE is active, data is synchronously parallel loaded into the shift registers from the 16 P inputs and no shifting takes place.

Pin numbers shown are for JT and NT packages.

maximum ratings recommended operating conditions, and electrical characteristics

See Table III, page 2-8.

TEXAS INSTRUMENTS
POST OFFICE BOX 225012 • DALLAS, TEXAS 75265

'HC673 timing requirements over recommended operating free-air temperature range (unless otherwise noted)

			V_{CC}	T_A = 25°C MIN	T_A = 25°C MAX	SN54HC673 MIN	SN54HC673 MAX	SN74HC673 MIN	SN74HC673 MAX	UNIT
f_{clock}	Clock frequency		2 V	0	5	0	3.3	0	4	MHz
			4.5 V	0	25	0	17	0	20	
			6 V	0	29	0	19	0	24	
t_w	Pulse duration, SH CLK or STRCLK high or low		2 V	100		150		125		ns
			4.5 V	20		30		25		
			6 V	17		26		21		
t_{su}	Setup time before SH CLK↓	SER/Q15, Y0 thru Y15	2 V	100		150		125		ns
			4.5 V	20		30		25		
			6 V	17		26		21		
		MODE, R/$\overline{W}$, $\overline{CS}$	2 V	175		265		220		ns
			4.5 V	35		53		44		
			6 V	30		45		37		
t_h	Hold time, SER/Q15, Y0 thru Y15, MODE, R/$\overline{W}$, $\overline{CS}$		2 V	0		0		0		ns
			4.5 V	0		0		0		
			6 V	0		0		0		

'HC674 timing requirements over recommended operating free-air temperature range (unless otherwise noted)

			V_{CC}	T_A = 25°C MIN	T_A = 25°C MAX	SN54HC674 MIN	SN54HC674 MAX	SN74HC674 MIN	SN74HC674 MAX	UNIT
f_{clock}	Clock frequency		2 V	0	5	0	3.3	0	4	MHz
			4.5 V	0	25	0	17	0	20	
			6 V	0	29	0	19	0	24	
t_w	Pulse duration, CLK high or low		2 V	100		150		125		ns
			4.5 V	20		30		25		
			6 V	17		26		21		
t_{su}	Setup time before CLK↓	SER/Q15, P0 thru P15	2 V	100		150		125		ns
			4.5 V	20		30		25		
			6 V	17		26		21		
		MODE, R/$\overline{W}$, $\overline{CS}$	2 V	175		265		220		ns
			4.5 V	35		53		44		
			6 V	30		45		37		
t_h	Hold time, SER/Q15, P0 thru P15, MODE, R/$\overline{W}$, $\overline{CS}$		2 V	0		0		0		ns
			4.5 V	0		0		0		
			6 V	0		0		0		

'HC673 switching characteristics over recommended operating free-air temperature range (unless otherwise noted), C_L = 50 pF (see Note 1)

PARAMETER	FROM (INPUT)	TO (OUTPUT)	V_{CC}	T_A = 25°C MIN	TYP	MAX	SN54HC673 MIN	MAX	SN74HC673 MIN	MAX	UNIT
f_{max}			2 V	5	7		3.3		4		MHz
			4.5 V	25	28		17		20		
			6 V	29	32		19		24		
t_{pd}	MODE/STRCLK	Y	2 V		66						ns
			4.5 V		23						
			6 V		20						
t_{pd}	$\overline{STRCLR}$	Y	2 V		57						ns
			4.5 V		19						
			6 V		16						
t_{pd}	SH CLK	SER/Q15	2 V		72						ns
			4.5 V		24						
			6 V		20						
t_{en}	$\overline{CS}$, R/$\overline{W}$	SER/Q15	2 V		66						ns
			4.5 V		23						
			6 V		20						
t_{dis}	$\overline{CS}$, R/$\overline{W}$	SER/Q15	2 V		66						ns
			4.5 V		23						
			6 V		20						

C_{pd}	Power dissipation capacitance	No load, T_A = 25°C	150 pF typ

4

ADVANCE INFORMATION

'HC674 switching characteristics over recommended operating free-air temperature range (unless otherwise noted), C_L = 50 pF (see Note 1)

PARAMETER	FROM (INPUT)	TO (OUTPUT)	V_{CC}	T_A = 25°C MIN	TYP	MAX	SN54HC674 MIN	MAX	SN74HC674 MIN	MAX	UNIT
f_{max}			2 V	5	7		3.3		4		MHz
			4.5 V	25	28		17		20		
			6 V	29	32		19		24		
t_{pd}	CLK	SER/Q15	2 V		72						ns
			4.5 V		24						
			6 V		20						
t_{en}	$\overline{CS}$, R/$\overline{W}$	SER/Q15	2 V		70						ns
			4.5 V		23						
			6 V		20						
t_{dis}	$\overline{CS}$, R/$\overline{W}$	SER/Q15	2 V		70						ns
			4.5 V		23						
			6 V		20						

C_{pd}	Power dissipation capacitance	No load, T_A = 25°C	150 pF typ

NOTE 1: For load circuit and voltage waveforms, see page 1-14.

TEXAS INSTRUMENTS
POST OFFICE BOX 225012 • DALLAS, TEXAS 75265

TYPES SN54HC677, SN54HC678, SN74HC677, SN74HC678 16-BIT ADDRESS COMPARATORS

D2833, MARCH 1984

- **'HC677 is a 16-Bit Address Comparator with Enable**
- **'HC678 is a 16-Bit Address Comparator with Latch**
- **Package Options Include Both Plastic and Ceramic Chip Carriers in Addition to Plastic and Ceramic DIPs**
- **Dependable Texas Instruments Quality and Reliability**

description

The 'HC677 and 'HC678 address comparators simplify addressing of memory boards and/or other peripheral devices. The four P inputs are normally hard wired with a preprogrammed address. An internal decoder determines what input information applied to the 16 A inputs must be low or high to cause a low state at the output (Y). For example, a positive-logic bit combination of 0111 (decimal 7) at the P input determines that inputs A1 through A7 must be low and that inputs A8 through A16 must be high to cause the output to go low. Equality of the address applied at the A inputs to the preprogrammed address is indicated by the output being low.

The 'HC677 features an enable input ($\overline{G}$). When $\overline{G}$ is low, the device is enabled. When $\overline{G}$ is high, the device is disabled and the output is high regardless of the A and P inputs. The 'HC678 features a transparent latch and a latch enable input (C). When C is high, the device is in the transparent mode. When C is low, the previous logic state of Y is latched.

The SN54HC677 and SN54HC678 are characterized for operation over the full military temperature range of −55°C to 125°C. The SN54HC677 and SN74HC678 are characterized for operation from −40°C to 85°C.

SN54HC677 . . . JT PACKAGE
SN74HC677 . . . JT OR NT PACKAGE
(TOP VIEW)

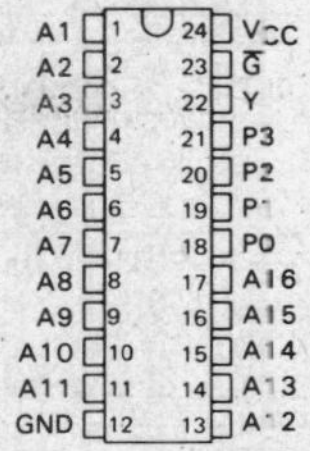

SN54HC677 . . . FH OR FK PACKAGE
SN74HC677 . . . FH OR FN PACKAGE
(TOP VIEW)

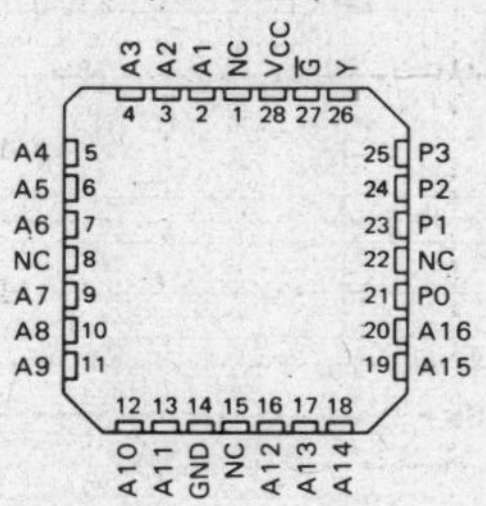

SN54HC678 . . . JT PACKAGE
SN74HC678 . . . JT OR NT PACKAGE
(TOP VIEW)

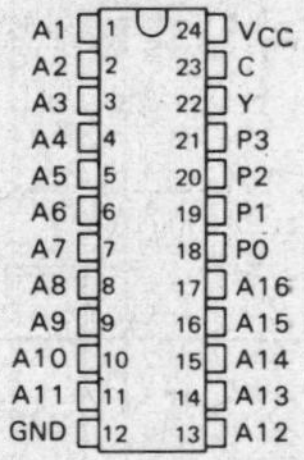

SN54HC678 . . . FH OR FK PACKAGE
SN74HC678 . . . FH OR FN PACKAGE
(TOP VIEW)

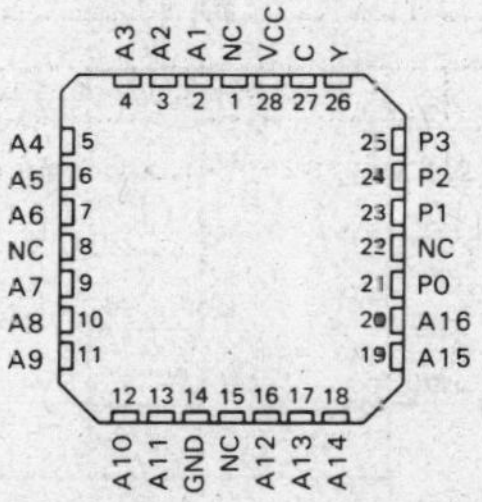

NC—No internal connection

ADVANCE INFORMATION
This document contains information on a new product. Specifications are subject to change without notice.

FUNCTION TABLE

'HC677	'HC678	INPUTS COMMON TO 'HC677 AND 'HC678																				OUTPUT
$\overline{G}$	C	P3	P2	P1	P0	A1	A2	A3	A4	A5	A6	A7	A8	A9	A10	A11	A12	A13	A14	A15	A16	Y
L	H	L	L	L	L	H	H	H	H	H	H	H	H	H	H	H	H	H	H	H	H	L
L	H	L	L	L	H	L	H	H	H	H	H	H	H	H	H	H	H	H	H	H	H	L
L	H	L	L	H	L	L	L	H	H	H	H	H	H	H	H	H	H	H	H	H	H	L
L	H	L	L	H	H	L	L	L	H	H	H	H	H	H	H	H	H	H	H	H	H	L
L	H	L	H	L	L	L	L	L	L	H	H	H	H	H	H	H	H	H	H	H	H	L
L	H	L	H	L	H	L	L	L	L	L	H	H	H	H	H	H	H	H	H	H	H	L
L	H	L	H	H	L	L	L	L	L	L	L	H	H	H	H	H	H	H	H	H	H	L
L	H	L	H	H	H	L	L	L	L	L	L	L	H	H	H	H	H	H	H	H	H	L
L	H	H	L	L	L	L	L	L	L	L	L	L	L	H	H	H	H	H	H	H	H	L
L	H	H	L	L	H	L	L	L	L	L	L	L	L	L	H	H	H	H	H	H	H	L
L	H	H	L	H	L	L	L	L	L	L	L	L	L	L	L	H	H	H	H	H	H	L
L	H	H	L	H	H	L	L	L	L	L	L	L	L	L	L	L	H	H	H	H	H	L
L	H	H	H	L	L	L	L	L	L	L	L	L	L	L	L	L	L	H	H	H	H	L
L	H	H	H	L	H	L	L	L	L	L	L	L	L	L	L	L	L	L	H	H	H	L
L	H	H	H	H	L	L	L	L	L	L	L	L	L	L	L	L	L	L	L	H	H	L
L	H	H	H	H	H	L	L	L	L	L	L	L	L	L	L	L	L	L	L	L	H	L
L	H	All other combinations																				H
H		'HC677: Any combination																				H
	L	'HC678: Any combination																				Latched

logic symbols

'HC677

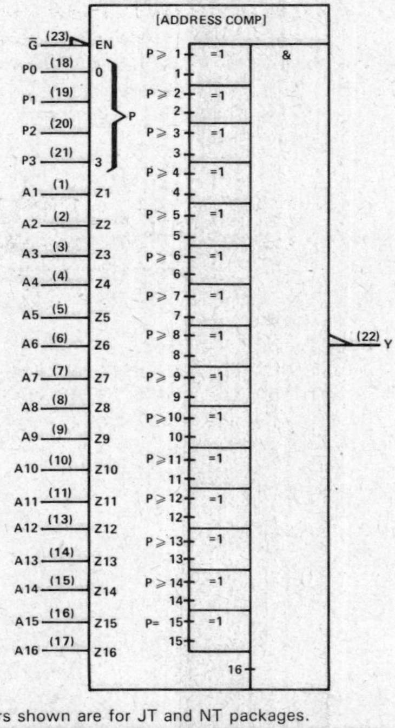

'HC678

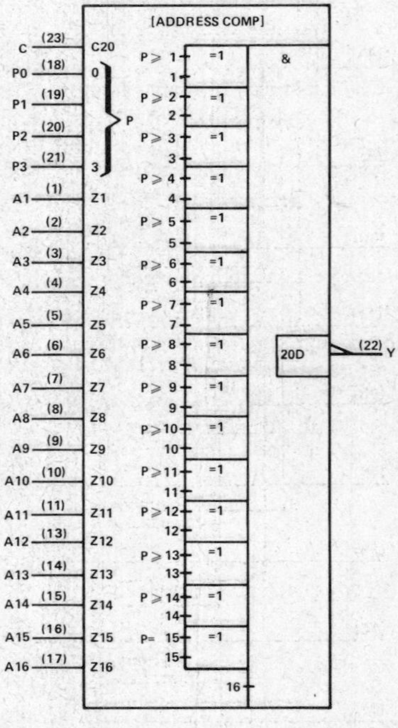

Pin numbers shown are for JT and NT packages.

TEXAS INSTRUMENTS
POST OFFICE BOX 225012 • DALLAS, TEXAS 75265

'HC677 logic diagram (positive logic)

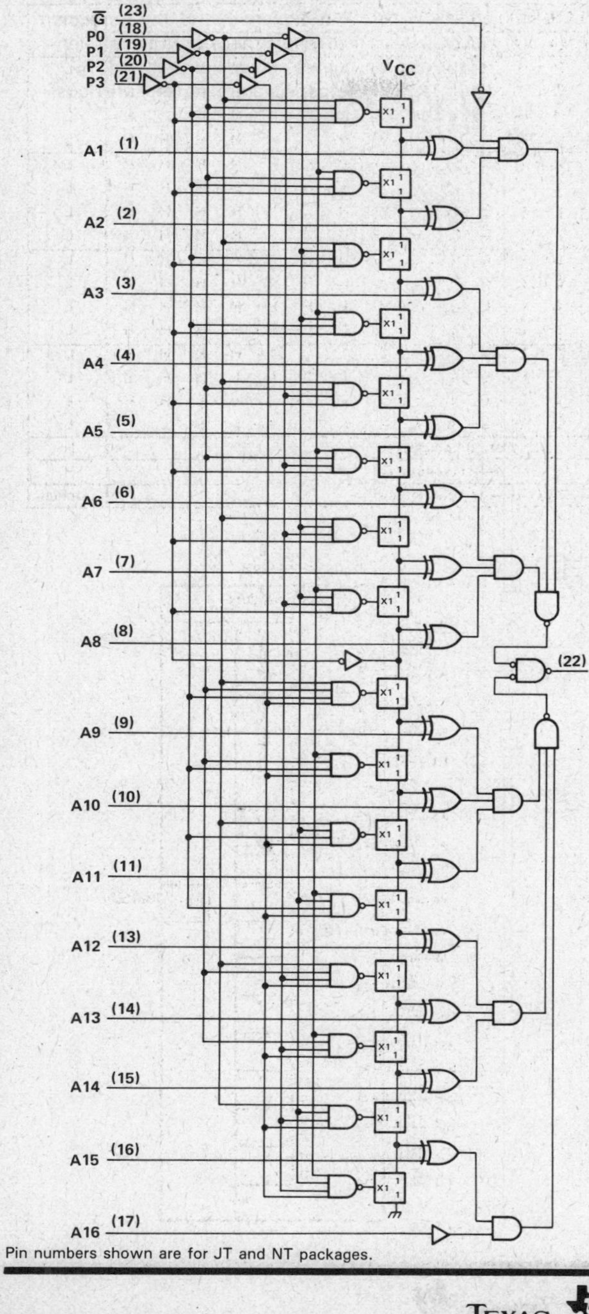

Pin numbers shown are for JT and NT packages.

In order to understand the implementation of this device, it is essential that the function of the vertical string of transmission gates be understood. A schematic of one of these gates is shown below. If the input to the transmission gate labeled "X1" is high, then the transmission path between the two ports labeled "1" is on. If the "X1" input is low, then the transmission path between the two ports labelec "1" is off. Only one of the 16 transmission gates can be off while the device is operating: which one is off is determined by inputs P0 through P3. The lines going from the string of transmission gates to the exclusive-OR gates located above the transmission gate that is off will be high. The lines going to the exclusive-OR gates located below that transmission gate will be low.

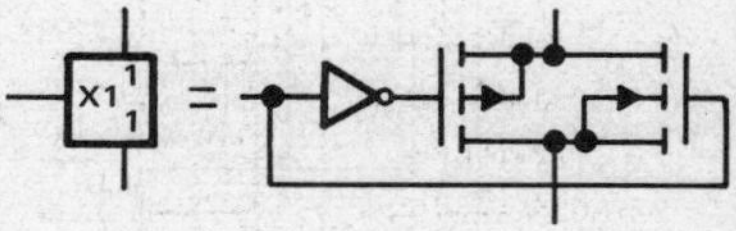

'HC678 logic diagram (positive logic)

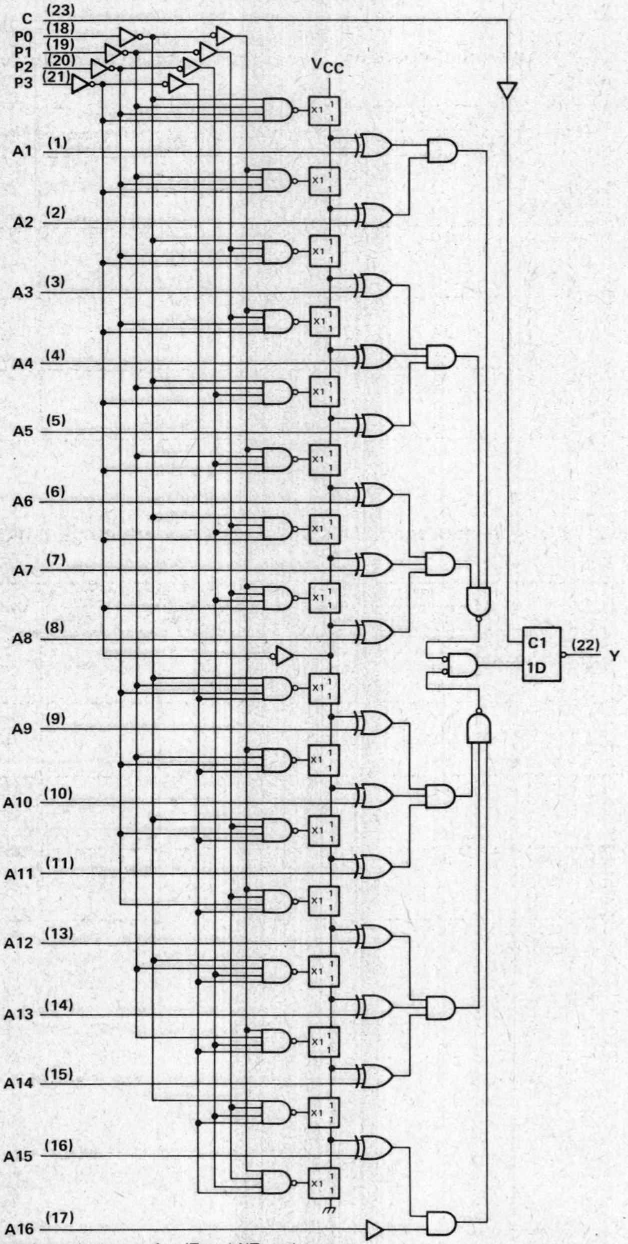

Pin numbers shown are for JT and NT packages.

An explanation of the function of the string of transmission gates appears with the 'HC677 logic diagram on the previous page.

maximum ratings, recommended operating conditions, and electrical characteristics

See Table IV, page 2-10.

'HC677 switching characteristics over recommended operating free-air temperature range (unless otherwise noted), C_L = 50 pF (see Note 1)

PARAMETER	FROM (INPUT)	TO (OUTPUT)	V_{CC}	T_A = 25°C			SN54HC677		SN74HC677		UNIT
				MIN	TYP	MAX	MIN	MAX	MIN	MAX	
t_{pd}	Any P	Y	2 V		160						ns
			4.5 V		32						
			6 V		27						
t_{pd}	Any A	Y	2 V		90						ns
			4.5 V		18						
			6 V		15						
t_{pd}	$\overline{G}$	Y	2 V		70						ns
			4.5 V		14						
			6 V		12						
t_t		Y	2 V		38						ns
			4.5 V		8						
			6 V		6						

'HC678 switching characteristics over recommended operating free-air temperature range (unless otherwise noted), C_L = 50 pF (see Note 1)

PARAMETER	FROM (INPUT)	TO (OUTPUT)	V_{CC}	T_A = 25°C			SN54HC678		SN74HC678		UNIT
				MIN	TYP	MAX	MIN	MAX	MIN	MAX	
t_{pd}	Any P	Y	2 V		165						ns
			4.5 V		33						
			6 V		28						
t_{pd}	Any A	Y	2 V		105						ns
			4.5 V		21						
			6 V		18						
t_{pd}	C	Y	2 V		75						ns
			4.5 V		15						
			6 V		13						
t_t		Y	2 V		38						ns
			4.5 V		8						
			6 V		6						

NOTE 1: For load circuit and voltage waveforms, see page 1-14.

TYPICAL APPLICATION INFORMATION

The 'HC677 and 'HC678 can be wired to recognize any one of 2^{16} addresses. The number of "lows" in the address determines the input pattern for the P inputs. Then those system address lines that are low in the address to be recognized are connected to the lowest numbered A inputs of the address comparator and the system address lines that are high are connected to the highest numbered A inputs.

For example, assume the comparator is to enable a device when the 16-bit system address is:

A15	A14	A13	A12	A11	A10	A9	A8	A7	A6	A5	A4	A3	A2	A1	A0
H	H	L	L	H	H	L	L	H	H	L	L	H	H	H	H

Since the address contains 6 lows and 10 highs, the following connections are made:

P3 to 0 V, P2 to V_{CC}, P1 to V_{CC}, and P0 to 0 V.
System address lines A13, A12, A9, A8, A5, and A4 to comparator inputs A1 through A6 in any convenient order.
The remaining ten system address lines to comparator inputs A7 through A16 in any convenient order.

The output provides an active-low enabling signal.

The following circuit is a modulo-N synchronous counter. The 'HC163 is connected to provide a low-level clear signal when N = $FEFF_{16}$.

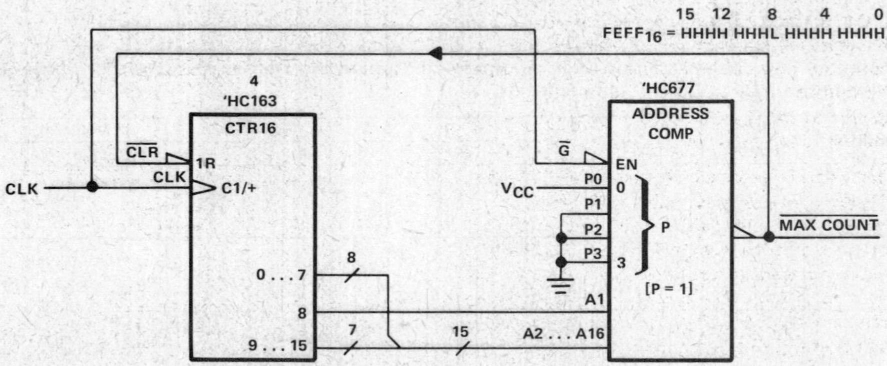

MODULO-N SYNCHRONOUS COUNTER

TYPES SN54HC679, SN54HC680, SN74HC679, SN74HC680 12-BIT ADDRESS COMPARATORS

D2833, MARCH 1984

- 'HC679 is a 12-Bit Address Comparator With Enable
- 'HC680 is a 12-Bit Address Comparator With Latch
- Package Options Include Both Plastic and Ceramic Chip Carriers in Addition to Plastic and Ceramic DIPs
- Dependable Texas Instruments Quality and Reliability

SN54HC679 . . . J PACKAGE
SN74HC679 . . . J OR N PACKAGE
(TOP VIEW)

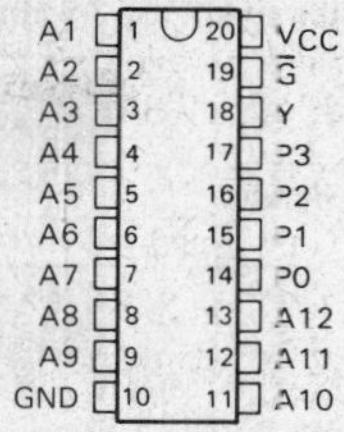

SN54HC679 . . . FH OR FK PACKAGE
SN74HC679 . . . FH OR FN PACKAGE
(TOP VIEW)

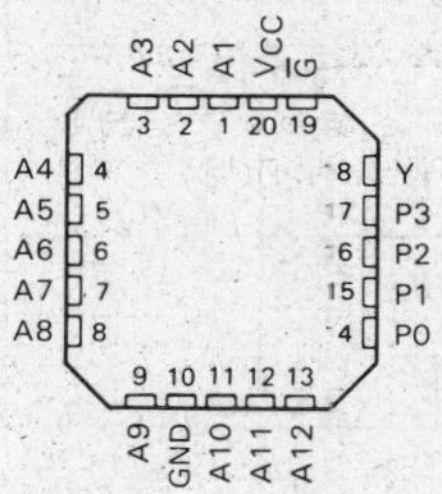

description

The 'HC679 and 'HC680 address comparators simplify addressing of memory boards and/or other peripheral devices. The four P inputs are normally hard wired with a preprogrammed address. An internal decoder determines what input information applied to the 12 A inputs must be low or high to cause a low state at the output (Y). For example, a positive-logic bit combination of 0111 (decimal 7) at the P input determines that inputs A1 through A7 must be low and that inputs A8 through A12 must be high to cause the output to go low. Equality of the address applied at the A inputs to the preprogrammed address is indicated by the output being low.

The 'HC679 features an enable input ($\overline{G}$). When $\overline{G}$ is low, the device is enabled. When $\overline{G}$ is high, the device is disabled and the output is high regardless of the A and P inputs. The 'HC680 features a transparent latch and a latch enable input (C). When C is high, the device is in the transparent mode. When C is low, the previous logical state of Y is latched.

The 'HC679 and 'HC680 are functionally unilaterally interchangeable with the ALSTTL counterparts, 'ALS679 and 'ALS680, in all cases of normal use as 12-bit address comparators. They differ in two respects. First, they may be programmed to recognize all A inputs low either by connecting all P inputs high (1111 = decimal 15), or by combination HHLL (1100 = 12), the latter option not being valid for the ALSTTL parts. Second, the combinations HHLH and HHHL (1101 = 13 and 1110 = 14) cannot be used (but are not needed) in address-comparator applications. These two combinations cause the outputs to be disabled (high).

SN54HC680 . . . J PACKAGE
SN74HC680 . . . J OR N PACKAGE
(TOP VIEW)

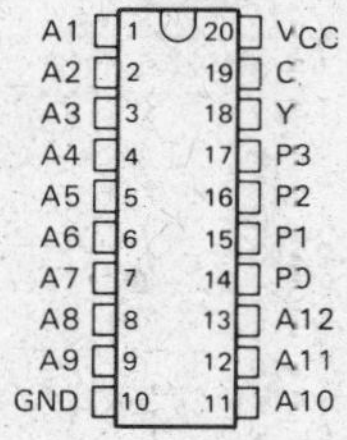

SN54HC680 . . . FH OR FK PACKAGE
SN74HC680 . . . FH OR FN PACKAGE
(TOP VIEW)

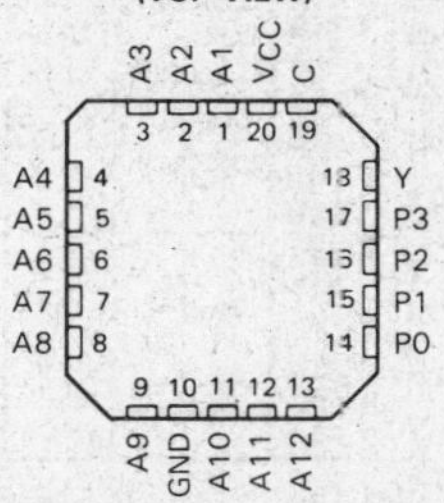

ADVANCE INFORMATION
This document contains information on a new product. Specifications are subject to change without notice.

description (continued)

The SN54HC679 and SN54HC680 are characterized for operation over the full military temperature range of −55°C to 125°C. The SN74HC679 and SN74HC680 are characterized for operation from −40°C to 85°C.

FUNCTION TABLE

'HC679	'HC680	INPUTS COMMON TO 'HC679 AND 'HC680																OUTPUT
$\overline{G}$	C	P3	P2	P1	P0	A1	A2	A3	A4	A5	A6	A7	A8	A9	A10	A11	A12	Y
L	H	L	L	L	L	H	H	H	H	H	H	H	H	H	H	H	H	L
L	H	L	L	L	H	L	H	H	H	H	H	H	H	H	H	H	H	L
L	H	L	L	H	L	L	L	H	H	H	H	H	H	H	H	H	H	L
L	H	L	L	H	H	L	L	L	H	H	H	H	H	H	H	H	H	L
L	H	L	H	L	L	L	L	L	L	H	H	H	H	H	H	H	H	L
L	H	L	H	L	H	L	L	L	L	L	H	H	H	H	H	H	H	L
L	H	L	H	H	L	L	L	L	L	L	L	H	H	H	H	H	H	L
L	H	L	H	H	H	L	L	L	L	L	L	L	H	H	H	H	H	L
L	H	H	L	L	L	L	L	L	L	L	L	L	L	H	H	H	H	L
L	H	H	L	L	H	L	L	L	L	L	L	L	L	L	H	H	H	L
L	H	H	L	H	L	L	L	L	L	L	L	L	L	L	L	H	H	L
L	H	H	L	H	H	L	L	L	L	L	L	L	L	L	L	L	H	L
L	H	H	H	L	L	L	L	L	L	L	L	L	L	L	L	L	L	L
L	H	H	H	L	H	X	X	X	X	X	X	X	X	X	X	X	X	H
L	H	H	H	H	L	X	X	X	X	X	X	X	X	X	X	X	X	H
L	H	H	H	H	H	L	L	L	L	L	L	L	L	L	L	L	L	L
L	H	All other combinations																H
H		'HC679: Any combination																H
	L	'HC680: Any combination																Latched

logic symbols

'HC679

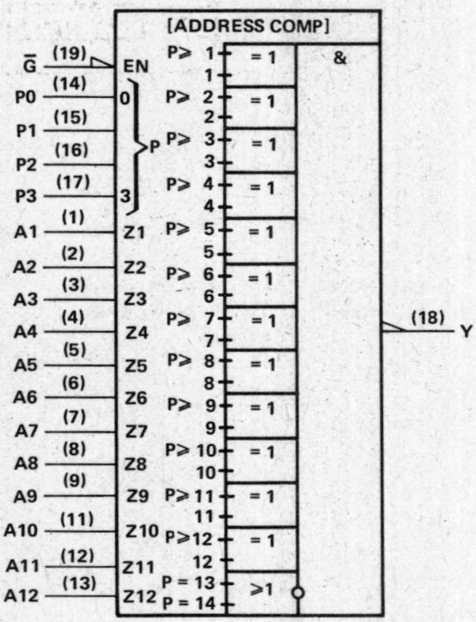

'HC680

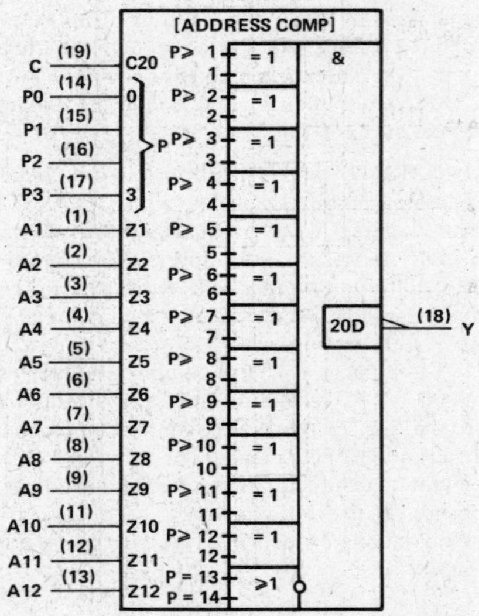

'HC679 logic diagram (positive logic)

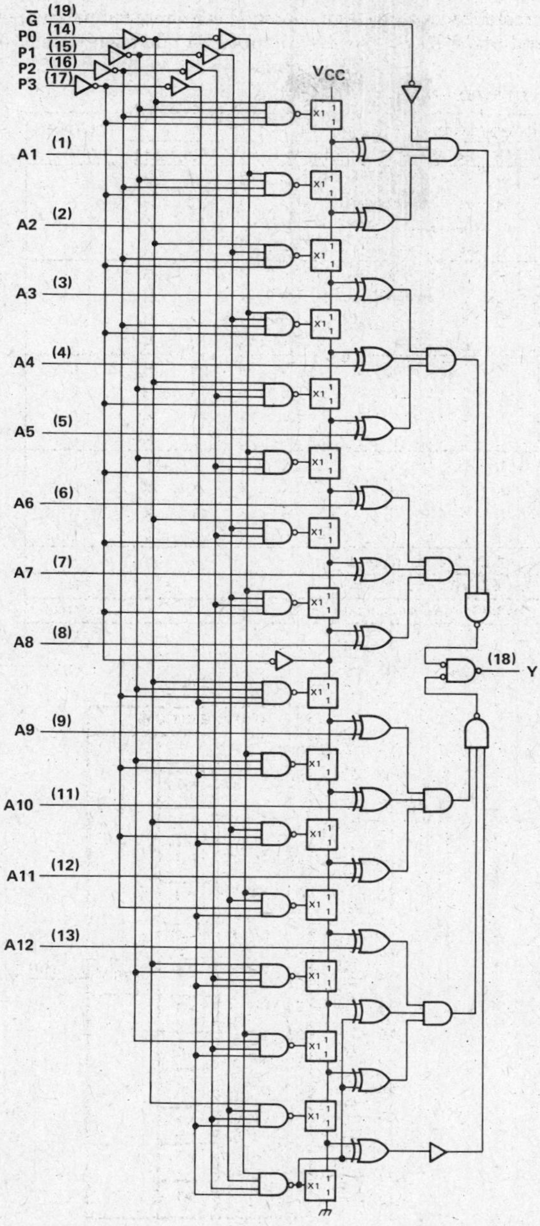

In order to understand the implementation of this device, it is essential that the function of the vertical string of transmission gates be understood. A schematic of one of these gates is shown below. If the input to the transmission gate labeled "X1" is high, then the transmission path between the two ports labeled "1" is on. If the "X1" input is low, then the transmission path between the two ports labeled "1" is off. Only one of the 16 transmission gates can be off while the device is operating; which one is off is determined by inputs P0 through P3. The lines going from the string of transmission gates to the Exclusive-OR gates located above the transmission gate that is off will be high. The lines going to the Exclusive-OR gates located below that transmission gate will be low.

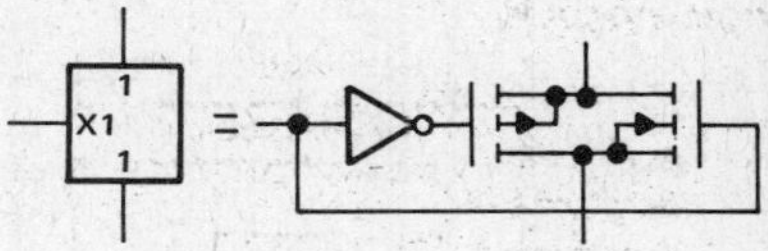

'HC680 logic diagram (positive logic)

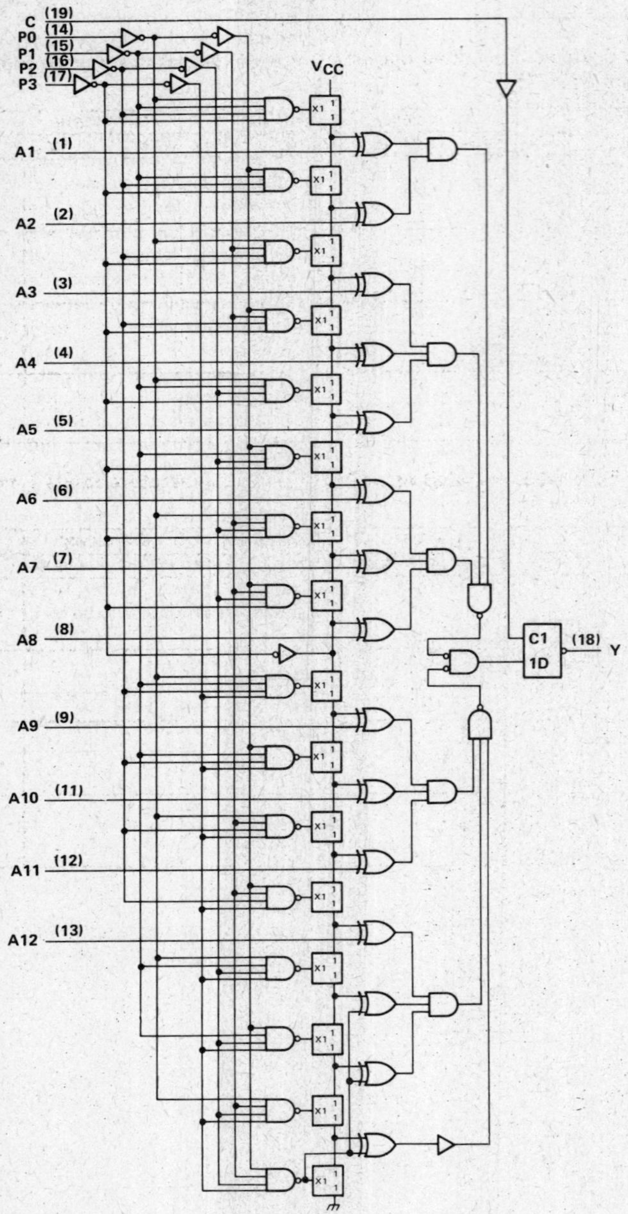

An explanation of the function of the string of transmission gates appears with the 'HC679 logic diagram on the previous page.

TEXAS INSTRUMENTS
POST OFFICE BOX 225012 • DALLAS, TEXAS 75265

maximum ratings, recommended operating conditions, and electrical characteristics

See Table IV, page 2-10.

'HC679 switching characteristics over recommended operating free-air temperature range (unless otherwise noted), C_L = 50 pF (see Note 1)

PARAMETER	FROM (INPUT)	TO (OUTPUT)	V_{CC}	T_A = 25°C MIN	TYP	MAX	SN54HC679 MIN	MAX	SN74HC679 MIN	MAX	UNIT
t_{pd}	Any P	Y	2 V		160						ns
			4.5 V		32						
			6 V		27						
t_{pd}	Any A	Y	2 V		90						ns
			4.5 V		18						
			6 V		15						
t_{pd}	$\overline{G}$	Y	2 V		70						ns
			4.5 V		14						
			6 V		12						
t_t		Y	2 V		38						ns
			4.5 V		8						
			6 V		6						

'HC680 switching characteristics over recommended operating free-air temperature range (unless otherwise noted), C_L = 50 pF (see Note 1)

PARAMETER	FROM (INPUT)	TO (OUTPUT)	V_{CC}	T_A = 25°C MIN	TYP	MAX	SN54HC680 MIN	MAX	SN74HC680 MIN	MAX	UNIT
t_{pd}	Any P	Y	2 V		165						ns
			4.5 V		33						
			6 V		28						
t_{pd}	Any A	Y	2 V		105						ns
			4.5 V		21						
			6 V		18						
t_{pd}	C	Y	2 V		75						ns
			4.5 V		15						
			6 V		13						
t_t		Y	2 V		38						ns
			4.5 V		8						
			6 V		6						

NOTE 1: For load circuit and voltage waveforms, see page 1-14.

4

ADVANCE INFORMATION

TYPICAL APPLICATION INFORMATION

The 'HC679 and 'HC680 can be wired to recognize any one of 2^{12} addresses. The number of "lows" in the address determines the input pattern for the P inputs. Then those system address lines that are low in the address to be recognized are connected to the lowest numbered A inputs of the address comparator and the system address lines that are high are connected to the highest numbered A inputs.

For example, assume the comparator is to enable a device when the 12-bit system address is:

A11	A10	A9	A8	A7	A6	A5	A4	A3	A2	A1	A0
H	H	L	L	H	H	L	L	H	H	H	H

Since the address contains 4 lows and 8 highs, the following connections are made.

P3 to 0 V, P2 to V_{CC}, P1 to 0 V, and P0 to 0 V.

System address lines A9, A8, A5, and A4 to comparator inputs A1 through A4 in any convenient order.

The remaining eight system address lines to comparator inputs A5 through A12 in any convenient order.

The output provides an active-low enabling signal.

The following circuit is a register bank decoder that examines the 14 most significant bits (A0 through A13) of a 20-bit address to select banks corresponding to the hex addresses 10000, 10040, 10080, and 100C0.

	0	4	8	12	16
10000_{16}	= LLLH	LLLL	LLLL	LLLL	LLLL
10040_{16}	= LLLH	LLLL	LLLL	LHLL	LLLL
10080_{16}	= LLLH	LLLL	LLLL	HLLL	LLLL
$100C0_{16}$	= LLLH	LLLL	LLLL	HHLL	LLLL

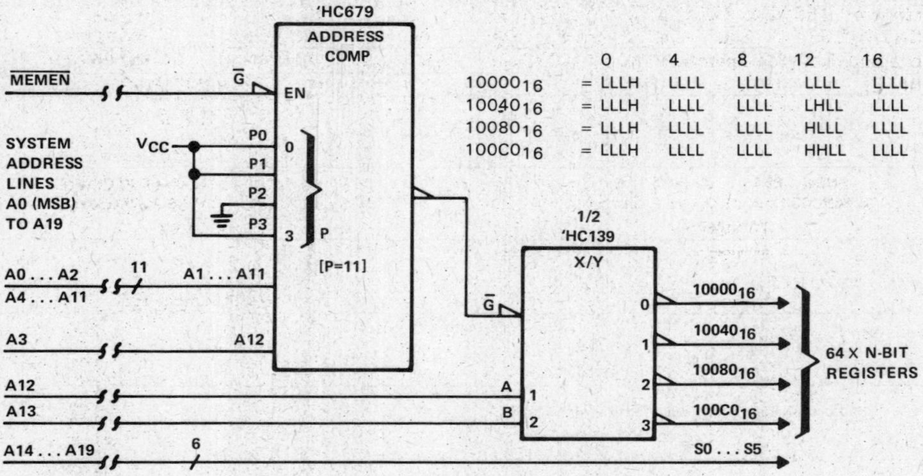

REGISTER BANK DECODER

TYPES SN54HC682, SN54HC684, SN54HC686 SN74HC682, SN74HC684, SN74HC686 8-BIT MAGNITUDE COMPARATORS

D2804, MARCH 1984

- **Compares Two 8-Bit Words**
- **'HC682 has 20-kΩ Pullup Resistors on the Q Inputs**
- **Package Options Include Both Plastic and Ceramic Chip Carriers in Addition to Plastic and Ceramic DIPs**
- **Dependable Texas Instruments Quality and Reliability**

SN54HC682, SN54HC684 . . . J PACKAGE
SN74HC682, SN54HC684 . . . J OR N PACKAGE
(TOP VIEW)

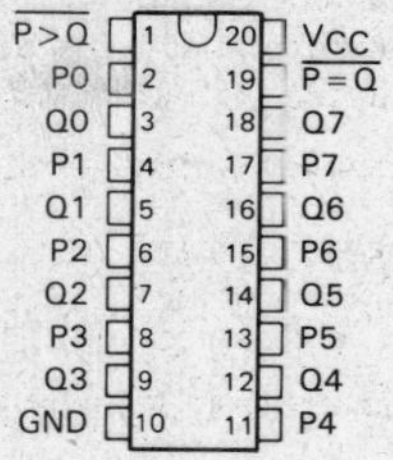

description

These magnitude comparators perform comparisons of two eight-bit binary or BCD words. All types provide $\overline{P=Q}$ and $\overline{P>Q}$ outputs. The 'HC682 features 20-kΩ pullup termination resistors on the Q inputs for analog or switch data.

The SN54HC682, SN54HC684, and SN54HC686 are characterized for operation over the full military temperature temperature range of −55°C to 125°C. The SN74HC682, SN74HC684, and SN74HC686 are characterized for operation from −40°C to 85°C.

SN54HC682, SN54HC684 . . . FH OR FK PACKAGE
SN74HC682, SN74HC684 . . . FH OR FN PACKAGE
(TOP VIEW)

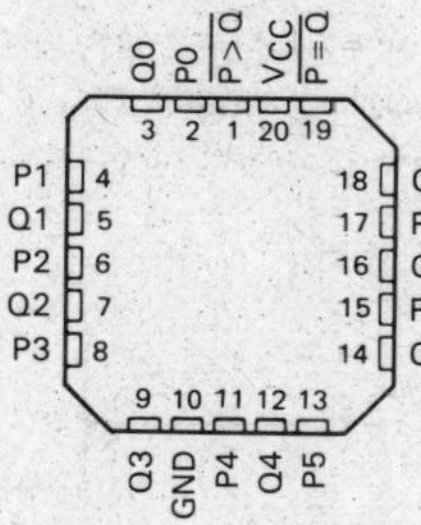

SN54HC686 . . . JT PACKAGE
SN74HC686 . . . JT OR NT PACKAGE
(TOP VIEW)

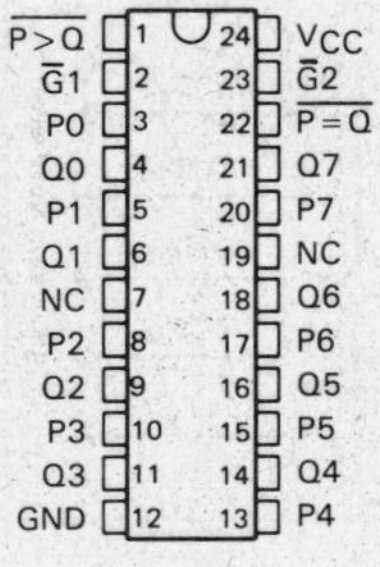

NC—No internal connection

SN54HC686 . . . FH OR FK PACKAGE
SN74HC686 . . . FH OR FN PACKAGE
(TOP VIEW)

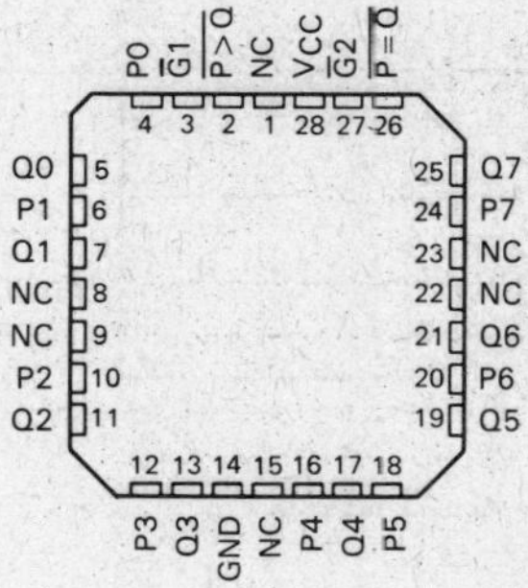

4

ADVANCE INFORMATION

FUNCTION TABLE

INPUTS			OUTPUTS	
DATA	ENABLES			
P, Q	$\overline{G}1$	$\overline{G}2$	$\overline{P=Q}$	$\overline{P>Q}$
P=Q	L	X	L	H
P>Q	X	L	H	L
P<Q	X	X	H	H
P=Q	H	X	H	H
P>Q	X	H	H	H
X	H	H	H	H

NOTES: 1. The last 3 lines of the function table apply only to the device having enable inputs, i.e., 'HC686.
2. The $\overline{P<Q}$ function can be generated by applying the $\overline{P=Q}$ and $\overline{P>Q}$ outputs to a 2-input NAND gate.

logic symbols

'HC682, 'HC684

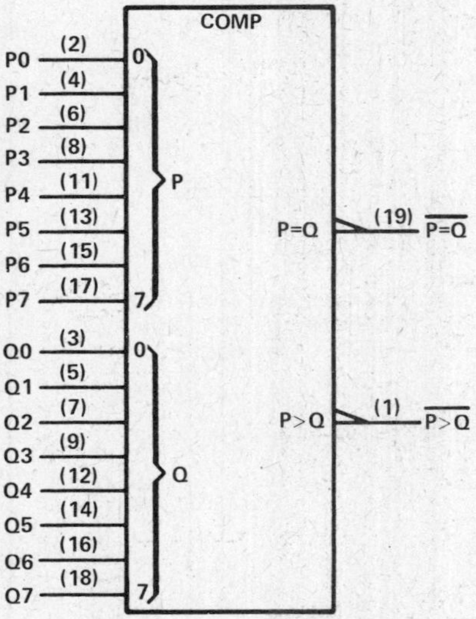

'HC682 has 20-kΩ pullup resistors on the Q inputs

'HC686

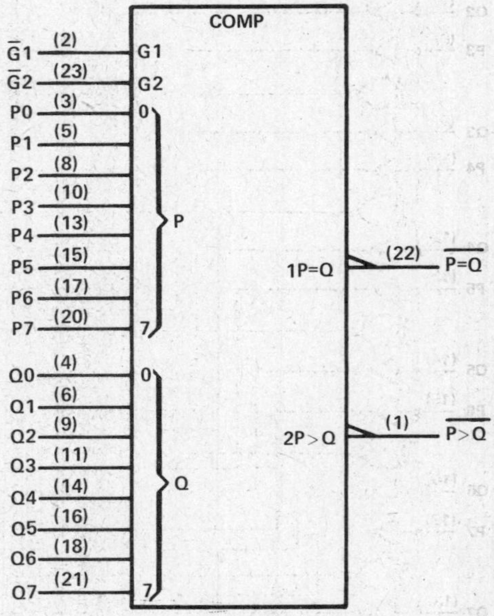

'HC686 pin numbers shown are for JT and NT packages.

'HC682, 'HC684 logic diagram (positive logic)

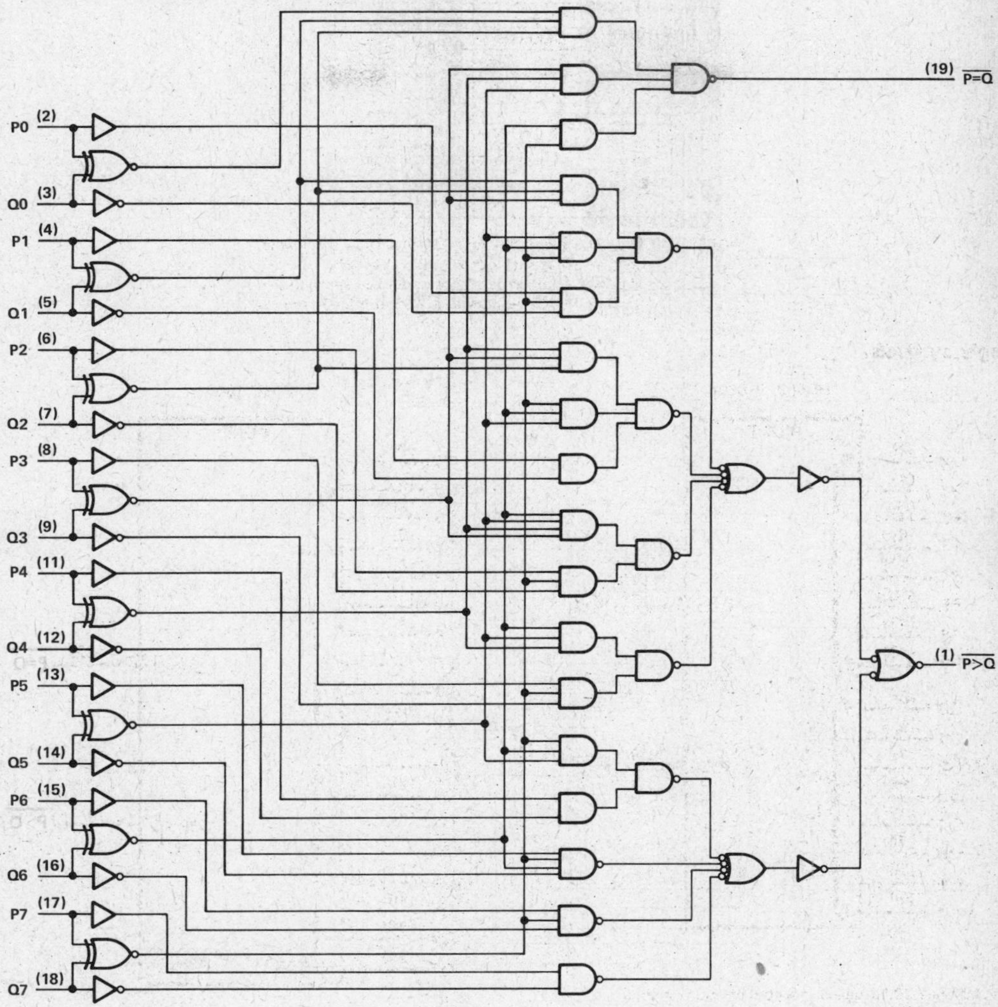

'HC686 logic diagram (positive logic)

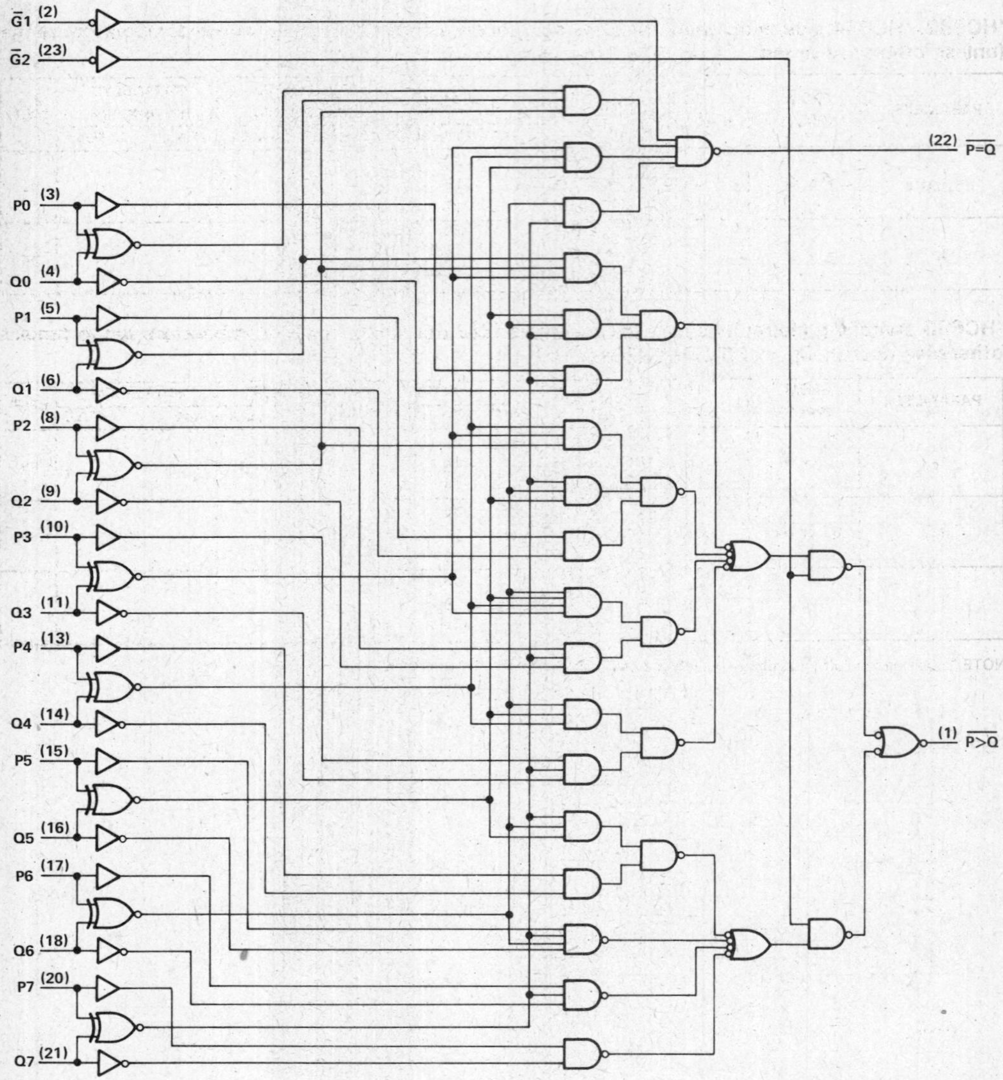

Pin numbers shown are for JT and NT packages.

TEXAS INSTRUMENTS
POST OFFICE BOX 225012 • DALLAS, TEXAS 75265

maximum ratings, recommended operating conditions, and electrical characteristics

See Table IV, page 2-10.

'HC682, 'HC684 switching characteristics over recommended operating free-air temperature range (unless otherwise noted), C_L = 50 pF (see Note 1)

PARAMETER	FROM (INPUT)	TO (OUTPUT)	V_{CC}	T_A = 25°C			SN54HC682 SN54HC684		SN74HC682 SN74HC684		UNIT
				MIN	TYP	MAX	MIN	MAX	MIN	MAX	
t_{pd}	P or Q	Any	2 V		80						ns
			4.5 V		16						
			6 V		14						
t_t		Any	2 V		38						ns
			4.5 V		8						
			6 V		6						

'HC686 switching characteristics over recommended operating free-air temperature range (unless otherwise noted), C_L = 50 pF (see Note 1)

PARAMETER	FROM (INPUT)	TO (OUTPUT)	V_{CC}	T_A = 25°C			SN54HC686		SN74HC686		UNIT
				MIN	TYP	MAX	MIN	MAX	MIN	MAX	
t_{pd}	P or Q	Any	2 V		80						ns
			4.5 V		16						
			6 V		14						
t_{pd}	$\overline{G}1$ or $\overline{G}2$	Any	2 V		55						ns
			4.5 V		11						
			6 V		9						
t_t		Any	2 V		38						ns
			4.5 V		8						
			6 V		6						

NOTE 1: For load circuit and voltage waveforms see page 1-14.

4 ADVANCE INFORMATION

TYPES SN54HC688, SN74HC688 8-BIT IDENTITY COMPARATORS

D2684, DECEMBER 1982—REVISED MARCH 1984

- Compares Two Eight-Bit Words
- Package Options Include Both Plastic and Ceramic Chip Carriers in Addition to Plastic and Ceramic DIPs
- Dependable Texas Instruments Quality and Reliability

SN54HC688 . . . J PACKAGE
SN74HC688 . . . J OR N PACKAGE
(TOP VIEW)

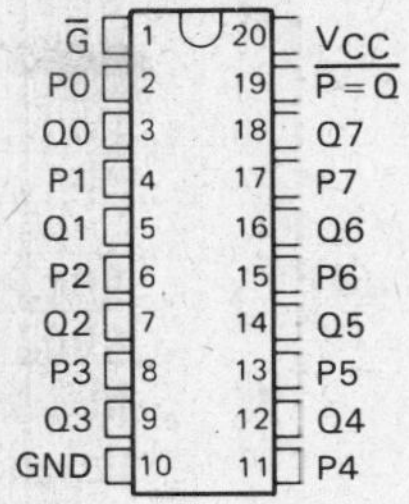

description

These identity comparators perform comparisons of two eight-bit binary or BCD words. An enable input (G̅) may be used to force the output to the high level.

The SN54HC688 is characterized for operation over the full military temperature range of −55°C to 125°C. The SN74HC688 is characterized for operation from −40°C to 85°C.

SN54HC688 . . . FH OR FK PACKAGE
SN74HC688 . . . FH OR FN PACKAGE
(TOP VIEW)

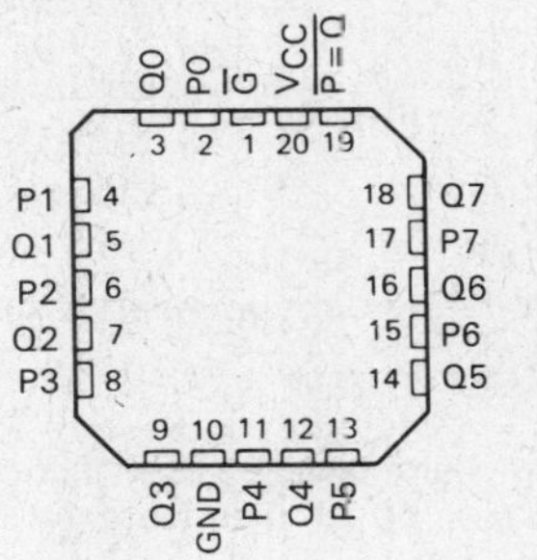

logic symbol

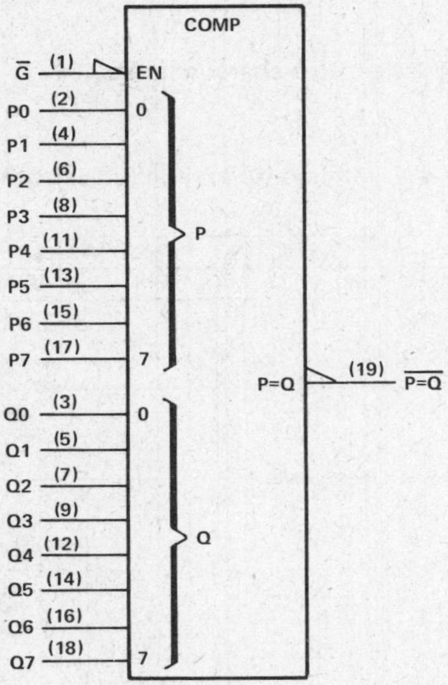

FUNCTION TABLE

INPUTS		OUTPUT
DATA P, Q	ENABLE G̅	P̅=̅Q̅
P = Q	L	L
P > Q	X	H
P < Q	X	H
X	H	H

4

ADVANCE INFORMATION

ADVANCE INFORMATION
This document contains information on a new product. Specifications are subject to change without notice.

logic diagram (positive logic)

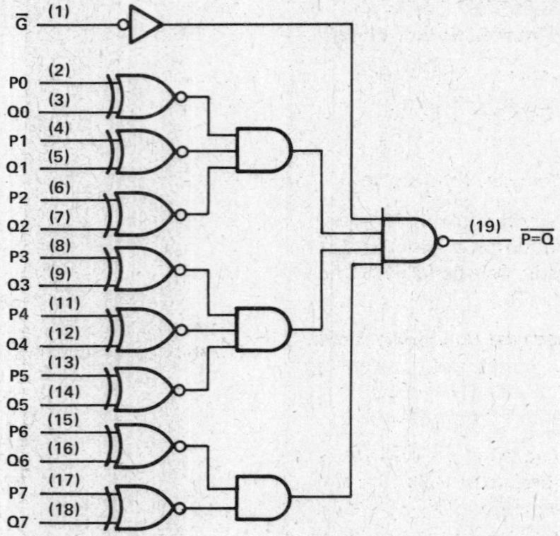

4

ADVANCE INFORMATION

maximum ratings, recommended operating conditions, and electrical characteristics

See Table IV, page 2-10.

switching characteristics over recommended operating free-air temperature range (unless otherwise noted), C_L = 50 pF (see Note 1)

PARAMETER	FROM (INPUT)	TO (OUTPUT)	V_{CC}	T_A = 25°C MIN	TYP	MAX	SN54HC688 MIN	MAX	SN74HC688 MIN	MAX	UNIT
t_{pd}	P	$\overline{P=Q}$	2 V		75						ns
			4.5 V		15						
			6 V		13						
t_{pd}	Q	$\overline{P=Q}$	2 V		75						ns
			4.5 V		15						
			6 V		13						
t_{pd}	$\overline{G}$	$\overline{P=Q}$	2 V		50						ns
			4.5 V		10						
			6 V		9						
t_t		Any	2 V		38						ns
			4.5 V		8						
			6 V		6						

NOTE 1: For load circuit and voltage waveforms, see page 1-14.

TYPES SN54HC690 THRU SN54HC693, SN74HC690 THRU SN74HC693 SYNCHRONOUS COUNTERS WITH OUTPUT REGISTERS AND MULTIPLEXED 3-STATE OUTPUTS

D2804, MARCH 1984

- **4-Bit Counters/Registers**
- **3-State Outputs Drive Bus Lines Directly**
- **'HC690 . . . Decade Counter, Direct Clear**
- **'HC691 . . . Binary Counter, Direct Clear**
- **'HC692 . . . Decade Counter, Synchronous Clear**
- **'HC693 . . . Binary Counter, Synchronous Clear**
- **Package Options Include Both Plastic and Ceramic Chip Carriers in Addition to Plastic and Ceramic DIPs**
- **Dependable Texas Instruments Quality and Reliability**

SN54HC690 THRU SN74HC693 . . . J PACKAGE
SN74HC690 THRU SN74HC693 . . . J OR N PACKAGE
(TOP VIEW)

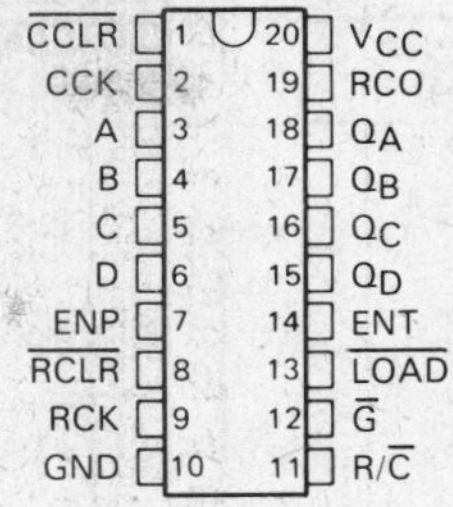

SN54HC690 THRU SN54HC693 . . . FH OR FK PACKAGE
SN74HC690 THRU SN74HC693 . . . FH OR FN PACKAGE
(TOP VIEW)

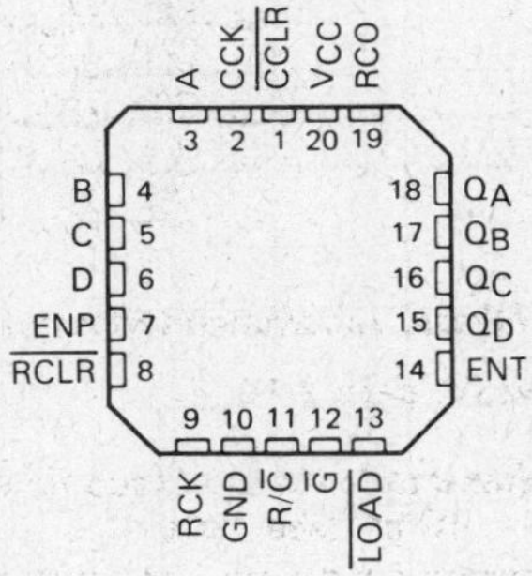

description

These devices incorporate synchronous up/down counters, four-bit D-type registers, and quadruple two-line to one-line multiplexers with three-state outputs in a single 20-pin package. The up/down counters are programmable from the data inputs and feature enable P (ENP) and enable T (ENT) inputs and a ripple-carry output (RCO) for easy expansion. The register/counter select input (R/$\overline{C}$) selects the counter when low and the register when high for the three-state outputs, Q_A, Q_B, Q_C, and Q_D.

Individual clock and clear inputs are provided for both the counter and the register. Both clock inputs are positive-edge triggered. The clear line is active low and is asynchronous on the 'HC690 and 'HC691, synchronous on the 'HC692 and 'HC693.

The SN54HC690 through SN54HC693 are characterized for operation over the full military range of −55°C to 125°C. The SN74HC690 through SN74HC693 are characterized for operation from −40°C to 85°C.

logic symbols

'HC690

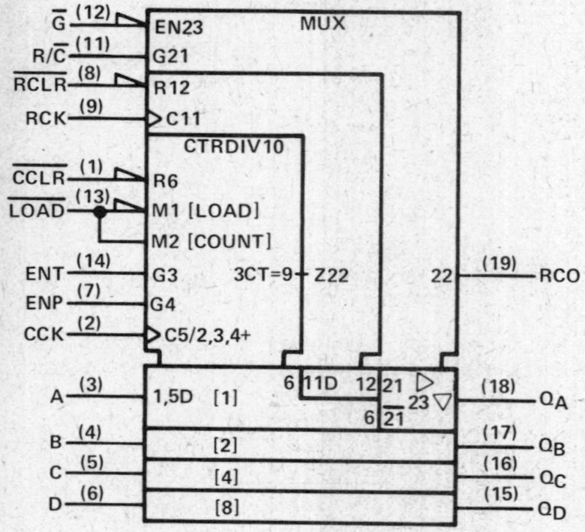

'HC691

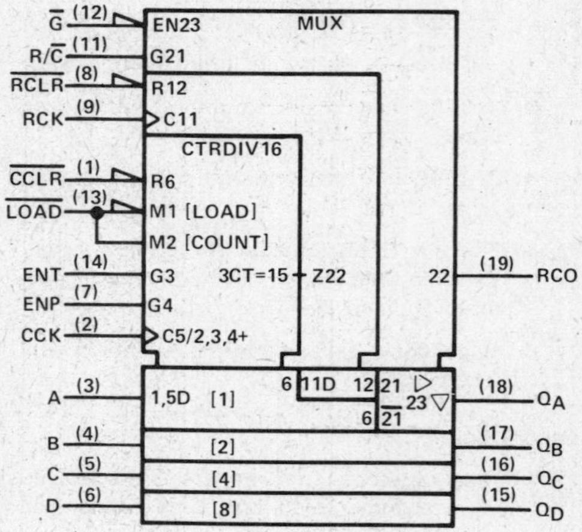

'HC692

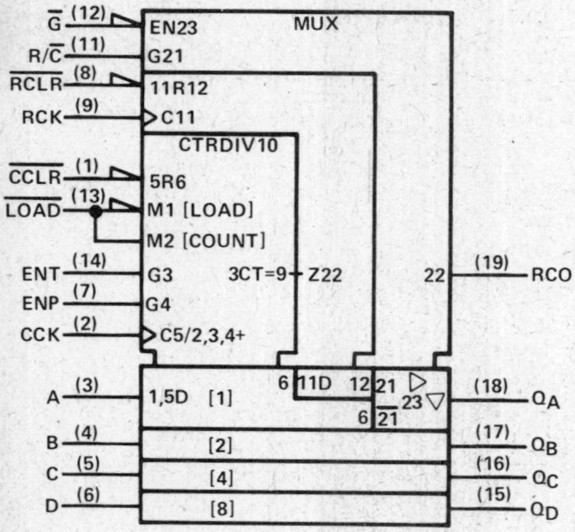

'HC693

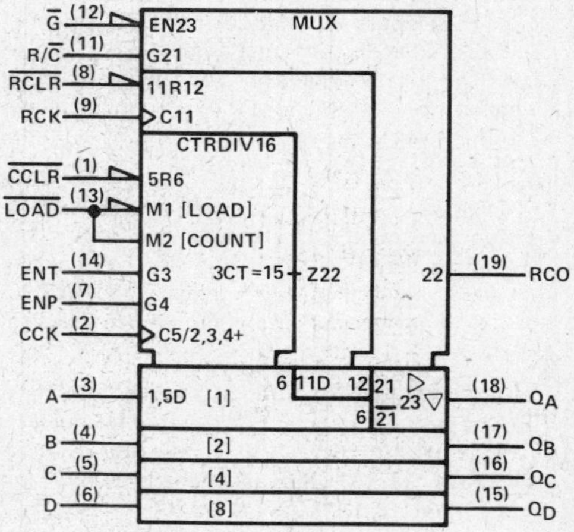

logic diagrams (positive logic)

'HC690

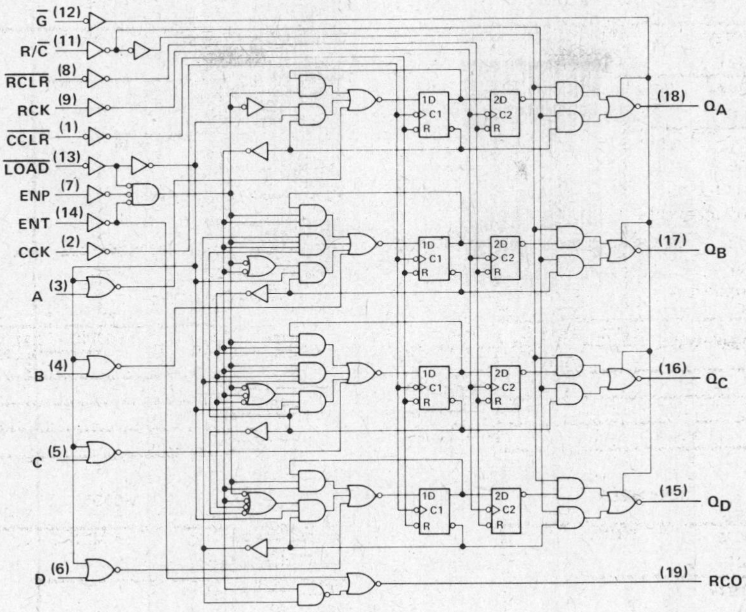

'HC691

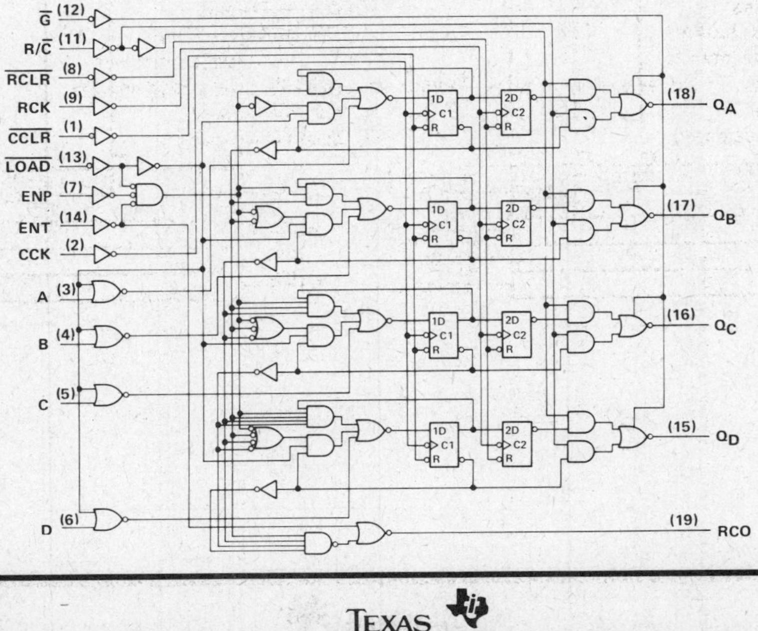

logic diagrams (positive logic)

'HC692

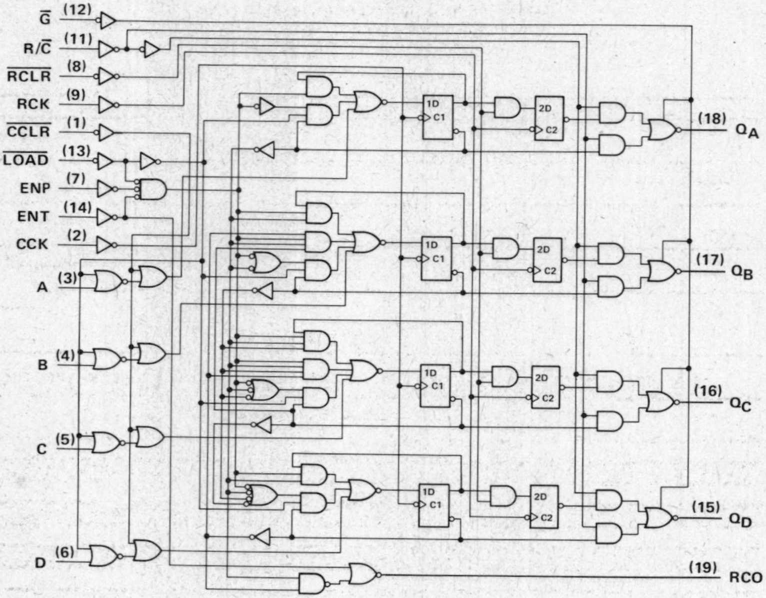

'HC693

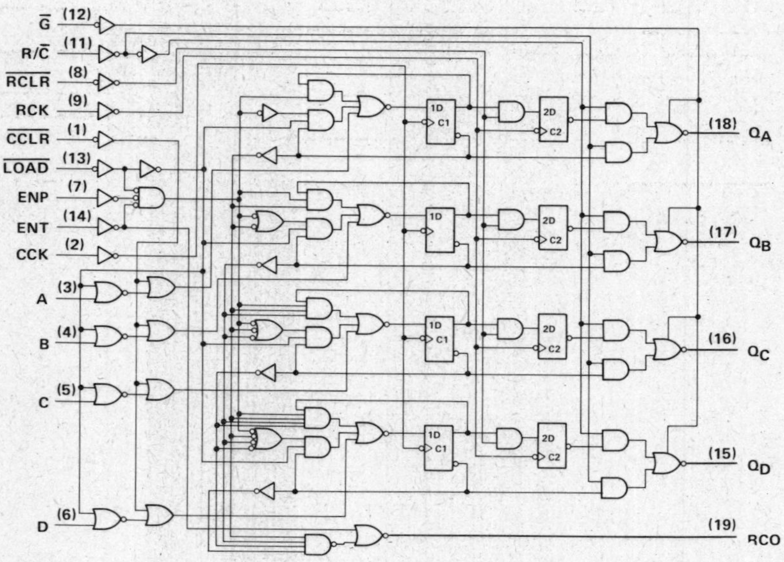

4

ADVANCE INFORMATION

typical operating sequences

'HC690 DECADE COUNTER, Asynchronous Clear
'HC692 DECADE COUNTER, Synchronous Clear

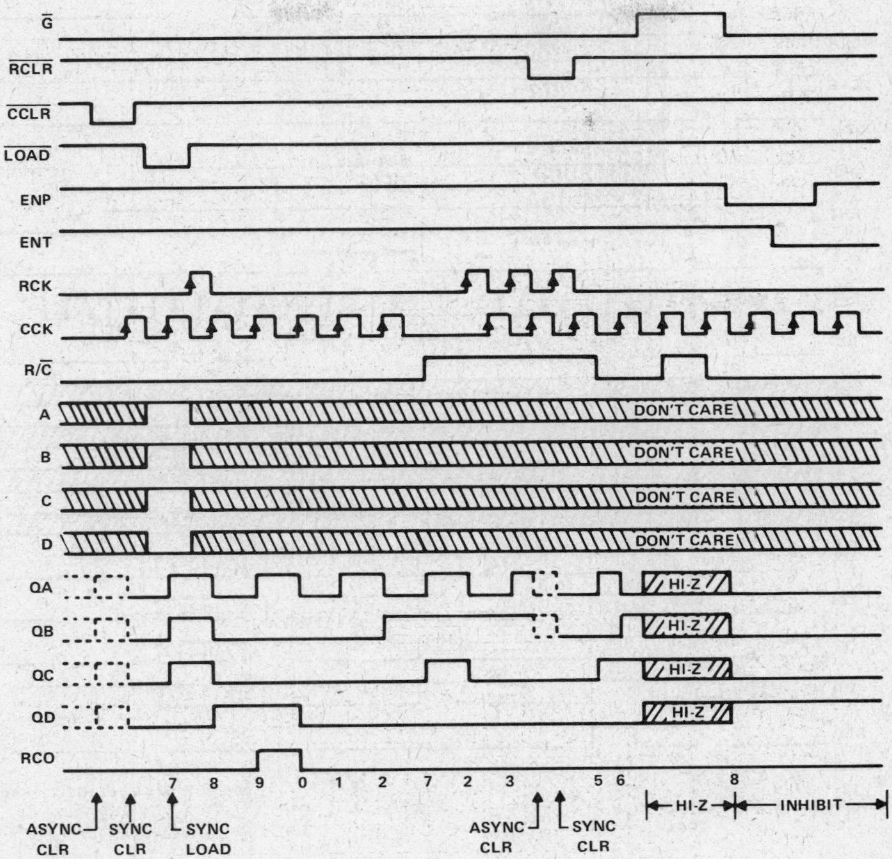

TEXAS INSTRUMENTS
POST OFFICE BOX 225012 • DALLAS, TEXAS 75265

typical operating sequences

'HC691 BINARY COUNTER, Asynchronous Clear
'HC693 BINARY COUNTER, Synchronous Clear

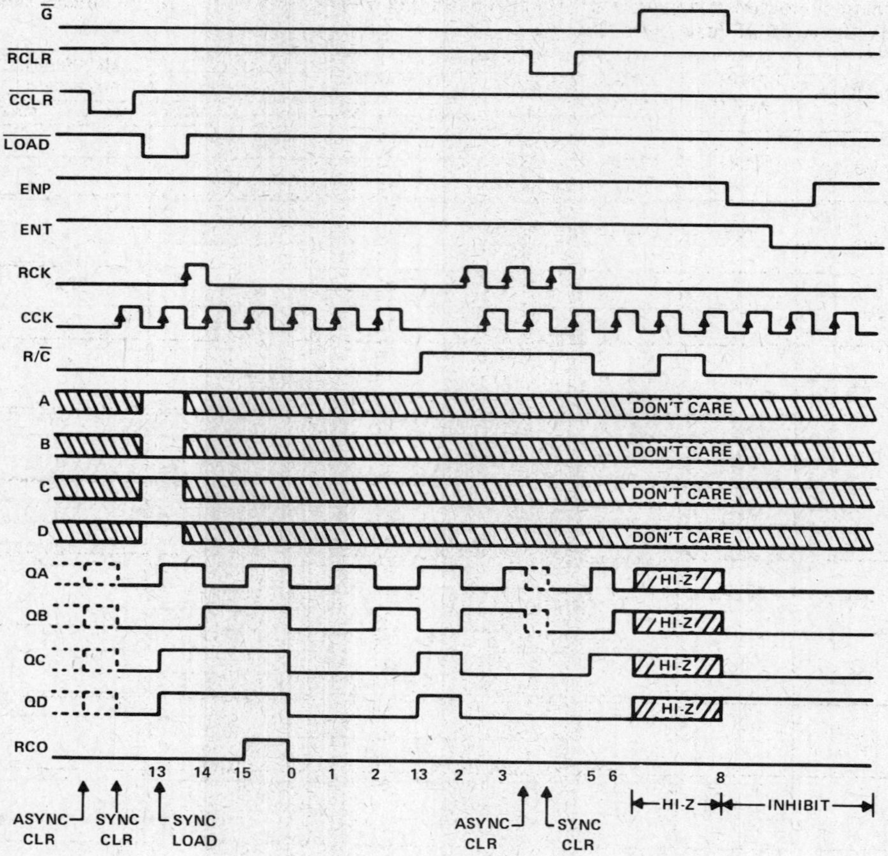

4

ADVANCE INFORMATION

maximum ratings, recommended operating conditions, and electrical characteristics

See Table III, page 2-8.

switching characteristics over recommended operating free-air temperature range (unless otherwise noted), C_L = 50 pF (see Note 1)

PARAMETER	FROM (INPUT)	TO (OUTPUT)	V_{CC}	T_A = 25°C MIN	TYP	MAX	SN54HC690 SN54HC691 MIN	MAX	SN74HC690 SN74HC691 MIN	MAX	UNIT
t_{pd}	CCK↑	RCO	2 V		105						ns
			4.5 V		21						
			6 V		18						
t_{pd}	ENT	RCO	2 V		50						ns
			4.5 V		10						
			6 V		9						
t_{pd}	CLK↑	Q	2 V		85						ns
			4.5 V		17						
			6 V		14						
t_{pd}	RCK↑	Q	2 V		75						ns
			4.5 V		15						
			6 V		13						
t_{PHL}	$\overline{CCLR}$↓	Q	2 V		85						ns
			4.5 V		17						
			6 V		14						
t_{PHL}	$\overline{CCLR}$↓	RCO	2 V		115						ns
			4.5 V		23						
			6 V		20						
t_{PHL}	$\overline{RCLR}$↓	Q	2 V		80						ns
			4.5 V		16						
			6 V		14						
t_{pd}	R/$\overline{C}$	Q	2 V		55						ns
			4.5 V		11						
			6 V		9						
t_{en}	$\overline{G}$↓	Q	2 V		50						ns
			4.5 V		10						
			6 V		9						
t_{dis}	$\overline{G}$↑	Q	2 V		80						ns
			4.5 V		16						
			6 V		14						
t_t		Any	2 V		38						ns
			4.5 V		8						
			6 V		6						

NOTE 1: For load circuit and voltage waveforms, see page 1-14.

4

ADVANCE INFORMATION

switching characteristics over recommended operating free-air temperature range (unless otherwise noted), C_L = 50 pF (see Note 1)

PARAMETER	FROM (INPUT)	TO (OUTPUT)	V_{CC}	T_A = 25°C MIN	TYP	MAX	SN54HC692 SN54HC693 MIN	MAX	SN74HC692 SN74HC693 MIN	MAX	UNIT
t_{pd}	CCK↑	RCO	2 V		105						ns
			4.5 V		21						
			6 V		18						
t_{pd}	ENT	RCO	2 V		50						ns
			4.5 V		10						
			6 V		9						
t_{pd}	CCK↑	Any Q	2 V		85						ns
			4.5 V		17						
			6 V		14						
t_{pd}	RCK↑	Any Q	2 V		80						ns
			4.5 V		16						
			6 V		14						
t_{pd}	R/$\overline{C}$	Any Q	2 V		55						ns
			4.5 V		11						
			6 V		9						
t_{en}	$\overline{G}$↓	Any Q	2 V		50						ns
			4.5 V		10						
			6 V		9						
t_{dis}	$\overline{G}$↑	Any Q	2 V		80						ns
			4.5 V		16						
			6 V		14						
t_t		Any	2 V		38						ns
			4.5 V		8						
			6 V		6						

NOTE 1: For load circuit and voltage waveforms, see page 1-14.

TEXAS INSTRUMENTS
POST OFFICE BOX 225012 • DALLAS, TEXAS 75265

TYPES SN54HC696 THRU SN54HC699, SN74HC696 THRU SN74HC699 SYNCHRONOUS UP/DOWN COUNTERS WITH OUTPUT REGISTERS AND MULTIPLEXED 3-STATE OUTPUTS

D2804, MARCH 1984

- **4-Bit Counters/Registers**
- **Multiplexed Outputs for Counter or Latched Data**
- **High-Current 3-State Outputs Drive Bus Lines Directly or up to 15 LSTTL Loads**
- **'HC696 . . . Decade Counter, Direct Clear**
  'HC697 . . . Binary Counter, Direct Clear
  'HC698 . . . Decade Counter, Synchronous Clear
  'HC699 . . . Binary Counter, Synchronous Clear
- **Package Options Include Both Plastic and Ceramic Chip Carriers in Addition to Plastic and Ceramic DIPs**
- **Dependable Texas Instruments Quality and Reliability**

SN54HC696 THRU SN54HC699 . . . J PACKAGE
SN74HC696 THRU SN74HC699 . . . J OR N PACKAGE
(TOP VIEW)

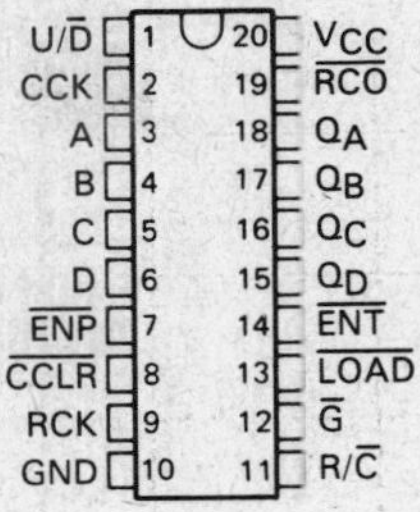

SN54HC696 THRU SN54HC699 . . . FH OR FK PACKAGE
SN74HC696 THRU SN74HC699 . . . FH OR FN PACKAGE
(TOP VIEW)

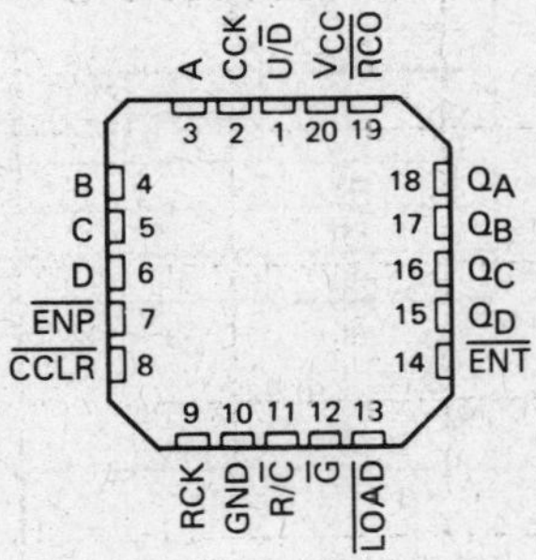

description

These high-speed CMOS devices incorporate synchronous up/down counters, four-bit D-type registers, and quadruple two-line to one-line multiplexers with three-state outputs in a single 20-pin package. The up/down counters are programmable from the data inputs and feature enable $\overline{P}$ ($\overline{ENP}$) and enable $\overline{T}$ ($\overline{ENT}$) and ripple-carry output (RCO) for easy expansion. The register/counter select input (R/$\overline{C}$) selects the counter when low and the register when high for the three-state outputs Q_A, Q_B, Q_C, and Q_D.

Both the counter clock (CCK) and register clock (RCK) are positive-edge triggered. The counter clear ($\overline{CCLR}$) is active low and asynchronous on the 'HC696 and 'HC697, synchronous on the 'HC698 and 'HC699.

The SN54HC696 through SN54HC699 are characterized for operation over the full military range of −55°C to 125°C. The SN74HC696 through SN74HC699 are characterized for operation from −40°C to 85°C.

ADVANCE INFORMATION
This document contains information on a new product. Specifications are subject to change without notice.

TYPES SN54HC696 THRU SN54HC699, SN74HC696 THRU SN74HC699 SYNCHRONOUS UP/DOWN COUNTERS WITH OUTPUT REGISTERS AND MULTIPLEXED 3-STATE OUTPUTS

logic symbols

'HC696

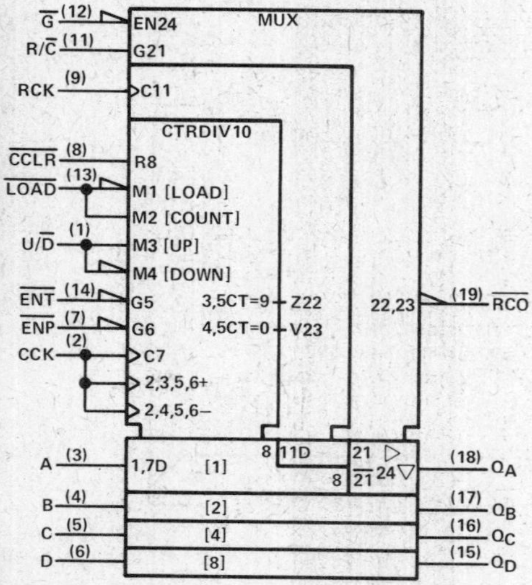

'HC697

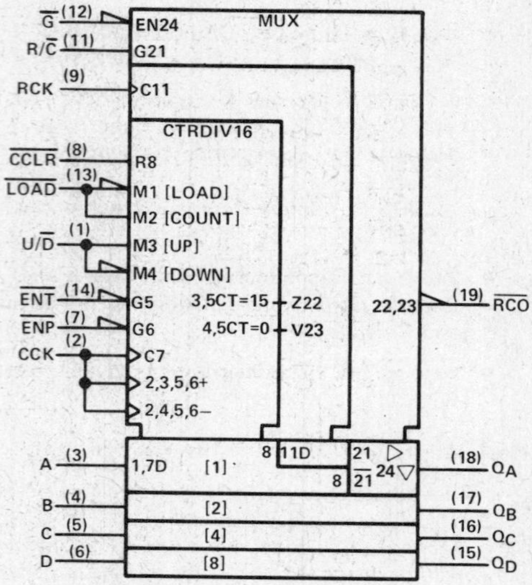

'HC698

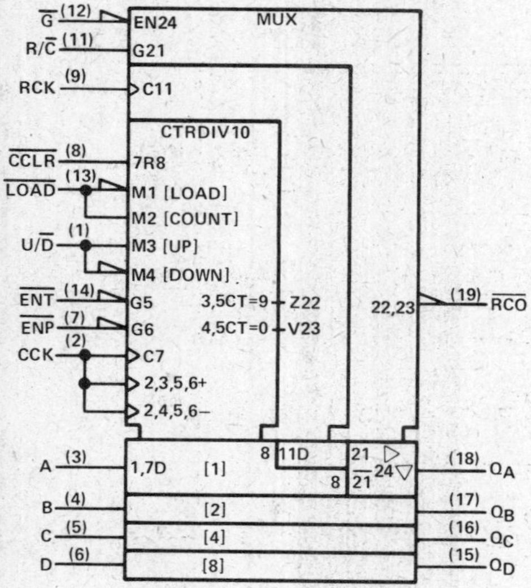

'HC699

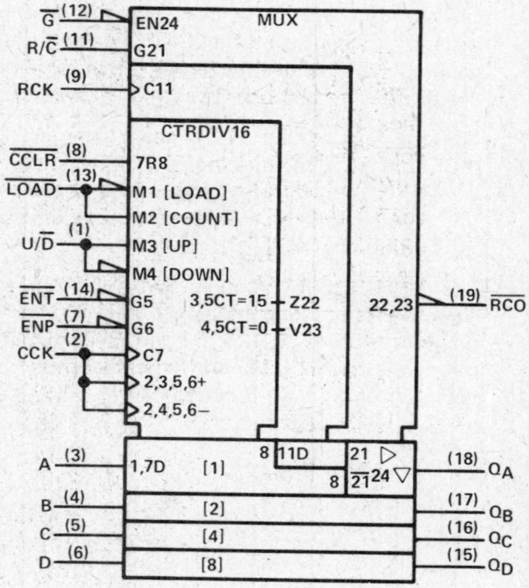

logic diagrams (positive logic)

'HC696

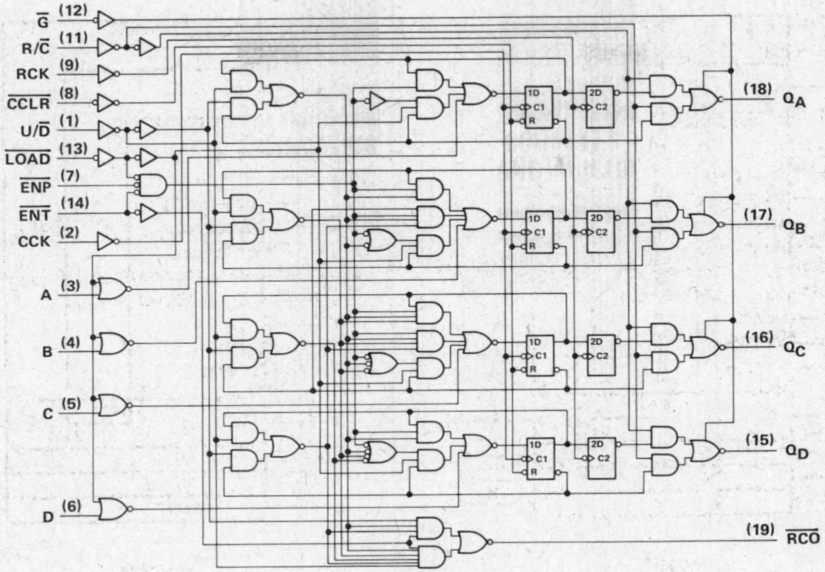

'HC697

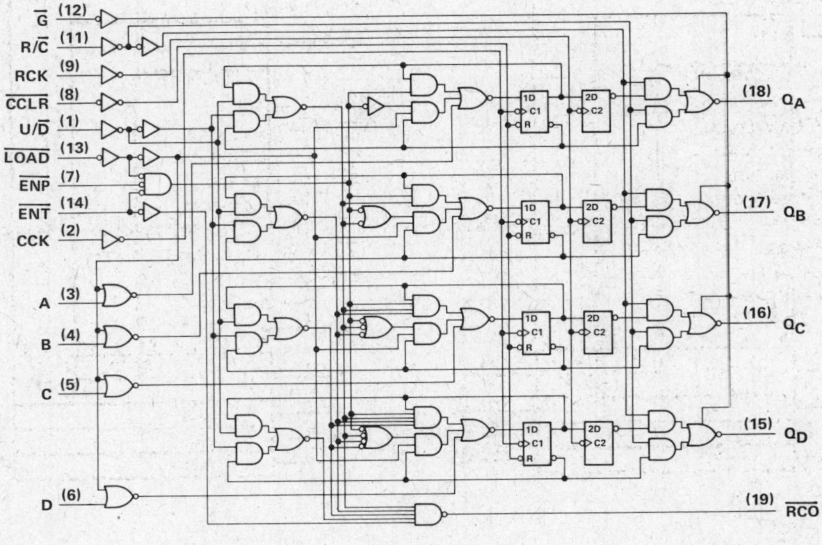

logic diagrams (positive logic)

'HC698

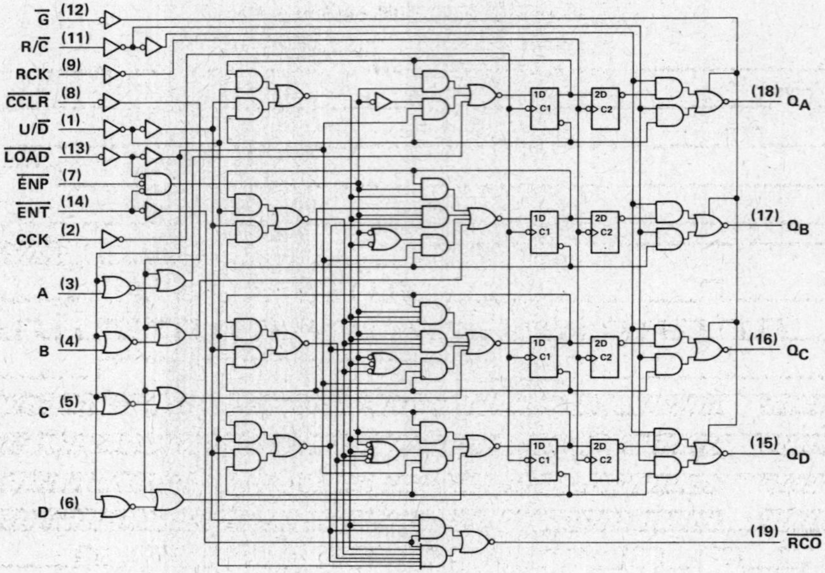

'HC699

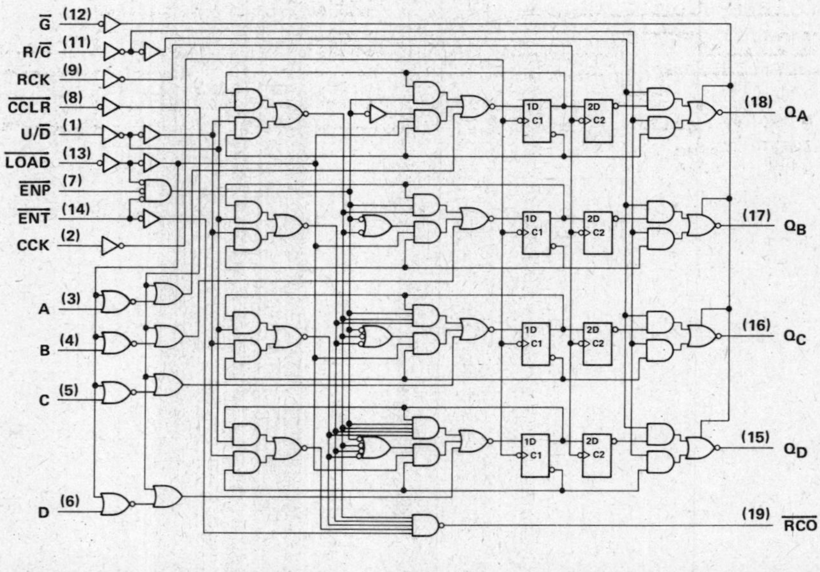

typical operating sequences

'HC696 DECADE COUNTER, Asynchronous Clear
'HC698 DECADE COUNTER, Synchronous Clear

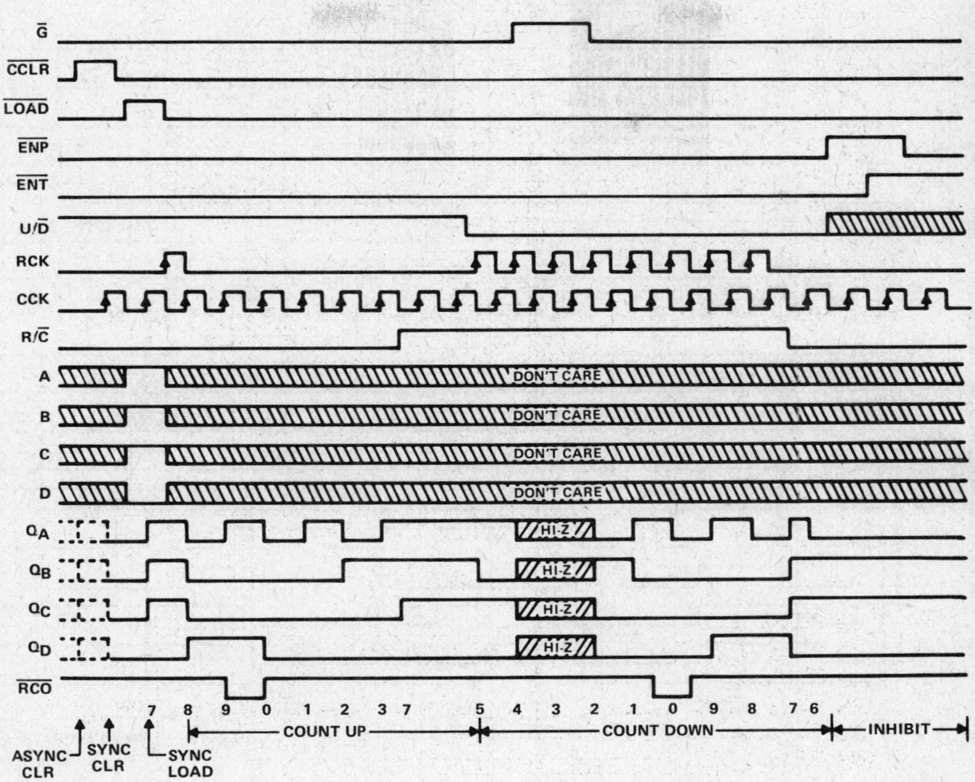

typical operating sequences

'HC697 BINARY COUNTER, Asynchronous Clear
'HC699 BINARY COUNTER, Synchronous Clear

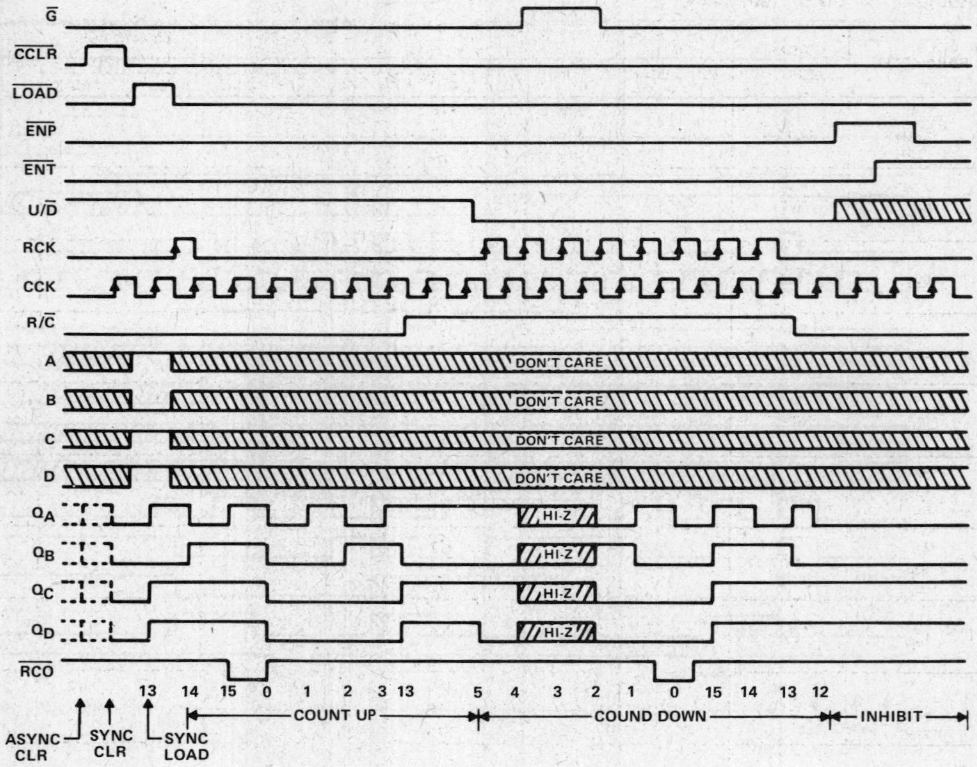

maximum ratings, recommended operating conditions, and electrical characteristics

See Table III, page 2-8.

switching characteristics over recommended operating free-air temperature range (unless otherwise noted), C_L = 50 pF (see Note 1)

PARAMETER	FROM (INPUT)	TO (OUTPUT)	V_{CC}	T_A = 25°C MIN	TYP	MAX	SN54HC696 SN54HC697 MIN	MAX	SN74HC696 SN74HC697 MIN	MAX	UNIT
t_{pd}	CCK↑	$\overline{RCO}$	2 V		115						ns
			4.5 V		23						
			6 V		20						
t_{pd}	$\overline{ENT}$	$\overline{RCO}$	2 V		55						ns
			4.5 V		11						
			6 V		9						
t_{pd}	$\overline{CCLR}$↓	$\overline{RCO}$	2 V		115						ns
			4.5 V		23						
			6 V		20						
t_{pd}	CCK↑	Any Q	2 V		85						ns
			4.5 V		17						
			6 V		14						
t_{pd}	RCK↑	Any Q	2 V		85						ns
			4.5 V		17						
			6 V		14						
t_{PHL}	$\overline{CCLR}$↓	Any Q	2 V		90						ns
			4.5 V		18						
			6 V		15						
t_{pd}	R/$\overline{C}$	Any Q	2 V		70						ns
			4.5 V		14						
			6 V		12						
t_{en}	$\overline{G}$↓	Any Q	2 V		70						ns
			4.5 V		14						
			6 V		12						
t_{dis}	$\overline{G}$↑	Any Q	2 V		95						ns
			4.5 V		19						
			6 V		16						
t_t		Any	2 V		38						ns
			4.5 V		8						
			6 V		6						

NOTE 1: For load circuits and voltage waveforms, see page 1-14.

4

ADVANCE INFORMATION

switching characteristics over recommended operating free-air temperature range (unless otherwise noted), C_L = 50 pF (see Note 1)

PARAMETER	FROM (INPUT)	TO (OUTPUT)	V_{CC}	T_A = 25°C MIN	TYP	MAX	SN54HC698 SN54HC699 MIN	MAX	SN74HC698 SN74HC699 MIN	MAX	UNIT
t_{pd}	CCK↑	$\overline{RCO}$	2 V		115						ns
			4.5 V		23						
			6 V		20						
t_{pd}	$\overline{ENT}$	$\overline{RCO}$	2 V		55						ns
			4.5 V		11						
			6 V		9						
t_{pd}	CCK↑	Any Q	2 V		95						ns
			4.5 V		19						
			6 V		16						
t_{pd}	RCK↑	Any Q	2 V		85						ns
			4.5 V		17						
			6 V		14						
t_{pd}	R/$\overline{C}$	Any Q	2 V		70						ns
			4.5 V		14						
			6 V		12						
t_{en}	$\overline{G}$↓	Any Q	2 V		70						ns
			4.5 V		14						
			6 V		12						
t_{dis}	$\overline{G}$↑	Any Q	2 V		95						ns
			4.5 V		19						
			6 V		16						
t_t		Any	2 V		38						ns
			4.5 V		8						
			6 V		6						

NOTE 1: For load circuits and voltage waveforms, see page 1-14.

D2684, DECEMBER 1982—REVISED MARCH 1984

- Carry-Out Output for Cascading
- Divide-by-N Counting
- DC Clock Input Circuit Allows Slow Rise Times
- Package Options Include Both Plastic and Ceramic Chip Carriers in Addition to Plastic and Ceramic DIPs
- Dependable Texas Instruments Quality and Reliability

SN54HC4017 . . . J PACKAGE
SN74HC4017 . . . J OR N PACKAGE
(TOP VIEW)

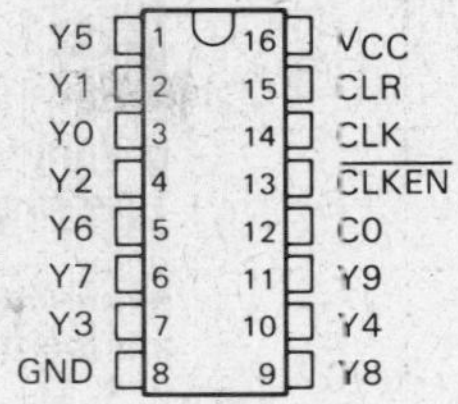

SN54HC4017 . . . FH OR FK PACKAGE
SN74HC4017 . . . FH OR FN PACKAGE
(TOP VIEW)

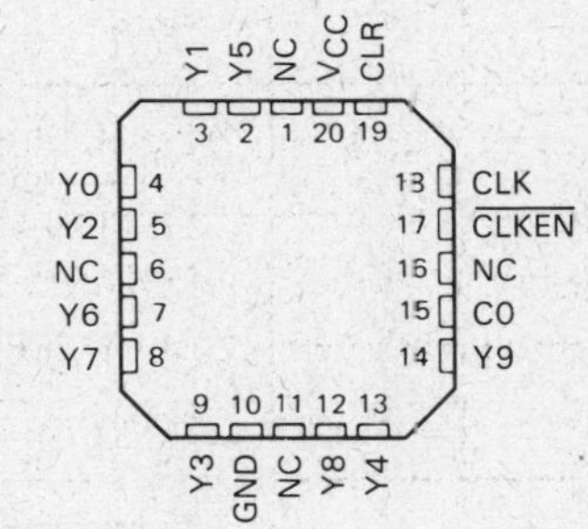

NC—No internal connection

description

The 'HC4017 is a 5-stage divide-by-10 Johnson counter with ten decoded outputs and a carry-out bit. High-speed operation and spike-free outputs are obtained by use of the Johnson decade counter configuration.

The ten decoded outputs are normally low and go high only at their respective decimal time periods. A high signal on CLR asynchronously clears the decade counter and sets the carry output and Y0 high. With $\overline{\text{CLKEN}}$ low, the count is advanced on a low-to-high transition at CLK. Alternatively, if CLK is high, the count is advanced on a high-to-low transition at $\overline{\text{CLKEN}}$. Each decoded output remains high for one full clock cycle. The carry output CO is high while Y0, Y1, Y2, Y3, or Y4 is high, then is low while Y5, Y6, Y7, Y8, or Y9 is high.

The SN54HC4017 is characterized for operation over the full military temperature range of −55°C to 125°C. The SN74HC4017 is characterized for operation from −40°C to 85°C.

logic symbol

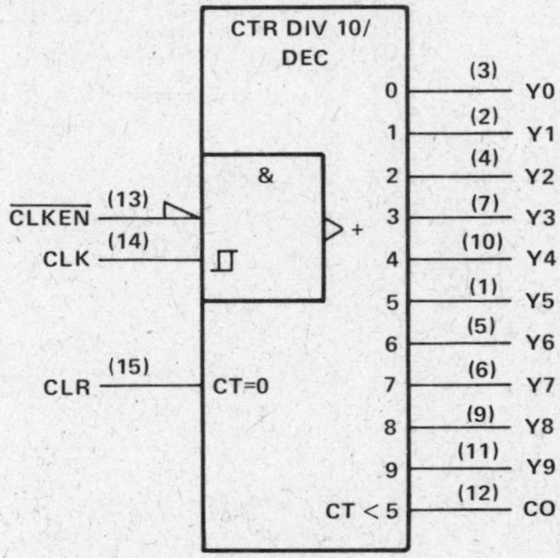

Pin numbers shown are for J and N packages.

4 ADVANCE INFORMATION

ADVANCE INFORMATION
This document contains information on a new product. Specifications are subject to change without notice.

logic diagram (positive logic)

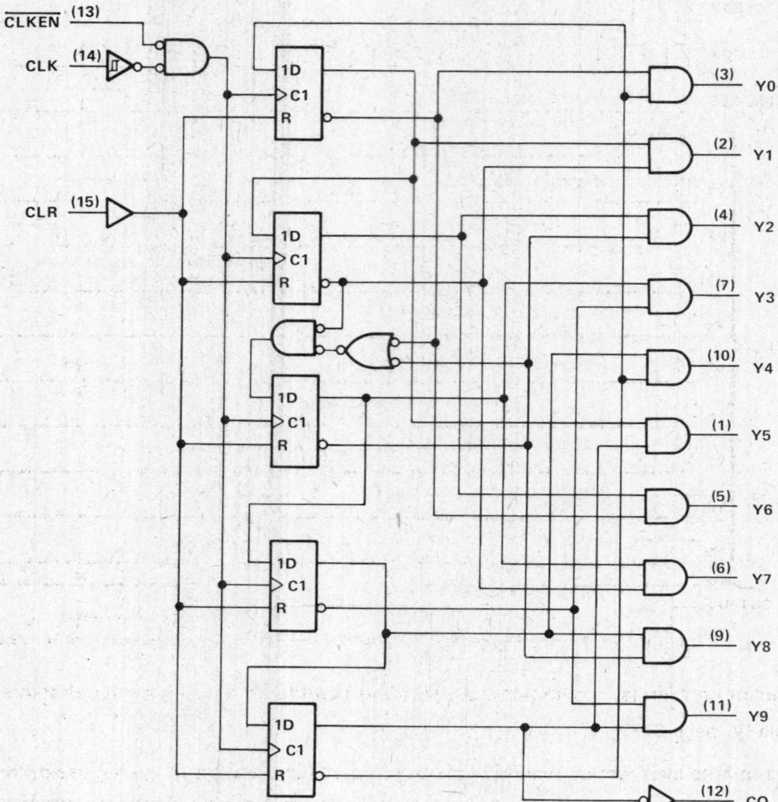

Pin numbers shown are for J and N packages.

4

ADVANCE INFORMATION

TEXAS INSTRUMENTS
POST OFFICE BOX 225012 • DALLAS, TEXAS 75265

typical clear, count, and inhibit sequences

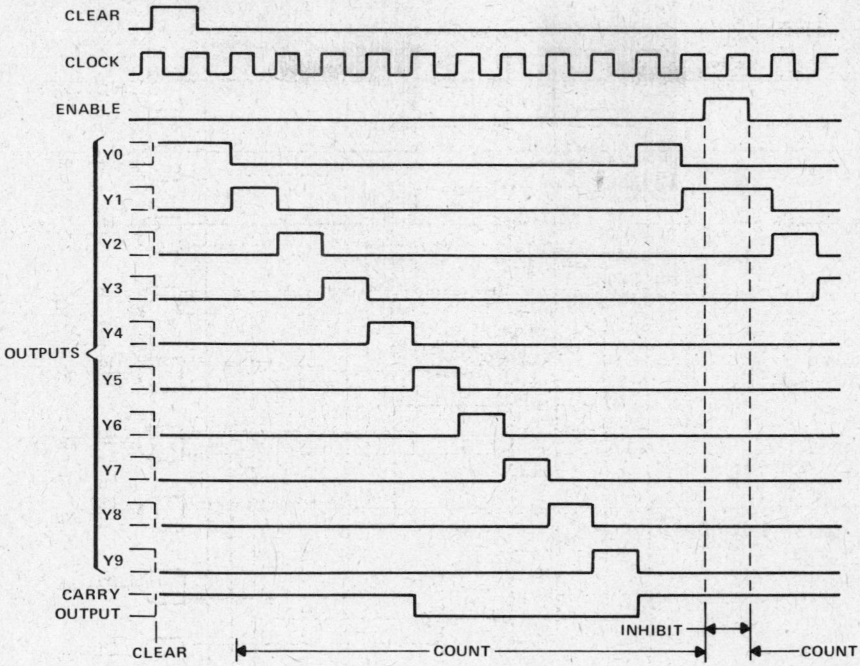

absolute maximum ratings, recommended operating conditions, and electrical characteristics

See Table IV, page 2-10.

timing requirements over recommended operating free-air temperature range (unless otherwise noted)

PARAMETER			V_{CC}	T_A = 25°C MIN	T_A = 25°C MAX	SN54HC4017 MIN	SN54HC4017 MAX	SN74HC4017 MIN	SN74HC4017 MAX	UNIT
f_{clock}	Clock frequency		2 V	0	6					MHz
			4.5 V	0	31					
			6 V	0	36					
t_w	Pulse duration	CLK high or low	2 V	80						ns
			4.5 V	16						
			6 V	14						
		CLR high	2 V	80						
			4.5 V	16						
			6 V	14						
t_{su}	Setup time, before CLK↑	$\overline{CLKEN}$ low	2 V	50						ns
			4.5 V	10						
			6 V	9						
		CLR high	2 V	50						
			4.5 V	10						
			6 V	9						

switching characteristics over recommended operating free-air temperature range (unless otherwise noted), C_L = 50 pF (see Note 1)

PARAMETER	FROM (INPUT)	TO (OUTPUT)	V_{CC}	T_A = 25°C MIN	T_A = 25°C TYP	T_A = 25°C MAX	SN54HC4017 MIN	SN54HC4017 MAX	SN74HC4017 MIN	SN74HC4017 MAX	UNIT
f_{max}			2 V	6	10						MHz
			4.5 V	31	50						
			6 V	36	55						
t_{pd}	CLK	Any Y	2 V		69						ns
			4.5 V		23						
			6 V		20						
t_{pd}	CLK	CO	2 V		69						ns
			4.5 V		23						
			6 V		20						
t_{pd}	CLR	Any Y	2 V		60						ns
			4.5 V		19						
			6 V		16						
t_{pd}	CLR	CO	2 V		60						ns
			4.5 V		19						
			6 V		16						
t_t		Any	2 V		28						ns
			4.5 V		8						
			6 V		6						

NOTE 1: For load circuit and voltage waveforms, see page 1-14.

TEXAS INSTRUMENTS
POST OFFICE BOX 225012 • DALLAS, TEXAS 75265

D2831, MARCH 1984

- Carry-Out Output for Cascading
- Divide-by-N Counting
- DC Clock Input Circuit Allows Slow Rise Times
- Package Options Include Both Plastic and Ceramic Chip Carriers in Addition to Plastic and Ceramic DIPs
- Dependable Texas Instruments Quality and Reliability

SN54HC4022 . . . J PACKAGE
SN74HC4022 . . . J OR N PACKAGE
(TOP VIEW)

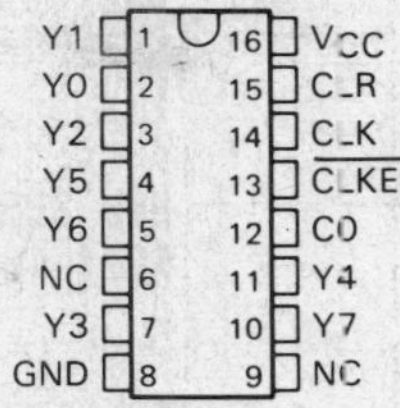

SN54HC4022 . . . FH OR FK PACKAGE
SN74HC4022 . . . FH OR FN PACKAGE
(TOP VIEW)

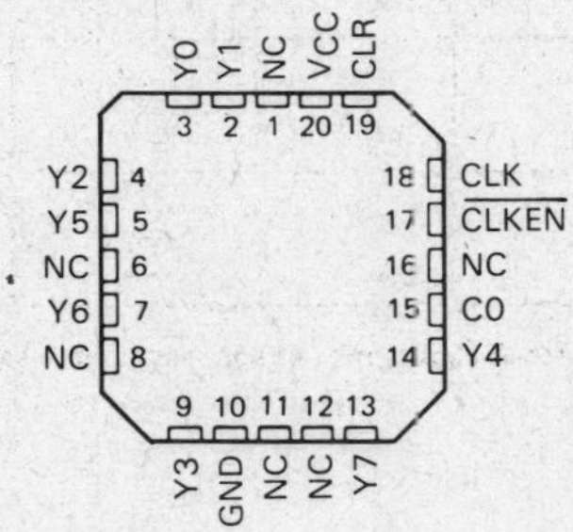

NC – No internal connection

description

The 'HC4022 is a four-stage divide-by-8 Johnson counter with eight decoded outputs and a carry-out bit. High-speed operation and spike-free outputs are obtained by use of the Johnson octal counter configuration.

The eight decoded outputs are normally low and go high only at their respective octal time periods. A high signal on CLR asynchronously clears the octal counter and sets the carry output and Y0 high. With $\overline{\text{CLKEN}}$ low, the count is advanced on a low-to-high transition at CLK. Alternatively, if CLK is high, the count is advanced on a high-to-low transition at $\overline{\text{CLKEN}}$. Each decoded output remains high for one full clock cycle. The carry output C0 is high while Y0, Y1, Y2, or Y3 is high, then is low while Y4, Y5, Y6, or Y7 is high.

The SN54HC4022 is characterized for operation over the full military temperature range of −55°C to 125°C. The SN74HC4022 is characterized for operation from −40°C to 85°C.

logic symbol

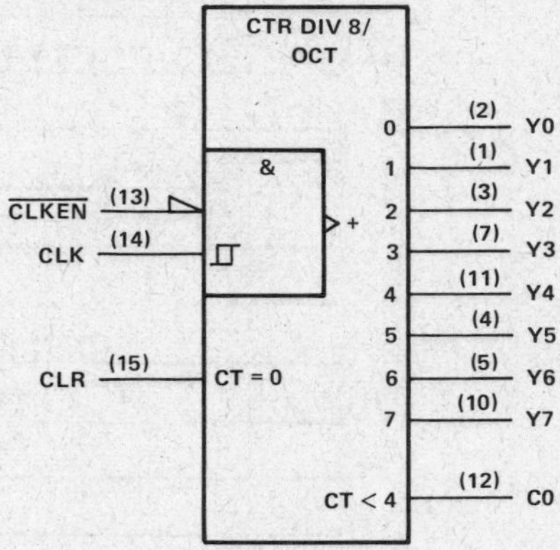

Pin numbers shown are for J and N packages.

logic diagram (positive logic)

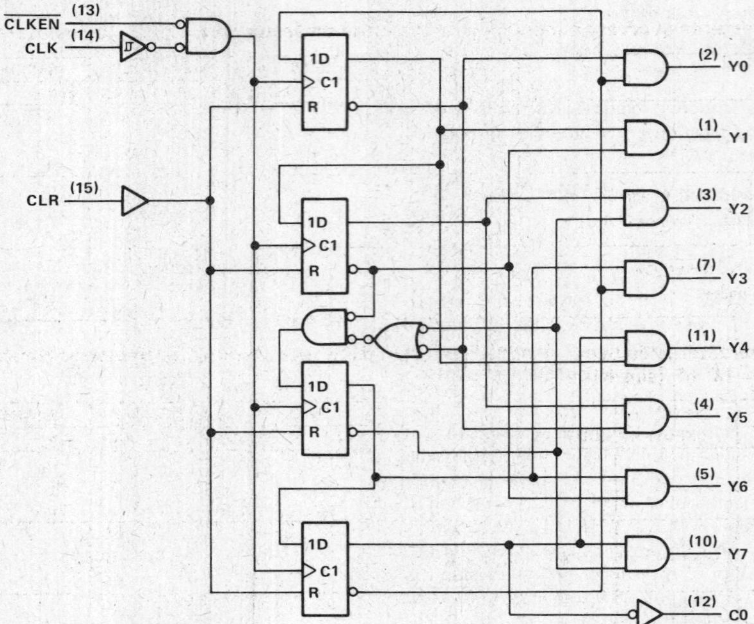

Pin numbers shown are for J and N packages.

typical clear, count, and inhibit sequences

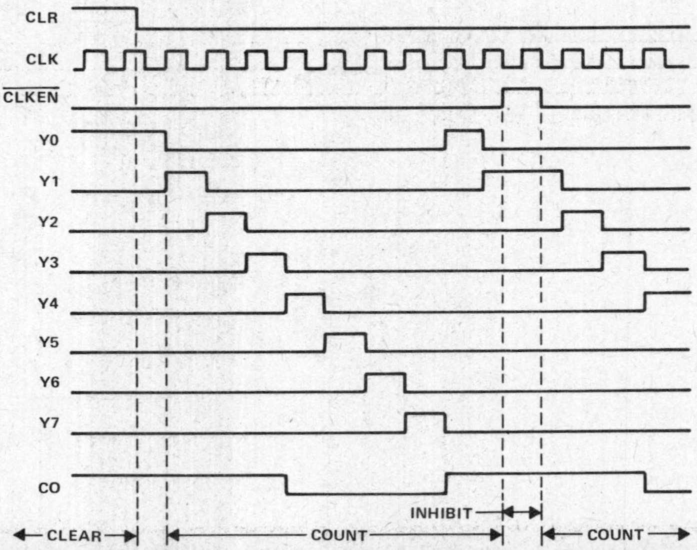

TEXAS INSTRUMENTS
POST OFFICE BOX 225012 • DALLAS, TEXAS 75265

maximum ratings, recommended operating conditions, and electrical characteristics

See Table IV, page 2-10.

timing requirements over recommended operating free-air temperature range (unless otherwise noted)

		V_{CC}	T_A = 25°C		SN54HC4022		SN74HC4022		UNIT
			MIN	MAX	MIN	MAX	MIN	MAX	
f_{clock}	Clock frequency	2 V	0	6	0	4.2	0	5	MHz
		4.5 V	0	31	0	21	0	25	
		6 V	0	36	0	25	0	29	
t_w	Pulse duration, CLK high or low, $\overline{\text{CLKEN}}$ high or low, or CLR high	2 V	80		120		100		ns
		4.5 V	16		24		20		
		6 V	14		20		17		
t_{su}	Setup time, $\overline{\text{CLKEN}}$ low or CLR inactive	2 V	50		75		65		ns
		4.5 V	10		15		13		
		6 V	9		13		11		

switching characteristics over recommended operating free-air temperature range (unless otherwise noted), C_L = 50 pF (see Note 1)

PARAMETER	FROM (INPUT)	TO (OUTPUT)	V_{CC}	T_A = 25°C			SN54HC4022		SN74HC4022		UNIT
				MIN	TYP	MAX	MIN	MAX	MIN	MAX	
f_{max}			2 V	6	10		4.2		5		MHz
			4.5 V	31	50		21		25		
			6 V	36	55		25		29		
t_{pd}	CLK or CLR	Any Y	2 V		70	230		345		290	ns
			4.5 V		24	46		69		58	
			6 V		20	39		59		49	
t_{pd}	CLK or CLR	CO	2 V		60	230		345		290	ns
			4.5 V		19	46		69		58	
			6 V		16	39		59		49	
t_t		Any	2 V		38	75		110		95	ns
			4.5 V		8	15		22		19	
			6 V		6	13		19		16	

C_{pd}	Power dissipation capacitance per counter	No load, T_A = 25°C	100 pF typ

NOTE 1: For load circuit and voltage waveforms, see page 1-14.

4

ADVANCE INFORMATION

TYPES SN54HC7022, SN74HC7022 OCTAL COUNTERS/DIVIDERS WITH POWER-UP CLEAR

D2804, MARCH 1984

- **Carry-Out Output for Cascading**
- **Divide-by-N Counting**
- **DC Clock Input Circuit Allows Slow Rise Times**
- **Power-Up Reset**
- **Pin-Out Compatible with 'HC4022**
- **Package Options Include Both Plastic and Ceramic Chip Carriers in Addition to Plastic and Ceramic DIPs**
- **Dependable Texas Instruments Quality and Reliability**

SN54HC7022 . . . J PACKAGE
SN74HC7022 . . . J OR N PACKAGE
(TOP VIEW)

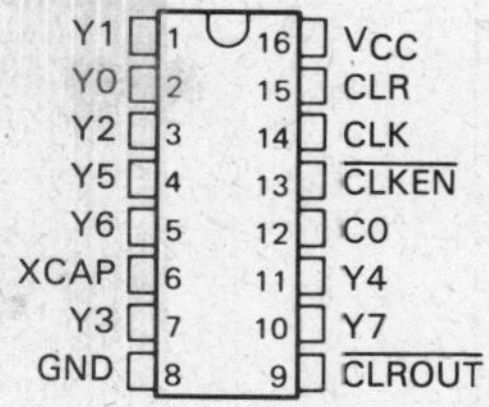

SN54HC7022 . . . FH OR FK PACKAGE
SN74HC7022 . . . FH OR FN PACKAGE
(TOP VIEW)

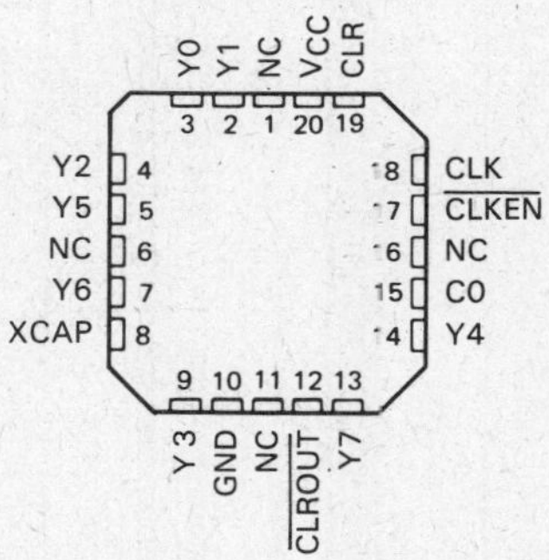

NC—No internal connection

description

The 'HC7022 is a four-stage divide-by-8 Johnson counter with eight decoded outputs and a carry-out bit. High-speed operation and spike-free outputs are obtained by use of the Johnson octal counter configuration.

The eight decoded outputs are normally low and go high only at their respective octal time periods. A high signal on CLR asynchronously clears the octal counter and sets the carry output and Y0 high. With $\overline{\text{CLKEN}}$ low, the count is advanced on a low-to-high transition at CLK. Alternatively, if CLK is high, the count is advanced on a high-to-low transition at $\overline{\text{CLKEN}}$. Each decoded output remains high for one full clock cycle. The carry output C0 is high while Y0, Y1, Y2, or Y3 is high, then is low while Y4, Y5, Y6, or Y7 is high.

This part is similar to the 'HC4022; the main difference is that it includes a power-up-clear circuit to reset the counter during the power-up of the device. The active-low open-drain clear output, $\overline{\text{CLROUT}}$, can be used to clear or reset external circuitry. The pulse duration of the power-up reset circuit can be controlled with an external capacitor C_{ext} connected to pin XCAP. If XCAP is connected to V_{CC}, the power-up reset function is bypassed.

The SN54HC7022 is characterized for operation over the full military temperature range of −55°C to 125°C. The SN74HC7022 is characterized for operation from −40°C to 85°C.

logic symbol

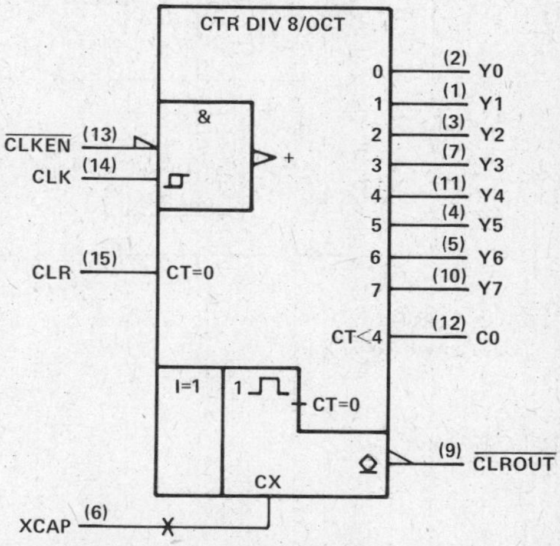

Pin numbers shown are for J and N packages.

ADVANCE INFORMATION
This document contains information on a new product. Specifications are subject to change without notice.

logic diagram (positive logic)

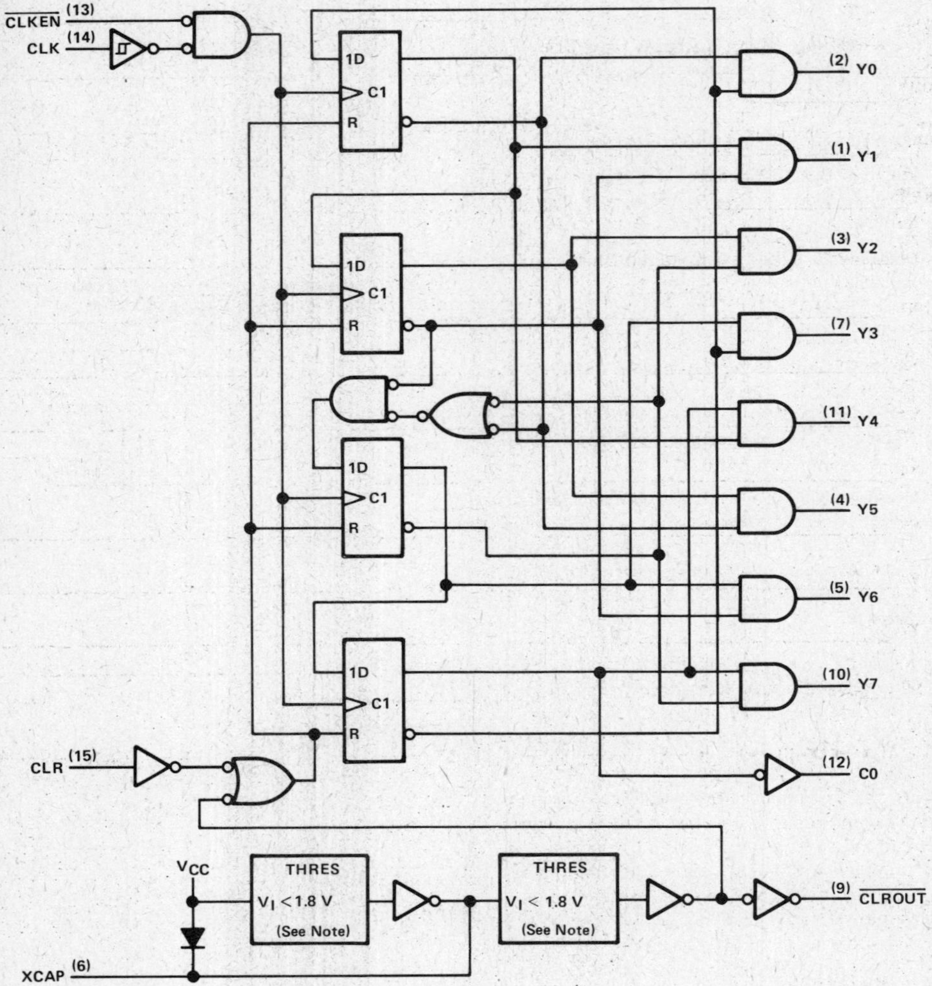

NOTE: The output of the each threshold detector is logically high until the input voltage exceeds the threshold level, typically 1.7 volts. Pin numbers shown are for J and N packages.

4

ADVANCE INFORMATION

TEXAS INSTRUMENTS
POST OFFICE BOX 225012 • DALLAS, TEXAS 75265

typical power-up clear, count, and inhibit sequences

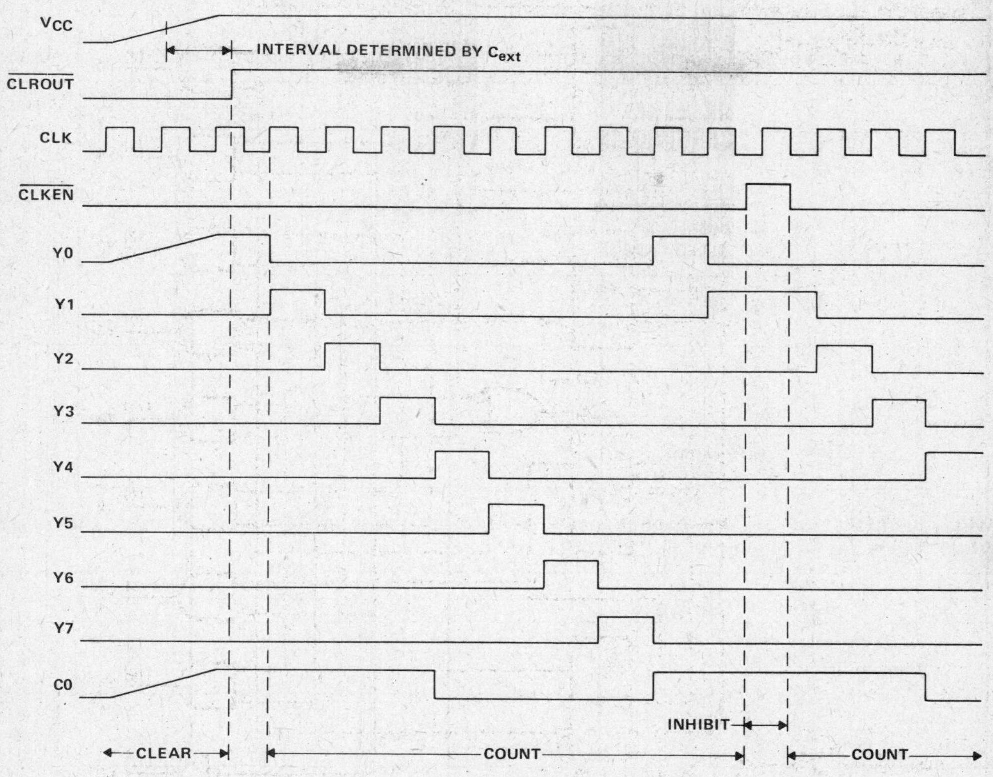

maximum ratings, recommended operating conditions, and electrical characteristics

See Table IV, page 2-10.

timing requirements over recommended operating free-air temperature range (unless otherwise noted)

			V_{CC}	T_A = 25°C		SN54HC7022		SN74HC7022		UNIT
				MIN	MAX	MIN	MAX	MIN	MAX	
f_{clock}	Clock frequency		2 V	0	5	0	3.3	0	4	MHz
			4.5 V	0	25	0	17	0	20	
			6 V	0	29	0	19	0	24	
t_w	Pulse duration	CLK high or low	2 V	100		150		125		ns
			4.5 V	20		30		25		
			6 V	17		26		21		
		CLR high	2 V	125		150		125		
			4.5 V	25		30		25		
			6 V	21		26		21		
t_{su}	Setup time, before CLK↑	$\overline{\text{CLKEN}}$ low	2 V	100		150		125		ns
			4.5 V	20		30		25		
			6 V	17		26		21		
		CLR inactive	2 V	100		150		125		
			4.5 V	20		30		25		
			6 V	17		26		21		

switching characteristics over recommended operating free-air temperature range (unless otherwise noted), C_L = 50 pF (see Note 1)

PARAMETER	FROM (INPUT)	TO (OUTPUT)	V_{CC}	T_A = 25°C			SN54HC7022		SN74HC7022		UNIT
				MIN	TYP	MAX	MIN	MAX	MIN	MAX	
f_{max}			2 V	5	10		3.3		4		MHz
			4.5 V	25	50		17		20		
			6 V	29	55		19		24		
t_{pd}	CLR	Any Y	2 V		60						ns
			4.5 V		19						
			6 V		16						
t_{pd}	CLK	CO	2 V		72						ns
			4.5 V		24						
			6 V		21						
t_{pd}	CLK	Any Y	2 V		69						ns
			4.5 V		23						
			6 V		20						
t_{pd}	CLR	CO	2 V		60						ns
			4.5 V		19						
			6 V		16						

NOTE 1: For load circuit and voltage waveforms, see page 1-14.

TEXAS INSTRUMENTS
POST OFFICE BOX 225012 • DALLAS, TEXAS 75265

TYPES SN54HC7074, SN74HC7074 6-SECTION MULTIFUNCTION (NAND, INVERT, NOR, FLIP-FLOP) CIRCUITS

D2831, MARCH 1984

- **Contains D-type Flip-Flops with Preset and Clear, NAND, NOR, and Inverter Gates**
- **Package Options Include Both Plastic and Ceramic Chip Carriers in Addition to Plastic and Ceramic DIPs**
- **Dependable Texas Instruments Quality and Reliability**

SN54HC7074 . . . JT PACKAGE
SN74HC7074 . . . JT OR NT PACKAGE
(TOP VIEW)

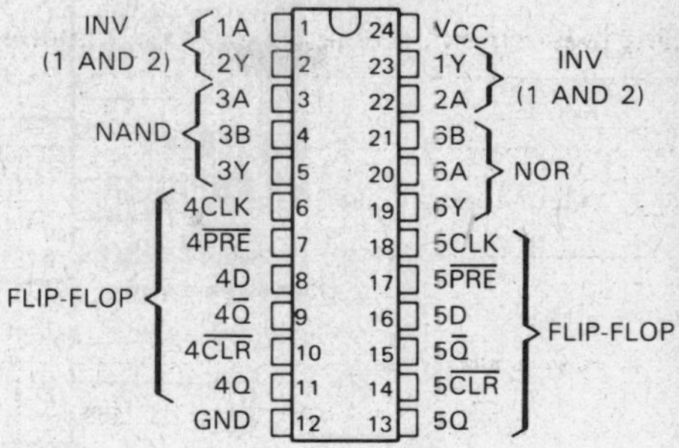

SN54HC7074 . . . FH OR FK PACKAGE
SN74HC7074 . . . FH OR FN PACKAGE
(TOP VIEW)

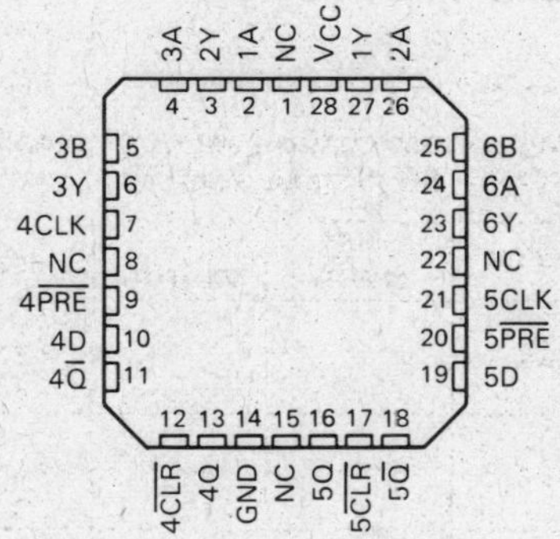

NC – No internal connection

description

The SN54HC7074 and SN74HC7074 are each comprised of the following sections:

Two inverters
One 3-input NOR gate
One 3-input NAND gate
Two D-type flip-flops

They perform the Boolean functions shown under the respective function table.

The D-type flip-flops are positive-edge-triggered and are functionally similar to the SN54HC74 and SN74HC74. A low level at the $\overline{PRE}$ or $\overline{CLR}$ inputs sets or resets the outputs regardless of the levels of the other inputs. When $\overline{PRE}$ and $\overline{CLR}$ are inactive (high), data at the D input meeting the setup time requirements are transferred to the outputs on the positive-going edge of the clock pulse. Clock triggering occurs at a voltage level and is not directly related to the rise time of the clock pulse. Following the hold time interval, data at the D input may be changed without affecting the levels at the outputs.

The SN54HC7074 is characterized for operation over the full military temperature range of −55°C to 125°C. The SN74HC7074 is characterized for operation from −40°C to 85°C.

logic symbol

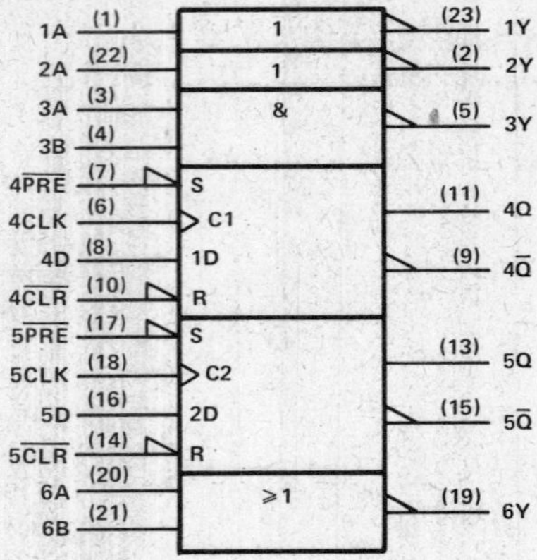

logic diagrams (positive logic)

INVERTERS

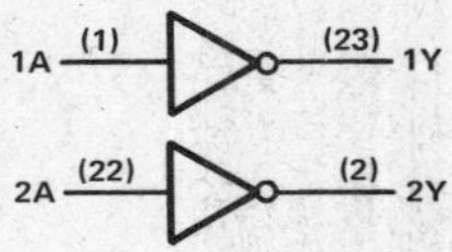

FUNCTION TABLE
(EACH INVERTER)

INPUT	OUTPUT
A	Y
H	L
L	H

positive logic: $Y = \overline{A}$

2-INPUT NAND GATE

FUNCTION TABLE

INPUTS		OUTPUT
A	B	Y
H	H	L
L	X	H
X	L	H

positive logic: $Y = \overline{A \cdot B}$ or $Y = \overline{A} + \overline{B}$

Pin numbers shown are for JT and NT packages.

4

ADVANCE INFORMATION

logic diagrams (positive logic)

D-TYPE FLIP-FLOP

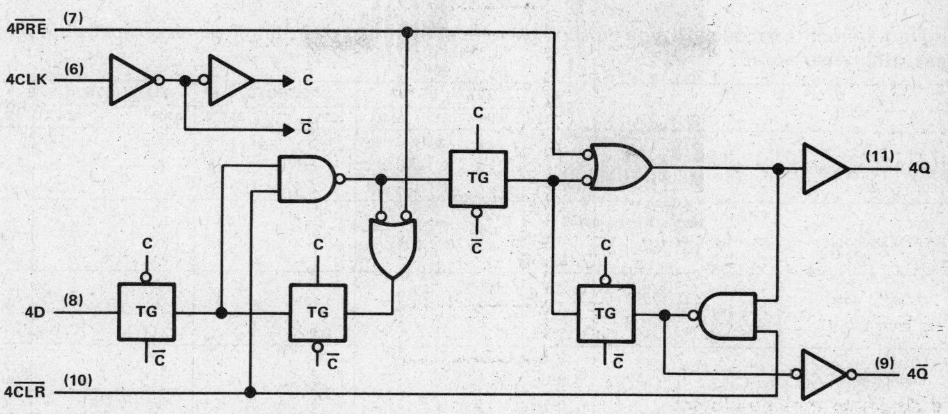

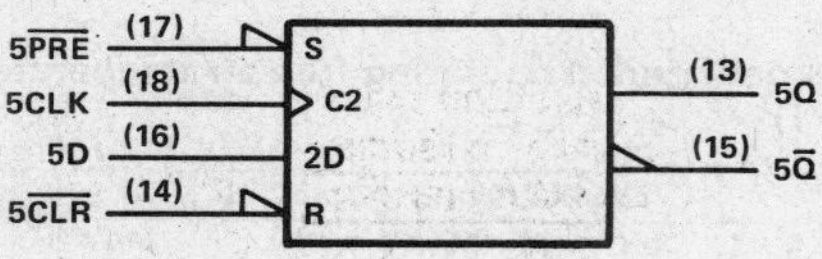

The detail above, and the composite logic symbol to the left, apply to both flip-flops.

FUNCTION TABLE
(EACH D FLIP-FLOP)

INPUTS				OUTPUTS	
$\overline{\text{PRE}}$	$\overline{\text{CLR}}$	CLK	D	Q	$\overline{\text{Q}}$
L	H	X	X	H	L
H	L	X	X	L	H
L	L	X	X	H*	H*
H	H	↑	H	H	L
H	H	↑	L	L	H
H	H	L	X	Q_0	$\overline{Q}_0$

*This configuration is nonstable; i.e., it will not persist when either $\overline{\text{PRE}}$ or $\overline{\text{CLR}}$ returns to the inactive (high) level.

2-INPUT NOR GATE

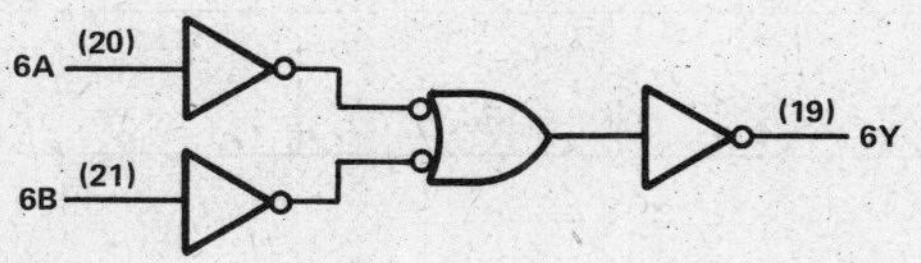

FUNCTION TABLE

INPUTS		OUTPUT
A	B	Y
H	X	L
X	H	L
L	L	H

positive logic: $Y = \overline{A+B}$ or $Y = \overline{A} \cdot \overline{B}$

Pin numbers shown are for JT and NT packages.

absolute maximum ratings, recommended operating conditions, electrical characteristics

For D-type flip-flops, see Table II, page 2-6. For gates and inverters, see Table I, page 2-4.

timing requirements for each D-type flip-flop over recommended operating free-air temperature range (unless otherwise noted)

			V_{CC}	T_A = 25°C MIN	T_A = 25°C MAX	SN54HC7074 MIN	SN54HC7074 MAX	SN74HC7074 MIN	SN74HC7074 MAX	UNIT
f_{clock}	Clock frequency		2 V	0	5	0	3.3	0	4	MHz
			4.5 V	0	25	0	17	0	20	
			6 V	0	29	0	20	0	24	
t_w	Pulse duration	$\overline{\text{PRE}}$ low, $\overline{\text{CLR}}$ low,	2 V	100		150		125		ns
		CLK high, or	4.5 V	20		30		25		
		CLK low	6 V	17		25		21		
t_{su}	Setup time before CLK↑	Data,	2 V	100		150		125		ns
		$\overline{\text{PRE}}$ high, or	4.5 V	20		30		25		
		$\overline{\text{CLR}}$ high	6 V	17		25		21		
t_h	Hold time, data after CLK↑		2 V	5		5		5		ns
			4.5 V	5		5		5		
			6 V	5		5		5		

switching characteristics for each D-type flip-flop over recommended operating free-air temperature range (unless otherwise noted), C_L = 50 pF (see Note 1)

PARAMETER	FROM (INPUT)	TO (OUTPUT)	V_{CC}	T_A = 25°C MIN	T_A = 25°C TYP	T_A = 25°C MAX	SN54HC7074 MIN	SN54HC7074 MAX	SN74HC7074 MIN	SN74HC7074 MAX	UNIT
f_{max}			2 V	5	10		3.3		4		MHz
			4.5 V	25	50		17		20		
			6 V	29	60		20		24		
t_{pd}	CLK	Q	2 V		45						ns
			4.5 V		15						
			6 V		13						
t_{pd}	CLK	$\overline{\text{Q}}$	2 V		45						ns
			4.5 V		15						
			6 V		13						

switching characteristics for gates and inverters over recommended operating free-air temperature range (unless otherwise noted), C_L = 50 pF (see Note 1)

PARAMETER	FROM (INPUT)	TO (OUTPUT)	V_{CC}	T_A = 25°C MIN	T_A = 25°C TYP	T_A = 25°C MAX	SN54HC7074 MIN	SN54HC7074 MAX	SN74HC7074 MIN	SN74HC7074 MAX	UNIT
t_{pd}	A or B	Y	2 V		30						ns
			4.5 V		10						
			6 V		9						
t_t		Y	2 V		38						ns
			4.5 V		8						
			6 V		6						

NOTE 1: For load circuit and voltage waveforms, see page 1-14.

TYPICAL APPLICATION DATA

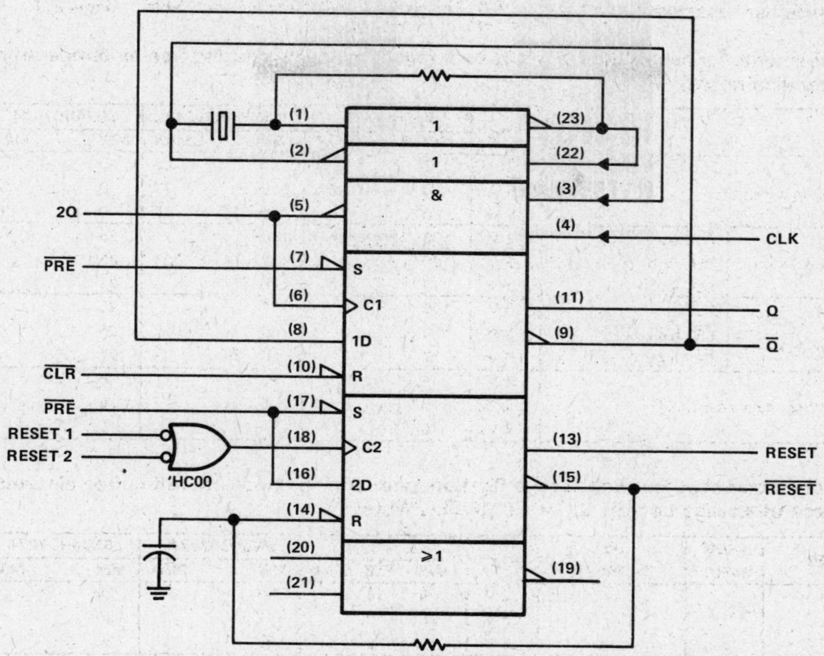

FIGURE 1. CLOCK AND RESET GENERATION FOR MICROPROCESSOR-BASED SYSTEM

TYPES SN54HC7340, SN74HC7340 OCTAL BUS DRIVERS WITH BIDIRECTIONAL REGISTERS

D2831, MARCH 1984

- **8-Bit Bus Drivers with Internal 8-Bit Register**
- **High-Current 3-State Outputs Can Drive up to 15 LSTTL Loads**
- **Package Options Include Both Plastic and Ceramic Chip Carriers in Addition to Plastic and Ceramic DIPs**
- **Dependable Texas Instruments Quality and Reliability**

SN54HC7340 . . . JT PACKAGE
SN74HC7340 . . . JT OR NT PACKAGE
(TOP VIEW)

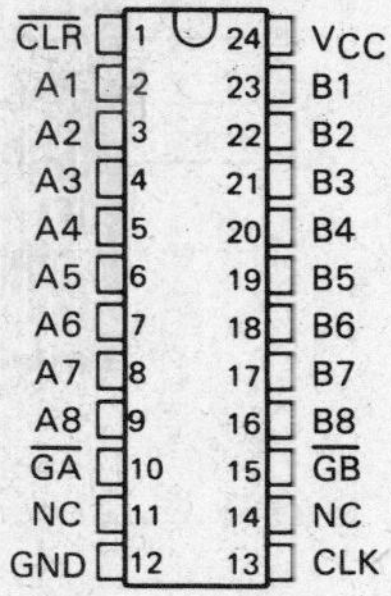

SN54HC7340 . . . FH OR FK PACKAGE
SN74HC7340 . . . FH OR FN PACKAGE
(TOP VIEW)

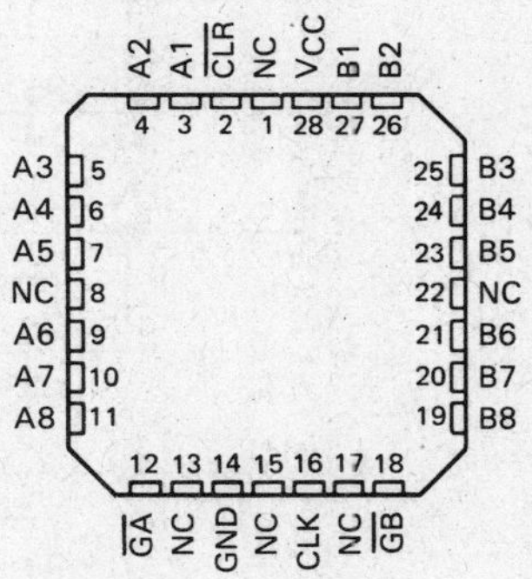

NC—No internal connection

description

These devices consist of bus driver circuits with three-state outputs, D-type flip-flops, clear, and control circuitry arranged for transmission of data directly from the A bus or from the internal register. The A bus is bidirectional and can be used either to load the internal register or to read its contents. Input data is loaded into the register on the low-to-high transition of the clock.

The SN54HC7340 is characterized for operation over the full military temperature range of −55 °C to 125 °C. The SN74HC7340 is characterized for operation from −40 °C to 85 °C.

logic symbol

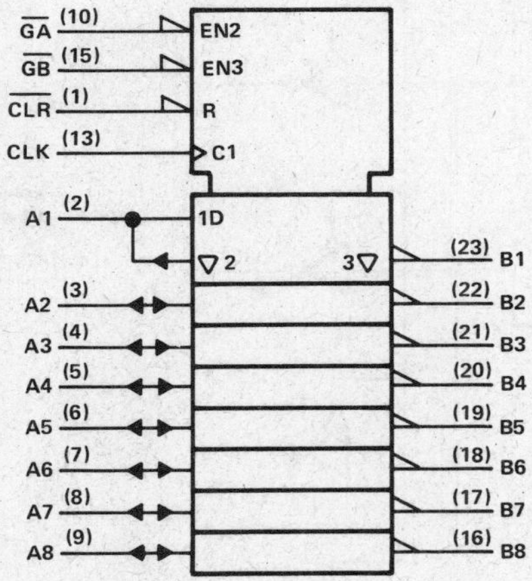

Pin numbers shown are for JT and NT packages.

FUNCTION TABLE

INPUTS				I/O	OUTPUT
$\overline{CLR}$	CLK	$\overline{GA}$	$\overline{GB}$	A	B
L	X	L	L	L	H
L	X	L	H	L	Z
L	X	H	L	Z	H
L	X	H	H	Z	Z
H	↑	L	L	Q_0	$\overline{Q_0}$
H	↑	L	H	Q_0	Z
H	↑	H	L	L	H
H	↑	H	L	H	L
H	↑	H	H	X	Z
H	L	L	L	Q_0	$\overline{Q_0}$
H	L	L	H	Q_0	Z
H	L	H	L	Z	$\overline{Q_0}$
H	L	H	H	Z	Z

logic diagram (positive logic)

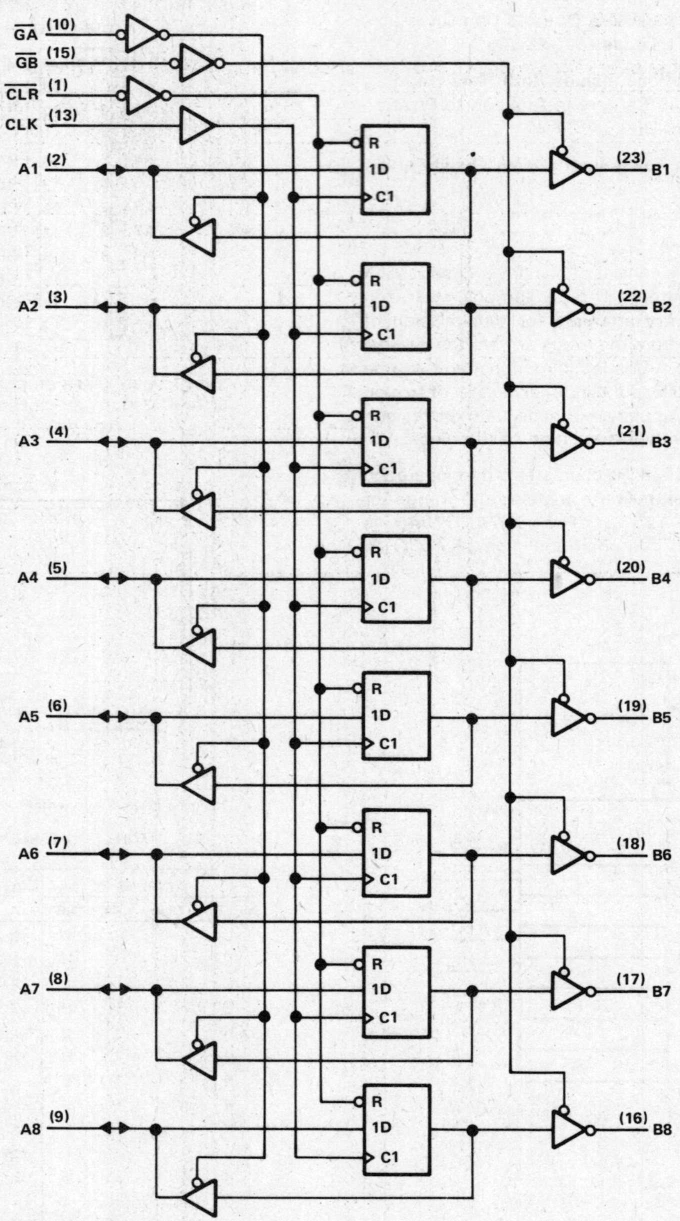

Pin numbers shown are for JT and NT packages.

4

ADVANCE INFORMATION

maximum ratings, recommended operating conditions, and electrical characteristics

See Table III, page 2-8.

timing requirements over recommended operating free-air temperature range (unless otherwise noted)

			V_{CC}	T_A = 25°C MIN	T_A = 25°C MAX	SN54HC7340 MIN	SN54HC7340 MAX	SN74HC7340 MIN	SN74HC7340 MAX	UNIT
f_{clock}	Clock frequency		2 V	0	5	0	3.3	0	4	MHz
			4.5 V	0	25	0	17	0	20	
			6 V	0	29	0	19	0	24	
t_w	Pulse duration	Clock high or low	2 V	100		150		125		ns
			4.5 V	20		30		25		
			6 V	17		26		21		
		Clear low	2 V	100		150		125		ns
			4.5 V	20		30		25		
			6 V	17		26		21		
t_{su}	Setup time before CLK↑	Data	2 V	100		150		125		ns
			4.5 V	20		30		25		
			6 V	17		26		21		
		Clear inactive	2 V	100		150		125		ns
			4.5 V	20		30		25		
			6 V	17		26		21		
t_h	Hold time, data after CLK↑		2 V	0		0		0		ns
			4.5 V	0		0		0		
			6 V	0		0		0		

switching characteristics over recommended operating free-air temperature range (unless otherwise noted), C_L = 50 pF (see Note 1)

PARAMETER	FROM (INPUT)	TO (OUTPUT)	V_{CC}	T_A = 25°C MIN	TYP	MAX	SN54HC7340 MIN	MAX	SN74HC7340 MIN	MAX	UNIT
f_{max}			2 V	5			3.3		4		MHz
			4.5 V	25			17		20		
			6 V	29			19		24		
t_{pd}	CLK	Any	2 V		72						ns
			4.5 V		24						
			6 V		21						
t_{PHL}	$\overline{CLR}$	Any	2 V		69						ns
			4.5 V		23						
			6 V		20						
t_{en}	$\overline{GA}$ or $\overline{GB}$	A or B	2 V		48						ns
			4.5 V		16						
			6 V		14						
t_{dis}	$\overline{GA}$ or $\overline{GB}$	A or B	2 V		51						ns
			4.5 V		17						
			6 V		15						
t_t		Any	2 V		28						ns
			4.5 V		8						
			6 V		6						

C_{pd}	Power dissipation capacitance	No load, T_A = 25°C	50 pF typ

switching characteristics over recommended operating free-air temperature range (unless otherwise noted), C_L = 150 pF (see Note 1)

PARAMETER	FROM (INPUT)	TO (OUTPUT)	V_{CC}	T_A = 25°C MIN	TYP	MAX	SN54HC7340 MIN	MAX	SN74HC7340 MIN	MAX	UNIT
t_{pd}	CLK	Any	2 V		93						ns
			4.5 V		31						
			6 V		27						
t_{PHL}	$\overline{CLR}$	Any	2 V		87						ns
			4.5 V		29						
			6 V		26						
t_{en}	$\overline{GA}$ or $\overline{GB}$	A or B	2 V		60						ns
			4.5 V		20						
			6 V		17						
t_{dis}	$\overline{GA}$ or $\overline{GB}$	A or B	2 V		63						ns
			4.5 V		21						
			6 V		18						
t_t		Any	2 V		60						ns
			4.5 V		17						
			6 V		14						

NOTE 1: For load circuit and voltage waveforms, see page 1-14.

PRODUCT PREVIEWS

5

PRODUCT PREVIEW

This section contains information on products under development in the formative or design phase. Characteristic data and other specifications are DESIGN GOALS. Texas Instruments reserves the right to change or discontinue these products without notice.

D2684, DECEMBER 1982—REVISED MARCH 1984

- D-C Triggered by Active-High or Active-Low Inputs
- Retriggerable for Very Long Output Pulses, Up to 100% Duty Cycle
- Overriding Clear Terminates Output Pulse
- Package Options Include Both Plastic and Ceramic Chip Carriers in Addition to Plastic and Ceramic DIPs
- Dependable Texas Instruments Quality and Reliability

SN54HC123 . . . J PACKAGE
SN74HC123 . . . J OR N PACKAGE
(TOP VIEW)

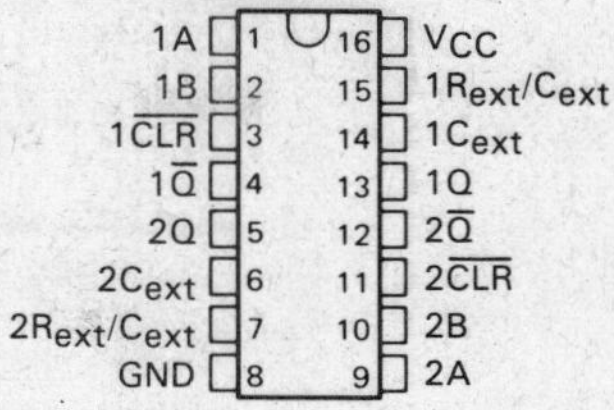

SN54HC123 . . . FH OR FK PACKAGE
SN74HC123 . . . FH OR FN PACKAGE
(TOP VIEW)

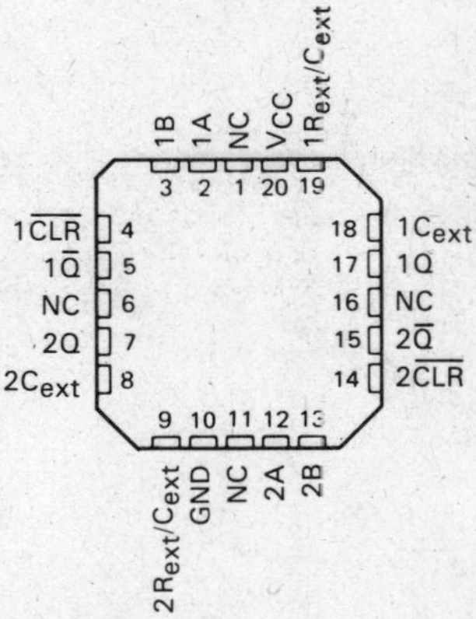

NC—No internal connection

description

These d-c triggered multivibrators feature output pulse duration control by three methods. The basic pulse duration is programmed by selection of external resistance and capacitance values. Once triggered, the basic pulse duration may be extended by retriggering the gated low-level-active (A) or high-level-active (B) inputs, or be reduced by use of the overriding clear. Figure 1 illustrates pulse control by retriggering and early clear.

The SN54HC123 is characterized for operation over the full military temperature range of −55 °C to 125 °C. The SN74HC123 is characterized for operation from −40 °C to 85 °C.

FUNCTION TABLE

INPUTS			OUTPUTS	
CLEAR	A	B	Q	$\overline{Q}$
L	X	X	L	H
X	H	X	L†	H†
X	X	L	L†	H†
H	L	↑	⊓	⊔
H	↓	H	⊓	⊔
↑	L	H	⊓	⊔

†The second and third lines each indicate the logic levels the outputs will take on after the completion of any pulse already started.

logic symbol

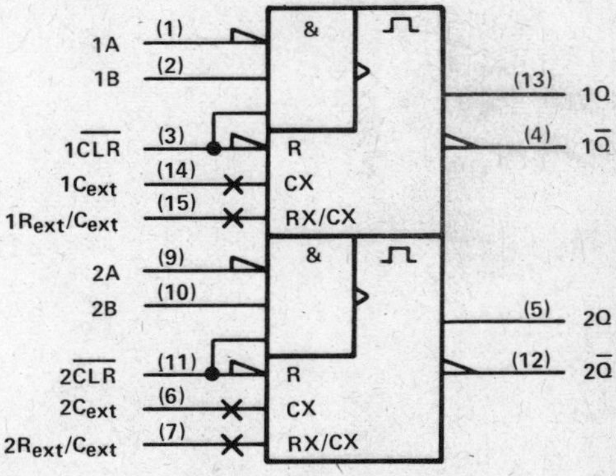

Pin numbers shown are for J and N packages.

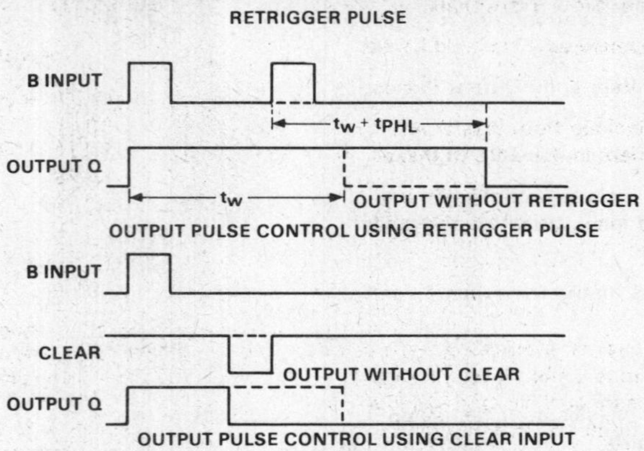

FIGURE 1 — TYPICAL INPUT/OUTPUT PULSES

maximum ratings, recommended operating conditions, and electrical characteristics

See Table IV, page 2-10.

TEXAS INSTRUMENTS
POST OFFICE BOX 225012 • DALLAS, TEXAS 75265

D2684, DECEMBER 1982—REVISED MARCH 1984

- Operation from Very Slow Transitions
- Temperature-Compensated Threshold Levels
- High Noise Immunity
- Package Options Include Both Plastic and Ceramic Chip Carriers in Addition to Plastic and Ceramic DIPs
- Dependable Texas Instruments Quality and Reliability

SN54HC132 . . . J PACKAGE
SN74HC132 . . . J OR N PACKAGE
(TOP VIEW)

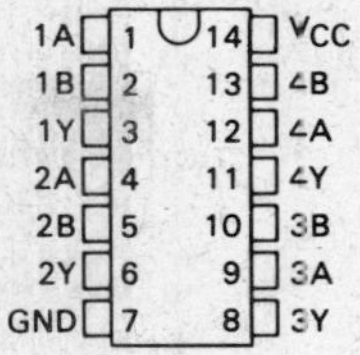

description

Each circuit functions as a NAND gate, but because of the Schmitt action, it has different input threshold levels for positive- and negative-going signals. It performs the Boolean function $Y = \overline{A \cdot B}$ or $Y = \overline{A} + \overline{B}$ in positive logic.

These circuits are temperature compensated and can be triggered from the slowest of input ramps and still give clean jitter-free output signals.

The SN54HC132 is characterized for operation over the full military temperature range of −55°C to 125°C. The SN74HC132 is characterized for operation from −40°C to 85°C.

SN54HC132 . . . FH OR FK PACKAGE
SN74HC132 . . . FH OR FN PACKAGE
(TOP VIEW)

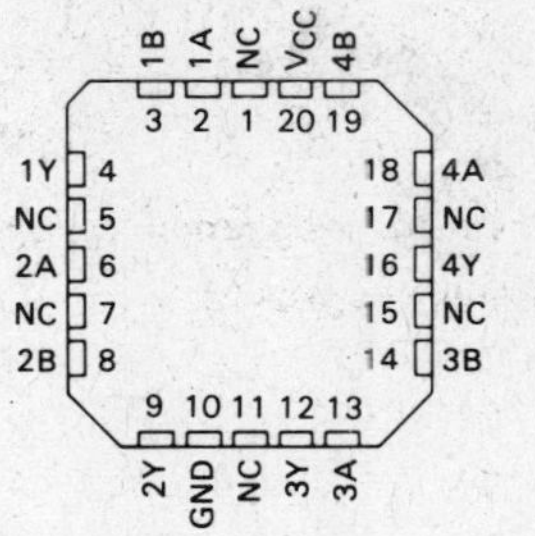

NC—No internal connection

logic symbol

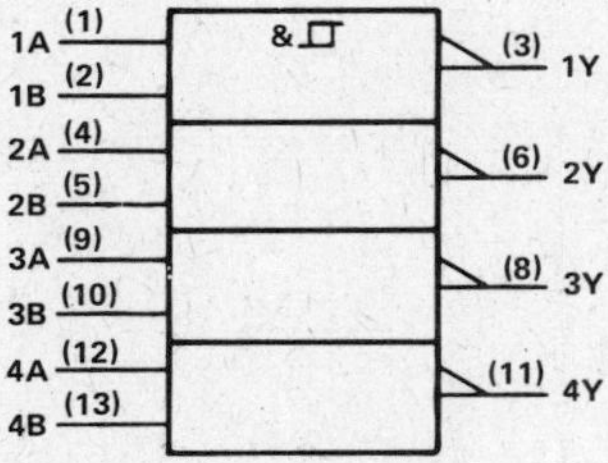

Pin numbers shown are for J and N packages.

FUNCTION TABLE (each gate)

INPUTS		OUTPUT
A	B	Y
H	H	L
L	X	H
X	L	H

maximum ratings, recommended operating conditions, and electrical characteristics

See Table I, page 2-4.

PRODUCT PREVIEWS 5

switching characteristics over recommended operating free-air temperature range (unless otherwise noted), C_L = 50 pF (see Note 1)

PARAMETER	FROM (INPUT)	TO (OUTPUT)	V_{CC}	T_A = 25°C MIN	TYP	MAX	SN54HC132 MIN	MAX	SN74HC132 MIN	MAX	UNIT
t_{pd}	A or B	Y	2 V		40						ns
			4.5 V		13						
			6 V		11						
t_t		Any	2 V		28						ns
			4.5 V		8						
			6 V		6						

NOTE 1: For load circuit and voltage waveforms, see page 1-14.

PRODUCT PREVIEWS

5

D2804, MARCH 1984

- **Offers Carry Functions in a Compatible Form for Direct Connections to the ALU**
- **Cascadable to Perform Look-Ahead Across n-Bit Adders**
- **Package Options Include Both Plastic and Ceramic Chip Carriers in Addition to Plastic and Ceramic DIPs**
- **Dependable Texas Instruments Quality and Reliability**

SN54HC182 . . . J PACKAGE
SN74HC182 . . . J OR N PACKAGE
(TOP VIEW)

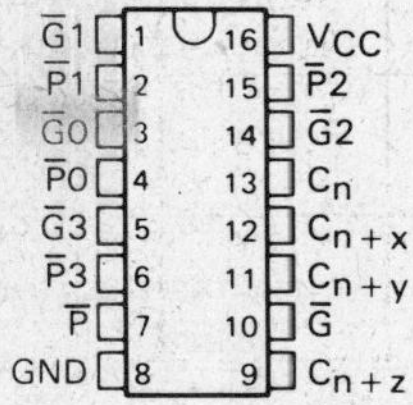

SN54HC182 . . . FH OR FK PACKAGE
SN74HC182 . . . FH OR FN PACKAGE
(TOP VIEW)

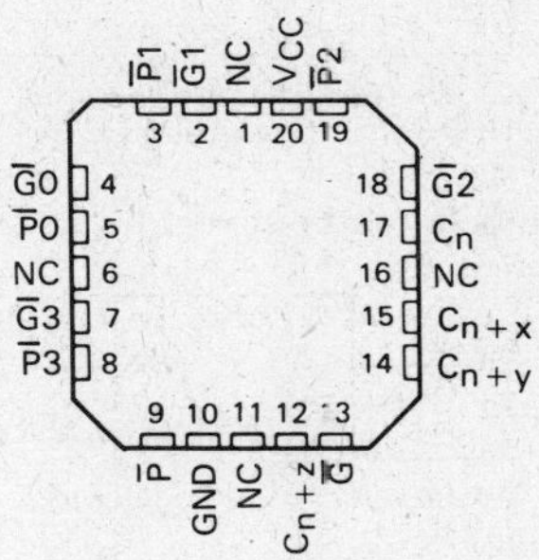

NC—No internal connection

PIN DESIGNATIONS

ALTERNATIVE	DESIGNATIONS[†]	FUNCTION
$\overline{G0}$, $\overline{G1}$, $\overline{G2}$, $\overline{G3}$	G0, G1, G2, G3	Carry Generate Inputs
$\overline{P0}$, $\overline{P1}$, $\overline{P2}$, $\overline{P3}$	P0, P1, P2, P3	Carry Propagate Inputs
C_n	$\overline{C}_n$	Carry Input
C_{n+x}, C_{n+y}, C_{n+z}	$\overline{C}_{n+x}$, $\overline{C}_{n+y}$, $\overline{C}_{n+z}$	Carry Outputs
$\overline{G}$	Y	Carry Generate Output
$\overline{P}$	X	Carry Propagate Output
V_{CC}		Supply Voltage
GND		Ground

[†] Interpretations are illustrated in connection with the Function Tables for the 'HC181 and 'HC881.

description

The 'HC182 look-ahead carry generators are capable of anticipating a carry across four binary adders or group of adders. They are cascadable to perform full look-ahead across n-bit adders.

This generator, when used in conjunction with the 'HC181 or 'HC881 Arithmetic Logic Unit ALU, provides high-speed carry look-ahead capability for any word length. The 'HC182 generates the look-ahead (anticipated carry) across a group of four ALUs. In addition, other carry look-ahead circuits may be employed to anticipate carry-across sections of four look-ahead packages up to n-bits.

The carry functions (inputs, outputs, generate, and propagate) of the look-ahead generators are implemented in the compatible forms for direct connections to the ALU. Reinterpretations of carry functions as explained on the 'HC181 and 'HC881 data sheet are also applicable to and compatible with the look-ahead generator. Logic equations for the 'HC182 are:

$$C_{n+x} = G0 + P0\,C_n$$
$$C_{n+y} = G1 + P1\,G0 + P1\,P0\,C_n$$
$$C_{n+z} = G2 + P2\,G1 + P2\,P1\,G0 + P2\,P1\,P0\,C_n$$
$$\overline{G} = \overline{G3 + P3\,G2 + P3\,P2\,G1 + P3\,P2\,P1\,G0}$$
$$\overline{P} = \overline{P3\,P2\,P1\,P0}$$

or

$$\overline{C}_{n+x} = \overline{Y0\,(X0 + C_n)}$$
$$\overline{C}_{n+y} = \overline{Y1\,[X1 + Y0\,(X0 + C_n)]}$$
$$\overline{C}_{n+z} = \overline{Y2\,\{X2 + Y1\,[X1 + Y0\,(X0 + C_n)]\}}$$
$$Y = Y3\,(X3 + Y2)\,(X3 + X2 + Y1)\,(X3 + X2 + X1 + Y0)$$
$$X = X3 + X2 + X1 + X0$$

maximum ratings recommended operating conditions, and electrical characteristics

See Table IV, page 2-10.

FUNCTION TABLE FOR $\overline{G}$ OUTPUT

INPUTS							OUTPUT
$\overline{G3}$	$\overline{G2}$	$\overline{G1}$	$\overline{G0}$	$\overline{P3}$	$\overline{P2}$	$\overline{P1}$	$\overline{G}$
L	X	X	X	X	X	X	L
X	L	X	X	L	X	X	L
X	X	L	X	L	L	X	L
X	X	X	L	L	L	L	L
All other combinations							H

FUNCTION TABLE FOR $\overline{P}$ OUTPUT

INPUTS				OUTPUT
$\overline{P3}$	$\overline{P2}$	$\overline{P1}$	$\overline{P0}$	$\overline{P}$
L	L	L	L	L
All other combinations				H

FUNCTION TABLE FOR C_{n+x} OUTPUT

INPUTS			OUTPUT
$\overline{G0}$	$\overline{P0}$	C_n	C_{n+x}
L	X	X	H
X	L	H	H
All other combinations			L

FUNCTION TABLE C_{n+y} OUTPUT

INPUTS					OUTPUT
$\overline{G1}$	$\overline{G0}$	$\overline{P1}$	$\overline{P0}$	C_n	C_{n+y}
L	X	X	X	X	H
X	L	L	X	X	H
X	X	L	L	H	H
All other combinations					L

FUNCTION TABLE FOR C_{n+z} OUTPUT

INPUTS							OUTPUT
$\overline{G2}$	$\overline{G1}$	$\overline{G0}$	$\overline{P2}$	$\overline{P1}$	$\overline{P0}$	C_n	C_{n+z}
L	X	X	X	X	X	X	H
X	L	X	L	X	X	X	H
X	X	L	L	L	X	X	H
X	X	X	L	L	L	H	H
All other combinations							L

H = High-level, L = Low-level, X = Irrelevant
Any inputs not shown in a given table are irrelevant with respect to that output.

logic symbols (alternatives)

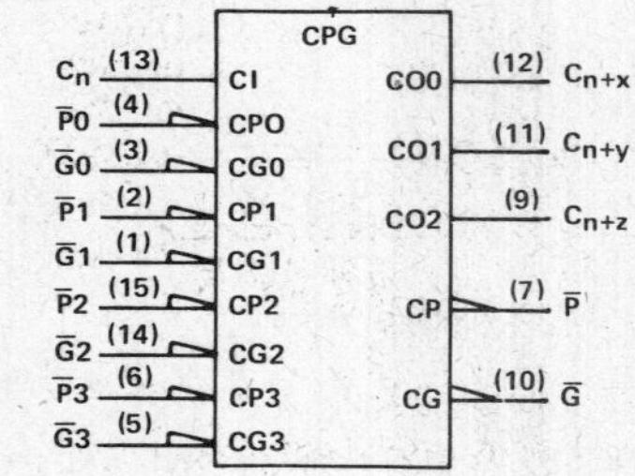

OR

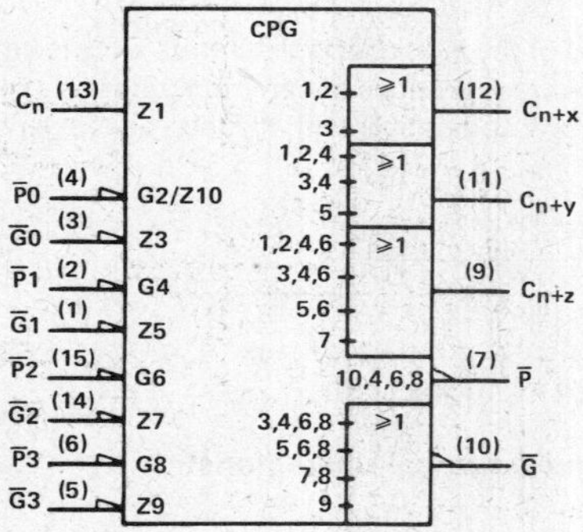

Pin numbers shown are for J and N packages only.

logic diagram (positive logic)

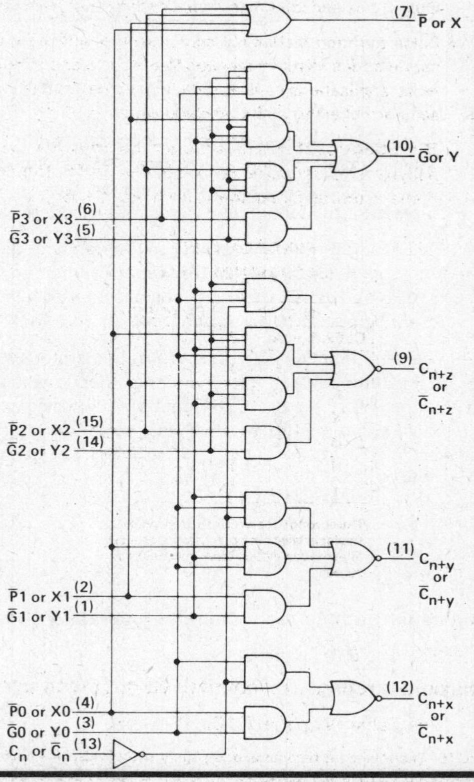

PRODUCT PREVIEWS

5

TYPES SN54HC221, SN74HC221 DUAL MONOSTABLE MULTIVIBRATORS WITH SCHMITT-TRIGGER INPUTS

D2684, DECEMBER 1982—REVISED MARCH 1984

- **Overriding Clear Terminates Output Pulse**
- **Package Options Include Both Plastic and Ceramic Chip Carriers in Addition to Plastic and Ceramic DIPs**
- **Dependable Texas Instruments Quality and Reliability**

SN54HC221 . . . J PACKAGE
SN74HC221 . . . J OR N PACKAGE
(TOP VIEW)

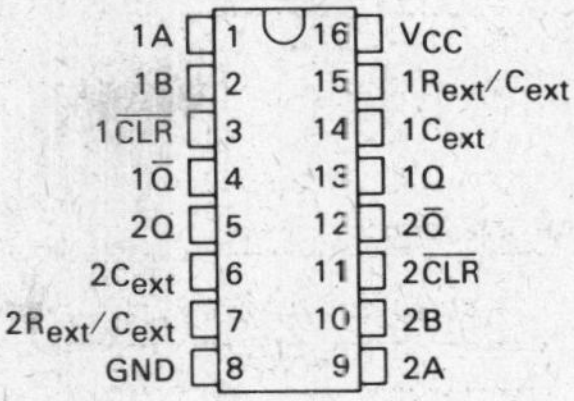

description

These devices are monolithic dual multivibrators featuring a negative-transition-triggered input and a positive-transition-triggered input either of which can be used as an inhibit input.

Pulse triggering occurs at a particular voltage level and is not directly related to the transition time of the input pulse. Schmitt-trigger input circuitry for the B input allows jitter-free triggering from inputs with slow transition rates.

Once fired, the outputs are independent of further transitions of the A and B inputs and are a function of the timing components, or the output pulses can be terminated by the overriding clear. Input pulses may be of any duration relative to the output pulse. Output rise and fall times are independent of pulse length.

Pulse duration stability is achieved through internal compensation and is virtually independent of V_{CC} and temperature. In most applications, pulse stability will be limited only by the accuracy of external timing components.

The SN54HC221 is characterized for operation over the full military temperature range of −55°C to 125°C. The SN74HC221 is characterized for operation from −40°C to 85°C.

SN54HC221 . . . FH OR FK PACKAGE
SN74HC221 . . . FH OR FN PACKAGE
(TOP VIEW)

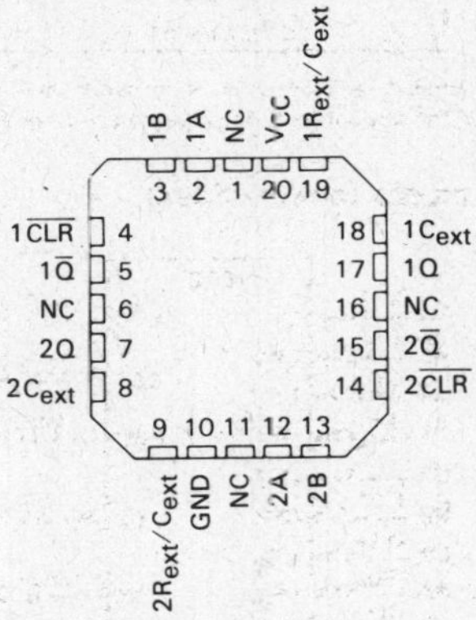

NC—No internal connection

FUNCTION TABLE (EACH MONOSTABLE)

INPUTS			OUTPUTS	
CLEAR	A	B	Q	Q̄
L	X	X	L	H
X	H	X	L†	H†
X	X	L	L†	H†
H	L	↑	⊓	⊔
H	↓	H	⊓	⊔
↑	L	H	⊓	⊔

†The second and third lines each indicate the logic levels the outputs will take on after the completion of any pulse already started.

logic symbol

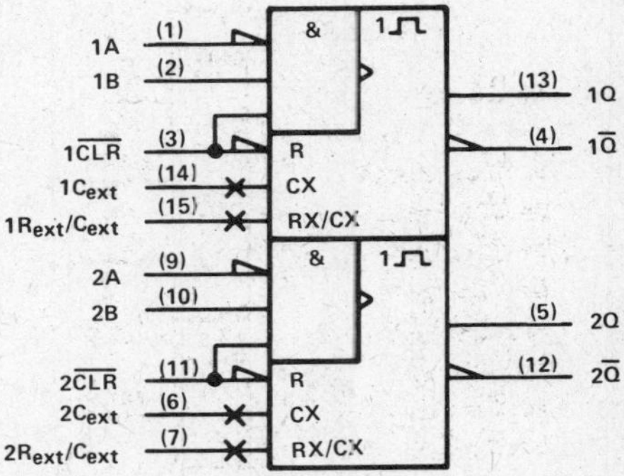

maximum ratings, recommended operating conditions, and electrical characteristics

See Table IV, page 2-10.

NOTE: The minimum recommended supply voltage for this device is 3 V.

PRODUCT PREVIEW

This document contains information on a product under development. Texas Instruments reserves the right to change or discontinue this product without notice.

TYPES SN54HC292, SN54HC294, SN74HC292, SN74HC294 PROGRAMMABLE FREQUENCY DIVIDERS/DIGITAL TIMERS

D2804, MARCH 1984

- Count Divider Chain
- Digitally Programmable from 2^2 to 2^{31} for 'HC292 or 2^{15} for 'HC294
- Usable Frequency Range from DC to 30 MHz
- Easily Expandable
- Applications
  - Frequency Division
  - Digital Timing
- Package Options Include Both Plastic and Ceramic Chip Carriers in Addition to Plastic and Ceramic DIPs
- Dependable Texas Instruments Quality and Reliability

SN54HC292 . . . J PACKAGE
SN74HC292 . . . J OR N PACKAGE
(TOP VIEW)

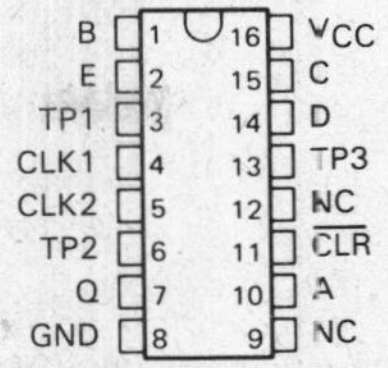

SN54HC292 . . . FH OR FK PACKAGE
SN74HC292 . . . FH OR FN PACKAGE
(TOP VEIW)

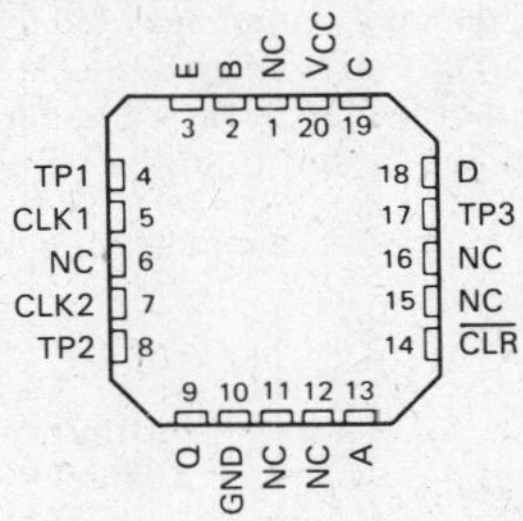

description

These are programmable frequency dividers/digital timers whose count modulo is under digital control of the inputs provided.

Both types feature an active-low clear input to initialize the state of all flip-flops. To facilitate incoming inspection, test points are provided (TP1, TP2, and TP3 on the 'HC292 and TP on the 'HC294). These test points are not intended to drive system loads. Both types feature two clock inputs; either one may be used for clock gating. (See the function table.)

A brief look at the digital timing capabilities of the 'HC292 will show that with a 1-MHz input frequency, programming for 2^{10} will give a period of 1.024 ms, and 2^{20} will give a period of 1.05 sec, 2^{26} will give a period of 1.12 min, and 2^{31} will give a period of 35.79 min.

These devices are easily cascadable giving limitless possibilities to timing delays that can be achieved.

The SN54HC292 and SN54HC294 are characterized for operation over the full military temperature range of −55°C to 125°C. The SN74HC292 and SN74HC294 are characterized for operation from −40°C to 85°C.

SN54HC294 . . . J PACKAGE
SN74HC294 . . . J OR N PACKAGE
(TOP VIEW)

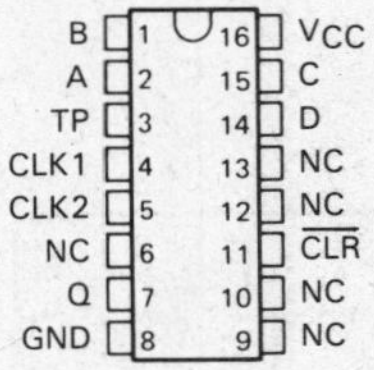

SN54HC294 . . . FH OR FK PACKAGE
SN74HC294 . . . FH OR FN PACKAGE
(TOP VIEW)

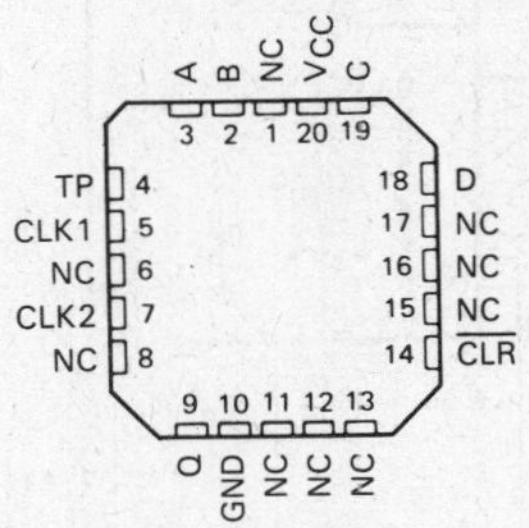

NC—No internal connection

PRODUCT PREVIEW

This document contains information on a product under development. Texas Instruments reserves the right to change or discontinue this product without notice.

'HC292, 'HC294 FUNCTION TABLE

CLEAR	CLK 1	CLK 2	Q OUTPUT MODE
L	X	X	Cleared to L
H	↑	L	Count
H	L	↑	Count
H	H	X	Inhibit
H	X	H	Inhibit

operation

The logic diagram shows that the count modulo is controlled by an X/Y decoder connected to the mode-control inputs of several flip-flops. These flip-flops with mode controls each have a ''D'' input connected to the parallel clock line and a ''T'' input driven by the preceding stage. The parallel clock frequency is always the input frequency divided by four.

The X/Y decoder output selected by the programming inputs goes low. While a mode control is low, the ''D'' input of that flip-flop is enabled, and the signal from the parallel clock line ($f_{in} \div 4$) is passed to the ''T'' input of the following stage. All the other mode controls are high enabling the ''T'' inputs and causing each flip-flop in turn to divide by two.

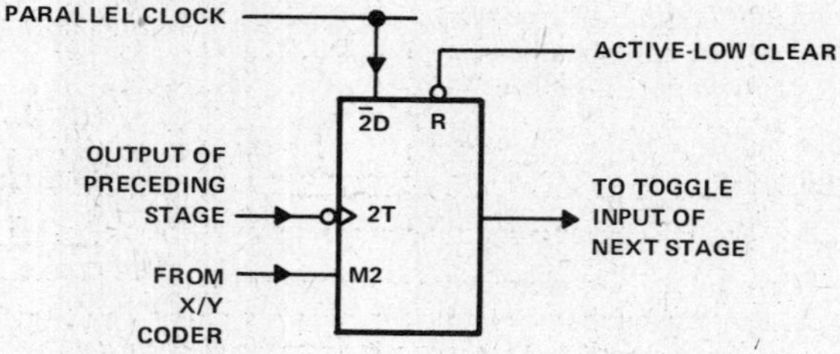

logic symbols

'HC292

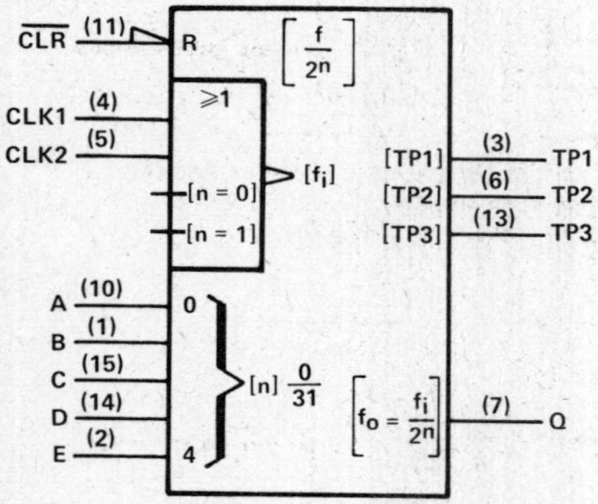

'HC294

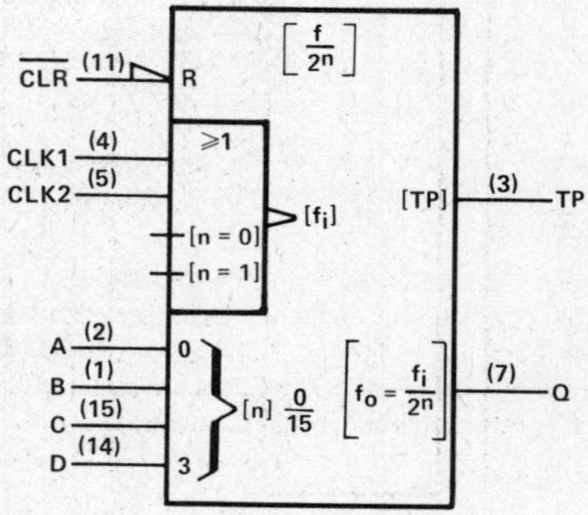

Pin numbers shown are for J and N packages.

'HC292 logic diagram (positive logic)

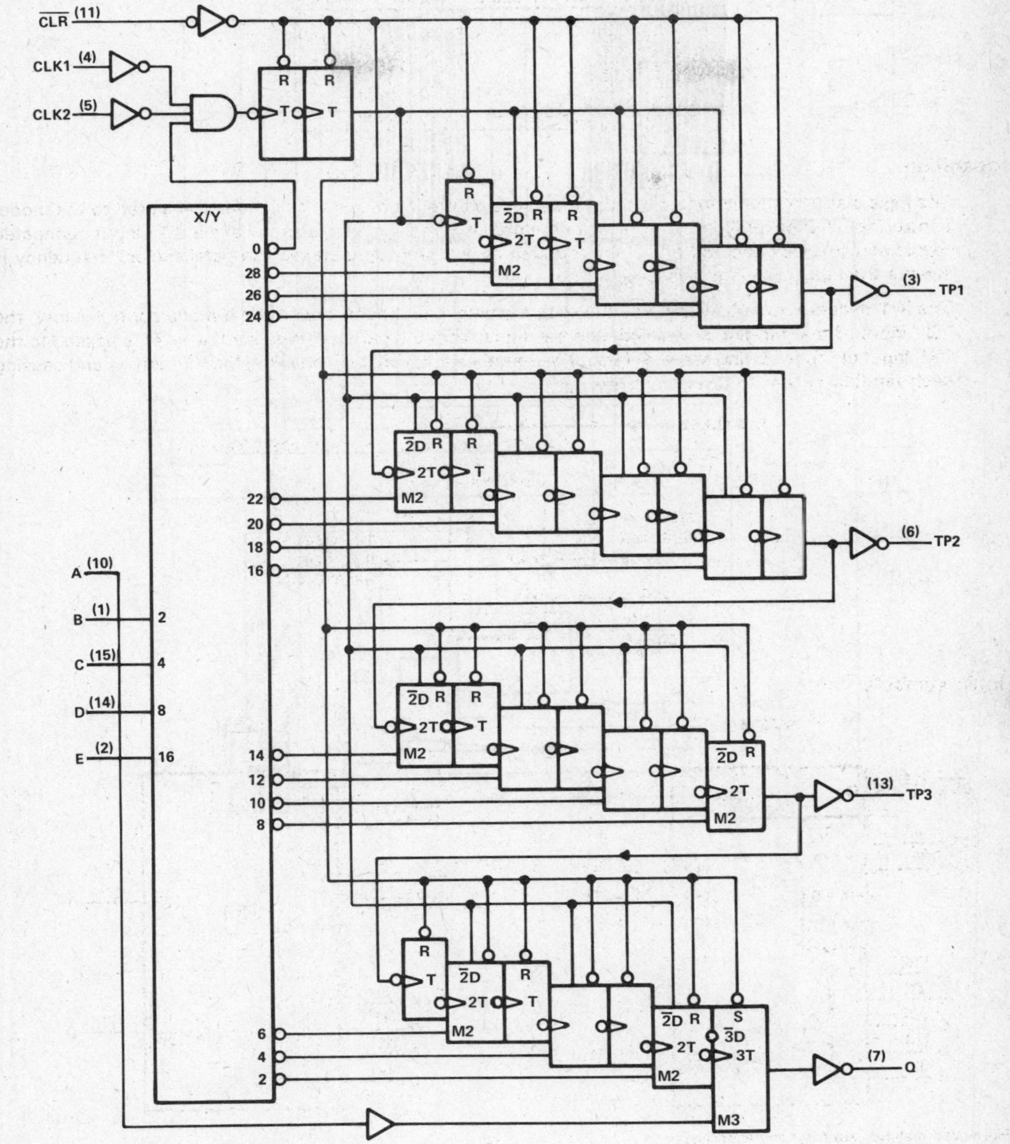

Pin numbers shown are for J and N packages.

'HC294 logic diagram (positive logic)

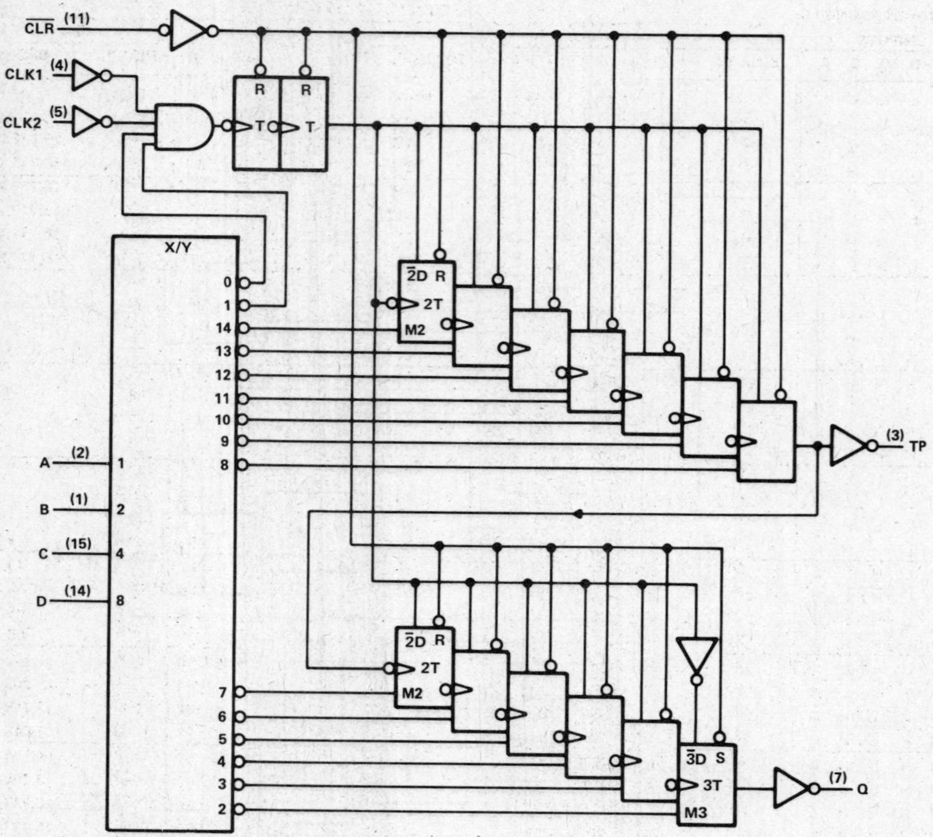

Pin numbers shown are for J and N packages.

TEXAS INSTRUMENTS
POST OFFICE BOX 225012 • DALLAS, TEXAS 75265

'HC292 FUNCTION TABLE

PROGRAMMING INPUTS					FREQUENCY DIVISION							
					Q		TP1		TP2		TP3	
E	D	C	B	A	BINARY	DECIMAL	BINARY	DECIMAL	BINARY	DECIMAL	BINARY	DECIMAL
L	L	L	L	L	Inhibit	Inhibit	Inhibit	Inhibit	Inhibit	Inhibit	Inhibit	Inhibit
L	L	L	L	H	Inhibit	Inhibit	Inhibit	Inhibit	Inhibit	Inhibit	Inhibit	Inhibit
L	L	L	H	L	2^{2}	4	2^{9}	512	2^{17}	131,072	2^{24}	16,777,216
L	L	L	H	H	2^{3}	8	2^{9}	512	2^{17}	131,072	2^{24}	16,777,216
L	L	H	L	L	2^{4}	16	2^{9}	512	2^{17}	131,072	2^{24}	16,777,216
L	L	H	L	H	2^{5}	32	2^{9}	512	2^{17}	131,072	2^{24}	16,777,216
L	L	H	H	L	2^{6}	64	2^{9}	512	2^{17}	131,072	2^{24}	16,777,216
L	L	H	H	H	2^{7}	128	2^{9}	512	2^{17}	131,072	2^{24}	16,777,216
L	H	L	L	L	2^{8}	256	2^{9}	512	2^{17}	131,072	2^{2}	4
L	H	L	L	H	2^{9}	512	2^{9}	512	2^{17}	131,072	2^{2}	4
L	H	L	H	L	2^{10}	1,024	2^{9}	512	2^{17}	131,072	2^{4}	16
L	H	L	H	H	2^{11}	2,048	2^{9}	512	2^{17}	131,072	2^{4}	16
L	H	H	L	L	2^{12}	4,096	2^{9}	512	2^{17}	131,072	2^{6}	64
L	H	H	L	H	2^{13}	8,192	2^{9}	512	2^{17}	131,072	2^{6}	64
L	H	H	H	L	2^{14}	16,384	2^{9}	512	Disabled Low		2^{8}	256
L	H	H	H	H	2^{15}	32,768	2^{9}	512	Disabled Low		2^{8}	256
H	L	L	L	L	2^{16}	65,536	2^{9}	512	2^{3}	8	2^{10}	1,024
H	L	L	L	H	2^{17}	131,072	2^{9}	512	2^{3}	8	2^{10}	1,024
H	L	L	H	L	2^{18}	262,144	2^{9}	512	2^{5}	32	2^{12}	4,096
H	L	L	H	H	2^{19}	524,288	2^{9}	512	2^{5}	32	2^{12}	4,096
H	L	H	L	L	2^{20}	1,048,576	2^{9}	512	2^{7}	128	2^{14}	16,384
H	L	H	L	H	2^{21}	2,097,152	2^{9}	512	2^{7}	128	2^{14}	16,384
H	L	H	H	L	2^{22}	4,194,304	Disabled Low		2^{9}	512	2^{16}	65,536
H	L	H	H	H	2^{23}	8,388,608	Disabled Low		2^{9}	512	2^{16}	65,536
H	H	L	L	L	2^{24}	16,777,216	2^{3}	8	2^{11}	2,048	2^{18}	262,144
H	H	L	L	H	2^{25}	33,554,432	2^{3}	8	2^{11}	2,048	2^{18}	262,144
H	H	L	H	L	2^{26}	67,108,864	2^{5}	32	2^{13}	8,192	2^{20}	1,048,576
H	H	L	H	H	2^{27}	134,217,728	2^{5}	32	2^{13}	8,192	2^{20}	1,048,576
H	H	H	L	L	2^{28}	268,435,456	2^{7}	128	2^{15}	32,768	2^{22}	4,194,304
H	H	H	L	H	2^{29}	536,870,912	2^{7}	128	2^{15}	32,768	2^{22}	4,194,304
H	H	H	H	L	2^{30}	1,073,741,824	2^{9}	512	2^{17}	131,072	2^{24}	16,777,216
H	H	H	H	H	2^{31}	2,147,483,648	2^{9}	512	2^{17}	131,072	2^{24}	16,777,216

maximum ratings, recommended operating conditions, and electrical characteristics

See Table IV, page 2-10.

'HC294 FUNCTION TABLE

PROGRAMMING INPUTS				FREQUENCY DIVISION			
				Q		TP	
D	C	B	A	BINARY	DECIMAL	BINARY	DECIMAL
L	L	L	L	Inhibit	Inhibit	Inhibit	Inhibit
L	L	L	H	Inhibit	Inhibit	Inhibit	Inhibit
L	L	H	L	2^2	4	2^9	512
L	L	H	H	2^3	8	2^9	512
L	H	L	L	2^4	16	2^9	512
L	H	L	H	2^5	32	2^9	512
L	H	H	L	2^6	64	2^9	512
L	H	H	H	2^7	128	Disabled Low	
H	L	L	L	2^8	256	2^2	4
H	L	L	H	2^9	512	2^3	8
H	L	H	L	2^{10}	1,024	2^4	16
H	L	H	H	2^{11}	2,048	2^5	32
H	H	L	L	2^{12}	4,096	2^6	64
H	H	L	H	2^{13}	8,192	2^7	128
H	H	H	L	2^{14}	16,384	2^8	256
H	H	H	H	2^{15}	32,768	2^9	512

maximum ratings, recommended operating conditions, and electrical characteristics

See Table IV, page 2-10.

TYPES SN54HC381, SN54HC382 SN74HC381, SN74HC382 ARITHMETIC LOGIC UNITS/FUNCTION GENERATORS

D2804, MARCH 1984

- Fully Parallel 4-Bit ALUs in 20-Pin Package
- Ideally Suited for High-Density Economical Processors
- 'HC381 Features $\overline{G}$ and $\overline{P}$ Outputs for Look-Ahead Carry Cascading
- 'HC382 Features Ripple Carry (C_{n+4}) and Overflow (OVR) Outputs
- Arithmetic and Logic Operations Selected Specifically to Simplify System Implementation:
  - A Minus B
  - B Minus A
  - A Plus B
  - and Five Other Functions
- Package Options Include Both Plastic and Ceramic Chip Carriers in Addition to Plastic and Ceramic DIPs
- Dependable Texas Instruments Quality and Reliability

SN54HC381 . . . J PACKAGE
SN74HC381 . . . J OR N PACKAGE
(TOP VIEW)

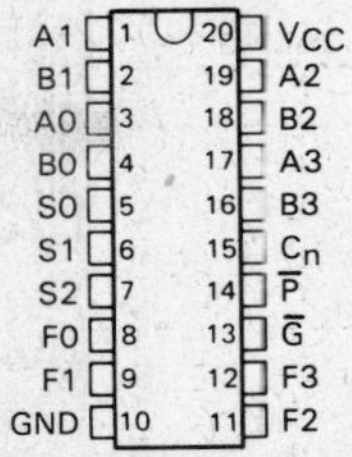

SN54HC381 . . . FH OR FK PACKAGE
SN74HC381 . . . FH OR FN PACKAGE
(TOP VIEW)

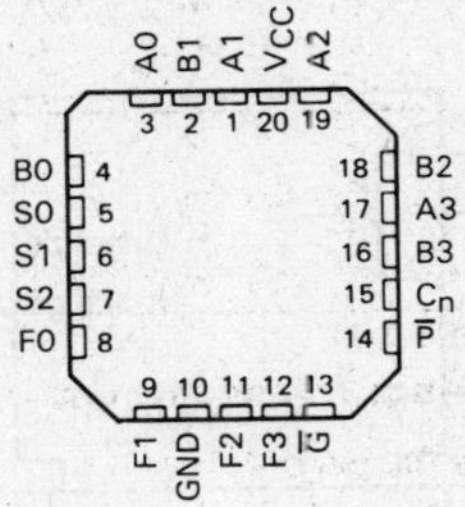

description

The 'HC381 and 'HC382 are arithmetic logic units (ALUs)/function generators that perform eight binary arithmetic/logic operations on two 4-bit words as shown in the function table. The Exclusive-OR, AND, or OR function of the two Boolean variables is provided without the use of external circuitry. Also the F outputs can be cleared (low) or preset (high) as desired. The 'HC381 provides two cascade outputs ($\overline{P}$ and $\overline{G}$) for expansion utilizing SN54HC182/SN74HC182 look-ahead carry generators. The 'HC382 provides a C_{n+4} output to ripple the carry to the C_n input of the next stage. The 'HC382 detects and indicates two's complement overflow condition via the OVR output. The overflow output is logically eqivalent to $C_{n+3} \oplus C_{n+4}$. When the 'HC382 is cascaded to handle word lengths longer than four bits in length, only the most significant overflow (OVR) output is used.

The SN54HC381 and SN54HC382 are characterized for operation over the full military temperature range of −55 °C to 125 °C. The SN74HC381 and SN74HC382 are characterized for operation from −40 °C to 85 °C.

SM54HC382 . . . J PACKAGE
SN74HC382 . . . J OR N PACKAGE
(TOP VIEW)

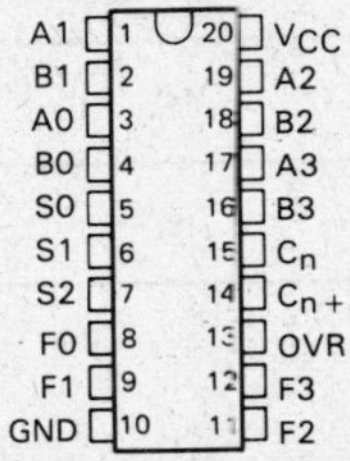

SN54HC382 . . . FH OR FK PACKAGE
SN74HC382 . . . FH OR FN PACKAGE
(TOP VIEW)

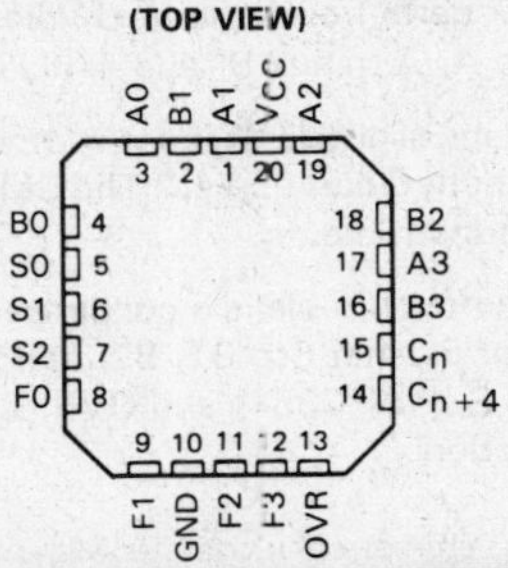

PRODUCT PREVIEW

This document contains information on a product under development. Texas Instruments reserves the right to change or discontinue this product without notice.

TEXAS INSTRUMENTS
POST OFFICE BOX 225012 • DALLAS, TEXAS 75265

FUNCTION TABLE

SELECTION			ARITHMETIC/LOGIC
S2	S1	S0	OPERATION
L	L	L	CLEAR
L	L	H	B MINUS A
L	H	L	A MINUS B
L	H	H	A PLUS B
H	L	L	A ⊕ B
H	L	H	A + B
H	H	L	AB
H	H	H	PRESET

H = high level, L = low level

logic symbols

'HC381

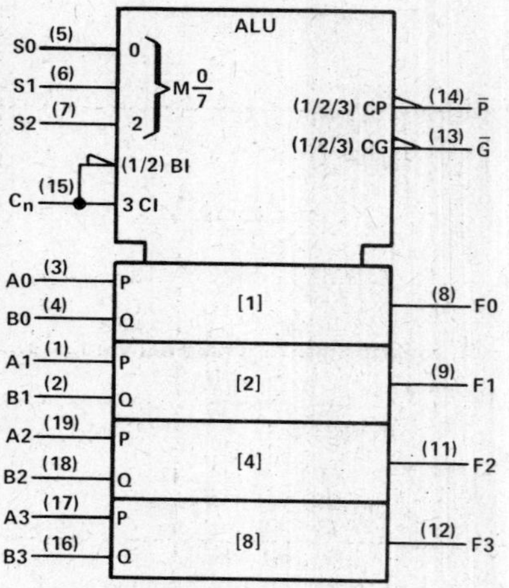

'HC382

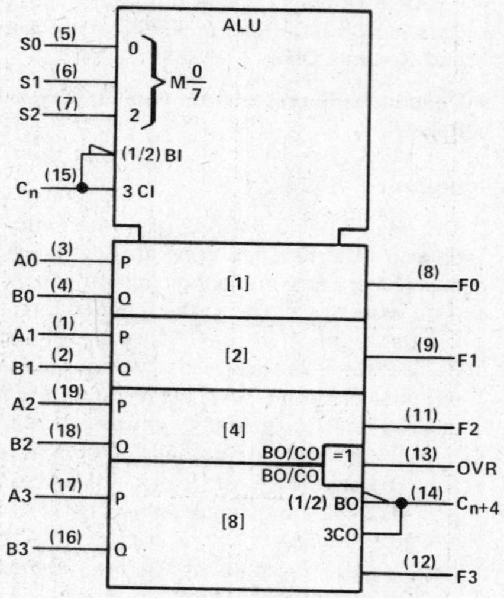

function table

Certain differences exist in the $\overline{G}$, $\overline{P}$ ('HC381) and OVR, C_{n+4} ('HC382) function table compared with similar parts from other technologies and other vendors. No differences exist in the arithmetic modes (B minus A, A minus B, and A plus B), where these outputs perform valuable cascade functions.

There are slight differences in other modes (CLEAR, A + B, A ⊕ B, AB, and PRESET) where these outputs are strictly "don't care." The CMOS implementation will be the same as for the LSTTL counterparts from Texas Instruments.

This function table is a condensed version and assumes for A_n that A0, A1, A2, and A3 inputs all agree and for B_n that B0, B1, B2, and B3 inputs all agree. This table is intended to point out the response of these $\overline{G}$, $\overline{P}$ ('HC381) and OVR, C_{n+4} ('HC382) outputs in all modes of operation to facilitate incoming inspection.

TEXAS INSTRUMENTS
POST OFFICE BOX 225012 • DALLAS, TEXAS 75265

FUNCTION TABLE

ARITHMETIC/LOGIC OPERATION	INPUTS						OUTPUTS				('HC381)		('HC382)	
	S2	S1	S0	C_n	A_n	B_n	F3	F2	F1	F0	$\overline{G}$	$\overline{P}$	OVR	C_{n+4}
Clear	L	L	L	X	X	X	L	L	L	L	H	H	L	L
B MINUS A	L	L	H	L	L	L	H	H	H	H	H	L	L	L
				L	L	H	H	H	H	L	L	H	L	H
				L	H	L	L	L	L	L	H	H	L	L
				L	H	H	H	H	H	H	H	L	L	L
				H	L	L	L	L	L	L	H	L	L	H
				H	L	H	H	H	H	H	L	H	L	H
				H	H	L	L	L	L	H	H	H	L	L
				H	H	H	L	L	L	L	H	L	L	H
A MINUS B	L	H	L	L	L	L	H	H	H	H	H	L	L	L
				L	L	H	L	L	L	L	H	H	L	L
				L	H	L	H	H	H	L	L	H	L	H
				L	H	H	H	H	H	H	H	L	L	L
				H	L	L	L	L	L	L	H	L	L	H
				H	L	H	L	L	L	H	H	H	L	L
				H	H	L	H	H	H	H	L	H	L	H
				H	H	H	L	L	L	L	H	L	L	H
A PLUS B	L	H	H	L	L	L	L	L	L	L	H	H	L	L
				L	L	H	H	H	H	H	H	L	L	L
				L	H	L	H	H	H	H	H	L	L	L
				L	H	H	H	H	H	L	L	H	L	H
				H	L	L	L	L	L	H	H	H	L	L
				H	L	H	L	L	L	L	H	L	L	H
				H	H	L	L	L	L	L	H	L	L	H
				H	H	H	H	H	H	H	L	H	L	H
A ⊕ B	H	L	L	X	L	L	L	L	L	L	H	H	L	L
				L	L	H	H	H	H	H	H	L	L	L
				H	L	H	H	H	H	H	H	L	H	H
				L	H	L	H	H	H	H	H	L	L	L
				H	H	L	H	H	H	H	H	L	H	H
				X	H	H	L	L	L	L	H	H	L	L
A + B	H	L	H	X	L	L	L	L	L	L	H	H	L	L
				L	L	H	H	H	H	H	H	L	L	L
				H	L	H	H	H	H	H	H	-	H	H
				L	H	L	H	H	H	H	H	L	L	L
				H	H	L	H	H	H	H	H	L	H	H
				L	H	H	H	H	H	H	H	L	L	L
				H	H	H	H	H	H	H	H	L	H	H
AB	H	H	L	X	L	L	L	L	L	L	H	H	L	L
				X	L	H	L	L	L	L	H	H	L	L
				X	H	L	L	L	L	L	H	H	L	L
				L	H	H	H	H	H	H	H	L	L	L
				H	H	H	H	H	H	H	H	L	H	H
PRESET	H	H	H	L	X	X	H	H	H	H	H	L	L	L
				H	X	X	H	H	H	H	H	L	H	H

maximum ratings, recommended operating conditions, and electrical characteristics

See Table IV, page 2-10.

'HC381 switching characteristics over recommended operating free-air temperature range (unless otherwise noted), C_L = 50 pF (see Note 1)

PARAMETER	FROM (INPUT)	TO (OUTPUT)	V_{CC}	T_A = 25°C MIN	T_A = 25°C TYP	T_A = 25°C MAX	SN54HC381 MIN	SN54HC381 MAX	SN74HC381 MIN	SN74HC381 MAX	UNIT
t_{pd}	C_n	Any F	2 V		36						ns
			4.5 V		12						
			6 V		10						
t_{pd}	A_i or B_i	F_i	2 V		36						ns
			4.5 V		12						
			6 V		10						
t_{pd}	Any A or B	$\overline{G}$ or $\overline{P}$	2 V		36						ns
			4.5 V		12						
			6 V		10						
t_{pd}	S0, S1, or S2	F_i	2 V		52						ns
			4.5 V		17						
			6 V		14						
t_{pd}	S0, S1, or S2	$\overline{G}$ or $\overline{P}$	2 V		50						ns
			4.5 V		17						
			6 V		14						

C_{pd}	Power dissipation capacitance	No load, T_A = 25°C	100 pF typ

'HC382 switching characteristics over recommended operating free-air temperature range (unless otherwise noted), C_L = 50 pF (see Note 1)

PARAMETER	FROM (INPUT)	TO (OUTPUT)	V_{CC}	T_A = 25°C MIN	T_A = 25°C TYP	T_A = 25°C MAX	SN54HC382 MIN	SN54HC382 MAX	SN74HC382 MIN	SN74HC382 MAX	UNIT
t_{pd}	C_n	Any F	2 V		36						ns
			4.5 V		12						
			6 V		10						
t_{pd}	C_n	OVR	2 V		36						ns
			4.5 V		12						
			6 V		10						
t_{pd}	C_n	C_{n+4}	2 V		33						ns
			4.5 V		11						
			6 V		10						
t_{pd}	A_i or B_i	F_i	2 V		36						ns
			4.5 V		12						
			6 V		10						
t_{pd}	Any A or B	C_{n+4}	2 V		39						ns
			4.5 V		13						
			6 V		11						
t_{pd}	Any A or B	OVR	2 V		42						ns
			4.5 V		14						
			6 V		12						
t_{pd}	S0, S1, or S2	F_i	2 V		52						ns
			4.5 V		17						
			6 V		14						
t_{pd}	S0, S1, or S2	C_{n+4} or OVR	2 V		60						ns
			4.5 V		20						
			6 V		17						

NOTE 1: For load circuit and voltage waveforms, see page 1-14.

TYPES SN54HC423, SN74HC423
DUAL RETRIGGERABLE MONOSTABLE MULTIVIBRATORS

D2684, DECEMBER 1982—REVISED MARCH 1984

- Retriggerable for Very Long Output Pulses, Up to 100% Duty Cycle
- Overriding Clear Terminates Output Pulse
- Package Options Include Both Plastic and Ceramic Chip Carriers in Addition to Plastic and Ceramic DIPs
- Dependable Texas Instruments Quality and Reliability

SN54HC423 . . . J PACKAGE
SN74HC423 . . . J OR N PACKAGE
(TOP VIEW)

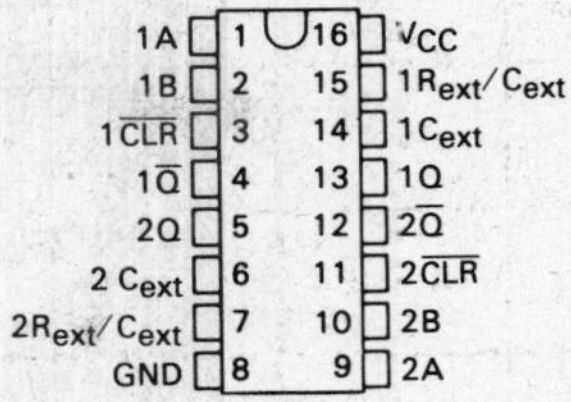

SN54HC423 . . . FH OR FK PACKAGE
SN74HC423 . . . FH OR FN PACKAGE
(TOP VIEW)

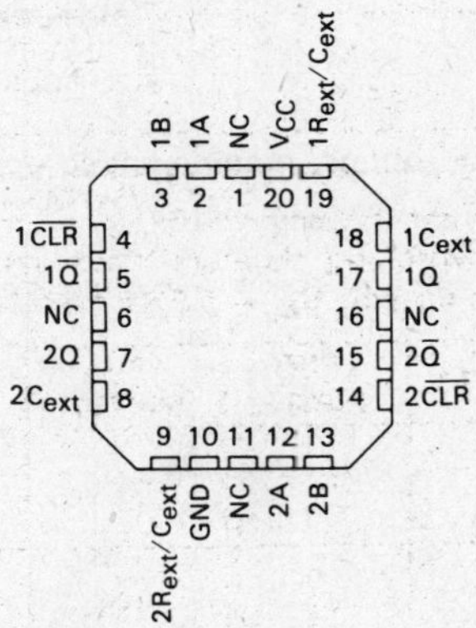

NC—No internal connection

description

These dc-triggered multivibrators feature output-pulse-duration control by two methods. The basic pulse duration is programmed by selection of external resistance and capacitance values. Once triggered, the basic pulse duration may be extended by retriggering the gated low-level-active (A) or high-level-active (B) inputs, or be reduced by use of the overriding clear. Figure 1 illustrates pulse control by retriggering and early clear.

The B input is a Schmitt trigger enabling jitter-free triggering from input signals with slow transition rates.

The SN54HC423 is characterized for operation over the full military temperature range of −55°C to 125°C. The SN74HC423 is characterized for operation from −40°C to 85°C.

FUNCTION TABLE

INPUTS			OUTPUTS	
CLEAR	A	B	Q	$\overline{Q}$
L	X	X	L	H
X	H	X	L*	H*
X	X	L	L*	H*
H	L	↑	⊓	⊔
H	↓	H	⊓	⊔

*These are the logic levels the outputs will take on after the completion of any pulse already started.

logic symbol

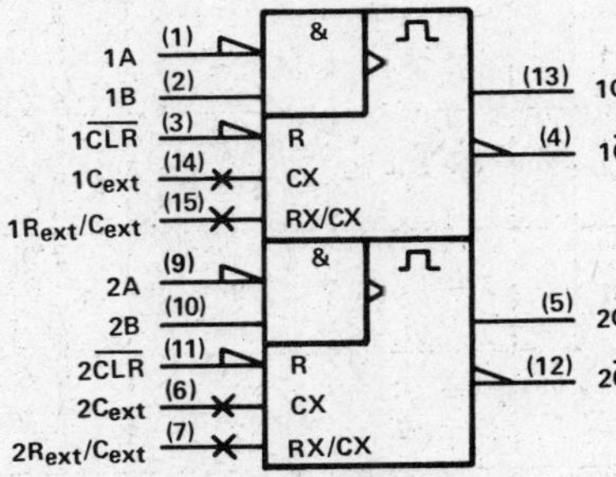

Pin numbers shown are for J and N packages.

TEXAS INSTRUMENTS
POST OFFICE BOX 225012 • DALLAS, TEXAS 75265

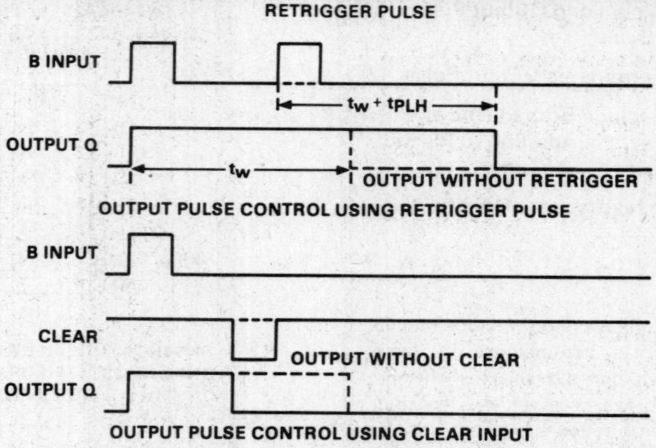

FIGURE 1—TYPICAL INPUT/OUTPUT PULSES

maximum ratings, recommended operating conditions, and electrical characteristics

See Table IV, page 2-10.

NOTE: The minimum recommended supply voltage for this device is 3 V.

TYPES SN54HCT534, SN74HCT534 OCTAL D-TYPE EDGE-TRIGGERED FLIP-FLOPS WITH 3-STATE OUTPUTS

D2804, MARCH 1984

- **Inputs are TTL-Voltage Compatible**
- **High-Current 3-State Inverting Outputs Can Drive up to 15 LSTTL Loads**
- **Package Options Include Both Plastic and Ceramic Chip Carriers in Addition to Plastic and Ceramic DIPs**
- **Dependable Texas Instruments Quality and Reliability**

SN54HCT534 . . . J PACKAGE
SN74HCT534 . . . J OR N PACKAGE
(TOP VIEW)

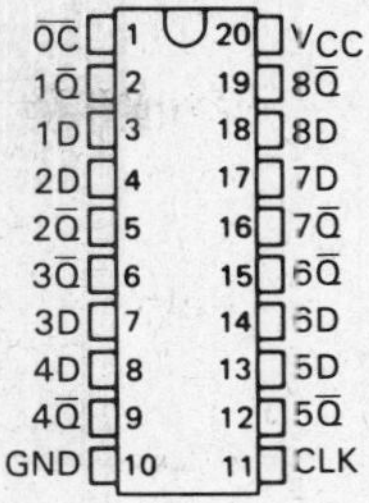

SN54HCT534 . . . FH OR FK PACKAGE
SN74HCT534 . . . FH OR FN PACKAGE
(TOP VIEW)

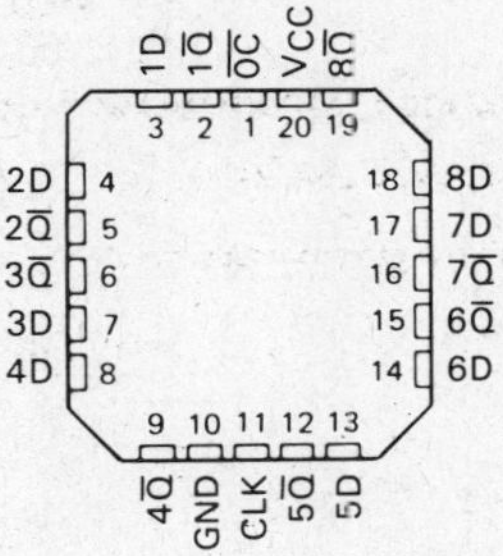

description

These 8-bit flip-flops feature three-state outputs designed specifically for driving highly capacitive or relatively low-impedance loads. They are particularly attractive for implementing buffer registers, I/O ports, bidirectional bus drivers, and working registers.

The eight flip-flops of the 'HCT534 are edge-triggered D-type flip-flops. On the positive transition of the clock, the $\overline{Q}$ outputs will be set to the complement of the logic states that were set up at the D inputs. The 'HCT534 is functionally equivalent to the 'HCT374 except for having inverted outputs.

An output-control input can be used to place the eight outputs in either a normal logic state (high or low logic levels) or a high-impedance state. In the high-impedance state the outputs neither load nor drive the bus lines significantly. The high-impedance third state and increased drive provide the capability to drive the bus lines in a bus-organized system without need for interface or pull-up components.

The output control does not affect the internal operation of the flip-flops. Old data can be retained or new data can be entered while the outputs are off.

The SN54HCT534 is characterized for operation over the full military temperature range of −55 °C to 125 °C. The SN74HCT534 is characterized for operation from −40 °C to 85 °C.

FUNCTION TABLE (EACH FLIP-FLOP)

INPUTS			OUTPUT
$\overline{OC}$	CLK	D	$\overline{Q}$
L	↑	H	L
L	↑	L	H
L	L	X	$\overline{Q}_0$
H	X	X	Z

TEXAS INSTRUMENTS
POST OFFICE BOX 225012 • DALLAS, TEXAS 75265

logic symbol

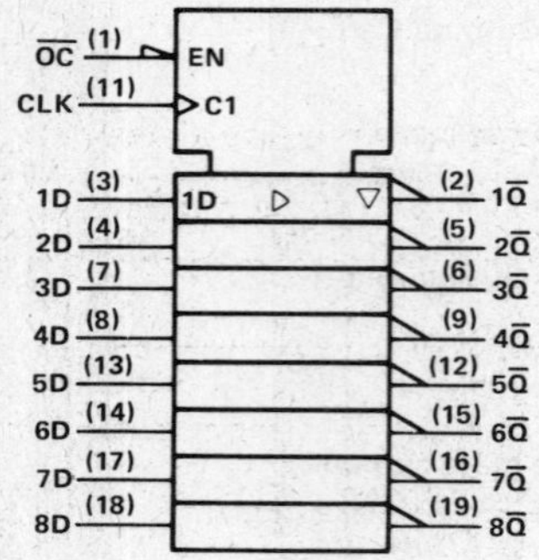

logic diagram (positive logic)

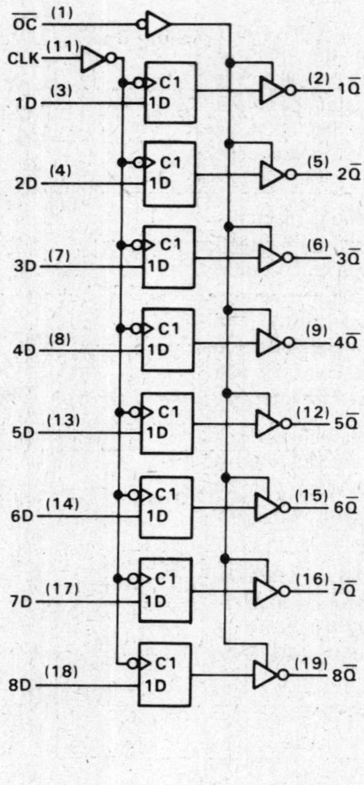

TEXAS INSTRUMENTS
POST OFFICE BOX 225012 • DALLAS, TEXAS 75265

maximum ratings, recommended operating conditions, and electrical characteristics

See Table VII, page 2-14.

timing requirements over recommended operating free-air temperature range (unless otherwise noted)

			V_{CC}	T_A = 25°C		SN54HCT534		SN74HCT534		UNIT
				MIN	MAX	MIN	MAX	MIN	MAX	
f_{clock}	Clock frequency		4.5 V	0	31	0	21	0	25	MHz
			5.5 V	0	36	0	23	0	28	
t_w	Pulse duration	CLK high	4.5 V	16		24		20		ns
			5.5 V	14		22		18		
		CLK low	4.5 V	16		24		20		ns
			5.5 V	14		22		18		
t_{su}	Setup time, data before CLK↑		4.5 V	20		30		25		ns
			5.5 V	17		27		23		
t_h	Hold time, data after CLK↑		4.5 V	5		5		5		ns
			5.5 V	5		5		5		

switching characteristics, over recommended operating free-air temperature range (unless otherwise noted), C_L = 50 pF (see Note 1)

PARAMETER	FROM (INPUT)	TO (OUTPUT)	V_{CC}	T_A = 25°C			SN54HCT534		SN74HCT534		UNIT
				MIN	TYP	MAX	MIN	MAX	MIN	MAX	
f_{max}			4.5 V	31	36		21		25		MHz
			5.5 V	36	40		23		28		
t_{pd}	CLK	Any	4.5 V		28	36		54		45	ns
			5.5 V		26	32		48		41	
t_{en}	$\overline{OC}$	Any	4.5 V		24	30		45		38	ns
			5.5 V		20	27		41		34	
t_{dis}	$\overline{OC}$	Any	4.5 V		22	30		45		38	ns
			5.5 V		20	27		41		34	
t_t		Any	4.5 V		10	12		18		15	ns
			5.5 V		9	11		16		14	

C_{pd}	Power dissipation capacitance per flip-flop	No load, T_A = 25°C	93 pF typ

switching characteristics, over recommended operating free-air temperature range (unless otherwise noted), C_L = 150 pF (see Note 1)

PARAMETER	FROM (INPUT)	TO (OUTPUT)	V_{CC}	T_A = 25°C			SN54HCT534		SN74HCT534		UNIT
				MIN	TYP	MAX	MIN	MAX	MIN	MAX	
t_{pd}	CLK	Any	4.5 V		38	53		80		66	ns
			5.5 V		36	47		71		60	
t_{en}	$\overline{OC}$	Any	4.5 V		30	47		71		59	ns
			5.5 V		27	39		59		49	
t_t		Any	4.5 V		18	42		63		53	ns
			5.5 V		16	38		57		48	

NOTE 1: For load circuit and voltage waveforms, see page 1-14.

D flip-flop signal conventions

It is TI practice to name the outputs and other inputs of a D-type flip-flop and to draw its logic symbol based on the assumption of true data (D) inputs. Then outputs that produce data in phase with the data inputs are called Q and those producing complementary data are called $\overline{Q}$. An input that causes a Q output to go high or a $\overline{Q}$ output to go low is called Preset; an input that causes a $\overline{Q}$ output to go high or a Q output to go low is called Clear. Bars are used over these pin names ($\overline{PRE}$ and $\overline{CLR}$) if they are active-low.

In some applications it may be advantageous to redesignate the data input $\overline{D}$. In that case all the other inputs and outputs should be renamed as shown below. Also shown are corresponding changes in the graphical symbol. Arbitrary pin numbers are shown in parentheses.

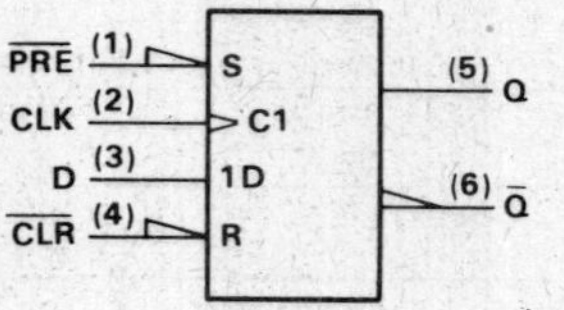

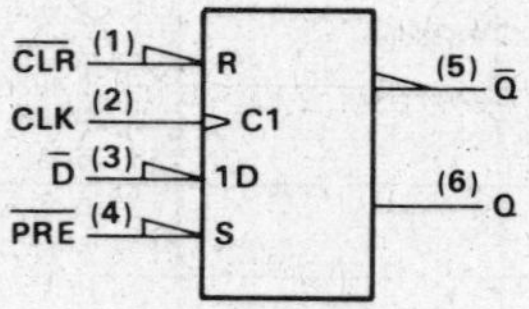

Notice that Q and $\overline{Q}$ exchange names, which causes Preset and Clear to do likewise. Also notice that the polarity indicators (⊵) on $\overline{PRE}$ and $\overline{CLR}$ remain since these inputs are still active-low, but that the presence or absence of the polarity indicator changes at $\overline{D}$, Q and $\overline{Q}$. Of course pin 5 ($\overline{Q}$) is still in phase with the data input $\overline{D}$, but now both are considered active-low.

TEXAS INSTRUMENTS
POST OFFICE BOX 225012 • DALLAS, TEXAS 75265

TYPES SN54HC592, SN54HC593, SN74HC592, SN74HC593 8-BIT BINARY COUNTERS WITH INPUT REGISTERS

D2684, DECEMBER 1982—REVISED MARCH 1984

- Parallel Register Inputs ('HC592)
- Parallel 3-State I/O: Register Inputs/ Counter Outputs ('HC593)
- Counter Has Direct Overriding Load and Clear
- High-Current Outputs Can Drive up to 15 LSTTL Loads ('HC593)
- Package Options Include Both Plastic and Ceramic Chip Carriers in Addition to Plastic and Ceramic DIPs
- Dependable Texas Instruments Quality and Reliability

SN54HC592 . . . J PACKAGE
SN74HC592 . . . J OR N PACKAGE
(TOP VIEW)

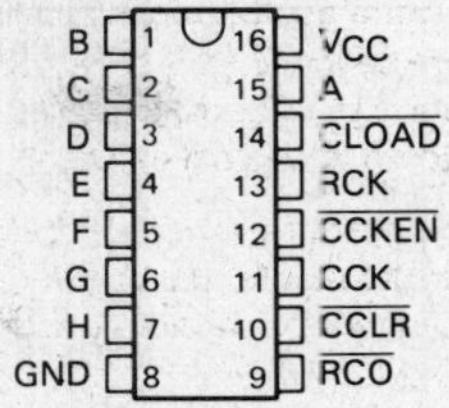

SN54HC592 . . . FH OR FK PACKAGE
SN74HC592 . . . FH OR FN PACKAGE
(TOP VIEW)

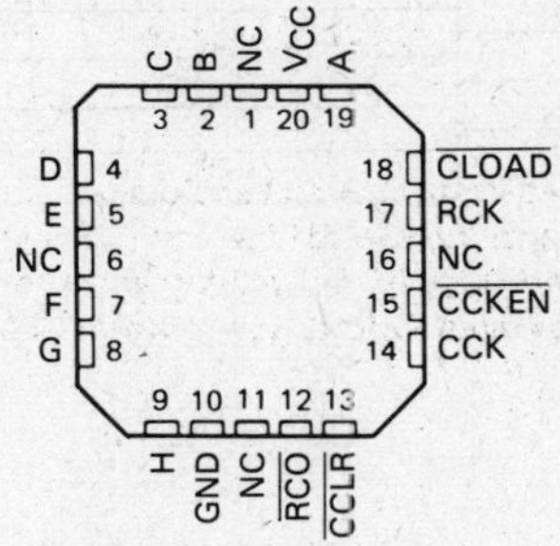

NC—No internal connection

description

The 'HC592 consists of a parallel input, 8-bit storage register feeding an 8-bit binary counter. Both the register and the counter have individual positive-edge-triggered clocks. In addition, the counter has direct load and clear functions. Expansion is easily accomplished by connecting $\overline{RCO}$ of the first stage to the count enable of the second stage, etc.

The 'HC593 has all the features of the 'HC592 plus 3-state I/O, which provides parallel counter outputs.

The SN54HC592 and SN54HC593 are characterized for operation over the full military temperature range of −55°C to 125°C. The SN74HC592 and SN74HC593 are characterized for operation from −40°C to 85°C.

SN54HC593 . . . J PACKAGE
SN74HC593 . . . J OR N PACKAGE
(TOP VIEW)

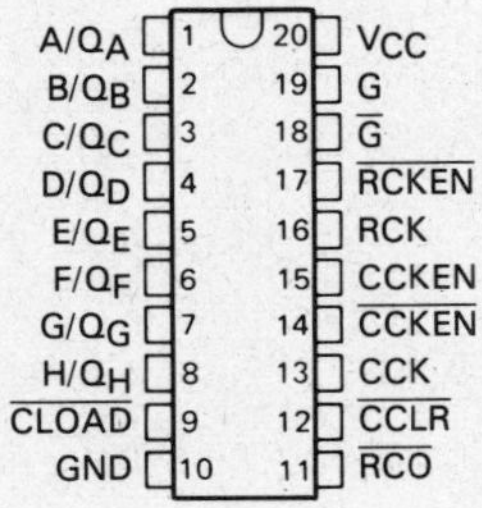

SN54HC593 . . . FH OR FK PACKAGE
SN74HC593 . . . FH OR FN PACKAGE
(TOP VIEW)

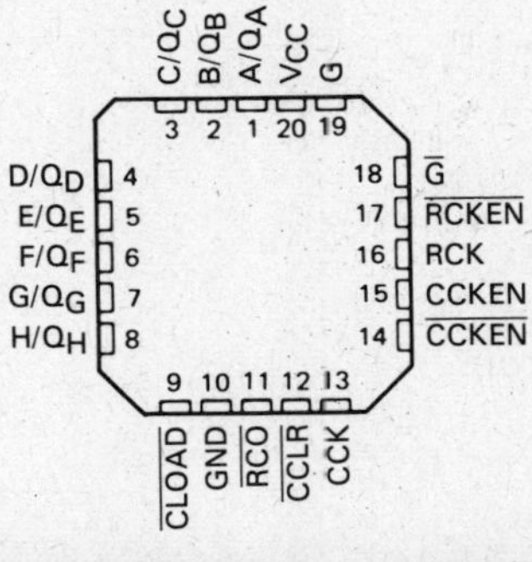

PRODUCT PREVIEW

This document contains information on a product under development. Texas Instruments reserves the right to change or discontinue this product without notice.

logic symbols

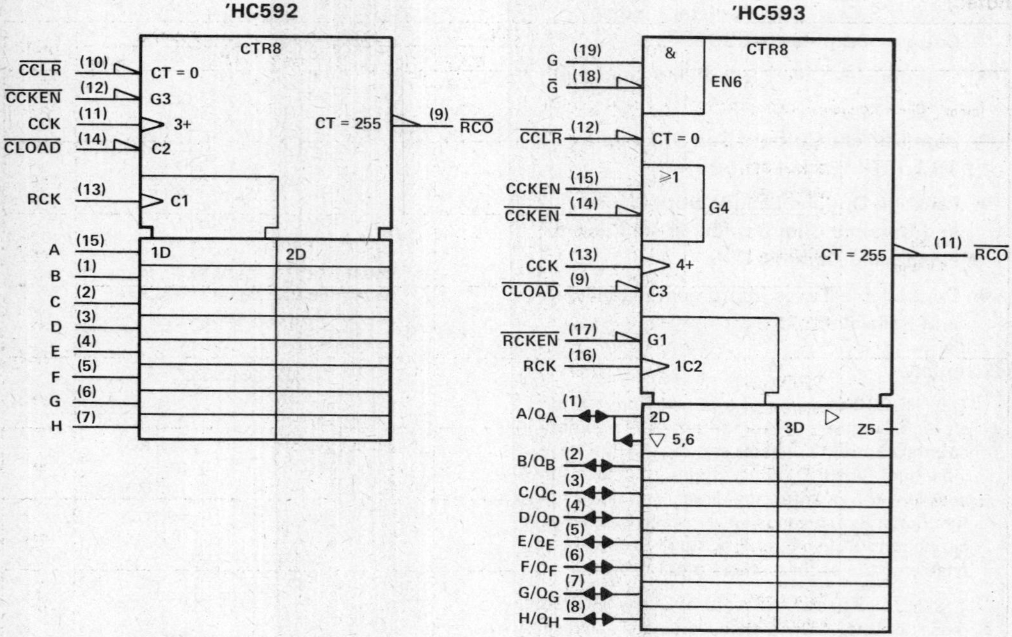

Pin numbers shown are for J and N packages.

maximum ratings, recommended operating conditions, and electrical characteristics

'HC592: See Table IV, page 2-10.
'HC593: See Table III, page 2-8.

TEXAS INSTRUMENTS
POST OFFICE BOX 225012 • DALLAS, TEXAS 75265

timing requirements over recommended operating free-air temperature range (unless otherwise noted)

		V_{CC}	T_A = 25°C		SN54HC'		SN74HC'		UNIT
			MIN	MAX	MIN	MAX	MIN	MAX	
f_{clock} Clock frequency, CCK or RCK		2 V	0	3.3					MHz
		4.5 V	0	17					
		6 V	0	19					
t_w Pulse duration	CCK or RCK high or low	2 V	150						ns
		4.5 V	30						
		6 V	26						
	$\overline{CCLR}$ low	2 V	125						ns
		4.5 V	25						
		6 V	21						
	$\overline{CLOAD}$ low	2 V	125						ns
		4.5 V	25						
		6 V	21						
t_{su} Setup time	$\overline{CCKEN}$ low before CCK	2 V	125						ns
		4.5 V	25						
		6 V	21						
	$\overline{CCLR}$ high (inactive) before CLK↑	2 V	125						ns
		4.5 V	25						
		6 V	21						
	RCK↑ before CCK↑ (see Note 1)	2 V	200						ns
		4.5 V	40						
		6 V	34						
	Data A thru H before RCK↑	2 V	125						ns
		4.5 V	25						
		6 V	21						
t_h Hold time		2 V	5						ns
		4.5 V	5						
		6 V	5						

NOTE 1: The RCK↑ to CCK↑ setup time ensures that the counter will see stable data from the register outputs.

'HC592 switching characteristics over recommended operating free-air temperature range (unless otherwise noted), C_L = 50 pF (see Note 2)

PARAMETER	FROM (INPUT)	TO (OUTPUT)	V_{CC}	T_A = 25°C			SN54HC592		SN74HC592		UNIT
				MIN	TYP	MAX	MIN	MAX	MIN	MAX	
f_{max}	CCK or RCK		2 V	3.3	8						MHz
			4.5 V	17	35						
			6 V	19	40						
t_{pd}	CCK↑	$\overline{RCO}$	2 V		75						ns
			4.5 V		25						
			6 V		21						
t_{pd}	$\overline{CLOAD}$↓	$\overline{RCO}$	2 V		75						ns
			4.5 V		25						
			6 V		21						
t_{PHL}	$\overline{CCLR}$↓	$\overline{RCO}$	2 V		85						ns
			4.5 V		28						
			6 V		24						
t_{pd}	RCK↑	$\overline{RCO}$	2 V		105						ns
			4.5 V		35						
			6 V		30						

NOTE 2: For load circuits and voltage waveforms, see page 1-14.

PRODUCT PREVIEWS

5

switching characteristics over recommended operating free-air temperature range (unless otherwise noted), C_L = 50 pF (see Note 1)

PARAMETER	FROM (INPUT)	TO (OUTPUT)	V_{CC}	T_A = 25°C MIN	TYP	MAX	SN54HC593 MIN	MAX	SN74HC593 MIN	MAX	UNIT
f_{max}	CCK or RCK		2 V	3.3	8						MHz
			4.5 V	17	35						
			6 V	19	40						
t_{pd}	CCK↑	Q	2 V		75						ns
			4.5 V		25						
			6 V		21						
t_{pd}	CCK↑	$\overline{RCO}$	2 V		75						ns
			4.5 V		25						
			6 V		21						
t_{pd}	$\overline{CLOAD}$↓	Q	2 V		75						ns
			4.5 V		25						
			6 V		21						
t_{pd}	$\overline{CLOAD}$↓	$\overline{RCO}$	2 V		75						ns
			4.5 V		25						
			6 V		21						
t_{pd}	RCK↑	$\overline{RCO}$	2 V		105						ns
			4.5 V		35						
			6 V		30						
t_{PHL}	$\overline{CCLR}$↓	Q	2 V		90						ns
			4.5 V		30						
			6 V		26						
t_{PHL}	$\overline{CCLR}$↓	$\overline{RCO}$	2 V		90						ns
			4.5 V		30						
			6 V		26						
t_{en}	G↑	Q	2 V		66						ns
			4.5 V		22						
			6 V		19						
t_{en}	$\overline{G}$↓	Q	2 V		75						ns
			4.5 V		25						
			6 V		21						
t_{dis}	G↓	Q	2 V		60						ns
			4.5 V		20						
			6 V		17						
t_{dis}	$\overline{G}$↑	Q	2 V		60						ns
			4.5 V		20						
			6 V		17						
t_t			2 V		28						ns
			4.5 V		8						
			6 V		6						

NOTE 1: For load circuit and voltage waveforms, see page 1-14.

D2684, DECEMBER 1982—REVISED MARCH 1984

TYPES SN54HC594, SN74HC594 8-BIT SHIFT REGISTERS WITH OUTPUT REGISTERS

- 8-Bit Serial-In, Parallel-Out Shift Registers With Storage
- Independent Direct-Overriding Clears On Shift and Storage Registers
- Independent Clocks for Both Shift and Storage Registers
- High-Current Outputs Can Drive up to 15 LSTTL Loads
- Package Options Include Both Plastic and Ceramic Chip Carriers in Addition to Plastic and Ceramic DIPs
- Dependable Texas Instruments Quality and Reliability

SN54HC594 . . . J PACKAGE
SN74HC594 . . . J OR N PACKAGE
(TOP VIEW)

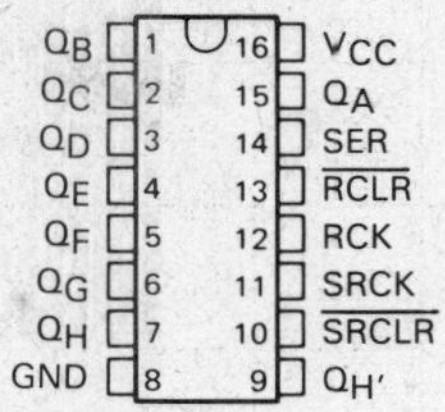

SN54HC594 . . . FH OR FK PACKAGE
SN74HC594 . . . FH OR FN PACKAGE
(TOP VIEW)

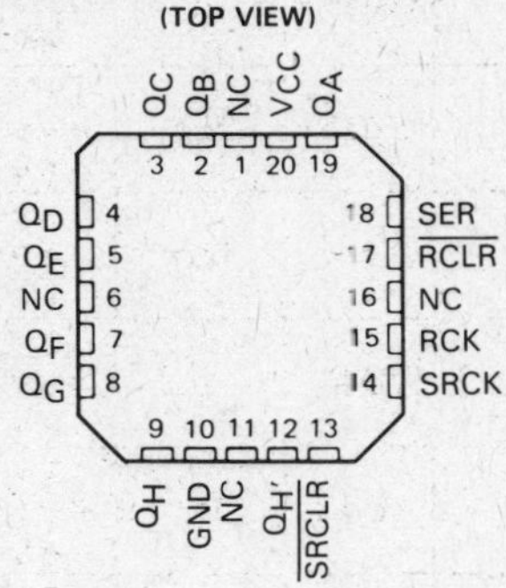

NC—No internal connection

description

These devices each contain an 8-bit serial-in, parallel-out shift register that feeds an 8-bit D-type storage register. Separate clocks and direct-overriding clears are provided on both the shift and storage registers. A serial output ($Q_{H'}$) is provided for cascading purposes.

Both the shift register and storage register clocks are positive-edge triggered. If the user wishes to connect both clocks together, the shift register will always be one clock pulse ahead of the storage register.

The parallel outputs (Q_A thru Q_H) have high-current capability; output $Q_{H'}$ is a standard output.

The SN54HC594 is characterized for operation over the full military temperature range of −55°C to 125°C. The SN74HC594 is characterized for operation from −40°C to 85°C.

logic symbol

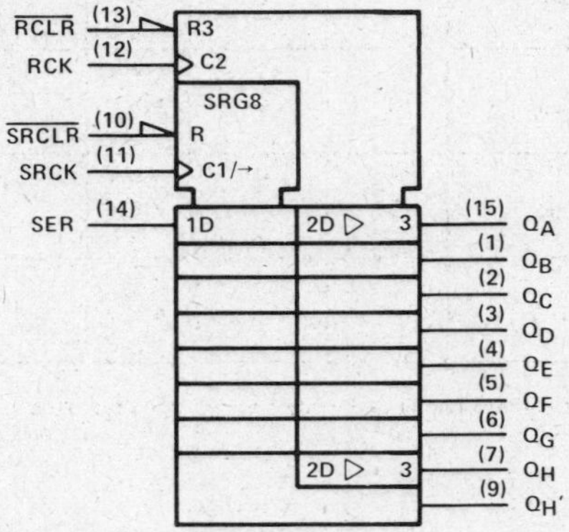

Pin numbers shown are for J and N packages.

maximum ratings, over recommended operating conditions, and electrical characteristics

See Table III, page 2-8.

PRODUCT PREVIEWS

5

PRODUCT PREVIEWS

5

D2684, DECEMBER 1982—REVISED MARCH 1984

- **8-Bit Serial-In, Parallel-Out Shift Registers with Storage**
- **High-Current 3-State Outputs Can Drive Up to 15 LSTTL Loads**
- **Shift Register Has Direct Clear**
- **Package Options Include Both Plastic and Ceramic Chip Carriers in Addition to Plastic and Ceramic DIPs**
- **Dependable Texas Instruments Quality and Reliability**

SN54HC595 . . . J PACKAGE
SN74HC595 . . . J OR N PACKAGE
(TOP VIEW)

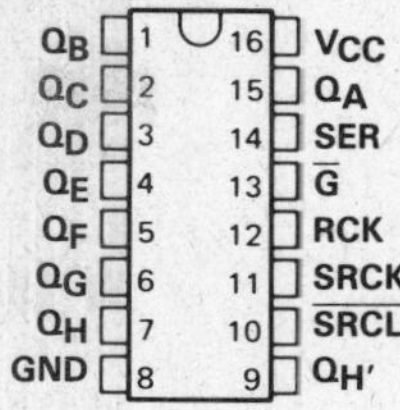

SN54HC595 . . . FH OR FK PACKAGE
SN74HC595 . . . FH OR FN PACKAGE
(TOP VIEW)

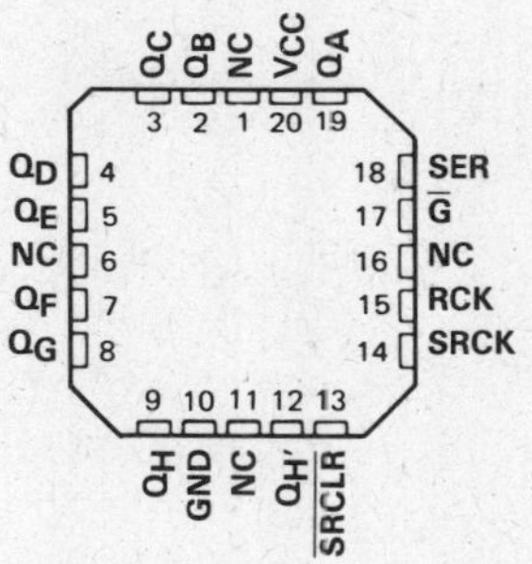

NC—No internal connection

description

These devices each contain an 8-bit serial-in, parallel-out shift register that feeds an 8-bit D-type storage register. The storage register has parallel 3-state outputs. Separate clocks are provided for both the shift register and the storage register. The shift register has a direct-overriding clear, serial input, and serial output pins for cascading.

Both the shift register and storage register clocks are positive-edge triggered. If the user wishes to connect both clocks together, the shift register state will always be one clock pulse ahead of the storage register.

The SN54HC595 is characterized for operation over the full military temperature range of −55°C to 125°C. The SN74HC595 is characterized for operation from −40°C to 85°C.

logic symbol

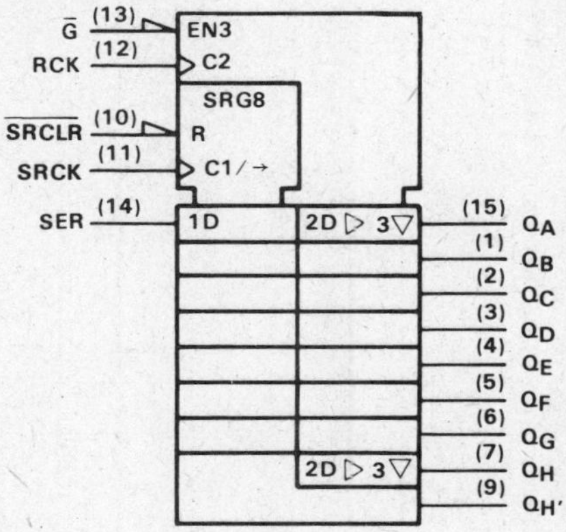

Pin numbers shown are for J and N packages.

maximum ratings, recommended operating conditions, and electrical characteristics

See Table III, page 2-8.

PRODUCT PREVIEW

This document contains information on a product under development. Texas Instruments reserves the right to change or discontinue this product without notice.

D2684, DECEMBER 1982—REVISED MARCH 1984

- **8-Bit Parallel Storage Register Inputs ('HC597)**
- **Parallel 3-State I/O; Storage Register Inputs, High-Current Shift Register Outputs Can Drive up to 15 LSTTL Loads ('HC598)**
- **Shift Register Has Direct Overriding Load and Clear**
- **Package Options Include Both Plastic and Ceramic Chip Carriers in Addition to Plastic and Ceramic DIPs**
- **Dependable Texas Instruments Quality and Reliability**

SN54HC597 . . . J PACKAGE
SN74HC597 . . . J OR N PACKAGE
(TOP VIEW)

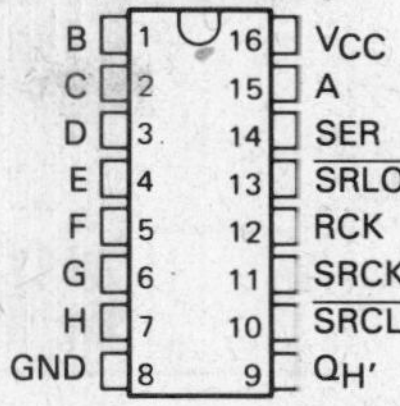

SN54HC597 . . . FH OR FK PACKAGE
SN74HC597 . . . FH OR FN PACKAGE
(TOP VIEW)

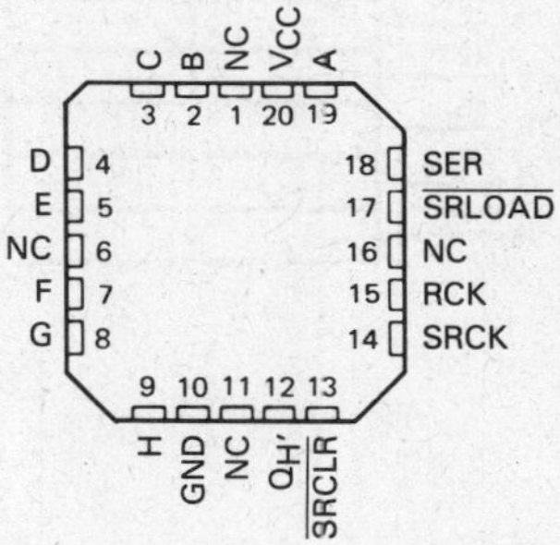

NC—No internal connection

description

The 'HC597 consists of an 8-bit storage latch feeding a parallel-in, serial-out 8-bit shift register. Both the storage register and shift register have positive edge-triggered clocks. The shift register also has direct load (from storage) and clear inputs.

The 'HC598 has all the features of the 'HC597 plus 3-state I/O ports that provide parallel shift register outputs. The 'HC598 also has multiplexed serial data inputs.

The SN54HC597 and SN54HC598 are characterized for operation over the full military temperature range of −55°C to 125°C. The SN74HC597 and SN74HC598 are characterized for operation from −40°C to 85°C.

SN54HC598 . . . J PACKAGE
SN74HC598 . . . J OR N PACKAGE
(TOP VIEW)

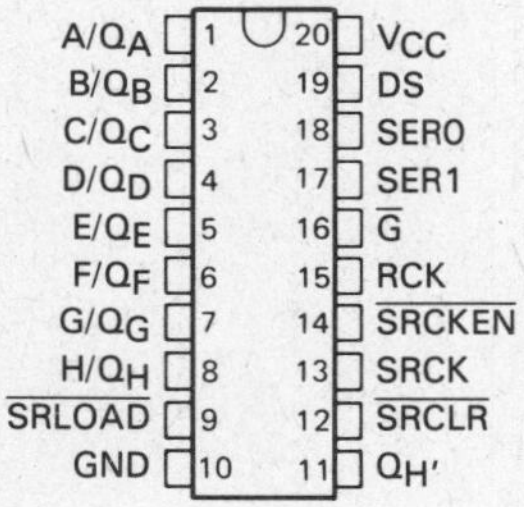

SN54HC598 . . . FH OR FK PACKAGE
SN74HC598 . . . FH OR FN PACKAGE
(TOP VIEW)

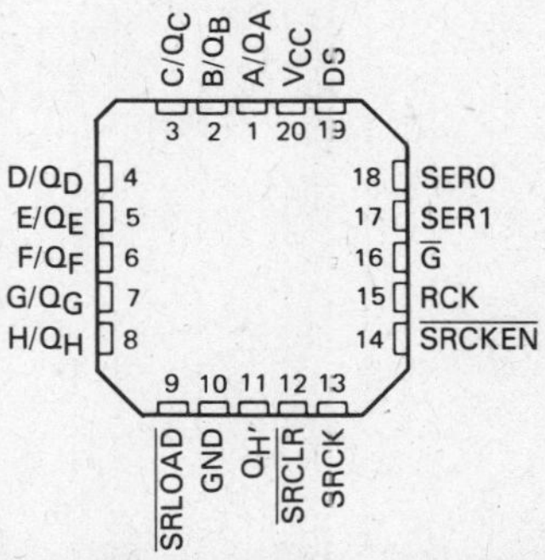

PRODUCT PREVIEW

This document contains information on a product under development. Texas Instruments reserves the right to change or discontinue this product without notice.

TEXAS INSTRUMENTS
POST OFFICE BOX 225012 • DALLAS, TEXAS 75265

logic symbols

'HC597

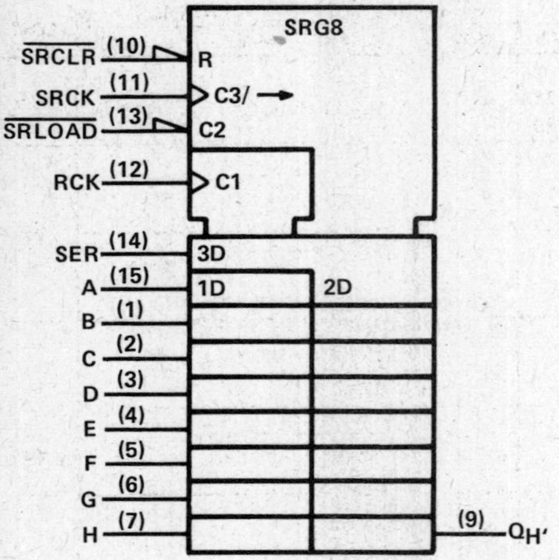

'HC598

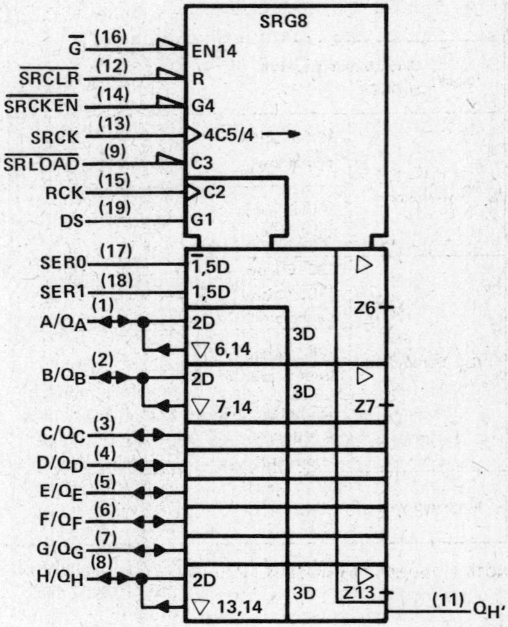

Pin numbers shown are for J and N packages.

maximum ratings, recommended operating conditions, and electrical characteristics

'HC597: See Table IV, page 2-10.
'HC598: See Table III, page 2-8.

TEXAS INSTRUMENTS
POST OFFICE BOX 225012 • DALLAS, TEXAS 75265

timing requirements over recommended operating free-air temperature range (unless otherwise noted)

			V_{CC}	T_A = 25°C MIN	T_A = 25°C MAX	SN54HC' MIN	SN54HC' MAX	SN74HC' MIN	SN74HC' MAX	UNIT
f_{clock}	Clock frequency, RCK or SRCK		2 V	0	5					MHz
			4.5 V	0	25					
			6 V	0	29					
t_w	Pulse duration	RCK or SRCK high or low	2 V	100						ns
			4.5 V	20						
			6 V	17						
		$\overline{SRCLR}$ low or $\overline{SRLOAD}$ low	2 V	100						
			4.5 V	20						
			6 V	17						
t_{su}	Setup time	$\overline{SRCKEN}$ low or $\overline{SRCLR}$ high (inactive) before SRCK↑	2 V	100						ns
			4.5 V	20						
			6 V	17						
		RCK↑ before SRCK↑ (see Note 1)	2 V	200						
			4.5 V	40						
			6 V	34						
		SER before SRCK↑ or A thru H before RCK↑	2 V	100						
			4.5 V	20						
			6 V	17						
t_h	Hold time		2 V	0						ns
			4.5 V	0						
			6 V	0						

NOTE 1: The RCK↑ before SRCK↑ setup time ensures that the shift register will see stable data coming from the input register.

'HC597 switching characteristics over recommended operating free-air temperature range (unless otherwise noted), C_L = 50 pF (see Note 2)

PARAMETER	FROM (INPUT)	TO (OUTPUT)	V_{CC}	T_A = 25°C MIN	T_A = 25°C TYP	T_A = 25°C MAX	SN54HC597 MIN	SN54HC597 MAX	SN74HC597 MIN	SN74HC597 MAX	UNIT
f_{max}	RCK or SRCK		2 V	5	8						MHz
			4.5 V	25	35						
			6 V	29	40						
t_{pd}	SRCK↑	$Q_{H'}$	2 V		75						ns
			4.5 V		25						
			6 V		21						
t_{pd}	$\overline{SRLOAD}$↓	$Q_{H'}$	2 V		75						ns
			4.5 V		25						
			6 V		21						
t_{PHL}	$\overline{SRCLR}$↓	$Q_{H'}$	2 V		60						ns
			4.5 V		20						
			6 V		17						
t_{pd}	RCK↑	$Q_{H'}$	2 V		60						ns
			4.5 V		20						
			6 V		17						

NOTE 2: For load circuits and voltage waveforms, see page 1-14.

'HC598 switching characteristics over recommended operating free-air temperature range (unless otherwise noted), C_L = 50 pF (see Note 1)

PARAMETER	FROM (INPUT)	TO (OUTPUT)	V_{CC}	T_A = 25°C MIN	TYP	MAX	SN54HC598 MIN	MAX	SN74HC598 MIN	MAX	UNIT
f_{max}	RCK or SRCK		2 V	5	8						MHz
			4.5 V	25	35						
			6 V	29	40						
t_{pd}	SRCK ↑	$Q_{H'}$	2 V		75						ns
			4.5 V		25						
			6 V		21						
t_{pd}	$\overline{SRLOAD}$ ↓	$Q_{H'}$	2 V		75						ns
			4.5 V		25						
			6 V		21						
t_{PHL}	$\overline{SRCLR}$ ↓	$Q_{H'}$	2 V		60						ns
			4.5 V		20						
			6 V		17						
t_{pd}	RCK ↑	$Q_{H'}$	2 V		60						ns
			4.5 V		20						
			6 V		17						
t_{pd}	SRCK ↑	Q_A thru Q_H	2 V		60						ns
			4.5 V		20						
			6 V		17						
t_{pd}	$\overline{SRLOAD}$ ↓	Q_A thru Q_H	2 V		60						ns
			4.5 V		20						
			6 V		17						
t_{PHL}	$\overline{SRCLR}$ ↓	Q_A thru Q_H	2 V		75						ns
			4.5 V		25						
			6 V		21						
t_{en}	$\overline{G}$ ↓	Q_A thru Q_H	2 V		66						ns
			4.5 V		22						
			6 V		19						
t_{dis}	$\overline{G}$ ↑	Q_A thru Q_H	2 V		60						ns
			4.5 V		20						
			6 V		17						

NOTE 1: For load circuit and voltage waveforms, see page 1-14.

TYPES SN54HC630, SN74HC630 16-BIT PARALLEL ERROR DETECTION AND CORRECTION CIRCUITS

D2804, MARCH 1984

- High-Current 3-State Outputs Can Drive up to 15 LSTTL Loads
- Detects and Corrects Single-Bit Errors
- Detects and Flags Dual-Bit Errors
- Dependable Texas Instruments Quality and Reliability

SN54HC630 . . . J PACKAGE
SN74HC630 . . . J OR N PACKAGE
(TOP VIEW)

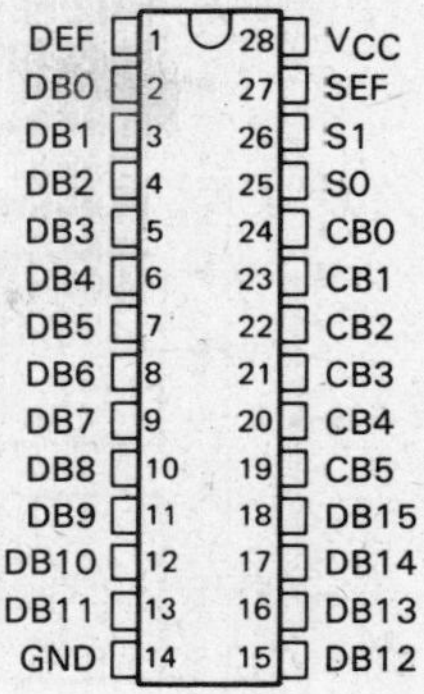

description

The 'HC630 device is a 16-bit parallel error detection and correction circuit (EDAC) in a 28-pin, 600-mil package. It uses a modified Hamming code to generate a 6-bit check word from a 16-bit data word. This check word is stored along with the data word during the memory write cycle. During the memory read cycle, the 22-bit words from memory are processed by the EDAC to determine if errors have occurred in memory.

Single-bit errors in the 16-bit data word are flagged and corrected.

Single-bit errors in the 6-bit check word are flagged, and the CPU sends the EDAC through the correction cycle even though the 16-bit word is not in error. The correction cycle will simply pass along the original 16-bit word in this case and produce error syndrome bits to pinpoint the error-generating location.

Dual-bit errors are flagged but not corrected. These dual errors may occur in any two bits of the 22-bit word from memory (two errors in the 16-bit data word, two errors in the 6-bit check word, or one error in each word).

The gross-error condition of all lows or all highs from memory will be detected. Otherwise, errors in three or more bits of the 22-bit word are beyond the capabilities of these devices to detect.

The SN54HC630 is characterized for operation over the full military temperature range of −55 °C to 125 °C. The SN74HC630 is characterized for operation from −40 °C to 85 °C.

CONTROL FUNCTION TABLE

MEMORY CYCLE	CONTROL		EDAC FUNCTION	DATA I/O	CHECK WORD I/O	ERROR FLAGS	
	S1	S0				SEF	DEF
WRITE	L	L	Generate Check Word	Input Data	Output Check Word	L	L
READ	L	H	Read Data & Check Word	Input Data	Input Check Word	L	L
READ	H	H	Latch & Flag Errors	Latch Data	Latch Check Word	Enabled	
READ	H	L	Correct Data Word & Generate Syndrome Bits	Output Corrected Data	Output Syndrome Bits	Enabled	

PRODUCT PREVIEW

This document contains information on a product under development. Texas Instruments reserves the right to change or discontinue this product without notice.

logic diagram

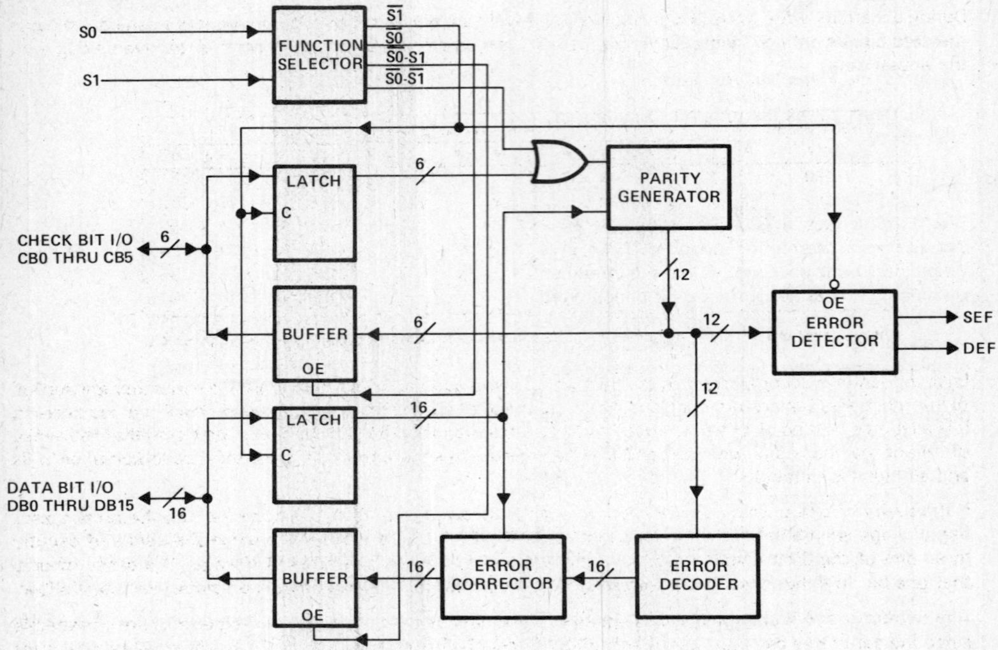

ERROR FUNCTION TABLE

TOTAL NUMBER OF ERRORS		ERROR FLAGS		DATE CORRECTION
16-BIT DATA	6-BIT CHECK WORD	SEF	DEF	
0	0	L	L	Not Applicable
1	0	H	L	Correction
0	1	H	L	Correction
1	1	H	H	Interrupt
2	0	H	H	Interrupt
0	2	H	H	Interrupt

In order to be able to determine whether the data from the memory is acceptable to use as presented to the bus, the EDAC must be strobed to enable the error flags and the flags will have to be tested for the zero condition.

The first case in the error function table represents the normal, no-error condition. The CPU sees lows on both flags. The next two cases of single-bit errors require data correction. Although the EDAC can discern the single check bit error and ignore it, the error flags are identical to the single error in the 16-bit data word. The CPU will ask for data correction in both cases. An interrupt condition to the CPU results in each of the last three cases, where dual errors occur.

TEXAS INSTRUMENTS
POST OFFICE BOX 225012 • DALLAS, TEXAS 75265

error detection and correction details

During a memory write cycle, six check bits (CB0-CB5) are generated by eight-input parity generators using the data bits as defined below. During a memory read cycle, the 6-bit check word is retrieved along with the actual data.

CHECK WORD	16-BIT DATA WORD															
BIT	0	1	2	3	4	5	6	7	8	9	10	11	12	13	14	15
CB0	X	X		X	X				X	X	X			X		
CB1	X		X	X		X	X		X			X			X	
CB2		X	X		X	X		X		X			X			X
CB3	X	X	X				X	X			X	X	X			
CB4				X	X	X	X	X						X	X	X
CB5									X	X	X	X	X	X	X	X

The six check bits are parity bits derived from the matrix of data bits as indicated by "x" for each bit.

Error detection is accomplished as the 6-bit check word and the 16-bit data word from memory are applied to internal parity generators/checkers. If the parity of all six groupings of data and check bits are correct, it is assumed that no error has occured and both error flags will be low. (It should be noted that the sense of two of the check bits, bits CB0 and CB1, is inverted to ensure that the gross-error condition of all lows and all highs is detected.)

If the parity of one or more of the check groups is incorrect, an error has occurred and the proper error flag or flags will be set high. Any single error in the 16-bit data word will change the sense of exactly three bits of the 6-bit check word. Any single error in the 6-bit check word changes the sense of only that one bit. In either case, the single error flag will be set high while the dual error flag will remain low.

Any two-bit error will change the sense of an even number of check bits. The two-bit error is not correctable since the parity tree can only identify single-bit errors. Both error flags are set high when any two-bit error is detected.

Three or more simultaneous bit errors can fool the EDAC into believing that no error, a correctable error, or an uncorrectable error has occurred and produce erroneous results in all three cases.

Error correction is accomplished by identifying the bad bit and inverting it. Identification of the erroneous bit is achieved by comparing the 16-bit data word and 6-bit check word from memory with the new check word with one (check word error) or three (data word error) inverted bits.

As the corrected word is made available on the data word I/O port, the check word I/O port presents a 6-bit syndrome error code. This syndrome code can be used to identify the bad memory chip.

ERROR SYNDROME TABLE

ERROR LOCATION	SYNDROME ERROR CODE					
	CB0	CB1	CB2	CB3	CB4	CB5
DB0	L	L	H	L	H	H
DB1	L	H	L	L	H	H
DB2	H	L	L	L	H	H
DB3	L	L	H	H	L	H
DB4	L	H	L	H	L	H
DB5	H	L	L	H	L	H
DB6	H	L	H	L	L	H
DB7	H	H	L	L	L	H
DB8	L	L	H	H	H	L
DB9	L	H	L	H	H	L
DB10	L	H	H	L	H	L
DB11	H	L	H	L	H	L
DB12	H	H	L	L	H	L
DB13	L	H	H	H	L	L
DB14	H	L	H	H	L	L
DB15	H	H	L	H	L	L
CB0	L	H	H	H	H	H
CB1	H	L	H	H	H	H
CB2	H	H	L	H	H	H
CB3	H	H	H	L	H	H
CB4	H	H	H	H	L	H
CB5	H	H	H	H	H	L
NO ERROR	H	H	H	H	H	H

maximum ratings, recommended operating conditions, and electrical characteristics

See Table III, page 2-8.

D2804, MARCH 1984

- High-Current 3-State Outputs Can Drive up to 15 LSTTL Loads
- Detects and Corrects Single-Bit Errors
- Detects and Flags Dual-Bit Errors
- Built-In Diagnostic Capability
- Fast Write and Read Cycle Processing Times
- Byte-Write Capability
- Dependable Texas Instruments Quality and Reliability

'HC632 . . . JD OR N PACKAGES
(TOP VIEW)

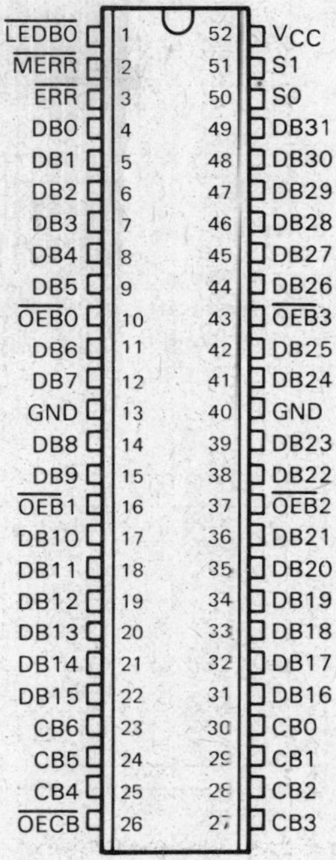

FOR CHIP CARRIER INFORMATION, CONTACT THE FACTORY.

description

The 'HC632 is a 32-bit parallel error detection and correction circuit (EDAC) in a 52-pin, 600-mil package. It uses a modified Hamming code to generate a 7-bit check word from a 32-bit data word. This check word is stored along with the data word during the memory write cycle. During the memory read cycle, the 39-bit words from memory are processed by the EDAC to determine if errors have occurred in memory.

Single-bit errors in the 32-bit data word are flagged and corrected.

Single-bit errors in the 7-bit check word are flagged, and the CPU sends the EDAC through the correction cycle even though the 32-bit data word is not in error. The correction cycle will simply pass along the original 32-bit data word in this case and produce error syndrome bits to pinpoint the error-generating location.

Dual-bit errors are flagged but not corrected. These errors may occur in any two bits of the 39-bit word from memory (two errors in the 32-bit data word, two errors in the 7-bit check word, or one error in each word). The gross-error condition of all lows or all highs from memory will be detected. Otherwise, errors in three or more bits of the 39-bit word are beyond the capabilities of these devices to detect.

Read-modify-write (byte-control) operations can be performed with the 'HC632 EDAC by using output latch enable, $\overline{LEDB0}$, and the individual $\overline{OEB0}$ thru $\overline{OEB3}$ byte control pins.

Diagnostics are performed on the EDAC by controls and internal paths that allow the user to read the contents of the DB and CB input latches. These will determine if the failure occured in memory or in the EDAC.

The SN54HC632 is characterized for operation over the full military temperature range of −55 °C to 125 °C. The SN74HC632 is characterized for operation from −40 °C to 85 °C.

PRODUCT PREVIEWS

5

PRODUCT PREVIEW

This document contains information on a product under development. Texas Instruments reserves the right to change or discontinue this product without notice.

TABLE I. WRITE CONTROL FUNCTION

MEMORY CYCLE	EDAC FUNCTION	CONTROL S1	CONTROL S0	DATA I/O	DB CONTROL $\overline{OEB0}$ THRU $\overline{OEB3}$	DB OUTPUT LATCH $\overline{LEDB0}$	CHECK I/O	CB CONTROL $\overline{OECB}$	ERROR FLAGS $\overline{ERR}$	ERROR FLAGS $\overline{MERR}$
Write	Generate check word	L	L	Input	H	X	Output check bits†	L	H	H

†See Table II for details on check bit generation.

memory write cycle details

During a memory write cycle, the check bits (CB0 thru CB6) are generated internally in the EDAC by seven 16-input parity generators using the 32-bit data word as defined in Table II. These seven check bits are stored in memory along with the original 32-bit data word. This 32-bit word will later be used in the memory read cycle for error detection.

TABLE II. PARITY ALGORITHM

CHECK WORD BIT	31	30	29	28	27	26	25	24	23	22	21	20	19	18	17	16	15	14	13	12	11	10	9	8	7	6	5	4	3	2	1	0
CB0	X		X	X		X					X		X	X	X			X			X		X	X	X	X		X				X
CB1				X		X		X		X		X		X	X	X				X		X		X		X		X		X	X	X
CB2	X		X			X	X		X			X	X			X	X		X			X	X		X			X	X			X
CB3			X	X	X				X	X	X				X	X			X	X	X				X	X	X				X	X
CB4	X	X							X	X	X	X	X	X			X	X							X	X	X	X	X	X		
CB5	X	X	X	X	X	X	X	X									X	X	X	X	X	X	X	X								
CB6	X	X	X	X	X	X	X	X																	X	X	X	X	X	X	X	X

(Column headers above: 32-BIT DATA WORD)

The seven check bits are parity bits derived from the matrix of data bits as indicated by "X" for each bit.

error detection and correction details

During a memory read cycle, the 7-bit check word is retrieved along with the actual data. In order to be able to determine whether the data from memory is acceptable to use as presented to the bus, the error flags must be tested to determine if they are at the high level.

The first case in Table III represents the normal, no-error conditions. The EDAC presents highs on both flags. The next two cases of single-bit errors give a high on $\overline{MERR}$ and a low on $\overline{ERR}$, which is the signal for a correctable error, and the EDAC should be sent through the correction cycle. The last three cases of double-bit errors will cause the EDAC to signal lows on both $\overline{ERR}$ and $\overline{MERR}$, which is the interrupt indication for the CPU.

TABLE III. ERROR FUNCTION

TOTAL NUMBER OF ERRORS 32-BIT DATA WORD	TOTAL NUMBER OF ERRORS 7-BIT CHECK WORD	ERROR FLAGS $\overline{ERR}$	ERROR FLAGS $\overline{MERR}$	DATA CORRECTION
0	0	H	H	Not applicable
1	0	L	H	Correction
0	1	L	H	Correction
1	1	L	L	Interrupt
2	0	L	L	Interrupt
0	2	L	L	Interrupt

PRODUCT PREVIEWS

5

TEXAS INSTRUMENTS
POST OFFICE BOX 225012 • DALLAS, TEXAS 75265

Error detection is accomplished as the 7-bit check word and the 32-bit data word from memory are applied to internal parity generators/checkers. If the parity of all seven groupings of data and check bits are correct, it is assumed that no error has occurred and both error flags will be high.

If the parity of one or more of the check groups is incorrect, an error has occurred and the proper error flag or flags will be set low. Any single error in the 32-bit data word will change the state of either three or five bits of the 7-bit check word. Any single error in the 7-bit check word changes the state of only that one bit. In either case, the single error flag ($\overline{\text{ERR}}$) will be set low while the dual error flag ($\overline{\text{MERR}}$) will remain high.

Any two-bit error will change the state of an even number of check bits. The two-bit error is not correctable since the parity tree can only identify single-bit errors. Both error flags are set low when any two-bit error is detected.

Three or more simultaneous bit errors can cause the EDAC to believe that no error, a correctable error, or an uncorrectable error has occurred and will produce erroneous results in all three cases. It should be noted that the gross-error conditions of all lows and all highs will be detected.

TABLE IV. READ, FLAG, AND CORRECT FUNCTION

MEMORY CYCLE	EDAC FUNCTION	CONTROL S1	CONTROL S0	DATA I/O	DB CONTROL $\overline{\text{OEB0}}$ THRU $\overline{\text{OEB3}}$	DB OUTPUT LATCH $\overline{\text{LEDB0}}$	CHECK I/O	CB CONTROL $\overline{\text{OECB}}$	ERROR FLAGS $\overline{\text{ERR}}$ $\overline{\text{MERR}}$
Read	Read & flag	H	L	Input	H	X	Input	H	Enabled†
Read	Latch input data & check bits	H	H	Latched input data	H	L	Latched input check word	H	Enabled†
Read	Output corrected data & syndrome bits	H	H	Output corrected data word	L	X	Output syndrome bits‡	L	Enabled†

†See Table III for error description.
‡See Table V for error location.

As the corrected word is made available on data I/O port (DB0 thru DB31), the check word I/O port (CB0 thru CB6) presents a 7-bit syndrome error code. This syndrome error code can be used to locate the bad memory chip. See Table V for syndrome decoding.

TABLE V. SYNDROME DECODING

6	5	4	3	2	1	0	ERROR
L	L	L	L	L	L	L	unc
L	L	L	L	L	L	H	2-bit
L	L	L	L	L	H	L	2-bit
L	L	L	L	L	H	H	unc
L	L	L	L	H	L	L	2-bit
L	L	L	L	H	L	H	unc
L	L	L	L	H	H	L	unc
L	L	L	L	H	H	H	2-bit
L	L	L	H	L	L	L	2-bit
L	L	L	H	L	L	H	unc
L	L	L	H	L	H	L	DB31
L	L	L	H	L	H	H	2-bit
L	L	L	H	H	L	L	unc
L	L	L	H	H	L	H	2-bit
L	L	L	H	H	H	L	2-bit
L	L	L	H	H	H	H	DB30
L	L	H	L	L	L	L	2-bit
L	L	H	L	L	L	H	unc
L	L	H	L	L	H	L	DB29
L	L	H	L	L	H	H	2-bit
L	L	H	L	H	L	L	DB28
L	L	H	L	H	L	H	2-bit
L	L	H	L	H	H	L	2-bit
L	L	H	L	H	H	H	DB27
L	L	H	H	L	L	L	DB26
L	L	H	H	L	L	H	2-bit
L	L	H	H	L	H	L	2-bit
L	L	H	H	L	H	H	DB25
L	L	H	H	H	L	L	2-bit
L	L	H	H	H	L	H	DB24
L	L	H	H	H	H	L	unc
L	L	H	H	H	H	H	2-bit

6	5	4	3	2	1	0	ERROR
L	H	L	L	L	L	L	2-bit
L	H	L	L	L	L	H	unc
L	H	L	L	L	H	L	DB7
L	H	L	L	L	H	H	2-bit
L	H	L	L	H	L	L	DB6
L	H	L	L	H	L	H	2-bit
L	H	L	L	H	H	L	2-bit
L	H	L	L	H	H	H	DB5
L	H	L	H	L	L	L	DB4
L	H	L	H	L	L	H	2-bit
L	H	L	H	L	H	L	2-bit
L	H	L	H	L	H	H	DB3
L	H	L	H	H	L	L	2-bit
L	H	L	H	H	L	H	DB2
L	H	L	H	H	H	L	unc
L	H	L	H	H	H	H	2-bit
L	H	H	L	L	L	L	DB0
L	H	H	L	L	L	H	2-bit
L	H	H	L	L	H	L	2-bit
L	H	H	L	L	H	H	unc
L	H	H	L	H	L	L	2-bit
L	H	H	L	H	L	H	DB1
L	H	H	L	H	H	L	unc
L	H	H	L	H	H	H	2-bit
L	H	H	H	L	L	L	2-bit
L	H	H	H	L	L	H	unc
L	H	H	H	L	H	L	unc
L	H	H	H	L	H	H	2-bit
L	H	H	H	H	L	L	unc
L	H	H	H	H	L	H	2-bit
L	H	H	H	H	H	L	2-bit
L	H	H	H	H	H	H	CB6

6	5	4	3	2	1	0	ERROR
H	L	L	L	L	L	L	2-bit
H	L	L	L	L	L	H	unc
H	L	L	L	L	H	L	unc
H	L	L	L	L	H	H	2-bit
H	L	L	L	H	L	L	unc
H	L	L	L	H	L	H	2-bit
H	L	L	L	H	H	L	2-bit
H	L	L	L	H	H	H	unc
H	L	L	H	L	L	L	unc
H	L	L	H	L	L	H	2-bit
H	L	L	H	L	H	L	2-bit
H	L	L	H	L	H	H	DB15
H	L	L	H	H	L	L	2-bit
H	L	L	H	H	L	H	unc
H	L	L	H	H	H	L	DB14
H	L	L	H	H	H	H	2-bit
H	L	H	L	L	L	L	unc
H	L	H	L	L	L	H	2-bit
H	L	H	L	L	H	L	2-bit
H	L	H	L	L	H	H	DB13
H	L	H	L	H	L	L	2-bit
H	L	H	L	H	L	H	DB12
H	L	H	L	H	H	L	DB11
H	L	H	L	H	H	H	2-bit
H	L	H	H	L	L	L	2-bit
H	L	H	H	L	L	H	DB10
H	L	H	H	L	H	L	DB9
H	L	H	H	L	H	H	2-bit
H	L	H	H	H	L	L	DB8
H	L	H	H	H	L	H	2-bit
H	L	H	H	H	H	L	2-bit
H	L	H	H	H	H	H	CB5

6	5	4	3	2	1	0	ERROR
H	H	L	L	L	L	L	unc
H	H	L	L	L	L	H	2-bit
H	H	L	L	L	H	L	2-bit
H	H	L	L	L	H	H	DB23
H	H	L	L	H	L	L	2-bit
H	H	L	L	H	L	H	DB22
H	H	L	L	H	H	L	DB21
H	H	L	L	H	H	H	2-bit
H	H	L	H	L	L	L	2-bit
H	H	L	H	L	L	H	DB20
H	H	L	H	L	H	L	DB19
H	H	L	H	L	H	H	2-bit
H	H	L	H	H	L	L	DB18
H	H	L	H	H	L	H	2-bit
H	H	L	H	H	H	L	2-bit
H	H	L	H	H	H	H	CB4
H	H	H	L	L	L	L	2-bit
H	H	H	L	L	L	H	DB16
H	H	H	L	L	H	L	unc
H	H	H	L	L	H	H	2-bit
H	H	H	L	H	L	L	DB17
H	H	H	L	H	L	H	2-bit
H	H	H	L	H	H	L	2-bit
H	H	H	L	H	H	H	CB3
H	H	H	H	L	L	L	unc
H	H	H	H	L	L	H	2-bit
H	H	H	H	L	H	L	2-bit
H	H	H	H	L	H	H	CB2
H	H	H	H	H	L	L	2-bit
H	H	H	H	H	L	H	CB1
H	H	H	H	H	H	L	CB0
H	H	H	H	H	H	H	none

(Column headers 6–0 are the SYNDROME BITS.)

CB X = error in check bit X
DB Y = error in data bit Y
2-bit = double-bit error
unc = uncorrectable multibit error

read-modify-write (byte control) operations

The 'HC632 is capable of byte-write operations. The 39-bit word from memory must first be latched into the DB and CB input latches. This is easily accomplished by switching from the read and flag mode (S1 = H,S0 = L) to the latch input mode (S1 = H,S0 = H). The EDAC will then make any corrections, if necessary, to the data word and place it at the input of the output data latch. This data word must then be latched into the output data latch by taking $\overline{\text{LEDBO}}$ from a low to a high.

Byte control can now be employed on the data word through the $\overline{\text{OEB0}}$ through $\overline{\text{OEB3}}$ controls. $\overline{\text{OEB0}}$ controls DB0—DB7 (byte 0), $\overline{\text{OEB1}}$ controls DB8—DB15 (byte 1), $\overline{\text{OEB2}}$ controls DB16—DB23 (byte 2), and $\overline{\text{OEB3}}$ controls DB24—DB31 (byte 3). Placing a high on the byte control will disable the output and the user can modify the byte. If a low is placed on the byte control, then the original byte is allowed to pass onto the data bus unchanged. If the original data word is altered through byte control, a new check word must be generated before it is written back into memory. This is easily accomplished by taking control S1 and S0 low. Table VI lists the read-modify-write functions.

TEXAS INSTRUMENTS
POST OFFICE BOX 225012 • DALLAS, TEXAS 75265

TABLE VI. READ-MODIFY-WRITE FUNCTION

MEMORY CYCLE	EDAC FUNCTION	CONTROL S1	CONTROL S0	BYTEn†	$\overline{\text{OEB}}$n†	DB OUTPUT LATCH $\overline{\text{LEDBO}}$	CHECK I/O	CB CONTROL	ERROR FLAG $\overline{\text{ERR}}$	ERROR FLAG $\overline{\text{MERR}}$
Read	Read & Flag	H	L	Input	H	X	Input	H	Enabled	
Read	Latch input data & check bits	H	H	Latched Input data	H	L	Latched input check word	H	Enabled	
Read	Latch corrected data word into output latch	H	H	Latched output data word	H	H	Hi-Z	H	Enabled	
							Output Syndrome bits	L		
Modify /write	Modify appropriate byte or bytes & generate new check word	L	L	Input modified BYTE0	H	H	Output check word	L	H	H
				Output unchanged BYTE0	L					

†$\overline{\text{OEB}}$0 controls DB0—DB7 (BYTE0), $\overline{\text{OEB}}$1 controls DB8—DB15 (BYTE1), $\overline{\text{OEB}}$2 controls DB16—DB23 (BYTE2), $\overline{\text{OEB}}$3 controls DB24—DB31 (BYTE3).

diagnostic operations

The 'HC632 is capable of diagnostics that will allow the user to determine whether the EDAC or the memory is failing. The diagnostic function tables will help the user to see the possibilities for diagnostic control.

In the diagnostic mode (S1 = L, S0 = H), the checkword is latched into the input latch while the data input latch remains transparent. This lets the user apply various data words against a fixed known checkword. If the user applies a diagnostic data word with an error in any bit location, the $\overline{\text{ERR}}$ flag should be low. If a diagnostic data word with two errors in any bit location is applied, the $\overline{\text{MERR}}$ flag should be low. After the checkword is latched into the input latch, it can be verified by taking $\overline{\text{OECB}}$ low. This outputs the latched checkword. The diagnostic data word can be latched into the output data latch and verified. By changing from the diagnostic mode (S1 = L, S0 = H) to the correction mode (S1 = H, S0 = H), the user can verify that the EDAC will correct the diagnostic data word. Also, the syndrome bits can be produced to verify that the EDAC pinpoints the error location. Table VII lists the diagnostic functions.

PRODUCT PREVIEWS

5

TABLE VII. DIAGNOSTIC FUNCTION

EDAC FUNCTION	CONTROL S1	CONTROL S0	DATA I/O	DB BYTE CONTROL $\overline{OEBn}$	DB OUTPUT LATCH $\overline{LEDBO}$	CHECK I/O	CB CONTROL $\overline{OECB}$	ERROR FLAGS $\overline{ERR}$	ERROR FLAGS $\overline{MERR}$
Read & flag	H	L	Input correct data word	H	X	Input correct check bits	H	H	H
Latch input check word while data input latch remains transparent	L	H	Input diagnostic data word†	H	L	Latched input check bits	H	Enabled	
Latch diagnostic data word into output latch	L	H	Input diagnostic data word†	H	H	Output latched check bits	L	Enabled	
						Hi-Z	H		
Latch diagnostic data word into input latch	H	H	Latched input diagnostic data word	H	H	Output syndrome bits	L	Enabled	
						Hi-Z	H		
Output diagnostic data word & syndrome bits	H	H	Output diagnostic data word	L	H	Output syndrome bits	L	Enabled	
						Hi-Z	H		
Output corrected diagnostic data word & output syndrome bits	H	H	Output corrected diagnostic data word	L	L	Output syndrome bits	L	Enabled	
						Hi-Z	H		

† Diagnostic data is a data word with an error in one bit location except when testing the $\overline{MERR}$ error flag. In this case, the diagnostic data word will contain errors in two bit locations.

TEXAS INSTRUMENTS
POST OFFICE BOX 225012 • DALLAS, TEXAS 75265

logic diagram (positive logic)

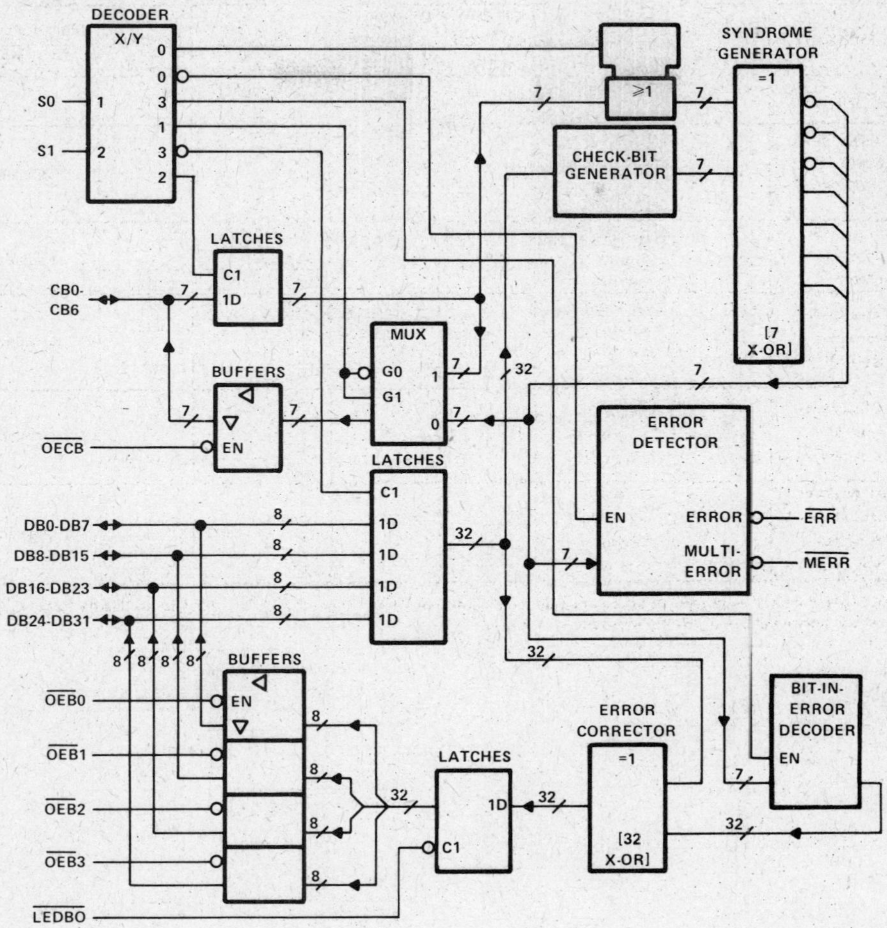

maximum ratings, recommended operating conditions, and electrical characteristics

See Table III, page 2-8.

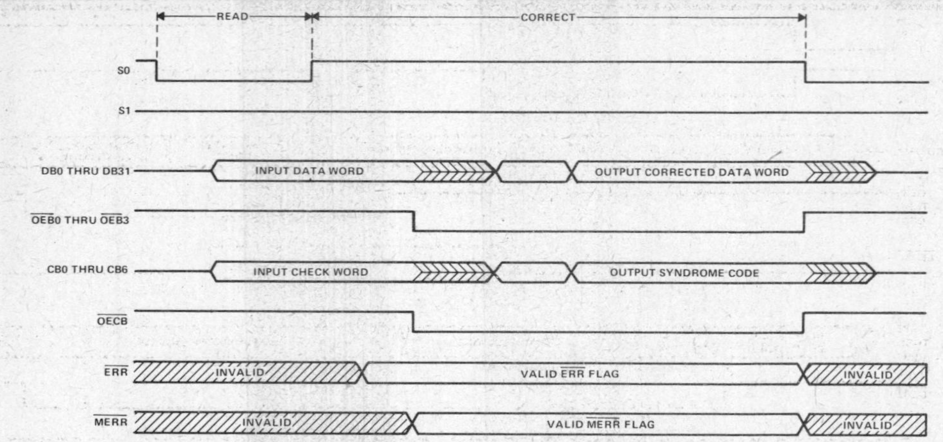

FIGURE 1-READ,FLAG,AND CORRECT MODE SWITCHING WAVEFORMS

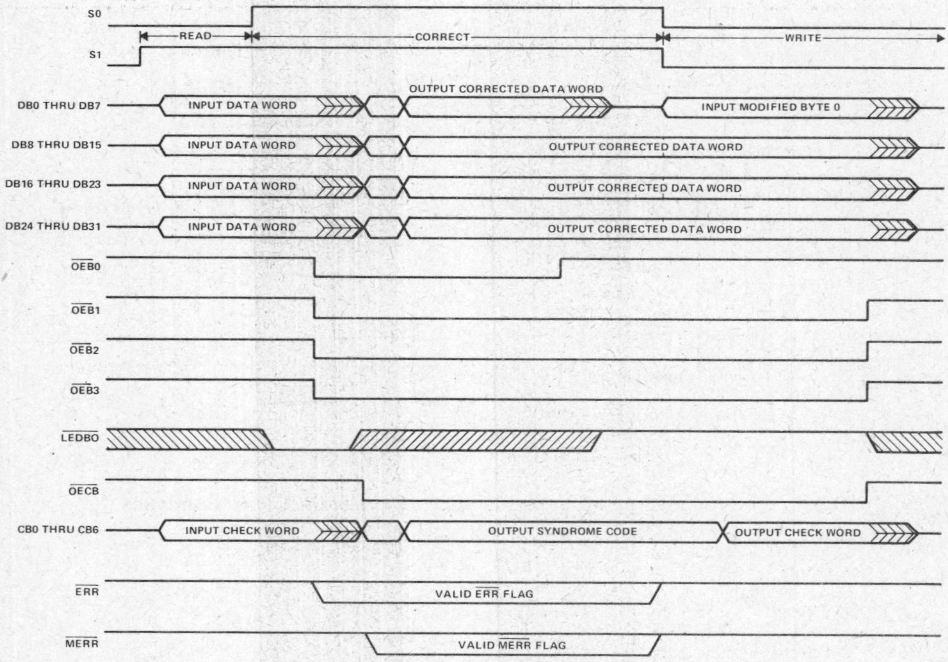

FIGURE 2-READ,CORRECT,MODIFY MODE SWITCHING WAVEFORMS

PRODUCT PREVIEWS

5

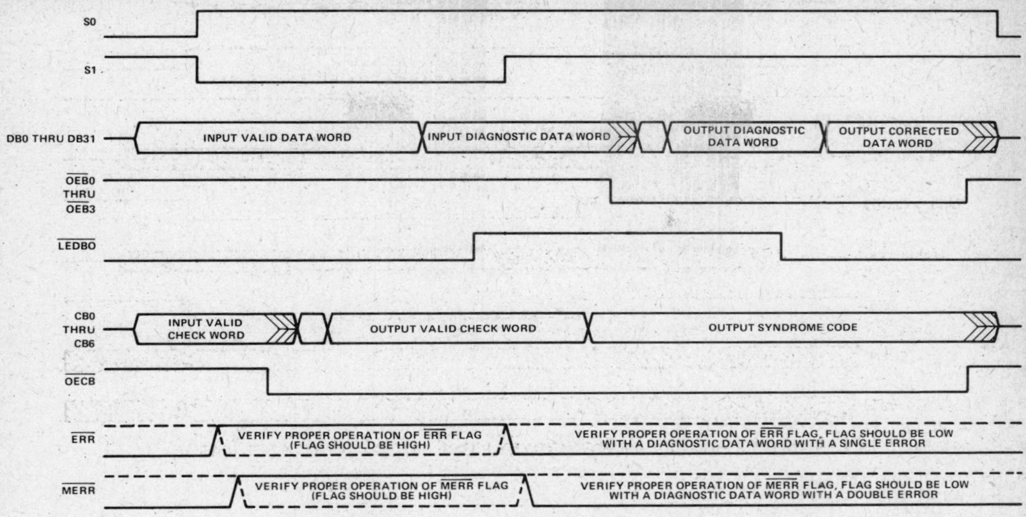

FIGURE 3-DIAGNOSTIC MODE SWITCHING WAVEFORM

TYPES SN54HC670, SN74HC670 4-BY-4 REGISTER FILES WITH 3-STATE OUTPUTS

D2804, MARCH 1984

- Separate Read/Write Addressing Permits Simultaneous Reading and Writing
- Organized as 4 Words of 4 Bits
- Expandable to 512 Words of n-Bits
- For Use as:
  - Scratch-Pad Memory
  - Buffer Storage Between Processors
  - Bit Storage in Fast Multiplication Designs
- High-Current 3-State Outputs Can Drive up to 15 LSTTL Loads
- Package Options Include Both Plastic and Ceramic Chip Carriers in Addition to Plastic and Ceramic DIPs
- Dependable Texas Instruments Quality and Reliability

SN54HC670 . . . J PACKAGE
SN74HC670 . . . J OR N PACKAGE
(TOP VIEW)

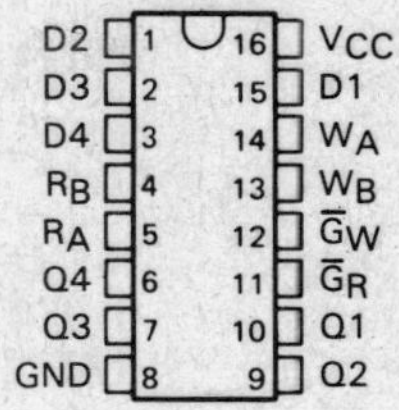

SN54HC670 . . . FH OR FK PACKAGE
SN74HC670 . . . FH OR FN PACKAGE
(TOP VIEW)

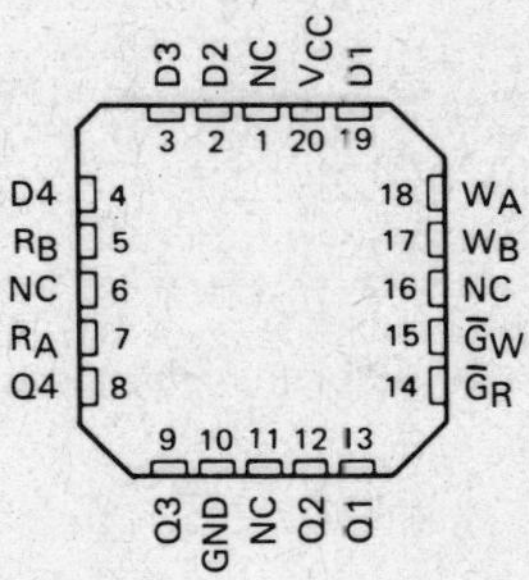

NC—No internal connections.

description

The SN54HC670 and SN74HC670 are 16-bit register files. The register file is organized as 4 words of 4 bits each and separate on-chip decoding is provided for addressing the four word locations to either write-in or retrieve data. This permits simultaneous writing into one location and reading from another word location.

Four data inputs are used to supply the 4-bit word to be stored. Location of the word is determined by the write-address inputs W_A and W_B in conjunction with a write-enable signal $\overline{G}_W$. Data applied at the inputs should be in its true form. That is, if a high-level signal is desired from the output, a high-level is applied at the data input for that particular bit location. When $\overline{G}_W$ is high, the data inputs are inhibited and their levels can cause no change in the information stored in the internal latches. When the read-enable input $\overline{G}_R$ is high, the data outputs are inhibited and go into the high-impedance state.

The individual address lines permit direct acquisition of data stored in any four of the latches. Four individual decoding gates are used to complete the address for reading a word. When the read address is made in conjunction with the read-enable signal, the word appears at the four outputs.

logic symbol

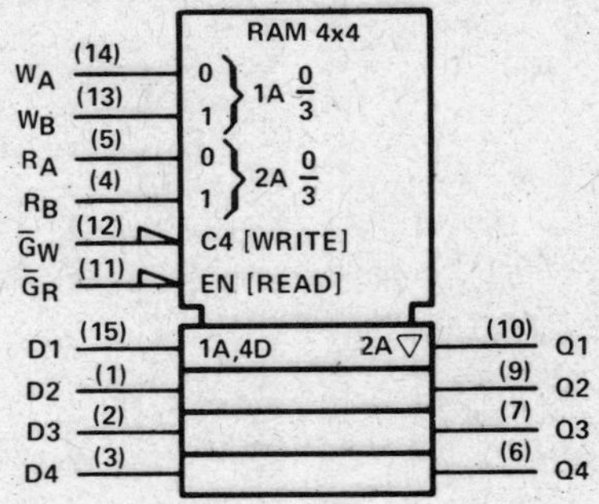

Pin numbers shown are for J and N packages.

TEXAS INSTRUMENTS
POST OFFICE BOX 225012 • DALLAS, TEXAS 75265

This arrangement—data-entry addressing separate from data-read addressing and individual sense line—eliminates recovery times, permits simultaneous reading and writing, and is limited in speed only by the write time and the read time. The register file has a nondestructive readout in that data is not lost when addressed.

The outputs are high-current, three-state outputs. These outputs may be bus connected for increasing the word capacity. Any number of these registers may be paralleled to provide n-bit word length.

The SN54HC670 is characterized for operation over the full military temperature range of −55 °C to 125 °C. The SN74HC670 is characterized for operation from −40 °C to 85 °C.

WRITE FUNCTION TABLE (SEE NOTES A, B, AND C)

WRITE INPUTS			WORD			
W_B	W_A	$\bar{G}_W$	0	1	2	3
L	L	L	Q = D	Q_0	Q_0	Q_0
L	H	L	Q_0	Q = D	Q_0	Q_0
H	L	L	Q_0	Q_0	Q = D	Q_0
H	H	L	Q_0	Q_0	Q_0	Q = D
X	X	H	Q_0	Q_0	Q_0	Q_0

READ FUNCTION TABLE (SEE NOTES A AND D)

READ INPUTS			OUTPUTS			
R_B	R_A	$\bar{G}_R$	Q1	Q2	Q3	Q4
L	L	L	W0B1	W0B2	W0B3	W0B4
L	H	L	W1B1	W1B2	W1B3	W1B4
H	L	L	W2B1	W2B2	W2B3	W2B4
H	H	L	W3B1	W3B2	W3B3	W3B4
X	X	H	Z	Z	Z	Z

NOTES: A. H = high level, L = low level, X = irrelevant, Z = high impedance (off)
B. (Q = D) = The four selected internal flip-flop outputs will assume the states applied to the four external data inputs.
C. Q_0 = the level of Q before the indicated input conditions were established.
D. W0B1 = The first bit of word 0, etc.

maximum ratings, recommended operating conditions, and electrical characteristics

See Table III, page 2-8.

PRODUCT PREVIEWS

5

TYPES SN54HC881, SN74HC881 ARITHMETIC LOGIC UNITS/FUNCTION GENERATORS

D2804, MARCH 1984

- Full Look-Ahead for High-Speed Operations on Long Words
- Arithmetic Operating Modes
  - Addition
  - Subtraction
  - Shift Operand A One Position
  - Magnitude Comparison
  - Plus Twelve Other Arithmetic Operations
- Logic Function Modes
  - Exclusive-OR
  - Comparator
  - AND, NAND, OR, NOR
  - Status Register Checks
  - Plus Ten Other Logic Operations
- Package Options Include Both Plastic and Ceramic Chip Carriers in Addition to Plastic and Ceramic DIPs
- Dependable Texas Instruments Quality and Reliability

SN54HC881 . . . JT PACKAGE
SN74HC881 . . . JT OR NT PACKAGE
(TOP VIEW)

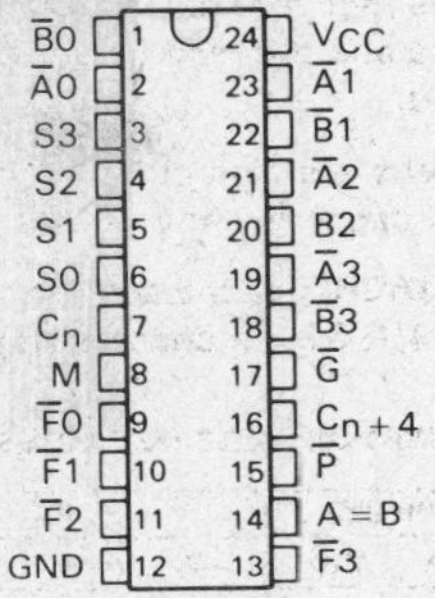

SN54HC881 . . . FH OR FK PACKAGE
SN74HC881 . . . FH OR FN PACKAGE
(TOP VIEW)

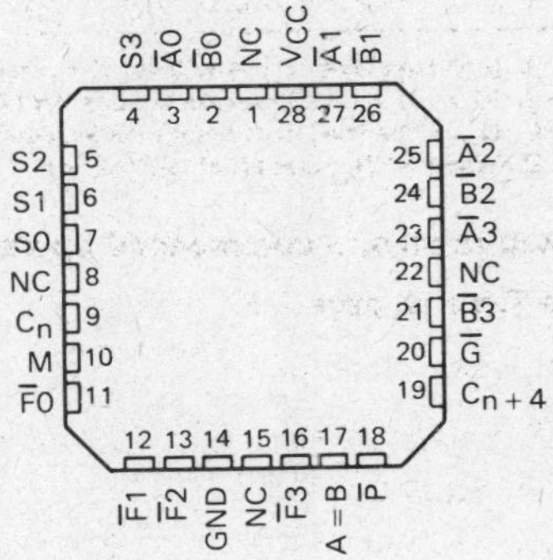

NC – No internal connection

logic symbol

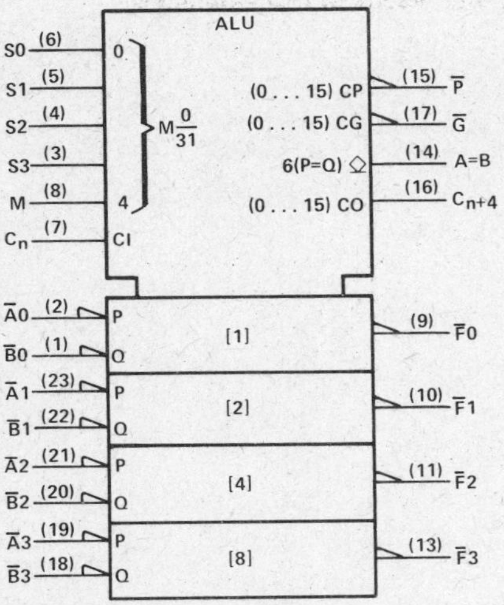

Pin numbers shown are for JT and NT packages.

For additional information on the SN54HC881 and SN74HC881, see page 3-149.

- Directly Compatible with the New 'HC181 and 'HC881 ALUs
- Capable of Anticipating the Carry Across a Group of Eight 4-Bit Binary Adders
- Cascadable to Perform Look-Ahead Across n-Bit Adders
- Package Options Include Both Plastic and Ceramic Chip Carriers in Addition to Plastic and Ceramic DIPs
- Dependable Texas Instruments Quality and Reliability

SN54HC882 . . . JT PACKAGE
SN74HC882 . . . JT OR NT PACKAGE
(TOP VIEW)

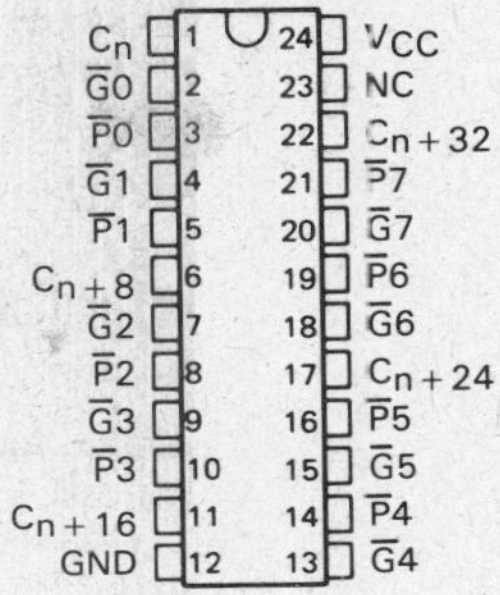

SN54HC882 . . . FH OR FK PACKAGE
SN74HC882 . . . FH OR FN PACKAGE
(TOP VIEW)

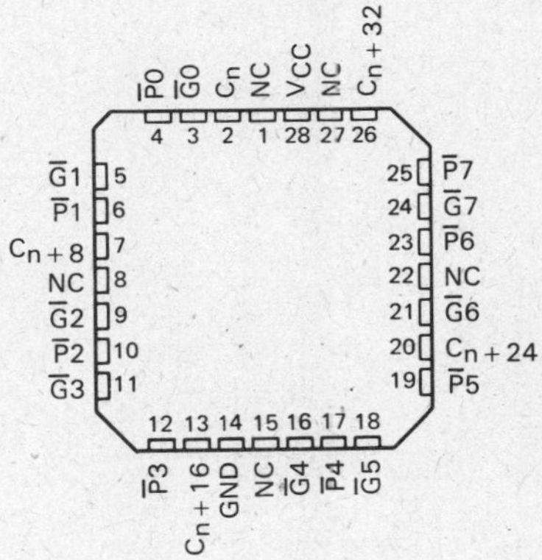

NC—No internal connection

description

The 'HC882 is a high-speed look-ahead carry generator capable of anticipating the carry across a group of eight 4-bit adders permitting the designer to implement look-ahead for a 32-bit ALU with a single package or, by cascading 'HC882's, full look-ahead is possible across n-bit adders.

The SN54HC882 is characterized for operation over the full military temperature range of −55°C to 125°C. The SN74HC882 is characterized for operation from −40°C to 85°C.

logic symbol

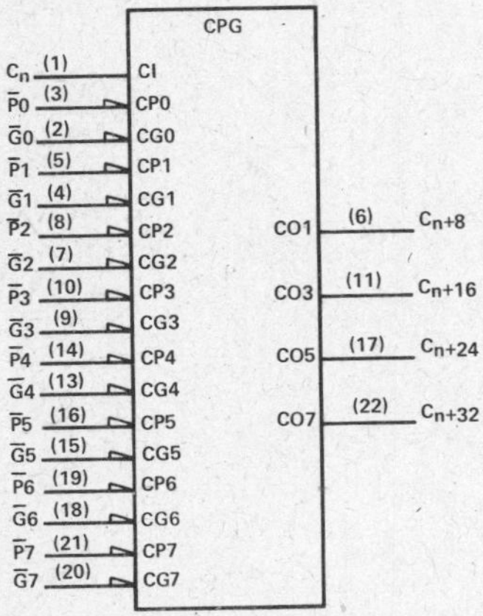

Pin numbers shown are for JT and NT packages.

LOGIC EQUATIONS

$$C_{n+8} = G1 + P1G0 + P1P0C_n$$

$$C_{n+16} = G3 + P3G2 + P3P2G1 + P3P2P1G0 + P3P2P1P0C_n$$

$$C_{n+24} = G5 + P5G4 + P5P4G3 + P5P4P3G2 + P5P4P3P2G1 + P5P4P3P2P1G0 + P5P4P3P2P1P0C_n$$

$$C_{n+32} = G7 + P7G6 + P7P6G5 + P7P6P5G4 + P7P6P5P4G3 + P7P6P5P4P3G2 + P7P6P5P4P3P2G1 + P7P6P5P4P3P2P1G0 + P7P6P5P4P3P2P1P0C_n$$

PRODUCT PREVIEW
This document contains information on a product under development. Texas Instruments reserves the right to change or discontinue this product without notice.

FUNCTION TABLE
FOR C_{n+32} OUTPUT

INPUTS																	OUTPUT
$\overline{G7}$	$\overline{G6}$	$\overline{G5}$	$\overline{G4}$	$\overline{G3}$	$\overline{G2}$	$\overline{G1}$	$\overline{G0}$	$\overline{P7}$	$\overline{P6}$	$\overline{P5}$	$\overline{P4}$	$\overline{P3}$	$\overline{P2}$	$\overline{P1}$	$\overline{P0}$	C_n	C_{n+32}
L	X	X	X	X	X	X	X	X	X	X	X	X	X	X	X	X	H
X	L	X	X	X	X	X	X	L	X	X	X	X	X	X	X	X	H
X	X	L	X	X	X	X	X	L	L	X	X	X	X	X	X	X	H
X	X	X	L	X	X	X	X	L	L	L	X	X	X	X	X	X	H
X	X	X	X	L	X	X	X	L	L	L	L	X	X	X	X	X	H
X	X	X	X	X	L	X	X	L	L	L	L	L	X	X	X	X	H
X	X	X	X	X	X	L	X	L	L	L	L	L	L	X	X	X	H
X	X	X	X	X	X	X	L	L	L	L	L	L	L	L	X	X	H
X	X	X	X	X	X	X	X	L	L	L	L	L	L	L	L	H	H
All other combinations																	L

FUNCTION TABLE
FOR C_{n+24} OUTPUT

INPUTS													OUTPUT
$\overline{G5}$	$\overline{G4}$	$\overline{G3}$	$\overline{G2}$	$\overline{G1}$	$\overline{G0}$	$\overline{P5}$	$\overline{P4}$	$\overline{P3}$	$\overline{P2}$	$\overline{P1}$	$\overline{P0}$	C_n	C_{n+24}
L	X	X	X	X	X	X	X	X	X	X	X	X	H
X	L	X	X	X	X	L	X	X	X	X	X	X	H
X	X	L	X	X	X	L	L	X	X	X	X	X	H
X	X	X	L	X	X	L	L	L	X	X	X	X	H
X	X	X	X	L	X	L	L	L	L	X	X	X	H
X	X	X	X	X	L	L	L	L	L	L	X	X	H
X	X	X	X	X	X	L	L	L	L	L	L	H	H
All other combinations													L

FUNCTION TABLE
FOR C_{n+16} OUTPUT

INPUTS									OUTPUT
$\overline{G3}$	$\overline{G2}$	$\overline{G1}$	$\overline{G0}$	$\overline{P3}$	$\overline{P2}$	$\overline{P1}$	$\overline{P0}$	C_n	C_{n+16}
L	X	X	X	X	X	X	X	X	H
X	L	X	X	L	X	X	X	X	H
X	X	L	X	L	L	X	X	X	H
X	X	X	L	L	L	L	X	X	H
X	X	X	X	L	L	L	L	H	H
All other combinations									L

FUNCTION TABLE
FOR C_{n+8} OUTPUT

INPUTS					OUTPUT
$\overline{G1}$	$\overline{G0}$	$\overline{P1}$	$\overline{P0}$	C_n	C_{n+8}
L	X	X	X	X	H
X	L	L	X	X	H
X	X	L	L	H	H
All other combinations					L

Any inputs not shown in a given table are irrelevant with respect to that output.

maximum ratings, recommended operating conditions, and electrical characteristics

See Table IV, page 2-10.

TYPES SN54HC4016, SN74HC4016 QUAD BILATERAL SWITCHES

D2831, MARCH 1984

- High Degree of Linearity
- Switches Can Transmit Signals in Either Direction at Frequencies up to 50 MHz
- Extremely Low Off-State Switch Current Resulting in Very High Effective Off-State Resistance
- High On/Off Output Voltage Ratio
- Extremely High Control-Input Impedance (Control Circuit Isolated from Switch Circuit)
- Low Crosstalk Between Switches
- Package Options Include Both Plastic and Ceramic Chip Carriers in Addition to Plastic and Ceramic DIPs
- Dependable Texas Instruments Quality and Reliability

SN54HC4016 . . . J PACKAGE
SN74HC4016 . . . J OR N PACKAGE
(TOP VIEW)

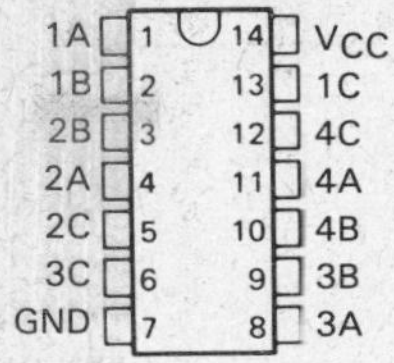

SN54HC4016 . . . FH OR FK PACKAGE
SN74HC4016 . . . FH OR FN PACKAGE
(TOP VIEW)

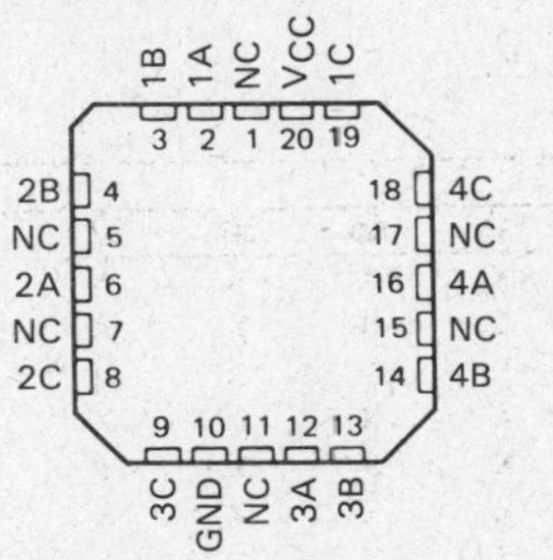

NC—No internal connection

description

The 'HC4016 is a quadruple bilateral switch for either digital or analog signals. Low power dissipation and high noise immunity allow the 'HC4016 to be used in many diverse environments.

Applications include digital switching and multiplexing, analog-to-digital and digital-to-analog conversion, digital control of frequency, impedance, phase, and analog-signal gain, signal gating, and as a squelch control, chopper, modulator, demodulator, or commutating switch.

The SN54HC4016 is characterized for operation over the full military temperature range of −55°C to 125°C. The SN74HC4016 is characterized for operation from −40°C to 85°C.

logic symbol

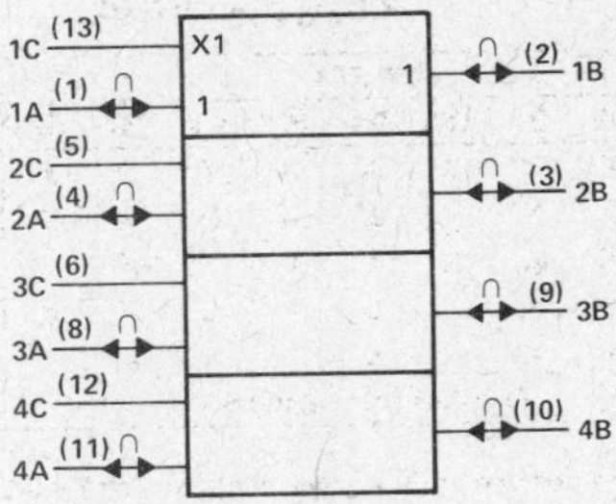

Pin numbers shown are for J and N packages.

logic diagram, each switch (positive logic)

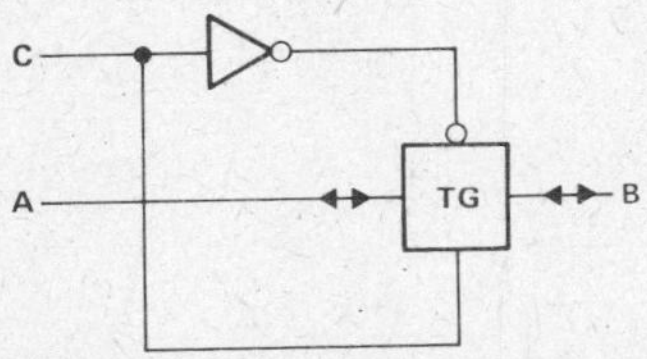

maximum ratings, recommended operating conditions, and electrical characteristics

See Table I, page 2-4.

PRODUCT PREVIEW

This document contains information on a product under development. Texas Instruments reserves the right to change or discontinue this product without notice.

TYPES SN54HC4049, SN54HC4050, SN74HC4049, SN74HC4050
HEX INVERTING AND NONINVERTING BUFFERS

D2831, MARCH 1984

- **Package Options Include Both Plastic and Ceramic DIPs**
- **Dependable Texas Instruments Quality and Reliability**

SN54HC4049, SN54HC4050 . . . J PACKAGE
SN74HC4049, SN74HC4050 . . . J OR N PACKAGE
(TOP VIEW)

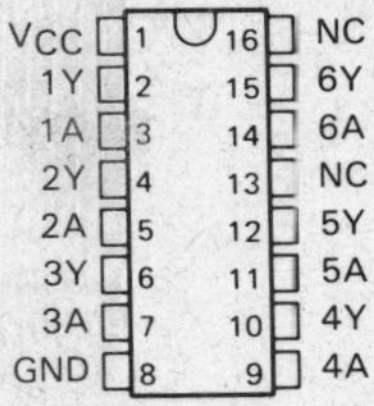

NC—No internal connection

Not available in chip carrier package with JEDEC-Standard pin-out. For chip carrier information, contact the factory.

description

The 'HC4049 and 'HC4050 hex inverting and noninverting buffers may be used as current sinks or source drivers, hex drivers, or high-to-low-logic-level (e.g., CMOS to TTL) converters. Logic-level conversion is accomplished using only one supply voltage (V_{CC}). The high-level input signal (V_{IH}) can exceed the V_{CC} supply voltage when this device is used for logic-level conversions.

Pin 16 of the 'HC4049 and 'HC4050 in the J or N package is not internally connected; therefore, any external connection to this pin does not affect circuit operation.

FUNCTION TABLE

INPUT A	OUTPUT Y 'HC4049	OUTPUT Y 'HC4050
H	L	H
L	H	L

logic symbols

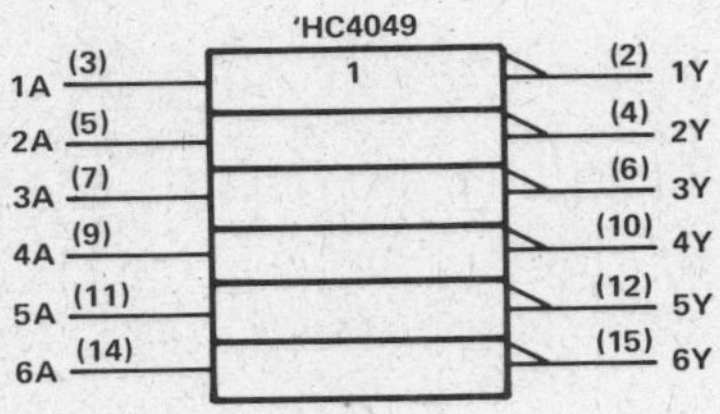

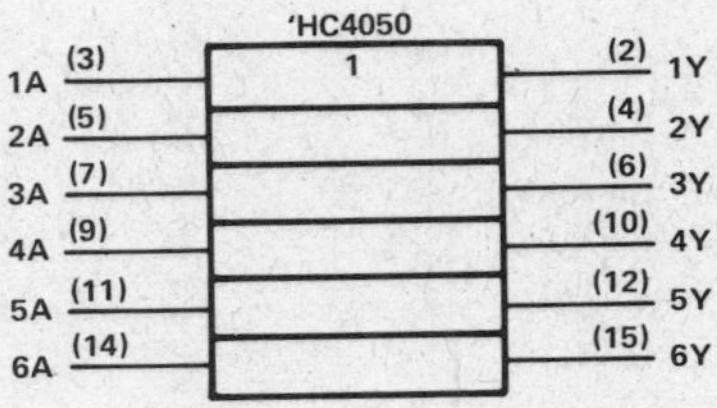

Pin numbers shown are for J and N packages.

TYPES SN54HC4051, SN54HC4052, SN54HC4053 SN74HC4051, SN74HC4052, SN74HC4053 ANALOG MULTIPLEXERS/DEMULTIPLEXERS

D2804, MARCH 1984

- Fast Switching
- Low Crosstalk Between Switches
- High On/Off Output Voltage Ratio
- Analog Supply Voltage Range ($V_{CC} - V_{EE}$) . . . 3 V to 12 V
- Digital Supply Voltage Range ($V_{CC} - GND$) . . . 2 V to 6 V
- Package Options Include Both Plastic and Ceramic Chip Carriers in Addition to Plastic and Ceramic DIPs
- Dependable Texas Instruments Quality and Reliability

SN54HC4051 . . . J PACKAGE
SN74HC4051 . . . J OR N PACKAGE
(TOP VIEW)

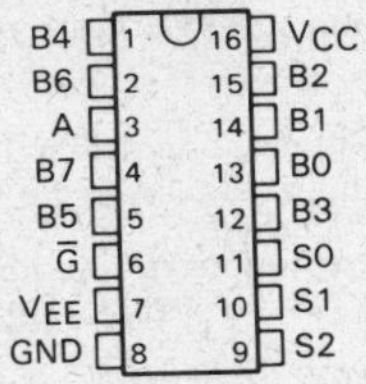

SN54HC4052 . . . J PACKAGE
SN74HC4052 . . . J OR N PACKAGE
(TOP VIEW)

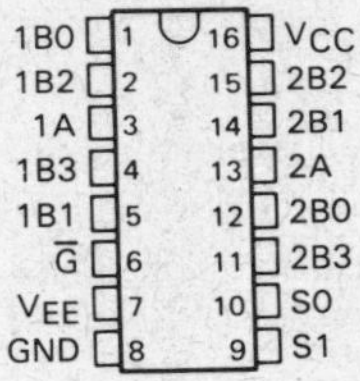

SN54HC4053 . . . J PACKAGE
SN74HC4053 . . . J OR N PACKAGE
(TOP VIEW)

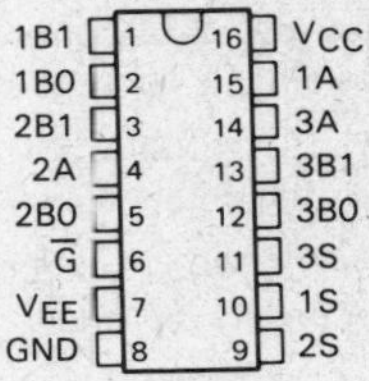

SN54HC4051 . . . FH OR FK PACKAGE
SN74HC4051 . . . FH OR FN PACKAGE
(TOP VIEW)

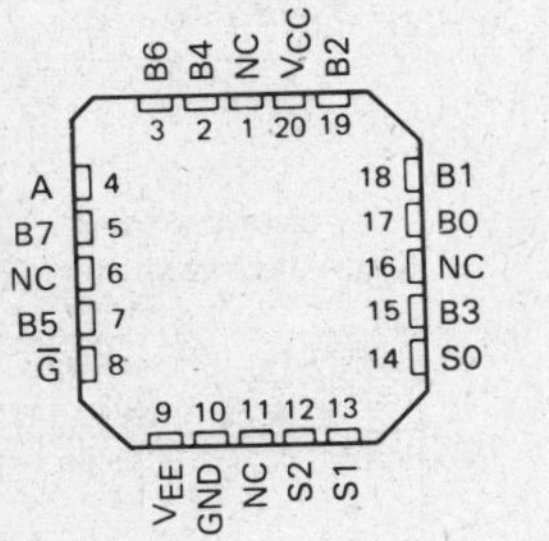

SN54HC4052 . . . FH OR FK PACKAGE
SN74HC4052 . . . FH OR FN PACKAGE
(TOP VIEW)

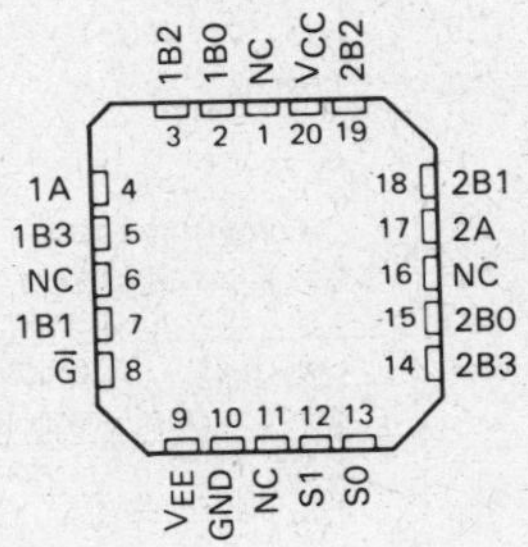

SN54HC4053 . . . FH OR FK PACKAGE
SN74HC4053 . . . FH OR FN PACKAGE
(TOP VIEW)

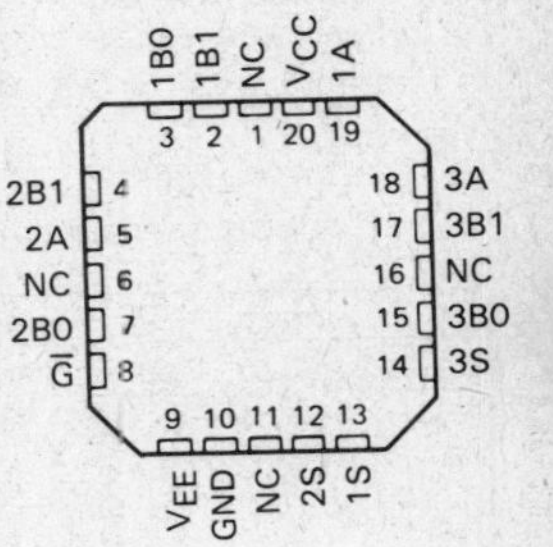

NC—No internal connection.

logic symbols

'HC4051

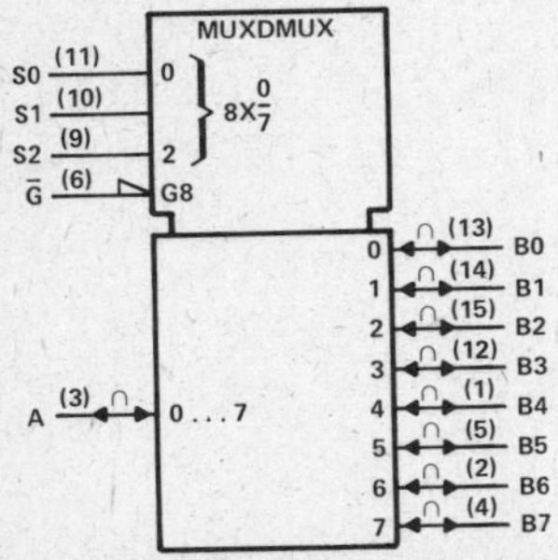

'HC4052

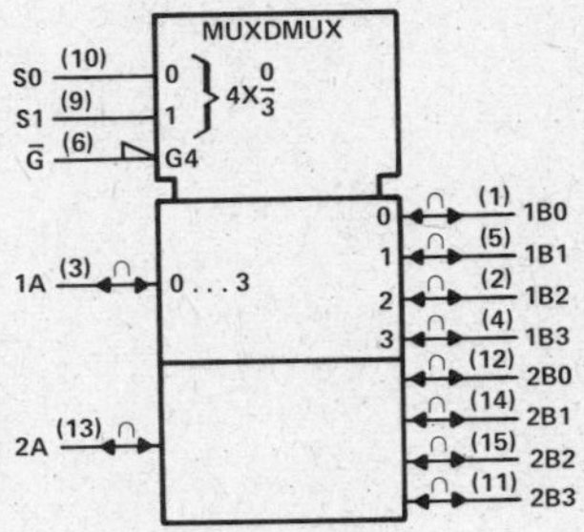

'HC4053

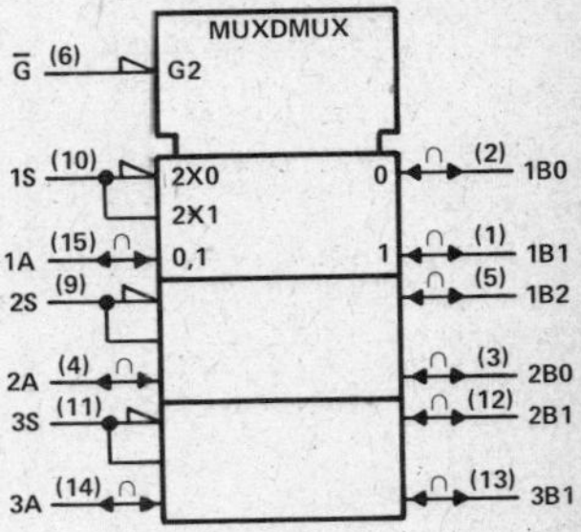

Pin numbers shown are for J and N packages.

PRODUCT PREVIEW

This document contains information on a product under development. Texas Instruments reserves the right to change or discontinue this product without notice.

TEXAS INSTRUMENTS

POST OFFICE BOX 225012 • DALLAS, TEXAS 75265

description

These devices are analog multiplexers/demultiplexers incorporating built-in level shifting. The level shifting allows a control input range of GND to V_{CC} for an analog signal range of V_{EE} to V_{CC}. Thus the common situation of positive digital signals controlling the multiplexing of both positive and negative analog signals can be accommodated.

These digitally controlled bilateral analog switches have low on-state impedance and very low off-state current. When the inhibit input terminal is high, all channels are off.

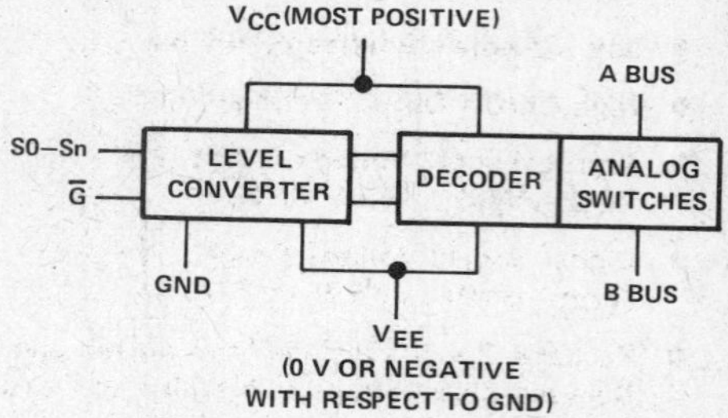

FIGURE 1. INTERNAL POWER SUPPLY CONNECTIONS

The 'HC4051 is a single eight-channel multiplexer/demultiplexer having three binary control inputs (S0, S1, and S2) and an enable input ($\overline{G}$). The three binary signals select one of eight channels to be turned on.

The 'HC4052 is a dual four-channel multiplexer/demultiplexer having two control inputs (S0 and S1) and an enable input ($\overline{G}$). The two binary signals select one of four channels in each of the two sections.

The 'HC4053 is a triple two-channel multiplexer/demultiplexer having three separate control inputs (1S, 2S, and 3S) and a common enable input ($\overline{G}$). Each S input independently selects one of two channels in one of the three sections.

The SN54HC4051, SN54HC4052, and SN54HC4053 are characterized for operation over the full military temperature range of −55°C to 125°C. The SN74HC4051, SN74HC4052, and SN74HC4053 are characterized for operation from −40°C to 85°C.

'HC4051
FUNCTION TABLE

INPUTS				CHANNEL
$\overline{G}$	S2	S1	S0	TURNED ON
H	X	X	X	None
L	L	L	L	0
L	L	L	H	1
L	L	H	L	2
L	L	H	H	3
L	H	L	L	4
L	H	L	H	5
L	H	H	L	6
L	H	H	H	7

'HC4052
FUNCTION TABLE
(EACH BILATERAL SWITCH)

INPUTS			CHANNEL
$\overline{G}$	S1	S0	TURNED ON
H	X	X	None
L	L	L	0
L	L	H	1
L	H	L	2
L	H	H	3

'HC4053
FUNCTION TABLE
(EACH BILATERAL SWITCH)

INPUTS		CHANNEL
$\overline{G}$	S	TURNED ON
H	X	None
L	L	0
L	H	1

TYPES SN54HC4316, SN74HC4316 QUAD BILATERAL SWITCHES

D2804, MARCH 1984

- Fast Switching Speeds
- Low Crosstalk Between Switches
- High On/Off Output Voltage Ratio
- Analog Supply Voltage Range ($V_{CC}-V_{EE}$) . . . 3 V to 12 V
- Digital Supply Voltage Range (V_{CC}−GND) . . . 2 V to 6 V
- Package Options Include Both Plastic and Ceramic Chip Carriers in Addition to Plastic and Ceramic DIPs
- Dependable Texas Instruments Quality and Reliability

SN54HC4316 . . . J PACKAGE
SN74HC4316 . . . J OR N PACKAGE
(TOP VIEW)

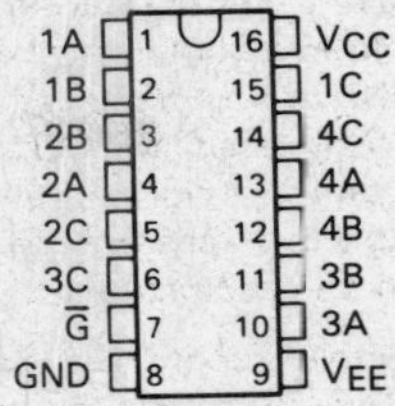

SN54HC4316 . . . FH OR FK PACKAGE
SN74HC4316 . . . FH OR FN PACKAGE
(TOP VIEW)

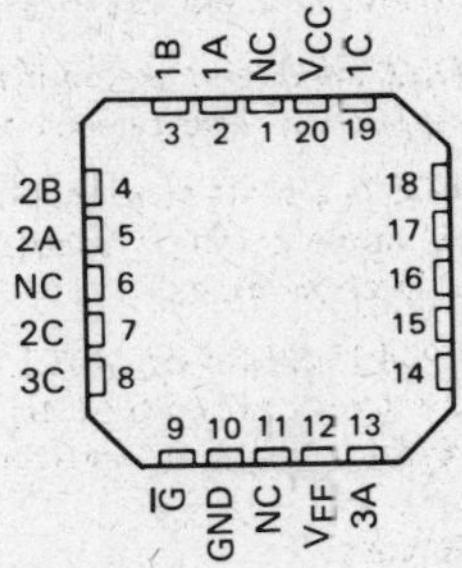

NC—No internal connection

description

The 'HC4316 is a quadruple bilateral switch. The switches can transmit analog or digital signals in either direction. The 'HC4316 offers high control input impedance and low crosstalk between switches.

Applications include digital switching and multiplexing analog-to-digital and digital-to-analog conversion; digital control of frequency, impedance, phase and analog-signal gain; and use as a squelch control, chopper, modulator, demodulator, or commutating switch.

The SN54HC4316 is characterized for operation over the full military temperature range of −55°C to 125°C. The SN74HC4316 is characterized for operation from −40°C to 85°C.

logic symbol

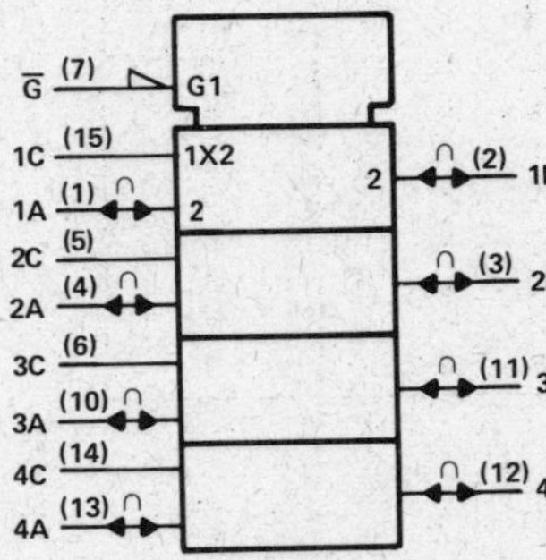

Pin numbers shown are for J and N packages.

logic diagram, each switch (positive logic)

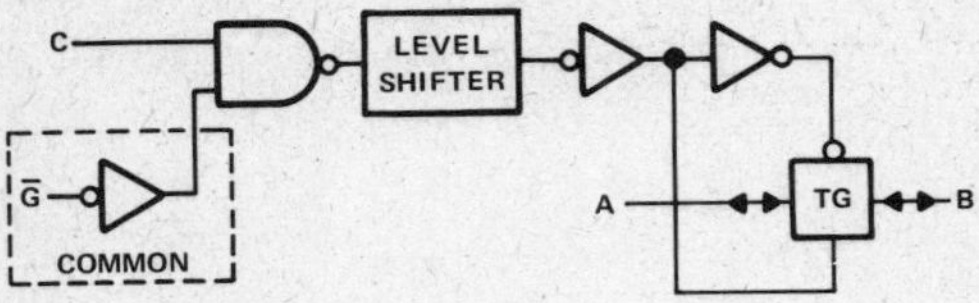

TEXAS INSTRUMENTS
POST OFFICE BOX 225012 • DALLAS, TEXAS 75265

TYPES SN54HC4511, SN74HC4511 BCD-TO-SEVEN-SEGMENT DECODERS/DRIVERS WITH LATCHED INPUTS

D2684, DECEMBER 1982—REVISED MARCH 1984

- Latch Storage of Code
- Blanking Input
- Lamp Test Provision
- Readout Blanking on All Illegal Input Combinations
- Package Options Include Both Plastic and Ceramic Chip Carriers in Addition to Plastic and Ceramic DIPs
- Dependable Texas Instruments Quality and Reliability

SN54HC4511 . . . J PACKAGE
SN74HC4511 . . . J OR N PACKAGE
(TOP VIEW)

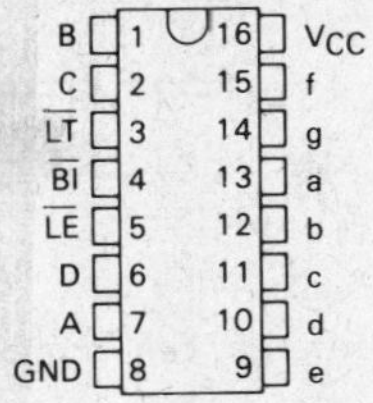

SN54HC4511 . . . FH OR FK PACKAGE
SN74HC4511 . . . FH OR FN PACKAGE
(TOP VIEW)

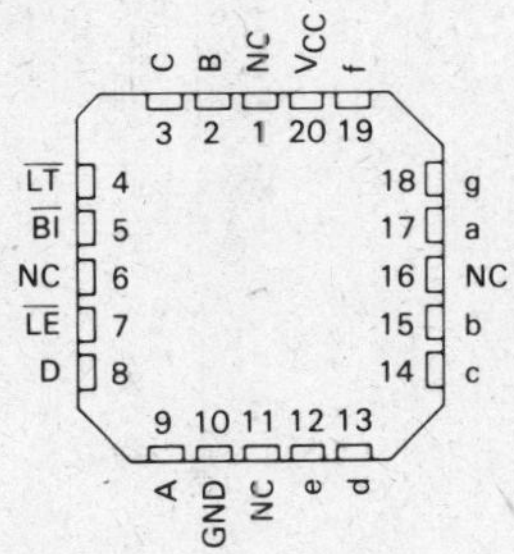

NC—No internal connection

description

The 'HC4511 provides the functions of a 4-bit storage latch, a BCD-to-seven-segment decoder, and an output driver. Lamp test ($\overline{LT}$), blanking ($\overline{BI}$), and latch enable ($\overline{LE}$) inputs are used to test the display, to turn off or pulse-modulate the brightness of the display, and to store a BCD code, respectively.

The SN54HC4511 is characterized for operation over the full military temperature range of −55°C to 125°C. The SN74HC4511 is characterized for operation from −40°C to 85°C.

FUNCTION TABLE

INPUTS							OUTPUTS							DISPLAY
$\overline{LE}$	$\overline{BI}$	$\overline{LT}$	D	C	B	A	a	b	c	d	e	f	g	
L	H	H	L	L	L	L	H	H	H	H	H	H	L	0
L	H	H	L	L	L	H	L	H	H	L	L	L	L	1
L	H	H	L	L	H	L	H	H	L	H	H	L	H	2
L	H	H	L	L	H	H	H	H	H	H	L	L	H	3
L	H	H	L	H	L	L	L	H	H	L	L	H	H	4
L	H	H	L	H	L	H	H	L	H	H	L	H	H	5
L	H	H	L	H	H	L	L	L	H	H	H	H	H	6
L	H	H	L	H	H	H	H	H	H	L	L	L	L	7
L	H	H	H	L	L	L	H	H	H	H	H	H	H	8
L	H	H	H	L	L	H	H	H	H	L	L	H	H	9
L	H	H	H	L	H	L	L	L	L	L	L	L	L	Blank
L	H	H	H	L	H	H	L	L	L	L	L	L	L	Blank
L	H	H	H	H	L	L	L	L	L	L	L	L	L	Blank
L	H	H	H	H	L	H	L	L	L	L	L	L	L	Blank
L	H	H	H	H	H	L	L	L	L	L	L	L	L	Blank
L	H	H	H	H	H	H	L	L	L	L	L	L	L	Blank
X	X	L	X	X	X	X	H	H	H	H	H	H	H	8
X	L	H	X	X	X	X	L	L	L	L	L	L	L	Blank
H	H	H	X	X	X	X	All outputs remain in state existing before $\overline{LE}$↑							

logic symbol

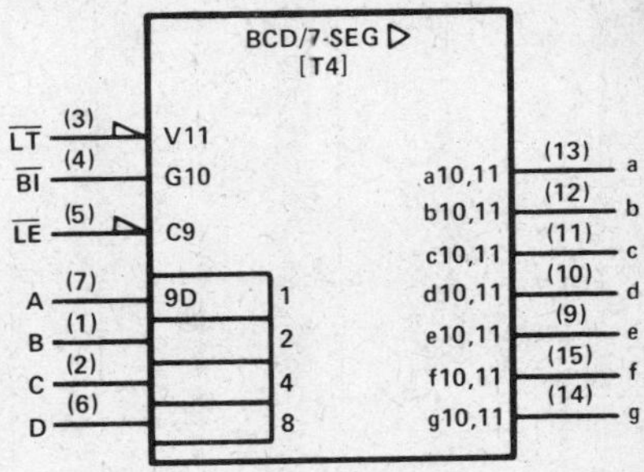

Pin numbers shown are for J and N packages.

PRODUCT PREVIEW

This document contains information on a product under development. Texas Instruments reserves the right to change or discontinue this product without notice.

TEXAS INSTRUMENTS
POST OFFICE BOX 225012 • DALLAS, TEXAS 75265

FONT TABLE T4 —
RESULTANT DISPLAYS USING 'HC4511

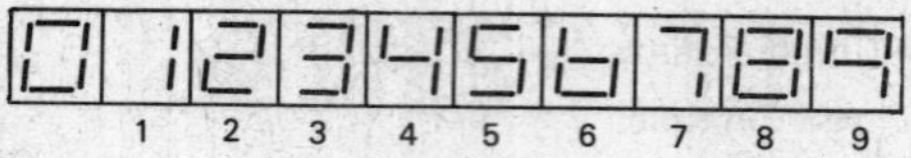

SEGMENT IDENTIFICATION

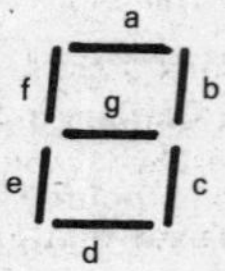

maximum ratings, recommended operating conditions, and electrical characteristics

See Table IV, page 2-10.

TEXAS INSTRUMENTS
POST OFFICE BOX 225012 • DALLAS, TEXAS 75265

D2684, DECEMBER 1982–REVISED MARCH 1984

- **Two Output Options:**
  'HC4514 Has Active-High Outputs
  'HC4515 Has Active-Low Outputs
- **Package Options Include Both Plastic and Ceramic Chip Carriers in Addition to Plastic and Ceramic DIPs**
- **Dependable Texas Instruments Quality and Reliability**

SN54HC' . . . JT PACKAGE
SN74HC' . . . JT OR NT PACKAGE
(TOP VIEW)

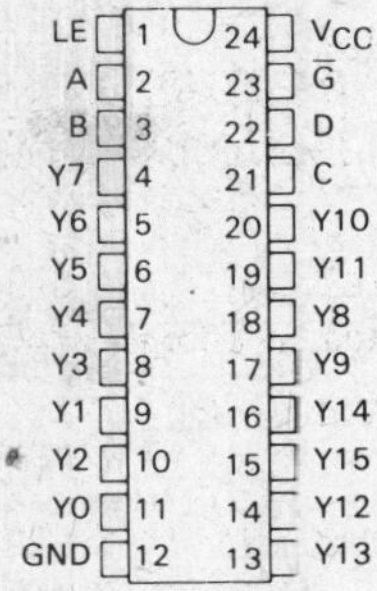

SN54HC' . . . FH OR FN PACKAGE
SN74HC' . . . FH OR FN PACKAGE
(TOP VIEW)

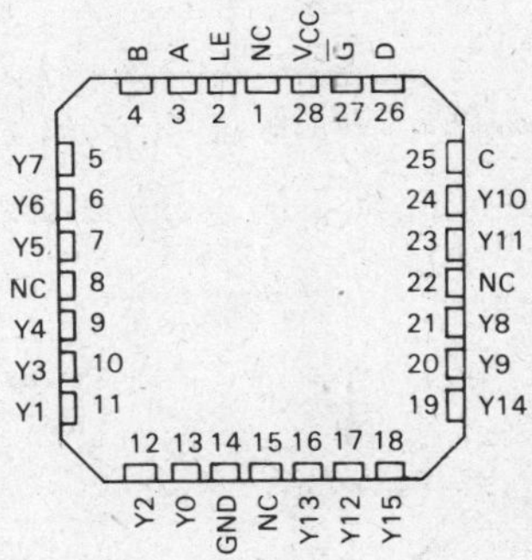

NC—No internal connection

description

These devices present two output options of a 4-line to 16-line decoder with latched inputs. The 'HC4514 presents a high level at the selected output. The 'HC4515 presents a low level at the selected output.

These devices consist of four storage latches with common latch enable (LE) and inhibit ($\overline{G}$) inputs. When a low signal is applied to the LE input, the input data is stored, decoded, and presented to the output. When $\overline{G}$ is high, all sixteen 'HC4514 outputs are at a low logic level, or all 'HC4515 outputs are at a high logic level.

The SN54HC4514 and the SN54HC4515 are characterized for operation over the full military temperature range of −55°C to 125°C. The SN74HC4514 and SN54HC4515 are characterized for operation from −40°C to 85°C.

FUNCTION TABLE

INPUTS						OUTPUT SELECTED	OUTPUTS	
LE	$\overline{G}$	D	C	B	A		'HC4514	'HC4515
H	L	L	L	L	L	0		
H	L	L	L	L	H	1		
H	L	L	L	H	L	2		
H	L	L	L	H	H	3		
H	L	L	H	L	L	4		
H	L	L	H	L	H	5	Selected	Selected
H	L	L	H	H	L	6	Output = H	Output = L
H	L	L	H	H	H	7	All others = L	All others = H
H	L	H	L	L	L	8		
H	L	H	L	L	H	9		
H	L	H	L	H	L	10		
H	L	H	L	H	H	11		
H	L	H	H	L	L	12		
H	L	H	H	L	H	13		
H	L	H	H	H	L	14		
H	L	H	H	H	H	15		
X	H	X	X	X	X		All = L	All = H
L	L	X	X	X	X	All outputs remain in state existing before LE↓		

5 PRODUCT PREVIEWS

PRODUCT PREVIEW

This document contains information on a product under development. Texas Instruments reserves the right to change or discontinue this product without notice.

TEXAS INSTRUMENTS
POST OFFICE BOX 225012 • DALLAS, TEXAS 75265

'HC4514 logic symbols (alternatives)

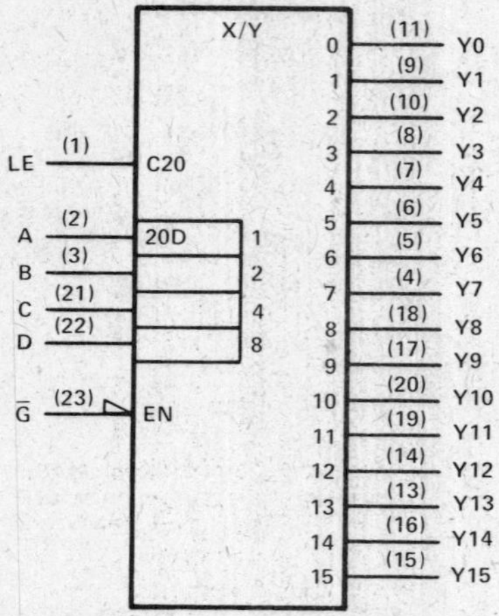

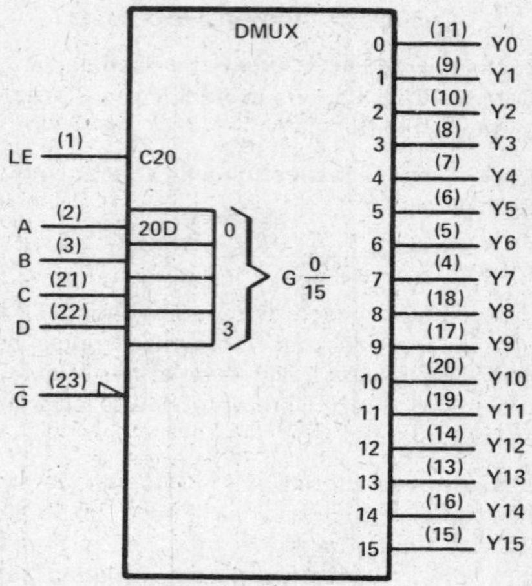

Pin numbers shown are for JT and NT packages.

'HC4515 logic symbols (alternatives)

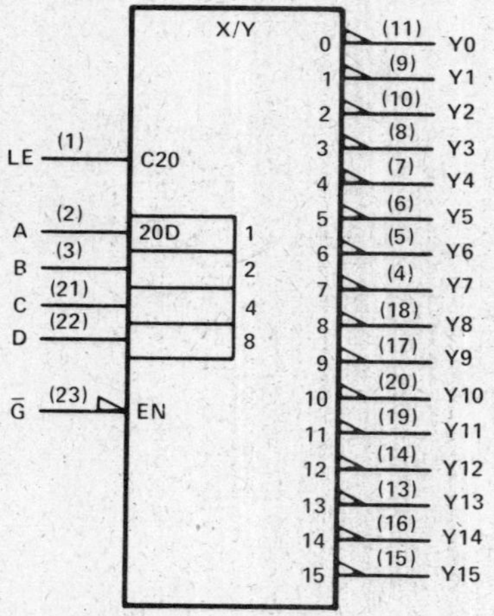

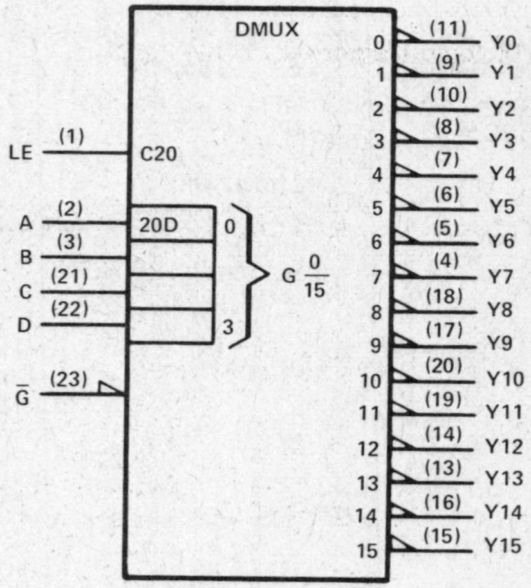

Pin numbers shown are for JT and NT packages.

'HC4514 logic diagram (positive logic)

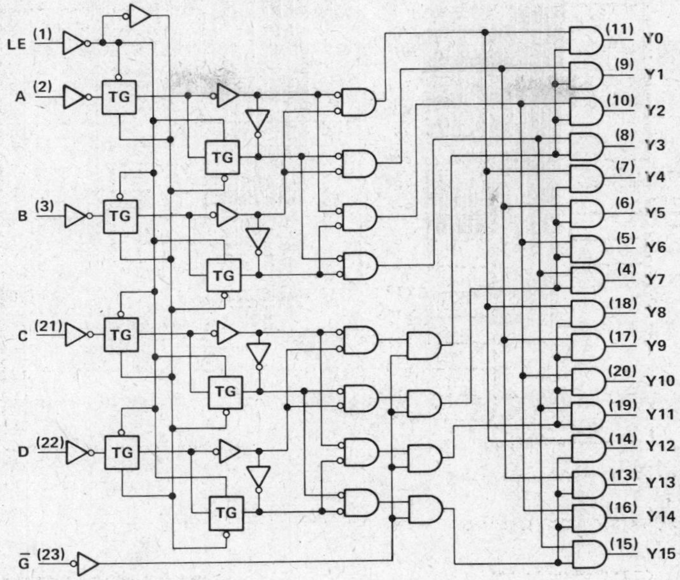

'HC4515 logic diagram (positive logic)

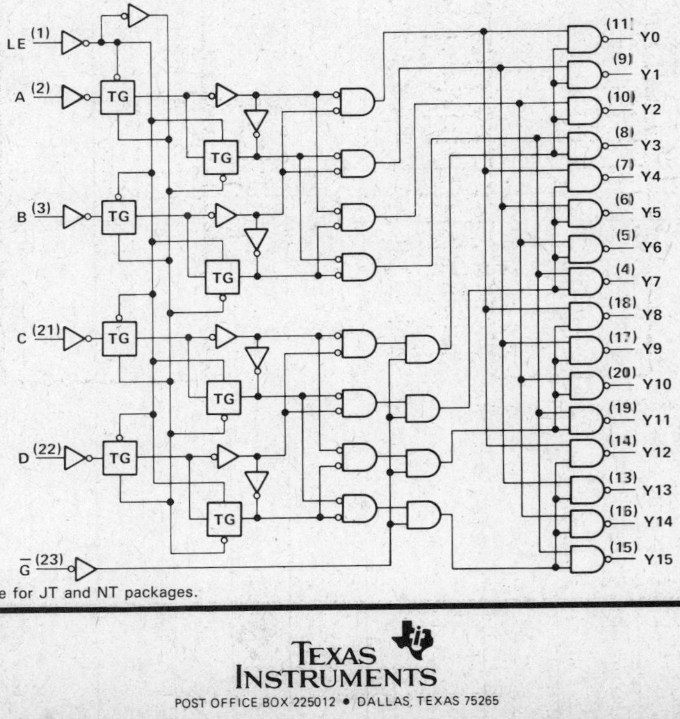

Pin numbers shown are for JT and NT packages.

absolute maximum ratings, recommended operating conditions, and electrical characteristics

See Table IV, page 2-10.

timing requirements over recommended operating free-air temperature range (unless otherwise noted)

PARAMETER		V_{CC}	T_A = 25°C MIN	T_A = 25°C MAX	SN54HC4514 SN54HC4515 MIN	SN54HC4514 SN54HC4515 MAX	SN74HC4514 SN74HC4515 MIN	SN74HC4514 SN74HC4515 MAX	UNIT
t_w	Pulse duration, LE high	2 V	40						ns
		4.5 V	8						
		6 V	7						
t_{su}	Setup time, A thru D before LE↓	2 V	50						ns
		4.5 V	10						
		6 V	9						
t_h	Hold time, A thru D before LE↓	2 V	0						ns
		4.5 V	0						
		6 V	0						

switching characteristics over recommended operating free-air temperature range (unless otherwise noted), C_L = 50 pF (see Note 1)

PARAMETER	FROM (INPUT)	TO (OUTPUT)	V_{CC}	T_A = 25°C MIN	T_A = 25°C TYP	T_A = 25°C MAX	SN54HC4514 SN54HC4515 MIN	SN54HC4514 SN54HC4515 MAX	SN74HC4514 SN74HC4515 MIN	SN74HC4514 SN74HC4515 MAX	UNIT
t_{pd}	A thru D	Any	2 V		72						ns
			4.5 V		24						
			6 V		21						
t_{pd}	LE	Any	2 V		78						ns
			4.5 V		26						
			6 V		22						
t_{pd}	$\overline{G}$	Any	2 V		60						ns
			4.5 V		20						
			6 V		17						
t_t		Any	2 V		38						ns
			4.5 V		8						
			6 V		6						

C_{pd}	Power dissipation capacitance	No load, T_A = 25°C	60 pF typ

NOTE 1: For load circuit and voltage waveforms, see page 1-14.

TYPES SN54HC7001, SN74HC7001 QUADRUPLE POSITIVE-AND GATES WITH SCHMITT-TRIGGER INPUTS

D2831, MARCH 1984

- Operation from Very Slow Transitions
- Temperature-Compensated Threshold Levels
- High Noise Immunity
- Same Pinouts as 'HC08
- Package Options Include Both Plastic and Ceramic Chip Carriers in Addition to Plastic and Ceramic DIPs
- Dependable Texas Instruments Quality and Reliability

SN54HC7001 . . . J PACKAGE
SN74HC7001 . . . J OR N PACKAGE
(TOP VIEW)

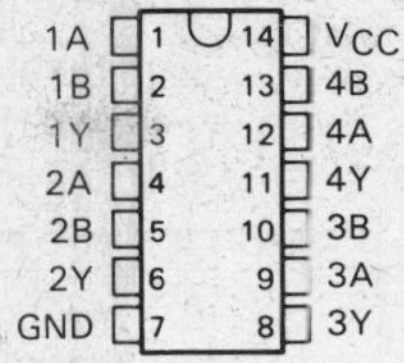

SN54HC7001 . . . FH OR FK PACKAGE
SN74HC7001 . . . FH OR FN PACKAGE
(TOP VIEW)

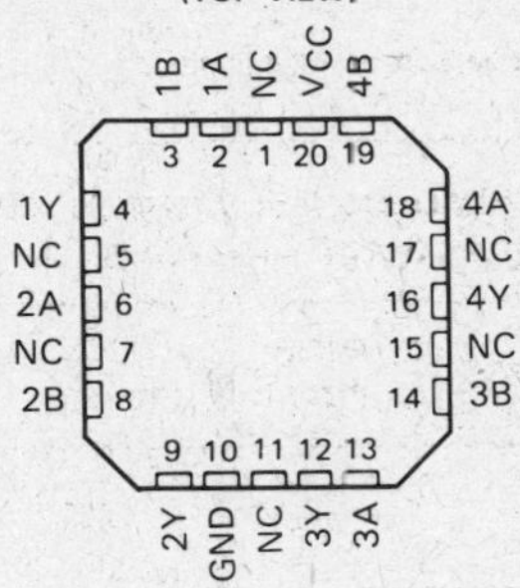

NC – No internal connection

description

Each circuit functions as a quadruple AND gate. They perform the Boolean function $Y = A \bullet B$ or $Y = \overline{\overline{A} + \overline{B}}$ in positive logic. However, because of the Schmitt action, the inputs have different input threshold levels for positive- and negative-going signals.

The circuits are temperature compensated and can be triggered from the slowest of input ramps and will still give clean jitter-free output signals.

The SN54HC7001 is characterized for operation over the full military temperature range of −55°C to 125°C. The SN74HC7001 is characterized for operation from −40°C to 85°C.

FUNCTION TABLE
(each gate)

INPUTS		OUTPUT
A	B	Y
H	H	H
L	X	L
X	L	L

logic symbol

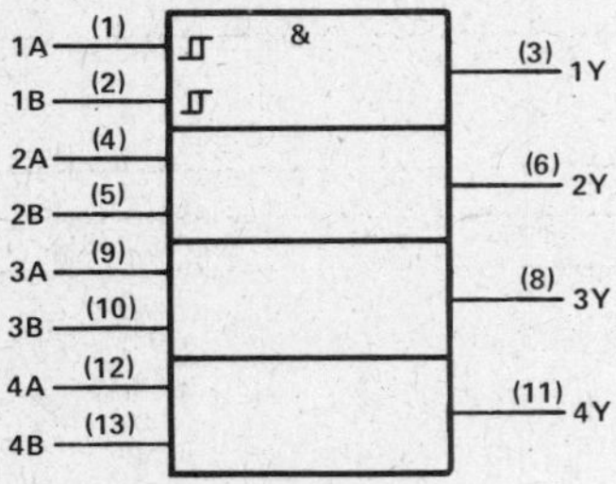

Pin numbers shown are for J and N packages.

logic diagram, each gate (positive logic)

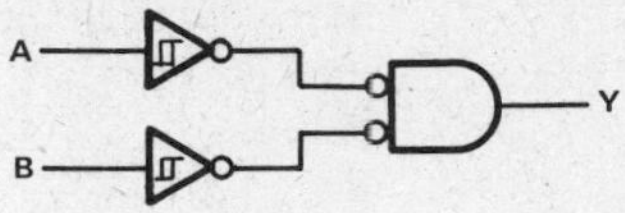

PRODUCT PREVIEWS

5

PRODUCT PREVIEW

This document contains information on a product under development. Texas Instruments reserves the right to change or discontinue this product without notice.

absolute maximum ratings, recommended operating conditions, and electrical characteristics

See Table I, page 2-4.

switching characteristics over recommended operating free-air temperature range (unless otherwise noted), C_L = 50 pF (see Note 1)

PARAMETER	FROM (INPUT)	TO (OUTPUT)	V_{CC}	T_A = 25°C MIN	TYP	MAX	SN54HC7001 MIN	MAX	SN74HC7001 MIN	MAX	UNIT
t_{pd}	A or B	Y	2 V		40						ns
			4.5 V		13						
			6 V		11						
t_t		Any	2 V		28						ns
			4.5 V		8						
			6 V		6						

NOTE 1: For load circuit and voltage waveforms, see page 1-14.

TYPES SN54HC7002, SN74HC7002
QUADRUPLE POSITIVE-NOR GATES WITH SCHMITT-TRIGGER INPUTS

D2831, MARCH 1984

- **Operation from Very Slow Transitions**
- **Temperature-Compensated Threshold Levels**
- **High Noise Immunity**
- **Same Pinouts as 'HC36**
- **Package Options Include Both Plastic and Ceramic Chip Carriers in Addition to Plastic and Ceramic DIPs**
- **Dependable Texas Instruments Quality and Reliability**

SN54HC7002 . . . J PACKAGE
SN74HC7002 . . . J OR N PACKAGE
(TOP VIEW)

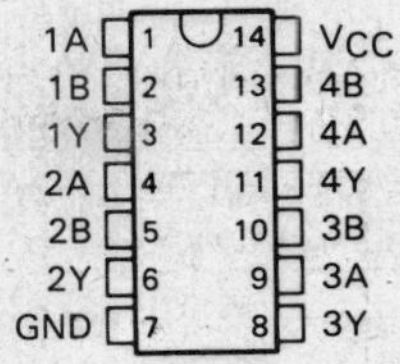

SN54HC7002 . . . FH OR FK PACKAGE
SN74HC7002 . . . FH OR FN PACKAGE
(TOP VIEW)

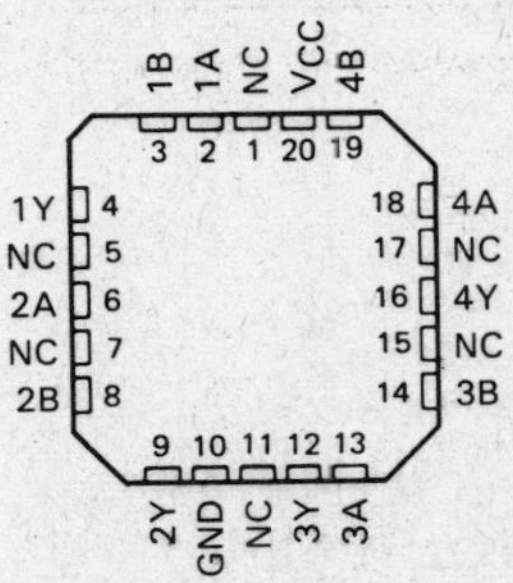

NC—No internal connection

description

Each circuit functions as a quadruple NOR gate. They perform the Boolean function $Y = \overline{A+B}$ or $Y = \overline{A} \cdot \overline{B}$ in positive logic. However, because of the Schmitt action, the inputs have different input threshold levels for positive- and negative-going signals.

The circuits are temperature compensated and can be triggered from the slowest of input ramps and will still give clean jitter-free output signals.

The SN54HC7002 is characterized for operation over the full military temperature range of −55°C to 125°C. The SN74HC7002 is characterized for operation from −40°C to 85°C.

FUNCTION TABLE

INPUTS		OUTPUT
A	B	Y
H	X	L
X	H	L
L	L	H

logic symbol

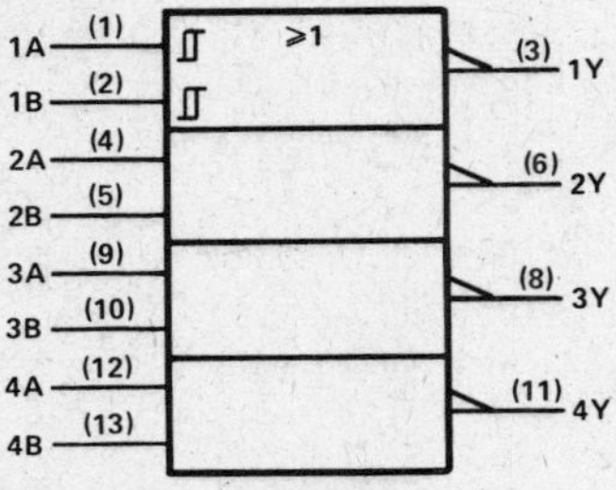

Pin numbers shown are for J and N packages.

logic diagram, each gate (positive logic)

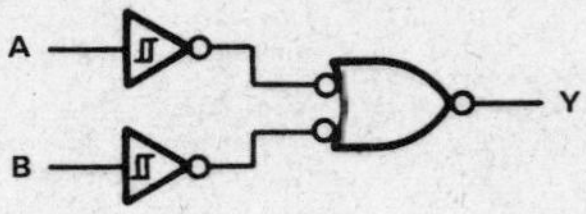

absolute maximum ratings, recommended operating conditions, and electrical characteristics

See Table I, page 2-4.

switching characteristics over recommended operating free-air temperature range (unless otherwise noted), C_L = 50 pF (see Note 1)

PARAMETER	FROM (INPUT)	TO (OUTPUT)	V_{CC}	T_A = 25°C			SN54HC7002		SN74HC7002		UNIT
				MIN	TYP	MAX	MIN	MAX	MIN	MAX	
t_{pd}	A or B	Y	2 V		40						ns
			4.5 V		13						
			6 V		11						
t_t		Any	2 V		28						ns
			4.5 V		8						
			6 V		6						

NOTE 1: For load circuit and voltage waveforms, see page 1-14.

TYPES SN54HC7003, SN74HC7003 QUADRUPLE POSITIVE-NAND GATES WITH SCHMITT-TRIGGER INPUTS AND OPEN-DRAIN OUTPUTS

D2831, MARCH 1984

- Operation from Very Slow Transitions
- Temperature-Compensated Threshold Levels
- High Noise Immunity
- Same Pinouts as 'HC03
- Package Options Include Both Plastic and Ceramic Chip Carriers in Addition to Plastic and Ceramic DIPs
- Dependable Texas Instruments Quality and Reliability

SN54HC7003 . . . J PACKAGE
SN74HC7003 . . . J OR N PACKAGE
(TOP VIEW)

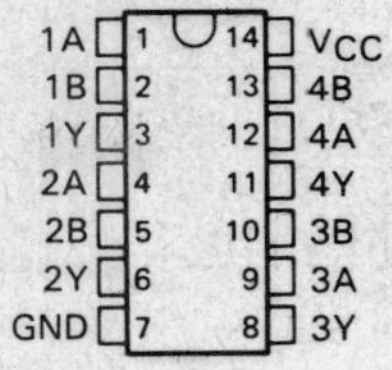

SN54HC7003 . . . FH OR FK PACKAGE
SN74HC7003 . . . FH OR FN PACKAGE
(TOP VIEW)

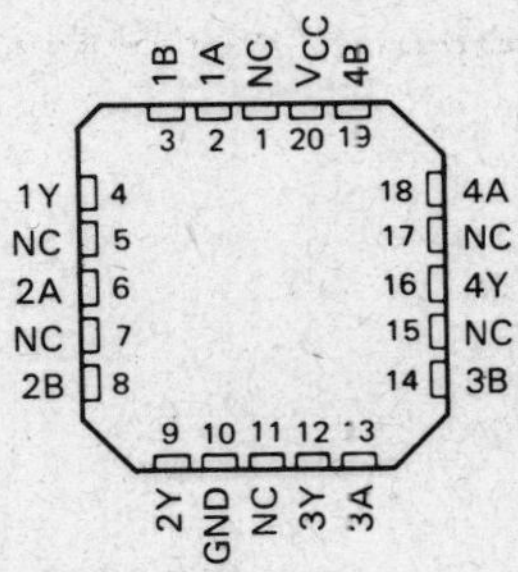

NC – No internal connection

description

Each circuit functions as a quadruple NAND gate. They perform the Boolean function $Y = \overline{A \cdot B}$ or $Y = \overline{A} + \overline{B}$ in positive logic. However, because of the Schmitt action, the inputs have different input threshold levels for positive- and negative-going signals.

The circuits are temperature compensated and can be triggered from the slowest of input ramps and will still give clean jitter-free output signals.

The SN54HC7003 is characterized for operation over the full military temperature range of −55 °C to 125 °C. The SN74HC7003 is characterized for operation from −40 °C to 85 °C.

FUNCTION TABLE (each gate)

INPUTS		OUTPUT
A	B	Y
H	H	L
L	X	H
X	L	H

logic symbol

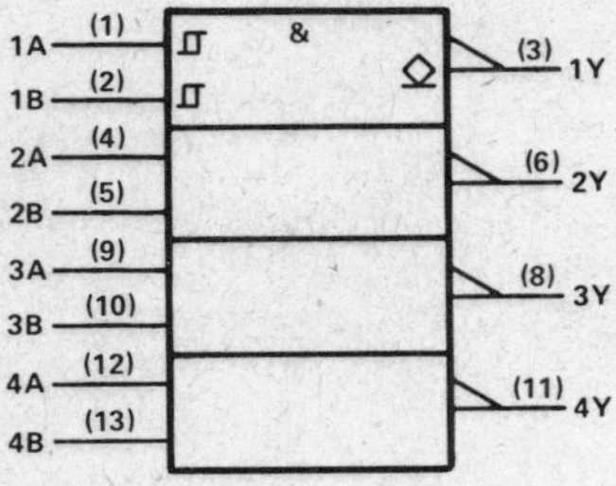

Pin numbers shown are for J and N packages.

logic diagram, each gate (positive logic)

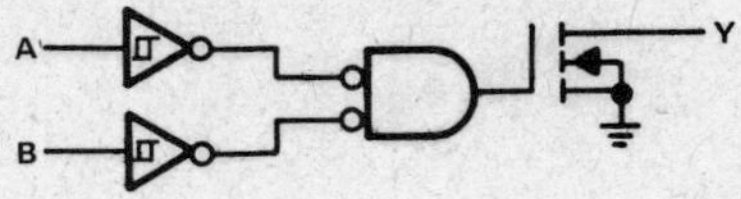

absolute maximum ratings, recommended operating conditions, and electrical characteristics

See Table I, page 2-4.

switching characteristics over recommended operating free-air temperature range (unless otherwise noted), C_L = 50 pF (see Note 1)

PARAMETER	FROM (INPUT)	TO (OUTPUT)	V_{CC}	T_A = 25°C MIN	TYP	MAX	SN54HC7003 MIN	MAX	SN74HC7003 MIN	MAX	UNIT
t_{pd}	A or B	Y	2 V		40						ns
			4.5 V		13						
			6 V		11						
t_t		Any	2 V		28						ns
			4.5 V		8						
			6 V		6						

NOTE 1: For load circuit and voltage waveforms, see page 1-14.

PRODUCT PREVIEWS

5

TYPES SN54HC7006, SN74HC7006 6-SECTION MULTIFUNCTION (NAND, INVERT, NOR) CIRCUITS

D2831, MARCH 1984

- Package Options Include Both Plastic and Ceramic Chip Carriers in Addition to Plastic and Ceramic DIPs
- Dependable Texas Instruments Quality and Reliability

SN54HC7006 . . . JT PACKAGE
SN74HC7006 . . . JT OR NT PACKAGE
(TOP VIEW)

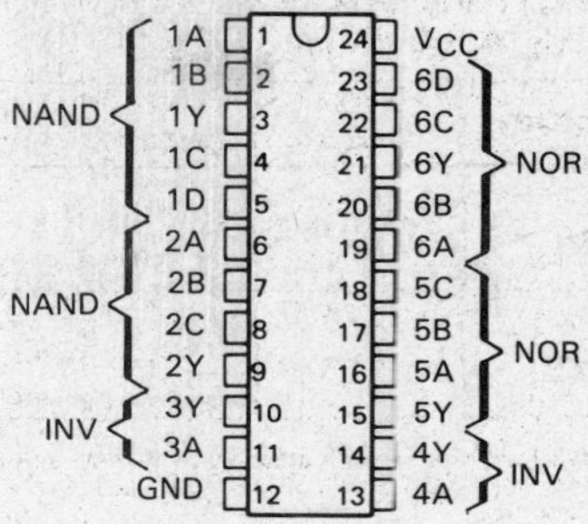

SN54HC7006 . . . FH OR FK PACKAGE
SN74HC7006 . . . FH OR FN PACKAGE
(TOP VIEW)

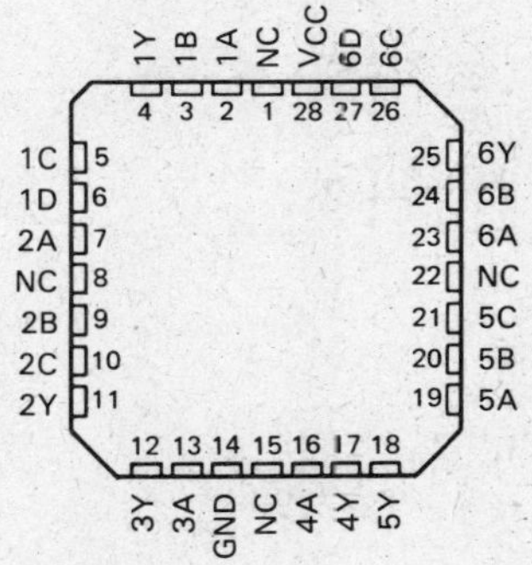

NC—No internal connection

description

The SN54HC7006 and SN74HC7006 are each comprised of the following sections:

One 3-input NAND gate
One 4-input NAND gate
One 3-input NOR gate
One 4-input NOR gate
Two inverters

They perform the Boolean functions shown under each function table.

The SN54HC7006 is characterized for operation over the full military temperature range of −55°C to 125°C. The SN74HC7006 is characterized for operation from −40°C to 85°C.

logic symbol

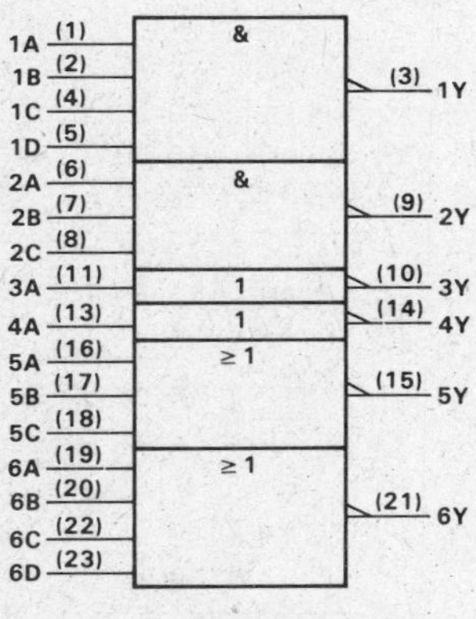

Pin numbers shown are for JT and NT packages.

logic diagrams (positive logic)

4-INPUT NAND GATE

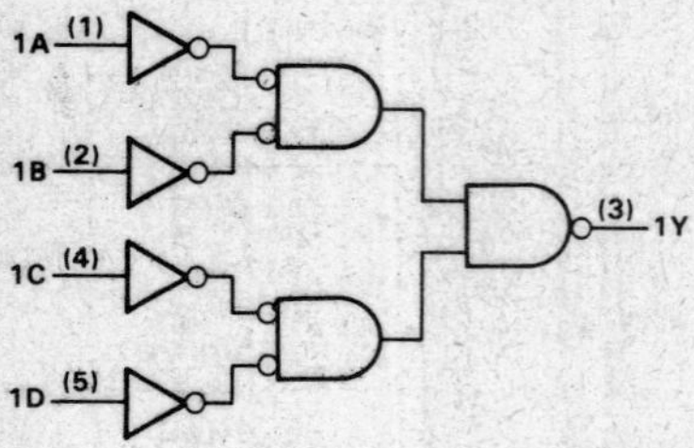

FUNCTION TABLE

INPUTS				OUTPUT
A	B	C	D	Y
H	H	H	H	L
L	X	X	X	H
X	L	X	X	H
X	X	L	X	H
X	X	X	L	H

positive logic: $Y = \overline{A \cdot B \cdot C \cdot D}$ or
$Y = \overline{A} + \overline{B} + \overline{C} + \overline{D}$

3-INPUT NAND GATE

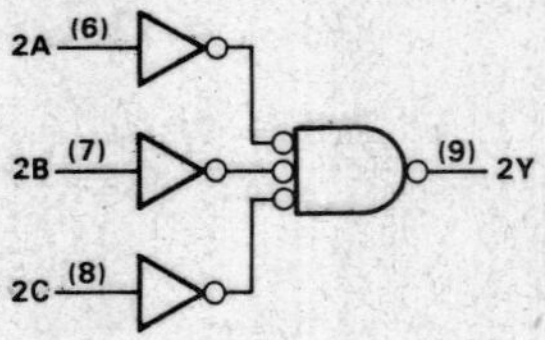

FUNCTION TABLE

INPUTS			OUTPUT
A	B	C	Y
H	H	H	L
L	X	X	H
X	L	X	H
X	X	L	H

positive logic: $Y = \overline{A \cdot B \cdot C}$ or
$Y = \overline{A} + \overline{B} + \overline{C}$

INVERTERS

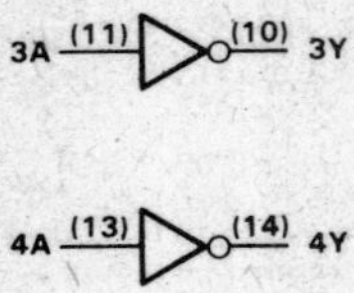

FUNCTION TABLE
(EACH INVERTER)

INPUT A	OUTPUT Y
H	L
L	H

positive logic: $Y = \overline{A}$

3-INPUT NOR GATE

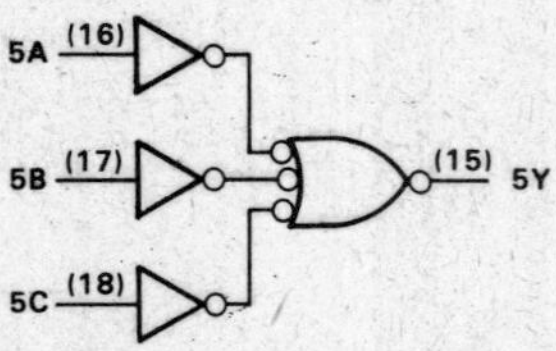

FUNCTION TABLE

INPUTS			OUTPUT
A	B	C	Y
H	X	X	L
X	H	X	L
X	X	H	L
L	L	L	H

positive logic: $Y = \overline{A + B + C}$ or
$Y = \overline{A} \cdot \overline{B} \cdot \overline{C}$

Pin numbers shown are for JT and NT packages.

logic diagram (positive logic)

4-INPUT NOR GATE

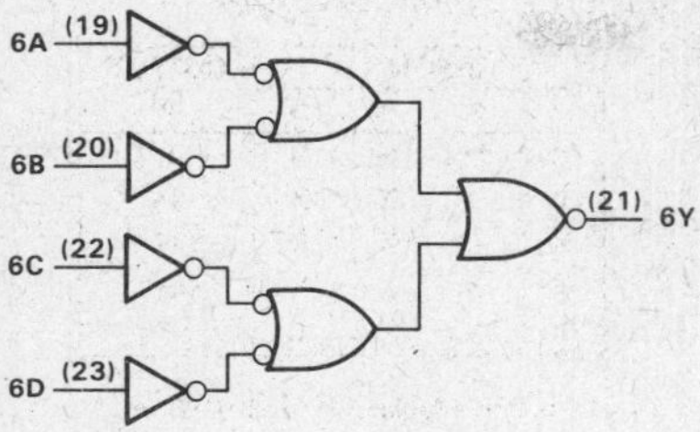

FUNCTION TABLE

INPUTS				OUTPUT
A	B	C	D	Y
H	X	X	X	L
X	H	X	X	L
X	X	H	X	L
X	X	X	H	L
L	L	L	L	H

positive logic: $Y = \overline{A+B+C+D}$ or
$Y = \overline{A}\cdot\overline{B}\cdot\overline{C}\cdot\overline{D}$

Pin numbers shown are for JT and NT packages.

absolute maximum ratings, recommended operating conditions, electrical characteristics

See Table I, page 2-4.

D2831, MARCH 1984

- Operation from Very Slow Transitions
- Temperature-Compensated Threshold Levels
- High Noise Immunity
- Same Pinouts as 'HC32
- Package Options Include Both Plastic and Ceramic Chip Carriers in Addition to Plastic and Ceramic DIPs
- Dependable Texas Instruments Quality and Reliability

SN54HC7032 . . . J PACKAGE
SN74HC7032 . . . J OR N PACKAGE
(TOP VIEW)

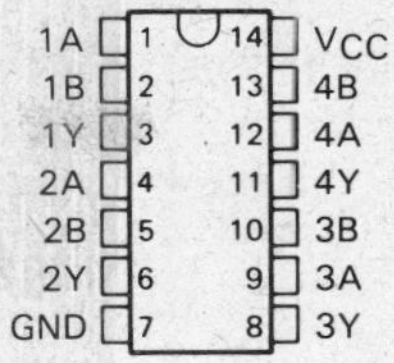

SN54HC7032 . . . FH OR FK PACKAGE
SN74HC7032 . . . FH OR FN PACKAGE
(TOP VIEW)

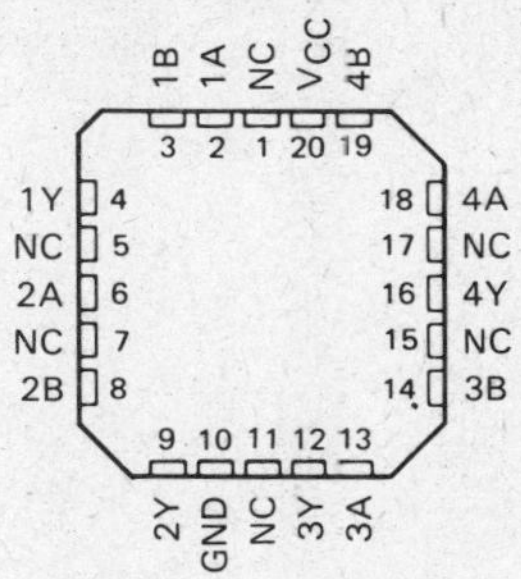

NC – No internal connection

description

Each circuit functions as a quadruple OR gate. They perform the Boolean function $Y = A + B$ or $Y = \overline{\overline{A} \cdot \overline{B}}$ in positive logic. However, because of the Schmitt action, the inputs have different input threshold levels for positive- and negative-going signals.

The circuits are temperature compensated and can be triggered from the slowest of input ramps and will still give clean jitter-free output signals.

The SN54HC7032 is characterized for operation over the full military temperature range of −55°C to 125°C. The SN74HC7032 is characterized for operation from −40°C to 85°C.

FUNCTION TABLE

INPUTS		OUTPUT
A	B	Y
H	X	H
X	H	H
L	L	L

logic symbol

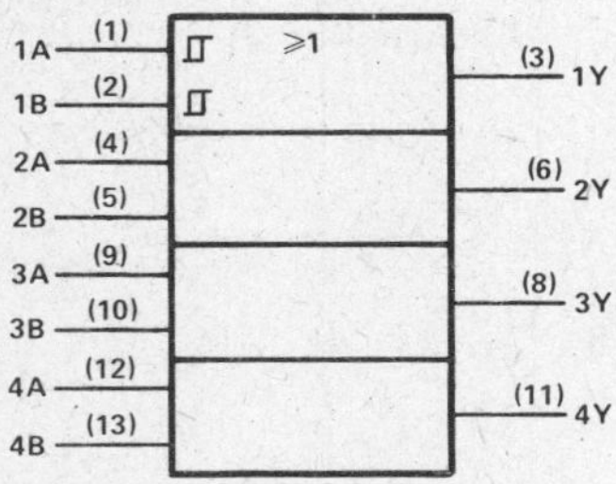

Pin numbers shown are for J and N packages.

logic diagram, each gate (positive logic)

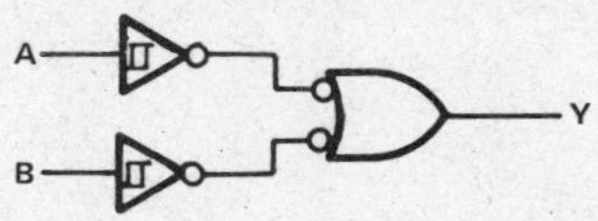

TEXAS INSTRUMENTS
POST OFFICE BOX 225012 • DALLAS, TEXAS 75265

absolute maximum ratings, recommended operating conditions, and electrical characteristics

See Table I, page 2-6.

switching characteristics over recommended free-air temperature range (unless otherwise noted), C_L = 50 pF (see Note 1)

PARAMETER	FROM (INPUT)	TO (OUTPUT)	V_{CC}	T_A = 25°C MIN	TYP	MAX	SN54HC7032 MIN	MAX	SN74HC7032 MIN	MAX	UNIT
t_{pd}	A or B	Y	2 V		40						ns
			4.5 V		13						
			6 V		11						
t_t		Any	2 V		28						ns
			4.5 V		8						
			6 V		6						

NOTE 1: For load circuit and voltage waveforms, see page 1-14.

TEXAS INSTRUMENTS
POST OFFICE BOX 225012 • DALLAS, TEXAS 75265

Explanation of Logic Symbols

F. A. Mann

Contents

If you have questions on this Explanation of Logic Symbols, please contact:

Texas Instruments Incorporated
F.A. Mann, MS 49
P.O. Box 225012
Dallas, Texas 75265

Telephone (214) 995-2867

IEEE Standards may be purchased from:

Institute of Electrical and Electronics Engineers, Inc.
IEEE Standards Office
345 East 47th Street
New York, N.Y. 10017

International Electrotechnical Comm ssion (IEC) publications may be purchased from:

American National Standards Institute, Inc.
1430 Broadway
New York, N.Y. 10018

List of Tables

List of Illustrations

1.0 INTRODUCTION

The International Electrotechnical Commission (IEC) has been developing a very powerful symbolic language that can show the relationship of each input of a digital logic circuit to each output without showing explicitly the internal logic. At the heart of the system is dependency notation, which will be explained in Section 4.

The system was introduced in the USA in a rudimentary form in IEEE/ANSI Standard Y32.14-1973. Lacking at that time a complete development of dependency notation, it offered little more than a substitution of rectangular shapes for the familiar distinctive shapes for representing the basic functions of AND, OR, negation, etc. This is no longer the case.

Internationally, Working Group 2 of IEC Technical Committee TC-3 has prepared a new document (Publication 617-12) that consolidates the original work started in the mid 1960's and published in 1972 (Publication 117-15) and the amendments and supplements that have followed. Similarly for the USA, IEEE Committee SCC 11.9 has revised the publication IEEE Std 91/ANSI Y32.14. Now numbered simply IEEE Std 91-1984, the IEEE standard contains all of the IEC work that has been approved, and also a small amount of material still under international consideration. Texas Instruments is participating in the work of both organizations and this document introduces new logic symbols in accordance with the new standards. When changes are made as the standards develop, future editions will take those changes into account.

The following explanation of the new symbolic language is necessarily brief and greatly condensed from what the standards publications will contain. This is not intended to be sufficient for those people who will be developing symbols for new devices. It is primarily intended to make possible the understanding of the symbols used in various data books and the comparison of the symbols with logic diagrams, functional block diagrams, and/or function tables will further help that understanding.

2.0 SYMBOL COMPOSITION

A symbol comprises an outline or a combination of outlines together with one or more qualifying symbols. The shape of the symbols is not significant. As shown in Figure 1, general qualifying symbols are used to tell exactly what logical operation is performed by the elements. Table I shows general qualifying symbols defined in the new standards. Input lines are placed on the left and output lines are placed on the right. When an exception is made to that convention, the direction of signal flow is indicated by an arrow as shown in Figure 11.

All outputs of a single, unsubdivided element always have identical internal logic states determined by the function of the element except when otherwise indicated by an associated qualifying symbol or label inside the element.

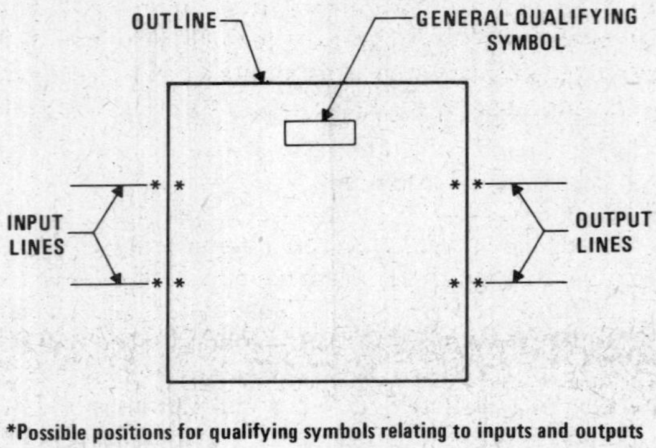

Figure 1. Symbol Composition

The outlines of elements may be abutted or embedded in which case the following conventions apply. There is no logic connection between the elements when the line common to their outlines is in the direction of signal flow. There is at least one logic connection between the elements when the line common to their outlines is perpendicular to the direction of signal flow. The number of logic connections between elements will be clarified by the use of qualifying symbols and this is discussed further under that topic. If no indications are shown on either side of the common line, it is assumed there is only one connection.

When a circuit has one or more inputs that are common to more than one element of the circuit, the common-control block may be used. This is the only distinctively shaped outline used in the IEC system. Figure 2 shows that unless otherwise qualified by dependency notation, an input to the common-control block is an input to each of the elements below the common-control block.

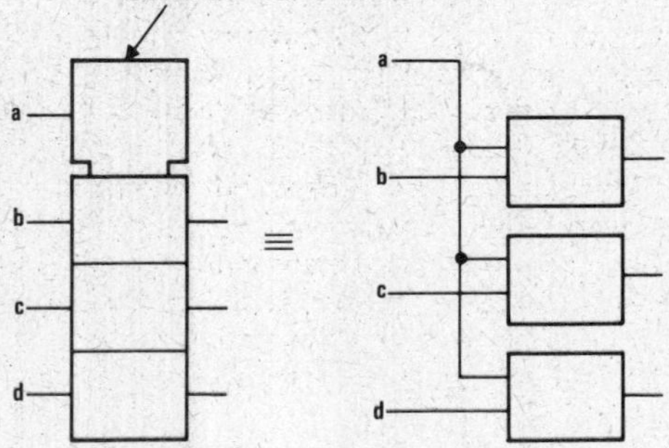

Figure 2. Common-Control Block

A common output depending on all elements of the array can be shown as the output of a common-output element. Its distinctive visual feature is the double line at its top. In addition the common-output element may have other inputs as shown in Figure 3. The function of the common-output element must be shown by use of a general qualifying symbol.

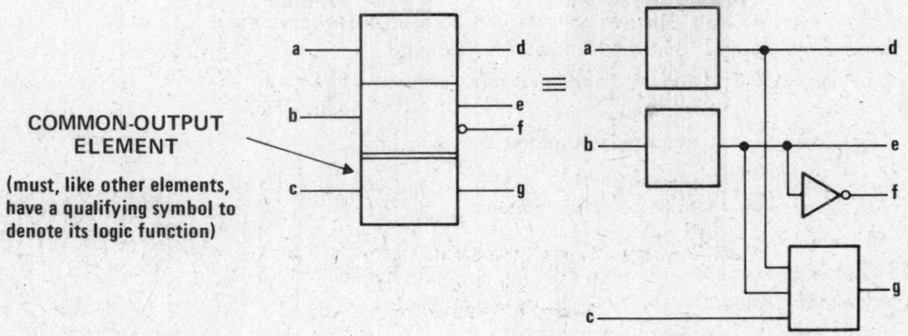

Figure 3. Common-Output Element

3.0 QUALIFYING SYMBOLS

3.1 General Qualifying Symbols

Table I shows general qualifying symbols defined by IEEE Standard 91. These characters are placed near the top center or the geometric center of a symbol or symbol element to define the basic function of the device represented by the symbol or of the element.

3.2 Qualifying Symbols for Inputs and Outputs

Qualifying symbols for inputs and outputs are shown in Table II and will be familiar to most users with the possible exception of the logic polarity and analog signal indicators. The older logic negation indicator means that the external 0 state produces the internal 1 state. The internal 1 state means the active state. Logic negation may be used in pure logic diagrams; in order to tie the external 1 and 0 logic states to the levels H (high) and L (low), a statement of whether positive logic (1 = H, 0 = L) or negative logic (1 = L, 0 = H) is being used is required or must be assumed. Logic polarity indicators eliminate the need for calling out the logic convention and are used in various data books in the symbology for actual devices. The presence of the triangular polarity indicator indicates that the L logic level will produce the internal 1 state (the active state) or that, in the case of an output, the internal 1 state will produce the external L level. Note how the active direction of transition for a dynamic input is indicated in positive logic, negative logic, and with polarity indication.

The internal connections between logic elements abutted together in a symbol may be indicated by the symbols shown in Table II. Each logic connection may be shown by the presence of qualifying symbols at one or both sides of the common line and if confusion can arise about the numbers of connections, use can be made of one of the internal connection symbols.

Table I. General Qualifying Symbols

SYMBOL	DESCRIPTION	CMOS EXAMPLE	TTL EXAMPLE
&	AND gate or function.	'HC00	SN7400
≥1	OR gate or function. The symbol was chosen to indicate that at least one active input is needed to activate the output.	'HC02	SN7402
=1	Exclusive OR. One and only one input must be active to activate the output.	'HC86	SN7486
=	Logic identity. All inputs must stand at the same state.	'HC86	SN74180
2k	An even number of inputs must be active.	'HC280	SN74180
2k+1	An odd number of inputs must be active.	'HC86	SN74ALS86
1	The one input must be active.	'HC04	SN7404
▷ or ◁	A buffer or element with more than usual output capability (symbol is oriented in the direction of signal flow).	'HC240	SN74S436
⎍	Schmitt trigger; element with hysteresis.	'HC132	SN74LS18
X/Y	Coder, code converter (DEC/BCD, BIN/OUT, BIN/7-SEG, etc.).	'HC42	SN74LS347
MUX	Multiplexer/data selector.	'HC151	SN74150
DMUX or DX	Demultiplexer.	'HC138	SN74138
Σ	Adder.	'HC283	SN74LS385
P−Q	Subtracter.	*	SN74LS385
CPG	Look-ahead carry generator	'HC182	SN74182
π	Multiplier.	*	SN74LS384
COMP	Magnitude comparator.	'HC85	SN74LS682
ALU	Arithmetic logic unit.	'HC181	SN74LS381
⎍	Retriggerable monostable.	'HC123	SN74LS422
1⎍	Nonretriggerable monostable (one-shot)	'HC221	SN74121
G ⎍⎍	Astable element. Showing waveform is optional.	*	SN74LS320
!G ⎍⎍	Synchronously starting astable.	*	SN74LS624
G! ⎍⎍	Astable element that stops with a completed pulse.	*	*
SRGm	Shift register. m = number of bits.	'HC164	SN74LS595
CTRm	Counter. m = number of bits; cycle length = 2^m.	'HC590	SN54LS590
CTR DIVm	Counter with cycle length = m.	'HC160	SN74LS668
RCTRm	Asynchronous (ripple-carry) counter; cycle length = 2^m.	'HC4020	*
ROM	Read-only memory.	*	SN74187
RAM	Random-access read/write memory.	'HC189	SN74170
FIFO	First-in, first-out memory.	*	SN74LS222
I=0	Element powers up cleared to 0 state.	*	SN74AS877
I=1	Element powers up set to 1 state.	'HC7022	SN74AS877
Φ	Highly complex function; "gray box" symbol with limited detail shown under special rules.	*	SN74LS608

*Not all of the general qualifying symbols have been used in TI's CMOS and TTL data books, but they are included here for the sake of completeness.

Table II. Qualifying Symbols for Inputs and Outputs

Logic negation at input. External 0 produces internal 1.

Logic negation at output. Internal 1 produces external 0.

Active-low input. Equivalent to —○ in positive logic.

Active-low output. Equivalent to ○— in positive logic.

Active-low input in the case of right-to-left signal flow.

Active-low output in the case of right-to-left signal flow.

Signal flow from right to left. If not otherwise indicated, signal flow is from left to right.

Bidirectional signal flow.

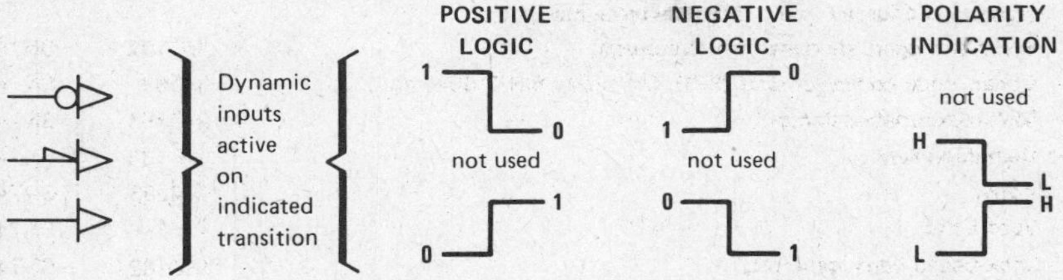

Nonlogic connection. A label inside the symbol will usually define the nature of this pin.

Input for analog signals (on a digital symbol) (see Figure 14).

Input for digital signals (on an analog symbol) (see Figure 14).

Internal connection. 1 state on left produces 1 state on right.

Negated internal connection. 1 state on left produces 0 state on right.

Dynamic internal connection. Transition from 0 to 1 on left produces transitory 1 state on right.

Internal input (virtual input). It always stands at its internal 1 state unless affected by an overriding dependency relationship.

Internal output (virtual output). Its effect on an internal input to which it is connected is indicated by dependency notation.

The internal (virtual) input is an input originating somewhere else in the circuit and is not connected directly to a terminal. The internal (virtual) output is likewise not connected directly to a terminal. The application of internal inputs and outputs requires an understanding of dependency notation, which is explained in Section 4.

Table III. Symbols Inside the Outline

Symbol	Meaning
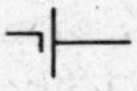	Postponed output (of a pulse-triggered flip-flop). The output changes when input initiating change (e.g., a C input) returns to its initial external state or level. See § 5.
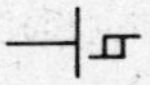	Bi-threshold input (input with hysteresis).
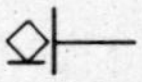	N-P-N open-collector or similar output that can supply a relatively low-impedance L level when not turned off. Requires external pull-up. Capable of positive-logic wired-AND connection. 
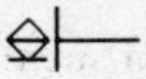	Passive-pull-up output is similar to N-P-N open-collector output but is supplemented with a built-in passive pull-up.
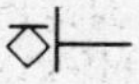	N-P-N open-emitter or similar output that can supply a relatively low-impedance H level when not turned off. Requires external pull-down. Capable of positive-logic wired-OR connection. 
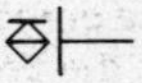	Passive-pull-down output is similar to N-P-N open-emitter output but is supplemented with a built-in passive pull-down.
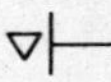	3-state output.
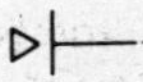	Output with more than usual output capability (symbol is oriented in the direction of signal flow).
EN	Enable input When at its internal 1-state, all outputs are enabled. When at its internal 0-state, open-collector and open-emitter outputs are off, three-state outputs are in the high-impedance state, and all other outputs (e.g., totem-poles) are at the internal 0-state.
J, K, R, S, T	Usual meanings associated with flip-flops (e.g., R = reset, T = toggle)
D	Data input to a storage element equivalent to: 
→m ←m	Shift right (left) inputs, m = 1, 2, 3, etc. If m = 1, it is usually not shown.
+m −m	Counting up (down) inputs, m = 1, 2, 3, etc. If m = 1, it is usually not shown.
	Binary grouping. m is highest power of 2.
CT = 15	The contents-setting input, when active, causes the content of a register to take on the indicated value.
CT = 9	The content output is active if the content of the register is as indicated.
	Input line grouping . . . indicates two or more terminals used to implement a single logic input. e.g., The paired expander inputs of SN7450. 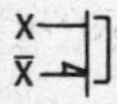
"1"	Fixed-state output always stands at its internal 1 state. For example, see SN74185.

In an array of elements, if the same general qualifying symbol and the same qualifying symbols associated with inputs and outputs would appear inside each of the elements of the array, these qualifying symbols are usually shown only in the first element. This is done to reduce clutter and to save time in recognition. Similarly, large identical elements that are subdivided into smaller elements may each be represented by an unsubdivided outline. The SN54HC242 or SN54LS440 symbol illustrates this principle.

3.3 Symbols Inside the Outline

Table III shows some symbols used inside the outline. Note particularly that open-collector (open-drain), open-emitter (open-source), and three-state outputs have cistinctive symbols. An EN input affects all the external outputs of the element in which it is placed, plus the external outputs of any elements shown to be influenced by that element. It has no effect on inputs. When an enable input affects only certain outputs, affects outputs located outside the indicated influence of the element in which the enable input is placed, and/or affects one or more inputs, a form of dependency notation will indicate this (see 4.10). The effects of the EN input on the various types of outputs are shown.

It is particularly important to note that a D input is always the data input of a storage element. At its internal 1 state, the D input sets the storage element to its 1 state, and at its internal 0 state it resets the storage element to its 0 state.

The binary grouping symbol will be explained more fully in Section 8. Binary-weighted inputs are arranged in order and the binary weights of the least-significant and the most-significant lines are indicated by numbers. In this document weights of input and output lines will be represented by powers of two usually only when the binary grouping symbol is used, otherwise decimal numbers will be used. The grouped inputs generate an internal number on which a mathematical function can be performed or that can be an identifying number for dependency notation (Figure 28). A frequent use is in addresses for memories.

Reversed in direction, the binary grouping symbol can be used with outputs. The concept is analogous to that for the inputs and the weighted outputs will indicate the internal number assumed to be developed within the circuit.

Other symbols are used inside the outlines in accordance with the IEC/IEEE standards but are not shown here. Generally these are associated with arithmetic operations and are self-explanatory.

When nonstandardized information is shown inside an outline, it is usually enclosed in square brackets [like these].

4.0 DEPENDENCY NOTATION

4.1 General Explanation

Dependency notation is the powerful tool that sets the IEC symbols apart from previous systems and makes compact, meaningful, symbols possible. It provides the means of denoting the relationship between inputs, outputs, or inputs and outputs without actually showing all the elements and interconnections involved. The information provided by dependency notation supplements that provided by the qualifying symbols for an element's function.

In the convention for the dependency notation, use will be made of the terms "affecting" and "affected." In cases where it is not evident which inputs must be considered as being the affecting or the affected ones (e.g., if they stand in an AND relationship), the choice may be made in any convenient way.

So far, eleven types of dependency have been defined and all of these are used in various TI data books. X dependency is used mainly with CMOS circuits. They are listed below in the order in which they are presented and are summarized in Table IV following 4.12.

Section	Dependency Type or Other Subject
4.2	G, AND
4.3	General Rules for Dependency Notation
4.4	V, OR
4.5	N, Negate (Exclusive-OR)
4.6	Z, Interconnection
4.7	X, Transmission
4.8	C, Control
4.9	S, Set and R, Reset
4.10	EN, Enable
4.11	M, Mode
4.12	A, Address

4.2 G (AND) Dependency

A common relationship between two signals is to have them ANDed together. This has traditionally been shown by explicitly drawing an AND gate with the signals connected to the inputs of the gate. The 1972 IEC publication and the 1973 IEEE/ANSI standard showed several ways to show this AND relationship using dependency notation. While ten other forms of dependency have since been defined, the ways to invoke AND dependency are now reduced to one.

In Figure 4 input **b** is ANDed with input **a** and the complement of **b** is ANDed with **c**. The letter G has been chosen to indicate AND relationships and is placed at input **b**, inside the symbol. A number considered appropriate by the symbol designer (1 has been used here) is placed after the letter G and also at each affected input. Note the bar over the 1 at input **c**.

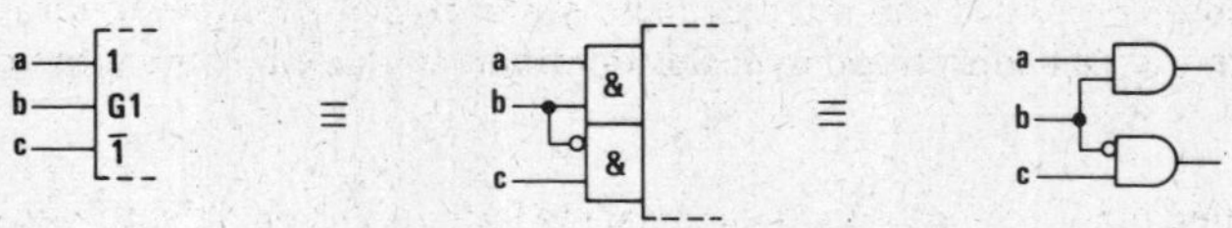

Figure 4. G Dependency Between Inputs

In Figure 5, output **b** affects input **a** with an AND relationship. The lower example shows that it is the internal logic state of **b**, unaffected by the negation sign, that is ANDed. Figure 6 shows input **a** to be ANDed with a dynamic input **b**.

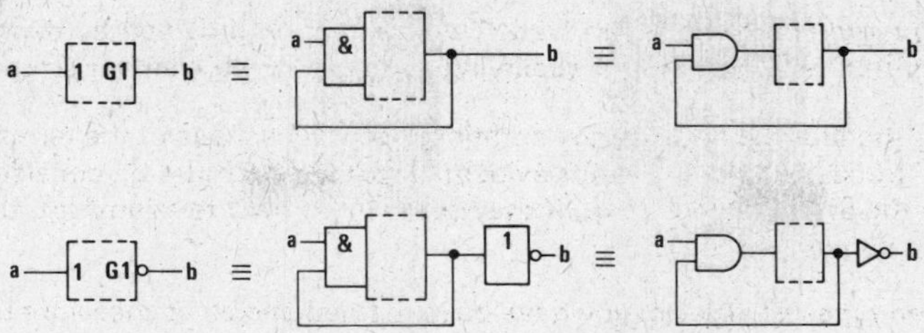

Figure 5. G Dependency Between Outputs and INputs

Figure 6. G Dependency with a Dynamic Input

The rules for G dependency can be summarized thus:

> When a G*m* input or output (*m* is a number) stands at its internal 1 state, all inputs and outputs affected by G*m* stand at their normally defined internal logic states. When the G*m* input or output stands at its 0 state, all inputs and outputs affected by G*m* stand at their internal 0 states.

4.3 Conventions for the Application of Dependency Notation in General

The rules for applying dependency relationships in general follow the same pattern as was illustrated for G dependency.

Application of dependency notation is accomplished by:

1) labeling the input (or output) *affecting* other inputs or outputs with the letter symbol indicating the relationship involved (e.g., G for AND) followed by an identifying number, appropriately chosen, and
2) labeling each input or output *affected* by that affecting input (or output) with that same number.

If it is the complement of the internal logic state of the affecting input or output that does the affecting, then a bar is placed over the identifying numbers at the affected inputs or outputs (Figure 4).

If two affecting inputs or outputs have the same letter and same identifying number, they stand in an OR relationship to each other (Figure 7).

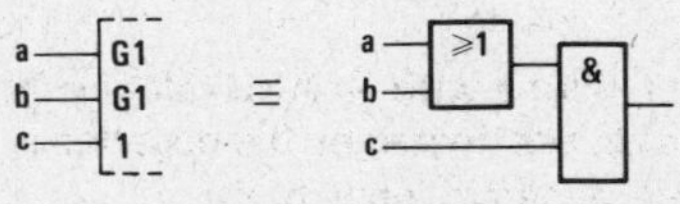

Figure 7. ORed Affecting Inputs

If the affected input or output requires a label to denote its function (e.g., "D"), this label will be *prefixed* by the identifying number of the affecting input (Figure 15).

If an input or output is affected by more than one affecting input, the identifying numbers of each of the affecting inputs will appear in the label of the affected one, separated by commas. The normal reading order of these numbers is the same as the sequence of the affecting relationships (Figure 15).

If the labels denoting the functions of affected inputs or outputs must be numbers (e.g., outputs of a coder), the identifying numbers to be associated with both affecting inputs and affected inputs or outputs will be replaced by another character selected to avoid ambiguity, e.g., Greek letters (Figure 8).

Figure 8. Substitution for Numbers

4.4 V (OR) Dependency

The symbol denoting OR dependency is the letter V (Figure 9).

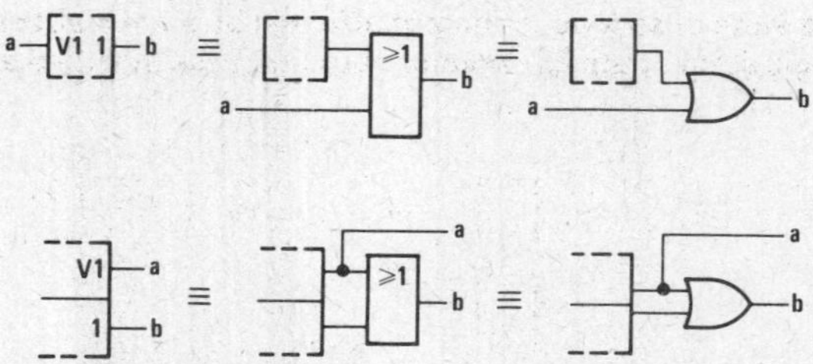

Figure 9. V (OR) Dependency

When a V*m* input or output stands at its internal 1 state, all inputs and outputs affected by V*m* stand at their internal 1 states. When the V*m* input or output stands at its internal 0 state, all inputs and outputs affected by V*m* stand at their normally defined internal logic states.

4.5 N (Negate) (Exclusive-OR) Dependency

The symbol denoting negate dependency is the letter N (Figure 10). Each input or output affected by an N*m* input or output stands in an Exclusive-OR relationship with the N*m* input or output.

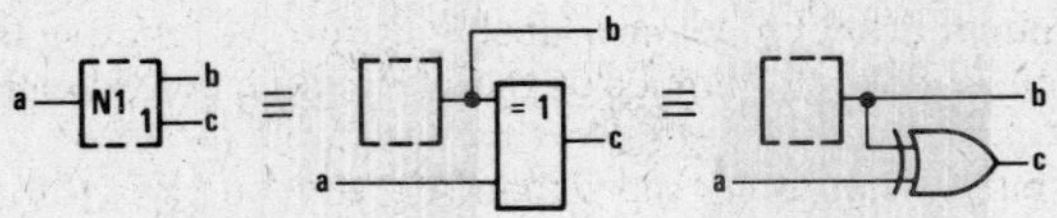

If a = 0, then c = b
If a = 1, then c = $\bar{b}$

Figure 10. N (Negate) (Exclusive-OR) Dependency

When an N*m* input or output stands at its internal 1 state, the internal logic state of each input and each output affected by N*m* is the complement of what it would otherwise be. When an N*m* input or output stands at its internal 0 state, all inputs and outputs affected by N*m* stand at their normally defined internal logic states.

4.6 Z (Interconnection) Dependency

The symbol denoting interconnection dependency is the letter Z.

Interconnection dependency is used to indicate the existence of internal logic connections between inputs, outputs, internal inputs, and/or internal outputs.

The internal logic state of an input or output affected by a Z*m* input or output will be the same as the internal logic state of the Z*m* input or output, unless modified by additional dependency notation (Figure 11).

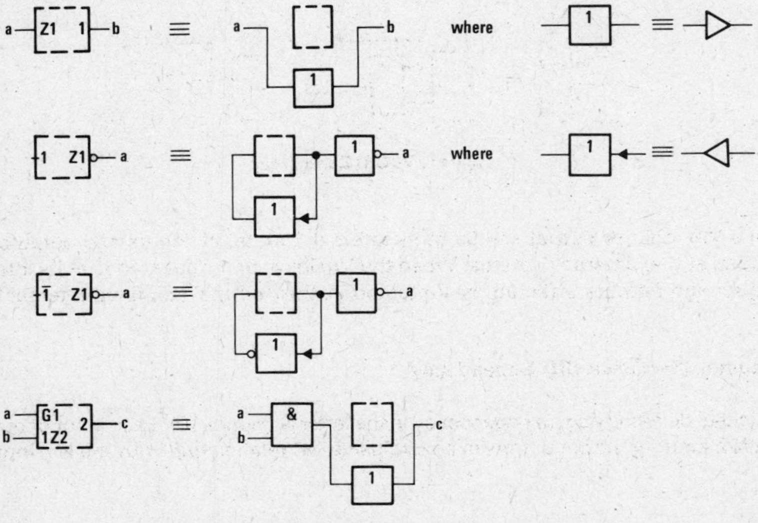

Figure 11. Z (Interconnection) Dependency

4.7 X (Transmission) Dependency

The symbol denoting transmission dependency is the letter X.

Transmission dependency is used to indicate controlled bidirectional connections between affected input/output ports (Figure 12).

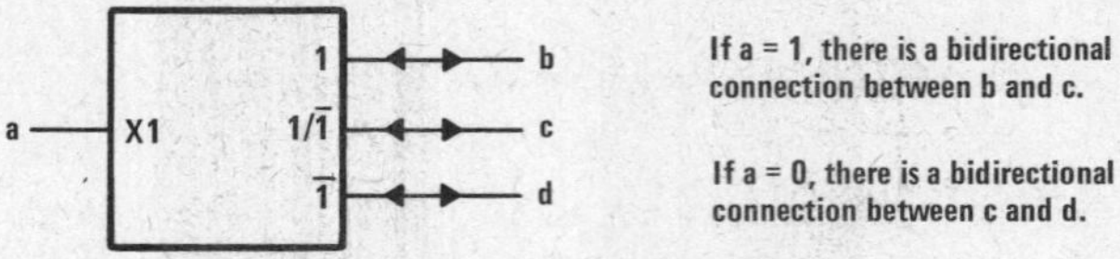

Figure 12. X (Transmission) Dependency

When an X*m* input or output stands at its internal 1 state, all input-output ports affected by this X*m* input or output are bidirectionally connected together and stand at the same internal logic state or analog signal level. When an X*m* input or output stands at its internal 0 state, the connection associated with this set of dependency notation does not exist.

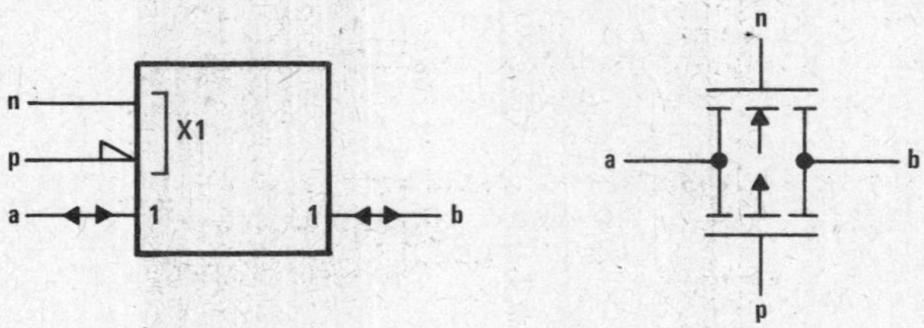

Figure 13. CMOS Transmission Gate Symbol and Schematic

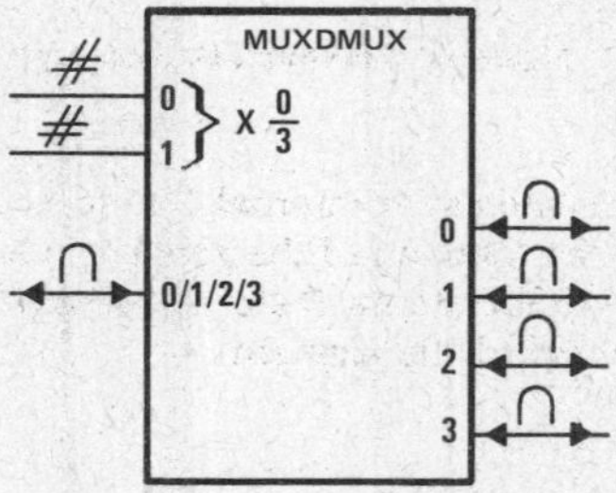

Figure 14. Analog Data Selector (Multiplexer/Demultiplexer)

Although the transmission paths represented by X dependency are inherently bidirectinal, use is not always made of this property. This is analogous to a piece of wire, which may be constrained to carry current in only one direction. If this is the case in a particular application, then the directional arrows shown in Figures 12, 13, and 14 would be omitted.

4.8 C (Control) Dependency

The symbol denoting control dependency is the letter C.

Control inputs are usually used to enable or disable the data (D, J, K, R, or S) inputs of storage elements. They may take on their internal 1 states (be active) either statically or dynamically. In the latter case the dynamic input symbol is used as shown in the third example of Figure 15.

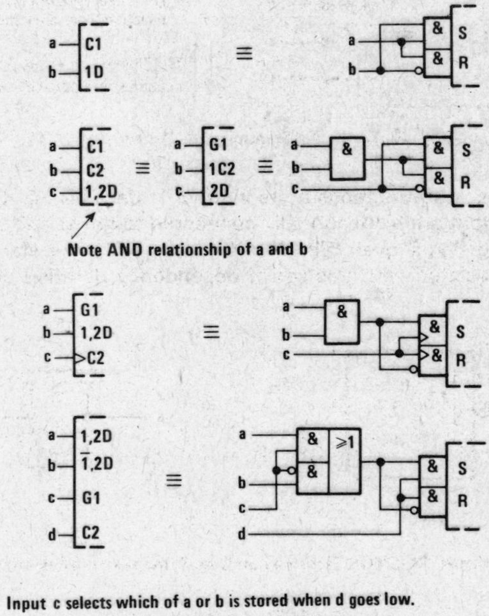

Figure 15. C (Control) Dependency

When a C*m* input or output stands at its internal 1 state, the inputs affected by C*m* have their normally defined effect on the function of the element, i.e., these inputs are enabled. When a C*m* input or output stands at its internal 0 state, the inputs affected by C*m* are disabled and have no effect on the function of the element.

4.9 S (Set) and R (Reset) Dependencies

The symbol denoting set dependency is the letter S. The symbol denoting reset dependency is the letter R.

Set and reset dependencies are used if it is necessary to specify the effect of the combination R = S = 1 on a bistable element. Case 1 in Figure 16 does not use S or R dependency.

When an S*m* input is at its internal 1 state, outputs affected by the S*m* input will react, regardless of the state of an R input, as they normally would react to the combination S = 1, R = 0. See cases 2, 4, and 5 in Figure 16.

When an R*m* input is at its internal 1 state, outputs affected by the R*m* input will react, regardless of the state of an S input, as they normally would react to the combination S = 0, R = 1. See cases 3, 4, and 5 in Figure 16.

When an S*m* or R*m* input is at its internal 0 state, it has no effect.

Note that the noncomplementary output patterns in cases 4 and 5 are only pseudo stable. The simultaneous return of the inputs to S = R = 0 produces an unforeseeable stable and complementary output pattern.

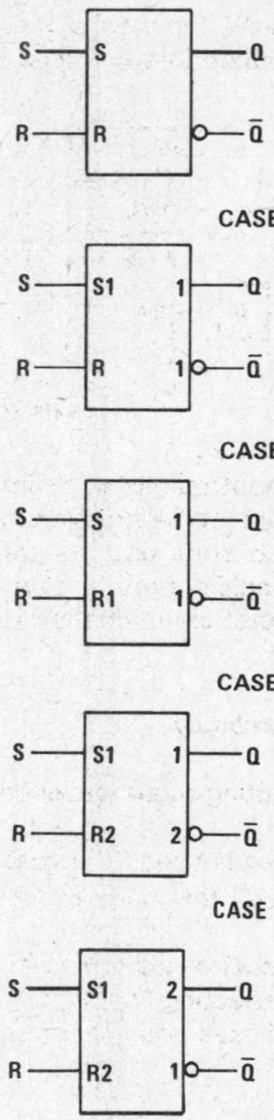

CASE 1

S	R	Q	$\bar{Q}$
0	0	nc	nc
0	1	0	1
1	0	1	0
1	1	?	?

CASE 2

S	R	Q	$\bar{Q}$
0	0	nc	nc
0	1	0	1
1	0	1	0
1	1	1	0

CASE 3

S	R	Q	$\bar{Q}$
0	0	nc	nc
0	1	0	1
1	0	1	0
1	1	0	1

CASE 4

S	R	Q	$\bar{Q}$
0	0	nc	nc
0	1	0	1
1	0	1	0
1	1	1	1

CASE 5

S	R	Q	$\bar{Q}$
0	0	nc	nc
0	1	0	1
1	0	1	0
1	1	0	0

0 = external 0 state   1 = external 1 state
nc = no change   ? = unspecified

Figure 16. S (Set) and R (Reset) Dependencies

4.10 EN (Enable) Dependency

The symbol denoting enable dependency is the combination of letters EN.

An EN*m* input has the same effect on outputs as an EN input, see 3.3, but it affects only those outputs labeled with the identifying number *m*. It also affects those inputs labeled with the identifying number *m*. By contrast, an EN input affects all outputs and no inputs. The effect of an EN*m* input on an affected input is identical to that of a C*m* input (Figure 17).

When an EN*m* input stands at its internal 1 state, the inputs affected by EN*m* have their normally defined effect on the function of the element and the outputs affected by this input stand at their normally defined internal logic states, i.e., these inputs and outputs are enabled.

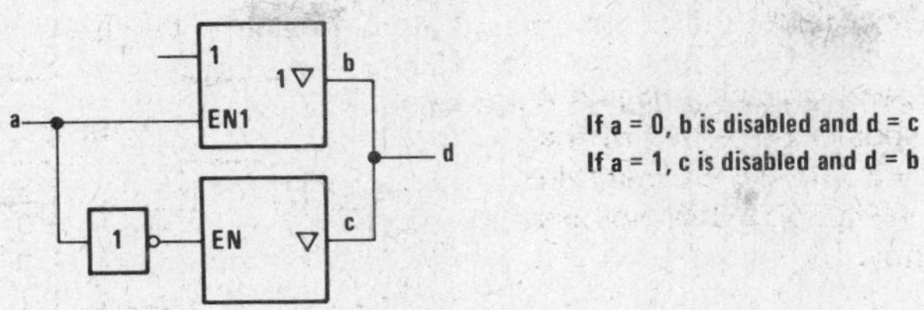

Figure 17. EN (Enable) Dependency

When an EN*m* input stands at its internal 0 state, the inputs affected by EN*m* are disabled and have no effect on the function of the element, and the outputs affected by EN*m* are also disabled. Open-collector outputs are turned off, three-state outputs stand at their normally defined internal logic states but externally exhibit high impedance, and all other outputs (e.g., totem-pole outputs) stand at their internal 0 states.

4.11 M (MODE) Dependency

The symbol denoting mode dependency is the letter M.

Mode dependency is used to indicate that the effects of particular inputs and outputs of an element depend on the mode in which the element is operating.

If an input or output has the same effect in different modes of operation, the identifying numbers of the relevant affecting M*m* inputs will appear in the label of that affected input or output between parentheses and separated by solidi (Figure 22).

4.11.1 M Dependency Affecting Inputs

M dependency affects inputs the same as C dependency. When an M*m* input or M*m* output stands at its internal 1 state, the inputs affected by this M*m* input or M*m* output have their normally defined effect on the function of the element, i.e., the inputs are enabled.

When an M*m* input or M*m* output stands at its internal 0 state, the inputs affected by this M*m* input or M*m* output have no effect on the function of the element. When an affected input has several sets of labels separated by solidi (e.g., C4/2→/3+), any set in which the identifying number of the M*m* input or M*m* output appears has no effect and is to be ignored. This represents disabling of some of the functions of a multifunction input.

The circuit in Figure 18 has two inputs, **b** and **c**, that control which one of four modes (0, 1, 2, or 3) will exist at any time. Inputs **d, e,** and **f** are D inputs subject to dynamic control (clocking) by the **a** input. The numbers 1 and 2 are in the series chosen to indicate the modes so inputs **e** and **f** are only enabled in mode 1 (for parallel loading) and input **d** is only enabled in mode 2 (for serial loading). Note that input **a** has three functions. It is the clock for entering data. In mode 2, it causes right shifting of data, which means a shift away from the control block. In mode 3, it causes the contents of the register to be incremented by one count.

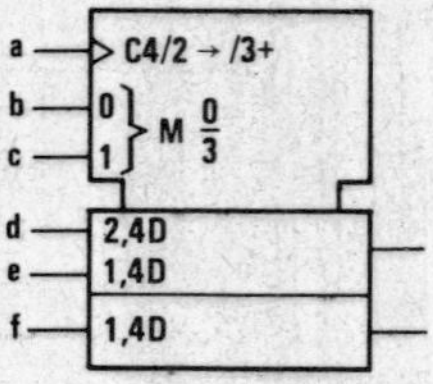

Note that all operations are synchronous.

In MODE 0 (b = 0, c = 0), the outputs remain at their existing states as none of the inputs has an effect.

In MODE 1 (b = 1, c = 0), parallel loading takes place thru inputs e and f.

In MODE 2 (b = 0, c = 1), shifting down and serial loading thru input d take place.

In MODE 3 (b = c = 1), counting up by increment of 1 per clock pulse takes place.

Figure 18. M (Mode) Dependency Affecting Inputs

4.11.2 M Dependency Affecting Outputs

When an M*m* input or M*m* output stands at its internal 1 state, the affected outputs stand at their normally defined internal logic states, i.e., the outputs are enabled.

When an M*m* input or M*m* output stands at its internal 0 state, at each affected output any set of labels containing the identifying number of that M*m* input or M*m* output has no effect and is to be ignored. When an output has several different sets of labels separated by solidi (e.g., 2,4/3,5), only those sets in which the identifying number of this M*m* input or M*m* output appears are to be ignored.

Figure 19 shows a symbol for a device whose output can behave like either a 3-state output or an open-collector output depending on the signal applied to input **a**. Mode 1 exists when input **a** stands at its internal 1 state and, in that case, the three-state symbol applies and the open-element symbol has no effect. When **a** = 0, mode 1 does not exist so the three-state symbol has no effect and the open-element symbol applies.

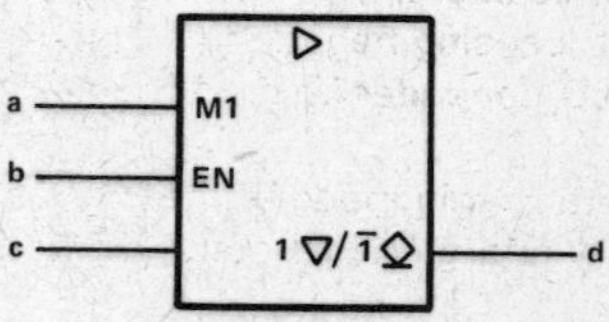

Figure 19. Type of Output Determined by Mode

In Figure 20, if input **a** stands at its internal 1 state establishing mode 1, output **b** will stand at its internal 1 state only when the content of the register equals 9. Since output **b** is located in the common-control block with no defined function outside of mode 1, the state of this output outside of mode 1 is not defined by the symbol.

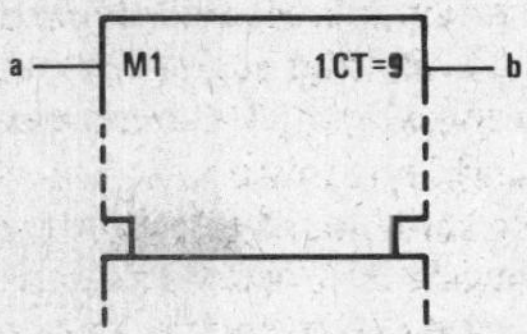

Figure 20. An Output of the Common-Control Block

In Figure 21, if input **a** stands at its internal 1 state establishing mode 1, output **b** will stand at its internal 1 state only when the content of the register equals 15. If input **a** stands at its internal 0 state, output **b** will stand at its internal 1 state only when the content of the register equals 0.

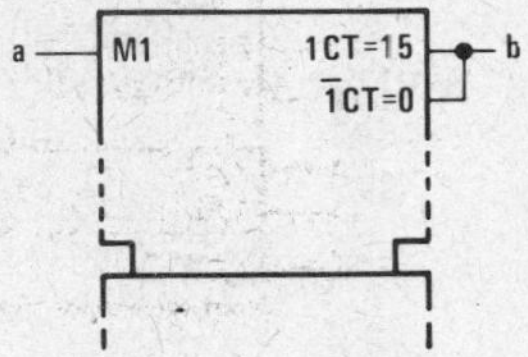

Figure 21. Determining an Output's Function

In Figure 22 inputs **a** and **b** are binary weighted to generate the numbers 0, 1, 2, or 3. This determines which one of the four modes exists.

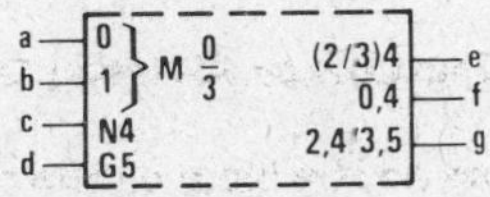

Figure 22. Dependent Relationships Affected by Mode

At output **e** the label set causing negation (if **c** = 1) is effective only in modes 2 and 3. In modes 0 and 1 this output stands at its normally defined state as if it had no labels. At output **f** the label set has effect when the mode is not 0 so output **e** is negated (if **c** = 1) in modes 1, 2, and 3. In mode 0 the label set has no effect so the output stands at its normally defined state. In this example $\overline{0}$,4 is equivalent to (1/2/3)4. At output **g** there are two label sets. The first set, causing negation (if **c** = 1), is effective only in mode 2. The second set, subjecting **g** to AND dependency on **d**, has effect only in mode 3.

Note that in mode 0 none of the dependency relationships has any effect on the outputs, so **e, f,** and **g** will all stand at the same state.

4.12 A (Address) Dependency

The symbol denoting address dependency is the letter A.

Address dependency provides a clear representation of those elements, particularly memories, that use address control inputs to select specified sections of a multildimensional arrays. Such a section of a memory array is usually called a word. The purpose of address dependency is to allow a symbolic presentation of the entire array. An input of the array shown at a particular element of this general section is common to the corresponding elements of all selected sections of the array. An output of the array shown at a particular element of this general section is the result of the OR function of the outputs of the corresponding elements of selected sections.

Inputs that are not affected by any affecting address input have their normally defined effect on all sections of the array, whereas inputs affected by an address input have their normally defined effect only on the section selected by that address input.

An affecting address input is labeled with the letter A followed by an identifying number that corresponds with the address of the particular section of the array selected by this input. Within the general section presented by the symbol, inputs and outputs affected by an A*m* input are labeled with the letter A, which stands for the identifying numbers, i.e., the addresses, of the particular sections.

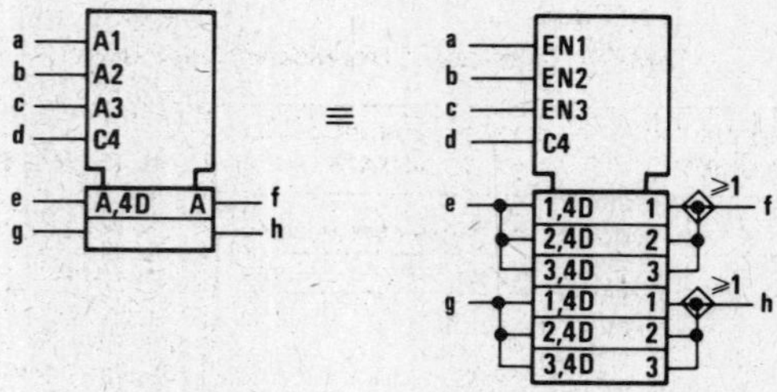

Figure 23. A (Address) Dependency

Figure 23 shows a 3-word by 2-bit memory having a separate address line for each word and uses EN dependency to explain the operation. To select word 1, input **a** is taken to its 1 state, which establishes mode 1. Data can now be clocked into the inputs marked "1,4D." Unless words 2 and 3 are also selected, data cannot be clocked in at the inputs marked "2,4D" and "3,4D." The outputs will be the OR functions of the selected outputs, i.e., only those enabled by the active EN functions.

The identifying numbers of affecting address inputs correspond with the addresses of the sections selected by these inputs. They need not necessarily differ from those of other affecting dependency-inputs (e.g., G, V, N, . . .), because in the general section presented by the symbol they are replaced by the letter A.

If there are several sets of affecting A*m* inputs for the purpose of independent and possibly simultaneous access to sections of the array, then the letter A is modified to 1A, 2A, ... Because they have access to the same sections of the array, these sets of A inputs may have the same identifying numbers. The symbols for 'HC170 or SN74LS170 make use of this.

Figure 24 is another illustration of the concept.

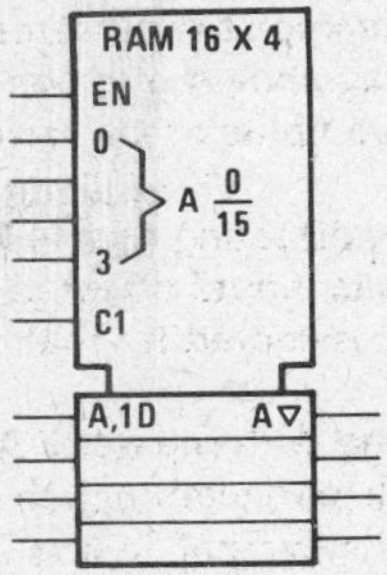

Figure 24. Array of 16 Sections of Four Transparent Latches with 3-State Outputs Comprising a 16-Word X 4-Bit Random-Access Memory

Table IV. Summary of Dependency Notation

TYPE OF DEPENDENCY	LETTER SYMBOL*	AFFECTING INPUT AT ITS 1-STATE	AFFECTING INPUT AT ITS 0-STATE
Address	A	Permits action (address selected)	Prevents action (address not selected)
Control	C	Permits action	Prevents action
Enable	EN	Permits action	Prevents action of inputs ◇outputs off ▽outputs at external high impedance, no change in internal logic state Other outputs at internal 0 state
AND	G	Permits action	Imposes 0 state
Mode	M	Permits action (mode selected)	Prevents action (mode not selected)
Negate (Ex-OR)	N	Complements state	No effect
Reset	R	Affected output reacts as it would to S = 0, R = 1	No effect
Set	S	Affected output reacts as it would to S = 1, R = 0	No effect
OR	V	Imposes 1 state	Permits action
Transmission	X	Bidirectional connection exists	Bidirectional connection does not exist
Interconnection	Z	Imposes 1 state	Imposes 0 state

*These letter symbols appear at the AFFECTING input (or output) and are followed by a number. Each input (or output) AFFECTED by that input is labeled with that same number. When the labels EN, R, and S appear at inputs without the following numbers, the descriptions above do not apply. The action of these inputs is described under "Symbols Inside the Outline," see 3.3.

5.0 BISTABLE ELEMENTS

The dynamic input symbol, the postponed output symbol, and dependency notation provide the tools to differentiate four main types of bistable elements and make synchronous and asynchronous inputs easily recognizable (Figure 25). The first column shows the essential distinguishing features; the other columns show examples.

Transparent latches have a level-operated control input. The D input is active as long as the C input is at its internal 1 state. The outputs respond immediately. Edge-triggered elements accept data from D, J, K, R, or S inputs on the active transition of C. Pulse-triggered elements require the setup of data before the start of the control pulse; the C input is considered static since the data must be maintained as long as C is at its 1 state. The output is postponed until C returns to its 0 state. The data-lock-out element is similar to the pulse-triggered version except that the C input is considered dynamic in that shortly after C goes through its active transition, the data inputs are disabled and data does not have to be held. However, the output is still postponed until the C input returns to its initial external level.

Notice that synchronous inputs can be readily recognized by their dependency labels (1D, 1J, 1K, 1S, 1R) compared to the asynchronous inputs (S, R), which are not dependent on the C inputs.

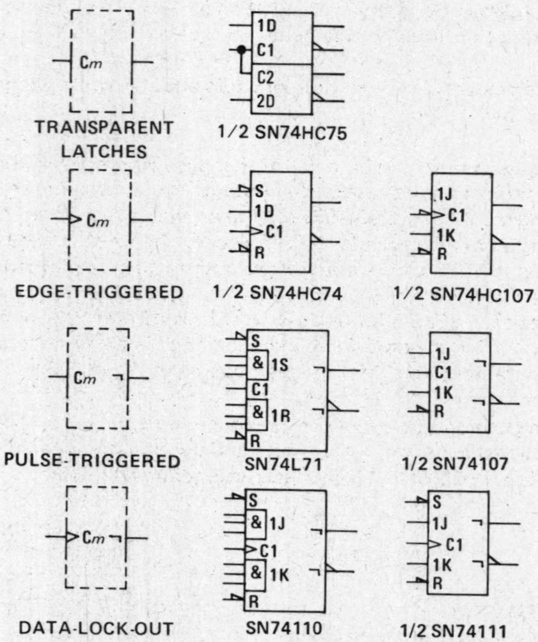

Figure 25. Four Types of Bistable Circuits

6.0 CODERS

The general symbol for a coder or code converter is shown in Figure 26. X and Y may be replaced by appropriate indications of the code used to represent the information at the inputs and at the outputs, respectively.

Figure 26. Coder General Symbol

Indication of code conversion is based on the following rule:

> Depending on the input code, the internal logic states of the inputs determine an internal value. This value is reproduced by the internal logic states of the outputs, depending on the output code.

The indication of the relationships between the internal logic states of the inputs and the internal value is accomplished by:

1) labeling the inputs with numbers. In this case the internal value equals the sum of the weights associated with those inputs that stand at their internal 1-state, or by
2) replacing X by an appropriate indication of the input code and labeling the inputs with characters that refer to this code.

The relationships between the internal value and the internal logic states of the outputs are indicated by:

1) labeling each output with a list of numbers representing those internal values that lead to the internal 1-state of that output. These numbers shall be separated by solidi as in Figure 27. This labeling may also be applied when Y is replaced by a letter denoting a type of dependency (see Section 7). If a continuous range of internal values produces the internal 1 state of an output, this can be indicated by two numbers that are inclusively the beginning and the end of the range, with these two numbers separated by three dots (e.g., 4 . . . 9 = 4/5/6/7/8/9) or by
2) replacing Y by an appropriate indiction of the output code and labeling the outputs with characters that refer to this code as in Figure 28.

Alternatively, the general symbol may be used together with an appropriate reference to a table in which the relationship between the inputs and outputs is indicated. This is a recommended way to symbolize a PROM after it has been programmed.

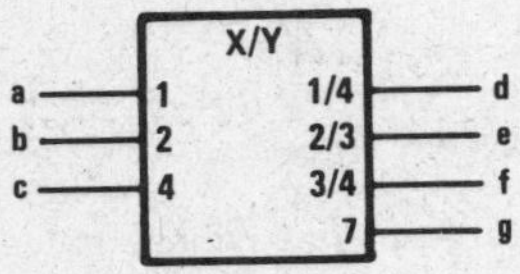

FUNCTION TABLE

INPUTS			OUTPUTS			
c	b	a	g	f	e	d
0	0	0	0	0	0	0
0	0	1	0	0	0	1
0	1	0	0	0	1	0
0	1	1	0	1	1	0
1	0	0	0	1	0	1
1	0	1	0	0	0	0
1	1	0	0	0	0	0
1	1	1	1	0	0	0

Figure 27. An X/Y Code Converter

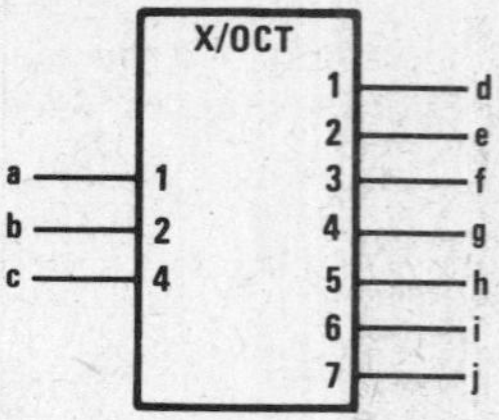

FUNCTION TABLE

INPUTS			OUTPUTS						
c	b	a	j	i	h	g	f	e	d
0	0	0	0	0	0	0	0	0	0
0	0	1	0	0	0	0	0	0	1
0	1	0	0	0	0	0	0	1	0
0	1	1	0	0	0	0	1	0	0
1	0	0	0	0	0	1	0	0	0
1	0	1	0	0	1	0	0	0	0
1	1	0	0	1	0	0	0	0	0
1	1	1	1	0	0	0	0	0	0

Figure 28. An X/Octal Code Converter

7.0 USE OF A CODER TO PRODUCE AFFECTING INPUTS

It often occurs that a set of affecting inputs for dependency notation is produced by decoding the signals on certain inputs to an element. In such a case use can be made of the symbol for a coder as an embedded symbol (Figure 29).

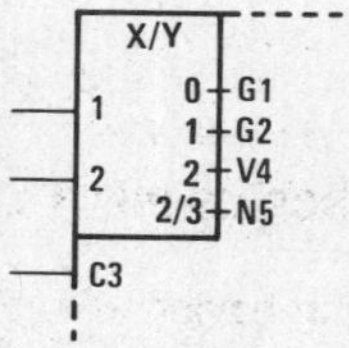

Figure 29. Producing Various Types of Dependencies

If all affecting inputs produced by a coder are of the same type and their identifying numbers shown at the outputs of the coder, Y (in the qualifying symbol X/Y) may be replaced by the letter denoting the type of dependency. The indications of the affecting inputs should then be omitted (Figure 30).

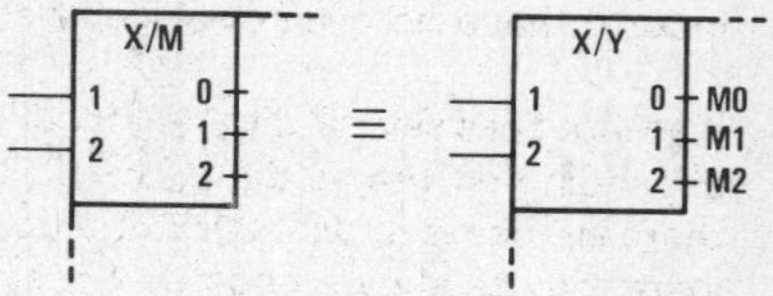

Figure 30. Producing One Type of Dependency

8.0 USE OF BINARY GROUPING TO PRODUCE AFFECTING INPUTS

If all affecting inputs produced by a coder are of the same type and have consecutive identifying numbers not necessarily corresponding with the numbers that would have been shown at the outputs of the coder, use can be made of the binary grouping symbol. k external lines effectively generate 2^k internal inputs. The bracket is followed by the letter denoting the type of dependency followed by m1/m2. The m1 is to be replaced by the smallest identifying number and the m2 by the largest one, as shown in Figure 31.

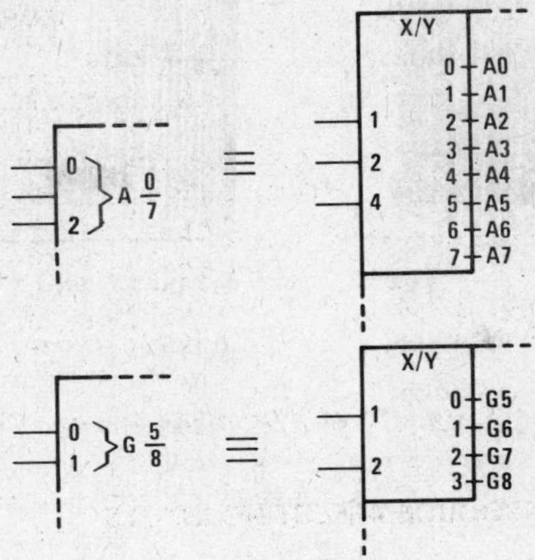

Figure 31. Use of the Binary Grouping Symbol

9.0 SEQUENCE OF INPUT LABELS

If an input having a single functional effect is affected by other inputs, the qualifying symbol (if there is any) for that functional effect is preceded by the labels corresponding to the affecting inputs. The left-to-right order of these preceding labels is the order in which the effects or modifications must be applied. The affected input has no functional effect on the element if the logic state of any one of the affecting inputs, considered separately, would cause the affected input to have no effect, regardless of the logic states of other affecting inputs.

If an input has several different functional effects or has several different sets of affecting inputs, depending on the mode of action, the input may be shown as often as required. However, there are cases in which this method of presentation is not advantageous. In those cases the input may be shown once with the different sets of labels separated by solidi (Figure 32). No meaning is attached to the order of these sets of labels. If one of the functional effects of an input is that of an unlabeled input to the element, a solidus will precede the first set of labels shown.

If all inputs of a combinational element are disabled (caused to have no effect on the function of the element), the internal logic states of the outputs of the element are not specified by the symbol. If all inputs of a sequential element are disabled, the content of this element is not changed and the outputs remain at their existing internal logic states.

Labels may be factored using algebraic techniques (Figure 33).

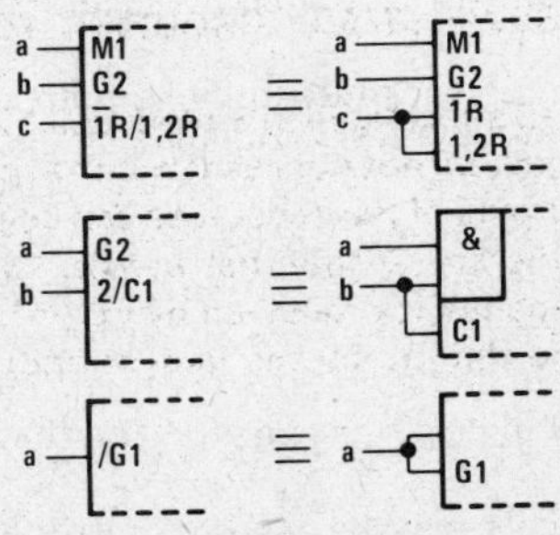

Figure 32. Input Labels

EXPLANATION OF LOGIC SYMBOLS

6

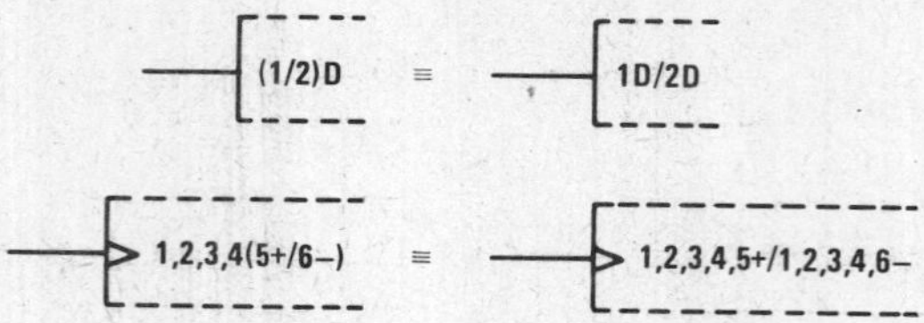

Figure 33. Factoring Input Labels

10.0 SEQUENCE OF OUTPUT LABELS

If an output has a number of different labels, regardless of whether they are identifying numbers of affecting inputs or outputs or not, these labels are shown in the following order:

1) If the postponed output symbol has to be shown, this comes first, if necessary preceded by the indications of the inputs to which it must be applied
2) Followed by the labels indicating modifications of the internal logic state of the output, such that the left-to-right order of these labels corresponds with the order in which their effects must be applied
3) Followed by the label indicating the effect of the output on inputs and other outputs of the element.

Symbols for open-circuit or three-state outputs, where applicable, are placed just inside the outside boundary of the symbol adjacent to the output line (Figure 34).

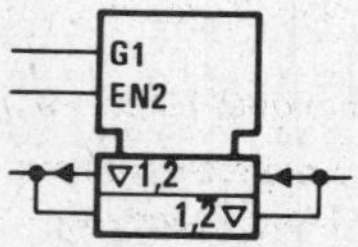

Figure 34. Placement of 3-State Symbols

If an output needs several different sets of labels that represent alternative functions (e.g., depending on the mode of action), these sets may be shown on different output lines that must be connected outside the outline. However, there are cases in which this method of presentation is not advantageous. In those cases the output may be shown once with the different sets of labels separated by solidi (Figure 35).

Two adjacent identifying numbers of affecting inputs in a set of labels that are not already separated by a nonnumeric character should be separated by a comma.

If a set of labels of an output not containing a solidus contains the identifying number of an affecting M*m* input standing at its internal 0 state, this set of labels has no effect on that output.

Labels may be factored using algebraic techniques (Figure 36).

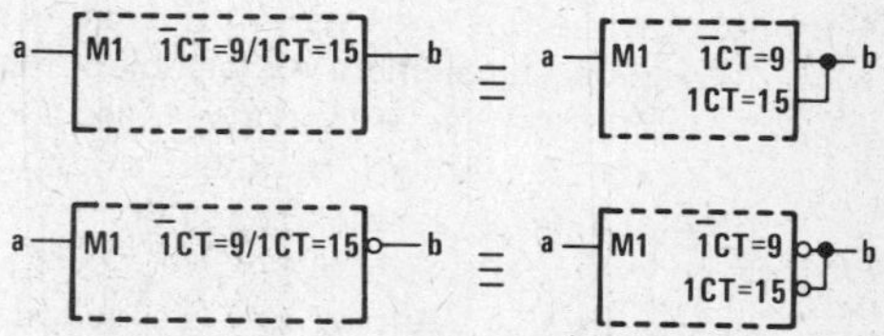

Figure 35. Output Labels

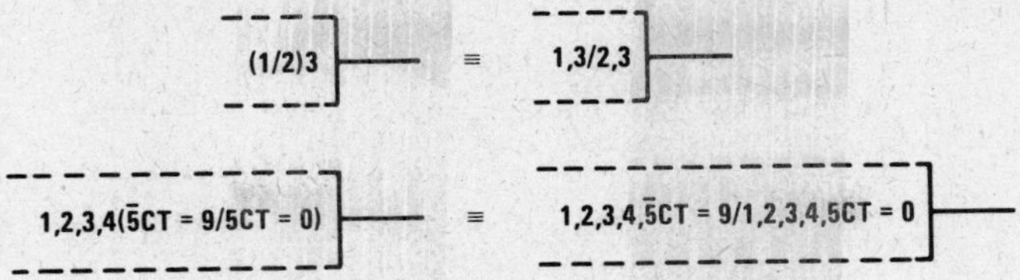

Figure 36. Factoring Output Labels

If you have questions on this Explanation of Logic Symbols, please contact:

Texas Instruments Incorporated
F.A. Mann, MS 49
P.O. Box 225012
Dallas, Texas 75265
Telephone (214) 995-2867

IEEE Standards may be purchased from:

Institute of Electrical and Electronics Engineers, Inc.
IEEE Standards Office
345 East 47th Street
New York, N.Y. 10017

International Electrotechnical Commission (IEC) publications may be purchased from:

American National Standards Institute, Inc.
1430 Broadway
New York, N.Y. 10018

DESIGNERS' INFORMATION

DESIGNERS' INFORMATION

CMOS Circuitry

The elementary CMOS building blocks are the inverter and the transmission gate. Each uses a complementary pair of one n-channel and one p-channel enhancement-type field-effect transistor. Figures 1 and 2 show these together with various logic symbols† used in this book to represent them.

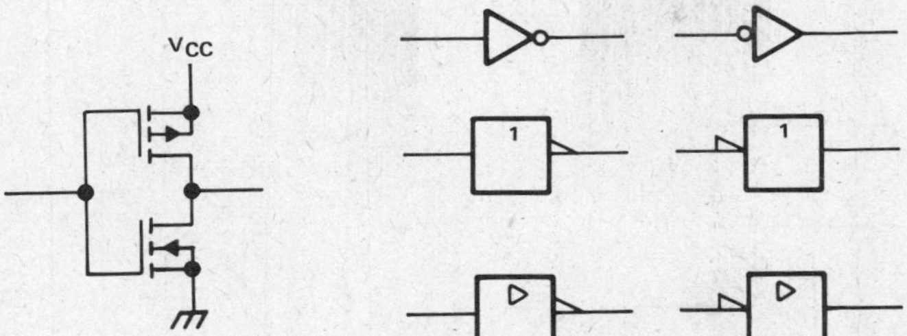

Figure 1. Inverters

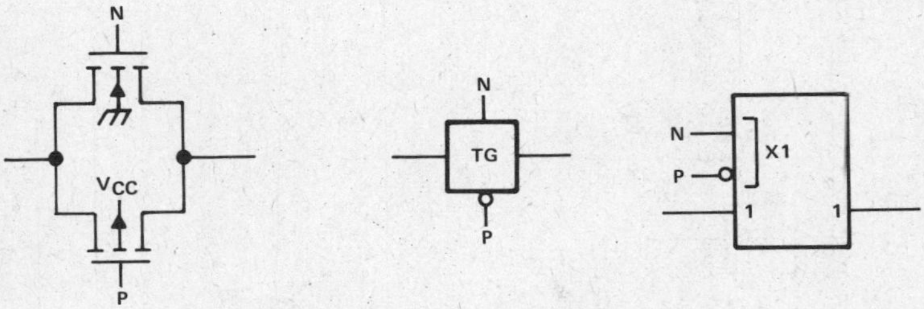

Figure 2. Transmission Gates

Logic gates are created by transistors added in parallel or series to the transistors making up the elementary inverter. Thus the simplest gates are inverting. See Figure 3. An odd number of additional inverters are sometimes added to the outputs of gates to make them noninverting. Basic CMOS gates usually have no more than three inputs. Arrays of gates are used when more than three signals are ANDed or ORed.

The Exclusive-OR or Exclusive-NOR gate is most easily implemented using two inverters and two transmission gates as shown in Figure 4. In complex chains of gates, the inverters may be made unnecessary by complementary signals being already available.

† The various logic symbols are equivalent. The distinctive-shape form of the inverter and gate symbols and the "TG" form of the transmission gate are usually used in the device logic diagrams. The logic inversion symbol (o) is shown at the input or the output, whichever maintains logical consistency with the driving output or the driven input, and this technique is used to indicate the true/complement levels of the signal as it progresses through the circuit. For example, see Figure 7 in this section. The rectangular forms of the inverter and gate symbols and the polarity indicator (⊵) replacing the inversion symbol are usually used in this book only in the device logic symbols. The ▷ indicates a high-current output.

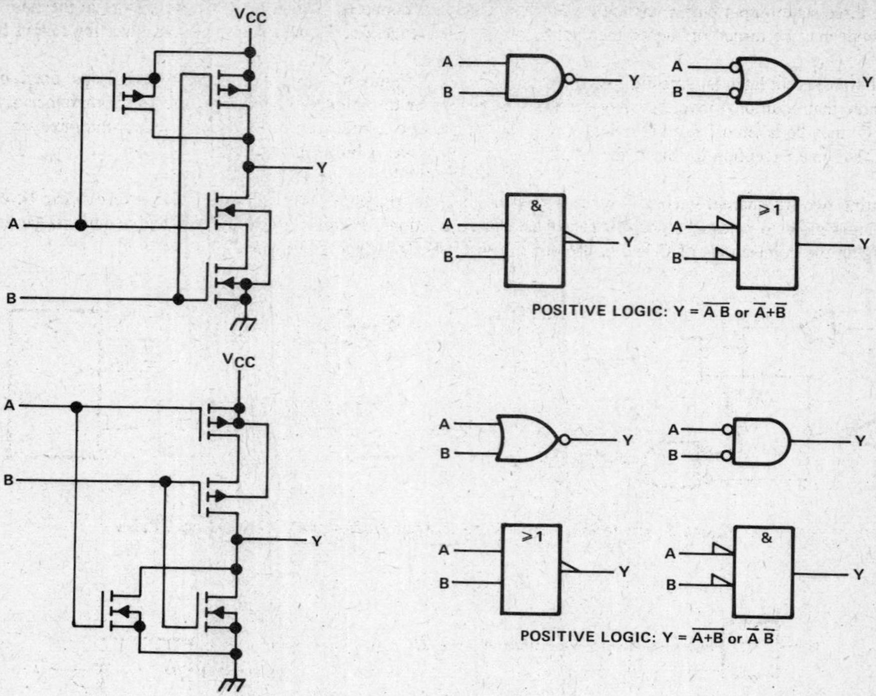

Figure 3. Gates

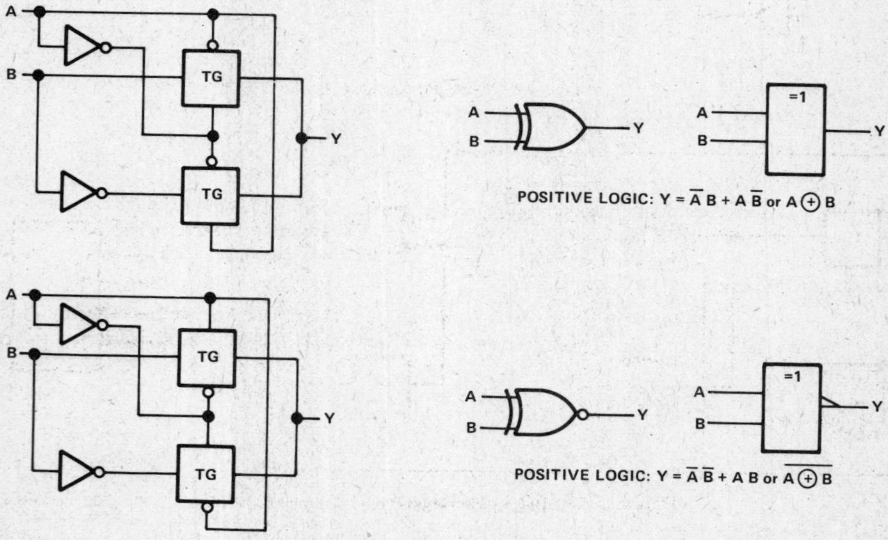

Figure 4. Exclusive-OR/NOR Gates

The three-state output buffer has logic elements in the gate connections to each of the transistors in the final inverter so that both may be turned off under the control of an enable function. Figure 5 illustrates an inverting output buffer.

The transparent latch is typically implemented as shown in Figure 6. This is the simplest form. Logic diagrams in this book show that additional inverters may be added as buffers or to optimize timing. The true and complementary outputs (Q and $\overline{Q}$) may be taken off at other points. Outputs brought out to terminals are always buffered to minimize any feedback effects. The one exception to this is the 'HCU device, which has unbuffered outputs.

Putting two transparent latches in series produces the edge-triggered D-type flip-flop. The inverters can be converted to two-input gates to provide asynchronous set and reset functions. Figure 7 illustrates a negative-edge-triggered circuit. Exchanging the connections of C and $\overline{C}$ produces a positive-edge-triggered version.

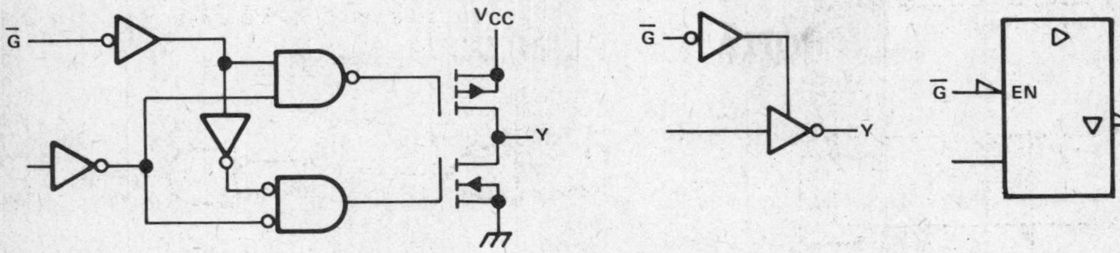

Figure 5. Inverting Three-State Output Buffer with Active-Low Enable

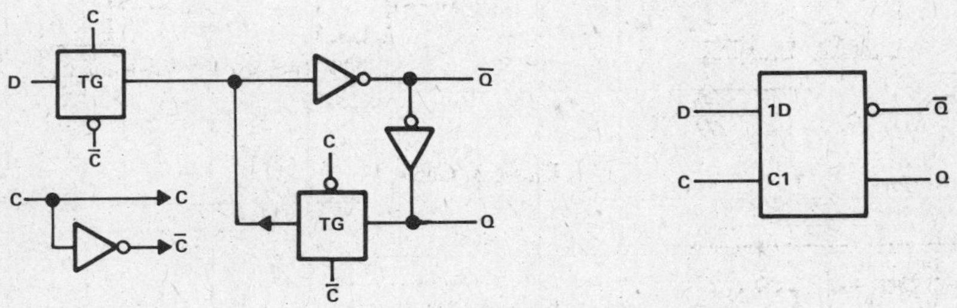

Figure 6. Transparent Latches

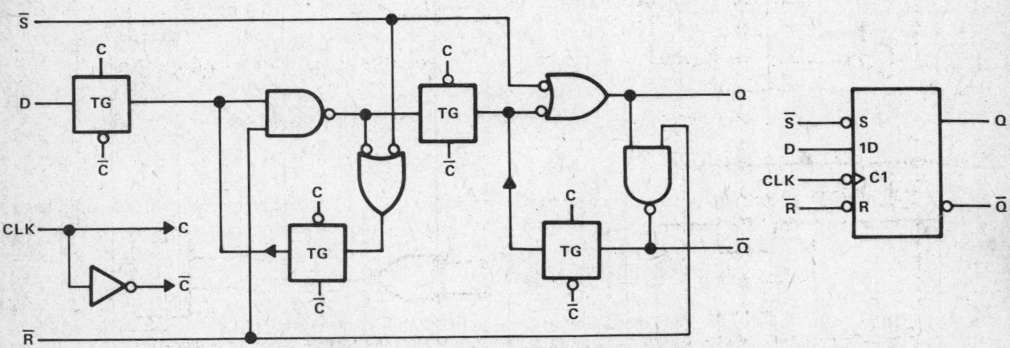

Figure 7. Negative-Edge-Triggered D-Type Flip-Flops

Detailed logic diagrams for flip-flops are given on the data sheets in this book when useful to illustrate special features such as synchronous clearing, J/K inputs, and toggle enabling.

In general the logic diagrams in this book have been simplified. They are believed to correctly indicate the logic implementation but should not be used to predict dynamic performance. Inverters existing in series may be combined or eliminated in the diagram as shown in Figure 8.

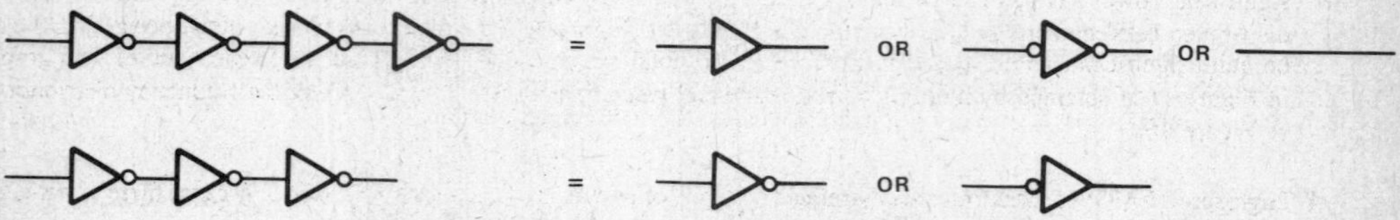

Figure 8. Simplification of Diagrams by Combining Inverters

High-Speed CMOS Characteristics

Table 1 compares the main characteristics of the high-speed CMOS family with those of standard TTL, LSTTL, STTL, ALSTTL, ASTTL, and metal-gate CMOS.

Table 1. Performance Comparison of High-Speed CMOS with Several Other Logic Families

TECHNOLOGY‡	SILICON-GATE CMOS	METAL GATE CMOS	STD TTL	LOW-POWER SCHOTTKY TTL	SCHOTTKY TTL	ADVANCED LOW-POWER SCHOTTKY TTL	ADVANCED SCHOTTKY TTL
Device series	SN74HC	4000	SN74	SN74LS	SN74AS	SN74ALS	SN74AS
Power dissipation per gate (mW)							
Static	0.0000025	0.001	10	2	19	1	8.5
At 100 kHz	0.17	0.1	10	2	19	1	8.5
Propagation delay time (ns) (C_L = 15 pF)	8	105	10	10	3	4	1.5
Maximum clock frequency (MHz) (C_L = 15 pF)	40	12	35	40	125	70	200
Speed/Power product (pJ) (at 100 kHz)	1.4	11	100	20	57	4	13
Minimum output drive (mA) (V_O = 0.4 V)							
Standard outputs	4	1.6	16	8	20	8	20
High-current outputs	6	1.6	48	24	64	24/48	48/64
Fan-out (LS loads)							
Standard outputs	10	4	40	20	50	20	50
High-current outputs	15	4	120	60	160	60/120	120/160
Maximum input current, I_{IL} (mA) (V_I = 0.4 V)	±0.001	−0.001	−1.6	−0.4	−2.0	−0.1	−0.5

‡ Family characteristics at 25°C, V_{CC} = 5 V; all values typical unless otherwise noted. This table is provided for broad comparisons only. Parameters for specific devices within a family may vary. For detailed comparisons, please consult the appropriate data book.

The major advantages of high-speed CMOS can be summarized as follows:

1. The high-speed CMOS family can operate at speeds comparable to LSTTL. The high-speed CMOS family has ac parameters guaranteed at a supply voltage of 2 V, 4.5 V, and 6 V over the full operating temperature range into a 50-pF load (also, 150 pF for high-current outputs). Note that at the higher operating frequencies, the power consumption is also comparable to LSTTL (Figure 9).

2. Figure 9 also shows that the high-speed CMOS family covers a wide range of applications: low power drain for low-speed systems, and a slightly higher drain for higher speed systems.

3. Minimum system power — only the gates that are switching contribute to system power consumption. This reduces the size of the power supply required, hence provides lower system cost and improved reliability through lower heat dissipation.

   As mentioned previously, the power consumption for an individual gate at the maximum speed is comparable to LSTTL. However in typical systems, only a fraction of the gates are switching at the clock frequency; therefore, significant power savings can be realized. On a system level where the individual gate switching frequencies are distributed between zero and the system clock frequency (Figure 10), the power saved with high-speed CMOS can be quite significant, as illustrated in Figure 11. The total system power is the area under each curve. The graph in Figure 11 is obtained by multiplying the individual gate characteristics (Figure 9) by the frequency distribution in Figure 10.

4. High-speed CMOS is ideal for battery-operated systems, or systems requiring battery back-up, because there is virtually no static power dissipation (Figure 9).

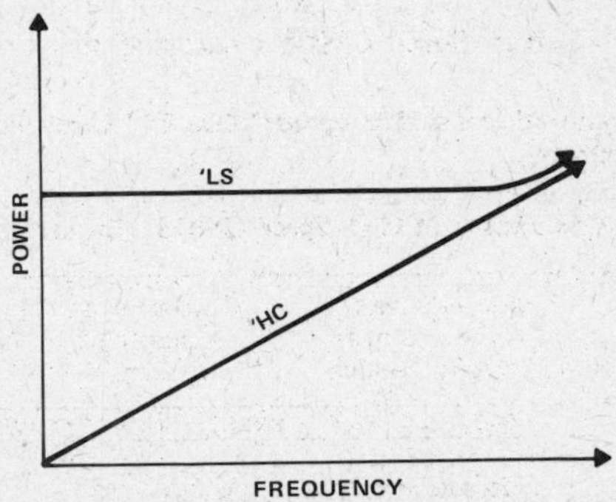

Figure 9. Power Consumed Versus Frequency for High-Speed CMOS Compared to LSTTL

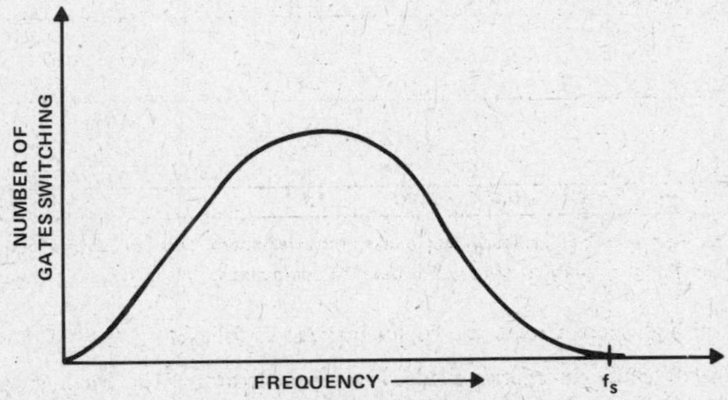

Figure 10. Typical Distribution of Switching Frequencies for Gates within a System with Maximum Clock Frequency, f_S

5. Improved noise immunity over bipolar devices is due to the rail-to-rail (V_{CC} to ground) output voltage swings. Figure 12 illustrates the noise immunity provided by the high-speed CMOS family as it compares to the LSTTL family. This noise immunity makes it ideal for high-noise environments. Minimum and maximum output voltages are guaranteed at 4 mA (6 mA for high-current devices). If the output currents exceed these limits, the noise immunity will be impaired. 'HCT devices have similar input noise margins to LSTTL because their inputs are TTL-voltage compatible. The outputs of 'HCT are the same as standard 'HC outputs.

6. High-speed CMOS devices can drive up to 10 LSTTL loads (15 LSTTL loads for high-current outputs) while maintaining good noise immunity. Although V_{OH}min and V_{OL}max are guaranteed for output currents up to 4 mA (6 mA for high-current outputs), currents up to ±25 mA (±35 mA for high-current outputs) can be obtained to drive LEDs or relays (see Driving LEDs and Relays in this section.)

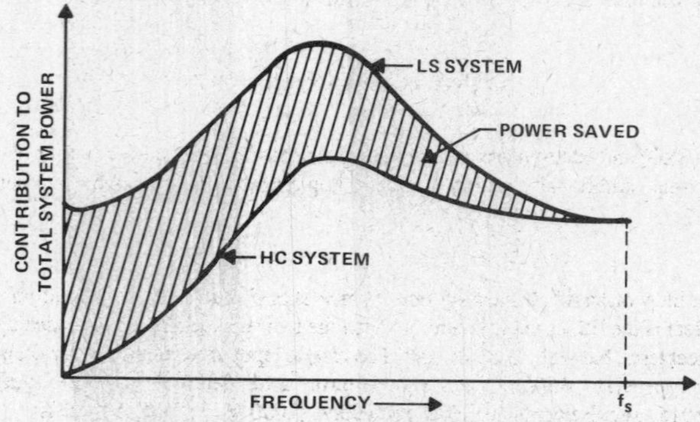

Figure 11. Contribution to Total Power by Gates Running at Frequencies from 0 to f_s

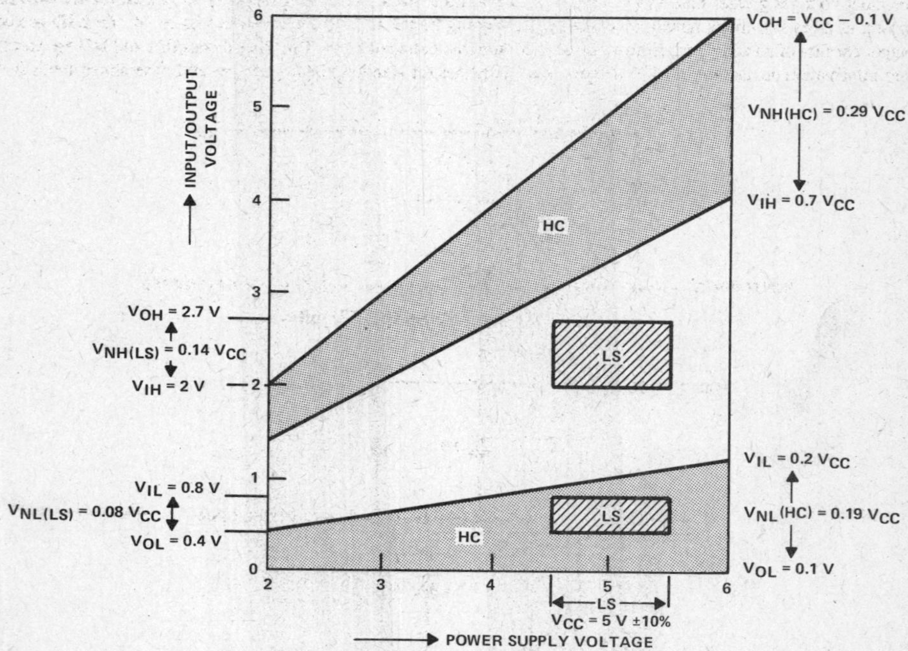

Figure 12. High-Speed CMOS and LS Noise Margins

7. High-speed CMOS devices are guaranteed over an extended temperature range:

SN54HC/HCT'	−55 °C to 125 °C	(military)
SN74HC/HCT'	−40 °C to 85 °C	(industrial)

All specified ac and dc characteristics are guaranteed over this range with the exception of Power Dissipation Capacitance (C_{pd}), which is specified as a typical value at 25 °C.

Protection Circuitry

Electrostatic discharge (ESD) and latch-up are two traditional causes of CMOS device failure. In order to protect HCMOS devices from ESD and latch-up, additional circuitry has been implemented on the inputs and outputs.

ESD PROTECTION

ESD occurs when a buildup of static charges on one surface arcs through a dielectric to another surface that has the opposite charge. The end effect is the ESD causes a short between the two surfaces. These damaged devices (walking-wounded) may still pass normal data sheet tests, but will eventually fail. The unique input protection circuitry designed by Texas Instruments provides immunity to typically 4500 V on the inputs and 3000 V on the outputs, which exceeds MIL-STD-883B, Method 3015, requirements for ESD protection (2000 V, 1.5 kΩ, 100 pF).

Figure 13 shows the circuitry implemented to provide protection for the input gates against ESD. The diode is forward biased for input voltages greater than V_{CC} + 0.5 V. The two transistors and resistor (actually one transistor diffused across a resistor) act as a resistor-diode network against negative-going transients. As illustrated in Figure 14, the ESD protection for the output consists of an additional diffused diode (D3) from the output to V_{CC}. The other diodes (D1 and D2) are parasitics. For further information on handling CMOS devices, see Guidelines for Handling ESDS Devices and Assemblies in this section.

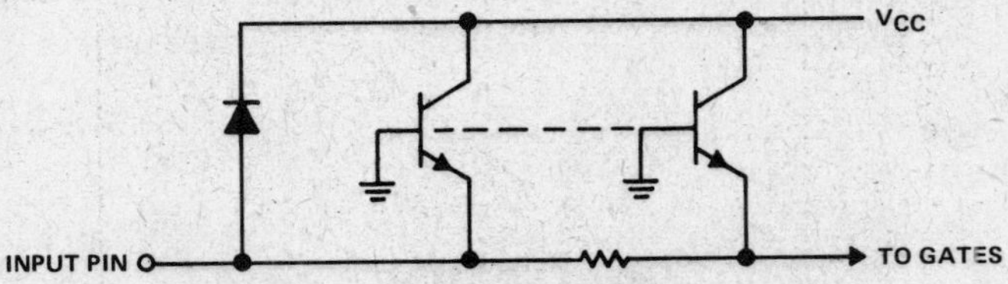

Figure 13. ESD Input Protection Circuitry

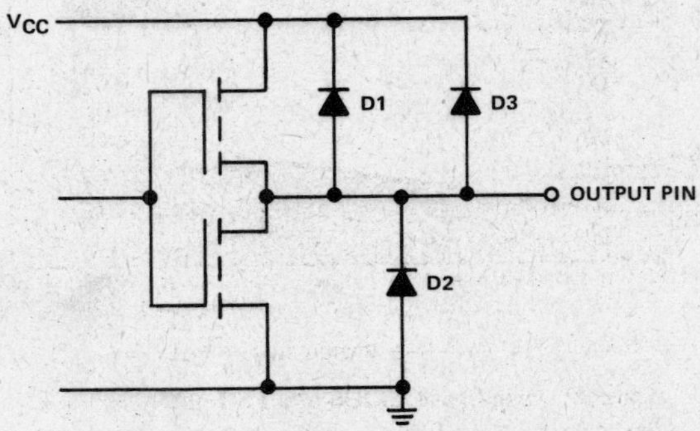

Figure 14. ESD Output Protection Circuitry. D1 and D2 are Parasitic Diodes

LATCH-UP PROTECTION

Internal to most all CMOS devices are two parasitic bipolar transistors; one p-n-p and one n-p-n. Figure 15 shows the cross section of a typical CMOS inverter with the parasitic bipolar transistors. Note that, as shown in Figure 16, these parasitic bipolar transistors are naturally configured as a thyristor or SCR. These transistors conduct when one or more of the p-n junctions become forward biased. When this happens, each parasitic transistor supplies the necessary base current for the other to remain in saturation. This is known as the ''latch-up'' condition and could possibly destroy the device if the supply current is not limited.

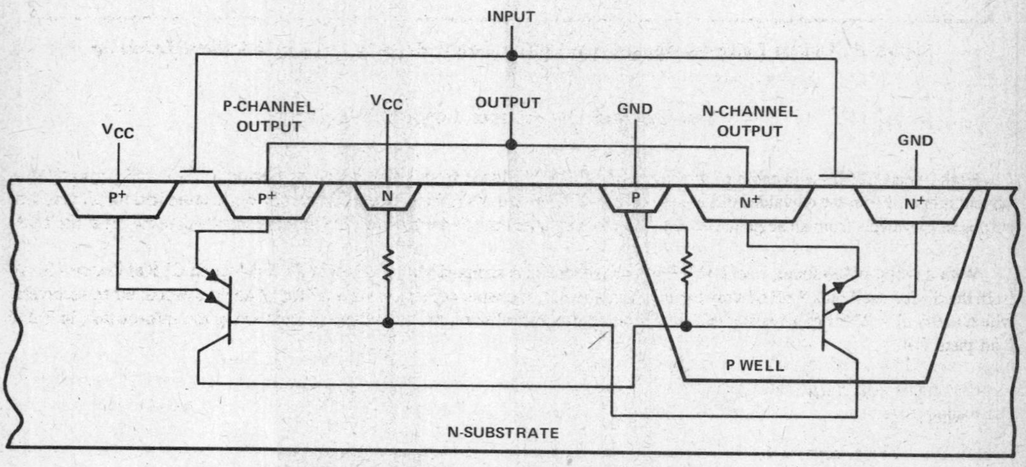

Figure 15. Parasitic Bipolar Transistors in CMOS

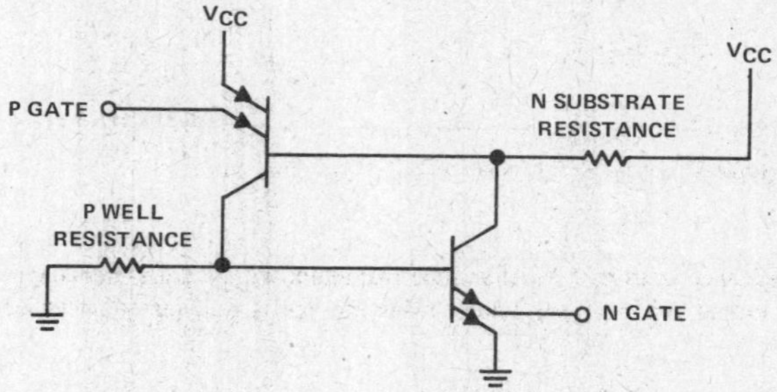

Figure 16. Schematic of Parasitic SCR — P Gate and N Gate Electrodes are Connected Together

A conventional thyristor is fired (turned on) by applying a voltage to the base of the n-p-n transistor, but the parasitic CMOS thyristor is fired by applying a voltage to the emitter of either transistor. One emitter of the p-n-p transistor is connected to an emitter of the n-p-n transistor, which is also the output of the CMOS gate. The other two emitters of the p-n-p and n-p-n transistors are connected to V_{CC} and ground, respectively. Therefore, to trigger the thyristor there must be a voltage greater than $V_{CC} + 0.5$ V or less than -0.5 V and there has to be sufficient current to cause the latch-up condition.

Latch-up cannot be completely eliminated! The alternative is to impede the thyristor from triggering. Texas Instruments has improved the circuit design by adding four additional diffusions or guard rings alternately connected to V_{CC} and ground as shown in Figure 17. The guard rings provide isolation between the device pins and any p-n junction that is not isolated by a transistor gate. All internal p-n junctions are separated by two guard rings. Tests have shown effective latch-up protection ranges from 450 mA to greater than 1 A at 25 °C, and typically greater than 250 mA at 125 °C.

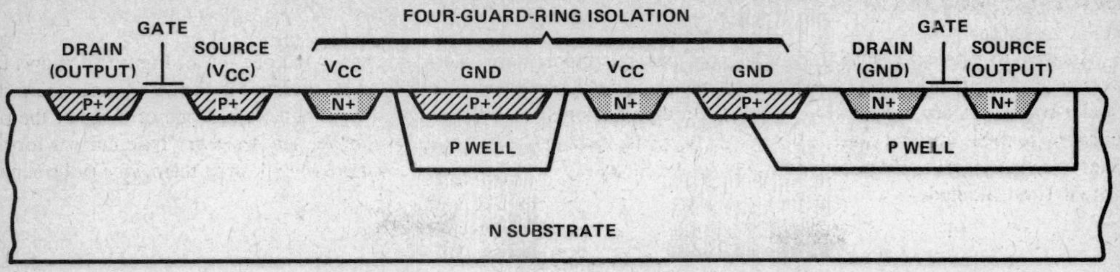

Figure 17. Unique Latch-Up Suppression Utilizes Guard Rings to Virtually Eliminate Latch-Up

Fan-Out and Capacitance Loading Effects

High-Speed CMOS is capable of driving up to 10 LSTTL loads from a single standard output, or 15 loads from a high-current output. From the dc values in Table I on page 2-4, the fan-out of high-speed CMOS devices is unlimited for all practical purposes. However, from an ac point of view, there is a definite limit to the fan-out. The limiting constraint is the input rise time.

With a worst-case model, about 15 pF of capacitance is associated with the input of a high-speed CMOS device (10 pF from the device itself plus 5 pF of stray capacitance; typically the input capacitance is 3 pF for all devices except the transceivers, which are 6 pF). The input resistance, r_I, can be approximated with the following equation using the information in Table I on page 2-4.

$$r_I = V_I/I_I$$

where

$$V_I = V_{CC} = 6\ V$$
$$I_I = 0.1\ nA$$

The output resistance can also be calculated from the values in Table I, page 2-4 and the following equation:

$$r_O = (V_{CC} - V_{OH})/I_{OH}$$

where

$$V_{CC} = 4.5\ V$$
$$V_{OH} = 4.3\ V\ \text{(typical)}$$
$$I_{OH} = 4\ mA$$

The calculated input resistance is about 60 MΩ and the maximum output resistance is approximately 50 Ω. Figure 18 shows the schematic of the output and the input models using the values previously determined.

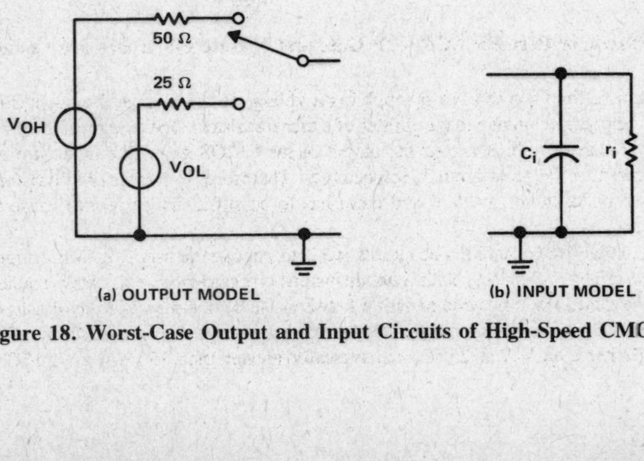

Figure 18. Worst-Case Output and Input Circuits of High-Speed CMOS

DESIGNERS' INFORMATION 7

For a fan-out of n high-speed CMOS devices, the input capacitance will be (n × 15) pF (capacitances are in parallel). When the driving device switches its output from the low level to the high level, the input capacitance of all devices in the fan-out must be charged up and reach $V_{IH}min$ within 500 ns (the recommended rise time). Therefore,

$$V_{IH}min = V_{OH}typ\,(1 - e^{-t/RC})$$

where

$R = 50\ \Omega$
$C = (15 \times n)$ pF
$t = 500$ ns
n = number of devices in the fan-out

Taking the natural log of both sides:

$$-t/RC = \ln(1 - V_{IH}min/V_{OH}typ)$$

Substituting in the appropriate values and solving for n indicates that the maximum fan-out of high-speed CMOS devices is approximately 505. Alternately, solving for t in terms of n shows that each high-speed CMOS device added to the fan-out will increase the propagation delay from input of the driving device to the input of the driven devices by about 0.989 ns. This corresponds to approximately 0.066 ns/pF of added delay. Table 2 contains typical values of fan-out and capacitive loading effects at different values of V_{CC}.

Table 2. Typical Fan-Out of High-Speed CMOS Devices and Propagation Delay per pF at Various Values of V_{CC}

V_{CC}	$V_{OH}min$	$V_{IH}min$	n	t_{pd}/pF
2 V	1.9 V	1.4 V	936	0.0667 ns
4.5 V	4.4 V	3.15 V	993	0.0629 ns
6 V	5.9 V	4.2 V	1004	0.0623 ns

NOTE:

$$n = \frac{-t/RC}{\ln\left[1 - \frac{V_{IH}min}{V_{OH}min}\right]}$$

where

$R = 50\ \Omega$
$C = 8$ pF
n = number of devices in the fan-out

$$t_{pd}/pF = \frac{500\ \text{ns}}{n \times 8\ \text{pF}}$$

Power Dissipation

The power dissipation of high-speed CMOS devices can be separated into three components: (1) quiescent power dissipation, P_Q; (2) transient power dissipation, P_T; and (3) capacitive power dissipation, P_C. The total power dissipation is the sum of the three components, $P_Q + P_T + P_C$.

The quiescent power is the product of V_{CC} and the quiescent current, I_{CC}. The quiescent current is the reverse current through the diodes that are reverse biased. This reverse current is generally very small (on the order of a few nA), which makes the quiescent power almost insignificant. However, for circuits that are in static conditions for long periods of time, the quiescent power becomes a factor to be considered.

The transient power is due to the current that flows only during the time the transistors are switching from one logic level to the other. During this time both transistors are partially on (one turning off, the other turning on), which produces a low-impedance path between V_{CC} and ground and results in a current spike. The rise (and fall) time of the input signal has a direct effect on the duration of the current spike. This is because the faster the input signal goes through the transition region, the less time both transistors are partially on. The transient power is dependent on the characteristics of the transistors, the switching frequency, and the rise time of the input signal. This component can be calculated using the following equation:

$$P_T = C_{pd} \times V_{CC}^2 \times f_i$$

where

C_{pd} = power dissipation capacitance (specified on each data sheet)
V_{CC} = supply voltage
f_i = input signal frequency

Additional capacitive power dissipation is caused by the charging and discharging of the external load capacitance and is dependent on the switching frequency. To calculate this power, the following equation may be used:

$$P_C = C_L \times V_{CC}^2 \times f_o$$

where

C_L = external (load) capacitance
V_{CC} = supply voltage
f_o = output signal frequency

'HCT POWER DISSIPATION

'HCT devices are primarily used to interface TTL output signals to high-speed CMOS inputs. To make the inputs of the 'HCT devices TTL-voltage compatible, the input transistor geometries were changed. This increased the power consumption compared to the equivalent 'HC device, however 'HCT still provides a considerable savings in power over TTL. The increase in power consumption is due to the fact that the TTL input levels cause both transistors in the transistor pair to be partially turned on. Included in the dc tables for 'HCT devices (Tables V through VIII in Section 2) is a parameter ΔI_{CC}, which enables the designer to compute how much additional current the 'HCT device draws per input when at a TTL voltage level.

Power Supply Decoupling

When an SN54HC/74HC gate switches, there is a brief period (on the order of a nanosecond) during which both transistors in the gate output buffer (Figure 19) are partially on. In this interval, the device draws a substantial supply current, producing a current spike on the V_{CC} and ground leads to the gate. This spike may exhibit di/dt as high as 5000 A/s. These spikes will react with the distributed inductance of the supply wiring to produce significant voltage transients on V_{CC} and ground unless adequate supply decoupling is provided. These transients, if allowed, will couple directly into the gate outputs, which in normal usage switch from rail-to-rail.

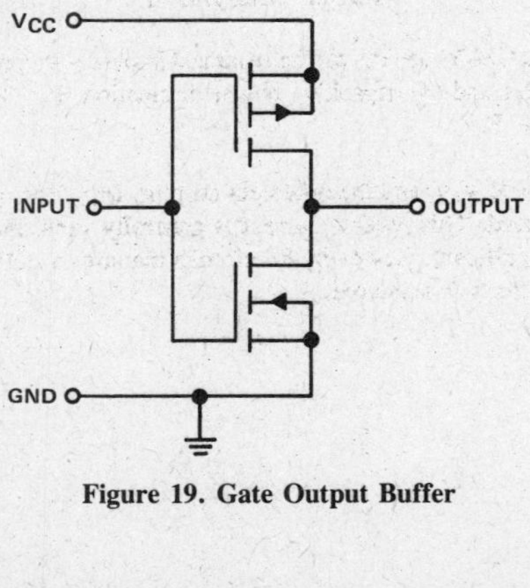

Figure 19. Gate Output Buffer

DECOUPLING PROCEDURE

Figure 20 illustrates a circuit for testing the effectiveness of decoupling. In this test circuit, the V_{CC} and ground connections consist of two parallel runs of one-eighth inch copper on a G-10 epoxy-glass circuit board. As a 0.01-μF decoupling capacitor between V_{CC} and ground is physically moved away from a driven gate in 1.5-inch increments, V_{CC} transients increase as shown in Figure 21.

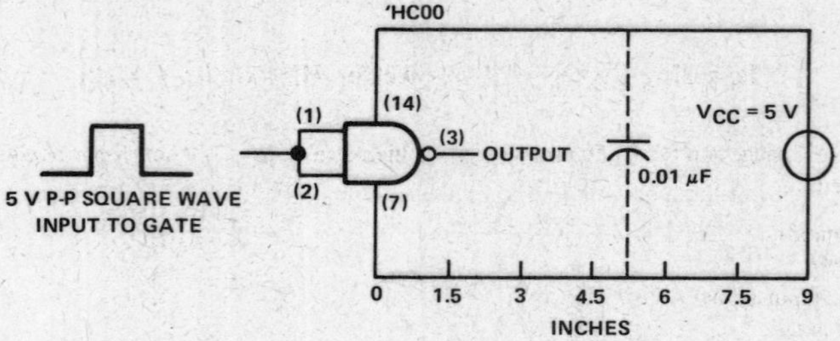

Figure 20. Test Circuit for Decoupling Effects

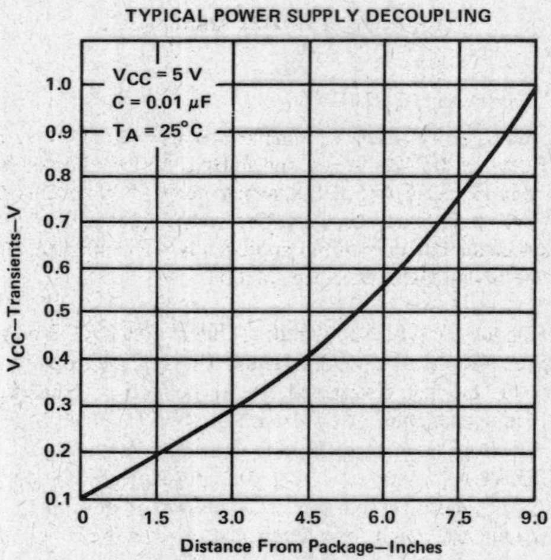

Figure 21. V_{CC} Transients vs Decoupling Capacitor Distance from DIP

The results indicate the importance of adequate decoupling, and illustrate the correct procedure for obtaining it. This procedure consists of locating decoupling capacitors as close as possible to the integrated circuit package, in order to maximize noise margins.

Connecting Unused Inputs

Unused inputs should be tied to V_{CC} or ground to prevent the input from floating. If left to float, the power consumption of the device will increase.

Matching

Another factor to consider when designing with high-speed CMOS is the V_{OH}min-to-V_I matching. This is important when the V_{OH}min of the driving device exceeds the V_{CC} + 0.5 V of the driven device. If this occurs, the ESD protection diode on the inputs will be forward biased. At this point, the driving device will attempt to ''power-up'' the driven device's power supply. No damage will occur to the driven device, provided the current flowing through the diode does not exceed 20 mA.

Powering Up/Down Sequence for High-Speed CMOS

To avoid any possible damage and reliability problems to the high-speed CMOS devices when applying power, the following steps should be followed:

1. Connect ground
2. Connect V_{CC}
3. Connect the input signal

When powering down a high-speed CMOS device, follow the above steps in reverse order.

High-Speed CMOS Interfacing

INTRODUCTION

The High-Speed CMOS logic family from Texas Instruments contains a broad spectrum of SSI/MSI functions. Within this family are TTL functions, HCT devices, HC4000 series, and an HCU device.[1] Entire CMOS systems may be implemented using this logic family. There is also a broad range of CMOS-system to non-CMOS-system interfaces that need to be considered. The design engineer will inevitably encounter these interfaces. To develop the necessary interfaces, a thorough understanding of data sheet parameters of both systems and an organized approach is recommended. This report uses basic examples to present one possible approach to the SN54/74HC interface solution.

There are two types of interfacing that must be considered: (1) interfacing CMOS system signals to non-CMOS systems and (2) interfacing non-CMOS system signals to CMOS systems. The first type requires an understanding of the CMOS output parameters and the non-CMOS input parameters and vice versa for the second type. In both cases, a model of the inputs and outputs of both systems may be useful.

GENERAL INTERFACING SOLUTION

An interfacing problem arises when the output logic levels and/or the current requirements of the driving system (or device) are different from the input logic levels and/or the current requirements of the driven system (or device). When determining the compatibility of the systems (or devices), the most important system/device parameters are V_{IH}, V_{IL}, V_{OH}, V_{OL}, I_{IH}, I_{IL}, I_{OH}, and V_{OL}.

Figure 22 is the voltage transfer characteristic of a typical unloaded inverter showing the various input and output voltage parameters. Loading the output of the inverter will tend to lower V_{OH} and raise V_{OL}. The tables of electrical characteristics specify minimum V_{OH} and maximum V_{OL} for various loads.

Noise Margin

There are two noise margins to be considered: the low-voltage noise margin and the high-voltage noise margin. The voltage difference between V_{IL}max of the driven system/device and V_{OL}max of the driving system/device is the low-voltage noise margin. The voltage difference between V_{OH}min of the driving system/device and V_{IH}min of the driven system/device is the high-voltage noise margin (Figure 23).

[1] HCT devices are explained later. The HC4000 series devices are pin-for-pin functionally compatible, but not electrically compatible, with the older metal-gate CMOS devices. The HCU device is unbuffered.

It is desirable to have the noise margin as large as possible and the uncertain region (the difference between $V_{IH}min$ and $V_{IL}max$) as small as possible. When an input voltage falls into the uncertain region, we do not know how the output in conjuction with other inputs driven by that output will respond. The problem with small noise margins is that any noise on the output of the driving system or device will cause the signal to fall into the uncertain region and possibly cause a bit error in the system. There are various sources of noise in digital systems. Three possible internal sources are inductive and resistive drops, capacitive coupling from another logic node, and mutual inductance with another logic node. Radio signals are possible external sources of noise.

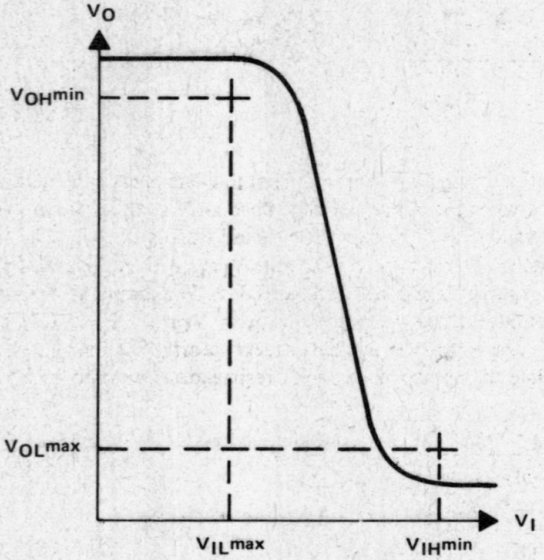

Figure 22. Voltage Transfer Characteristic of a Typical Inverter

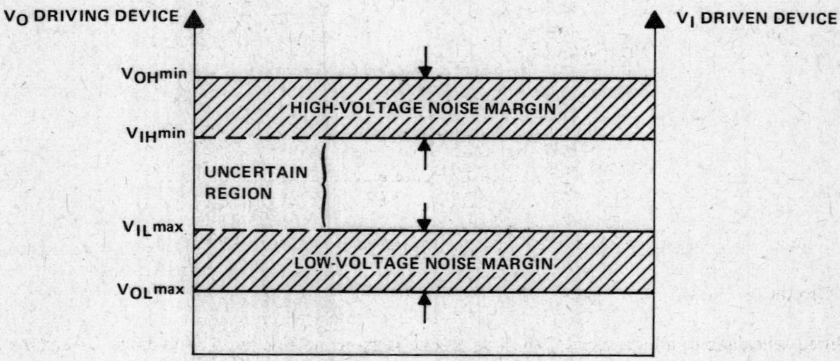

Figure 23. Noise Margins

As an aid for interfacing between the various TTL families, the eight parameters previously defined are shown in Table 3. The values are for V_{CC} = 5 V and T_A = 25 °C (worst-case device parameters — the device will perform at least this well). All currents are designated positive when flowing into the device.

Table 3. Worst-Case Values of Primary Interfacing Parameters

PARAMETER	74HCMOS	74TTL	74LSTTL	74ASTTL	74ALSTTL
V_{IH}min	3.5 V	2 V	2 V	2 V	2 V
V_{IL}max	1 V	0.8 V	0.8 V	0.8 V	0.8 V
V_{OH}min	4.9 V	2.4 V	2.7 V	2.7 V	2.7 V
V_{OL}max	0.1 V	0.4 V	0.4 V	0.4 V	0.4 V
I_{IH}max	1 μA	40 μA	20 μA	200 μA	20 μA
I_{IL}max	−1 μA	−1.6 mA	−400 μA	−2 mA	−100 μA
I_{OH}max	−4 mA	−400 μA	−400 μA	−2 mA	−400 μA
I_{OL}max	4 mA	16 mA	8 mA	20 mA	4 mA

Driving Gate Output Model

Figure 24 shows the model of a driving gate derived from the data sheet specifications. $V_{OH(nl)}$ (nl = no load) is the high-level output voltage expected when the output gate is unloaded. $V_{OL(nl)}$ is the low-level output voltage expected when the output gate is unloaded. The values for these two voltages are usually not given on the data sheets. As a rule of thumb for MOS devices, the output switches between the power rails $V_{OH(nl)} = V_{CC}$ and $V_{OL(nl)} =$ GND; for bipolar devices (e.g., the TTL Family) $V_{OL(nl)}$ is about V_{CC}(sat) or about 0.3 V. Within the TTL family $V_{OH(nl)}$ varies. Standard TTL has a $V_{OH(nl)}$ within two base-emitter drops of V_{CC}($V_{OH(nl)} = V_{CC} - 1.2$ V); LSTTL has a $V_{OH(nl)}$ within one base-emitter drop of V_{CC}($V_{OH(nl)} = V_{CC} - 0.6$ V). The data sheets specify V_{OH}max and V_{OL}max at a nonzero I_{OH} and I_{OL}, respectively. Therefore to calculate the approximate series resistances, the following two equations may be used:

$$R_{OH} = \frac{|V_{OH(nl)} - V_{OH}\text{min}|}{I_{OH}}$$

$$R_{OL} = \frac{|V_{OL(nl)} - V_{OL}\text{max}|}{I_{OL}}$$

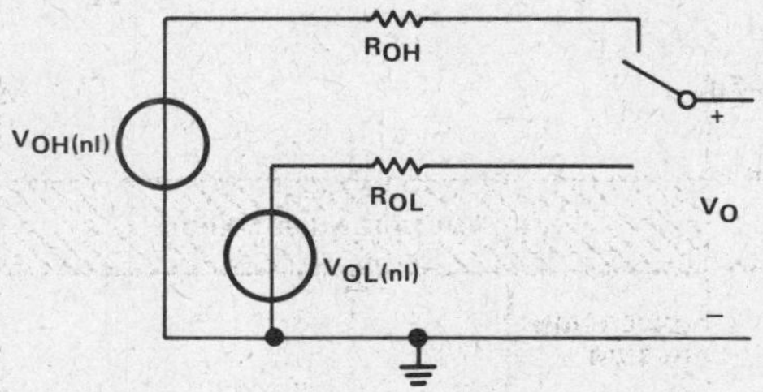

Figure 24. Output Model of a Driving Gate

Input Gate Circuit

A simplified schematic of a high-speed CMOS input gate is shown in Figure 25. The diode D1 and the transistors Q1 and Q2 provide static discharge and input transient clamping for the device. Any inputs higher than V_{CC} + 0.5 V or lower than −0.5 V will clamp the input. The capacitors C1 and C2 represent the parasitic capacitances present at the gate input. The data sheet specifices that the input capacitance (C1 + C2) will not exceed 10 pF (typical is about 5 pF). The input capacitance is split between V_{CC} and ground of the device and provides a feedback path between V_{CC} and the input. If the input is driven by a high-impedance source, then any transient noise on V_{CC} may be coupled back into the input.

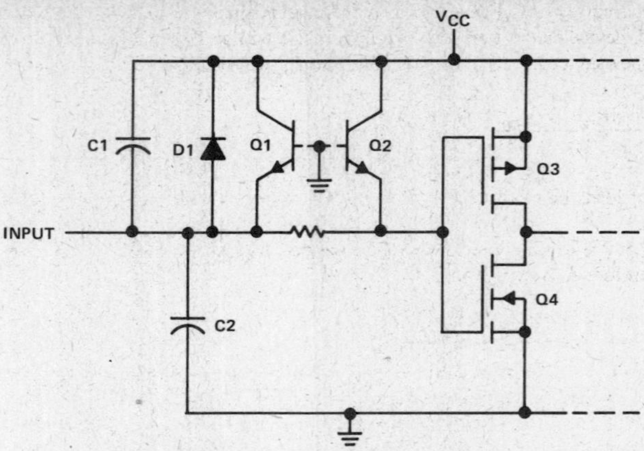

Figure 25. SN54/74HC Input Gate

CMOS-to-STANDARD-TTL INTERFACE

CMOS devices can drive TTL loads with no additional interfacing required. The output voltages of CMOS devices are compatible with the input voltage requirements of TTL devices. The input current requirements of the TTL devices does place a strict limitation on the number of TTL devices that CMOS devices can drive from a single output (the fan-out).

Figure 26 is a schematic of a CMOS output gate driving a TTL input gate. When the CMOS gate drives the emitter of Q3 low, a current will flow into the CMOS gate from R1 and the emitter of the TTL gate. The maximum guaranteed current that the CMOS device can sink is 4 mA. However, the device can sink up to 25 mA, but the output voltage is not guaranteed above 4 mA. Therefore, the maximum TTL fan-out that a device can drive without exceeding the specified limit is two (I_{IL} for TTL is −1.6 mA).

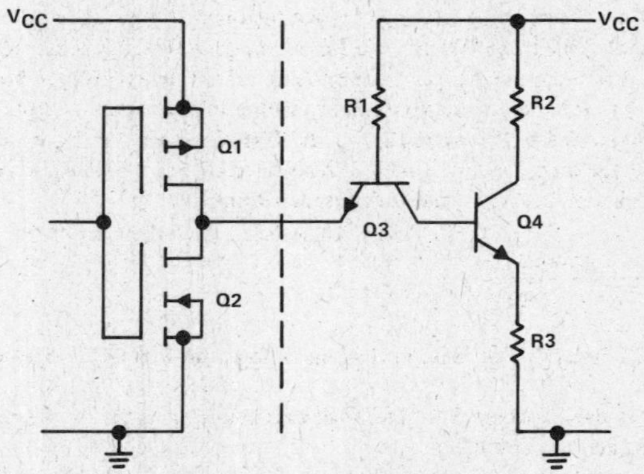

Figure 26. SN54/74HC to TTL Interface

STANDARD TTL-to-CMOS INTERFACE

The interface for TTL driving CMOS is not as simple as the CMOS-to-Standard-TTL interface. Taking the voltage levels from Table 3, it can be seen they are not compatible as far as V_{OH}min of the TTL device and V_{IH}min of the CMOS device. Figure 27 shows the schematic of TTL to CMOS interface. The pull-up resistor R_P eliminates the voltage incompatibility.

The lower limit of the pull-up resistor is determined by the current-sinking capability of the driving device (TTL for this interface). When the TTL device output goes low, Q3 (Figure 27) will be required to sink a current of $(V_{CC}\text{-}V_{OL}max)/R_P$ in addition to the sum of the output currents of the driven devices I_{IL} worse case. All of this is shown in the following equation:

$$R_P min = \frac{V_{CC} - V_{OL}max\ (TTL)}{I_{OL}(TTL) + n\ I_{IL}(load)}$$

where n is the number of loads being driven, and V_{CC} is the voltage applied to the pull-up resistor.

Example: An SN74LS00 is driving three SN74HC00 devices. $V_{CC}min = 4.75$ V, $V_{OL}max = 0.4$ V, $I_{OL} = 8$ mA, $I_{IL} = 1\ \mu A$, n = 3, therefore $R_P min = 543\ \Omega$.

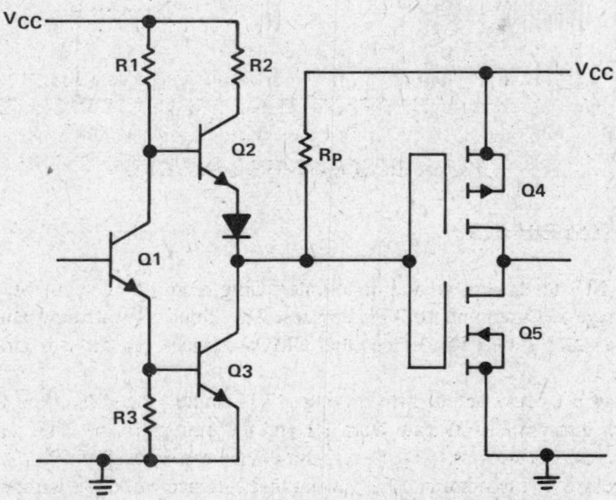

Figure 27. TTL to SN54/74HC Interface with a Pull-Up Resistor

The upper limit of the pull-up resistor is determined by two factors: (1) the total input capacitance of the loads and (2) the total high-level input currents of the loads. When the TTL output goes high, Q_2 is turned off due to the pull-up resistor. Therefore, all the current that flows into the devices that are being driven flows through the pull-up resistor R_P. The input voltage of the CMOS devices will therefore rise exponentially with a time constant of $R_P Ci$ ($C_i = 10$ pF max). The time constant cannot exceed the 500-ns rise time requirement of the CMOS device. Along with this limitation, the total input currents must not cause the voltage drop across the pull-up resistor to exceed $V_{IH}min$ for the CMOS devices. Bringing all this into play, the following equation may be used to determine $R_P max$.

$$R_P max = \frac{V_{CC} - V_{IH}min\ (load)}{|n\ I_{IH}(load) - I_{OH}(driver)|}$$

where n is the number of loads being driven, and V_{CC} is the voltage applied to the pull-up resistor.

Example: An SN74LS00 is driving three SN74HC00 devices. $V_{CC} = 5.25$ V, $V_{IH}min = 3.675$ V, $I_{IH} = 1\ \mu A$, $I_{OH} = 0$, n = 3, therefore $R_P max = 525$ kΩ.

However, if the rise time is calculated using this value of Rpmax, the recommended 500 ns will be exceeded.

From the relationship:

$$V_{IH}min = V_{CC}max\ (1 - e^{-t/R_P C_i})$$

with

$$V_{IH}min = 3.675\ \text{V and } V_{CC}max = 5.25\ \text{V}$$

then

$$R_P = \frac{t}{1.2\ C_i} = 13.8\ \text{k}\Omega \text{ for } t = 500\ \text{ns and } C_i = 30\ \text{pF}$$

Generally, this rise-time constraint is the limiting factor on the upper limit of the pull-up resistor.

CMOS-to-LSTTL INTERFACE

The interface of CMOS to LSTTL is very similar to the interface of CMOS to TTL. Figure 28 shows a schematic of the interface. As can be seen, there is no pull-up resistor required. When the LSTTL input is pulled low, the current will flow through R1 and D2 into the CMOS output. In the worst-case condition, this current is about 0.4 mA. Because the CMOS output parameter I_{OL} specifies a 4-mA current sink for the device, the maximum LSTTL fan-out is ten.

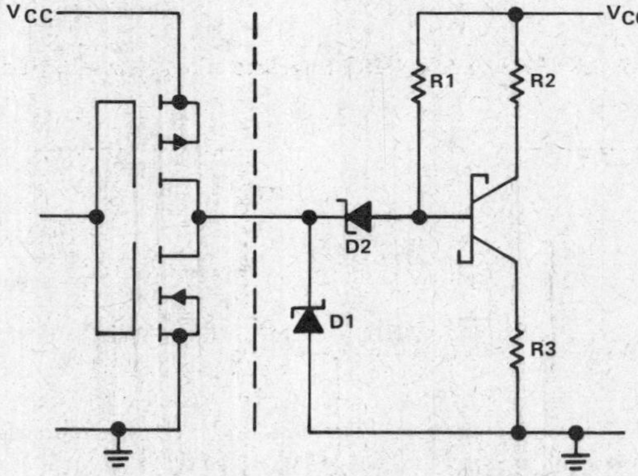

Figure 28. SN54/74HC to LSTTL Interface

LSTTL-to-CMOS INTERFACE

For an LSTTL device to drive a CMOS device, a pull-up resistor must be used because the $V_{OH}min$ of the LSTTL is less than the specified $V_{IH}min$ of the CMOS device. Figure 29 shows the schematic of the LSTTL/CMOS interface. The upper and lower limits of the pull-up resistor are determined in the same method as the TTL/CMOS interface. Remember the upper limit of the pull-up resistor is limited by the input currents and the input capacitance.

CMOS-to-ALSTTL INTERFACE

The output logic level of CMOS devices are completely compatible with the input logic levels of ALSTTL devices. The interface structure with ALSTTL is shown in Figure 30. As with the other CMOS-to-TTL interfaces, there is no pull-up resistor required. The fan-out of ALSTTL devices is determined by the amount of current that flows through Q3 into the

CMOS device, and the amount of current the CMOS device can sink. When the input of the ALSTTL device is low, there is 0.1 mA flowing though Q2. The maximum current that the CMOS device can sink (according to the parameters) is 4 mA. This corresponds to a ALSTTL fan-out of 40.

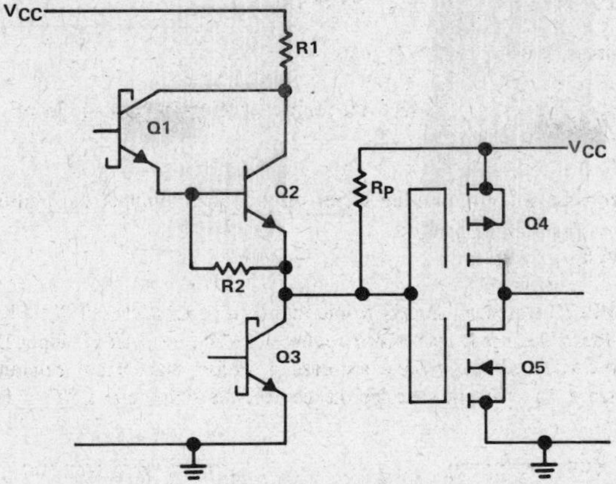

Figure 29. LSTTL to SN54/74HC Interface with a Pull-Up Resistor

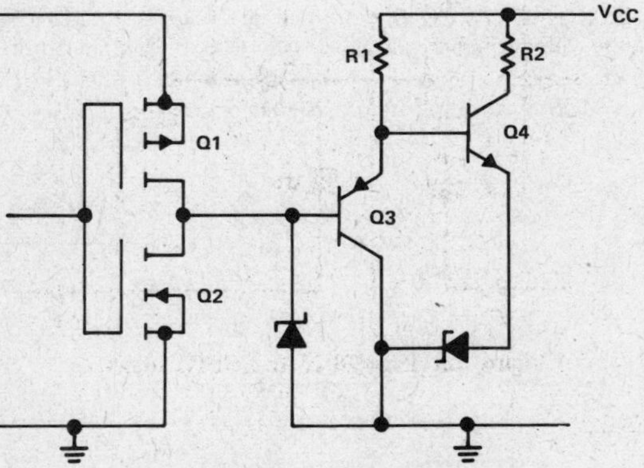

Figure 30. SN54/74HC to ALSTTL Interface

ALSTTL-to-CMOS INTERFACE

The high-level output voltage of ALSTTL devices is incompatible with the required high-level input voltage of CMOS devices. Because of this incompatibility, a pull-up resistor is required to make the two voltage levels compatible. The method of determining the upper and lower limits of the pull-up resistor is the same as the other two TTL-to-CMOS interfaces. Figure 31 shows a schematic of the interface.

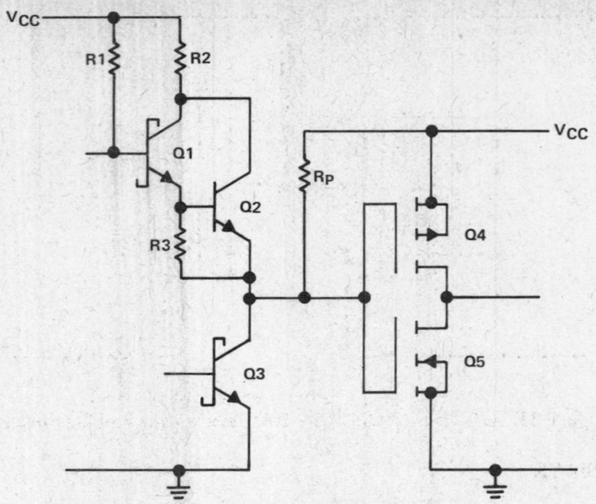

Figure 31. Interface with a Pull-Up Resistor

CMOS-to-ASTTL INTERFACE

As in the case of the other CMOS-to-TTL interfaces, no pull-up resistor is required (Figure 32) because the input voltage levels of ASTTL are compatible with the output voltage levels of CMOS. The fan-out of ASTTL devices is limited by the low-level input current (I_{IL}) of ASTTL and the current sinking capability of CMOS (I_{OL}). I_{IL} for the ASTTL is 2 mA, and the current sink limit of CMOS is 4 mA. Therefore, the fan-out is two ASTTL devices.

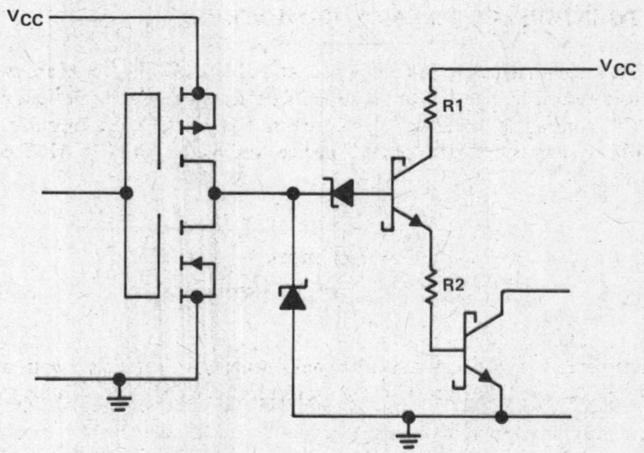

Figure 32. SN54/74HC to ASTTL Interface

ASTTL-to-CMOS INTERFACE

Not all the ouput logic levels of ASTTL are compatible with the input logic levels of CMOS. Table 3 shows there is incompatibility between the V_{OH} of ASTTL and V_{IH} of CMOS. As with other TTL-to-CMOS interfaces, a pull-up resistor is required (Figure 33). The appropriate value of the pull-up resistor is determined by the same procedure previously explained.

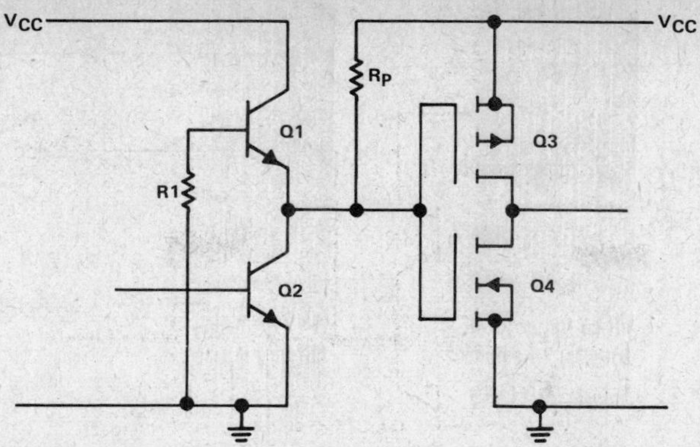

Figure 33. ASTTL to SN54/74HC Interface with a Pull-Up Resistor

CMOS-to-NMOS INTERFACE

NMOS is used extensively in large-scale-integration products such as microprocessors, microcomputers, and memories. The logic levels of NMOS are usually TTL-compatible. CMOS devices can drive NMOS devices with no pull-up resistors. The input impedance of NMOS is very high, which is similar to the input impedance of CMOS.

NMOS-to-CMOS INTERFACE

A pull-up resistor may be necessary when an NMOS device drives a CMOS device. The method of determining the value range of the pull-up resistor is the same as the method described previously for TTL. A quick look at NMOS output parameters and CMOS input parameters will determine if a pull-up resistor will be required.

USING HCT DEVICES TO INTERFACE TO CMOS FROM TTL

To interface from a TTL system (standard TTL, LSTTL, ASTTL, ALSTTL), there are two methods: (1) the use of pull-up resistors (as previously described) and (2) the use of HCT devices. Using HCT devices is by far the easier method. The HCT device inputs are TTL compatible, while the outputs are both TTL and CMOS compatible. Therefore, all the interface requires is to connect the TTL system output into the HCT device, and the output of the HCT device can then be used for the input of the CMOS system.

Oscillators

RC OSCILLATORS

Simple oscillator circuits using a minimum number of components can be designed with high-speed CMOS devices, e.g., two 'HC04, 'HCU04, 'HC00, or 'HC02 gates. These oscillators generate a period of approximately 1.8 RC seconds (Figure 34).

CRYSTAL-CONTROLLED OSCILLATORS

A crystal or ceramic resonator may be used to set the oscillator period (Figure 35). The value of the resistor, typically 100 kΩ, may require special selection to ensure oscillation at the desired fundamental resonator frequency. The capacitor, typically 100 pF, is required to dampen parasitic oscillations in the 30-MHz to 50-MHz range.

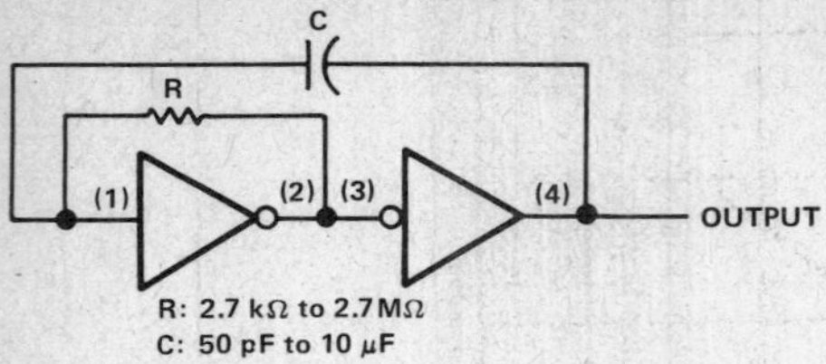

Figure 34. Simple RC Oscillator Using Two 'HC04 Gates

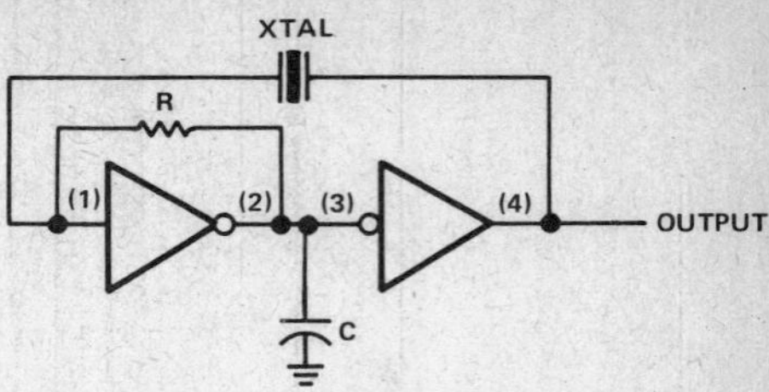

Figure 35. Oscillator Circuit Using a Crystal to Set the Period

VOLTAGE-CONTROLLED OSCILLATORS

Voltage-controlled oscillators (VCOs) can also be designed using a minimal number of components. Figure 36 shows a VCO using NAND and inverter gates. This VCO design exploits the phenomena of the slight variations in the propagation delay of an 'HC gate with changes in the supply voltage. The 'HC00 is connected as a three-stage ring oscillator with a buffer. As the control (supply) voltage V_C is varied, the ring oscillator's frequency changes according to the following:

$$f_{out} \approx 5.8 \times V_C$$

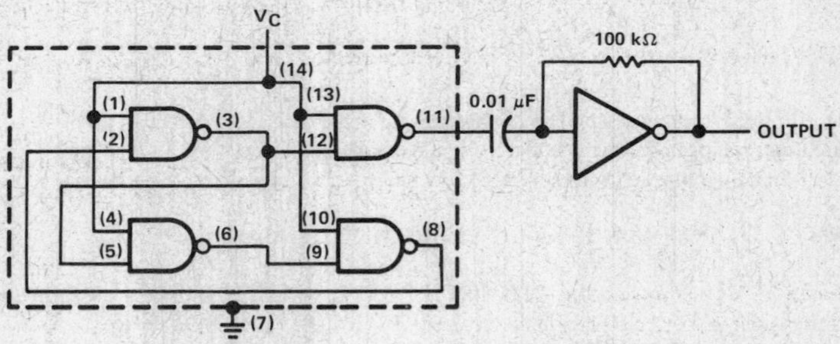

Figure 36. Voltage-Controlled Oscillator (VCO)

The inverter, which is powered by a separate voltage source, serves to restore the oscillator output voltage to 5 V peak-to-peak. This function is required, because the 'HC00 switches from rail-to-rail (as do all HC devices). The magnitude of the oscillator output voltage is thus dependent on V_C. The 100-kΩ resistor across the inverter provides bias such that operation will be within the linear operating region of the gate. The capacitor serves to ac-couple the oscillator to the inverter.

The VCO output is linear for control voltages in the range of 1.5 to 4.5 V (Figure 37).

To prevent oscillator "bleed-through" onto the V_{CC} line, adequate decoupling of the 'HC device power supply is required.

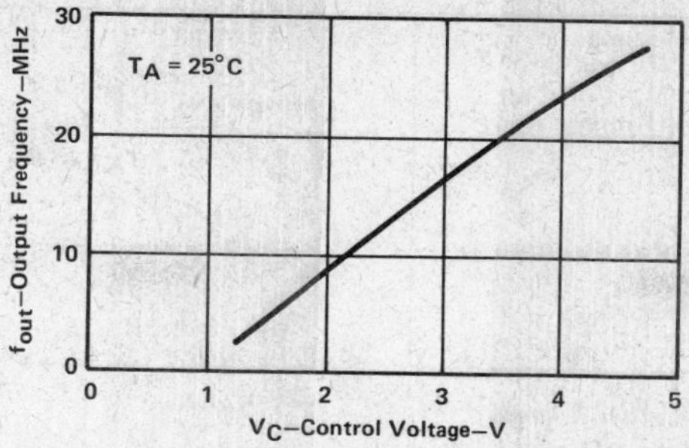

Figure 37. VCO Output Frequency vs Input Voltage

Drivers for LEDs and Relays

INTRODUCTION

SN54/74HC devices are capable of sinking or sourcing up to 25 mA (35 mA for high-current devices) per gate. As the device sinks or sources more current, V_{OH}min or V_{OL}max levels will begin to fall or rise respectively.

Because of these characteristics, SN54/74HC devices can be used to drive LEDs and relays.

DRIVING LEDs

Figure 38 shows an 'HC04 driving a TIL221 gallium phosphide light-emitting diode. The resistor performs the function of current limiter. The luminous intensity of the LED depends on the amount of forward current.

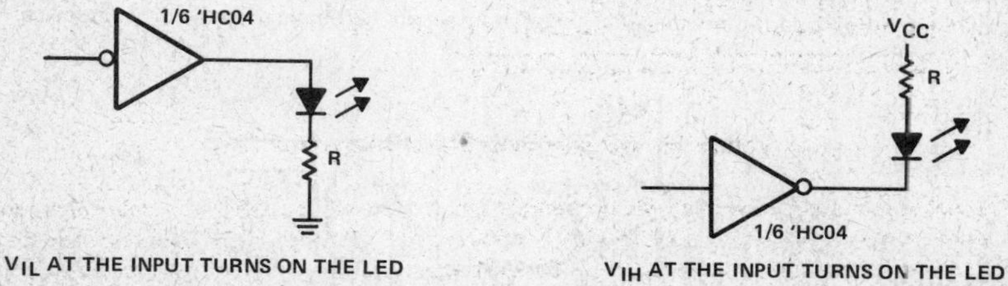

Figure 38. 'HC04 Driving a LED

Example: Using 10 mA forward current and 2.2 V forward voltage, the value of the current-limiting resistor can be calculated using the following equations:

$$\text{[for Figure 38(a)]} \quad R = \frac{V_{OH} - 2.2\ \text{V}}{10\ \text{mA}}$$

$$\text{[for Figure 38(b)]} \quad R = \frac{V_{CC} - 2.2\ \text{V} - V_{OL}}{10\ \text{mA}}$$

It should be noted that as used here, V_{OH} and V_{OL} are not the V_{OH}min and V_{OL}max specified in the data book. Figures 39 and 40 show typical values for V_{OH} and V_{OL} for an 'HC00.

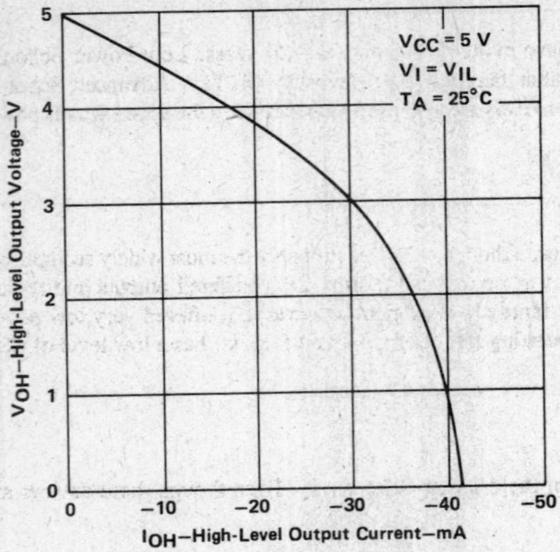

Figure 39. Typical Values for V_{OH}

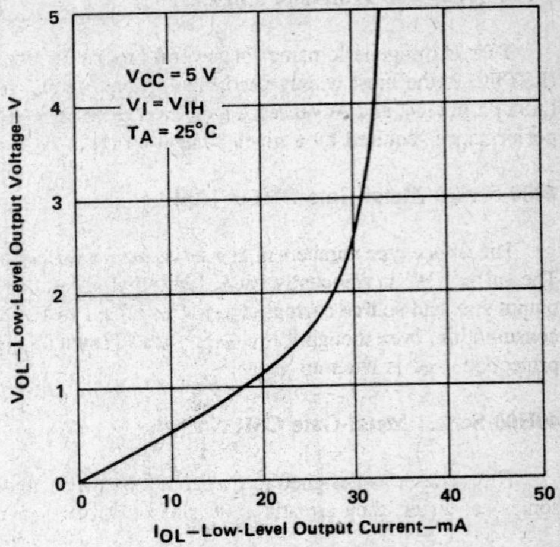

Figure 40. Typical Values for V_{OL}

DRIVING RELAYS

Multiple gates can be connected in parallel to increase the current sinking or sourcing capability of SN54/74HC devices. Figure 41 shows two 'HC04 gates connected in parallel for relay driver application.

Precautions should be taken to prevent one gate from "hogging" the current. Small resistors (typically 50 Ω) in series with the output gate will limit the possibility of "current hogging" by any one gate.

In all applications in which the SN54/74HC output is required to source or sink substantial current (6 mA to 25 mA), particular attention should be paid to providing adequate power supply decoupling for the driving device.

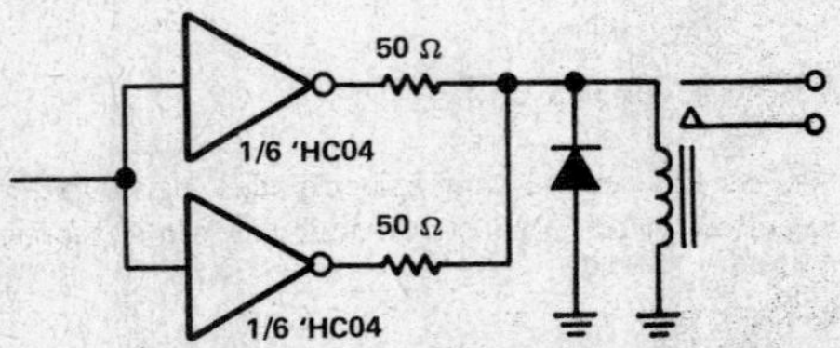

Figure 41. SN54/74HC04 Gates Connected in Parallel to Drive a Relay

SN54HC/SN74HC Interchangeability Guide

INTRODUCTION

The following has been prepared as a guide to interchanging devices from other logic families, both bipolar and CMOS, with those from the SN54HC/SN74HC family. This is not intended to be a comprehensive guide since interchangeability can depend on many factors, and only careful data sheet comparisons can provide definitive answers. The considerations listed below are based upon information accumulated in answering a large number of inquiries in this area.

First, a brief review is given on each logic technology, and second, discussion is given on the various aspects involved in attempting to interchange that technology with the SN54HC/SN74HC family.

TTL: Transistor-Transistor Logic

TTL is the generic name for several bipolar families that have evolved over the past 20 years. Low-Power Schottky (LSTTL) is the most widely used bipolar logic family today. Other families, e.g., Schottky (STTL), Advanced Schottky (ASTTL or AS), and Advanced Low-Power Schottky (ALSTTL or ALS) are also used, depending on the speed versus power performance required by a given system design.

4000 Series: Metal-Gate CMOS Logic

The device type numbers in this series have a variety of prefixes, although "CD" is probably the most widely recognized. The suffix "B" is frequently used, indicating an improvement over the original family, i.e., buffered outputs and typical output sink and source current capabilities of ± 1 mA. This logic family became popular because it offered very low power consumption, even though it is slower than TTL with a typical operating frequency of about 5 MHz, has a low level of ESD protection, and is latch-up prone.

40H00 Series: Metal-Gate CMOS Logic

This series was designed to overcome the speed limitations of the original 4000 family. Even though these devices are somewhat faster, they are still slow when compared to LSTTL.

74C00 Series: Metal-Gate CMOS Logic

The distinguishing feature of this family is that the pinouts correspond to those of TTL, making interchangeability easier. The devices, however, exhibit many of the same speed/power limitations as those of the 4000 series. The fan-out is typically higher than the 4000 series, however, with typical output sink and source capabilities of ± 1.75 mA.

74SC00 Series: Silicon-Gate CMOS Logic

This series was the forerunner to the SN54HC/SN74HC family, or more closely, to the SN54HCT/SN74HCT family. The 74SC family was designed to overcome many of the 4000 series deficiencies, particularly the slower speed and the lower drive capability.

Note: The "SC" designation should not be confused with that of Texas Instruments new Standard Cell family (SN54SC/SN74SC series).

INTERCHANGEABILITY CONSIDERATIONS

Listed below are the highlights of benefits derived from replacing other logic families with SN54HC/SN74HC; also listed are important considerations that may affect the feasibility or desirability of such replacement. All comparisons are by necessity general in nature.

LSTTL

Considerations:

1. SN54HC/SN74HC high-level input voltage is not TTL-compatible. In a mixed family system (LS output driving HC input) it will be necessary to use SN54HCT/SN74HCT, pull-up resistors, or level shifters.
2. SN54HC/SN74HC has less drive capability than some LSTTL functions.
3. LSTTL open-collector outputs have higher breakdowns than SN54HC/SN74HC open-drain equivalent functions.

HCMOS advantages:

1. Lower system power consumption
2. Improved noise immunity
3. Wider supply voltage range.

Other TTL Families

Considerations:

1. SN54HC/SN74HC high-level input voltage is not TTL-compatible. In a mixed family system (TTL output driving HC input) it will be necessary to use SN54HCT/SN74HCT, pull-up resistors, or level shifters.
2. SN54HC/SN74HC has less drive capability than some TTL functions.
3. TTL open-collector outputs have higher breakdowns than SN54HC/SN74HC open-drain equivalent functions.
4. Some of the TTL families offer greater operating speed, e.g., STTL, AS, and ALS.

HCMOS advantages:

1. Lower system power consumption
2. Improved noise immunity
3. Wider supply voltage range.

4000 Series and 74C00 Series

Considerations:

1. Although most applications use a 5-V supply, these older families operate in the 3-V to 15-V range.
2. SN54HC/SN74HC must be operated with a supply voltage in the 2-V to 6-V range.

HCMOS advantages:

1. Higher frequency of operation
2. Improved ESD protection and latch-up performance
3. Higher output drive capability.

40H00 Series

Considerations:

1. Although most applications use a 5-V supply, this family will operate in the 2-V to 8-V range.
2. SN54HC/SN74HC must be operated with a supply voltage in the 2-V to 6-V range.

HCMOS advantages:

1. Higher frequency of operation
2. Improved ESD protection and latch-up performance
3. Higher output drive capability
4. Multiple-sourced family.

As a quick reference guide, Table 4 shows highlights of interchanging other logic families with high-speed CMOS.

CONCLUSION

Within the constraints given above, the SN54HC/SN74HC family can be regarded as pin-for-pin equivalents to the other logic families. The rapidly-expanding SN54HC/SN74HC family is ideally suited for system upgrading, system shrinking, or especially, new system design.

Table 4. Highlights of Interchangeability

	TTL FAMILY (TTL, LSTTL, STTL, ALS, AS)	METAL-GATE CMOS
Power	HCMOS offers lower system power consumption than any of the TTL families.	Power consumption of HCMOS is less than metal-gate CMOS.
Speed	HCMOS operating speed is comparable to LSTTL. Some TTL families (STTL, AS, and ALS) offer greater operating speed.	HCMOS operating speed is much faster than metal-gate CMOS.
Input Voltage	The V_{IH}min of HCMOS is not compatible with the V_{OH}min of TTL. In a mixed family system, it is necessary to use 'HCT devices, pull-up resistors, or level shifters.	HCMOS input voltage levels are compatible with metal-gate CMOS outputs only when the power supply voltage for the metal-gate CMOS devices is between 2 V and 6 V.
Output Voltage	The output voltages of HCMOS are TTL-compatible.	HCMOS output voltage levels are compatible with metal-gate CMOS inputs only when the power supply voltage for the metal-gate CMOS devices is between 2 V and 6 V.
Drive Capability	The output current capability of HCMOS is not as large as the TTL family.	HCMOS has a higher current drive capability.
Fan-out (LS devices)	HCMOS has a smaller fan-out to LS devices than the TTL family.	HCMOS has a higher fan-out to LS cevices.
Supply Voltage	HCMOS has a wide operating supply voltage range (2 V to 6 V).	Operating supply range of metal-gate is larger than HCMOS (from 3 V to 15 V).
ESD and Latch-Up	TTL family devices are not as vulnerable to ESD and latch-up damage.	HCMOS has an improved protection circuitry against ESD and latch-up.

Guidelines for Handling Electrostatic-Discharge Sensitive (ESDS) Devices and Assemblies

SCOPE

This specification establishes the requirements for methods and materials used to protect electronic parts, devices, and assemblies (items) susceptible to damage or degradation from electrostatic discharge (ESD). The electrostatic charges referred to in this specification are generated and stored on surfaces of ordinary plastics, most common textile garments, ungrounded people's bodies, and many other commonly unnoticed static generators. The passage of these charges through an electrostatic-sensitive part may result in catastrophic failure or performance degradation of the part.

The part types for which these requirements are applicable include, but are not limited to, those listed:

1) All metal-oxide semiconductor (MOS) devices, e.g., CMOS, PMOS, etc.
2) Junction field-effect transistors (JFET)
3) Bipolar digital and linear circuits
4) Op Amps, monolithic microcircuits with MOS compensating networks, on-board MOS capacitors, or other MOS elements
5) Hybrid microcircuits and assemblies containing any of the types of devices listed
6) Printed circuit boards and any other type of assembly containing static-sensitive devices.

Definitions

1. Antistatic material: ESD protective material having a surface resistivity between 10^9 and 10^{14} Ω/square.
2. Static dissipative material: ESD protective material having surface resistivity between 10^5 and 10^9 Ω/square.
3. Conductive material: ESD protective material having a surface resistivity of 10^5 Ω/square maximum.

4. Electrostatic discharge (ESD): A transfer of electrostatic charge between bodies at different electrostatic potentials caused by direct contact or induced by an electrostatic field.
5. Surface resistivity: An inverse measure of the conductivity of a material and is the resistance of unit length and unit width of a surface. Note: Surface resistivity of a material is numerically equal to the surface resistance between two electrodes forming opposite sides of a square. The size of the square is immaterial. Surface resistivity applies to both surface and volume conductive materials and has the dimension of Ω/square.
6. Volume resistivity: Also referred to as bulk resistivity. It is normally determined by measuring the resistance (R) of a square of material (surface resistivity) and multiplying this value by the thickness (T).
7. Ionizer: A blower that generates positive and negative ions, either by electrostatic means or by means of a radioactive energy source, in an airstream, and distributes a layer of low velocity ionized air over a work area to neutralize static charges.
8. Close proximity: For the purpose of this specification, is 6 inches or less.

Device Sensitivity per Test Circuit of Method 3015, MIL-STD-883

Devices are categorized according to their susceptibility to damage resulting from electrostatic discharge (ESD), and the type packaging required to adequately protect them.

1) Device electrostatic sensitivity:

Category	ESD Sensitivity (V)	Minimum Protective Packaging
A	20-2000	Antistatic Magazine & Conductive Bag/Box
B	> 2000	Antistatic Magazine & Antistatic Bag

2) Devices are to be categorized by their sensitivity
3) Devices are to be protected from ESD damage from receipt at incoming inspection through assembly, test and shipment of completed equipment.

APPLICABLE REFERENCE DOCUMENTS

The following reference documents (of latest issue) can provide additional information on ESD controls.

1) MIL-M-38510 Microcircuits, General Specification
2) MIL-STD-883 Test Methods and Procedures for Microelectronics
3) MIL-S-19491 Semiconductor Devices, Packaging of
4) MIL-M-55565 Microcircuits, Packaging of
5) DOD-HDBK-263 Electrostatic Discharge Control Handbook for Protection
6) DOD-STD-1686 Electrostatic Discharge Control Program
7) NAVSEA SE 003-11-TRN-010 Electrostatic Discharge Training Manual

FACILITIES FOR STATIC-FREE WORK STATION

The minimum acceptable static-free work station shall consist of the work surface covered with an ESD protective material attached to ground through a 1 MΩ ±10% resistor, an attached grounding wrist strap with integral 1 MΩ ±10% resistor for each operator, and air ionizer(s) of sufficient capacity for each operator. The wrist strap shall be connected to the ESD protective material. Ground shall utilize the standard building earth ground, refer to Figure 42. Conductive floor tile along with conductive shoes may be used in lieu of the conductive wrist straps. The Site Safety Engineer must review and approve all electrical connections at the static-free work station prior to its implementation.

Air ionizers shall be positioned so that the devices at the static-free work stations are within a 4-foot arc measured by a vertical line from the face of the ionizer and 45 degrees on each side of this line.

General grounding requirements are to be in accordance with Table 5.

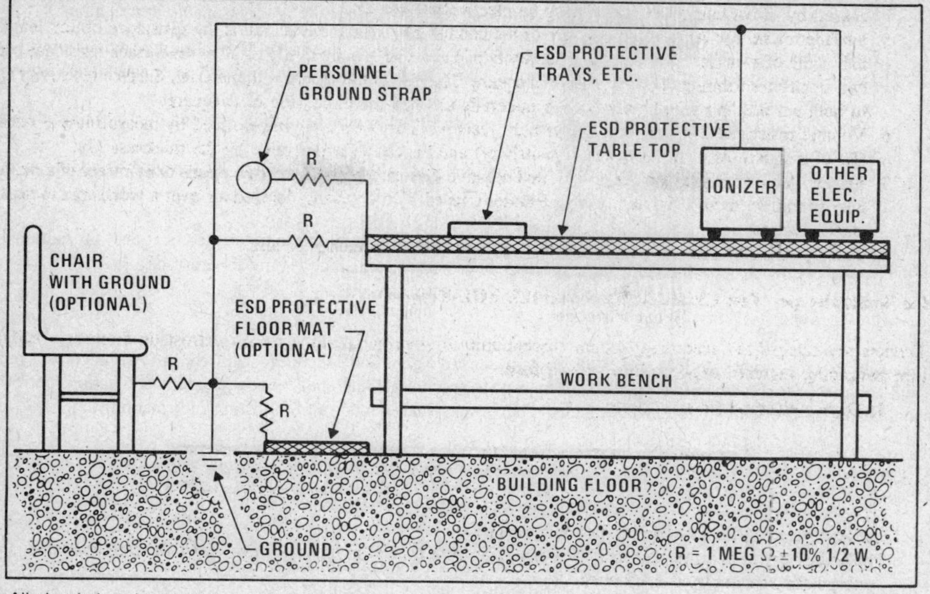

All electrical equipment sitting on the conductive table top must be hard grounded but must be isolated from the conductive table top.

NOTE: Earth ground is not computer ground or RF ground or any other limited type ground.

Figure 42. Static-Free Work Station

Table 5. General Grounding Requirements

	TREATED WITH ANTISTATIC SOLUTION OR MADE OF CONDUCTIVE MATERIAL	GROUNDED TO COMMON POINT
Handling Equipment/Handtools	X	
Metal Parts of Fixtures and Tools/Storage Racks		X
Handling Trays/Tubes	X	
Soldering Irons/Bath		X
Table Tops/Floor Mats	X	X
Personnel		X Using Wrist Strap*

*With 1 MΩ ± 10% resistor

Usage of Antistatic Solution in Areas to Control the Generation of Static Charges

The use of antistatic chemicals (antistats) should be a supplemental part of an overall organized ESD program. Any antistatic chemical application shall be considered as a means to reduce or eliminate static charge generation on nonconductive materials in the manufacturing or storage areas.

The application of any antistatic chemical in a clean room of class 10,000 or less shall not be permitted. Accordingly, any user of antistatic solutions must consider the following precautions:

1. Do not apply antistatic spray or solutions in any form to energized electrical parts, assemblies, panels, or equipment.
2. Do not perform antistatic chemical applications in any area when bare chips, raw parts, packages, and/or personnel are exposed to spray mists and evaporation vapors.

The need for initial application and frequency of reapplication can only be established through routine electrostatic voltage measurements using an electrostatic voltmeter. The following durability schedule is a reasonable expectation.

1) Soft surfaces (carpet, fabric seats, foam padding, etc.): each 6 months or after cleaning, by spraying.
2) Hard abused surfaces (floors, table tops, tools, etc.): each week (or day for heavy use) and after cleaning, by wiping or mopping.
3) Hard unabused surfaces (cabinets, walls, fixtures, etc.): each 6 months or annually and after cleaning, by wiping or spraying.
4) Company-furnished and maintained clothing and smocks: after each cleaning, by spraying or adding antistatic concentrate to final rinse water when cleaned.

The use of antistatic chemicals, their application, and compliance with all appropriate specifications, precautions, and requirements shall be the responsibility of the Area Supervisor where antistatic chemicals are used.

ESD Labels and Signs in Work Areas

ESD caution signs at work stations and labels on static-sensitive parts and containers shall be consistent in color, symbols, class, and voltage sensitivity identification, and appropriate instructions. Signs shall be posted at all work stations performing any handling operations with static-sensitive items. These signs shall contain the following information.

> **CAUTION**
>
> **STATIC CAN DAMAGE COMPONENTS**
>
> Do not handle ESDS items unless grounding wrist strap is properly worn and grounded. Do not let clothing or plain plastic materials contact or come in close proximity to ESDS items.

Labels shall be affixed to all containers containing static-sensitive items at a place readily visible and proper for the intended purpose. Additionally, labels must be consistently placed on containers and packages at a standard location to eliminate mishandling. Use only QC accepted and approved signs and labels to identify static-sensitive products and work areas. The use of ESD signs and labels, and their information content shall be the responsibility of the Area Supervisor to assure consistency and compatibility throughout the static-sensitive routing.

Relative Humidity Control

Since relative humidity has a significant impact on the generation of static electricity, when possible, the work area should be maintained within the following relative humidity ranges: incoming/assembly/test/storage 50%-65% (ref. Ashrae, 55—74), within ±5% to avoid static voltage monitor variations.

PREPARATION FOR WORKING AT STATIC-FREE WORK STATION

A work station with a conductive work surface connected to ground through a 1 MΩ ±10% resistor, a grounding wrist strap with the ground wire connected to the conductive work surface, and an ionizer constitute a static-free work station (Figure 42). An operator is properly grounded when the wrist strap is in snug (no slack) contact with the bare skin, usually positioned on the left wrist for a right-handed operator. The wrist strap must be worn the entire time an operator is at a static-free work station. The operator should first touch the grounded bench top before handling static-sensitive items. This precaution should be observed in addition to wearing the grouding wrist strap. If possible, operators should avoid touching leads or contacts even though grounded.

> **CAUTION**
>
> Personnel shall never be attached to ground without the presence of the 1 MΩ ±10% series resistor in the ground wire.

An operator's clothing should never make contact or come in close proximity with static sensitive items. They must be especially careful to prevent any static-sensitive items (being handled) from touching their clothing. Long sleeves must be rolled up or covered with antistatic sleeve protector banded to the bare wrist which shall "cage" the sleeve at least as far up as the elbow. Only antistatic finger cots may be used when handling static-sensitive items.

Any person not properly prepared, while at or near the work station, shall not touch or come in close proximity with any static-sensitive items. It is the responsibility of the operator and the Area Supervisor to ensure that the static-free work area is clear of unnecessary static hazards, including such personal items as plastic coated cups or wrappers, plastic cosmetic bottles or boxes, combs, tissue boxes, cigarette packages, and vinyl or plastic purses. All work-related items, including information sheets, fluid containers, tools, and parts carriers must be those approved for use at the static-free work station.

GENERAL HANDLING PROCEDURES AND REQUIREMENTS

1. All static-sensitive items must be received in an antistatic/conductive container and must not be removed from the container except at static-free work station. All protective folders or envelopes holding documentation (lot travelers, etc.) shall be made of nonstatic-generating material.
2. Each packing (outermost) container and package (internal or intermediate) shall have a bright yellow warning label attached, stating the following information or equivalent:

   The warning label shall be legible and easily readable to normal vision at a distance of 3 feet.
3. Static-sensitive items are to remain in their protective containers except when actually in work at the static-free station.
4. Before removing the items from their protective container, the operator should place the container on the conductive grounded bench top and make sure the wrist strap fits snugly around the wrist and is properly plugged into the ground receptacle, then touch hands to the conductive bench top.
5. All operations on the items should be performed with the items in contact with the grounded bench top as much as possible. Do not allow conductive magazine to touch hard grounded test gear on bench top.
6. Ordinary plastic solder-suckers and other plastic assembly aids shall not be used.
7. In cases where it is impossible or impractical to ground the operator with a wrist strap, a conductive shoe strap may be used along with conductive tile/mats.
8. When the operator moves from any other place to the static-free station, the start-up procedure shall be the same as in PREPARATION FOR WORKING AT STATIC-FREE WORK STATION.
9. The ionizer shall be in operation prior to presenting any static-sensitive items to the static-free station, and shall be in operation during the entire time period the items are at the station.
10. "Plastic snow" polystyrene foam, "peanuts," or other high-dielectric materials shall never come in contact with or be used around electrostatic sensitive items, unless they have been treated with an antistat (as evidenced by pink color and generation of less than ± 100 volts).
11. Static-sensitive items shall not be transported or stored in trays, tote boxes, vials, or similar containers made of untreated plastic material unless items are protectively packaged in conductive material.

PACKAGING REQUIREMENTS

Packaging of static-sensitive items is to be in accordance with Device Sensitivity, item 1). The use of tape and plain plastic bags is prohibited. All outer and inner containers are to be marked as outlined in GENERAL HANDLING PROCEDURES AND REQUIREMENTS, item 2, and conductive magazines/boxes may be used in lieu of conductive bags.

SPECIFIC HANDLING PROCEDURES FOR STATIC-SENSITIVE ITEMS

Stockroom Operations

1. Containers of static-sensitive items are not to be accepted into stock unless adequately identified as containing static-sensitive items.
2. Items may be removed from the protective container (magazine/bag, etc.) for the purpose of subdividing for order issue only by a properly grounded operator at an approved static-free station as defined in FACILITIES FOR and PREPARATION FOR WORKING AT STATIC-FREE WORK STATION.
3. All subdivided lots must be carefully repackaged in protective containers (magazine/bag, etc.) prior to removal from the static-free work station and labeled to indicate that the package(s) contain static-sensitive items. If it is suspected that a static-sensitive item is not adequately protected, do not transfer it to another container, return it to the originator for disposition unless the originator is a Customer. In that case, the QC Engineer should contact the Customer and negotiate an appropriate disposition.
4. It is the responsibility of the Stockroom Supervisor to ensure that all personnel assigned to this operation are familiar with handling procedures as outlined in this specification. A copy of this specification is to be posted in the vicinity so that it is accessible to the operators. Stock handlers and all others who might have occasion to move stock are to be instructed to avoid direct contact with unprotected static-sensitive items.

Module and Subassembly Operations

1. Static-sensitive items are not to be received from a stockroom, kitting, or machine insertion area unless received in approved static-protective packaging, and properly labeled to indicate that its contents are static sensitive.
2. All single station, progressive line manual assembly operators, and visual inspectors prior to wave soldering operations are to be properly grounded with a grounding wrist strap when handling static-sensitive items.
3. Progressive lines used as single stations where operators will be working on a mix of boards, both static-sensitive and nonstatic-sensitive, will require that all operators working on the line be properly grounded. This is necessary to accommodate the sliding of static-sensitive boards along the assembly bench or across positions not engaged in the assembly of this type board.
4. It is the responsibility of the Area Supervisor to ensure that all personnel handling static-sensitive items are familiar with this procedure and fully aware of the damage or degradation of these units in the event of noncompliance. A periodic inspection should be made using an electrostatic voltmeter to assure that the static-free stations are in proper working order and to ensure that operators are wearing grounding wrist straps properly (snugly in contact with bare skin).

Soldering and Lead-Forming Operations

1. All soldering machines, conveyors, cleaning machines, and equipment shall be electrically grounded to ensure that they are at the same ground potential as the grounded operators working on their stations. No machine surfaces exposed to static-sensitive items are to be above the ground potential.
2. All processing equipment shall be grounded, including all loading and unloading stations, that is, the stations before and after each piece of processing equipment.
3. All nonmetallic, static-generating components in the handling systems shall be treated to ensure protection from static.
4. All stations shall be identified by posting signs as outlined in **ESD Labels and Signs in Work Areas.**
5. Operators are to be properly grounded with a grounding wrist strap during any handling, loading, unloading, inspection, rework, or proximity to static-sensitive items.
6. Unloading operators working at a grounded station shall place static-sensitive items into approved static-protective bags or containers.
7. All manual soldering, repair, and touch-up work stations on the solder line are to be static protected. Operators are to wear grounding wrist straps when working on static-sensitive items. Only grounded-tip soldering/desoldering irons are allowed when working on static-sensitive items.
8. It is the responsibility of the Area Spervisor to ensure that all personnel handling static-sensitive items are familiar with this procedure and fully aware of the damage or degradation of these units in the event of noncompliance. A periodic inspection should be made using an electrostatic voltmeter to assure that the static-free stations are in proper working order and to ensure that operators are wearing grounding wrist straps properly (comfortably snug in contact with bare skin).

Electrical Testing Operations

1. All electrical test stations shall be static protected. Operators shall be properly grounded when working on these items.
2. Reused antistatic magazines must be monitored for maintenance of antistatic characteristics.
3. Devices should be in an antistatic/conductive environment except at the moment when actually under test.
4. Devices should not be inserted into or removed from circuits or tester with the power on or with signals applied to inputs to prevent transient voltages from causing permanent damage.
5. All unused input leads should be biased if possible.
6. Device or module repairs must be performed at static-free stations with the operator attached to a grounding wrist strap. Grounded-tip soldering irons shall be used when working on static-sensitive items.
7. Static-sensitive items shall be handled through all electrical inspections in static protective containers. Removal of the items from the protective containers shall be done at a static-free work station as discussed in **PREPARATION FOR WORKING AT A STATIC-FREE WORK STATION**. The units must be returned to the containers before leaving the station.
8. All such items shall be shipped with an ESD warning label affixed as listed.
9. It is the responsibility of the Area Supervisor to ensure that all personnel handling static-sensitive items are familiar with this procedure and fully aware of the damage or possible degradation of these units in the event of noncompliance. A periodic inspection should be made using an electrostatic voltmeter to assure that the static-free stations are in proper working order and to ensure that operators are wearing grounding straps properly (snugly in contact with bare skin).

Packing Operations

1. Static-sensitive items are not to be accepted into the packing area unless they are contained in a static-protected bag or conductive container.
2. A static-sensitive item delivered to the packer within an approved container or bag and found to be in order regarding identification shall be packed in the standard shipping carton or other regular packaging material. Containers are to be labeled in accordance with **GENERAL HANDLING PROCEDURES AND REQUIREMENTS,** item 2.
3. Any void-fillers shall be made of an approved antistatic material.

Burn-In Operations

1. Burn-in board loading and unloading of static-sensitive items shall be done at a static-free station.
2. Shorting clips/shorted connectors shall be installed on the board plug-in tab prior to loading any units into the board sockets. The clip/connector shall be taken off just prior to plugging the board into the oven connector. The clip/connector shall be installed immediately upon removal of the board from the oven connector. Installation and removal of the clip/connector shall be done by a properly grounded operator.
3. All automatic or semiautomatic loading and unloading equipment shall be properly electrically grounded.
4. It is the responsibility of the Area Supervisor to ensure that all personnel handling static-sensitive items are familiar with this procedure and fully aware of the damage or possible degradation of these units in the event of noncompliance. A periodic inspection should be made using an electrostatic voltmeter to assure that the static-free stations are in proper working order and to ensure that operators are wearing grounding straps properly (snugly in contact with bare skin).

CUSTOMER RETURNED ITEM HANDLING PROCEDURE

Receipt of ESDS-labeled items is to be done at a static-free work station and handled in accordance with applicable sections within this guideline.

QUALITY CONTROL PROVISIONS

Sampling

Each manufacturing, stockroom, and testing operation handling ESDS devices will be audited a minimum of once each quarter for compliance with all terms of this specification by the responsible process control or QRA organization. Ground continuity and the presence of uncontrolled static voltages are considered critical and shall be checked more frequently as specified below.

Ground Continuity (minimum of once a week).

Ground connections (grounding wrist strap, ground wires on cords, etc.) shall be checked for electrical continuity. The presence of a 1 MΩ ±10% resistor in the ground connections between both the operator wrist straps to the work surface and the work surface to ground connector must be verified.

Grounded Conditions (minimum of once a week).

A visual inspection shall be made to determine full compliance with this specification at static-free work stations during handling of static-sensitive items, including operator being grounded as required, static-sensitive items not being handled in unprotected or unauthorized areas, and no static-generating materials at the grounded work station.

Sleeve Protectors (minimum of once a week).

A visual check shall be made to determine that each operator wearing loose-fitting or long-sleeved clothing either has sleeves properly rolled or covered with sleeve protectors properly grounded to the bare skin at the wrist.

Static Voltage Levels (minimum of once a week).

In addition to the visual inspections, a sample inspection using an electrostatic voltmeter will be used to check for uncontrolled electrostatic voltages at or near electrostatic-controlled work stations.

Conductive Floor Tiles (minimum of once a month).

Conductive floors must have a resistance of not less than 25 kΩ from any point on the tile to earth ground. Also, resistance from any point-to-point on the tile floor 3 feet apart shall be not less than 25 kΩ. The test methods to be used are ASTM-F-150-72 and NFPA 56.

Records

Written records must be kept of all these QC audits.

TRAINING

Training is applicable for all areas where individuals come in contact with ESDS (category A) devices. It is the responsibility of each Area Supervisor to make sure that his/her people receive ESD training initially and every 12 months thereafter to maintain proficiency. Training should include static fundamentals, a review of applicable parts of this specification, and actual applications in the work area.

ORDERING INSTRUCTIONS

Electrical characteristics presented in this data book, unless otherwise noted, apply for circuit type(s) listed in the page heading regardless of package. The availability of a circuit function in a particular package is denoted by an alphabetical reference above the pin-connection diagram(s). These alphabetical references refer to mechanical outline drawings shown in this section.

Factory orders for circuits described in this catalog should include a four-part type number as explained in the following example.

EXAMPLE **SN** **54HC02** **J** **-00**

1. Prefix → SN

MUST CONTAIN TWO TO FOUR LETTERS

SN	Standard Prefix
SNJ	MIL-STD-883 Processed and Screened per JEDEC Standard 101
JANB	MIL-M-38510 Processed

2. Unique Circuit Description → 54HC02

MUST CONTAIN SIX TO NINE CHARACTERS

Examples:

54HC00
74HC74
74HCT620
74HC4002

3. Package → J

MUST CONTAIN ONE OR TWO LETTERS

J, JT, N, NT (Dual-in-line packages)†

FH, FK, or FN (Chip carriers)

(From pin-connection diagram on individual data sheet)

4. Instructions (Dash No.) → -00

MUST CONTAIN TWO NUMBERS

– 00 No special instructions
– 10 Solder-dipped leads (N and NT packages only)

† These circuits in dual-in-line packages are shipped in one of the carriers shown below. Unless a specific method of shipment is specified by the customer (with possible additional costs), circuits will be shipped in the most practical carrier. Please contact your TI sales representative for the method that will best suit your particular needs.

Dual-in-line (J, JT, N, NT)

– Slide Magazines
– A-Channel Plastic Tubing
– Barnes Carrier (N only)
– Sectioned Cardboard Box
– Individual Plastic Box

FH and FK ceramic chip carrier packages

Both versions of these hermetically sealed chip carrier packages have ceramic bases. The FH package is an all-ceramic package with a glass seal. The FK package has a three-layer base with a metal lid and braze seal.

The packages are intended for surface mounting on solder lands on 1,27 (0.050) centers. Terminals require no additional cleaning or processing when used in soldered assembly.

FH and FK package terminal assignments conform to JEDEC Standards 1 and 2.

FH AND FK CERAMIC CHIP CARRIER PACKAGES

(28-terminal package shown)

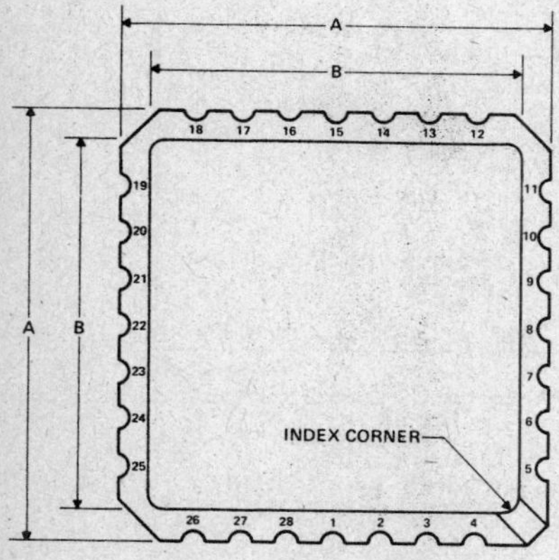

CERAMIC CHIP CARRIERS

JEDEC OUTLINE DESIGNATION*	NO. OF TERMINALS	A MIN	A MAX	B MIN	B MAX
MS004CB	20	8,69 (0.342)	9,09 (0.358)	7,80 (0.307)	9,09 (0.358)
MS004CC	28	11,23 (0.442)	11,63 (0.458)	10,31 (0.406)	11,63 (0.458)

*All dimensions and notes for the specified JEDEC outline apply.

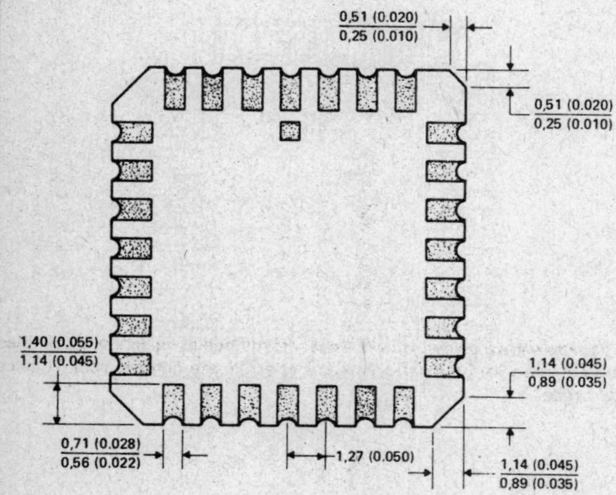

ALL LINEAR DIMENSIONS ARE IN MILLIMETERS AND PARENTHETICALLY IN INCHES

FN plastic chip carrier package

Each of these chip carrier packages consists of a circuit mounted on a lead frame and encapsulated within an electrically nonconductive plastic compound. The compound withstands soldering temperatures with no deformation, and circuit performance characteristics remain stable when the devices are operated in high-humidity conditions. The packages are intended for surface mounting on solder lands on 1,27 (0.050) centers. Leads require no additional cleaning or processing when used in soldered assembly.

FN PLASTIC CHIP CARRIER PACKAGE

(28-terminal package shown)

NO. OF TERMINALS	A MIN	A MAX	B MIN	B MAX	C MIN	C MAX
20	9,70 (0.382)	10,03 (0.395)	8,89 (0.350)	9,04 (0.356)	8,08 (0.318)	8,38 (0.330)
28	12,24 (0.482)	12,57 (0.495)	11,43 (0.450)	11,58 (0.456)	10,62 (0.418)	10,92 (0.430)

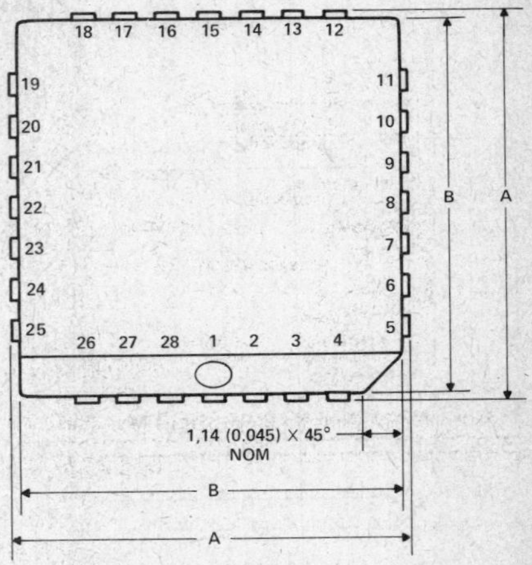

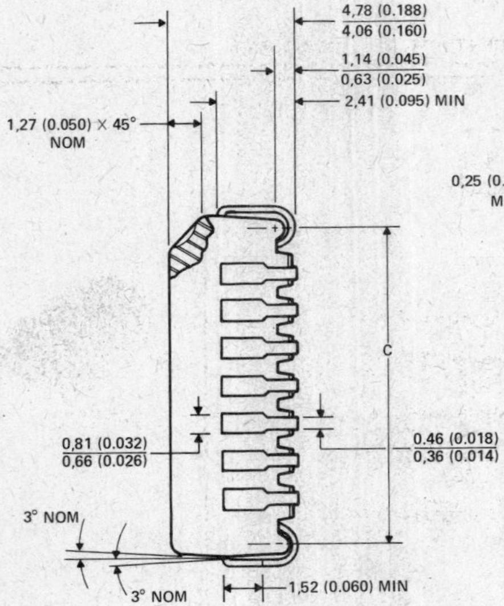

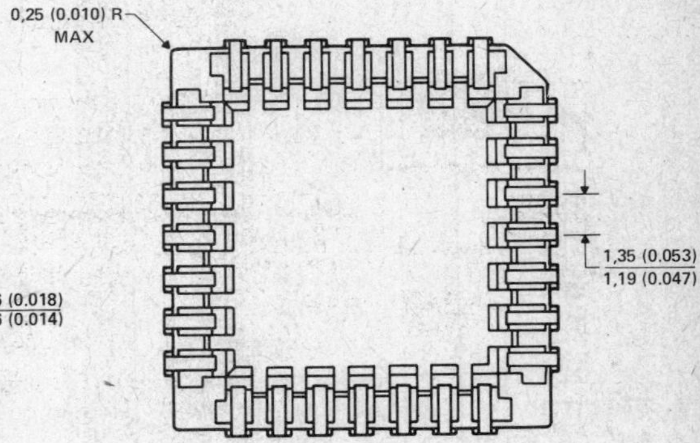

ALL LINEAR DIMENSIONS ARE IN MILLIMETERS AND PARENTHETICALLY IN INCHES

MECHANICAL DATA

8

J ceramic packages (including JT packages)

Each of these hermetically sealed dual-in-line packages consists of a ceramic base, ceramic cap, and a lead frame. Hermetic sealing is accomplished with glass. Once the leads are compressed and inserted, sufficient tension is provided to secure the package in the board during soldering. Tin-plated (''bright-dipped'') leads require no additional cleaning or processing when used in soldered assembly.

14-PIN J CERAMIC

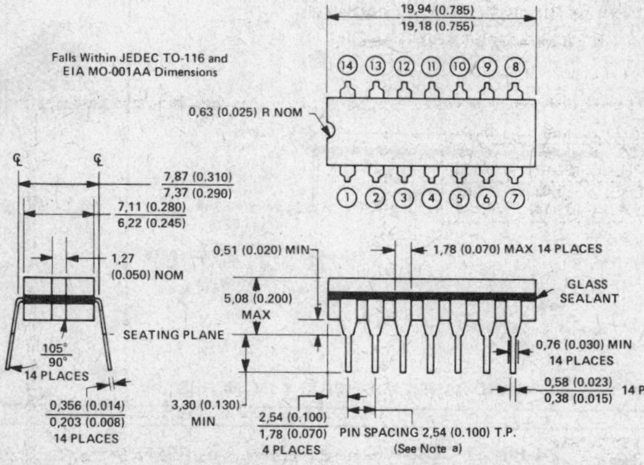

ALL LINEAR DIMENSIONS ARE IN MILLIMETERS AND PARENTHETICALLY IN INCHES

16-PIN J CERAMIC

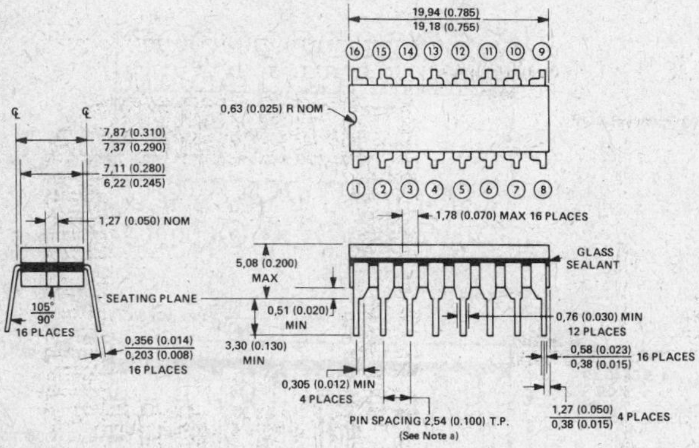

ALL LINEAR DIMENSIONS ARE IN MILLIMETERS AND PARENTHETICALLY IN INCHES

NOTE A: Each pin centerline is located within 0,25 (0.010) of its true longitudinal position.

J ceramic dual-in-line packages (continued)

20-PIN J CERAMIC

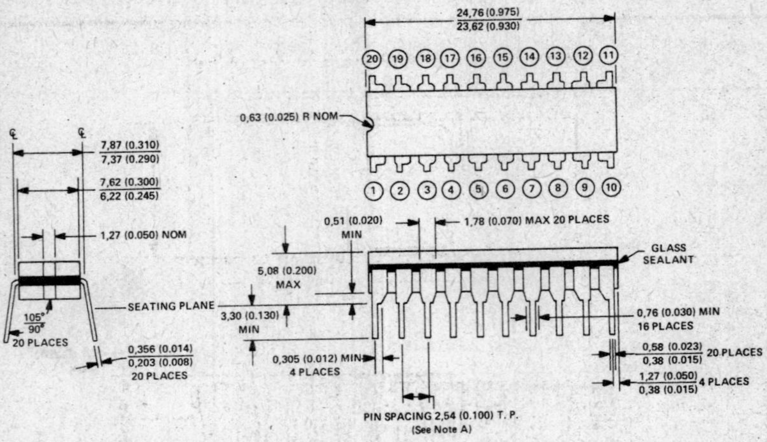

ALL LINEAR DIMENSIONS ARE IN MILLIMETERS AND PARENTHETICALLY IN INCHES

24-PIN JT CERAMIC

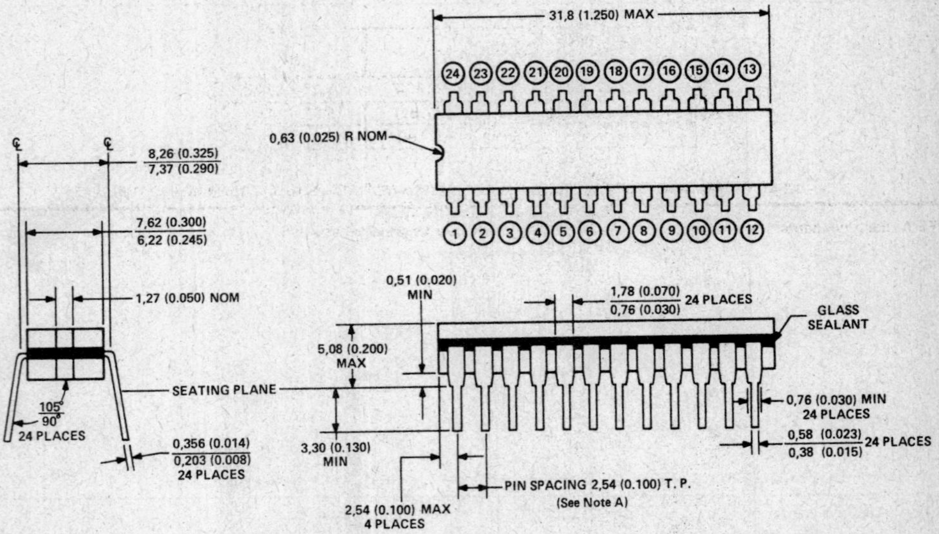

ALL LINEAR DIMENSIONS ARE IN MILLIMETERS AND PARENTHETICALLY IN INCHES

NOTE A: Each pin centerline is located within 0,25 (0.010) of its true longitudinal position.

JD ceramic dual-in-line packages — side-braze

This is a hermetically sealed ceramic package with a metal cap and side-brazed tin-plated leads.

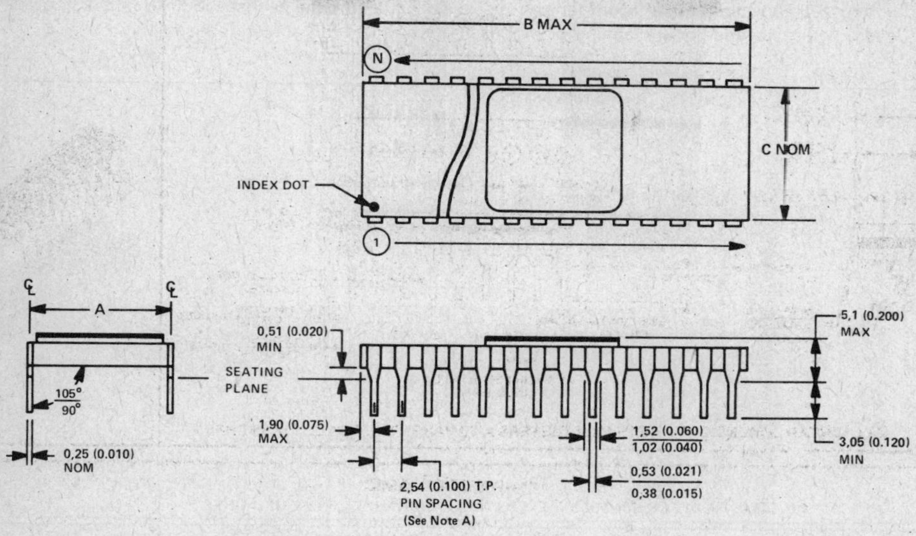

PINS / DIM	52	28
A ± 0,25 (0.010)	15,24 (0.600)	15,24 (0.600)
B MAX	67,3 (2.65)	35,94 (1.415)
C NOM	15,0 (0.590)	15,11 (0.595)

ALL LINEAR DIMENSIONS ARE IN MILLIMETERS AND PARENTHETICALLY IN INCHES

NOTE A: Each pin centerline is located within 0,25 (0.010) of its true longitudinal position.

N plastic packages (including NT package)

Each of these dual-in-line packages consists of a circuit mounted on a lead frame and encapsulated within a plastic compound. The compound will withstand soldering temperature with no deformation and circuit performance characteristics remain stable when operated in high-humidity conditions. Once the leads are compressed and inserted, sufficient tension is provided to secure the package in the board during soldering. Leads require no additional cleaning or processing when used in soldered assembly.

14-PIN N PLASTIC

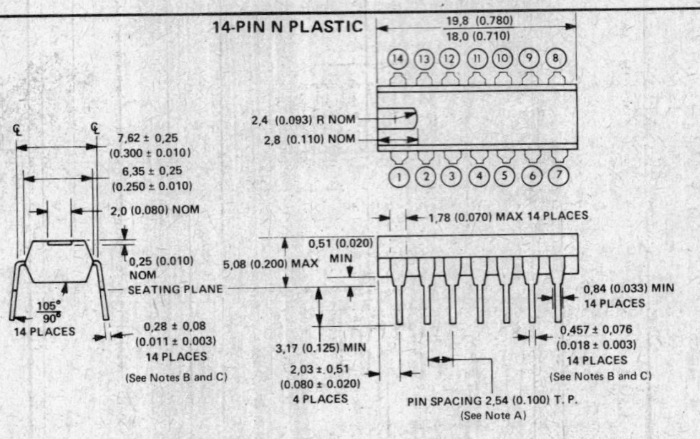

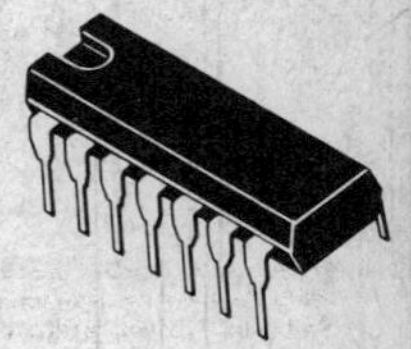

Falls Within JEDEC TO-116 and EIA MO-001AA Dimensions

ALL LINEAR DIMENSIONS ARE IN MILLIMETERS AND PARENTHETICALLY IN INCHES

16-PIN N PLASTIC

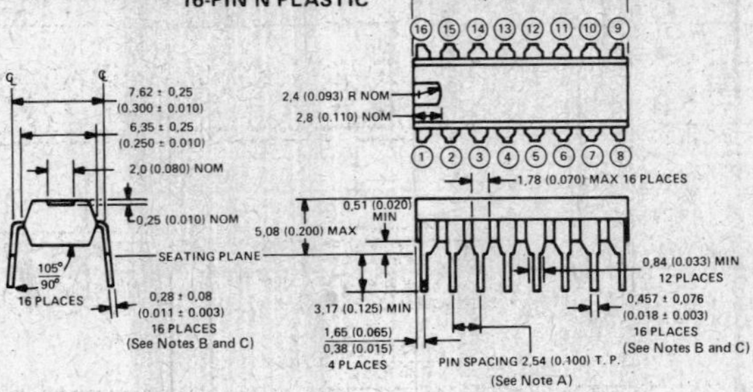

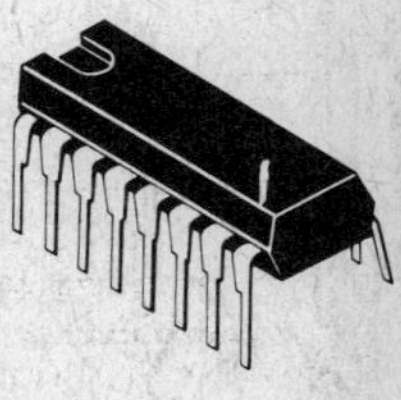

ALTERNATE SIDE VIEW

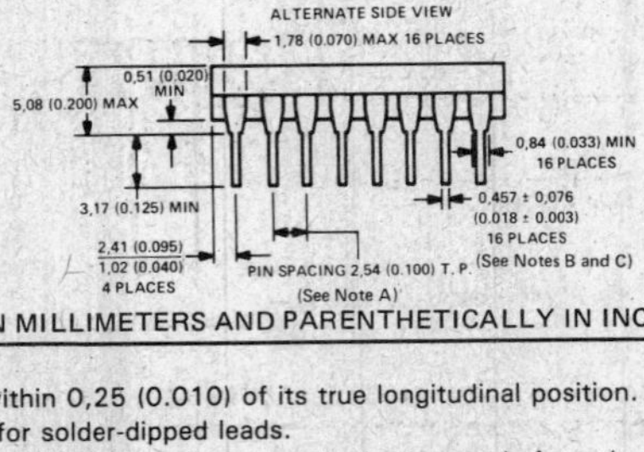

Parts may be supplied in accordance with the alternate side view at the option of TI plants located in Europe. In this case, the overall length of the package is 22,1 (0.870) max.

ALL LINEAR DIMENSIONS ARE IN MILLIMETERS AND PARENTHETICALLY IN INCHES

NOTES: A. Each pin centerline is located within 0,25 (0.010) of its true longitudinal position.
B. This dimension does not apply for solder-dipped leads.
C. When solder-dipped leads are specified, dipped areas of the lead extends from the lead tip to at least 0,51 (0.020) above seating plane.

N plastic dual-in-line packages (continued)

20-PIN N PACKAGE

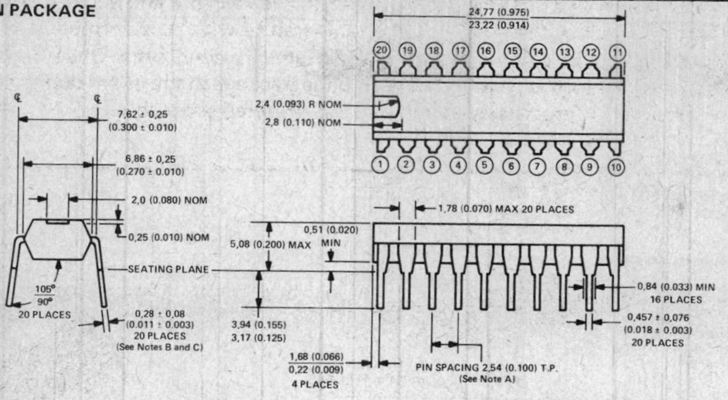

ALTERNATE SIDE VIEW

Parts may be supplied in accordance with the alternate side view at the option of TI plants located in Europe. In this case, the overall length of the package is 26,7 (1.050) max.

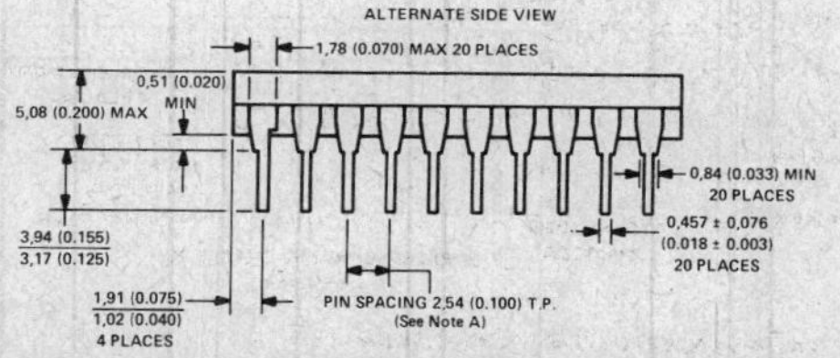

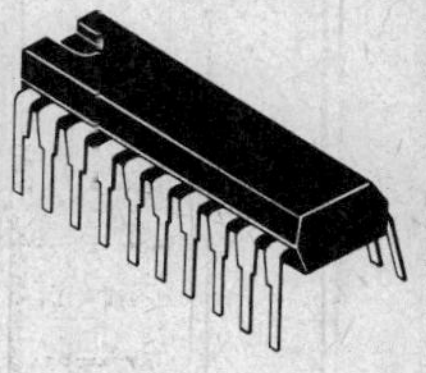

ALL LINEAR DIMENSIONS ARE IN MILLIMETERS AND PARENTHETICALLY IN INCHES

24-PIN NT PLASTIC

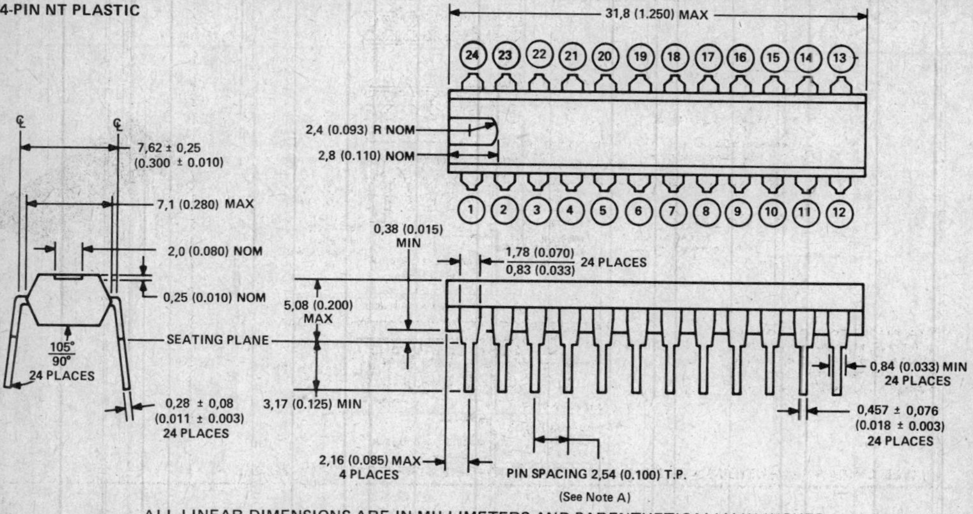

ALL LINEAR DIMENSIONS ARE IN MILLIMETERS AND PARENTHETICALLY IN INCHES

NOTES: A. Each pin centerline is located within 0,25 (0.010) of its true longitudinal position.
B. This dimension does not apply for solder-dipped leads.
C. When solder-dipped leads are specified, dipped areas of the lead extends from the lead tip to at least 0,51 (0.020) above seating plane.

N plastic dual-in-line packages (continued)

28-PIN N PLASTIC

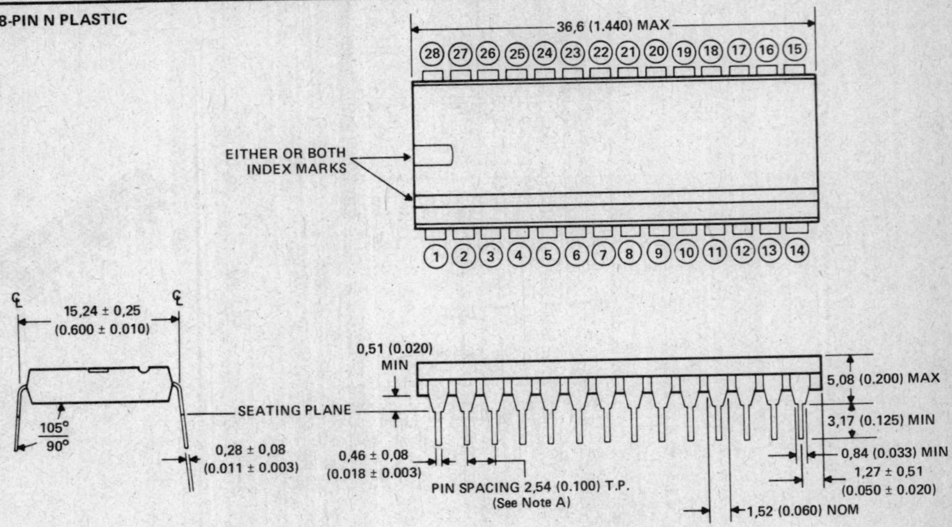

ALL LINEAR DIMENSIONS ARE IN MILLIMETERS AND PARENTHETICALLY IN INCHES

NOTE A: Each pin centerline is located within 0,25 (0.010) of its true longitudinal position.

52-PIN N PLASTIC

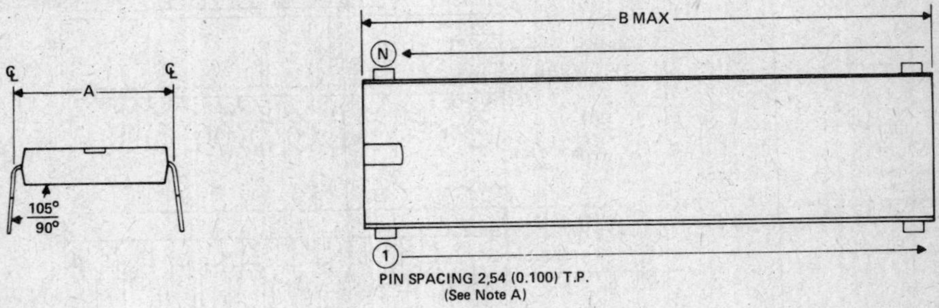

DIM \ PINS	52
A ± 0,25 (0.010)	15,24 (0.600)
B MAX	67,3 (2.65)

ALL LINEAR DIMENSIONS ARE IN MILLIMETERS AND PARENTHETICALLY IN INCHES

NOTE A: Each pin centerline is located within 0,25 (0.010) of its true longitudinal position.

TEXAS INSTRUMENTS
POST OFFICE BOX 225012 • DALLAS, TEXAS 75265

NOTES

NOTES

NOTES

NOTES

NOTES

Texas Instruments Semiconductor Technical Literature

TTL Data Book, Vol. 1, 1984, 336 pages.
Product guide for all TI TTL devices, functional indexes, alphanumeric index, and general information.

TTL Data Book, Vol. 2, 1984, 1,000 pages.
Detailed specifications and application information on the TI family of Low-power Schottky (LS), Schottky (S), and standard TTL logic devices.

TTL Data Book, Vol. 3, 1984, 792 pages.
Detailed specifications and application information on the TI family of Advanced Low-power Schottky (ALS) and Advanced Schottky (AS) logic devices.

TTL Data Book, Vol. 4, 1984, 416 pages.
Detailed specifications and application information on the TI family of bipolar field-programmable logic (FPL), programmable read-only memories (PROM), random-access memories (RAM), microprocessors, and support circuits.

High-speed CMOS Logic Data Book, 1984, 580 pages.
Detailed specifications and application information on the TI family of High-speed CMOS logic devices. Includes product selection guide, glossary, and alphanumeric index.

Linear Circuits Data Book, 1984, 820 pages.
Detailed specifications on operational amplifiers, voltage comparators, voltage regulators, data-acquisition devices, a/d converters, timers, switches, amplifiers, and special functions. Includes LinCMOS™ functions. Contains product guide, interchangeability guide, glossary, and alphanumeric index.

Interface Circuits Data Book, 1981, 700 pages.
Includes specifications and applications information on TTL logic interface circuits, as well as product profiles on the line drivers/receivers and peripheral drivers.

Optoelectronics Data Book, 1983, 480 pages.
Contains more than 300 device types representing traditional optoelectronics (IREDs, LEDs, detectors, couplers, and displays), special components (avalanche, photodiodes, and transimpedance amplifiers), fiber optic components (sources, detectors, and interconnecting cables), and new image sensors (linear and arrays).

MOS Memory Data Book, 1984, 456 pages.
Detailed specifications on dynamic RAMs, static RAMs, EPROMs, ROMs, cache address comparators, and memory controllers. Contains product guide, interchangeability guide, glossary, and alphanumeric index. Also, chapters on testing and reliability.

TMS7000 Family Data Manual, 1983, 350 pages.
Detailed specifications and application information on TI's family of microprogrammable 8-bit microcomputers. Includes architecture description, device operation, instruction set, electrical characteristics, and mechanical data. TMS7000 microcomputers include versions in CMOS and SMOS and with on-board UART.

TMSxxxxx Microcomputer Data Manuals
These manuals contain detailed specifications and application information on specific TMSxxxxx microcomputers and peripherals. Include architecture description, device operation, instruction set, electrical characteristics, and mechanical data.

Assembly Language Programmer's Guides.
TMS32010, 1983, 160 pages.
TMS99000, 1983, 322 pages.
TMS7000, 1983, 160 pages.
Include general programming information, assembly instructions, assembler directives, assembler output, and application notes.

TMS32010 User's Guide, 1984, 400 pages.
Detailed application information on the TMS32010 Digital Signal Processor. Detailed reference manual on use of the TMS320 instruction set. Data sheets included.

Fundamentals of Microcomputer Design, 1982, 584 pages.
University textbook. Subjects include microprocessors, software, instruction sets, microcomputer programming, high-level languages, hardware features, microcomputer memory, and I/O design. A design example is included.

Understanding Series™ Books
The Understanding Series books form a library written for anyone who wants to learn quickly and easily about today's technology, its impact on our world, and its application in our lives. Each book is written in bright, clear, down-to-earth language and focuses on one aspect of what's new in today's electronics. Engineering concepts and theory are explained using simple arithmetic. Technical terms are explained in layman's language. Ideal for self-paced, individualized instruction. Currently 13 different titles in the series.

Write for current availability and prices to:
Texas Instruments
Information Publishing Center
P.O. Box 225012 MS-54
Dallas, TX 75265

April 1984

TI Sales Offices

ALABAMA: Huntsville (205) 837-7530.

ARIZONA: Phoenix (602) 995-1007.

CALIFORNIA: Irvine (714) 660-1200;
Sacramento (916) 929-1521;
San Diego (619) 278-9601;
Santa Clara (408) 980-9000;
Torrance (213) 217-7010;
Woodland Hills (213) 704-7759.

COLORADO: Aurora (303) 695-2800.

CONNECTICUT: Wallingford (203) 269-0074.

FLORIDA: Ft. Lauderdale (305) 973-8502;
Maitland (305) 660-4600; **Tampa** (813) 870-6420.

GEORGIA: Atlanta (404) 452-4600.

ILLINOIS: Arlington Heights (312) 640-3000.

INDIANA: Ft. Wayne (219) 424-5174;
Indianapolis (317) 248-8555.

IOWA: Cedar Rapids (319) 395-9550.

MARYLAND: Baltimore (301) 944-8600.

MASSACHUSETTS: Waltham (617) 895-9100.

MICHIGAN: Farmington Hills (313) 553-1500.

MINNESOTA: Eden Prairie (612) 828-9300.

MISSOURI: Kansas City (816) 523-2500;
St. Louis (314) 569-7600.

NEW JERSEY: Iselin (201) 750-1050.

NEW MEXICO: Albuquerque (505) 345-2555.

NEW YORK: East Syracuse (315) 463-9291;
Endicott (607) 754-3900; **Melville** (516) 454-6600;
Pittsford (716) 385-6770;
Poughkeepsie (914) 473-2900.

NORTH CAROLINA: Charlotte (704) 527-0930;
Raleigh (919) 876-2725.

OHIO: Beachwood (216) 464-6100;
Dayton (513) 258-3877.

OKLAHOMA: Tulsa (918) 250-0633.

OREGON: Beaverton (503) 643-6758.

PENNSYLVANIA: Ft. Washington (215) 643-6450;
Coraopolis (412) 771-8550.

TEXAS: Austin (512) 250-7655;
Houston (713) 778-6592; **Richardson** (214) 680-5082;
San Antonio (512) 496-1779.

UTAH: Murray (801) 266-8972.

VIRGINIA: Fairfax (703) 849-1400.

WASHINGTON: Redmond (206) 881-3080.

WISCONSIN: Brookfield (414) 785-7140.

CANADA: Nepean, Ontario (613) 726-1970;
Richmond Hill, Ontario (416) 884-9181;
St. Laurent, Quebec (514) 334-3635.

G

TI Regional Technology Centers

CALIFORNIA: Irvine (714) 660-8140,
Hotline: (714) 660-8164; **Santa Clara** (408) 748-2220,
Hotline: (408) 980-0305.

GEORGIA: Atlanta (404) 452-4682,
Hotline: (404) 452-4686.

ILLINOIS: Arlington Heights (312) 640-2909,
Hotline: (312) 228-6008.

MASSACHUSETTS: Waltham (617) 890-6671,
Hotline: (617) 890-4271.

TEXAS: Richardson (214) 680-5066,
Hotline: (214) 680-5096.

TI Distributors

TI AUTHORIZED DISTRIBUTORS IN USA

Arrow Electronics
Diplomat Electronics
ESCO Electronics
General Radio Supply Company
Graham Electronics
Hall-Mark Electronics
Harrison Equipment Co.
International Electronics
JACO Electronics
Kieruiff Electronics
LCOMP, Incorporated
Marshall Industries
Milgray Electronics
Newark Electronics
Rochester Radio Supply
Time Electronics
R.V. Weatherford Co.
Wyle Laboratories

TI AUTHORIZED DISTRIBUTORS IN CANADA

CESCO Electronics, Inc.
Future Electronics
ITT Components
L.A. Varah, Ltd.

ALABAMA: Arrow (205) 882-2730;
Kierulff (205) 883-6070; Marshall (205) 881-9235.

ARIZONA: Phoenix: Arrow (602) 968-4800;
Kierulff (602) 243-4101; Marshall (602) 968-6181;
Wyle (602) 249-2232; **Tucson:** Kierulff (602) 624-9986.

CALIFORNIA: Los Angeles/Orange County:
Arrow (213) 701-7500, (714) 838-5422;
Kierulff (213) 725-0325, (714) 731-5711;
Marshall (213) 999-5001, (213) 442-7204,
(714) 556-6400; R.V. Weatherford (714) 634-9600,
(213) 849-3451, (714) 623-1261; Wyle (213) 322-8100,
(714) 863-9953; **Sacramento:** Arrow (916) 925-7456;
Wyle (916) 638-5282; **San Diego:**
Arrow (619) 565-4800; Kierulff (619) 278-2112;
Marshall (619) 578-9600; Wyle (619) 565-9171;
San Francisco Bay Area: Arrow (408) 745-6600;
Kierulff (415) 968-6292; Marshall (408) 732-1100;
Wyle (408) 727-2500; **Santa Barbara:**
R.V. Weatherford (805) 965-8551.

COLORADO: Arrow (303) 696-1111;
Kierulff (303) 790-4444; Wyle (303) 457-9953.

CONNECTICUT: Arrow (203) 265-7741;
Diplomat (203) 797-9674; Kierulff (203) 265-1115;
Marshall (203) 265-3822; Milgray (203) 795-0714.

FLORIDA: Ft. Lauderdale: Arrow (305) 776-7790;
Diplomat (305) 974-8700; Kierulff (305) 486-4004;
Orlando: Arrow (305) 725-1480;
Milgray (305) 647-5747; **Tampa:**
Diplomat (813) 443-4514; Kierulff (813) 576-1966.

GEORGIA: Arrow (404) 449-8252;
Kierulff (404) 447-5252; Marshall (404) 923-5750.

TEXAS INSTRUMENTS

Creating useful products
and services for you.

ILLINOIS: Arrow (312) 397-3440;
Diplomat (312) 595-1000; Hall-Mark (312) 860-3800;
Kierulff (312) 640-0200; Newark (312) 638-4411.

INDIANA: Indianapolis: Arrow (317) 243-9353;
Graham (317) 634-8202; **Ft. Wayne:**
Graham (219) 423-3422.

IOWA: Arrow (319) 395-7230.

KANSAS: Kansas City: Hall-Mark (913) 888-4747;
Wichita: LCOMP (316) 265-9507.

MARYLAND: Arrow (301) 247-5200;
Diplomat (301) 995-1226; Kierulff (301) 636-5800;
Milgray (301) 468-6400.

MASSACHUSETTS: Arrow (617) 933-8130;
Diplomat (617) 935-6611; Kierulff (617) 667-8331;
Marshall (617) 272-8200; Time (617) 935-8080.

MICHIGAN: Detroit: Arrow (313) 971-8220;
Marshall (313) 525-5850; Newark (313) 967-0600;
Grand Rapids: Arrow (616) 243-0912.

MINNESOTA: Arrow (612) 830-1800;
Hall-Mark (612) 854-3223; Kierulff (612) 941-7500.

MISSOURI: Kansas City: LCOMP (816) 221-2400;
St. Louis: Arrow (314) 567-6888
Hall-Mark (314) 291-5350; Kierulff (314) 739-0855.

NEW HAMPSHIRE: Arrow (603) 668-6968.

NEW JERSEY: Arrow (201) 575-5300, (609) 596-8000;
Diplomat (201) 785-1830;
General Radio (609) 964-8560; Kierulff (201) 575-6750;
Marshall (201) 882-0320, (609) 234-9100;
Milgray (609) 983-5010.

NEW MEXICO: Arrow (505) 243-4566;
International Electronics (505) 345-8127.

NEW YORK: Long Island: Arrow (516) 231-1000;
Diplomat (516) 454-6334; JACO (516) 273-5500;
Marshall (516) 273-2424; Milgray (516) 420-9800;
Rochester: Arrow (716) 275-0300;
Marshall (716) 235-7620;
Rochester Radio Supply (716) 454-7800; **Syracuse:**
Arrow (315) 652-1000; Diplomat (315) 652-5000;
Marshall (607) 754-1570.

NORTH CAROLINA: Arrow (919) 876-3132,
(919) 725-8711; Kierulff (919) 872-8410.

OHIO: Cincinnati: Graham (513) 772-1661;
Hall-Mark (513) 563-5980; **Cleveland:**
Arrow (216) 248-3990; Hall-Mark (216) 349-4632;
Kierulff (216) 587-6558; Marshall (216) 248-1788;
Columbus: Hall-Mark (614) 891-4555;
Dayton: Arrow (513) 435-5563;
ESCO (513) 226-1133; Marshall (513) 236-8088.

OKLAHOMA: Arrow (918) 665-7700;
Hall-Mark (918) 665-3200; Kierulff (918) 252-7537.

OREGON: Arrow (503) 684-7690;
Kierulff (503) 641-9150; Wyle (503) 640-6000.

PENNSYLVANIA: Arrow (412) 856-7000,
(215) 928-1800; General Radio (215) 922-7037.

TEXAS: Austin: Arrow (512) 835-4180;
Hall-Mark (512) 258-8848; Kierulff (512) 835-2090;
Wyle (512) 244-3091; **Dallas:** Arrow (214) 386-7500;
Hall-Mark (214) 341-1147;
International Electronics (214) 233-9323;
Kierulff (214) 343-2400; Wyle (214) 235-9953;
El Paso: International Electronics (915) 598-3406;
Houston: Arrow (713) 530-4700;
Hall-Mark (713) 781-6100;
Harrison Equipment (713) 879-2600;
Kierulff (713) 530-7030.

UTAH: Diplomat (801) 486-4134;
Kierulff (801) 973-6913; Wyle (801) 974-9953.

VIRGINIA: Arrow (804) 282-0413.

WASHINGTON: Arrow (206) 643-4800;
Kierulff (206) 575-4420; Wyle (206) 453-8300.

WISCONSIN: Arrow (414) 764-6600;
Hall-Mark (414) 761-3000; Kierulff (414) 784-8160.

CANADA: Calgary: Future (403) 486-0974;
Varah (403) 230-1235; **Hamilton:** Varah (416) 561-9311;
Montreal: CESCO (514) 735-5511;
Future (514) 694-7710;
ITT Components (514) 735-1177; **Ottawa:**
CESCO (613) 226-6903; Future (613) 820-8313; ITT
Components (613) 226-7406; Varah (613) 592-4728;
Quebec City: CESCO (418) 687-4231; **Toronto:**
CESCO (416) 661-0220; Future (416) 663-5563;
ITT Components (416) 630-7971; **Vancouver:**
Future (604) 438-5545; Varah (604) 873-3211;
ITT Components (604) 270-7805; **Winnipeg:**
Varah (204) 633-6190.

BG

TI Worldwide Sales Offices

ALABAMA: Huntsville: 500 Wynn Drive, Suite 514, Huntsville, AL 35805, (205) 837-7530.

ARIZONA: Phoenix: 8102 N. 23rd Ave., Suite 1B, Phoenix, AZ 85021, (602) 995-1007.

CALIFORNIA: Irvine: 17891 Cartwright Rd., Irvine, CA 92714, (714) 660-1200; **Sacramento:** 1900 Point West Way, Suite 171, Sacramento, CA 95815, (916) 929-1521; **San Diego:** 4333 View Ridge Ave., Suite B., San Diego, CA 92123, (619) 278-9601; **Santa Clara:** 5353 Betsy Ross Dr., Santa Clara, CA 95054, (408) 980-9000; **Torrance:** 19505 Hamilton St., Bldg. A, Suite 1, Torrance, CA 90502, (213) 217-7010; **Woodland Hills:** 21220 Erwin St., Woodland Hills, CA 91367, (213) 704-7759.

COLORADO: Aurora: 1400 S. Potomac Ave., Suite 101, Aurora, CO 80012, (303) 695-2800.

CONNECTICUT: Wallingford: 9 Barnes Industrial Park Rd., Barnes Industrial Park, Wallingford, CT 06492, (203) 269-0074.

FLORIDA: Ft. Lauderdale: 2765 N.W. 62nd St., Ft. Lauderdale, FL 33309, (305) 973-8502; **Maitland:** 2601 Maitland Center Parkway, Maitland, FL 32751, (305) 660-4600; **Tampa:** 5010 W. Kennedy Blvd., Suite 101, Tampa, FL 33609, (813) 870-6420.

GEORGIA: Atlanta: 3300 Northeast Expy., Building 9, Atlanta, GA 30341, (404) 452-4600.

ILLINOIS: Arlington Heights: 515 W. Algonquin, Arlington Heights, IL 60005, (312) 640-3000.

INDIANA: Ft. Wayne: 2020 Inwood Dr., Ft. Wayne, IN 46815, (219) 424-5174; **Indianapolis:** 2346 S. Lynhurst, Suite J-400, Indianapolis, IN 46241, (317) 248-8555.

IOWA: Cedar Rapids: 373 Collins Rd. NE, Suite 200, Cedar Rapids, IA 52402, (319) 395-9550.

MARYLAND: Baltimore: 1 Rutherford Pl., 7133 Rutherford Rd., Baltimore, MD 21207, (301) 944-8600.

MASSACHUSETTS: Waltham: 504 Totten Pond Rd., Waltham, MA 02154, (617) 895-9100.

MICHIGAN: Farmington Hills: 33737 W. 12 Mile Rd., Farmington Hills, MI 48018, (313) 553-1500.

MINNESOTA: Eden Prairie: 11000 W. 78th St., Eden Prairie, MN 55344 (612) 828-9300.

MISSOURI: Kansas City: 8080 Ward Pkwy., Kansas City, MO 64114, (816) 523-2500; **St. Louis:** 11861 Westline Industrial Drive, St. Louis, MO 63141, (314) 569-7600.

NEW JERSEY: Iselin: 485E U.S. Route 1 South, Iselin, NJ 08830 (201) 750-1050

NEW MEXICO: Albuquerque: 5907 Alice NSE, Suite E., Albuquerque, NM 87110, (505) 395-2555.

NEW YORK: East Syracuse: 6700 Old Collamer Rd., East Syracuse, NY 13057, (315) 463-9291; **Endicott:** 112 Nanticoke Ave., P.O. Box 618, Endicott, NY 13760, (607) 754-3900; **Melville:** 1 Huntington Quadrangle, Suite 3C10, P.O. Box 2936, Melville, NY 11747, (516) 454-6600; **Pittsford:** 2851 Clover St., Pittsford, NY 14534, (716) 385-6770; **Poughkeepsie:** 385 South Rd., Poughkeepsie, NY 12601, (914) 473-2900.

NORTH CAROLINA: Charlotte: 8 Woodlawn Green, Woodlawn Rd., Charlotte, NC 28210, (704) 527-0930; **Raleigh:** 2809 Highwoods Blvd., Suite 100, Raleigh, NC 27625, (919) 876-2725.

OHIO: Beachwood: 23408 Commerce Park Rd., Beachwood, OH 44122, (216) 464-6100; **Dayton:** Kingsley Bldg., 4124 Linden Ave., Dayton, OH 45432, (513) 258-3877.

OKLAHOMA: Tulsa: 7615 East 63rd Place, 3 Memorial Place, Tulsa, OK 74133, (918) 250-0633.

OREGON: Beaverton: 6700 SW 105th St., Suite 110, Beaverton, OR 97005, (503) 643-6758.

PENNSYLVANIA: Ft. Washington: 260 New York Dr., Ft. Washington, PA 19034, (215) 643-6450; **Coraopolis:** 420 Rouser Rd., 3 Airport Office Park, Coraopolis, PA 15108, (412) 771-8550.

TEXAS: Austin: 12501 Research Blvd., P.O. Box 2909, Austin, TX 78723, (512) 250-7655; **Richardson:** 1001 E. Campbell Rd., Richardson, TX 75080, (214) 680-5082; **Houston:** 9100 Southwest Frwy., Suite 237, Houston, TX 77036, (713) 778-6592; **San Antonio:** 1000 Central Parkway South, San Antonio, TX 78232, (512) 496-1779.

UTAH: Murray: 5201 South Green SE, Suite 200, Murray, UT 84107, (801) 266-8972.

VIRGINIA: Fairfax: 3001 Prosperity, Fairfax, VA 22031, (703) 849-1400.

WASHINGTON: Redmond: 2723 152nd Ave., N.E. Bldg. 6, Redmond, WA 98052, (206) 881-3080.

WISCONSIN: Brookfield: 450 N. Sunny Slope, Suite 150, Brookfield, WI 53005, (414) 785-7140.

CANADA: Nepean: 301 Moodie Drive, Mallorn Center, Nepean, Ontario, Canada, K2H9C4, (613) 726-1970. **Richmond Hill:** 280 Centre St. E., Richmond Hill L4C1B1, Ontario, Canada (416) 884-9181; **St. Laurent:** Ville St. Laurent Quebec, 9460 Trans Canada Hwy., St. Laurent, Quebec, Canada H4S1R7, (514) 334-3635. G

ARGENTINA: Texas Instruments Argentina S.A.I.C.F.: Esmeralda 130, 15th Floor, 1035 Buenos Aires, Argentina, 394-2963.

AUSTRALIA (& NEW ZEALAND): Texas Instruments Australia Ltd.: 6-10 Talavera Rd., North Ryde (Sydney), New South Wales, Australia 2113, 02 + 887-1122; 5th Floor, 418 St. Kilda Road, Melbourne, Victoria, Australia 3004, 03 + 267-4677; 171 Philip Highway, Elizabeth, South Australia 5112, 08 + 255-2066.

AUSTRIA: Texas Instruments Ges.m.b.H.: Industriestrabe B/16, A-2345 Brunn/Gebirge, 2236-846210.

BELGIUM: Texas Instruments N.V. Belgium S.A.: Mercure Centre, Raketstraat 100, Rue de la Fusee, 1130 Brussels, Belgium, 02/720.80.00.

BRAZIL: Texas Instruments Electronicos do Brasil Ltda.: Av. Faria, Lima, 2003, 20 0 Andar—Pinheiros, Cep-01451 Sao Paulo, Brazil, 815-6166.

DENMARK: Texas Instruments A/S, Mairelundvej 46E, DK-2730 Herlev, Denmark, 2 - 91 74 00.

FINLAND: Texas Instruments Finland OY: PL 56, 00510 Helsinki 51, Finland, (90) 7013133.

FRANCE: Texas Instruments France: Headquarters and Prod. Plant, BP 05, 06270 Villeneuve-Loubet, (93) 20-01-01; Paris Office, BP 67 8-10 Avenue Morane-Saulnier, 78141 Velizy-Villacoublay, (3) 946-97-12; Lyon Sales Office, L'Oree D'Ecully, Batiment B, Chemin de la Forestiere, 69130 Ecully, (7) 833-04-40; Strasbourg Sales Office, Le Sebastopol 3, Quai Kleber, 67055 Strasbourg Cedex, (88) 22-12-66; Rennes, 23-25 Rue du Puits Mauger, 35100 Rennes, (99) 31-54-86; Toulouse Sales Office, Le Peripole—2, Chemin du Pigeonnier de la Cepiere, 31100 Toulouse, (61) 44-18-19; Marseille Sales Office, Noilly Paradis—146 Rue Paradis, 13006 Marseille, (91) 37-25-30.

GERMANY: Texas Instruments Deutschland GmbH: Haggerty- strasse 1, D-8050 Freising, 08161-801; Kurfuerstendamm 195/196, D-1000 Berlin 15, 030-8827365; III, Hagen 43/Kibbelstrasse, D-4300 Essen, 0201-24250; Frankfurter Allee 6-8, D-6236 Eschborm 1, 06196-43074; Hamburger Strasse 11, D-2000 Hamburg 76, 040-2201154, Kirchhorsterstrasse 2, D-3000 Hannover 51, 0511-648021; Arabellastrasse 15, D-8000 Muenchen 81, 089-92341; Maybachstrasse 11, D-7302 Ostfildern 2/Nellingen, 0711-34030.

HONG KONG (+ PEOPLES REPUBLIC OF CHINA): Texas Instruments Asia Ltd., 8th Floor, World Shipping Ctr., Harbour City, 7 Canton Rd., Kowloon, Hong Kong, 3 + 722-1223.

IRELAND: Texas Instruments (Ireland) Limited: Brewery Rd., Stillorgan, County Dublin, Eire, 01 831269.

ITALY: Texas Instruments Semiconduttori Italia Spa: Viale Delle Scienze, 1, 02015 Cittaducale (Rieti), Italy, 0746 694.1; Via Salaria KM 24 (Palazzo Cosma), Monterotondo Scalo (Rome), Italy, 06 9004395; Viale Europa, 38-44, 20093 Cologno Monzese (Milano), 02 2532541; Corso Svizzera, 185, 10100 Torino, Italy, 011 774545; Via J. Barozzi 6, 45100 Bologna, Italy, 051 355851.

JAPAN: Texas Instruments Asia Ltd.: 4F Aoyama Fuji Bldg., 6-12, Kita Aoyama 3-Chome, Minato-ku, Tokyo, Japan 107, 03-498-2111; Osaka Branch, 5F, Nissho Iwai Bldg., 30 Imabashi 3- Chome, Higashi-ku, Osaka, Japan 541, 06-204-1881; Nagoya Branch, 7F Daini Toyota West Bldg., 10-27, Meieki 4-Chome, Nakamura-ku Nagoya, Japan 450, 052-583-8691.

KOREA: Texas Instruments Supply Co.: 3rd Floor, Samon Bldg., Yuksam-Dong, Gangnam-ku, 135 Seoul, Korea, 02 + 462-8661678-39.

MEXICO: Texas Instruments de Mexico S.A.: Poniente 116, No. 489, Colonia Vallejo, Mexico, D.F. 02300, 567-9200.

MIDDLE EAST: Texas Instruments: No. 13, 1st Floor Mannai Bldg., Diplomatic Area, Manama, P.O. Box 26335, Bahrain, Arabian Gulf, 973 - 72 46 81.

NETHERLANDS: Texas Instruments Holland B.V., P.O. Box 12995, (Bullewijk) 1100 AZ Amsterdam, Zuid-Oost, Holland (020) 5602911.

NORWAY: Texas Instruments Norway A/S: PB106, Refstad 131, Oslo 1, Norway, (2) 155090.

PHILIPPINES: Texas Instruments Asia Ltd.: 14th Floor, Ba- Lepanto Bldg., 8747 Paseo de Roxas, Makati, Metro Manila, Philippines, 882465.

PORTUGAL: Texas Instruments Equipamento Electronico (Portugal), Lda.: Rua Eng. Frederico Ulrich, 2650 Moreira Da Maia, 4470 Maia, Portugal, 2-948-4532.

SINGAPORE (+ INDIA, INDONESIA, MALAYSIA, THAILAND): Texas Instruments Asia Ltd.: P.O. Box 138, Unit #02-08, Block 6, Kolam Ayer Industrial Est., Kallang Sector, Singapore 1334, Republic of Singapore, 747-2255.

SPAIN: Texas Instruments Espana, S.A.: C/Jose Lazaro Galdiano No. 6, Madrid 16, 1/458.14.58. C/Balmes, 89 Barcelona-8, 25360 00/253 29 02.

SWEDEN: Texas Instruments International Trade Corporation (Sverigefilialen): Box 39103, 10054 Stockholm, Sweden, 08 - 235480.

SWITZERLAND: Texas Instruments, Inc., Reidstrasse 6, CH-8953 Dietikon (Zuerich) Switzerland, 1-740 2220.

TAIWAN: Texas Instruments Supply Co.: 10th Floor, Fu- Shing Bldg., 71 Sung-Kiang Road, Taipei, Taiwan, Republic of China, 02 + 521-9321.

UNITED KINGDOM: Texas Instruments Limited: Manton Lane, Bedford, MK41 7PA, England, 0234 67466; St. James House, Wellington Road North, Stockport, SK4 2RT, England, 061 442-8448. BG

TEXAS INSTRUMENTS

Creating useful products
and services for you